Human Anatomy Laboratory Manual

WITH CAT DISSECTIONS

Seventh Edition

Elaine N. Marieb, R.N., Ph.D.
Holyoke Community College

Susan J. Mitchell, Ph.D.
Onondaga Community College

Lori A. Smith, Ph.D.
American River College

PEARSON

Boston Columbus Indianapolis New York San Francisco Upper Saddle River
Amsterdam Cape Town Dubai London Madrid Milan Munich Paris Montréal Toronto
Delhi Mexico City São Paulo Sydney Hong Kong Seoul Singapore Taipei Tokyo

Editor-in-Chief: Serina Beauparlant
Associate Project Editor: Lisa Damerel
Director of Development: Barbara Yien
Marketing Manager: Derek Perrigo
Senior Managing Editor: Deborah Cogan
Production and Design Manager: Michele Mangelli
Production Supervisor: Janet Vail
Copyeditor: Sally Peyrefitte
Art Coordinator: Jean Lake
Interior Designer: tani hasegawa
Cover Designer: Riezebos Holzbaur Design Group
Compositor: Cenveo® Publisher Services
Senior Manufacturing Buyer: Stacey Weinberger

Cover image: Colored X ray of a lateral view of the neck showing the cervical vertebrae during forward bending of the head. The seven cervical neck vertebrae are the smallest, lightest vertebrae; they support the head and neck and aid in flexion, extension, and rotation of the head. The hyoid bone of the neck is seen inferior to the mandible (jawbone). SPL/Photo Researchers, Inc.

Credits and acknowledgments borrowed from other sources and reproduced, with permission, in this textbook appear on the appropriate page within the text or on page 545.

The Authors and Publisher believe that the lab experiments described in this publication, when conducted in conformity with the safety precautions described herein and according to the school's laboratory safety procedures, are reasonably safe for the student to whom this manual is directed. Nonetheless, many of the described experiments are accompanied by some degree of risk, including human error, the failure or misuses of laboratory or electrical equipment, mismeasurement, chemical spills, and exposure to sharp objects, heat, bodily fluids, blood, or other biologics. The Authors and Publisher disclaim any liability arising from such risks in connection with any of the experiments contained in this manual. If students have any questions or problems with materials, procedures, or instructions on any experiment, they should always ask their instructor for help before proceeding.

PEARSON

www.pearsonhighered.com

ISBN 10: 0-321-88418-3
ISBN 13: 978-0-321-88418-3

1 2 3 4 5 6 7 8 9 10—WBC—17 16 15 14 13

Contents

Preface for the Instructor vii
A Word to the Student xi

THE HUMAN BODY: AN ORIENTATION

EXERCISE 1 The Language of Anatomy 1
Activity 1 Locating Body Regions 5
Activity 2 Observing Sectioned Specimens 5
Activity 3 Practicing Using Correct Anatomical Terminology 7
Activity 4 Identifying Organs in the Abdominopelvic Cavity 8
Activity 5 Locating Abdominal Surface Regions 9
Group Challenge The Language of Anatomy 10
Review Sheet The Language of Anatomy 11

EXERCISE 2 Organ Systems Overview 15
Dissection and Identification: The Organ Systems of the Rat 16
Activity 1 Observing External Structures 17
Activity 2 Examining the Oral Cavity 17
Activity 3 Opening the Ventral Body Cavity 17
Activity 4 Examining the Ventral Body Cavity 18
Activity 5 Examining the Human Torso Model 23
Group Challenge Odd Organ Out 24
Review Sheet Organ Systems Overview 25

THE MICROSCOPE AND ITS USES

EXERCISE 3 The Microscope 27
Activity 1 Identifying the Parts of a Microscope 28
Activity 2 Viewing Objects Through the Microscope 30
Activity 3 Estimating the Diameter of the Microscope Field 32
Activity 4 Perceiving Depth 33
Activity 5 Preparing and Observing a Wet Mount 33
Review Sheet The Microscope 35

THE CELL

EXERCISE 4 The Cell: Anatomy and Division 39
Activity 1 Identifying Parts of a Cell 40
Activity 2 Identifying Components of a Plasma Membrane 41
Activity 3 Locating Organelles 41
Activity 4 Examining the Cell Model 44
Activity 5 Observing Various Cell Structures 44
Activity 6 Identifying the Mitotic Stages 48
Review Sheet The Cell: Anatomy and Division 49

HISTOLOGY: BASIC TISSUES OF THE BODY

EXERCISE 5 Classification of Tissues 53
Activity 1 Examining Epithelial Tissue Under the Microscope 54
Activity 2 Examining Connective Tissue Under the Microscope 61
Activity 3 Examining Nervous Tissue Under the Microscope 67
Activity 4 Examining Muscle Tissue Under the Microscope 68
Review Sheet Classification of Tissues 71

THE INTEGUMENTARY SYSTEM

EXERCISE 6 The Integumentary System 77
Activity 1 Locating Structures on a Skin Model 78
Activity 2 Differentiating Sebaceous and Sweat Glands Microscopically 82
Activity 3 Plotting the Distribution of Sweat Glands 82
Activity 4 Identifying Nail Structures 83
Activity 5 Comparing Hairy and Relatively Hair-Free Skin Microscopically 84
Activity 6 Taking and Identifying Inked Fingerprints 85
Review Sheet The Integumentary System 87

THE SKELETAL SYSTEM

EXERCISE 7 Overview of the Skeleton: Classification and Structure of Bones and Cartilages 91
Activity 1 Examining and Classifying Bones 93
Activity 2 Examining a Long Bone 93
Activity 3 Examining the Effects of Heat and Hydrochloric Acid on Bones 96
Activity 4 Examining the Microscopic Structure of Compact Bone 96
Activity 5 Examining the Osteogenic Epiphyseal Plate 98
Review Sheet Overview of the Skeleton: Classification and Structure of Bones and Cartilages 101

EXERCISE 8 The Axial Skeleton 105
Activity 1 Identifying the Bones of the Skull 106
Activity 2 Palpating Skull Markings 114
Group Challenge Odd Bone Out 115
Activity 3 Examining Spinal Curvatures 116

Activity 4 Examining Vertebral Structure 120
Activity 5 Examining the Relationship Between Ribs and Vertebrae 121
Activity 6 Examining a Fetal Skull 122
Review Sheet The Axial Skeleton 125

EXERCISE 9 The Appendicular Skeleton 133
Activity 1 Examining and Identifying Bones of the Appendicular Skeleton 134
Activity 2 Palpating the Surface Anatomy of the Pectoral Girdle and Upper Limb 138
Activity 3 Observing Pelvic Articulations 139
Activity 4 Comparing Male and Female Pelves 142
Activity 5 Palpating the Surface Anatomy of the Pelvic Girdle and Lower Limb 144
Activity 6 Constructing a Skeleton 144
Review Sheet The Appendicular Skeleton 145

EXERCISE 10 Articulations and Body Movements 153
Activity 1 Identifying Fibrous Joints 155
Activity 2 Identifying Cartilaginous Joints 155
Activity 3 Examining Synovial Joint Structure 156
Activity 4 Demonstrating the Importance of Friction-Reducing Structures 156
Activity 5 Demonstrating Movements of Synovial Joints 157
Activity 6 Demonstrating Actions at the Hip Joint 158
Activity 7 Demonstrating Actions at the Knee Joint 160
Activity 8 Examining the Action at the Temporomandibular Joint 161
Activity 9 Demonstrating Actions at the Shoulder Joint 161
Group Challenge Articulations: "Simon Says" 163
Review Sheet Articulations and Body Movements 167

THE MUSCULAR SYSTEM

EXERCISE 11 Microscopic Anatomy and Organization of Skeletal Muscle 171
Activity 1 Examining Skeletal Muscle Cell Anatomy 174
Activity 2 Observing the Histologic Structure of a Skeletal Muscle 174
Activity 3 Studying the Structure of a Neuromuscular Junction 176
Review Sheet Microscopic Anatomy and Organization of Skeletal Muscle 177

EXERCISE 12 Gross Anatomy of the Muscular System 181
Activity 1 Identifying Muscles of the Head and Neck 182
Activity 2 Identifying Muscles of the Trunk 191
Activity 3 Identifying Muscles of the Upper Limb 197
Activity 4 Identifying Muscles of the Lower Limb 204
Activity 5 Review of Human Musculature 204
Activity 6 Making a Muscle Painting 204
Group Challenge Muscle IDs 215
Dissection and Identification: Cat Muscles 215
Activity 7 Preparing the Cat for Dissection 215
Activity 8 Dissecting Trunk and Neck Muscles 217
Activity 9 Dissecting Forelimb Muscles 222
Activity 10 Dissecting Hindlimb Muscles 225
Review Sheet Gross Anatomy of the Muscular System 231

THE NERVOUS SYSTEM

EXERCISE 13 Histology of Nervous Tissue 239
Activity 1 Identifying Parts of a Neuron 242
Activity 2 Studying the Microscopic Structure of Selected Neurons 244
Activity 3 Examining the Microscopic Structure of a Nerve 245
Activity 4 Studying the Structure of Selected Sensory Receptors 247
Review Sheet Histology of Nervous Tissue 249

EXERCISE 14 Gross Anatomy of the Brain and Cranial Nerves 253
Activity 1 Identifying External Brain Structures 254
Activity 2 Identifying Internal Brain Structures 257
Activity 3 Identifying and Testing the Cranial Nerves 267
Dissection: The Sheep Brain 267
Group Challenge Odd (Cranial) Nerve Out 272
Review Sheet Gross Anatomy of the Brain and Cranial Nerves 273

EXERCISE 15 The Spinal Cord and Spinal Nerves 279
Activity 1 Identifying Structures of the Spinal Cord 280
Activity 2 Identifying Spinal Cord Tracts 283
Dissection: Spinal Cord 283
Activity 3 Identifying the Major Nerve Plexuses and Peripheral Nerves 291
Dissection and Identification: Cat Spinal Nerves 291
Activity 4 Dissecting Nerves of the Brachial Plexus 291
Activity 5 Dissecting Nerves of the Lumbosacral Plexus 291
Review Sheet The Spinal Cord and Spinal Nerves 295

EXERCISE 16 The Autonomic Nervous System 299
Activity 1 Locating the Sympathetic Trunk (Chain) 302
Activity 2 Comparing Sympathetic and Parasympathetic Effects 302
Review Sheet The Autonomic Nervous System 303

EXERCISE 17 **Special Senses: Anatomy of the Visual System 305**
Activity 1 Identifying Accessory Eye Structures 306
Activity 2 Identifying Internal Structures of the Eye 310
Activity 3 Studying the Microscopic Anatomy of the Retina 310
Activity 4 Predicting the Effects of Visual Pathway Lesions 310
Dissection: The Cow (Sheep) Eye 310
Review Sheet Special Senses: Anatomy of the Visual System 313

EXERCISE 18 **Special Senses: Visual Tests and Experiments 317**
Activity 1 Demonstrating the Blind Spot 317
Activity 2 Determining Near Point of Accommodation 318
Activity 3 Testing Visual Acuity 319
Activity 4 Testing for Astigmatism 320
Activity 5 Testing for Color Blindness 320
Activity 6 Testing for Depth Perception 320
Activity 7 Conducting an Ophthalmoscopic Examination 322
Review Sheet Special Senses: Visual Tests and Experiments 323

EXERCISE 19 **Special Senses: Hearing and Equilibrium 327**
Activity 1 Identifying Structures of the Ear 328
Activity 2 Examining the Ear with an Otoscope (Optional) 329
Activity 3 Examining the Microscopic Structure of the Cochlea 330
Activity 4 Conducting Laboratory Tests of Hearing 331
Activity 5 Examining the Microscopic Structure of the Crista Ampullaris 333
Activity 6 Conducting Laboratory Tests on Equilibrium 334
Review Sheet Special Senses: Hearing and Equilibrium 337

EXERCISE 20 **Special Senses: Olfaction and Taste 341**
Activity 1 Microscopic Examination of the Olfactory Epithelium 343
Activity 2 Microscopic Examination of the Taste Buds 344
Activity 3 Stimulating Taste Buds 344
Activity 4 Examining the Effect of Olfactory Stimulation 344
Activity 5 Demonstrating Olfactory Adaptation 344
Activity 6 Examining the Combined Effects of Smell, Texture, and Temperature on Taste 345
Review Sheet Special Senses: Olfaction and Taste 347

THE ENDOCRINE SYSTEM

EXERCISE 21 **Functional Anatomy of the Endocrine Glands 349**
Activity 1 Identifying the Endocrine Organs 353
Dissection and Identification: Selected Endocrine Organs of the Cat 353
Activity 2 Opening the Ventral Body Cavity 353
Activity 3 Identifying Organs 354
Activity 4 Examining the Microscopic Structure of Endocrine Glands 355
Group Challenge Odd Hormone Out 358
Review Sheet Functional Anatomy of the Endocrine Glands 359

THE CIRCULATORY SYSTEM

EXERCISE 22 **Blood 365**
Activity 1 Determining the Physical Characteristics of Plasma 367
Activity 2 Examining the Formed Elements of Blood Microscopically 368
Activity 3 Conducting a Differential White Blood Cell Count 372
Activity 4 Determining the Hematocrit 372
Activity 5 Determining Hemoglobin Concentration 374
Activity 6 Determining Coagulation Time 376
Activity 7 Typing for ABO and Rh Blood Groups 377
Activity 8 Observing Demonstration Slides 378
Activity 9 Measuring Plasma Cholesterol Concentration 378
Review Sheet Blood 379

EXERCISE 23 **Anatomy of the Heart 383**
Activity 1 Using the Heart Model to Study Heart Anatomy 386
Activity 2 Tracing the Path of Blood Through the Heart 386
Activity 3 Using the Heart Model to Study Cardiac Circulation 388
Dissection: The Sheep Heart 389
Activity 4 Examining Cardiac Muscle Tissue Anatomy 392
Review Sheet Anatomy of the Heart 393

EXERCISE 24 **Anatomy of Blood Vessels 397**
Activity 1 Examining the Microscopic Structure of Arteries and Veins 400
Activity 2 Locating Arteries on an Anatomical Chart or Model 405
Activity 3 Identifying the Systemic Veins 409
Activity 4 Identifying Vessels of the Pulmonary Circulation 410
Activity 5 Tracing the Hepatic Portal Circulation 411
Activity 6 Tracing the Pathway of Fetal Blood Flow 411

Dissection and Identification: The Blood Vessels of the Cat 413

Activity 7 Opening the Ventral Body Cavity and Preliminary Organ Identification 413

Activity 8 Identifying the Blood Vessels 413

Review Sheet Anatomy of Blood Vessels 421

EXERCISE 25 **The Lymphatic System and Immune Response 429**

Activity 1 Identifying the Organs of the Lymphatic System 430

Activity 2 Studying the Microscopic Anatomy of a Lymph Node, the Spleen, and a Tonsil 432

Dissection and Identification: The Main Lymphatic Ducts of the Cat 434

Activity 3 Identifying the Main Lymphatic Ducts of the Cat 434

Group Challenge Compare and Contrast Lymphoid Organs and Tissues 435

Review Sheet The Lymphatic System and Immune Response 437

THE RESPIRATORY SYSTEM

EXERCISE 26 **Anatomy of the Respiratory System 441**

Activity 1 Identifying Respiratory System Organs 447

Activity 2 Examining Prepared Slides of Trachea and Lung Tissue 447

Activity 3 Demonstrating Lung Inflation in a Sheep Pluck 447

Dissection and Identification: The Respiratory System of the Cat 448

Activity 4 Identifying Organs of the Respiratory System of the Cat 448

Activity 5 Observing Lung Tissue Microscopically 450

Review Sheet Anatomy of the Respiratory System 451

THE DIGESTIVE SYSTEM

EXERCISE 27 **Anatomy of the Digestive System 457**

Activity 1 Identifying Organs of the Alimentary Canal 458

Activity 2 Studying the Histologic Structure of Selected Digestive System Organs 463

Activity 3 Observing the Histologic Structure of the Small Intestine 466

Activity 4 Examining the Histologic Structure of the Large Intestine 468

Activity 5 Identifying Types of Teeth 469

Activity 6 Studying Microscopic Anatomy of the Tooth 470

Activity 7 Examining Salivary Gland Tissue 470

Activity 8 Examining the Histology of the Pancreas 471

Activity 9 Examining the Histology of the Liver 473

Dissection and Identification: The Digestive System of the Cat 473

Activity 10 Exposing and Viewing the Salivary Glands and Oral Cavity Structures 474

Activity 11 Identifying Alimentary Canal Organs 475

Review Sheet Anatomy of the Digestive System 479

THE URINARY SYSTEM

EXERCISE 28 **Anatomy of the Urinary System 485**

Activity 1 Identifying Urinary System Organs 486

Dissection: Gross Internal Anatomy of the Pig or Sheep Kidney 486

Activity 2 Studying Nephron Structure 492

Activity 3 Studying Bladder Structure 493

Dissection and Identification: The Urinary System of the Cat 493

Activity 4 Identifying Organs of the Urinary System 493

Group Challenge Urinary System Sequencing 496

Review Sheet Anatomy of the Urinary System 497

THE REPRODUCTIVE SYSTEM

EXERCISE 29 **Anatomy of the Reproductive System 503**

Activity 1 Identifying Male Reproductive Organs 506

Activity 2 Conducting a Microscopic Study of Selected Male Reproductive Organs 507

Activity 3 Identifying Female Reproductive Organs 509

Activity 4 Conducting a Microscopic Study of Selected Female Reproductive Organs 511

Dissection and Identification: The Reproductive System of the Cat 513

Activity 5 Identifying Organs of the Male Reproductive System 513

Activity 6 Identifying Organs of the Female Reproductive System 515

Review Sheet Anatomy of the Reproductive System 517

SURFACE ANATOMY

EXERCISE 30 **Surface Anatomy Roundup 525**

Activity 1 Palpating Landmarks of the Head 527

Activity 2 Palpating Landmarks of the Neck 528

Activity 3 Palpating Landmarks of the Trunk 529

Activity 4 Palpating Landmarks of the Abdomen 532

Activity 5 Palpating Landmarks of the Upper Limb 534

Activity 6 Palpating Landmarks of the Lower Limb 536

Review Sheet Surface Anatomy Roundup 541

Appendix The Metric System 543

Credits 545

Index 547

Preface for the Instructor

The philosophy behind the revision of this manual mirrors that of all earlier editions. It reflects a still-developing sensibility for the way teachers teach and students learn, engendered by years of teaching the subject and by listening to the suggestions of other instructors as well as those of students enrolled in multifaceted healthcare programs. *Human Anatomy Laboratory Manual with Cat Dissections* was originally developed to facilitate and enrich the laboratory experience for both teachers and students. This edition retains those same goals.

This manual, intended for students in introductory human anatomy courses, presents a wide range of anatomical laboratory experiences for students concentrating in nursing, physical therapy, dental hygiene, pharmacology, respiratory therapy, and health and physical education, as well as biology and premedical programs. This manual studies anatomy of the human specimen in particular, but the cat and isolated animal organs are also used in the dissection experiments.

Basic Approach and Features

The organization and scope of this laboratory manual lend themselves to use in the one-term human anatomy course. The variety of anatomical studies enables instructors to gear their courses to specific academic programs or to their own teaching preferences. Although the main textbook, *Human Anatomy,* Seventh Edition (Elaine N. Marieb, Patricia Brady Wilhelm, and Jon Mallatt, Pearson Education © 2014), provided the impetus for this revision, the laboratory manual, as in previous editions, is based largely on exercises developed for use independent of any textbook. It contains all the background discussion and terminology necessary to perform all manipulations effectively and eliminates the need for students to bring a textbook into the laboratory.

The laboratory manual is comprehensive and balanced enough to be flexible and is carefully written so that students can successfully complete each of its 30 exercises with little supervision. Exercises covering anatomical terminology and an orientation to the human body (Exercises 1 and 2) together provide the necessary tools for studying the various body systems. Exercises on the microscope, the cell, and tissues (Exercises 3 through 5) lay the groundwork for a study of each body system from the cellular to the organ level. Other exercises explore the anatomy of organ systems in detail (Exercises 6 though 29), and students also can gain experience with the clinically valuable study of surface anatomy (Exercise 30).

Homeostasis is continually emphasized as a requirement for optimal health. Pathological conditions are viewed as a loss of homeostasis; these discussions can be recognized by the homeostatic imbalance logo within the descriptive material of each exercise. This holistic approach encourages an integrated understanding of the human body.

Features

- Each exercise begins with learning objectives.
- Key terms appear in boldface print, and each term is defined when introduced.
- Illustrations are large and of exceptional quality. Full-color photographs and drawings highlight and differentiate important structures and focus student attention on them.
- Body structures are studied from simple to complex levels, and physiological experiments allow ample opportunity for student observation and manipulation.
- All laboratory instructions and procedures incorporate the latest precautions as recommended by the Centers for Disease Control and Prevention (CDC); these are reinforced by the laboratory safety procedures described inside the front cover and in the front section of the *Instructor's Guide.* These procedures can be easily photocopied and posted in the lab.
- Laboratory Review Sheets at the end of each exercise require students to label diagrams and answer multiple-choice, short-answer, and essay questions. We have strived to achieve an acceptable balance between questions that require students to recognize structures and those that ask students to explain important concepts.
- Three icons alert students to special features or instructions:

The **dissection scissors icon** appears at the beginning of activities that entail the dissection of the cat as well as isolated animal organs.

The **homeostatic imbalance icon** appears where a clinical disorder is described to indicate what happens when there is a structural abnormality or physiological malfunction, that is, a loss of homeostasis.

A **safety icon** notifies students that they are to observe specific safety precautions when they are using certain equipment or conducting particular lab procedures. For example, when handling body fluids such as blood, urine, and saliva, they are to wear gloves.

Unique Approach to Anatomy

In this revision, we have continued to try to respond to reviewers' and users' feedback concerning trends that are having an impact on the anatomy and physiology laboratory experience:

- The increased use of multimedia in the laboratory
- The ever-increasing computer literacy of our students
- The ongoing search for good pedagogy and effective use of laboratory time
- The continued importance of visual learning for today's student
- The need to reinforce writing, computation, and critical-thinking skills across the curriculum

The specific changes implemented to address these trends are described next.

New to the Seventh Edition

Improved Organization

We have integrated important parts of two short exercises from the previous edition into other appropriate exercises. Serous membranes are now more elaborately discussed in The Language of Anatomy (Exercise 1), and coverage of the fetal skull has been integrated into the axial skeleton exercise (Exercise 8). Also, the spinal cord and spinal nerves and the autonomic nervous system, which in the previous edition were covered in one long exercise, are now covered in two exercises. Similarly, vision is now covered in two exercises: Special Senses: Anatomy of the Visual System, and Special Senses: Visual Tests and Experiments. As a result of these changes, some exercise numbers are different from the previous edition. (Please consult the Table of Contents for number changes to Exercises 7 through 18. The numbering of Exercises 1 through 6 and 19 through 30 remain the same in this edition as in the previous edition of this lab manual.)

Group Challenge Activities

New to this edition, Group Challenge activities ask students to find the relationships between anatomical structures and to use that information to understand anatomy and physiology at a deeper level. These activities, which are designed to teach critical-thinking skills, include the following:

The Language of Anatomy (Exercise 1, p. 10)

Odd Organ Out (Exercise 2, p. 24)

Odd Bone Out (Exercise 8, p. 115)

Articulations: "Simon Says" (Exercise 10, p. 165)

Muscle IDs (Exercise 12, p. 215)

Odd (Cranial) Nerve Out (Exercise 14, p. 272)

Odd Hormone Out (Exercise 21, p. 358)

Compare and Contrast Lymphoid Organs and Tissues (Exercise 25, p. 436)

Urinary System Sequencing (Exercise 28, p. 500)

Updated Pre-lab Quizzes

The pre-lab quizzes at the beginning of each exercise motivate your students to prepare for lab by asking them basic information they should know before doing the lab. These have been updated and written in groups of either five or ten questions to facilitate grading.

Updated Art Program

Many new and improved histology images have been added to this edition. Also included are several new cadaver photos. These additions complement our art program's use of three-dimensional, realistic styles with dramatic views and perspectives, and rich, vibrant colors. The art includes key anatomy figures that are rendered with detail, depth, and a clear focus on key anatomical structures.

Customization Options

For information on creating a custom version of this manual, visit www.pearsonlearningsolutions.com/, or contact your Pearson sales representative for details.

Supplements

Instructor's Guide

The *Instructor's Guide* that accompanies the *Human Anatomy Laboratory Manual* contains a wealth of information, including answers to the updated pre-lab quizzes. The guide includes help for anticipating pitfalls and problem areas, directions for lab setup, a complete materials list for each lab, and answers to the pre-lab quiz, activity, Group Challenge, and Review Sheet questions. The probable in-class time required for each lab is indicated by an hourglass icon. (ISBN 0-321-88439-6)

Anatomy Atlas

The Bassett Atlas of Human Anatomy (ISBN 0-805-30118-6)

Multimedia

Human Anatomy & Physiology Videos

The following videos reinforce many of the concepts covered in this lab manual, and are now available on a single DVD. Please contact your Pearson sales representative for more information about the *Instructor's Review Copy for Cadaver Dissection Video Series for Human Anatomy and Physiology* (ISBN 0-321-85920-0).

- *Human Musculature* by Rose Leigh Vines and Allan Hinderstein
- *The Human Nervous System: The Brain and Cranial Nerves* by Rose Leigh Vines, Rosalee Carter, and University Media Services, California State University, Sacramento
- *The Human Nervous System: The Spinal Cord and Spinal Nerves* by Rose Leigh Vines, Rosalee Carter, and University Media Services, California State University, Sacramento
- *The Human Cardiovascular System: The Heart* by Rose Leigh Vines and University Media Services, California State University, Sacramento
- *The Human Cardiovascular System: The Blood Vessels* by Rose Leigh Vines and University Media Services, California State University, Sacramento
- *The Human Respiratory System* by Rose Leigh Vines, Ann Motekaitis, and University Media Services, California State University, Sacramento
- *The Human Digestive System* by Rose Leigh Vines, Ann Motekaitis, and University Media Services, California State University, Sacramento
- *The Human Urinary System* by Rose Leigh Vines, Ann Motekaitis, and University Media Services, California State University, Sacramento
- *The Human Reproductive Systems* by Rose Leigh Vines, Ann Motekaitis, and University Media Services, California State University, Sacramento

Practice Anatomy Lab™ 3.0

(ISBN 0-321-68211-4)

Practice Anatomy Lab™ (PAL) 3.0 is an indispensable virtual anatomy study and practice tool that gives students 24/7 access to the most widely used laboratory specimens including human cadaver, cat, and fetal pig as well as anatomical models and histological images that are used in the laboratory.

PAL 3.0 features:

- **An interactive cadaver** that allows students to peel back layers of the human cadaver and view hundreds of brand-new dissection photographs specifically commissioned for this version.
- **Interactive histology** that allows students to view the same tissue slide at varying magnifications.
- **Quizzes that give students more opportunity for practice.** Each time the student takes a quiz or lab practical exam, a new set of questions is generated.
- **Integration of nerves, arteries, and veins** across body systems.
- **Integrated muscle animations** of the origin, insertion, action, and innervations of key muscles.
- **Rotatable bones** help students appreciate the three-dimensionality of bone structures.

PAL 3.0 is available in the Study Area of MasteringA&P®. Students may also purchase a 12-month subscription to PAL 3.0 online at www.practiceanatomylab.com.

The **PAL 3.0 DVD** can be packaged with the lab manual for no additional cost. It can also be purchased separately.

The **Instructor Resource DVD for PAL 3.0** provides instructors access to all images in PAL in PowerPoint® and JPEG formats. PowerPoint slides include images with editable labels and leader lines, as well as embedded links to relevant 3D anatomy animations and bone rotations.

The **PAL 3.0 Test Bank** includes more than 4,000 customizable multiple-choice quiz and fill-in-the-blank lab practical questions in TestGen® format. The Test Bank is available for download in the Instructor Resource Center and is also fully assignable in the MasteringA&P® Item Library.

Contact your Pearson sales representative for more information on these titles, or visit our web site at www.pearsonhighered.com

Acknowledgments

Many thanks to the Pearson Education editorial team: Serina Beauparlant, Editor-in-Chief; and Lisa Damerel, Associate Project Editor. Thanks also to Stacey Weinberger, Senior Manufacturing Buyer; and Derek Perrigo, Marketing Manager. Kudos as usual to Michele Mangelli and her production team: to Janet Vail, our production editor for this project; to Jean Lake, our art coordinator; to Yvo Riezebos for a beautiful cover; to tani hasegawa for the wonderful interior design; and to our incredibly conscientious copyeditor, Sally Peyrefitte.

As always, we invite users of this edition to send us their comments and suggestions for subsequent editions.

Elaine N. Marieb,
Susan J. Mitchell,
and Lori A. Smith
Pearson Education
1301 Sansome Street
San Francisco, CA 94111

A Word to the Student

We hope you will enjoy your laboratory experiences. As with any unfamiliar experience, it really helps to know in advance what you can expect and what will be expected of you.

Laboratory Activities

The laboratory exercises in this manual are designed to help you gain a broad understanding of anatomy. You can anticipate examining models, dissecting a specimen and isolated animal organs, and using a microscope to look at tissue slides (anatomical approaches). You will also investigate a limited selection of physiological phenomena (conduct visual tests, plot the distribution of sweat glands, and so forth) to make your anatomy studies more meaningful.

Icons/Visual Mnemonics

We have tried to make this manual easy for you to use, and to this end two colored section heads and three different icons (visual mnemonics) are used throughout:

The **Dissection** head is purple and is accompanied by the **dissection scissors icon** at the beginning of activities that require you to dissect the cat as well as isolated animal organs.

The **Activity** head is blue. Because most exercises provide some explanatory background before the experiment(s), this visual cue alerts you that your lab involvement is imminent.

The **homeostatic imbalance icon** appears where a clinical disorder is described to indicate what happens when there is a structural abnormality or physiological malfunction, that is, a loss of homeostasis.

The **safety icon** notifies you that you are to observe specific safety precautions when using certain equipment or conducting particular lab procedures (for example, using a hood when you are working with ether, or wearing gloves when you are handling body fluids, such as blood, urine, or saliva).

Hints for Success in the Laboratory

Most students can use helpful hints and guidelines to ensure that they have successful lab experiences.

1. Perhaps the best bit of advice is to attend all your scheduled labs and to participate in all the assigned exercises. Learning is an *active* process.

2. *Before* going to lab, complete the pre-lab quiz, and scan the scheduled lab exercise and the questions in the Review Sheet at the end of the exercise.

3. Be on time. Most instructors explain what the lab is about, pitfalls to avoid, and the sequence or format to be followed at the beginning of the lab session. If you are late, you will miss this information and also risk annoying the instructor.

4. Review your lab notes after completing the lab session to help you focus on and remember the important concepts.

5. Keep your work area clean and neat. This reduces confusion and accidents.

6. Assume that all lab chemicals and equipment are sources of potential danger to you. Follow directions for equipment use, and observe the laboratory safety guidelines provided inside the front cover of this manual.

7. Keep in mind the real value of the laboratory experience—a place for you to observe, manipulate, and experience hands-on activities that will dramatically enhance your understanding of the lecture presentations.

Supplements

Practice Anatomy Lab 3.0 Lab Guide

by Ruth Heisler, Nora Hebert, Jett Chinn, Karen Krabbenhoft, and Olga Malakhova

Written to accompany PAL™ 3.0, the new *Practice Anatomy Lab 3.0 Lab Guide* contains lab exercises that direct you to select images and features in PAL 3.0, and then assess your understanding by completing labeling, matching, short answer, and fill-in-the-blank questions. Exercises cover three key lab specimens in PAL 3.0—human cadaver, anatomical models, and histology.

without PAL 3.0 DVD (ISBN 0-321-84025-9)

with PAL 3.0 DVD (ISBN 0-321-85767-4)

The Anatomy Coloring Book, Fourth Edition

by Wynn Kapit and Lawrence M. Elson

For more than 35 years, *The Anatomy Coloring Book* has been the #1 best-selling human anatomy coloring book! A useful tool for anyone with an interest in learning anatomical structures, this concisely written text features precise, extraordinary hand-drawn figures that were crafted especially for easy coloring and interactive study. Organized according to body systems, each of the 162 spreads featured in this book includes an ingenious color-key system where anatomical terminology is linked to detailed illustrations of the structures of the body.

The *Fourth Edition* features user-friendly two-page spreads with enlarged art, clearer, more concise text descriptions, and new boldface headings that make this classic coloring book accessible to a wider range of learners.

(ISBN 0-321-83201-9)

We really hope that you enjoy your anatomy laboratories and that this lab manual makes learning about intricate structures and functions of the human body a fun and rewarding process. We're always open to constructive criticism and suggestions for improvement in future editions. If you have any, please write to us in care of Pearson Education.

Elaine N. Marieb,
Susan J. Mitchell,
and Lori A. Smith
Pearson Education
1301 Sansome Street
San Francisco, CA 94111

EXERCISE 1

The Language of Anatomy

MATERIALS

- ☐ Human torso model (dissectible)
- ☐ Human skeleton
- ☐ Demonstration: sectioned and labeled kidneys [three separate kidneys uncut or cut so that (a) entire, (b) transverse sectional, and (c) longitudinal sectional views are visible]
- ☐ Gelatin-spaghetti molds
- ☐ Scalpel

OBJECTIVES

1. Describe the anatomical position, and explain its importance.
2. Use proper anatomical terminology to describe body regions, orientation and direction, and body planes.
3. Name the body cavities, and indicate the important organs in each cavity.
4. Name and describe the serous membranes of the ventral body cavities.
5. Identify the abdominopelvic quadrants and regions on a torso model or image.

PRE-LAB QUIZ

1. Circle True or False. In the anatomical position, the body is lying down.
2. Circle the correct underlined term. With regard to surface anatomy, abdominal / axial refers to the structures along the center line of the body.
3. The term *superficial* refers to a structure that is:
   a. attached near the trunk of the body
   b. toward or at the body surface
   c. toward the head
   d. toward the midline
4. The ________ plane runs longitudinally and divides the body into right and left parts.
   a. frontal
   b. sagittal
   c. transverse
   d. ventral
5. Circle the correct underlined terms. The dorsal body cavity can be divided into the cranial / thoracic cavity, which contains the brain, and the sural / vertebral cavity, which contains the spinal cord.

Most of us are naturally curious about our bodies. This curiosity is particularly evident in infants, who are fascinated with their own waving hands or their mother's nose. Unlike the infant, however, the student of anatomy must learn to observe and identify the dissectible body structures formally.

A student new to any science is often overwhelmed at first by the terminology used in that subject. The study of anatomy is no exception. But without this specialized terminology, confusion is inevitable. For example, what do *over, on top of, superficial to, above,* and *behind* mean in reference to the human body? Anatomists have an accepted set of reference terms that are universally understood. These allow body structures to be located and identified with a minimum of words and a high degree of clarity.

This exercise presents some of the most important anatomical terminology used to describe the body and introduces you to basic concepts of **gross anatomy,** the study of body structures visible to the naked eye.

Anatomical Position

When anatomists or doctors refer to specific areas of the human body, the picture they keep in mind is a universally accepted standard position called the **anatomical position.** It is essential to understand this position because much of the body terminology used in this book refers to this body positioning, regardless of the position the body happens to be in. In the anatomical position, the human body is erect, with the feet only slightly apart, head and toes pointed forward, and arms hanging at the sides with palms facing forward **(Figure 1.1)**.

☐ Assume the anatomical position, and notice that it is not particularly comfortable. The hands are held unnaturally forward rather than with the palms toward the thighs.

Check the box when you have completed this task.

Surface Anatomy

Body surfaces provide a wealth of visible landmarks for study of the body (Figure 1.1).

Axial: Relating to head, neck, and trunk, the axis of the body

Appendicular: Relating to limbs and their attachments to the axis

Anterior Body Landmarks

Note the following regions **(Figure 1.2a)**:

Abdominal: Anterior body trunk region inferior to the ribs

Acromial: Point of the shoulder

Antebrachial: Forearm

Antecubital: Anterior surface of the elbow

Axillary: Armpit

Brachial: Arm

Buccal: Cheek

Carpal: Wrist

Cephalic: Head

Cervical: Neck region

Coxal: Hip

Crural: Leg

Digital: Fingers or toes

Femoral: Thigh

Fibular (peroneal): Side of the leg

Frontal: Forehead

Hallux: Great toe

Inguinal: Groin area

Mammary: Breast region

Manus: Hand

Mental: Chin

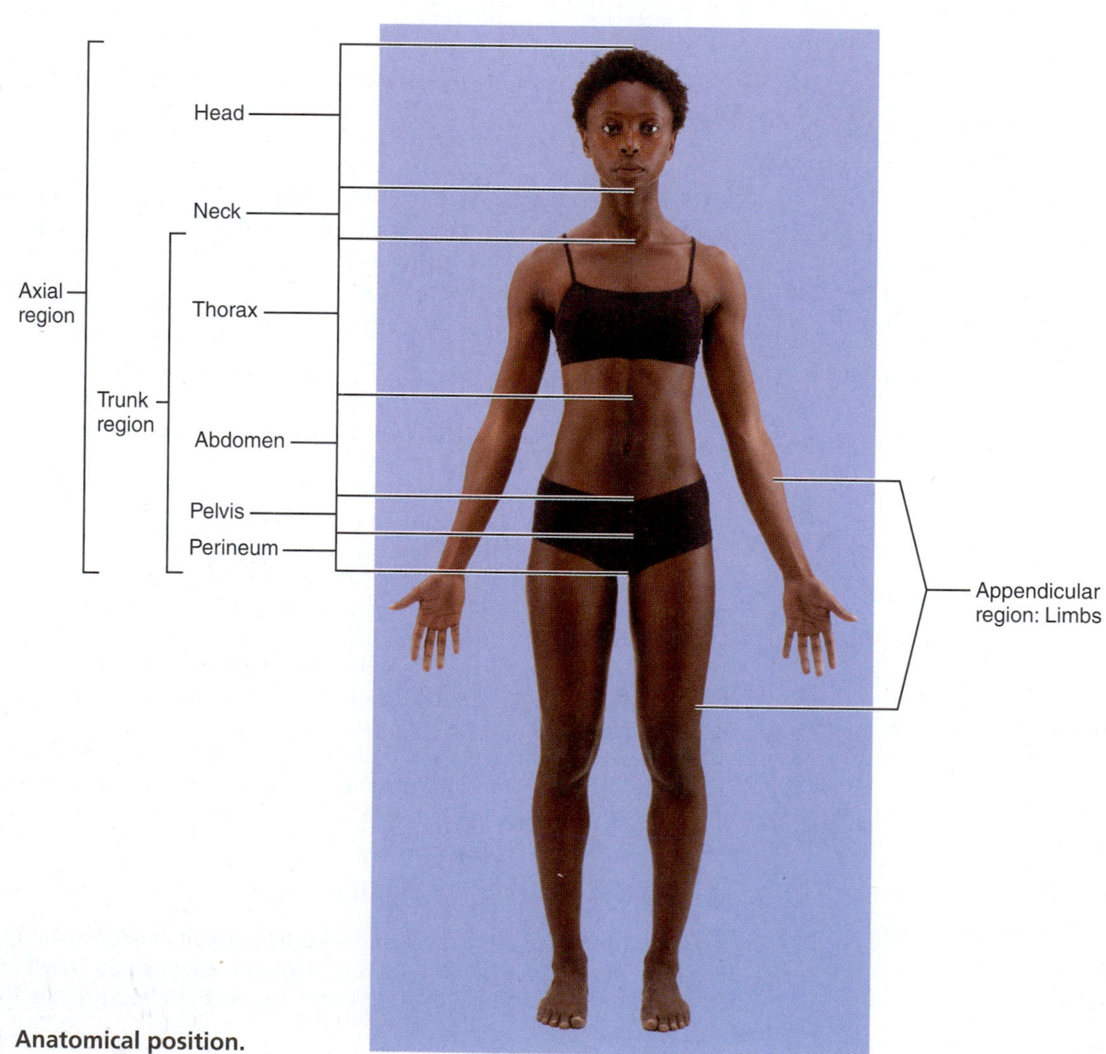

Figure 1.1 Anatomical position.

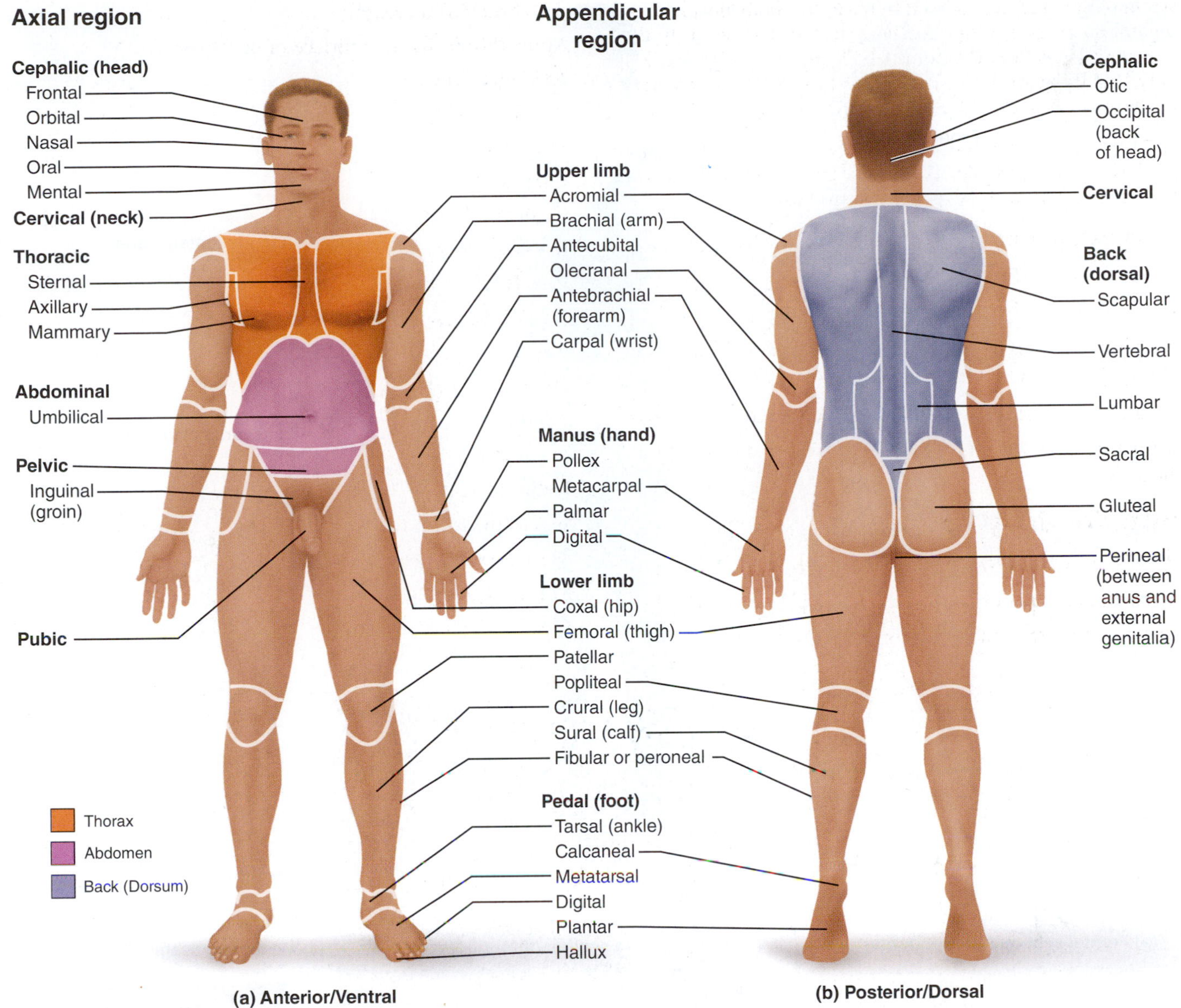

Figure 1.2 Surface anatomy. (a) Anatomical position. **(b)** Heels are raised to illustrate the plantar surface of the foot.

Nasal: Nose
Oral: Mouth
Orbital: Bony eye socket (orbit)
Palmar: Palm of the hand
Patellar: Anterior knee (kneecap) region
Pedal: Foot
Pelvic: Pelvis region
Pollex: Thumb
Pubic: Genital region
Sternal: Region of the breastbone
Tarsal: Ankle
Thoracic: Chest
Umbilical: Navel

Posterior Body Landmarks

Note the following body surface regions (Figure 1.2b):

Acromial: Point of the shoulder
Brachial: Arm
Calcaneal: Heel of the foot
Cephalic: Head
Dorsal: Back
Femoral: Thigh
Gluteal: Buttocks or rump
Lumbar: Area of the back between the ribs and hips; the loin
Manus: Hand
Occipital: Posterior aspect of the head or base of the skull
Olecranal: Posterior aspect of the elbow

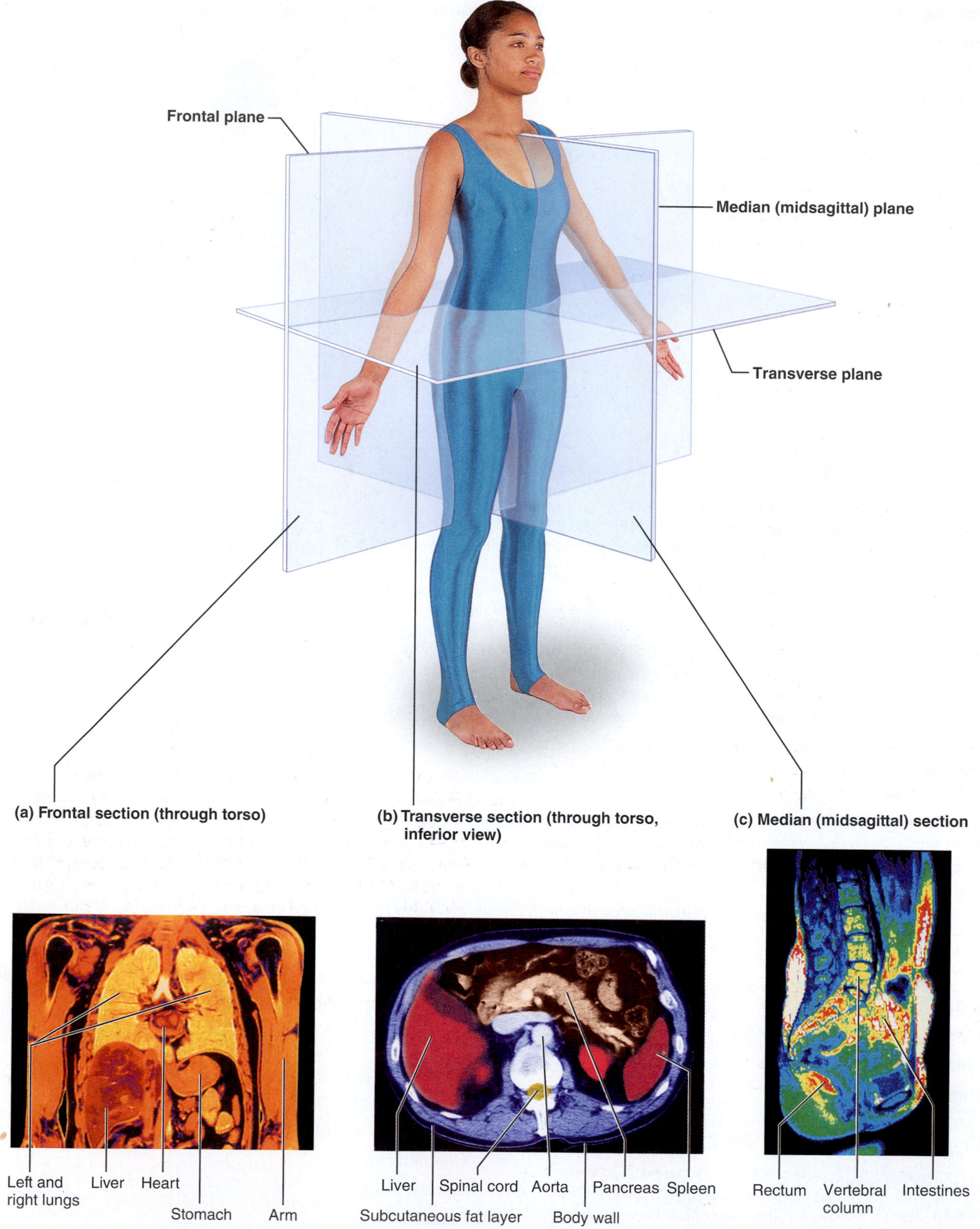

Figure 1.3 Planes of the body, with corresponding magnetic resonance imaging (MRI) scans.

Otic: Ear

Pedal: Foot

Perineal: Region between the anus and external genitalia

Plantar: Sole of the foot

Popliteal: Back of the knee

Sacral: Region between the hips (overlying the sacrum)

Scapular: Scapula or shoulder blade area

Sural: Calf or posterior surface of the leg

Vertebral: Area of the spinal column

ACTIVITY 1

Locating Body Regions

Before continuing, locate the anterior and posterior body landmarks on yourself, your lab partner, and a human torso model. ■

Body Planes and Sections

The body is three-dimensional, and to observe its internal structures, it is often helpful and necessary to make use of a **section,** or cut. When the section is made through the body wall or through an organ, it is made along an imaginary surface or line called a **plane.** Anatomists commonly refer to three planes **(Figure 1.3)**, or sections, that lie at right angles to one another.

Sagittal plane: A plane that runs longitudinally and divides the body into right and left parts is referred to as a sagittal plane. If it divides the body into equal parts, right down the midline of the body, it is called a **median,** or **midsagittal, plane.**

Frontal plane: Sometimes called a **coronal plane,** the frontal plane is a longitudinal plane that divides the body (or an organ) into anterior and posterior parts.

Transverse plane: A transverse plane runs horizontally, dividing the body into superior and inferior parts. When organs are sectioned along the transverse plane, the sections are commonly called **cross sections.**

On microscope slides, the abbreviation for a longitudinal section (sagittal or frontal) is l.s. Cross sections are abbreviated x.s. or c.s.

A sagittal or frontal plane section of any nonspherical object, be it a banana or a body organ, provides quite a different view from a transverse section **(Figure 1.4)**.

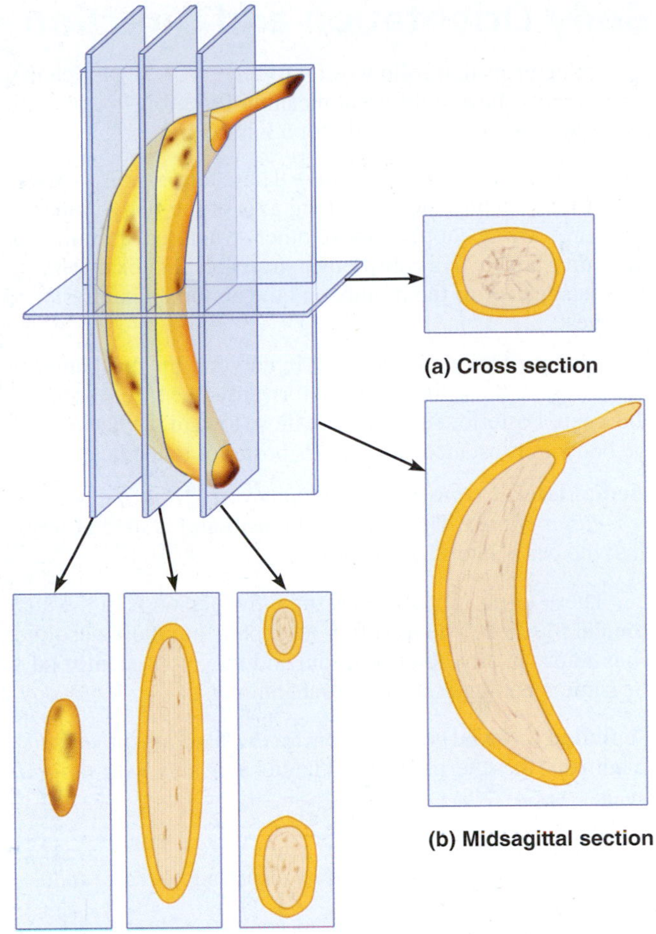

Figure 1.4 Objects can look odd when viewed in section. This banana has been sectioned in three different planes **(a–c)**, and only in one of these planes **(b)** is it easily recognized as a banana. To recognize human organs in section, one must anticipate how the organs will look when cut that way. If one cannot recognize a sectioned organ, it is possible to reconstruct its shape from a series of successive cuts, as from the three serial sections in **(c)**.

ACTIVITY 2

Observing Sectioned Specimens

1. Go to the demonstration area and observe the transversely and longitudinally cut organ specimens (kidneys). Pay close attention to the different structural details in the samples because you will need to draw these views in the Review Sheet at the end of this exercise.

2. After completing instruction 1, obtain a gelatin-spaghetti mold and a scalpel, and bring them to your laboratory bench. (Essentially, this is just cooked spaghetti added to warm gelatin, which is then allowed to gel.)

3. Cut through the gelatin-spaghetti mold along any plane, and examine the cut surfaces. You should see spaghetti strands that have been cut transversely (x.s.), some cut longitudinally, and some cut obliquely.

4. Draw the appearance of each of these spaghetti sections below, and verify the accuracy of your section identifications with your instructor.

Transverse cut Longitudinal cut Oblique cut ■

Body Orientation and Direction

Study the terms that follow (refer to **Figure 1.5**). Notice that certain terms have a different meaning for a four-legged animal (quadruped) than they do for a human (biped).

Superior/inferior *(above/below):* These terms refer to placement of a structure along the long axis of the body. Superior structures always appear above other structures, and inferior structures are always below other structures. For example, the nose is superior to the mouth, and the abdomen is inferior to the chest.

Anterior/posterior *(front/back):* In humans, the most anterior structures are those that are most forward—the face, chest, and abdomen. Posterior structures are those toward the backside of the body. For instance, the spine is posterior to the heart.

Medial/lateral *(toward the midline/away from the midline or median plane):* The sternum (breastbone) is medial to the ribs; the ear is lateral to the nose.

These terms of position assume the person is in the anatomical position. The next four term pairs are more absolute. They apply in any body position, and they consistently have the same meaning in all vertebrate animals.

Cephalad (cranial)/caudal *(toward the head/toward the tail):* In humans, these terms are used interchangeably with *superior* and *inferior,* but in four-legged animals they are the same as *anterior* and *posterior,* respectively.

Dorsal/ventral *(backside/belly side):* These terms are used chiefly in discussing the comparative anatomy of animals, assuming the animal is standing. *Dorsum* is a Latin word meaning "back." Thus, *dorsal* refers to the animal's back or the *back*side of any other structures; for example, the posterior surface of the human leg is its dorsal surface. The term *ventral* derives from the Latin term *venter,* meaning "belly," and always refers to the belly side of animals. In humans, the terms *ventral* and *dorsal* are used interchangeably with the terms *anterior* and *posterior,* but in four-legged animals, *ventral* and *dorsal* are the same as *inferior* and *superior,* respectively.

Proximal/distal *(nearer the trunk or attached end/farther from the trunk or point of attachment):* These terms are used primarily to locate various areas of the body limbs. For example, the fingers are distal to the elbow; the knee is proximal to the toes. However, these terms may also be used to indicate regions (closer to or farther from the head) of internal tubular organs.

Superficial (external)/deep (internal) *(toward or at the body surface/away from the body surface):* These terms locate body organs according to their relative closeness to the body surface. For example, the skin is superficial to the skeletal muscles, and the lungs are deep to the rib cage.

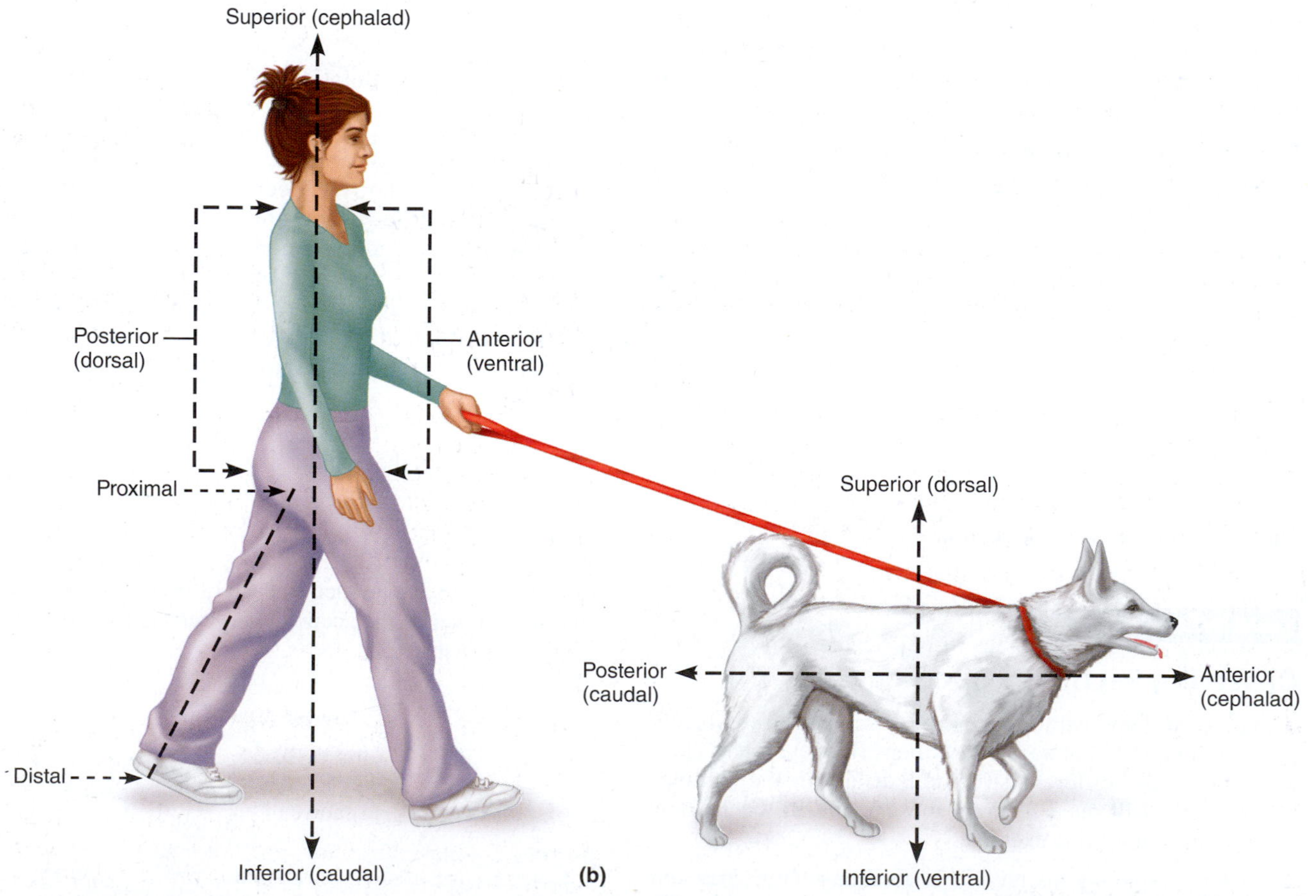

Figure 1.5 Anatomical terminology describing body orientation and direction. **(a)** With reference to a human. **(b)** With reference to a four-legged animal.

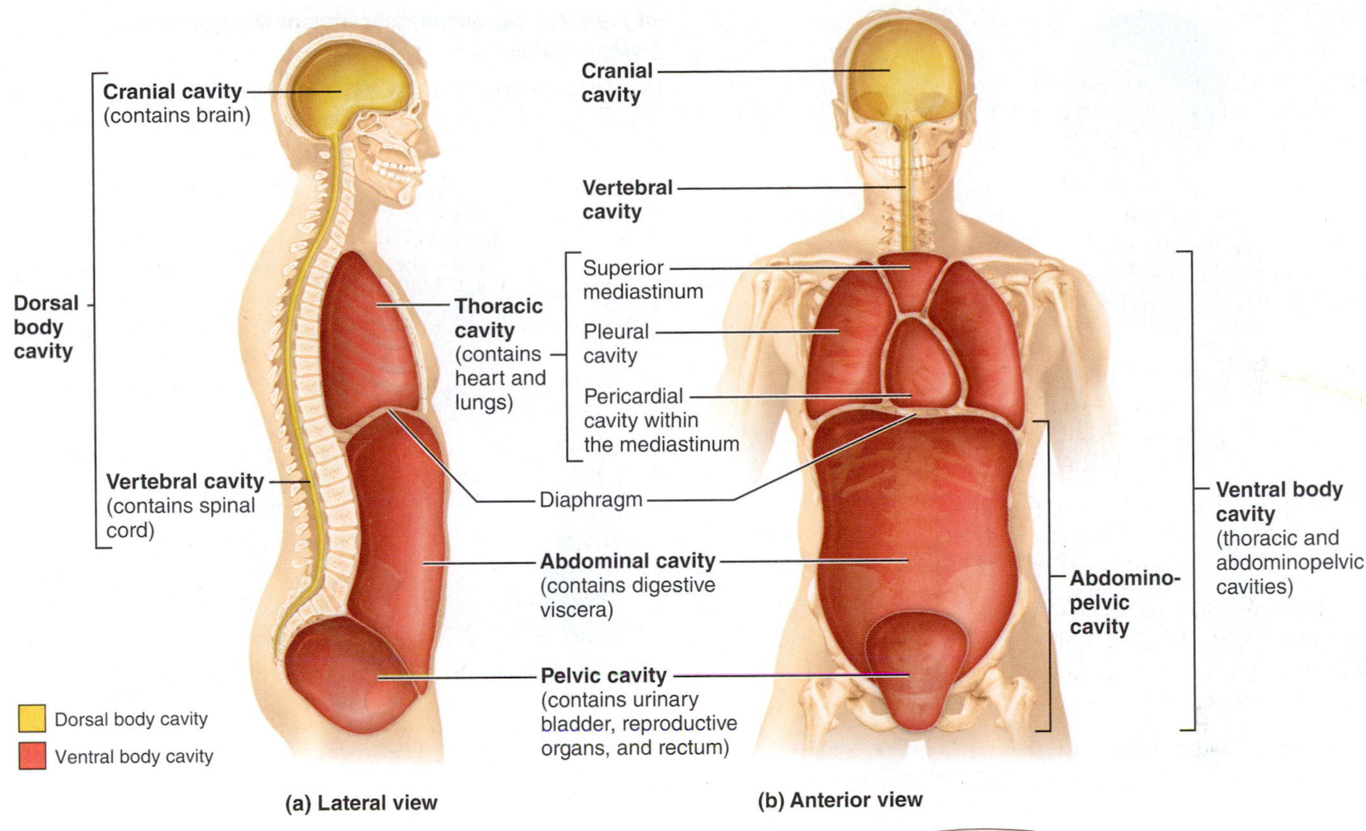

Figure 1.6 **Dorsal and ventral body cavities and their subdivisions.**

ACTIVITY 3

Practicing Using Correct Anatomical Terminology

Before continuing, use a human torso model, a human skeleton, or your own body to specify the relationship between the following structures when the body is in the anatomical position.

1. The wrist is proximal to the hand.
2. The trachea (windpipe) is anterior to the spine.
3. The brain is superior to the spinal cord.
4. The kidneys are posterior to the liver.
5. The nose is medial (or anterior) to the cheekbones.
6. The thumb is lateral to the ring finger.
7. The thorax is superior to the abdomen.
8. The skin is superficial to the skeleton. ■

Body Cavities

The axial portion of the body has two large cavities that provide different degrees of protection to the organs within them **(Figure 1.6)**.

Dorsal Body Cavity

The dorsal body cavity can be subdivided into two cavities. The **cranial cavity,** within the rigid skull, contains the brain. The **vertebral** (or **spinal**) **cavity,** which is within the bony vertebral column, protects the delicate spinal cord. Because the spinal cord is a continuation of the brain, these cavities are continuous with each other.

Ventral Body Cavity

Like the dorsal cavity, the ventral body cavity is subdivided. The superior **thoracic cavity** is separated from the rest of the ventral cavity by the dome-shaped diaphragm. The heart and lungs, located in the thoracic cavity, are protected by the bony rib cage. The cavity inferior to the diaphragm is often referred to as the **abdominopelvic cavity.** Although there is no further physical separation of the ventral cavity, some describe the abdominopelvic cavity as two areas, a superior **abdominal cavity** (the area that houses the stomach, intestines, liver, and other organs) and an inferior **pelvic cavity** (the region that is partially enclosed by the bony pelvis and contains the reproductive organs, bladder, and rectum). The abdominal and pelvic cavities are not continuous with each other in a straight plane; the pelvic cavity is tipped forward (Figure 1.6).

Serous Membranes of the Ventral Body Cavity

The walls of the ventral body cavity and the outer surfaces of the organs it contains are covered with an exceedingly thin, double-layered membrane called the **serosa,** or **serous membrane.** The part of the membrane lining the cavity walls is referred to as the **parietal serosa,** and it is continuous with a similar membrane, the **visceral serosa,** covering the external surface of the organs within the cavity. These membranes produce a thin lubricating fluid that allows the visceral organs to

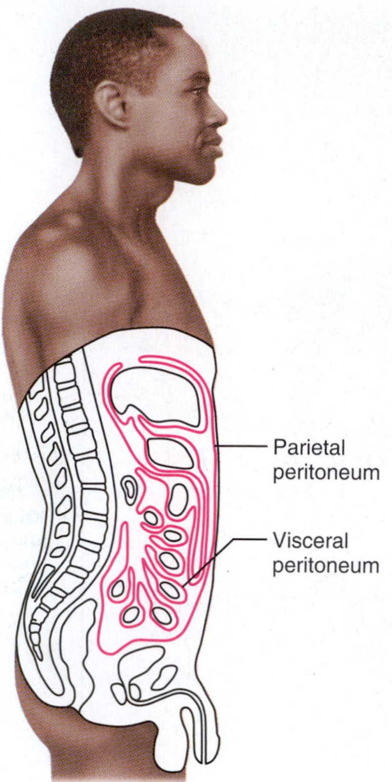

Figure 1.7 Serous membranes of the ventral body cavities.

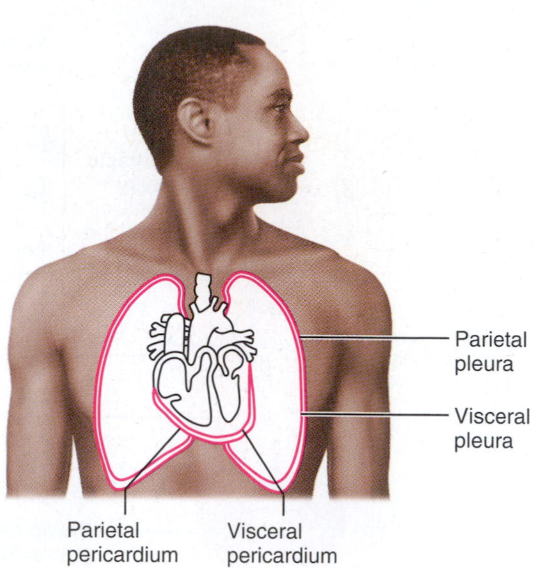

slide over one another or to rub against the body wall without friction. Serous membranes also compartmentalize the various organs. This helps prevent infection in one organ from spreading to others.

The specific names of the serous membranes depend on the structures they surround. The serosa lining the abdominal cavity and covering its organs is the **peritoneum,** that enclosing the lungs is the **pleura,** and that around the heart is the **pericardium** (Figure 1.7).

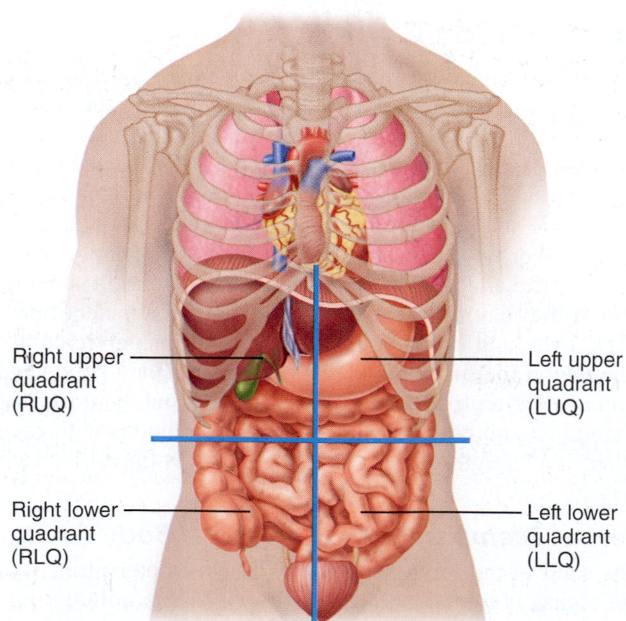

Figure 1.8 Abdominopelvic quadrants. Superficial organs are shown in each quadrant.

Abdominopelvic Quadrants and Regions

Because the abdominopelvic cavity is quite large and contains many organs, it is helpful to divide it up into smaller areas for discussion or study.

Most physicians and nurses use a scheme that divides the abdominal surface and the abdominopelvic cavity into four approximately equal regions called **quadrants.** These quadrants are named according to their relative position—that is, *right upper quadrant, right lower quadrant, left upper quadrant,* and *left lower quadrant* (Figure 1.8). (Note that the terms *left* and *right* refer to the left and right side of the body in Figure 1.8, not the left and right side of the art on the page). The left and right of the body viewed are referred to as **anatomical left** and **right.**

ACTIVITY 4

Identifying Organs in the Abdominopelvic Cavity

Examine the torso model to respond to the following directions and questions.

Name two organs found in the left upper quadrant.

______________________ and ______________________

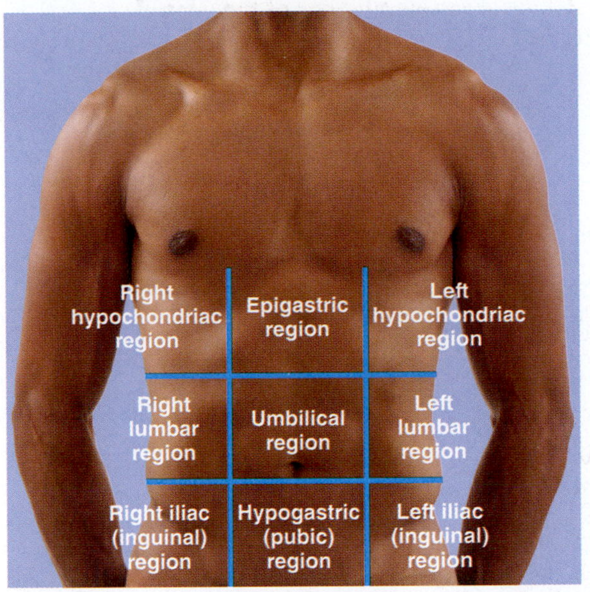

(a)

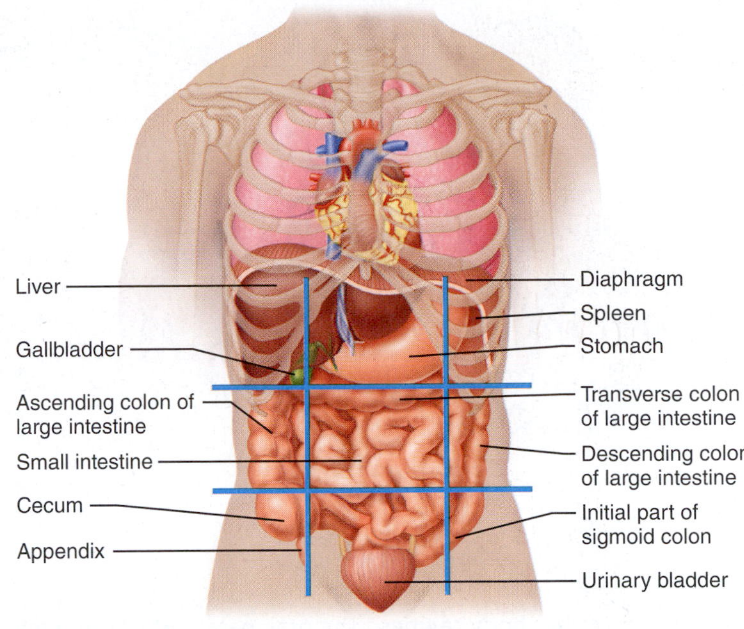

(b)

Figure 1.9 Abdominopelvic regions. Nine regions are delineated by four planes. **(a)** The superior horizontal plane is just inferior to the ribs; the inferior horizontal plane is at the superior aspect of the hip bones. The vertical planes are just medial to the nipples. **(b)** Superficial organs are shown in each region.

Name two organs found in the right lower quadrant.

____________________ and ____________________

Which organ (Figure 1.8) is divided into identical halves by the median plane? ________________

A different scheme commonly used by anatomists divides the abdominal surface and abdominopelvic cavity into nine separate regions by four planes **(Figure 1.9a)**. Although the names of these nine regions are unfamiliar to you now, with a little patience and study they will become easier to remember. As you read through the descriptions of these nine regions and locate them, also note the organs the regions contain (Figure 1.9b).

Umbilical region: The centermost region, which includes the umbilicus (navel).

Epigastric region: Immediately superior to the umbilical region; overlies most of the stomach.

Hypogastric (pubic) region: Immediately inferior to the umbilical region; encompasses the pubic area.

Iliac or **inguinal regions:** Lateral to the hypogastric region and overlying the superior parts of the hip bones.

Lumbar regions: Between the ribs and the flaring portions of the hip bones; lateral to the umbilical region.

Hypochondriac regions: Flanking the epigastric region laterally and overlying the lower ribs.

ACTIVITY 5

Locating Abdominal Surface Regions

Locate the regions of the abdominal surface on a human torso model and on yourself before continuing.

Other Body Cavities

Besides the large, closed body cavities, there are several types of smaller body cavities **(Figure 1.10)**. Many of these are in the head, and most open to the body exterior.

Oral cavity: The oral cavity, commonly called the *mouth*, contains the tongue and teeth. It is continuous with the rest of the digestive tube, which opens to the exterior at the anus.

Nasal cavity: Located within and posterior to the nose, the nasal cavity is part of the passages of the respiratory system.

Orbital cavities: The orbital cavities (orbits) in the skull house the eyes and present them in an anterior position.

Middle ear cavities: Each middle ear cavity lies just medial to an eardrum and is carved into the bony skull. These cavities contain tiny bones that transmit sound vibrations to the hearing receptor in the inner ears.

Synovial cavities: Synovial cavities are joint cavities—they are enclosed within fibrous capsules that surround the freely movable joints of the body, such as those between the vertebrae and the knee and hip joints. Like the serous membranes of the ventral body cavity, membranes lining the synovial cavities secrete a lubricating fluid that reduces friction as the enclosed structures move across one another.

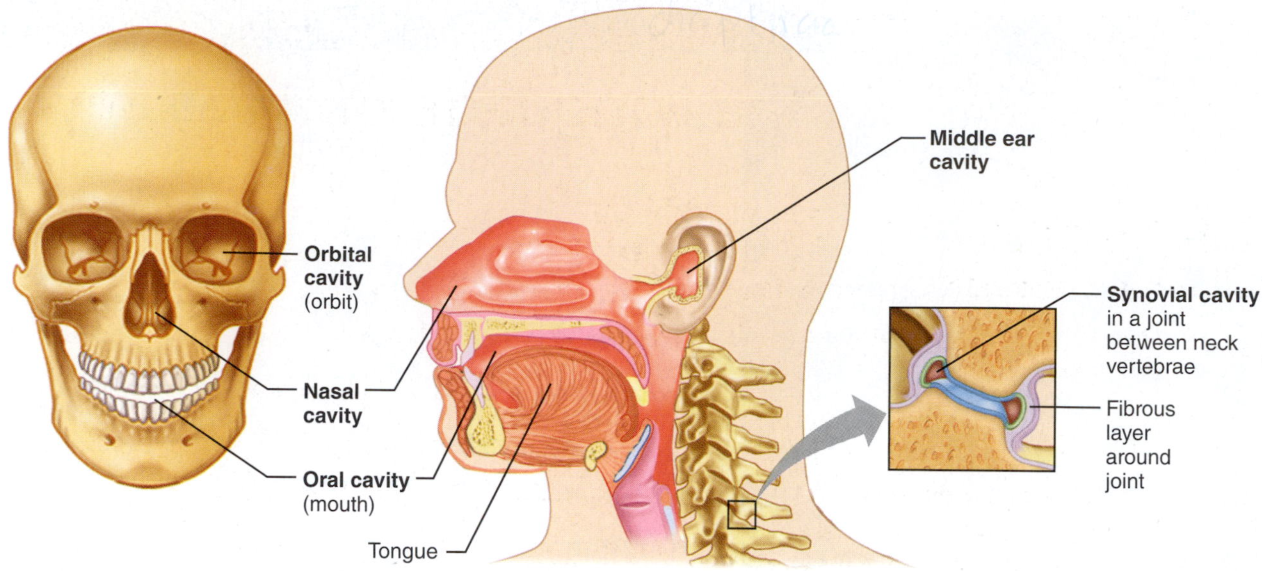

Figure 1.10 Other body cavities. The oral, nasal, orbital, and middle ear cavities are located in the head and open to the body exterior. Synovial cavities are found in joints between many bones, such as the vertebrae of the spine, and at the knee, shoulder, and hip.

GROUP CHALLENGE

The Language of Anatomy

Working in small groups, complete the tasks described below. Work together, but refrain from using a figure or other reference to answer the questions. As usual, assume that the human body is in the anatomical position.

1. Arrange the following terms from superior to inferior: cervical, coxal, crural, femoral, lumbar, mental, nasal, plantar, sternal, and tarsal. ____________

____

____

____

2. Arrange the following terms from proximal to distal: antebrachial, antecubital, brachial, carpal, digital, and palmar.

____

____

____

3. Arrange the following terms from medial to lateral: acromial, axillary, buccal, otic, pollex, and umbilical.

____

____

____

4. Arrange the following terms from distal to proximal: calcaneal, femoral, hallux, plantar, popliteal, and sural.

____

____

____

5. Name a plane that you could use to section a four-legged chair and still be able to sit in the chair without falling over. ____________

6. Name the abdominopelvic region that is both medial and inferior to the right lumbar region. ____________

7. Name the type of inflammation (think "-itis") that is typically accompanied by pain in the lower right quadrant.

____

Name ______________________

Lab Time/Date ______________________

EXERCISE 1

REVIEW SHEET

The Language of Anatomy

Surface Anatomy

1. Match each of the numbered descriptions with the related term in the key.

Key: a. buccal b. calcaneal c. cephalic d. digital e. patellar f. scapular

buccal 1. cheek

digital 2. the fingers

scapular 3. shoulder blade region

patellar 4. anterior aspect of knee

calcaneal 5. heel of foot

cephalic 6. the head

2. Indicate the following body areas on the accompanying diagram by placing the correct key letter at the end of each line.

Key:

a. abdominal
b. antecubital
c. brachial
d. cervical
e. crural
f. femoral
g. fibular
h. gluteal
i. lumbar
j. occipital
k. oral
l. popliteal
m. pubic
n. sural
o. thoracic
p. umbilical

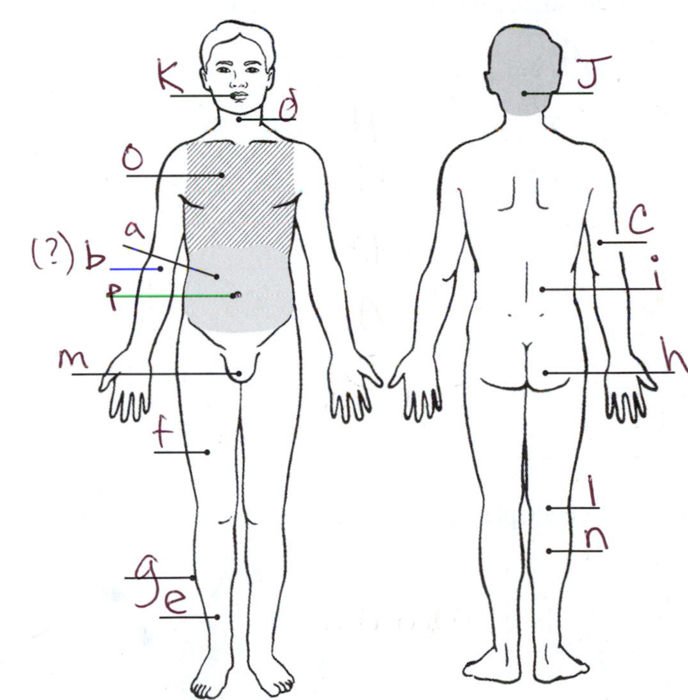

3. Classify each of the terms in the key of question 2 above into one of the large body regions indicated below. Insert the appropriate key letters on the answer blanks.

______________________ 1. appendicular

______________________ 2. axial

Body Orientation, Direction, Planes, and Sections

4. Describe completely the standard human anatomical position. ________________________________

____

5. Define *section*. ____

6. Several incomplete statements are listed below. Correctly complete each statement by choosing the appropriate anatomical term from the key. Record the key letters and/or terms on the correspondingly numbered blanks below. Some terms are used more than once.

Key:			
a. anterior	d. inferior	g. posterior	j. superior
b. distal	e. lateral	h. proximal	k. transverse
c. frontal	f. medial	i. sagittal	

In the anatomical position, the face and palms are on the __1__ body surface; the buttocks and shoulder blades are on the __2__ body surface; and the top of the head is the most __3__ part of the body. The ears are __4__ and __5__ to the shoulders and __6__ to the nose. The heart is __7__ to the vertebral column (spine) and __8__ to the lungs. The elbow is __9__ to the fingers but __10__ to the shoulder. The abdominopelvic cavity is __11__ to the thoracic cavity and __12__ to the spinal cavity. In humans, the dorsal surface can also be called the __13__ surface; however, in quadruped animals, the dorsal surface is the __14__ surface.

If an incision cuts the heart into right and left parts, the section is a __15__ section; but if the heart is cut so that superior and inferior portions result, the section is a __16__ section. You are told to cut a dissection animal along two planes so that both kidneys are observable in each section. The two sections that will always meet this requirement are the __17__ and __18__ sections. A section that demonstrates the continuity between the spinal and cranial cavities is a __19__ section.

1. anterior A
2. G
3. J
4. H and
5. F
6. E
7. A
8. F
9. H
10. B
11. D
12. A
13. D
14. ?
15. I
16. K
17. I
18. C
19. I ?

7. Correctly identify each of the body planes by inserting the appropriate term for each on the answer line below the drawing.

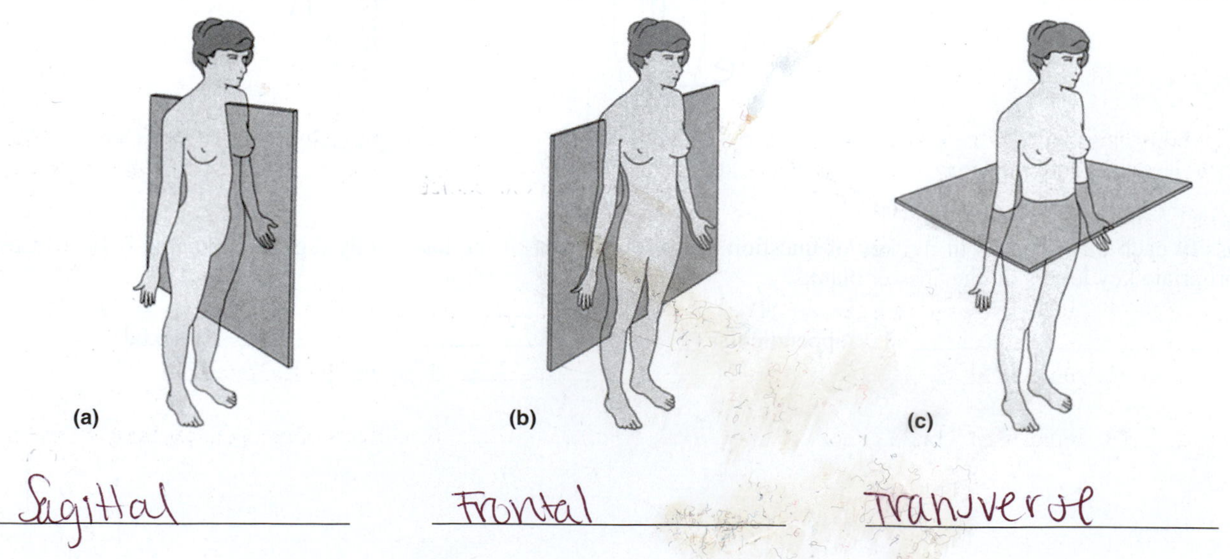

8. Draw a kidney as it appears when sectioned in each of the three different planes.

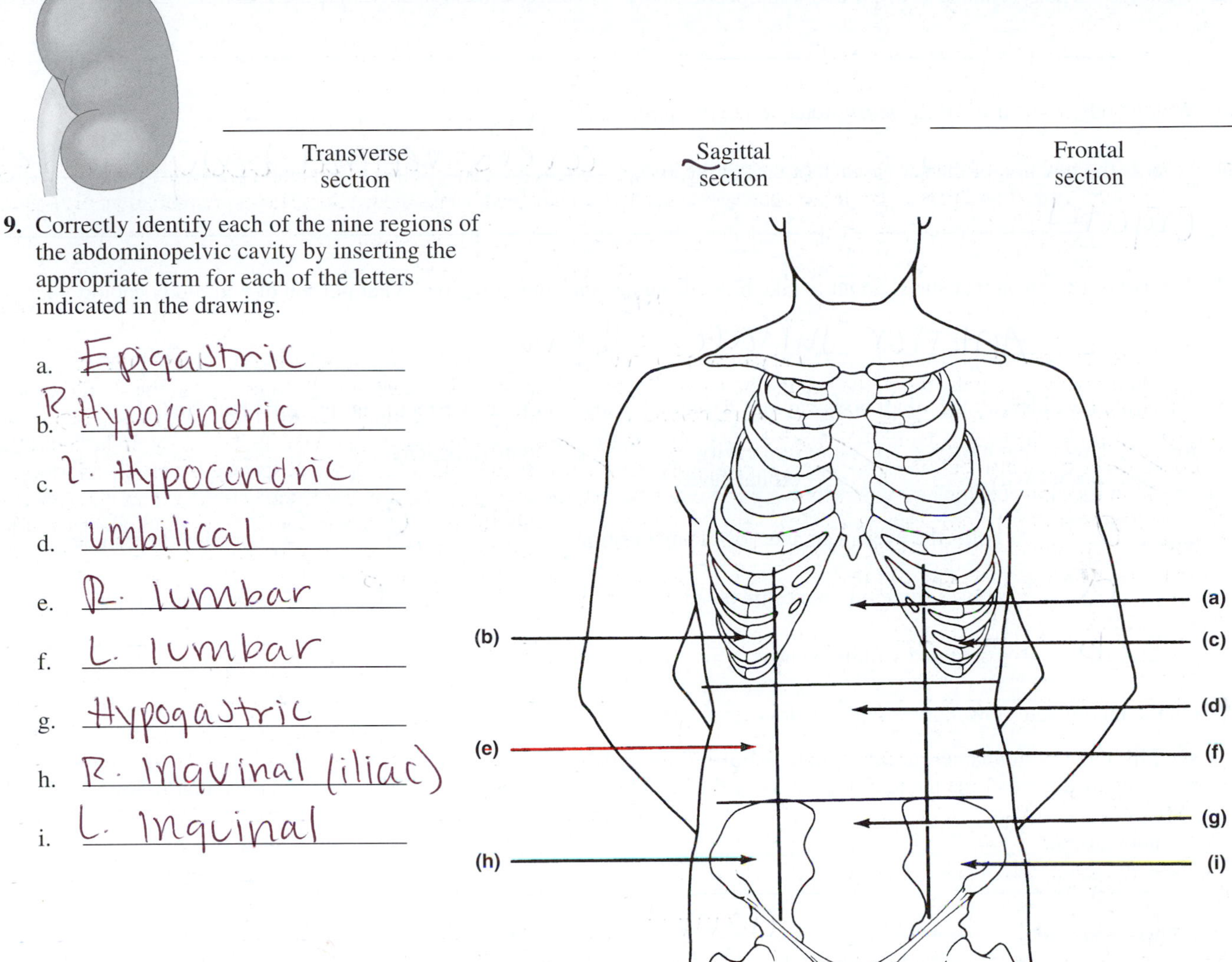

9. Correctly identify each of the nine regions of the abdominopelvic cavity by inserting the appropriate term for each of the letters indicated in the drawing.

a. Epigastric
b. R. Hypocondric
c. L. Hypocondric
d. Umbilical
e. R. lumbar
f. L. lumbar
g. Hypogastric
h. R. Inguinal (iliac)
i. L. Inguinal

Body Cavities

10. Which body cavity would have to be opened for the following types of surgeries or procedures? (Use the key to find the correct choice, and write the letter in same-numbered blank. More than one choice may apply for some surgeries/procedures.)

Key: a. abdominopelvic c. dorsal e. thoracic
b. cranial d. spinal f. ventral

e 1. surgery to remove a cancerous lung lobe
a 2. removal of the uterus, or womb
b 3. removal of a brain tumor
a (?) 4. appendectomy
a, f 5. stomach ulcer operation
(?) 6. delivery of preoperative "saddle" anesthesia

11. Name the muscle that subdivides the ventral body cavity. diaphragm

12. What are the bony landmarks of the abdominopelvic cavity? Abdominal, pelvic

13. Which body cavity affords the *least* protection to its internal structures?

14. What is the function of the serous membranes of the body? covers ventral body cavities organs

15. A nurse informs you that she is about to take blood from the antecubital region. What portion of your body should you present to her? Anterior surface elbow

16. Using the key, identify the small body cavities described below. Write the correct letter in each blank line.

Key: a. middle ear cavity | b. nasal cavity | c. oral cavity | d. orbital cavity | e. synovial cavity

d 1. holds the eyes in an anterior-facing position

a 2. houses three tiny bones involved in hearing

b 3. contained within the nose

c 4. contains the tongue

e 5. surrounds a joint

17. On the incomplete flowchart provided below:

- Fill in the cavity names as appropriate to boxes 3 through 8.
- Then, using either the name of the cavity or the box numbers, identify the descriptions in the list that follows.

Body cavities

1 Dorsal body cavity → 3 cranial cavity (superior); 4 vertebral cavity (inferior)

2 Ventral body cavity → 5 Thoracic cavity (superior); 6 Abdominopelvic cavity (inferior)

6 → 7 Abdominal cavity (superior); 8 Pelvic cavity (inferior)

1 a. contained within the skull and vertebral column

8 b. houses female reproductive organs

3 c. the most protective body cavity

6 d. its name means "belly"

5 e. contains the heart

7 f. contains the small intestine

2, 5 g. bounded by the ribs

? h. its walls are muscular

EXERCISE 2

Organ Systems Overview

MATERIALS

- ☐ Freshly killed or preserved rat (predissected by instructor as a demonstration; or for student dissection, one rat for every two to four students) or predissected human cadaver
- ☐ Dissection trays
- ☐ Twine or large dissecting pins
- ☐ Scissors
- ☐ Probes
- ☐ Forceps
- ☐ Disposable gloves
- ☐ Human torso model (dissectible)

OBJECTIVES

1. Name the human organ systems, and indicate the major functions of each system.
2. List several major organs of each system, and identify them in a dissected rat, human cadaver or cadaver image, or dissectible human torso model.
3. Name the correct organ system for each organ studied in the laboratory.

PRE-LAB QUIZ

1. Name the structural and functional unit of all living things. ______________
2. The small intestine is an example of a(n) ______________, because it is composed of two or more tissue types that perform a particular function for the body.
   a. epithelial tissue
   b. muscle tissue
   c. organ
   d. organ system
3. The ______________________ system is responsible for maintaining homeostasis of the body via rapid communication.
4. The kidneys are part of the ______________________ system.
5. The thin muscle that separates the thoracic and abdominal cavities is the ______________________.

The basic unit or building block of all living things is the **cell.** Cells fall into four different categories according to their structures and functions. Each of these corresponds to one of the four tissue types: epithelial, muscular, nervous, and connective. A **tissue** is a group of cells that are similar in structure and function. An **organ** is a structure composed of two or more tissue types that performs a specific function for the body. For example, the small intestine, which digests and absorbs nutrients, is made up of all four tissue types.

An **organ system** is a group of organs that act together to perform a particular body function. For example, the organs of the digestive system work together to break down foods and absorb the end products into the bloodstream to provide nutrients and fuel for all the body's cells. In all, there are 11 organ systems **(Table 2.1)**. The lymphatic system also encompasses a *functional system* called the immune system, which is composed of an army of mobile *cells* that act to protect the body from foreign substances.

Read through this summary of the body's organ systems before beginning your rat dissection or examination of the predissected human cadaver. If a human cadaver is not available, photographs provided in this exercise (Figures 2.3 through 2.6) will serve as a partial replacement.

Table 2.1 Overview of Organ Systems of the Body

Organ system	Major component organs	Function
Integumentary (Skin)	Epidermal and dermal regions; cutaneous sense organs and glands	• Protects deeper organs from mechanical, chemical, and bacterial injury, and from drying out • Excretes salts and urea • Aids in regulation of body temperature • Produces vitamin D
Skeletal	Bones, cartilages, tendons, ligaments, and joints	• Supports the body and protects internal organs • Provides levers for muscular action • Cavities provide a site for blood cell formation
Muscular	Muscles attached to the skeleton	• Primary function is to contract or shorten; in doing so, skeletal muscles allow locomotion (running, walking, etc.), grasping and manipulation of the environment, and facial expression • Generates heat
Nervous	Brain, spinal cord, nerves, and sensory receptors	• Allows body to detect changes in its internal and external environment and to respond to such information by activating appropriate muscles or glands • Helps maintain homeostasis of the body via rapid communication
Endocrine	Pituitary, thymus, thyroid, parathyroid, adrenal, and pineal glands; ovaries, testes, and pancreas	• Helps maintain body homeostasis, promotes growth and development; produces chemical messengers called hormones that travel in the blood to exert their effect(s) on various target organs of the body
Cardiovascular	Heart, blood vessels, and blood	• Primarily a transport system that carries blood containing oxygen, carbon dioxide, nutrients, wastes, ions, hormones, and other substances to and from the tissue cells where exchanges are made; blood is propelled through the blood vessels by the pumping action of the heart • Antibodies and other protein molecules in the blood protect the body
Lymphatic/immune	Lymphatic vessels, lymph nodes, spleen, thymus, tonsils, and scattered collection of lymphoid tissue	• Picks up fluid leaked from the blood vessels and returns it to the blood • Cleanses blood of pathogens and other debris • Houses lymphocytes that act via the immune response to protect the body from foreign substances
Respiratory	Nasal passages, pharynx, larynx, trachea, bronchi, and lungs	• Keeps the blood continuously supplied with oxygen while removing carbon dioxide • Contributes to the acid-base balance of the blood via its carbonic acid–bicarbonate buffer system
Digestive	Oral cavity, esophagus, stomach, small and large intestines, and accessory structures including teeth, salivary glands, liver, and pancreas	• Breaks down ingested foods to minute particles, which can be absorbed into the blood for delivery to the body cells • Undigested residue removed from the body as feces
Urinary	Kidneys, ureters, bladder, and urethra	• Rids the body of nitrogen-containing wastes, including urea, uric acid, and ammonia, which result from the breakdown of proteins and nucleic acids • Maintains water, electrolyte, and acid-base balance of blood
Reproductive	Male: testes, prostate, scrotum, penis, and duct system, which carries sperm to the body exterior	• Provides germ cells called sperm for producing offspring
	Female: ovaries, uterine tubes, uterus, mammary glands, and vagina	• Provides germ cells called eggs; the female uterus houses the developing fetus until birth; mammary glands provide nutrition for the infant

DISSECTION AND IDENTIFICATION

The Organ Systems of the Rat

Many of the external and internal structures of the rat are quite similar in structure and function to those of the human, so a study of the gross anatomy of the rat should help you understand our own physical structure. The following instructions include directions for dissecting and observing a rat. In addition, instructions for observing organs (Activity 4, "Examining the Ventral Body Cavity," page 18) also apply to superficial observations of a previously dissected human cadaver. The general instructions for observing external structures also apply to human cadaver observations. The photographs (Figures 2.3 through 2.6) will provide visual aids.

Note that four of the organ systems listed in the table (Table 2.1) (integumentary, skeletal, muscular, and nervous) will not be studied at this time because they require microscopic study or more detailed dissection. ■

ACTIVITY 1

Observing External Structures

1. If your instructor has provided a predissected rat, go to the demonstration area to make your observations. Alternatively, if you and/or members of your group will be dissecting the specimen, obtain a preserved or freshly killed rat, a dissecting tray, dissecting pins or twine, scissors, probe, forceps, and disposable gloves. Bring these items to your laboratory bench.

If a predissected human cadaver is available, obtain a probe, forceps, and disposable gloves before going to the demonstration area.

2. Don the gloves before beginning your observations. This precaution is particularly important when handling freshly killed animals, which may harbor internal parasites.

3. Observe the major divisions of the body—head, trunk, and extremities. If you are examining a rat, compare these divisions to those of humans. ■

ACTIVITY 2

Examining the Oral Cavity

Examine the structures of the oral cavity. Identify the teeth and tongue. Observe the extent of the hard palate (the portion underlain by bone) and the soft palate (immediately posterior to the hard palate, with no bony support). Notice that the posterior end of the oral cavity leads into the throat, or pharynx, a passageway used by both the digestive and respiratory systems. ■

ACTIVITY 3

Opening the Ventral Body Cavity

1. Pin the animal to the wax of the dissecting tray by placing its dorsal side down and securing its extremities to the wax with large dissecting pins **(Figure 2.1a)**.

If the dissecting tray is not waxed, you will need to secure the animal with twine as follows. (Some may prefer this method in any case.) Obtain the roll of twine. Make a loop knot around one upper limb, pass the twine under the tray, and secure the opposing limb. Repeat for the lower extremities.

2. Lift the abdominal skin with a forceps, and cut through it with the scissors (Figure 2.1b). Close the scissor blades, and insert them flat under the cut skin. Moving in a cephalad direction, open and close the blades to loosen the skin from the underlying connective tissue and muscle. Now cut the skin along the body midline, from the pubic region to the lower jaw (Figure 2.1c, page 18). Finally, make a lateral cut about halfway down the ventral surface of each limb. Complete the job of freeing the skin with the scissor tips, and pin the flaps to the tray (Figure 2.1d). The underlying tissue that is now exposed is the skeletal musculature of the body wall and limbs. It allows voluntary body movement. Notice that the muscles are packaged in sheets of pearly white connective tissue (fascia), which protect the muscles and bind them together.

3. Carefully cut through the muscles of the abdominal wall in the pubic region, avoiding the underlying organs. Remember, to *dissect* means "to separate"—not mutilate! Now, hold and lift the muscle layer with a forceps and cut through the muscle layer from the pubic region to the bottom of the rib cage. Make two lateral cuts at the base of the rib

(a)

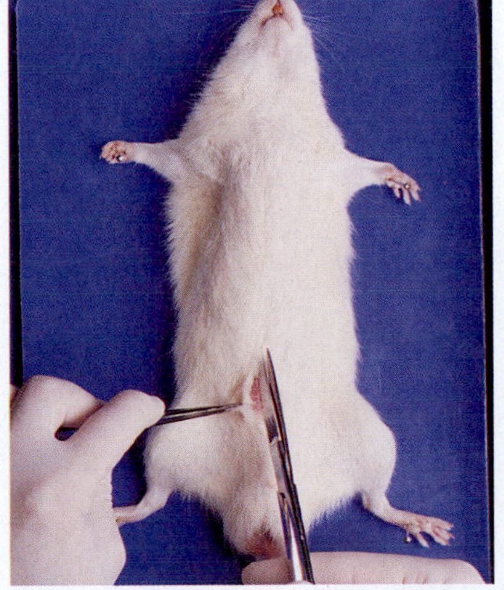

(b)

Figure 2.1 Rat dissection: Securing for dissection and the initial incision. (a) Securing the rat to the dissection tray with dissecting pins. **(b)** Using scissors to make the incision on the median line of the abdominal region.

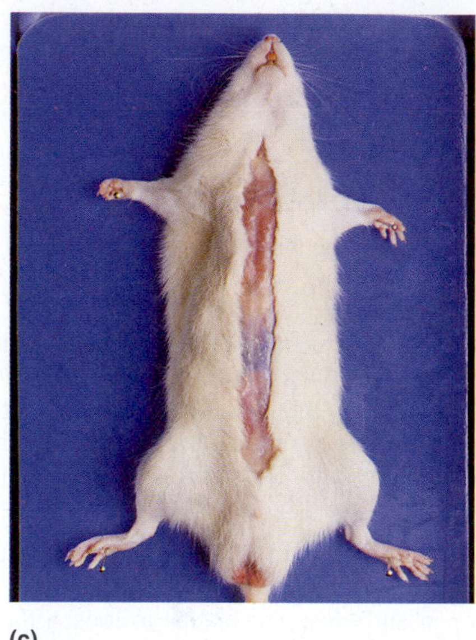

(c)

(d)

Figure 2.1 *(continued)* **Rat dissection: Securing for dissection and the initial incision. (c)** Completed incision from the pelvic region to the lower jaw. **(d)** Reflection (folding back) of the skin to expose the underlying muscles.

cage **(Figure 2.2)**. A thin membrane attached to the inferior boundary of the rib cage should be obvious; this is the **diaphragm,** which separates the thoracic and abdominal cavities. Cut the diaphragm where it attaches to the ventral ribs to loosen the rib cage. Cut through the rib cage on either side. You can now lift the ribs to view the contents of the thoracic cavity. Cut across the flap at the level of the neck, and remove it. ■

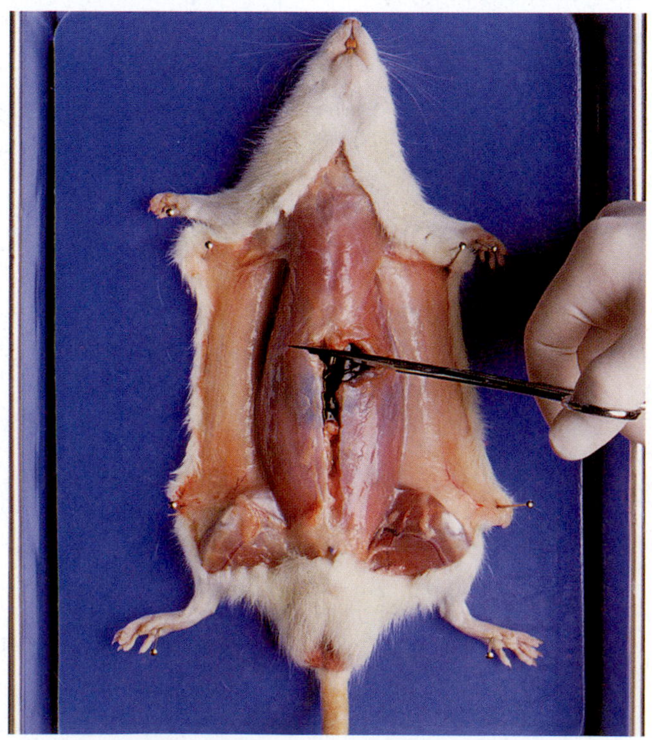

Figure 2.2 Rat dissection: Making lateral cuts at the base of the rib cage.

ACTIVITY 4

Examining the Ventral Body Cavity

1. Starting with the most superficial structures and working deeper, examine the structures of the thoracic cavity. (Refer to **Figure 2.3** as you work.) Choose the appropriate view depending on whether you are examining a rat (a) or a human cadaver (b).

Thymus: An irregular mass of glandular tissue overlying the heart (not illustrated in the human cadaver photograph).

With the probe, push the thymus to the side to view the heart.

Heart: Medial oval structure enclosed within the pericardium (serous membrane sac).

Lungs: Lateral to the heart on either side.

Now observe the throat region to identify the trachea.

Trachea: Tubelike “windpipe” running medially down the throat; part of the respiratory system.

Follow the trachea into the thoracic cavity; notice where it divides into two branches. These are the bronchi.

Bronchi: Two passageways that plunge laterally into the tissue of the two lungs.

To expose the esophagus, push the trachea to one side.

Esophagus: A food chute; the part of the digestive system that transports food from the pharynx (throat) to the stomach.

Diaphragm: A thin muscle attached to the inferior boundary of the rib cage; separates the thoracic and abdominal cavities.

Follow the esophagus through the diaphragm to its junction with the stomach.

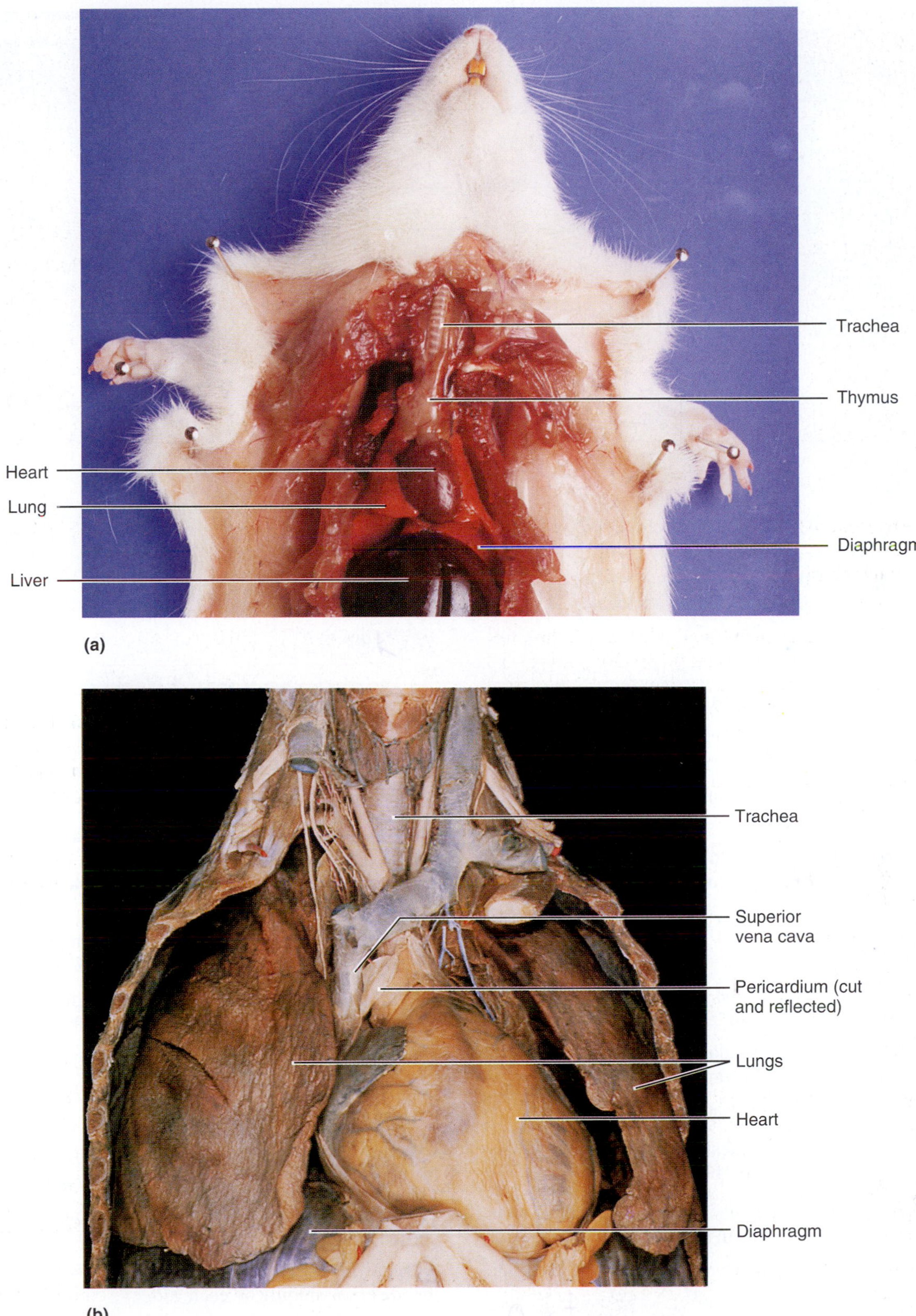

Figure 2.3 Superficial organs of the thoracic cavity. (a) Dissected rat. **(b)** Human cadaver.

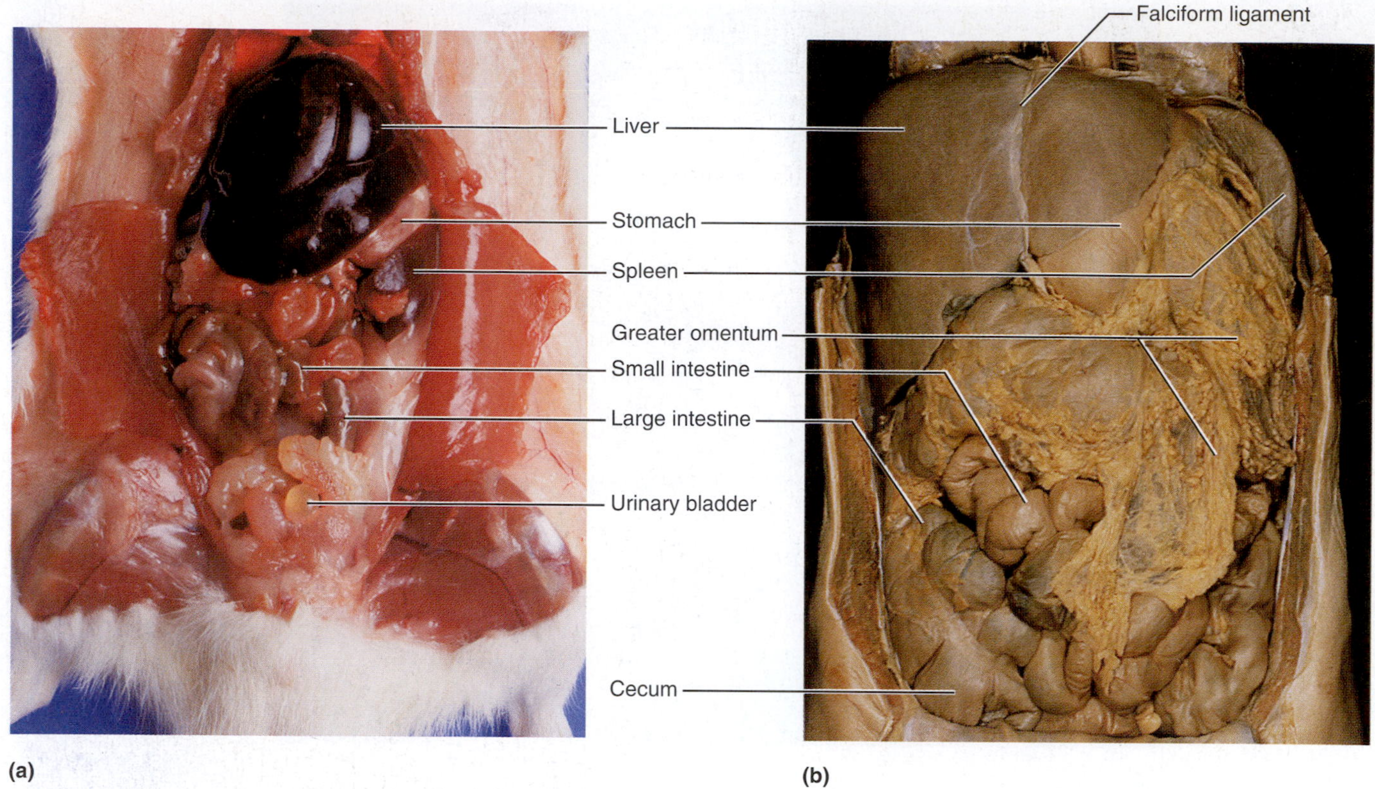

Figure 2.4 Abdominal organs. (a) Dissected rat, superficial view. **(b)** Human cadaver, superficial view.

Stomach: A curved organ important in food digestion and temporary food storage.

2. Examine the superficial structures of the abdominopelvic cavity. Lift the **greater omentum,** an extension of the peritoneum that covers the abdominal viscera. Continuing from the stomach, trace the rest of the digestive tract **(Figure 2.4)**.

Small intestine: Connected to the stomach and ending just before the saclike cecum.

Large intestine: A large muscular tube connected to the small intestine and ending at the anus.

Cecum: The initial portion of the large intestine.

Follow the course of the large intestine to the rectum, which is partially covered by the urinary bladder.

Rectum: Terminal part of the large intestine; continuous with the anal canal (not visible in this dissection).

Anus: The opening of the digestive tract (through the anal canal) to the exterior.

Now lift the small intestine with the forceps to view the mesentery.

Mesentery: An apronlike serous membrane; suspends many of the digestive organs in the abdominal cavity. Notice that it is heavily invested with blood vessels and, more likely than not, riddled with large fat deposits.

Locate the remaining abdominal structures.

Pancreas: A diffuse gland; rests dorsal to and in the mesentery between the first portion of the small intestine and the stomach. You will need to lift the stomach to view the pancreas.

Spleen: A dark red organ curving around the left lateral side of the stomach; considered part of the lymphatic system and often called the red blood cell "graveyard."

Liver: Large and brownish red; the most superior organ in the abdominal cavity, directly beneath the diaphragm.

3. To locate the deeper structures of the abdominopelvic cavity, move the stomach and the intestines to one side with the probe.

Examine the posterior wall of the abdominal cavity to locate the two kidneys **(Figure 2.5)**.

Kidneys: Bean-shaped organs; retroperitoneal (behind the peritoneum).

Adrenal glands: Large endocrine glands that sit on top of the superior margin of each kidney; considered part of the endocrine system.

Carefully strip away part of the peritoneum with forceps, and attempt to follow the course of one of the ureters to the bladder.

Ureter: Tube running from the indented region of a kidney to the urinary bladder.

Urinary bladder: The sac that serves as a reservoir for urine.

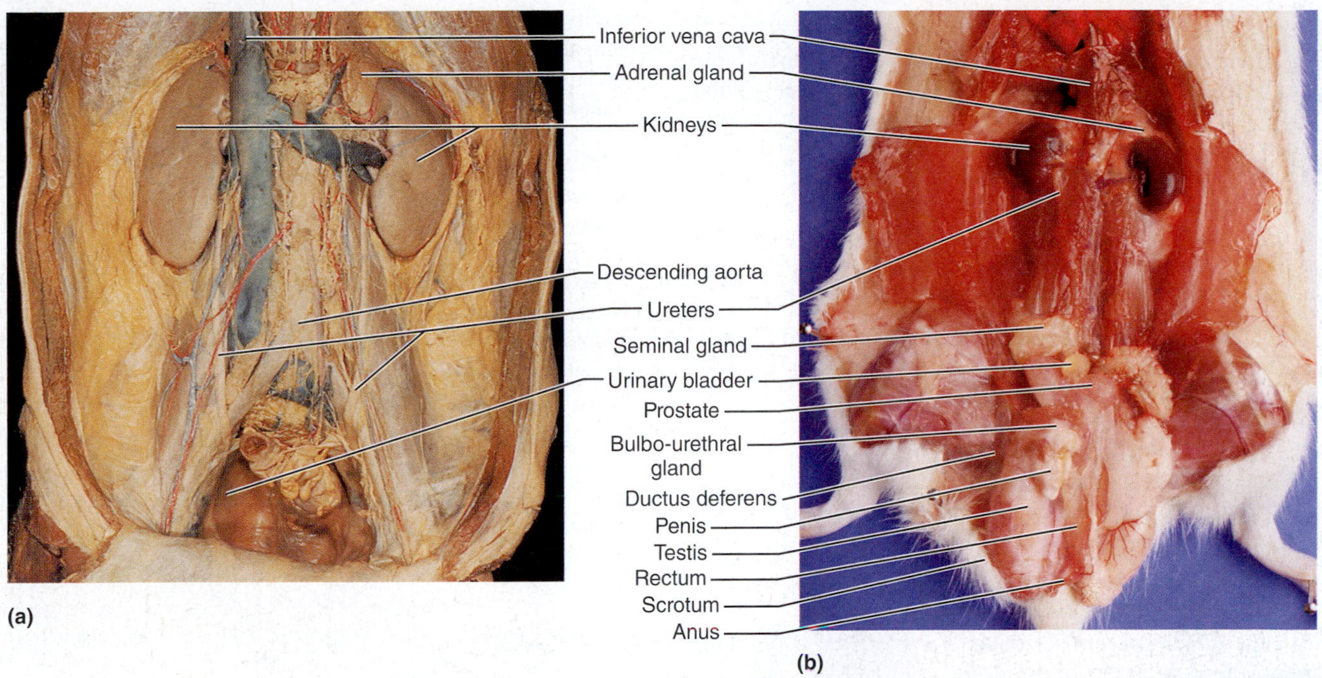

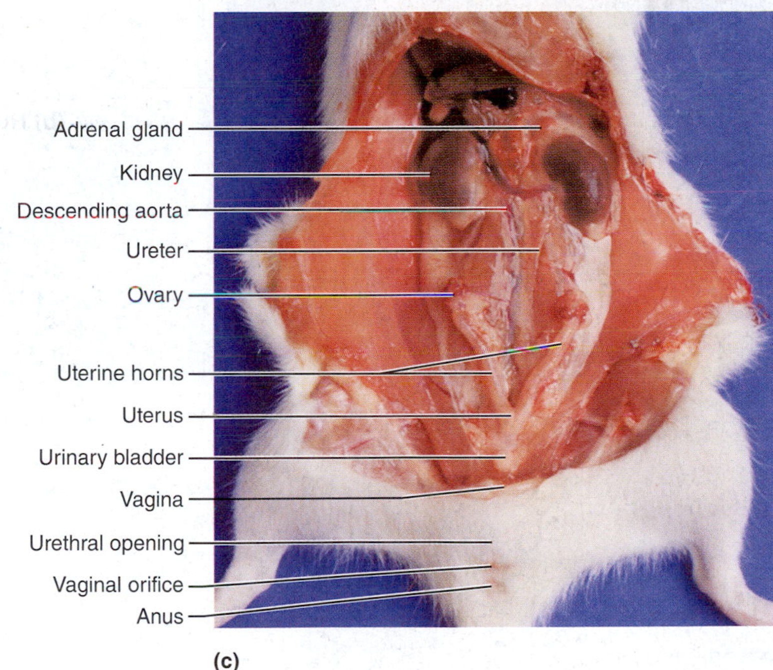

Figure 2.5 Deep structures of the abdominopelvic cavity. (a) Human cadaver. **(b)** Dissected male rat. (Some reproductive structures also shown.) **(c)** Dissected female rat. (Some reproductive structures also shown.)

4. In the midline of the body cavity lying between the kidneys are the two principal abdominal blood vessels. Identify each.

Inferior vena cava: The large vein that returns blood to the heart from the lower body regions.

Descending aorta: Deep to the inferior vena cava; the largest artery of the body; carries blood away from the heart down the midline of the body.

5. You will perform only a brief examination of reproductive organs. If you are working with a rat, first determine whether the animal is a male or female. Observe the ventral body surface beneath the tail. If a saclike scrotum and an opening for the anus are visible, the animal is a male. If three body openings—urethral, vaginal, and anal—are present, it is a female.

Male Animal

Make a shallow incision into the **scrotum.** Loosen and lift out one oval **testis.** Exert a gentle pull on the testis to identify the slender **ductus deferens,** or **vas deferens,** which carries sperm from the testis superiorly into the abdominal cavity and joins with the urethra. The urethra runs through the penis of the male and carries both urine and sperm out of the body. Identify the **penis,** extending from the bladder to the ventral body wall. You may see other glands of the male rat's reproductive system (Figure 2.5b), but you don't need to identify them at this time.

Female Animal

Inspect the pelvic cavity to identify the Y-shaped **uterus** lying against the dorsal body wall and beneath the bladder (Figure 2.5c). Follow one of the uterine horns superiorly to identify an **ovary,** a small oval structure at the end of the uterine horn. (The rat uterus is quite different from the uterus of a human female, which is a single-chambered organ about the size and shape of a pear.) The inferior undivided part of the rat uterus is continuous with the **vagina,** which leads to the body exterior. Identify the **vaginal orifice** (external vaginal opening).

If you are working with a human cadaver, proceed as indicated next.

Male Cadaver

Make a shallow incision into the **scrotum (Figure 2.6a)**. Loosen and lift out the oval **testis.** Exert a gentle pull on the testis to identify the slender **ductus (vas) deferens,** which carries sperm from the testis superiorly into the abdominal cavity (Figure 2.6b) and joins with the urethra. The urethra runs through the penis of the male and carries both urine and sperm out of the body. Identify the **penis,** extending from the bladder to the ventral body wall.

Female Cadaver

Inspect the pelvic cavity to identify the pear-shaped **uterus** lying against the dorsal body wall and superior to the bladder. Follow one of the **uterine tubes** superiorly to identify an **ovary,** a small oval structure at the end of the uterine tube (Figure 2.6c). The inferior part of the uterus is continuous with the **vagina,** which leads to the body exterior. Identify the **vaginal orifice** (external vaginal opening).

6. When you have finished your observations, rewrap or store the dissection animal or cadaver according to your instructor's directions. Wash the dissecting tools and equipment with laboratory detergent. Dispose of the gloves. Then wash and dry your hands before continuing with the examination of the human torso model. ■

Figure 2.6 Human reproductive organs. (a) Male external genitalia. **(b)** Sagittal section of the male pelvis. **(c)** Sagittal section of the female pelvis.

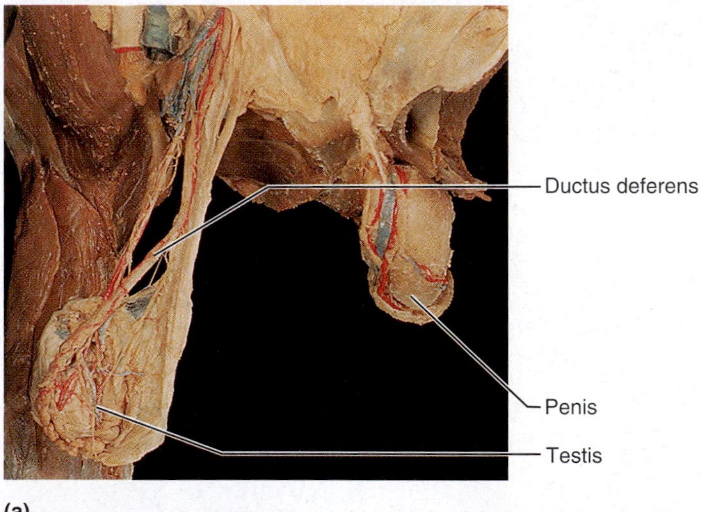

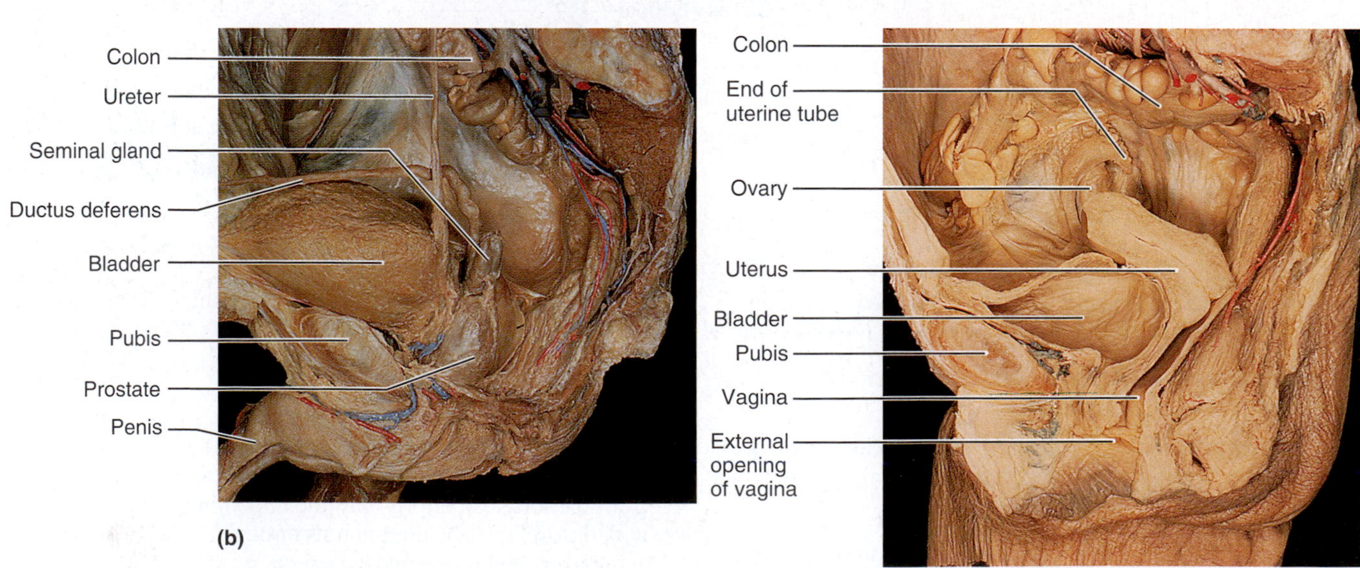

ACTIVITY 5

Examining the Human Torso Model

1. Examine a human torso model to identify the organs listed next to the photograph of the human torso model **(Figure 2.7)**. (If a torso model is not available, Figure 2.7 may be used for this part of the exercise). Some model organs will have to be removed to see the deeper organs.

2. Using the terms to the right of the figure (Figure 2.7), label each organ supplied with a leader line in the figure (Figure 2.7).

3. Place each of the organs listed in the correct body cavity or cavities. For organs found in the abdominopelvic cavity, also indicate which quadrant they occupy.

Dorsal body cavity ______________________________

Thoracic cavity ______________________________

____

Abdominopelvic cavity ______________________________

____

____

____

____

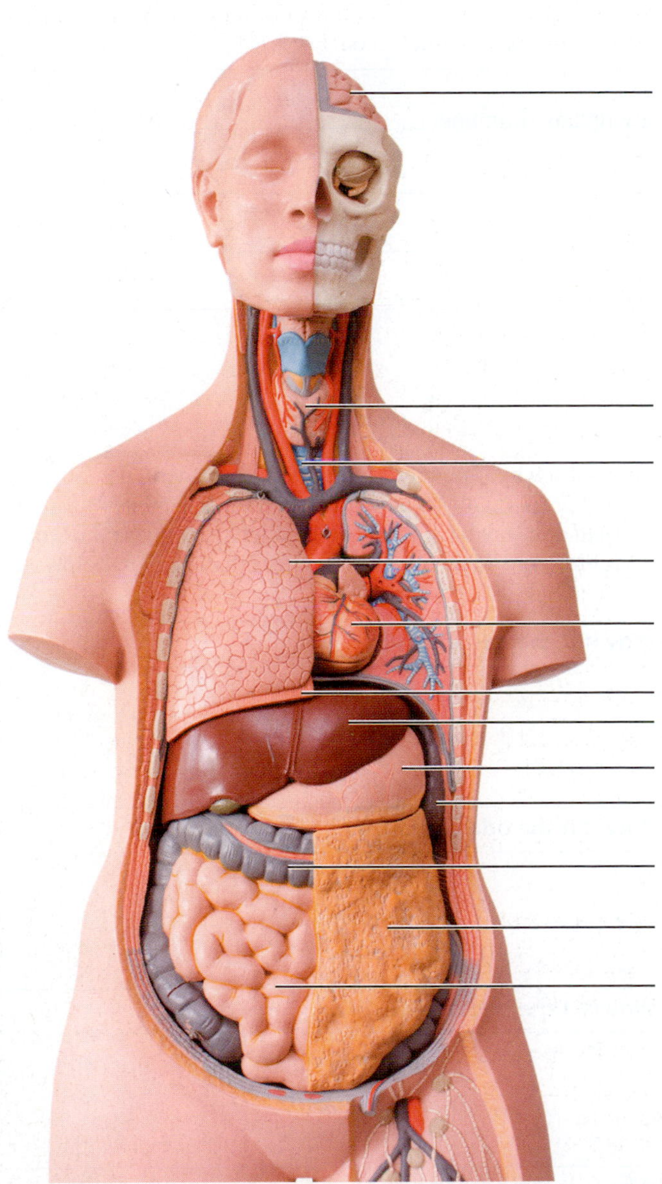

Adrenal gland
Aortic arch
Brain
Bronchi
Descending aorta
Diaphragm
Esophagus
Greater omentum
Heart
Inferior vena cava
Kidneys
Large intestine
Liver
Lungs
Pancreas
Rectum
Small intestine
Spinal cord
Spleen
Stomach
Thyroid gland
Trachea
Ureters
Urinary bladder

Figure 2.7 **Human torso model.**

4. Determine which organs are found in each abdominopelvic region, and record below.

Umbilical region: ______________________

Epigastric region: ______________________

Hypogastric region: ______________________

Right iliac region: ______________________

Left iliac region: ______________________

Right lumbar region: ______________________

Left lumbar region: ______________________

Right hypochondriac region: ______________________

Left hypochondriac region: ______________________

Now, assign each of the organs just identified to one of the organ system categories listed below.

Digestive: ______________________

____

Urinary: ______________________

____

Cardiovascular: ______________________

____

Endocrine: ______________________

____

Reproductive: ______________________

____

Respiratory: ______________________

____

Lymphatic/immune: ______________________

____

Nervous: ______________________

____

GROUP CHALLENGE

Odd Organ Out

Each box below contains four organs. One of the listed organs in each case does *not* share a characteristic that the other three do. Circle the organ that doesn't belong with the others, and explain why it is singled out. What characteristic is it missing? Sometimes there may be multiple reasons why the organ doesn't belong with the others. Include as many as you can think of, but make sure the organ does not have the key characteristic(s). Use the table (Table 2.1) and the pictures in your lab manual to help you select and justify your answer.

1. Which is the "odd organ"?	Why is it the odd one out?
Stomach Teeth Small intestine Oral cavity	
2. Which is the "odd organ"?	**Why is it the odd one out?**
Thyroid gland Thymus Spleen Lymph nodes	
3. Which is the "odd organ"?	**Why is it the odd one out?**
Ovaries Prostate gland Uterus Uterine tubes	
4. Which is the "odd organ"?	**Why is it the odd one out?**
Stomach Small intestine Esophagus Large intestine	

Name ______________________

Lab Time/Date ______________________

EXERCISE 2

REVIEW SHEET

Organ Systems Overview

1. Use the key below to indicate which body systems perform the following functions. (Some body systems are used more than once.) Then, circle the organ systems (in the key) that are present in all subdivisions of the ventral body cavity.

Key: a. cardiovascular d. integumentary g. nervous j. skeletal
b. digestive e. lymphatic/immune h. reproductive k. urinary
c. endocrine f. muscular i. respiratory

______________ 1. rids the body of nitrogen-containing wastes

______________ 2. is affected by removal of the thyroid gland

______________ 3. provides support and levers on which the muscular system acts

______________ 4. includes the heart

______________ 5. protects underlying organs from drying out and from mechanical damage

______________ 6. protects the body; destroys bacteria and tumor cells

______________ 7. breaks down ingested food into its building blocks

______________ 8. removes carbon dioxide from the blood

______________ 9. delivers oxygen and nutrients to the tissues

______________ 10. moves the limbs; facilitates facial expression

______________ 11. conserves body water or eliminates excesses

______________ and ______________ 12. facilitate conception and childbearing

______________ 13. controls the body by means of chemical molecules called hormones

______________ 14. is damaged when you cut your finger or get a severe sunburn

2. Using the key above, choose the *organ system* to which each of the following sets of organs or body structures belongs.

______________ 1. thymus, spleen, lymphatic vessels

______________ 2. bones, cartilages, tendons

______________ 3. pancreas, pituitary, adrenal glands

______________ 4. trachea, bronchi, lungs

______________ 5. epidermis, dermis, cutaneous sense organs

______________ 6. testis, ductus deferens, urethra

______________ 7. esophagus, large intestine, rectum

______________ 8. muscles of the thigh, postural muscles

3. Using the key below, place the following organs in their proper body cavity. Letters may be used more than once.

Key: a. abdominopelvic b. cranial c. spinal d. thoracic

______	1. stomach	______	4. liver	______	7. heart
______	2. esophagus	______	5. spinal cord	______	8. trachea
______	3. large intestine	______	6. urinary bladder	______	9. rectum

4. Using the organs listed in question 3 above, record, by number, which would be found in the abdominal regions listed below.

______	1. hypogastric region	______	4. epigastric region
______	2. right lumbar region	______	5. left iliac region
______	3. umbilical region	______	6. left hypochondriac region

5. The levels of organization of a living body are as follows: chemicals, ______, ______, ______, ______, and organism.

6. Define *organ.* ______

______

7. Using the terms provided, correctly identify all of the body organs indicated with leader lines in the drawings below. Then name the organ systems by entering the name of each on the answer blank below each drawing.

Key: blood vessels, brain, heart, kidney, nerves, sensory receptor, spinal cord, ureter, urethra, urinary bladder

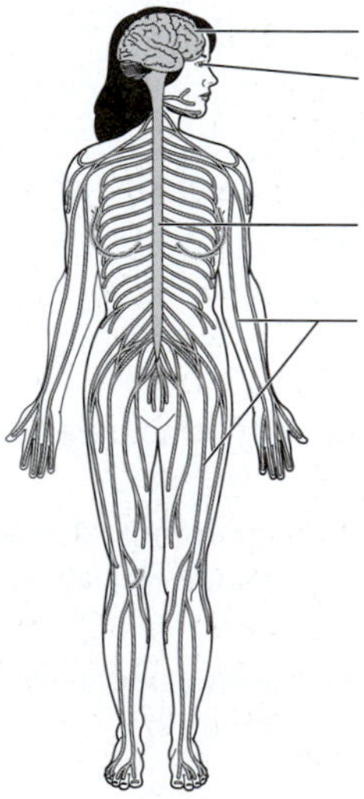

a. 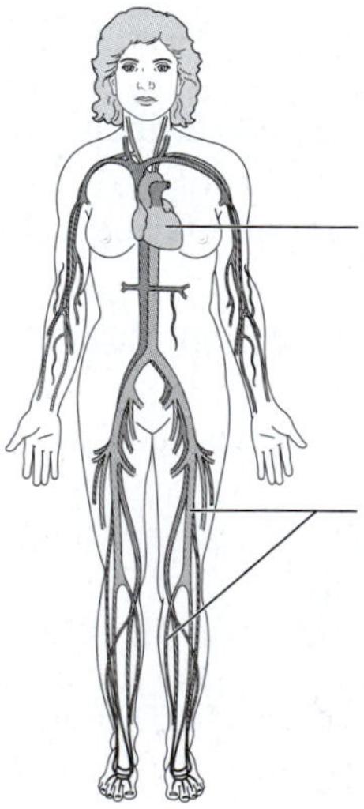 ______ b. 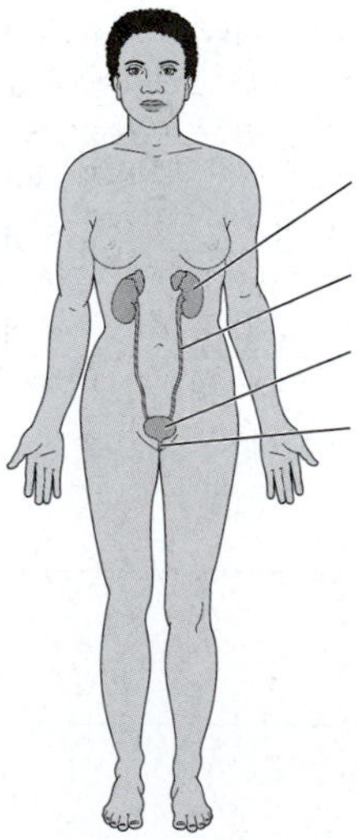 ______ c. ______

8. Why is it helpful to study the external and internal structures of the rat? ______

______

3

EXERCISE

The Microscope

MATERIALS

- ☐ Compound microscope
- ☐ Millimeter ruler
- ☐ Prepared slides of the letter *e* or newsprint
- ☐ Immersion oil
- ☐ Lens paper
- ☐ Prepared slide of grid ruled in millimeters
- ☐ Prepared slide of three crossed colored threads
- ☐ Clean microscope slide and coverslip
- ☐ Toothpicks (flat-tipped)
- ☐ Physiological saline in a dropper bottle
- ☐ Iodine or dilute methylene blue stain in a dropper bottle
- ☐ Filter paper or paper towels
- ☐ Beaker containing fresh 10% household bleach solution for wet mount disposal
- ☐ Disposable autoclave bag
- ☐ Prepared slide of cheek epithelial cells

Note to the Instructor: The slides and coverslips used for viewing cheek cells are to be soaked for 2 hours (or longer) in 10% bleach solution and then drained. The slides and disposable autoclave bag containing coverslips, lens paper, and used toothpicks are to be autoclaved for 15 min at 121°C and 15 pounds pressure to ensure sterility. After autoclaving, the disposable autoclave bag may be discarded in any disposal facility and the slides and glassware washed with laboratory detergent and prepared for use. These instructions also apply to any bloodstained glassware or disposable items used in other experimental procedures.

OBJECTIVES

1. Identify the parts of the microscope, and list the function of each part.
2. Describe and demonstrate the proper techniques for care of the microscope.
3. Demonstrate proper focusing technique.
4. Define *total magnification, resolution, parfocal, field, depth of field,* and *working distance*.
5. Measure the field size for one objective lens, calculate it for all the other objective lenses, and estimate the size of objects in each field.
6. Discuss the general relationships among magnification, working distance, and field size.

PRE-LAB QUIZ

1. The microscope slide rests on the ________________ while being viewed.
   a. base b. condenser c. iris d. stage
2. Your lab microscope is *parfocal*. What does this mean?
   a. The specimen is clearly in focus at this depth.
   b. The slide should be almost in focus when you change to higher magnifications.
   c. You can easily discriminate two close objects as separate.
3. If the ocular lens magnifies a specimen 10×, and the objective lens magnifies the specimen 35×, what is the total magnification being used to observe the specimen? ________________
4. How do you clean the lenses of your microscope?
   a. with a paper towel b. with soap and water
   c. with special lens paper and cleaner
5. Circle True or False. You should always start your observation of specimens with the oil-immersion lens.

With the invention of the microscope, biologists gained a valuable tool to observe and study structures, such as cells, that are too small to be seen by the unaided eye. The information gained helped in establishing many of the theories basic to the understanding of biological sciences. This exercise will familiarize you with the workhorse of microscopes—the compound microscope—and provide you with the necessary instructions for its proper use.

Care and Structure of the Compound Microscope

The **compound microscope** is a precision instrument and should always be handled with care. *At all times you must observe the following rules for its transport, cleaning, use, and storage:*

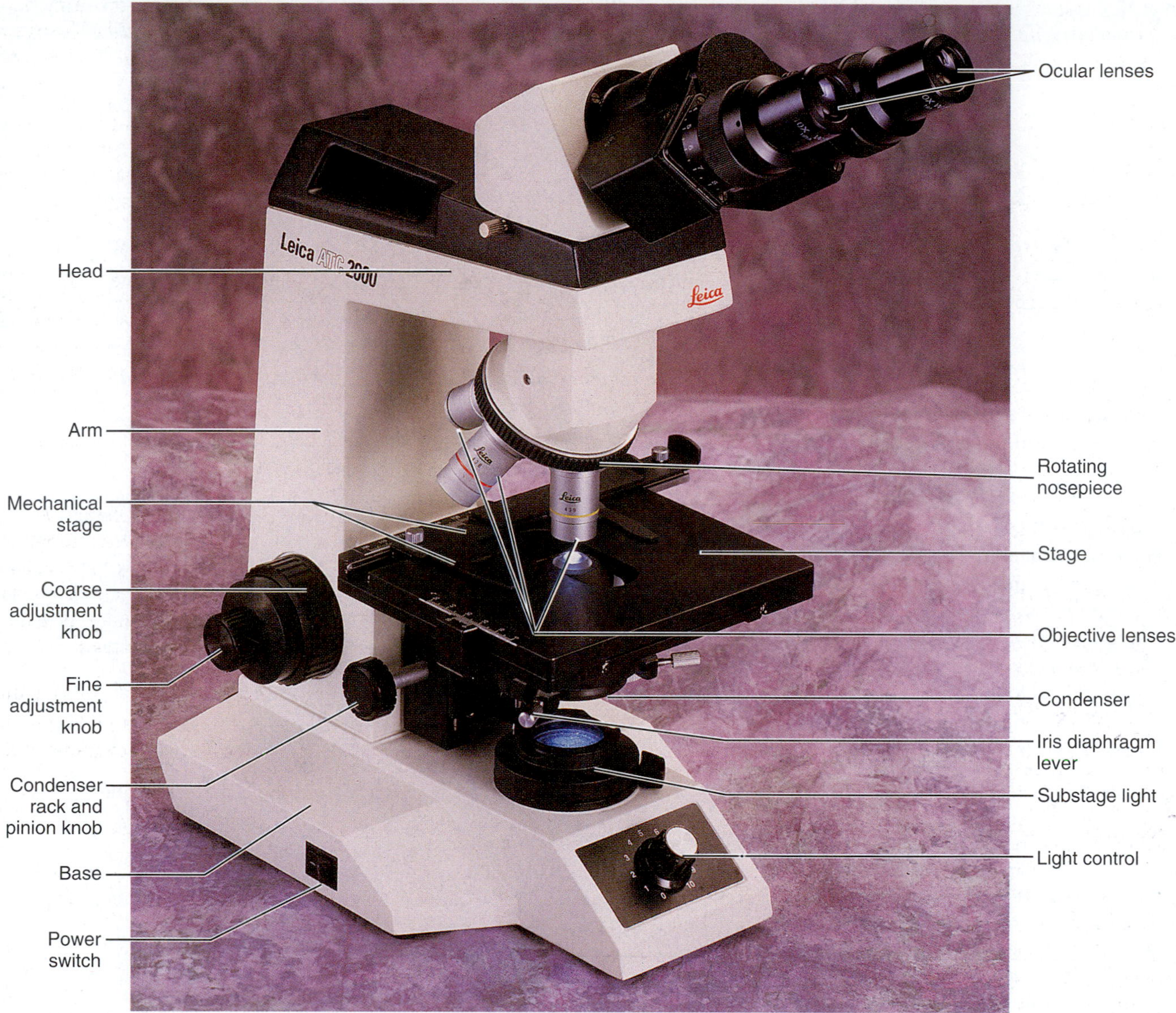

Figure 3.1 **Compound microscope and its parts.**

- When transporting the microscope, hold it in an upright position with one hand on its arm and the other supporting its base. Avoid swinging the instrument during its transport and jarring the instrument when setting it down.
- Use only special grit-free lens paper to clean the lenses. Use a circular motion to wipe the lenses, and clean all lenses before and after use.
- Always begin the focusing process with the lowest-power objective lens in position, changing to the higher-power lenses as necessary.
- Use the coarse adjustment knob only with the lowest-power lens.
- Always use a coverslip with wet mount preparations.
- Before putting the microscope in the storage cabinet, remove the slide from the stage, rotate the lowest-power objective lens into position, wrap the cord neatly around the base, and replace the dust cover or return the microscope to the appropriate storage area.
- Never remove any parts from the microscope; inform your instructor of any mechanical problems that arise.

ACTIVITY 1

Identifying the Parts of a Microscope

1. Using the proper transport technique, obtain a microscope and bring it to the laboratory bench.

- Record the number of your microscope in the **Summary Chart** (on page 31).

Compare your microscope with the illustration **(Figure 3.1)**, and identify the following microscope parts:

Base: Supports the microscope. (**Note:** Some microscopes are provided with an inclination joint, which allows the instrument to be tilted backward for viewing dry preparations.)

Substage light or **mirror:** Located in the base. In microscopes with a substage light source, the light passes directly upward through the microscope: light controls are located on the microscope base. If a mirror is used, light must be reflected from a separate free-standing lamp.

Stage: The platform the slide rests on while being viewed. The stage has a hole in it to permit light to pass through both it and the specimen. Some microscopes have a stage equipped with *spring clips;* others have a clamp-type *mechanical stage* (Figure 3.1). Both hold the slide in position for viewing; in addition, the mechanical stage has two adjustable knobs that control precise movement of the specimen.

Condenser: Small substage lens that concentrates the light on the specimen. The condenser may have a rack and pinion knob that raises and lowers the condenser to vary light delivery. Generally, the best position for the condenser is close to the inferior surface of the stage.

Iris diaphragm lever: Arm attached to the base of the condenser that regulates the amount of light passing through the condenser. The iris diaphragm permits the best possible contrast when viewing the specimen.

Coarse adjustment knob: Used to focus on the specimen.

Fine adjustment knob: Used for precise focusing once coarse focusing has been completed.

Head or **body tube:** Supports the objective lens system, which is mounted on a movable nosepiece, and the ocular lens or lenses.

Arm: Vertical portion of the microscope connecting the base and head.

Ocular (or *eyepiece*): Depending on the microscope, there are one or two lenses at the superior end of the head or body tube. Observations are made through the ocular(s). An ocular lens has a magnification of 10×; it increases the apparent size of the object by ten times, or ten diameters. If your microscope has a **pointer** to indicate a specific area of the viewed specimen, it is attached to one ocular and can be positioned by rotating the ocular lens.

Nosepiece: Rotating mechanism at the base of the head. Generally carries three or four objective lenses and permits sequential positioning of these lenses over the light beam passing through the hole in the stage. Use the nosepiece to change the objective lenses. Do not directly grab the lenses.

Objective lenses: Adjustable lens system that permits the use of a **scanning lens,** a **low-power lens,** a **high-power lens,** or an **oil immersion lens.** The objective lenses have different magnifying and resolving powers.

2. Examine the objective lenses carefully; note their relative lengths and the numbers inscribed on their sides. On many microscopes, the scanning lens, with a magnification between 4× and 5×, is the shortest lens. If there is no scanning lens, the low-power objective lens is the shortest and typically has a magnification of 10×. The high-power objective lens is of intermediate length and has a magnification range from 40× to 50×, depending on the microscope. The oil immersion objective lens is usually the longest of the objective lenses and has a magnifying power of 95× to 100×. Some microscopes lack the oil immersion lens.

- Record the magnification of each objective lens of your microscope in the first row of the Summary Chart (page 31). Also, cross out the column relating to a lens that your microscope does not have. Plan on using the same microscope for all microscopic studies.

3. Rotate the lowest-power objective lens until it clicks into position, and turn the coarse adjustment knob about 180 degrees. Notice how far the stage (or objective lens) travels during this adjustment. Move the fine adjustment knob 180 degrees, noting again the distance that the stage (or the objective lens) moves. ■

Magnification and Resolution

The microscope is an instrument of magnification. In the compound microscope, magnification is achieved through the interplay of two lenses—the ocular lens and the objective lens. The objective lens magnifies the specimen to produce a **real image** that is projected to the ocular. This real image is magnified by the ocular lens to produce the **virtual image** seen by your eye **(Figure 3.2)**.

The **total magnification (TM)** of any specimen being viewed is equal to the power of the ocular lens multiplied by the power of the objective lens being used. For example, if the

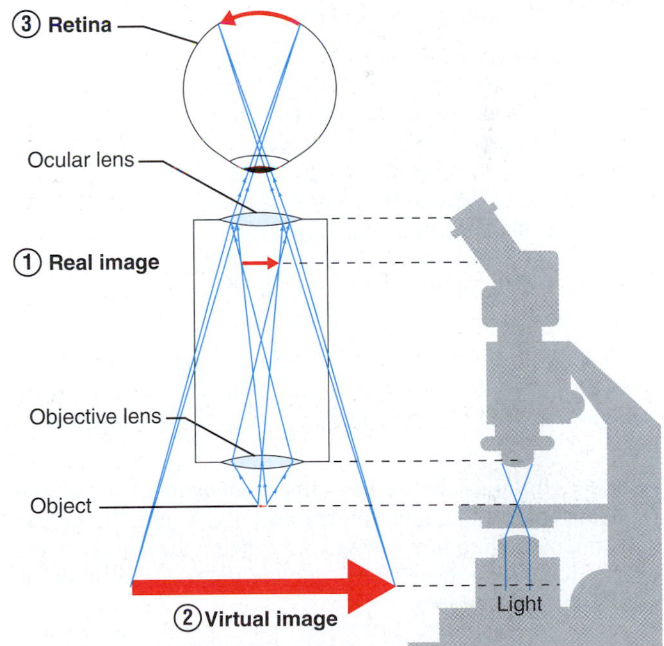

Figure 3.2 Image formation in light microscopy. **Step 1:** The objective lens magnifies the object, forming the real image. **Step 2:** The ocular lens magnifies the real image, forming the virtual image. **Step 3:** The virtual image passes through the lens of the eye and is focused on the retina.

ocular lens magnifies 10× and the objective lens magnifies 45×, the total magnification is 450× (or 10 × 45).

- Determine the total magnification you can achieve with each of the objectives on your microscope, and record the figures on the third row of the Summary Chart.

The compound light microscope has certain limitations. Although the level of magnification is almost limitless, the **resolution** (or resolving power), that is, the ability to discriminate two close objects as separate, is not. The human eye can resolve objects about 100 μm apart, but the compound microscope has a resolution of 0.2 μm under ideal conditions. Objects closer than 0.2 μm are seen as a single fused image.

Resolving power is determined by the amount and physical properties of the visible light that enters the microscope. In general, the more light delivered to the objective lens, the greater the resolution. The size of the objective lens aperture (opening) decreases with increasing magnification, allowing less light to enter the objective. Thus, you will probably find it necessary to increase the light intensity at the higher magnifications.

ACTIVITY 2

Viewing Objects Through the Microscope

1. Obtain a millimeter ruler, a prepared slide of the letter *e* or newsprint, a dropper bottle of immersion oil, and some lens paper. Adjust the condenser to its highest position, and switch on the light source of your microscope. If the light source is not built into the base, use the curved surface of the mirror to reflect the light up into the microscope.

2. Secure the slide on the stage so that you can read the slide label and the letter *e* is centered over the light beam passing through the stage. If you are using a microscope with spring clips, make sure the slide is secured at both ends. If your microscope has a mechanical stage, open the jaws of its slide holder by using the control lever, typically located at the rear left corner of the mechanical stage. Insert the slide squarely within the confines of the slide holder. Check to see that the slide is resting on the stage, not on the mechanical stage frame, before releasing the control lever.

3. With your lowest-power (scanning or low-power) objective lens in position over the stage, use the coarse adjustment knob to bring the objective lens and stage as close together as possible.

4. Look through the ocular lens, and using the iris diaphragm, adjust the light for comfort. Now use the coarse adjustment knob to focus slowly away from the *e* until it is as clearly focused as possible. Complete the focusing with the fine adjustment knob.

5. Sketch the letter *e* in the circle on the Summary Chart just as it appears in the **field** (the area you see through the microscope).

How far is the bottom of the objective lens from the specimen? In other words, what is the **working distance**? Use a millimeter ruler to make this measurement.

Record the working distance in the Summary Chart.

How has the apparent orientation of the *e* changed top to bottom, right to left, and so on?

____

____

6. Move the slide slowly away from you on the stage as you view it through the ocular lens. In what direction does the image move?

____

Move the slide to the left. In what direction does the image move?

____

At first this change in orientation may confuse you, but with practice you will learn to move the slide in the desired direction with no problem.

7. Today most good laboratory microscopes are **parfocal;** that is, the slide should be in focus (or nearly so) at the higher magnifications once you have properly focused. *Without touching the focusing knobs,* increase the magnification by rotating the next higher magnification lens into position over the stage. Make sure it clicks into position. Using the fine adjustment only, sharpen the focus. If you are unable to focus with a new lens, your microscope is not parfocal. Do not try to force the lens into position. Consult your instructor. Note the decrease in working distance. As you can see, focusing with the coarse adjustment knob could drive the objective lens through the slide, breaking the slide and possibly damaging the lens. Sketch the letter *e* in the Summary Chart. What new details become clear?

____

____

As best you can, measure the distance between the objective and the slide.

Record the working distance in the Summary Chart.

Is the image larger or smaller? ______________

Approximately how much of the letter *e* is visible now?

____

Is the field larger or smaller? ______________

Why is it necessary to center your object (or the portion of the slide you wish to view) before changing to a higher power?

____

Move the iris diaphragm lever while observing the field. What happens?

____

Summary Chart for Microscope # ____________

	Scanning	Low power	High power	Oil immersion
Magnification of objective lens	________ ×	________ ×	________ ×	________ ×
Magnification of ocular lens	10 ×	10 ×	10 ×	10 ×
Total magnification	________ ×	________ ×	________ ×	________ ×
Working distance	________ mm	________ mm	________ mm	________ mm
Detail observed Letter e	◯	◯	◯	◯
Field size (diameter)	___ mm ___µm	___ mm ___ µm	___ mm ___ µm	___ mm ___ µm

Is it more desirable to increase *or* decrease the light when changing to a higher magnification?

____________Why? ____________________________

____

8. If you have just been using the low-power objective, repeat the steps given in direction 7 using the high-power objective lens. What new details become clear?

____

____

Record the working distance in the Summary Chart.

9. Without touching the focusing knob, rotate the high-power lens out of position so that the area of the slide over the opening in the stage is unobstructed. Place a drop of immersion oil over the *e* on the slide and rotate the oil immersion lens into position. Set the condenser at its highest point (closest to the stage), and open the diaphragm fully. Adjust the fine focus and fine-tune the light for the best possible resolution.

Note: If for some reason the specimen does not come into view after you adjust the fine focus, do not go back to the 40× lens to recenter. You do not want oil from the oil immersion lens to cloud the 40× lens. Turn the revolving nosepiece in the other direction to the low-power lens, and recenter and refocus the object. Then move the immersion lens back into position, again avoiding the 40× lens. Sketch the letter *e* in the Summary Chart. What new details become clear?

____

____

Is the field again decreased in size? ____________

As best you can, estimate the working distance, and record it in the Summary Chart. Is the working distance less *or* greater than it was when the high-power lens was focused?

____

Compare your observations on the relative working distances of the objective lenses with the illustration **(Figure 3.3)**. Explain why it is desirable to begin the focusing process in the lowest power.

____

____

____

10. Rotate the oil immersion lens slightly to the side and remove the slide. Clean the oil immersion lens carefully with lens paper, and then clean the slide in the same manner with a fresh piece of lens paper. ■

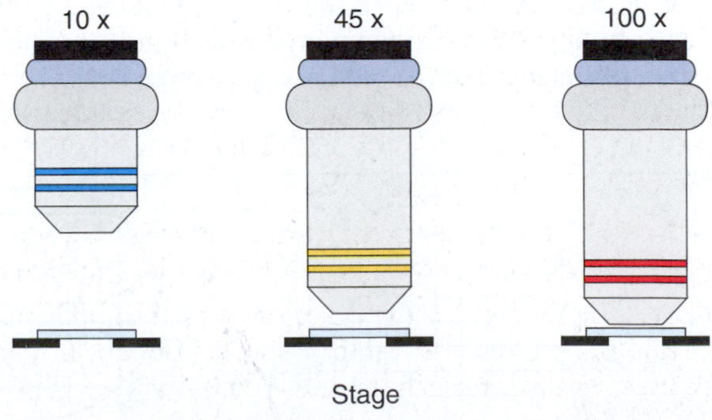

Figure 3.3 Relative working distances of the 10×, 45×, and 100× objectives.

Table 3.1 Comparison of Metric Units of Length

Metric unit	Abbreviation	Equivalent
Meter	m	(about 39.3 in.)
Centimeter	cm	10^{-2} m
Millimeter	mm	10^{-3} m
Micrometer (or micron)	μm (μ)	10^{-6} m
Nanometer (or millimicrometer or millimicron)	nm (mμ)	10^{-9} m
Ångstrom	Å	10^{-10} m

The Microscope Field

By this time you should know that the size of the microscope field decreases with increasing magnification. For future microscope work, it will be useful to determine the diameter of each of the microscope fields. This information will allow you to make a fairly accurate estimate of the size of the objects you view in any field. For example, if you have calculated the field diameter to be 4 mm and the object being observed extends across half this diameter, you can estimate the length of the object to be approximately 2 mm.

Microscopic specimens are usually measured in micrometers and millimeters, both units of the metric system. You can get an idea of the relationship and meaning of these units from the table **(Table 3.1)**. (A more detailed treatment appears in Appendix A.)

ACTIVITY 3

Estimating the Diameter of the Microscope Field

1. Obtain a grid slide (a slide prepared with graph paper ruled in millimeters). Each of the squares in the grid is 1 mm on each side. Use your lowest-power objective to bring the grid lines into focus.

2. Move the slide so that one grid line touches the edge of the field on one side, and then count the number of squares you can see across the diameter of the field. If you can see only part of a square, as in the accompanying diagram, estimate the part of a millimeter that the partial square represents.

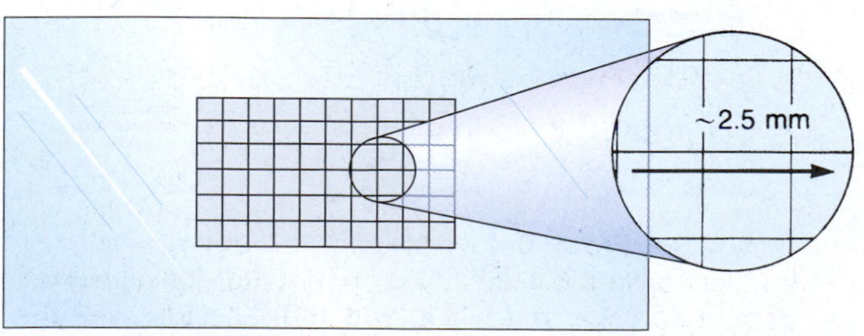

Record this figure in the appropriate space marked "field size" on the Summary Chart (page 31). (If you have been using the scanning lens, repeat the procedure with the low-power objective lens.)

Complete the chart by computing the approximate diameter of the high-power and oil immersion fields. The general formula for calculating the unknown field diameter is as follows:

$$\text{Diameter of field } A \times \text{total magnification of field } A = \text{diameter of field } B \times \text{total magnification of field } B$$

where *A* represents the known or measured field and *B* represents the unknown field. This can be simplified to

$$\text{Diameter of field } B = \frac{\text{diameter of field } A \times \text{total magnification of field } A}{\text{total magnification of field } B}$$

For example, if the diameter of the low-power field (field *A*) is 2 mm and the total magnification is 50×, you would compute the diameter of the high-power field (field *B*) with a total magnification of 100× as follows:

$$\text{Field diameter } B = (2 \text{ mm} \times 50)/100$$

$$\text{Field diameter } B = 1 \text{ mm}$$

3. Estimate the length (longest dimension) of the following microscopic objects. *Base your calculations on the field sizes you have determined for your microscope.*

a. Object seen in low-power field:

approximate length:

_______ mm

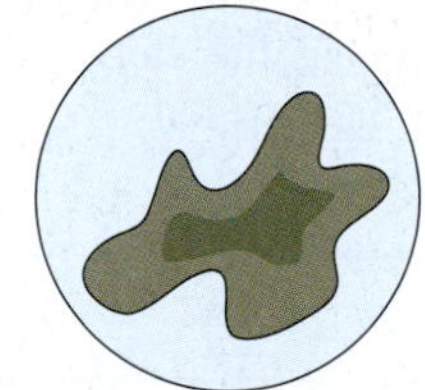

b. Object seen in high-power field:

approximate length:

_______ mm

or _______ μm

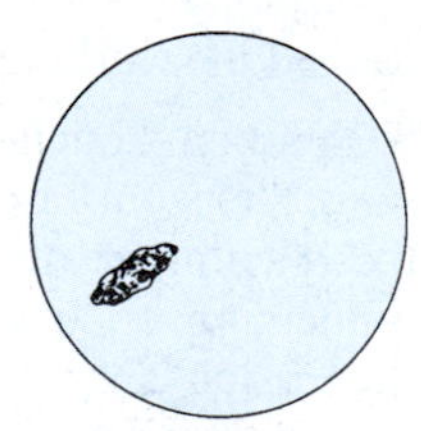

c. Object seen in oil immersion field:

approximate length:

_______ μm

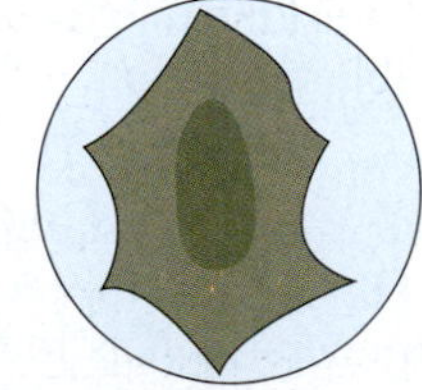

4. If an object viewed with the oil immersion lens looked as it does in the field depicted just below, could you determine its approximate size from this view?

If not, then how could you determine it? ______________________

____

____

____

Perceiving Depth

Any microscopic specimen has depth as well as length and width; it is rare indeed to view a tissue slide with just one layer of cells. Normally you can see two or three cell thicknesses. Therefore, it is important to learn how to determine relative depth with your microscope. In microscope work the **depth of field** (the thickness of the plane that is clearly in focus) is greater at lower magnifications. As magnification increases, the depth of field decreases.

ACTIVITY 4

Perceiving Depth

1. Obtain a slide with colored crossed threads. Focusing at low magnification, locate the point where the three threads cross each other.

2. Use the iris diaphragm lever to greatly reduce the light, thus increasing the contrast. Focus down with the coarse adjustment until the threads are out of focus, then slowly focus upward again, noting which thread comes into clear focus first. (You will see two or even all three threads, so you must be very careful in determining which one first comes into clear focus.) Observe: As you rotate the adjustment knob forward (away from you), does the stage rise or fall? If the stage rises, then the first clearly focused thread is the top one; the last clearly focused thread is the bottom one.

If the stage descends, how is the order affected? ______________

____

Record your observations as to which color of thread is uppermost, in the middle, or lowest:

Top thread ______________________________________

Middle thread ____________________________________

Bottom thread ____________________________________

Viewing Cells Under the Microscope

There are various ways to prepare cells for viewing under a microscope. Cells and tissues can look very different with different stains and preparation techniques. One method of preparation is to mix the cells in physiological saline (called a wet mount) and stain them with methylene blue stain.

If you are not instructed to prepare your own wet mount, obtain a prepared slide of epithelial cells to make the observations in step 10 of Activity 5.

ACTIVITY 5

Preparing and Observing a Wet Mount

1. Obtain the following: a clean microscope slide and coverslip, two flat-tipped toothpicks, a dropper bottle of physiological saline, a dropper bottle of iodine or methylene blue stain, and filter paper (or paper towels). Handle only your own slides throughout the procedure.

2. Place a drop of physiological saline in the center of the slide. Using the flat end of the toothpick, *gently* scrape the inner lining of your cheek. Transfer your cheek scrapings to the slide by agitating the end of the toothpick in the drop of saline **(Figure 3.4a)**.

Immediately discard the used toothpick in the disposable autoclave bag provided at the supplies area.

3. Add a tiny drop of the iodine or methylene blue stain to the preparation. (These epithelial cells are nearly transparent and thus difficult to see without the stain, which colors the nuclei of the cells and makes them look much darker than the cytoplasm.) Stir with a clean toothpick.

Immediately discard the used toothpick in the disposable autoclave bag provided at the supplies area.

4. Hold the coverslip with your fingertips so that its bottom edge touches one side of the fluid drop (Figure 3.4b), then

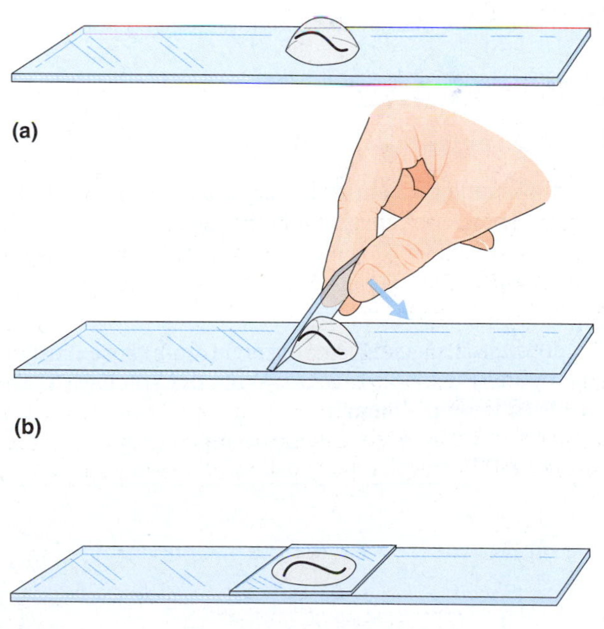

Figure 3.4 Procedure for preparing a wet mount. **(a)** The object is placed in a drop of water (or saline) on a clean slide, **(b)** a coverslip is held at a 45° angle with the fingertips, and **(c)** it is lowered carefully over the water and the object.

carefully lower the coverslip onto the preparation (Figure 3.4c). *Do not just drop the coverslip,* or you will trap large air bubbles under it, which will obscure the cells. *Always use a coverslip with a wet mount* to prevent soiling the lens if you should misfocus.

5. Examine your preparation carefully. The coverslip should be tight against the slide. If there is excess fluid around its edges, you will need to remove it. Obtain a piece of filter paper, fold it in half, and use the folded edge to absorb the excess fluid. You may use a twist of paper towel as an alternative.

 Before continuing, discard the filter paper or paper towel in the disposable autoclave bag.

6. Place the slide on the stage, and locate the cells in low power. You will probably want to dim the light with the iris diaphragm to provide more contrast for viewing the lightly stained cells. Furthermore, a wet mount will dry out quickly in bright light because a bright light source is hot.

7. Cheek epithelial cells are very thin, six-sided cells. In the cheek, they provide a smooth, tilelike lining **(Figure 3.5)**. Move to high power to examine the cells more closely.

8. Make a sketch of the epithelial cells that you observe.

Use information on your Summary Chart (page 31) to estimate the diameter of cheek epithelial cells.

______ μm

Why do *your* cheek cells look different from those illustrated in the figure (Figure 3.5)? (*Hint:* What did you have to *do* to your cheek to obtain them?)

____

____

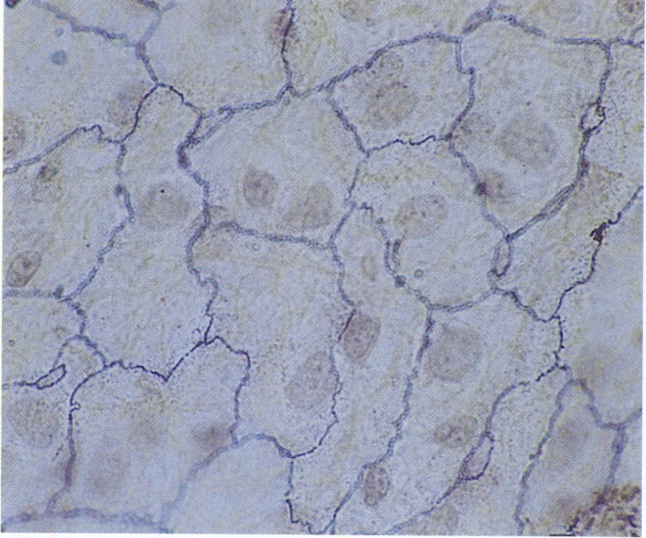

Figure 3.5 Epithelial cells of the cheek cavity (surface view, 630×).

! 9. When you complete your observations of the wet mount, dispose of your wet mount preparation in the beaker of bleach solution, and put the coverslips in an autoclave bag.

10. Obtain a prepared slide of cheek epithelial cells, and view them under the microscope.

Estimate the diameter of one of these cheek epithelial cells using information from the Summary Chart (page 31).

______ μm

Why are these cells more similar to those seen in the figure (Figure 3.5) and easier to measure than those of the wet mount?

____

____

11. Before leaving the laboratory, make sure all other materials are properly discarded or returned to the appropriate laboratory station. Clean the microscope lenses, and put the dust cover on the microscope before you return it to the storage cabinet. ■

Name ______________________

Lab Time/Date ______________________

EXERCISE 3

REVIEW SHEET

The Microscope

Care and Structure of the Compound Microscope

1. Label all indicated parts of the microscope.

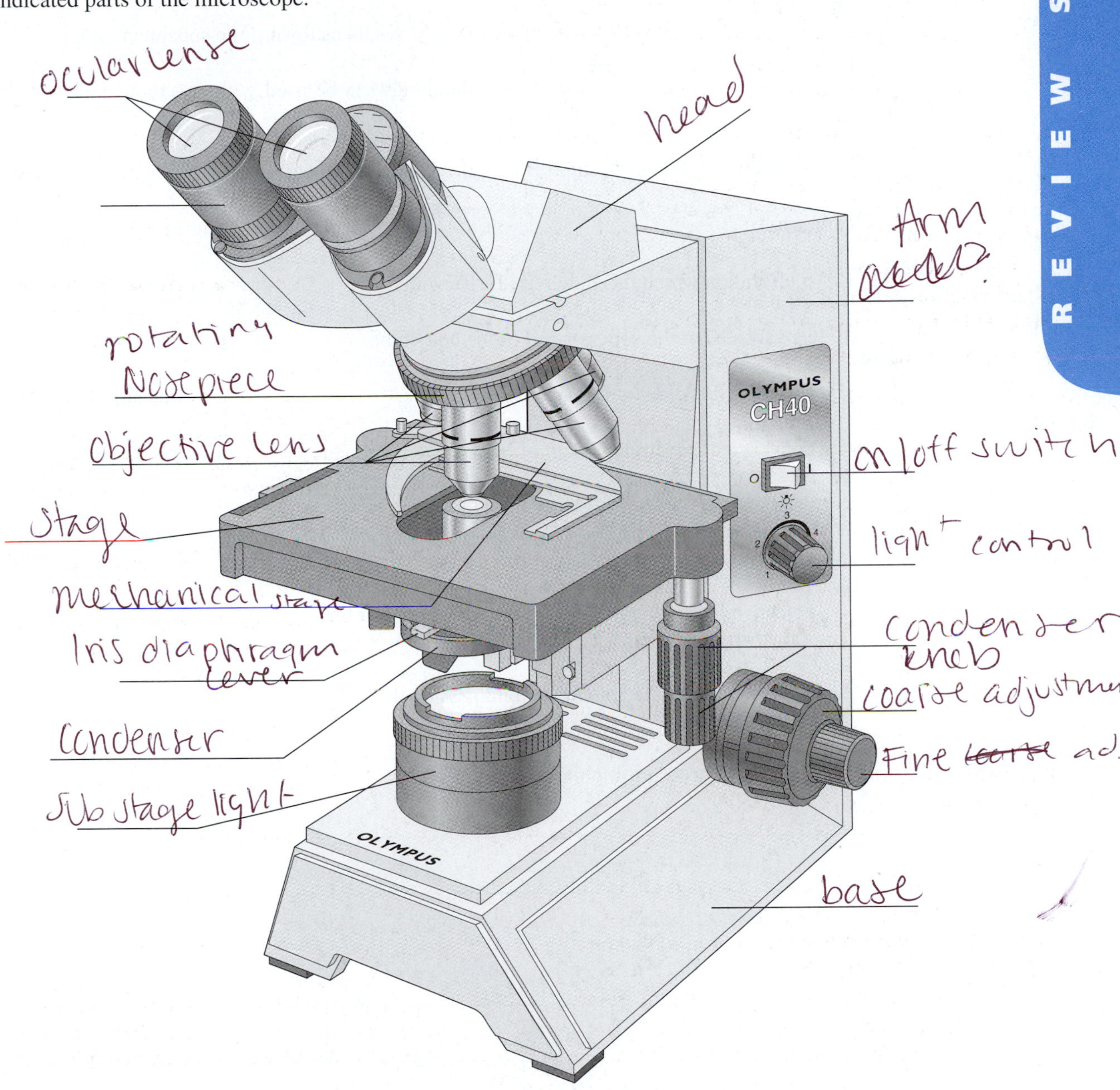

2. Determine whether each of the following statements is true or false. If it is true, write *T* on the answer blank. If it is false, correct the statement by writing on the blank the proper word or phrase to replace the one that is underlined.

________________ 1. The microscope lens may be cleaned with any soft tissue.

________________ 2. The microscope should be stored with the oil immersion lens in position over the stage.

________________ 3. When beginning to focus, the lowest-power lens should be used.

________________ 4. When focusing, always focus toward the specimen.

________________ 5. A coverslip should always be used with wet mounts and the high-power and oil lenses.

3. Match the microscope structures given in column B with the statements in column A that identify or describe them.

	Column A	Column B
________	1. platform on which the slide rests for viewing	a. coarse adjustment knob
________	2. lens located at the superior end of the body tube	b. condenser
________	3. secure(s) the slide to the stage	c. fine adjustment knob
________	4. delivers a concentrated beam of light to the specimen	d. iris diaphragm
________	5. used for precise focusing once initial focusing has been done	e. mechanical stage or spring clips
		f. movable nosepiece
________	6. carries the objective lenses; rotates so that the different objective lenses can be brought into position over the specimen	g. objective lenses
		h. ocular
________	7. used to increase the amount of light passing through the specimen	i. stage

4. Explain the proper technique for transporting the microscope.

____

____

5. Define the following terms.

real image: ____

____

resolution: ____

____

Viewing Objects Through the Microscope

6. Complete, or respond to, the following statements:

__________ 1. The distance from the bottom of the objective lens in use to the specimen is called the ______.

__________ 2. Assume there is an object on the left side of the field that you want to bring to the center (that is, toward the apparent right). In what direction would you move your slide?

__________ 3. The area of the specimen seen when looking through the microscope is the ______.

__________ 4. If a microscope has a 10× ocular and the total magnification at a particular time is 950×, the objective lens in use at that time is ______×.

__________ 5. Why should the light be dimmed when you are looking at living (nearly transparent) cells?

__________ 6. After focusing in low power, you find that you need to use only the fine adjustment to focus the specimen at the higher powers. The microscope is therefore said to be ______.

__________ 7. You are using a 10× ocular and a 15× objective. If the field size is 1.5 mm, the approximate field size with a 30× objective is ______ mm.

__________ 8. If the size of the high-power field is 1.2 mm, an object that occupies approximately a third of that field has an estimated diameter of ______ mm.

7. You have been asked to prepare a slide with the letter *k* on it (as shown below). In the circle below, draw the *k* as seen in the low-power field.

k

8. The numbers for the field sizes below are too large to represent the typical compound microscope lens system, but the relationships depicted are accurate. Figure out the magnification of fields 1 and 3, and the field size of 2. (Hint: Use your ruler.)

5 mm	______ mm	0.5 mm
1. →○←	2. →○←	3. →∘←
______×	100×	______×

9. Say you are observing an object in the low-power field. When you switch to high power, it is no longer in your field of view.

Why might this occur? ____

____

What should you have done initially to prevent this from happening? ____

____

10. Do the following factors increase or decrease as one moves to higher magnifications with the microscope?

resolution: ______________ amount of light needed: ______________

working distance: ______________ depth of field: ______________

11. A student has the high-dry lens in position and appears to be intently observing the specimen. The instructor, noting a working distance of about 1 cm, knows the student isn't actually seeing the specimen.

How so? ______________

12. Describe the proper procedure for preparing a wet mount.

____

____

____

13. Indicate the probable cause of the following situations arising during use of a microscope.

a. Only half of the field is illuminated: ______________

____

b. Field does not change as mechanical stage is moved: ______________

____

EXERCISE 4

The Cell: Anatomy and Division

MATERIALS

- ☐ Three-dimensional model of the "composite" animal cell or laboratory chart of cell anatomy
- ☐ Compound microscope
- ☐ Prepared slides of simple squamous epithelium ($AgNO_3$ stain), teased smooth muscle (l.s.), human blood cell smear, and sperm
- ☐ Video of mitosis
- ☐ Three-dimensional models of mitotic stages
- ☐ Prepared slides of whitefish blastulas
- ☐ Chenille sticks (pipe cleaners), two different colors, cut into 3" pieces, 8 pieces per group

Note to the Instructor: See directions for handling wet mount preparations and disposable supplies (pages 33–34, Exercise 3). For suggestions on video of mitosis, see Instructor's Guide.

OBJECTIVES

1. Define *cell, organelle,* and *inclusion.*
2. Identify on a cell model or diagram the following cellular regions, and list the major function of each region: nucleus, cytoplasm, and plasma membrane.
3. Identify the cytoplasmic organelles, and discuss their structure and function.
4. Compare and contrast specialized cells with the concept of the "generalized cell."
5. Define *interphase, mitosis,* and *cytokinesis.*
6. List the stages of mitosis, and describe the key events of each stage.
7. Identify the mitotic phases on slides or appropriate diagrams.
8. Explain the importance of mitotic cell division, and describe its product.

PRE-LAB QUIZ

1. Define *cell.* ____

   ____
2. When a cell is not dividing, the DNA is loosely spread throughout the nucleus in a threadlike form called:
   a. chromatin
   b. chromosomes
   c. cytosol
   d. ribosomes
3. The plasma membrane not only provides a protective boundary for the cell but also determines which substances enter or exit the cell. We call this characteristic:
   a. diffusion
   b. membrane potential
   c. osmosis
   d. selective permeability
4. Proteins are assembled on these organelles. ________________
5. Because these organelles are responsible for providing most of the ATP the cell needs, they are often referred to as the "powerhouses" of the cell. They are the:
   a. centrioles
   b. lysosomes
   c. mitochondria
   d. ribosomes
6. Circle the correct underlined term. During cytokinesis / interphase the cell grows and performs its usual activities.
7. Circle True or False. The end product of mitosis is four genetically identical daughter nuclei.
8. How many stages of mitosis are there? ________________
9. DNA replication occurs during:
   a. cytokinesis
   b. interphase
   c. metaphase
   d. prophase
10. Circle True or False. All animal cells have a cell wall.

The **cell,** the structural and functional unit of all living things, is a complex entity. The cells of the human body are highly diverse, and their differences in size, shape, and internal composition reflect their specific roles in the body. Nonetheless, cells do have many common anatomical features, and all cells must carry out certain functions to sustain life. For example, all cells can maintain their boundaries, metabolize, digest nutrients and dispose of wastes, grow and reproduce, move, and respond to a stimulus. Most of these functions are considered in detail in later exercises. This exercise focuses on structural similarities found in many cells and illustrated by a "composite," or "generalized," cell. The only function considered here is cell reproduction (cell division).

Anatomy of the Composite Cell

In general, all animal cells have three major regions, or parts, that can readily be identified with a light microscope: the **nucleus,** the **plasma membrane,** and the **cytoplasm.** The nucleus is typically a round or oval structure near the center of the cell. It is surrounded by cytoplasm, which in turn is enclosed by the plasma membrane. Since the advent of the electron microscope, even smaller cell structures—organelles—have been identified. A diagrammatic representation of the composite cell can help you begin to identify these five structures **(Figure 4.1a)**. An electron microscope can provide a detailed view of the cellular structure, particularly that of the nucleus (Figure 4.1b).

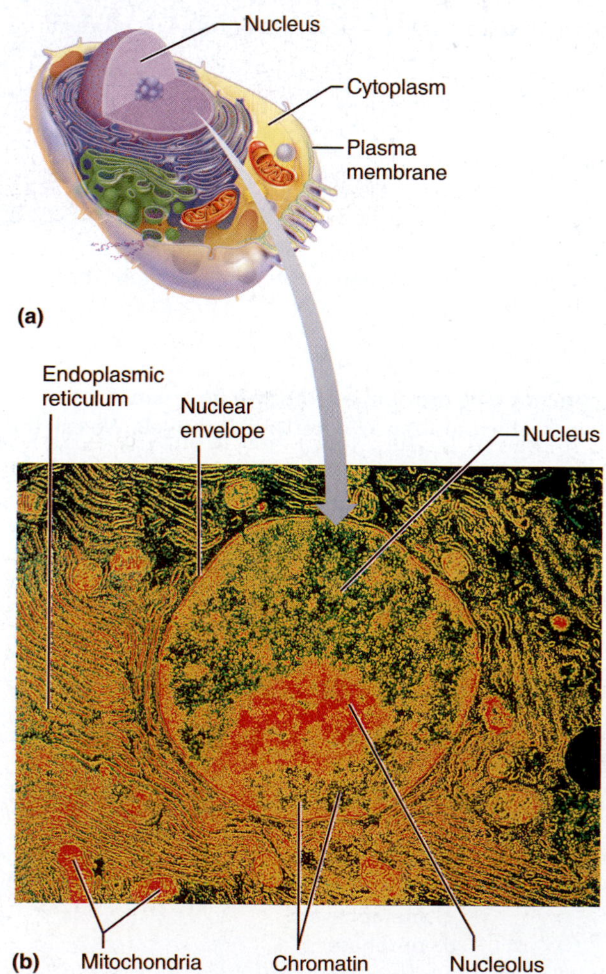

Figure 4.1 Anatomy of the composite animal cell. **(a)** Diagram. **(b)** Transmission electron micrograph (5500×).

Nucleus

The nucleus contains the genetic material, DNA, sections of which are called *genes*. Often described as the control center of the cell, the nucleus is necessary for cell reproduction. A cell that has lost or ejected its nucleus is literally programmed to die.

When the cell is not dividing, the genetic material is loosely dispersed throughout the nucleus in a threadlike form called **chromatin.** When the cell is in the process of dividing to form daughter cells, the chromatin coils and condenses, forming dense, darkly staining rodlike bodies called **chromosomes**—much in the way a stretched spring becomes shorter and thicker when it is released. Carefully note the appearance of the nucleus—it is somewhat nondescript when a cell is healthy. When the nucleus appears dark and the chromatin becomes clumped, this is an indication that the cell is dying and undergoing degeneration.

The nucleus also contains one or more small round bodies, called **nucleoli,** composed primarily of proteins and ribonucleic acid (RNA). The nucleoli are assembly sites for ribosomal particles that are particularly abundant in the cytoplasm. Ribosomes are the actual protein-synthesizing "factories."

The nucleus is bound by a double-layered porous membrane, the **nuclear envelope.** The nuclear envelope is similar in composition to other cellular membranes, but it is distinguished by its large **nuclear pores.** Nuclear pores are spanned by protein complexes that regulate what passes through, and they permit easy passage of protein and RNA molecules.

ACTIVITY 1

Identifying Parts of a Cell

As able, identify the nuclear envelope, chromatin, nucleolus, and the nuclear pores (Figure 4.1a and b and Figure 4.3). ■

Plasma Membrane

The plasma membrane separates cell contents from the surrounding environment. Its main structural building blocks are phospholipids (fats) and globular protein molecules. Some of the externally facing proteins and lipids have sugar (carbohydrate) side chains attached to them that are important in cellular interactions **(Figure 4.2)**. Described in terms of the fluid mosaic model, the membrane is a bilayer of phospholipid molecules in which the protein molecules float. Occasional cholesterol molecules dispersed in the fluid phospholipid bilayer help stabilize it.

Besides providing a protective barrier for the cell, the plasma membrane plays an active role in determining which substances may enter or leave the cell and in what quantity. Because of its molecular composition, the plasma membrane is selective about what passes through it. It allows nutrients to enter the cell but keeps out undesirable substances. By the

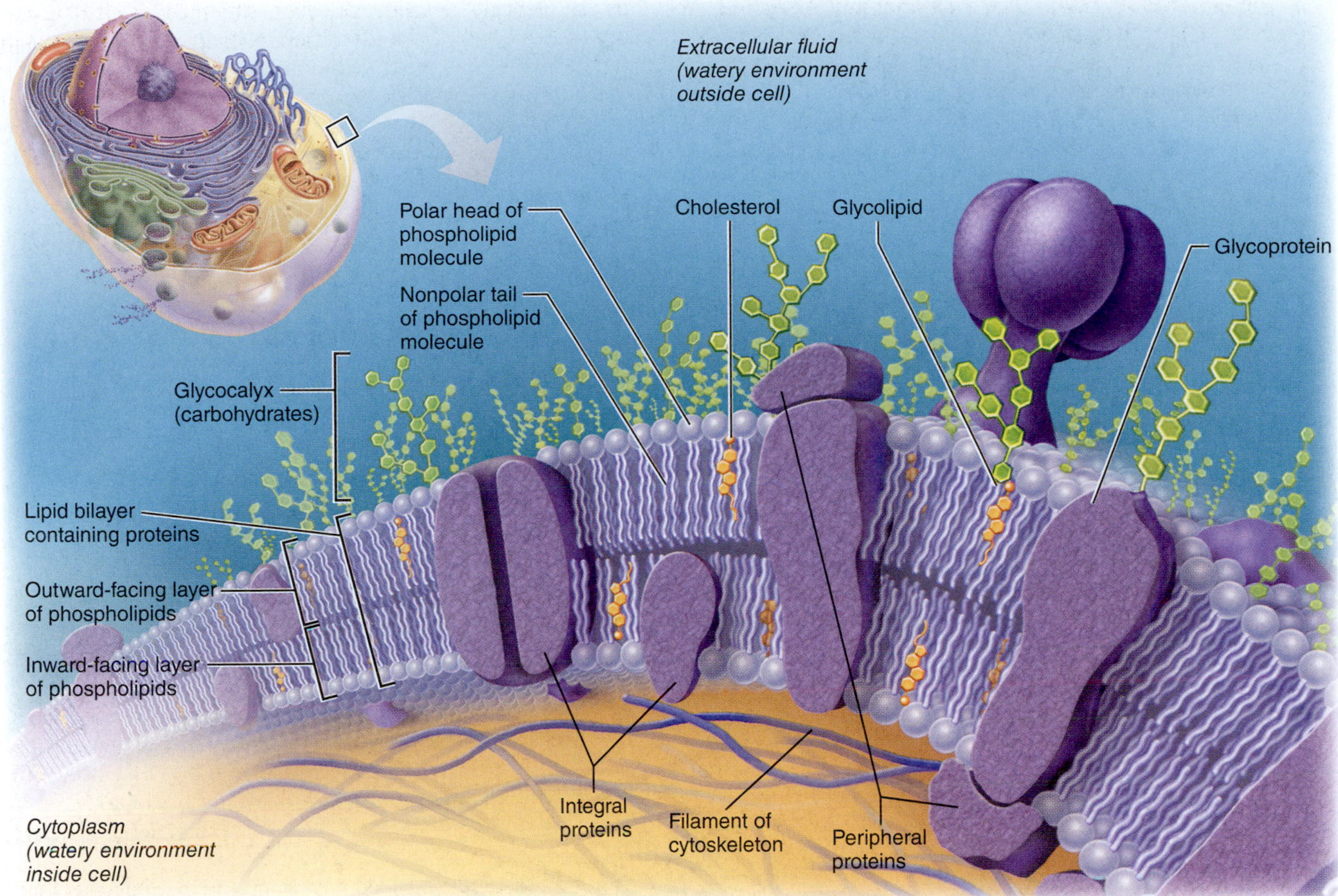

Figure 4.2 Structural details of the plasma membrane.

same token, valuable cell proteins and other substances are kept within the cell, and excreta or wastes pass to the exterior. This property is known as **selective permeability.** Transport through the plasma membrane occurs in two basic ways. In *active transport*, the cell must provide energy in the form of adenosine triphosphate, or ATP, to power the transport process. In *passive transport*, the transport process is driven by concentration or pressure differences.

Additionally, the plasma membrane maintains a resting potential that is essential to normal functioning of excitable cells, such as neurons and muscle cells, and plays a vital role in cell signaling and in cell-to-cell interactions. In some cells the membrane is thrown into minute fingerlike projections or folds called **microvilli** (Figure 4.3). Microvilli greatly increase the surface area of the cell available for absorption or passage of materials and for the binding of signaling molecules.

ACTIVITY 2

Identifying Components of a Plasma Membrane

Identify the phospholipid and protein portions of the plasma membrane (Figure 4.2). Also locate the sugar (*glyco* = carbohydrate) side chains and cholesterol molecules. Identify the microvilli (Figure 4.3). ■

Cytoplasm and Organelles

The cytoplasm consists of the cell contents outside the nucleus. It is the major site of most activities carried out by the cell. Suspended in the **cytosol**, the fluid cytoplasmic material, are many small structures called **organelles** (literally, "small organs"). The organelles are the metabolic machinery of the cell, and they are highly organized to carry out specific functions for the cell as a whole. The organelles include the ribosomes, endoplasmic reticulum, Golgi apparatus, lysosomes, peroxisomes, mitochondria, cytoskeletal elements, and centrioles.

ACTIVITY 3

Locating Organelles

Each organelle type is summarized in the table (Table 4.1) and described briefly below. Read through this material and then, as best you can, locate the organelles (see Figures 4.1b and 4.3). ■

- **Ribosomes** are densely staining, roughly spherical bodies composed of RNA and protein. They are the actual sites of protein synthesis. They are seen floating free in the cytoplasm or attached to a membranous structure. When they are attached, the whole ribosome-membrane complex is called the *rough endoplasmic reticulum*.

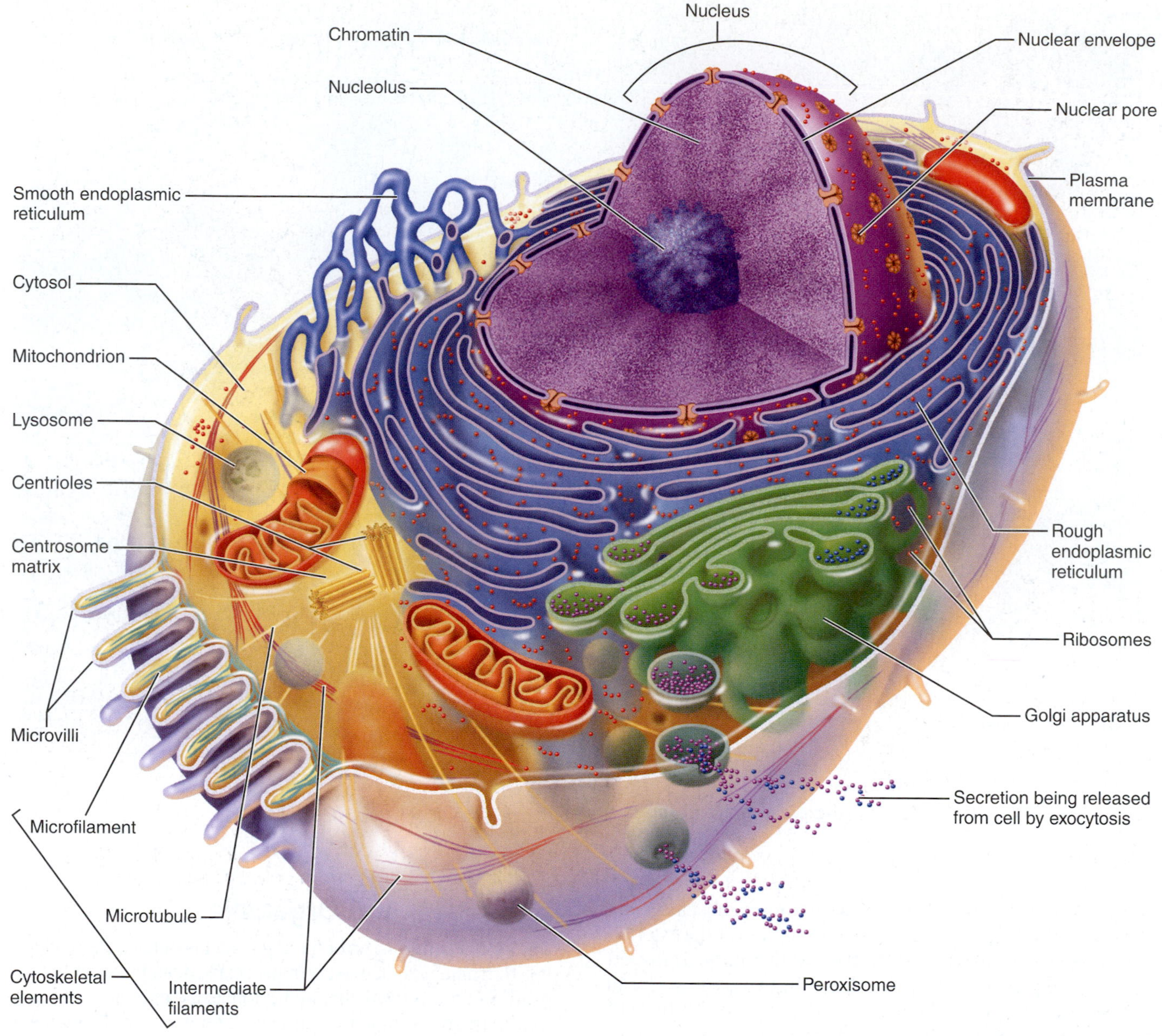

Figure 4.3 Structure of the generalized cell. No cell is exactly like this one, but this composite illustrates features common to many human cells. Note that not all organelles are drawn to the same scale in this illustration.

• The **endoplasmic reticulum (ER)** is a highly folded system of membranous tubules and cisterns (sacs) that extends throughout the cytoplasm (see Figure 4.3). The ER is continuous with the nuclear envelope, forming a system of channels for the transport of cellular substances (primarily proteins) from one part of the cell to another. The ER exists in two forms: **rough ER** and **smooth ER.** A particular cell may have both or only one, depending on its specific functions. The rough ER is studded with ribosomes. Its cisterns modify and store the newly formed proteins and dispatch them to other areas of the cell. The external face of the rough ER is involved in phospholipid and cholesterol synthesis. The amount of rough ER is closely correlated with the amount of protein a cell manufactures and is especially abundant in cells that make protein products for export—for example, the pancreas cells that produce digestive enzymes destined for the small intestine.

The smooth ER does not participate in protein synthesis but is present in conspicuous amounts in cells that produce steroid-based hormones—for example, the interstitial endocrine cells of the testes, which produce testosterone. Smooth ER is also abundant in cells that are active in lipid metabolism and drug detoxification activities—liver cells, for instance.

• The **Golgi apparatus** is a stack of flattened sacs with bulbous ends and associated membranous vesicles that is generally found close to the nucleus. Within its cisterns, the proteins delivered to it by transport vesicles from the

Table 4.1 Cytoplasmic Organelles

Organelle		Location and function
Ribosomes		Tiny spherical bodies composed of RNA and protein; floating free or attached to a membranous structure (the rough ER) in the cytoplasm. Actual sites of protein synthesis.
Endoplasmic reticulum (ER)		Membranous system of tubules that extends throughout the cytoplasm. Two varieties: Rough ER is studded with ribosomes (tubules of the rough ER provide an area for storage and transport of the proteins made on the ribosomes to other cell areas; external face synthesizes phospholipids and cholesterol). Smooth ER has no function in protein synthesis; rather, it is a site of steroid and lipid synthesis, lipid metabolism, and drug detoxification.
Golgi apparatus		Stack of flattened sacs with bulbous ends and associated small vesicles; found close to the nucleus. Plays a role in packaging proteins or other substances for export from the cell or incorporation into the plasma membrane and in packaging lysosomal enzymes.
Lysosomes	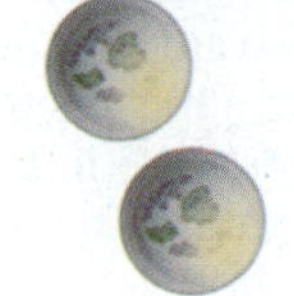	Various-sized membranous sacs containing digestive enzymes (acid hydrolases). Function to digest worn-out cell organelles and foreign substances that enter the cell; have the capacity to destroy the cell entirely if they are ruptured.
Peroxisomes		Small lysosome-like membranous sacs containing oxidase enzymes that detoxify alcohol, hydrogen peroxide, and other harmful chemicals.
Mitochondria		Generally rod-shaped bodies with a double-membrane wall; inner membrane is thrown into folds, or cristae. Contain enzymes that oxidize foodstuffs to produce cellular energy (ATP); often referred to as "powerhouses of the cell."
Centrioles		Paired, cylindrical bodies lie at right angles to each other, close to the nucleus. Direct the formation of the mitotic spindle during cell division; form the bases of cilia and flagella.
Cytoskeletal elements: microfilaments, intermediate filaments, and microtubules		Provide cellular support; function in intracellular transport. Microfilaments are formed largely of actin, a contractile protein, and thus are important in cell mobility (particularly in muscle cells). Intermediate filaments are stable elements composed of a variety of proteins and resist mechanical forces acting on cells. Microtubules form the internal structure of the centrioles and help determine cell shape.

rough ER are modified, segregated, and packaged into membranous vesicles that ultimately (1) are incorporated into the plasma membrane, (2) become secretory vesicles that release their contents from the cell, or (3) become lysosomes.

- **Lysosomes,** which appear in various sizes, are membrane-bound sacs containing an array of powerful digestive enzymes. A product of the packaging activities of the Golgi apparatus, the lysosomes contain *acid hydrolases,* enzymes capable of digesting worn-out cell structures and foreign substances that enter the cell via vesicle formation through phagocytosis or endocytosis. Because they have the capacity of total cell destruction, the lysosomes are often referred to as the "suicide sacs" of the cell.

- **Peroxisomes,** like lysosomes, are enzyme-containing sacs. However, their *oxidases* have a different task. Using oxygen, they detoxify a number of harmful substances, the most important of which are free radicals. Peroxisomes are particularly abundant in kidney and liver cells—cells that are actively involved in detoxification.

- **Mitochondria** are generally rod-shaped bodies with a double-membrane wall; the inner membrane is thrown into folds, or *cristae.* Oxidative enzymes on or within the mitochondria catalyze the reactions of the citric acid cycle and the electron transport chain (collectively called aerobic cellular respiration), in which the end products of food digestion are broken down to produce energy. The released energy is captured in the bonds of ATP molecules, which are then transported out of the mitochondria to provide a ready energy supply to power the cell. Every living cell requires a constant supply of ATP for its many activities. Because the mitochondria provide the bulk of this ATP, they are referred to as the "powerhouses" of the cell.

- **Cytoskeletal elements** ramify throughout the cytoplasm, forming an internal scaffolding called the *cytoskeleton* that supports and moves substances within the cell. The **microtubules** are slender tubules formed of proteins called *tubulins.* Most microtubules radiate from a region of cytoplasm near the nucleus called the *centrosome,* and they have the ability to aggregate and then disaggregate spontaneously. Microtubules organize the cytoskeleton and form the spindle during cell division. They also transport substances down the length of elongated cells (such as neurons), suspend organelles, and help maintain cell shape by providing rigidity to the soft cellular substance. The stable **intermediate filaments** are proteinaceous cytoskeletal elements that act as internal guy-wires to resist mechanical (pulling) forces acting on cells. **Microfilaments,** ribbon or cordlike elements, are formed of contractile proteins, primarily *actin.* Because of their ability to shorten and then relax to assume a more elongated form, these are important in cell mobility and are conspicuous in muscle cells that are specialized to contract. A cross-linked network of microfilaments called the *terminal web* braces and strengthens the internal face of the plasma membrane.

  The cytoskeletal structures are changeable and microscopic. With the exception of the microtubules of the mitotic spindle, which are obvious during cell division (see pages 46–47), and the microfilaments of skeletal muscle cells, they are rarely seen, even in electron micrographs (and they are not visible in Figure 4.1b). However, special stains can reveal the plentiful supply of these important structures.

- The paired **centrioles** lie close to the nucleus within the centrosome in cells capable of reproducing themselves. They are rod-shaped bodies that lie at right angles to each other. Internally each centriole is composed of nine triplets of microtubules. During cell division, the centrosome complex that contains the centrioles directs the formation of the mitotic spindle. Centrioles also form the basis for cell projections called cilia and flagella.

The cell cytoplasm contains various other substances and structures, including stored foods (glycogen granules and lipid droplets), pigment granules, crystals of various types, water vacuoles, and ingested foreign materials. However, these are not part of the active metabolic machinery of the cell and are therefore called **inclusions.**

ACTIVITY 4

Examining the Cell Model

Once you have located all of these structures in the figure (Figure 4.3), examine the cell model (or cell chart) to repeat and reinforce your identifications. ■

Differences and Similarities in Cell Structure

ACTIVITY 5

Observing Various Cell Structures

1. Obtain a compound microscope and prepared slides of simple squamous epithelium, smooth muscle cells (teased), human blood, and sperm.

2. Observe each slide under the microscope, carefully noting similarities and differences in the cells. See photomicrographs of simple squamous epithelium (Figure 3.5 in Exercise 3) and teased smooth muscle (Figure 5.7c in Exercise 5). You'll need the oil immersion lens to observe blood and sperm. Distinguish the limits of the individual cells, and notice the shape and position of the nucleus in each case. When you look at the human blood smear, direct your attention to the red blood cells, the pink-stained cells that are most numerous. The color photomicrographs illustrating a blood smear (Figure 22.3 in Exercise 22) and sperm (Figure 29.3 in Exercise 29) may be helpful in this cell structure study. Sketch your observations in the circles provided (page 45).

3. Measure the length and/or diameter of each cell, and record below the appropriate sketch.

4. How do these four cell types differ in shape and size?

____

____

____

____

How might cell shape affect cell function?

____

____

____

____

____

Which cells have visible projections?

____

How do these projections relate to the function of these cells?

____

____

Do any of these cells lack a plasma membrane? ____________

Simple squamous epithelium

Diameter ________

Sperm cells

Length ________

Human red blood cells

Diameter ________

Teased smooth muscle cells

Length ________

A nucleus? ________________________

In the cells with a nucleus, can you discern nucleoli?

____

____

Were you able to observe any of the organelles in these cells?

________ Why or why not? ________

____

____

____

____

Cell Division: Mitosis and Cytokinesis

A cell's *life cycle* is the series of changes it goes through from the time it is formed until it reproduces itself. It consists of two stages—**interphase,** the longer period during which the cell grows and carries out its usual activities **(Figure 4.4a)**, and **cell division,** when the cell reproduces itself by dividing. In an interphase cell about to divide, the genetic material (DNA) is copied exactly. Once this important event has occurred, cell division ensues.

Cell division in all cells other than bacteria consists of two events, mitosis and cytokinesis. **Mitosis** is the division of the copied DNA of the mother cell to two daughter cells. **Cytokinesis** is the division of the cytoplasm. It begins when mitosis is nearly complete. Although mitosis is usually accompanied by cytokinesis, in some instances cytoplasmic division does not occur, leading to the formation of binucleate or multinucleate cells. This is relatively common in the human liver.

The product of **mitosis** is two daughter nuclei that are genetically identical to the mother nucleus. This distinguishes mitosis from **meiosis,** a specialized type of nuclear division that occurs only in the reproductive organs (testes or ovaries). Meiosis, which yields four daughter nuclei that differ genetically in composition from the mother nucleus, is used only for producing gametes (eggs and sperm) for sexual reproduction. The function of cell division, including mitosis and cytokinesis in the body, is to increase the number of cells for growth and repair while maintaining their genetic heritage.

The phases of mitosis include **prophase, metaphase, anaphase,** and **telophase.** The detailed events of interphase, mitosis, and cytokinesis are described and illustrated in the figure (Figure 4.4).

Mitosis is essentially the same in all animal cells, but depending on the tissue, it takes from 5 minutes to several hours to complete. In most cells, centriole replication occurs during interphase of the next cell cycle.

At the end of cell division, two daughter cells exist—each with a smaller cytoplasmic mass than the mother cell but genetically identical to it. The daughter cells grow and carry out the normal spectrum of metabolic processes until it is their turn to divide.

Cell division is extremely important during the body's growth period. Most cells divide until puberty, when normal body size is achieved and overall body growth ceases.

(Text continues on page 48.)

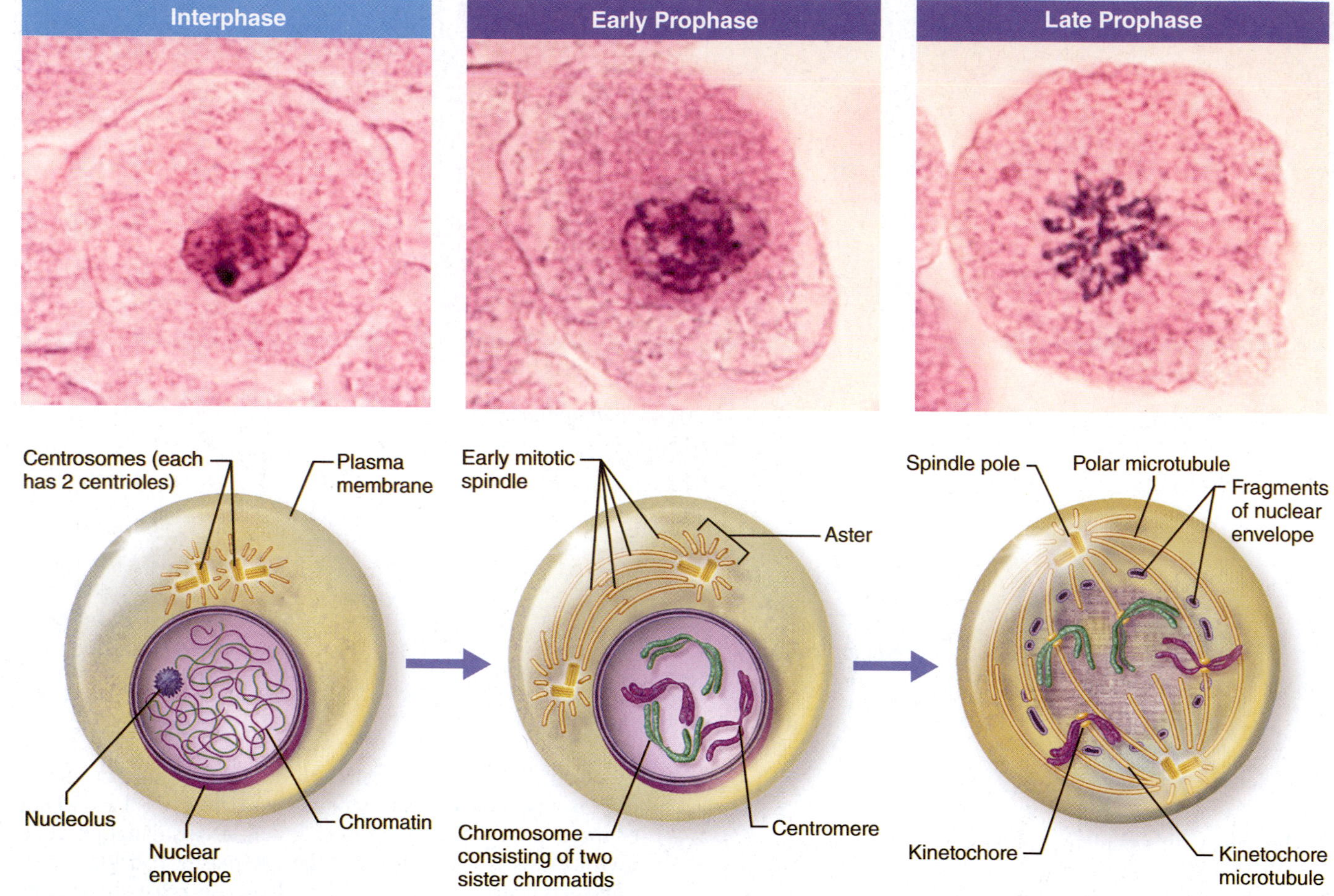

Interphase

Interphase is the period of a cell's life when it carries out its normal metabolic activities and grows. Interphase is not part of mitosis.

- During interphase, the DNA-containing material is in the form of chromatin. The nuclear envelope and one or more nucleoli are intact and visible.
- There are three distinct periods of interphase:
  - G_1: The centrioles begin replicating.
  - S: DNA is replicated.
  - G_2: Final preparations for mitosis are completed and centrioles finish replicating.

Prophase—first phase of mitosis

Early Prophase

- The chromatin condenses, forming barlike chromosomes.
- Each duplicated chromosome consists of two identical threads, called **sister chromatids**, held together at the **centromere**. (Later when the chromatids separate, each will be a new chromosome.)
- As the chromosomes appear, the nucleoli disappear, and the two centrosomes separate from one another.
- The centrosomes act as focal points for growth of a microtubule assembly called the **mitotic spindle**. As the microtubules lengthen, they propel the centrosomes toward opposite ends (poles) of the cell.
- Microtubule arrays called **asters** ("stars") extend from the centrosome matrix.

Late Prophase

- The nuclear envelope breaks up, allowing the spindle to interact with the chromosomes.
- Some of the growing spindle microtubules attach to **kinetochores**, special protein structures at each chromosome's centromere. Such microtubules are called **kinetochore microtubules**.
- The remaining spindle microtubules (not attached to any chromosomes) are called **polar microtubules**. The microtubules slide past each other, forcing the poles apart.
- The kinetochore microtubules pull on each chromosome from both poles in a tug-of-war that ultimately draws the chromosomes to the center, or equator, of the cell.

Figure 4.4 The interphase cell and the events of cell division. The cells shown are from an early embryo of a whitefish. Photomicrographs are above; corresponding diagrams are below. (Photomicrographs approximately 1530×.)

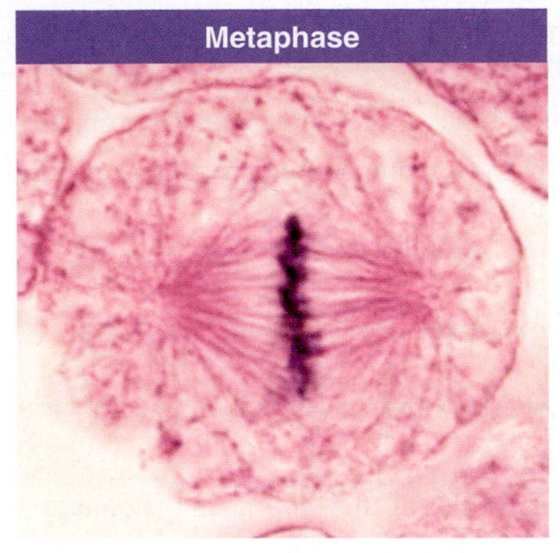

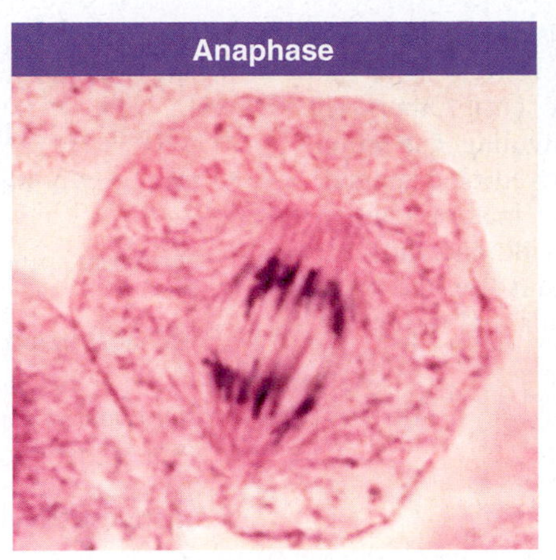

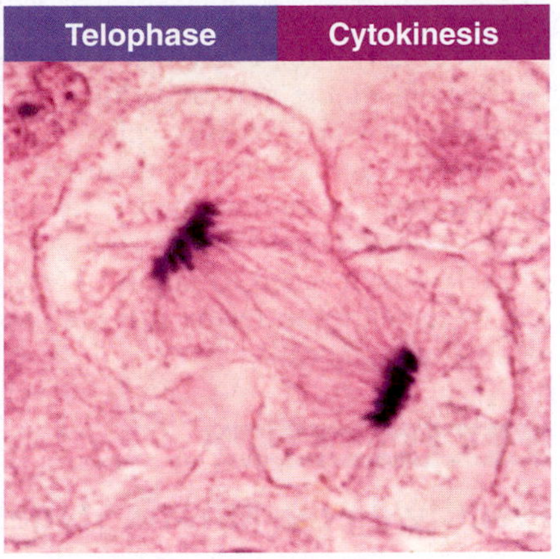

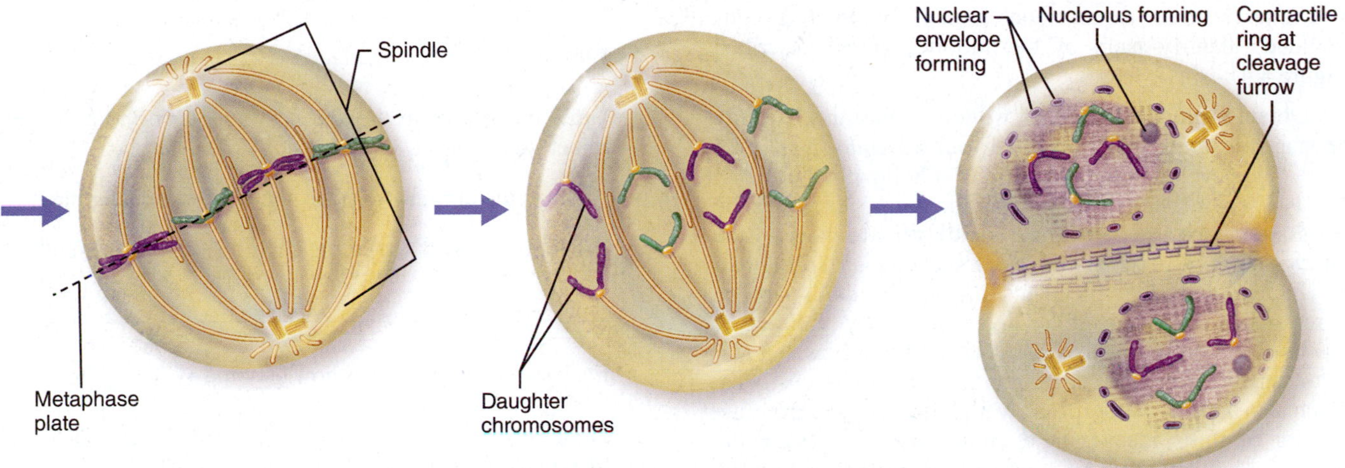

Metaphase—second phase of mitosis

- The two centrosomes are at opposite poles of the cell.
- The chromosomes cluster at the midline of the cell, with their centromeres precisely aligned at the **equator** of the spindle. This imaginary plane midway between the poles is called the **metaphase plate**.
- Enzymes act to separate the chromatids from each other.

Anaphase—third phase of mitosis

The shortest phase of mitosis, anaphase begins abruptly as the centromeres of the chromosomes split simultaneously. Each chromatid now becomes a chromosome in its own right.

- The kinetochore microtubules, moved along by motor proteins in the kinetochores, gradually pull each chromosome toward the pole it faces.
- At the same time, the polar microtubules slide past each other, lengthen, and push the two poles of the cell apart.
- The moving chromosomes look V shaped. The centromeres lead the way, and the chromosomal "arms" dangle behind them.
- Moving and separating the chromosomes is helped by the fact that the chromosomes are short, compact bodies. Diffuse threads of chromatin would trail, tangle, and break, resulting in imprecise "parceling out" to the daughter cells.

Telophase—final phase of mitosis

Telophase begins as soon as chromosomal movement stops. This final phase is like prophase in reverse.

- The identical sets of chromosomes at the opposite poles of the cell uncoil and resume their threadlike chromatin form.
- A new nuclear envelope forms around each chromatin mass, nucleoli reappear within the nuclei, and the spindle breaks down and disappears.
- Mitosis is now ended. The cell, for just a brief period, is binucleate (has two nuclei) and each new nucleus is identical to the original mother nucleus.

Cytokinesis—division of cytoplasm

Cytokinesis begins during late anaphase and continues through and beyond telophase. A contractile ring of actin microfilaments forms the **cleavage furrow** and pinches the cell apart.

Figure 4.4 *(continued)*

After this time in life, only certain cells carry out cell division routinely—for example, cells subjected to abrasion (epithelium of the skin and lining of the gut). Other cell populations—such as liver cells—stop dividing but retain this ability should some of them be removed or damaged. Skeletal muscle, cardiac muscle, and most mature neurons almost completely lose this ability to divide and thus are severely handicapped by injury. Throughout life, the body retains its ability to repair cuts and wounds and to replace some of its aged cells.

ACTIVITY 6

Identifying the Mitotic Stages

1. Watch a video presentation of mitosis (if available).

2. Using the three-dimensional models of dividing cells provided, identify each of the mitotic phases described (Figure 4.4).

3. Obtain a prepared slide of whitefish blastulas to study the stages of mitosis. The cells of each *blastula* (a stage of embryonic development consisting of a hollow ball of cells) are at approximately the same mitotic stage, so it may be necessary to observe more than one blastula to view all the mitotic stages. A good analogy for a blastula is a soccer ball in which each leather piece making up the ball's surface represents an embryonic cell. The exceptionally high rate of mitosis observed in this tissue is typical of embryos, but if it occurs in specialized tissues it can indicate cancerous cells, which also have an extraordinarily high mitotic rate. Examine the slide carefully, identifying the four mitotic phases and the process of cytokinesis. Compare your observations with the figure (Figure 4.4), and verify your identifications with your instructor. ■

"Chenille Stick" Mitosis

1. Obtain a total of 8 chenille sticks, each measuring 3 inches: four of one color and four of another color (e.g., four green and four purple).

2. Assemble the chenille sticks into a total of four chromosomes (each with two sister chromatids) by twisting two sticks of the same color together at the center with a single twist.

What does the twist at the center represent? ____________

3. Arrange the chromosomes as they appear in early prophase.

Name the structure that assembles during this phase.

____

Draw early prophase in the space provided on your Review Sheet (question 11, page 52).

4. Arrange the chromosomes as they appear in late prophase.

What structure on the chromosome centromere do the growing spindle microtubules attach to? ____________

What structure is now present as fragments? ____________

Draw late prophase in the space provided on your Review Sheet (question 11, page 52).

5. Arrange the chromosomes as they appear in metaphase.

What is the name of the imaginary plane that the chromosomes align along? ______________________________

Draw metaphase in the space provided on your Review Sheet (question 11, page 52).

6. Arrange the chromosomes as they appear in anaphase.

What does untwisting of the chenille sticks represent?

____

Each sister chromatid has now become a ____________.

Draw anaphase in the space provided on your Review Sheet (question 11, page 52).

7. Arrange the chromosomes as they appear in telophase.

Briefly list four reasons why telophase is like the reverse of prophase.

____

____

____

____

Draw telophase in the space provided on your Review Sheet (question 11, page 52).

Name ______________________

Lab Time/Date ______________________

EXERCISE 4

REVIEW SHEET

The Cell: Anatomy and Division

Anatomy of the Composite Cell

1. Define the following terms:

organelle: ______________________

____

cell: ______________________

2. Although cells have differences that reflect their specific functions in the body, what functions do they have in common?

____

3. Identify the following cell parts:

______________________ 1. external boundary of cell; regulates flow of materials into and out of the cell; site of cell signaling

______________________ 2. contains digestive enzymes of many varieties; "suicide sac" of the cell

______________________ 3. scattered throughout the cell; major site of ATP synthesis

______________________ 4. slender extensions of the plasma membrane that increase its surface area

______________________ 5. stored glycogen granules, crystals, pigments, and so on

______________________ 6. membranous system consisting of flattened sacs and vesicles; packages proteins for export

______________________ 7. control center of the cell; necessary for cell division and cell life

______________________ 8. two rod-shaped bodies near the nucleus; associated with the formation of the mitotic spindle

______________________ 9. dense, darkly staining nuclear body; packaging site for ribosomes

______________________ 10. contractile elements of the cytoskeleton

______________________ 11. membranous system; involved in intracellular transport of proteins and synthesis of membrane lipids

______________________ 12. attached to membrane systems or scattered in the cytoplasm; site of protein synthesis

______________________ 13. threadlike structures in the nucleus; contain genetic material (DNA)

______________________ 14. site of free-radical detoxification

4. In the following diagram, label all parts provided with a leader line.

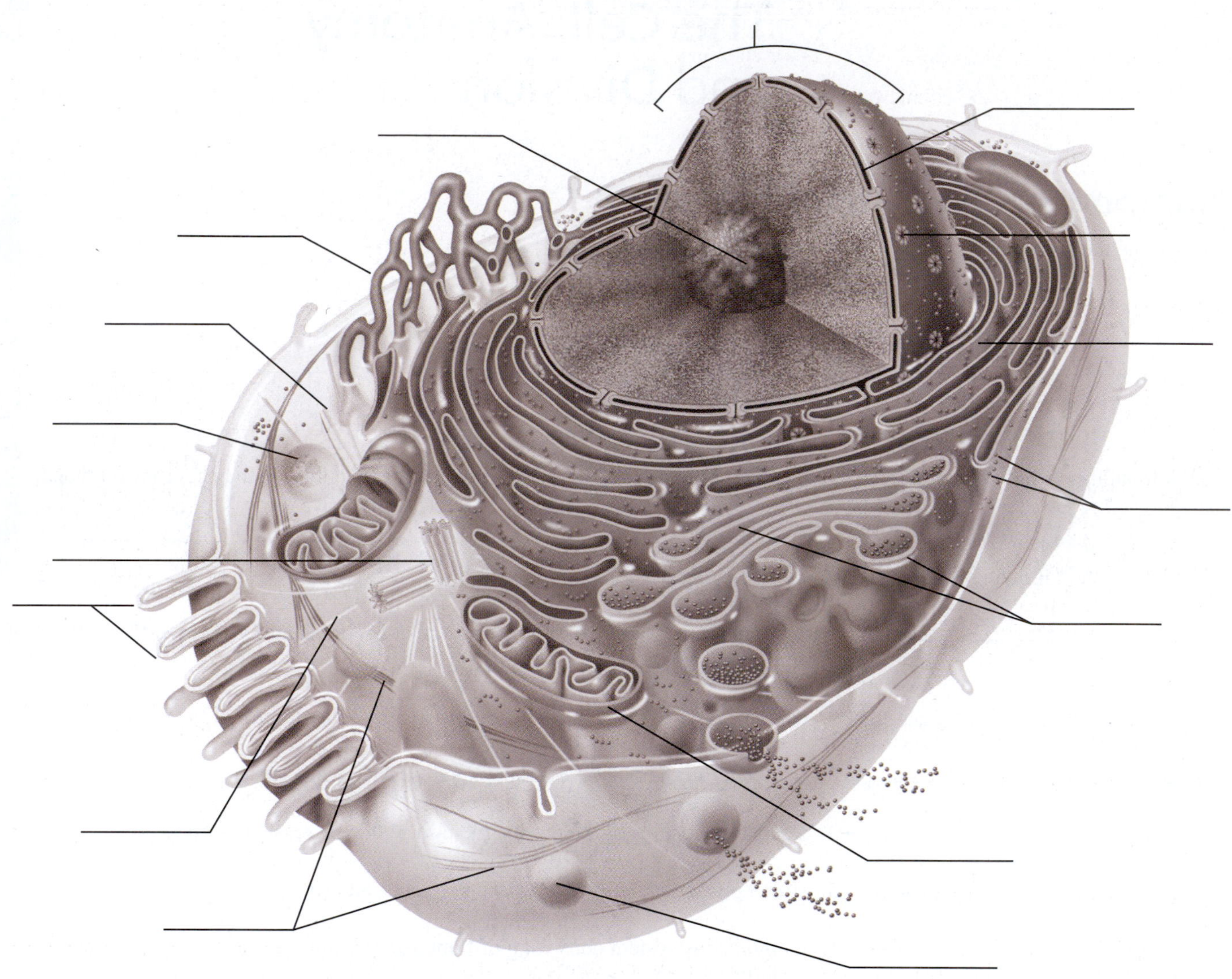

Differences and Similarities in Cell Structure

5. For each of the following cell types, list (a) *one* important structural characteristic observed in the laboratory, and (b) the function that the structure complements or ensures.

squamous epithelium a. ______________________________

b. ______________________________

sperm a. ______________________________

b. ______________________________

smooth muscle a. ______________________________

b. ______________________________

red blood cells a. ______________________________

b. ______________________________

6. What is the significance of the red blood cell being anucleate (without a nucleus)? ______________________

____

Did it ever have a nucleus? __________ If so, when? ______________________ (Use an appropriate reference.)

7. Of the four cells observed microscopically (squamous epithelial cells, red blood cells, smooth muscle cells, and sperm),

which has the smallest diameter? ______________________ Which is the longest? ______________________

Cell Division: Mitosis and Cytokinesis

8. Identify the three phases of mitosis in the following photomicrographs.

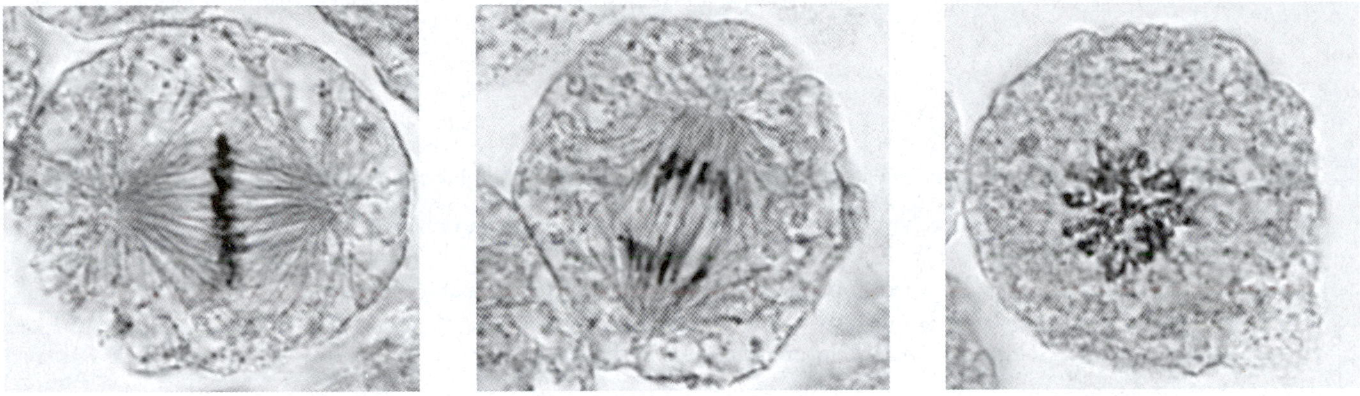

a. ______________________ b. ______________________ c. ______________________

9. What is the importance of mitotic cell division? ______________________

____

10. Complete the following statements:

Division of the __1__ is referred to as mitosis. Cytokinesis is division of the __2__. The major structural difference between chromatin and chromosomes is that the latter is __3__. Chromosomes attach to the spindle fibers by undivided structures called __4__. If a cell undergoes mitosis but not cytokinesis, the product is __5__. The structure that acts as a scaffolding for chromosomal attachment and movement is called the __6__. __7__ is the period of cell life when the cell is not involved in division. Two cell populations in the body that do not routinely undergo cell division are __8__ and __9__.

1. ______________________
2. ______________________
3. ______________________
4. ______________________
5. ______________________
6. ______________________
7. ______________________
8. ______________________
9. ______________________

11. Draw the phases of mitosis for a cell with a chromosome number of 4.

12. Using the key, categorize each of the events described below according to the phase in which it occurs.

Key: a. anaphase b. interphase c. metaphase d. prophase e. telophase

__________ 1. Chromatin coils and condenses, forming chromosomes.

__________ 2. The chromosomes are V-shaped.

__________ 3. The nuclear membrane re-forms.

__________ 4. Chromosomes stop moving toward the poles.

__________ 5. Chromosomes line up in the center of the cell.

__________ 6. The nuclear membrane fragments.

__________ 7. The mitotic spindle forms.

__________ 8. DNA synthesis occurs.

__________ 9. Centrioles replicate.

__________ 10. Chromosomes first appear to be duplex structures.

__________ 11. Chromosomal centromeres are attached to the kinetochore fibers.

__________ 12. Cleavage furrow forms.

__________ and __________ 13. The nuclear membrane(s) is absent.

13. What is the physical advantage of the chromatin coiling and condensing to form short chromosomes at the onset of mitosis?

____

____

EXERCISE 5

Classification of Tissues

MATERIALS

- ☐ Compound microscope
- ☐ Immersion oil
- ☐ Prepared slides of simple squamous, simple cuboidal, simple columnar, stratified squamous (nonkeratinized), stratified cuboidal, stratified columnar, pseudostratified ciliated columnar, and transitional epithelium
- ☐ Prepared slides of mesenchyme; of adipose, areolar, reticular, and dense (regular, irregular, and elastic) connective tissues; of hyaline and elastic cartilage; of fibrocartilage; of bone (x.s.); and of blood
- ☐ Prepared slide of nervous tissue (spinal cord smear)
- ☐ Prepared slides of skeletal, cardiac, and smooth muscle (l.s.)

OBJECTIVES

1. Name the four primary tissue types in the human body, and state a general function of each type.
2. Name the major subcategories of the primary tissue types, and identify the tissues of each subcategory microscopically or in an appropriate image.
3. State the locations of the various tissues in the body.
4. List the general function and structural characteristics of each of the tissues studied.

PRE-LAB QUIZ

1. Groups of cells that are anatomically similar and share a function are called:
   a. organ systems   c. organs
   b. organisms   d. tissues
2. How many primary tissue types are found in the human body? ___________
3. Circle True or False. Endocrine and exocrine glands are classified as epithelium because they usually develop from epithelial membranes.
4. Epithelial tissues can be classified according to cell shape. _______________ epithelial cells are scalelike and flattened.
   a. Columnar   c. Squamous
   b. Cuboidal   d. Transitional
5. All connective tissue is derived from an embryonic tissue known as:
   a. cartilage   c. mesenchyme
   b. ground substance   d. reticular
6. All the following are examples of connective tissue *except:*
   a. bones   c. neurons
   b. ligaments   d. tendons
7. Circle True or False. Blood is a type of connective tissue.
8. Circle the correct underlined term. Of the two major cell types found in nervous tissue, neurons / neuroglial cells are highly specialized to generate and conduct electrical signals.
9. How many basic types of muscle tissue are there? ___________________
10. This type of muscle tissue is found in the walls of hollow organs. It has no striations, and its cells are spindle shaped. It is:
    a. cardiac muscle
    b. skeletal muscle
    c. smooth muscle

Cells are the building blocks of life and the all-inclusive functional units of unicellular organisms. However, in higher organisms, cells do not usually operate as isolated, independent entities. In humans and other multicellular organisms, cells depend on one another and cooperate to maintain homeostasis in the body.

With a few exceptions, even the most complex animal starts out as a single cell, the fertilized egg, which divides almost endlessly. The trillions of cells that result become specialized for a particular function; some become supportive bone, others the transparent lens of the eye, still others skin cells, and so on. Thus a division of labor exists, with certain groups of cells highly specialized to perform functions that benefit the organism as a whole. Cell specialization carries with it certain hazards, because when a small specific group of cells is indispensable, any inability to function on its part can paralyze or destroy the entire body.

Groups of cells that are similar in structure and function are called **tissues.** The four primary tissue types—epithelial tissue, connective tissue, nervous tissue, and muscle—have distinctive structures, patterns, and functions. The four primary tissues are further divided into subcategories, as described shortly.

To perform specific body functions, the tissues are organized into **organs** such as the heart, kidneys, and lungs. Most organs contain several representatives of the primary tissues, and the arrangement of these tissues determines the organ's structure and function. Thus **histology,** the study of tissues, complements a study of gross anatomy and provides the structural basis for a study of organ physiology.

The main objective of this exercise is to familiarize you with the major similarities and dissimilarities of the primary tissues, so that when the tissue composition of an organ is described, you will be able to more easily understand (and perhaps even predict) the organ's major function. This exercise focuses chiefly on epithelial tissue and some types of connective tissue. (Muscle, nervous tissue, and bone, a connective tissue, are covered in more depth in later exercises.)

Epithelial Tissue

Epithelial tissue, or an **epithelium,** is a sheet of cells that covers a body surface or lines a body cavity. It occurs in the body as (1) covering and lining epithelium and (2) glandular epithelium. Covering and lining epithelium forms the outer layer of the skin and lines body cavities that open to the outside. It covers the walls and organs of the closed body cavity. Because glands almost invariably develop from epithelial sheets, glands are classified as epithelium.

Epithelial functions include protection, absorption, filtration, excretion, secretion, and sensory reception. For example, the epithelium covering the body surface protects against bacterial invasion and chemical damage; that lining the respiratory tract is ciliated to sweep dust and other foreign particles away from the lungs. Epithelium specialized to absorb substances lines the stomach and small intestine. In the kidney tubules, the epithelium absorbs, secretes, and filters. Secretion is a specialty of the glands.

The following characteristics distinguish epithelial tissues from other types:

- **Polarity.** The membranes always have one free surface, called the *apical surface,* and typically that surface is significantly different from the *basal surface.*
- **Specialized contacts.** Cells fit closely together to form membranes, or sheets of cells, and are bound together by specialized junctions.
- **Supported by connective tissue.** The cells are attached to and supported by an adhesive **basement membrane,** which is an amorphous material secreted partly by the epithelial cells *(basal lamina)* and connective tissue cells *(reticular lamina)* that lie adjacent to each other.
- **Avascular but innervated.** Epithelial tissues are supplied by nerves but have no blood supply of their own (are avascular). Instead they depend on diffusion of nutrients from the underlying connective tissue. Glandular epithelia, however, are very vascular.
- **Regeneration.** If well nourished, epithelial cells can easily regenerate themselves. This is an important characteristic because many epithelia are subjected to a good deal of friction.

The covering and lining epithelia are classified according to two criteria—arrangement or relative number of layers and cell shape **(Figure 5.1)**. On the basis of arrangement, there are **simple** epithelia, consisting of one layer of cells attached to the basement membrane, and **stratified** epithelia, consisting of two or more layers of cells. The general types based on shape are **squamous** (scalelike), **cuboidal** (cubelike), and **columnar** (column-shaped) epithelial cells. The terms denoting shape and arrangement of the epithelial cells are combined to describe the epithelium fully. *Stratified epithelia are named according to the cells at the apical surface of the epithelial sheet,* not those resting on the basement membrane.

There are, in addition, two less easily categorized types of epithelia. **Pseudostratified epithelium** is actually a simple columnar epithelium (one layer of cells), but because its cells vary in height and their nuclei lie at different levels above the basement membrane, it gives the false appearance of being stratified. This epithelium is often ciliated. **Transitional epithelium** is a rather peculiar stratified squamous epithelium formed of rounded, or "plump," cells with the ability to slide over one another to allow the organ to be stretched. Transitional epithelium is found only in urinary system organs subjected to periodic distension, such as the bladder. The superficial cells are flattened (like true squamous cells) when the organ is distended and rounded when the organ is empty.

Epithelial cells forming glands are highly specialized to remove materials from the blood and to manufacture them into new materials, which they then secrete. There are two types of glands: exocrine and endocrine glands **(Figure 5.2)**. **Endocrine glands** lose their surface connection (duct) as they develop; thus they are referred to as ductless glands. They secrete hormones into the extracellular fluid, and from there the hormones enter the blood or the lymphatic vessels that weave through the glands. **Exocrine glands** retain their ducts, and their secretions empty through these ducts, either to a body surface or into body cavities. The exocrine glands include the sweat and oil glands, liver, and pancreas. Glands are discussed in conjunction with the organ systems to which their products are functionally related.

The most common types of epithelia, their characteristic locations in the body, and their functions are described in the illustrations on the next several pages **(Figure 5.3)**.

ACTIVITY 1

Examining Epithelial Tissue Under the Microscope

Obtain slides of simple squamous, simple cuboidal, simple columnar, pseudostratified ciliated columnar, stratified squamous

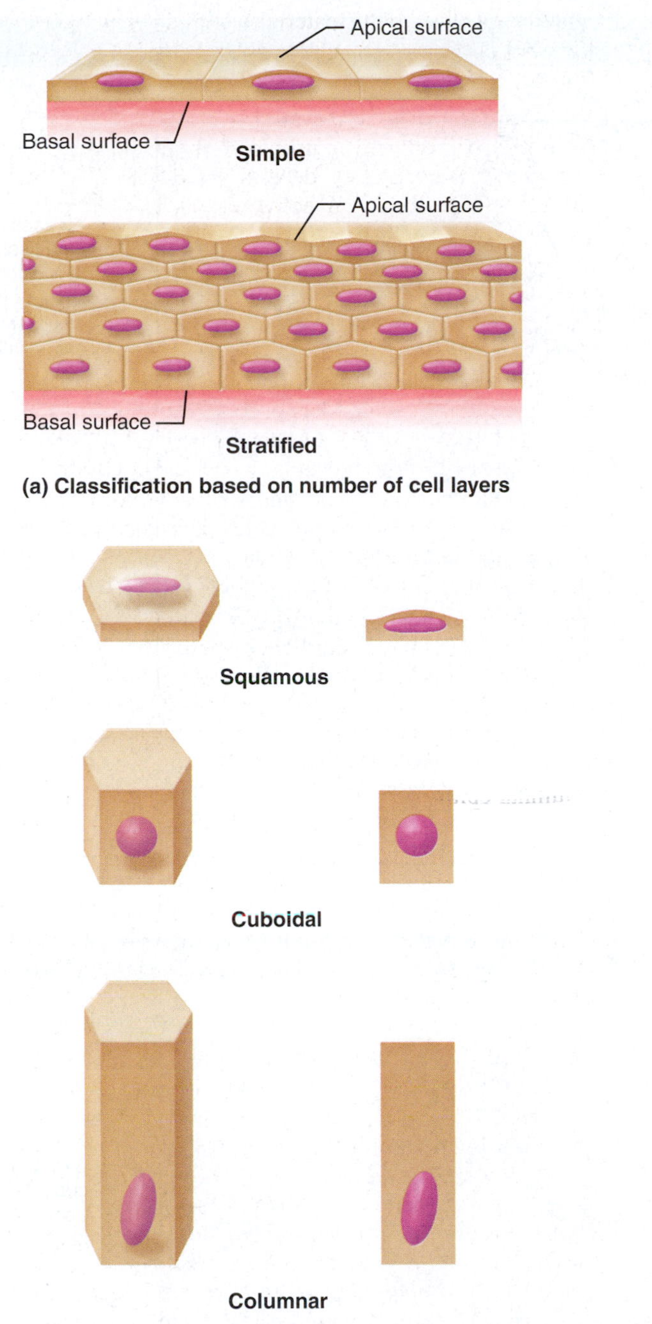

Figure 5.1 Classification of epithelia. For each category in (b), a whole cell is shown on the left and a longitudinal section on the right.

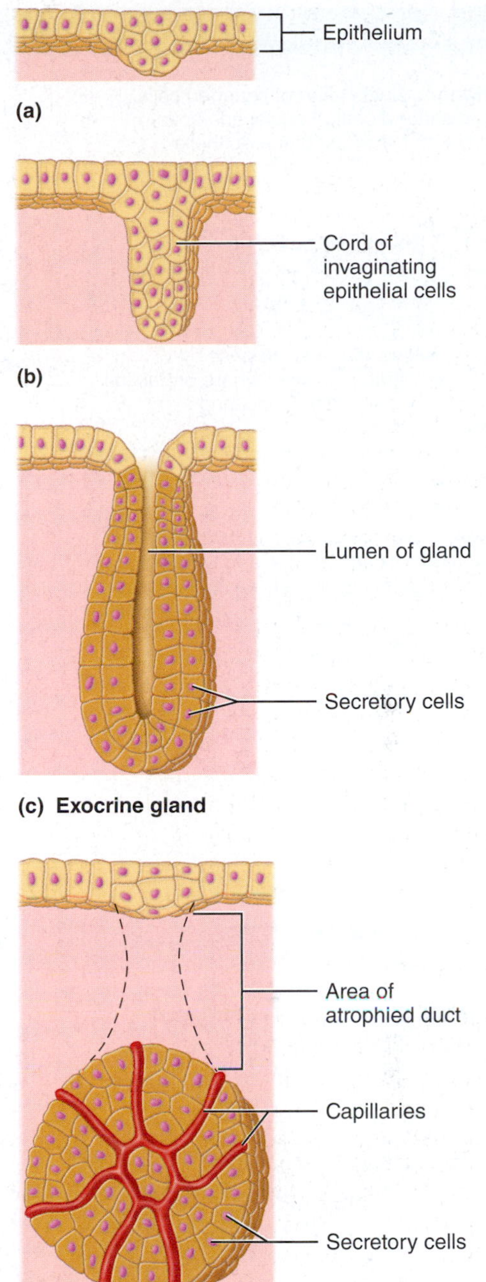

Figure 5.2 Formation of endocrine and exocrine glands from epithelial sheets. (a) Epithelial cells grow and push into the underlying tissue. **(b)** A cord of epithelial cells forms. **(c)** In an exocrine gland, a lumen (cavity) forms. The inner cells form the duct, and the outer cells produce the secretion. **(d)** In a forming endocrine gland, the connecting duct cells atrophy, leaving the secretory cells with no connection to the epithelial surface. However, they do become heavily invested with blood and lymphatic vessels that receive the secretions.

(nonkeratinized), stratified cuboidal, stratified columnar, and transitional epithelia. Examine each carefully, and notice how the epithelial cells fit closely together to form intact sheets of cells, a necessity for a tissue that forms linings or covering membranes. Scan each epithelial type for modifications for specific functions, such as cilia (motile cell projections that help to move substances along the cell surface), and microvilli, which increase the surface area for absorption. Also be alert for goblet cells, which secrete lubricating mucus. Compare your observations with the descriptions and photomicrographs (Figure 5.3).

While you are working, check the questions in the Review Sheet at the end of this exercise. A number of the questions there refer to some of the observations you are asked to make during your microscopic study. ■

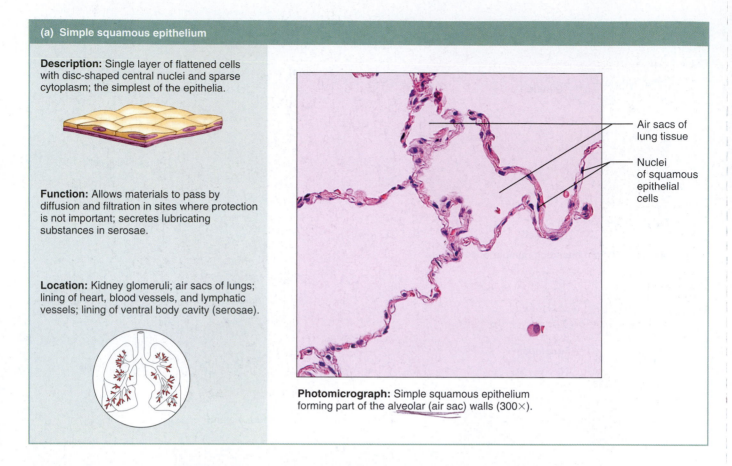

(a) Simple squamous epithelium

Description: Single layer of flattened cells with disc-shaped central nuclei and sparse cytoplasm; the simplest of the epithelia.

Function: Allows materials to pass by diffusion and filtration in sites where protection is not important; secretes lubricating substances in serosae.

Location: Kidney glomeruli; air sacs of lungs; lining of heart, blood vessels, and lymphatic vessels; lining of ventral body cavity (serosae).

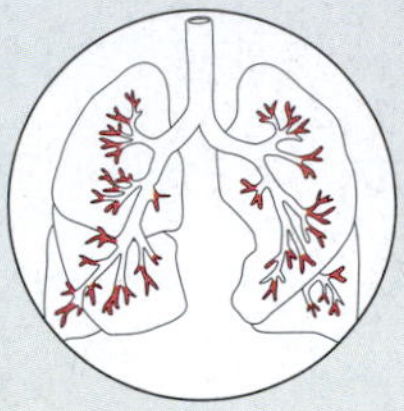

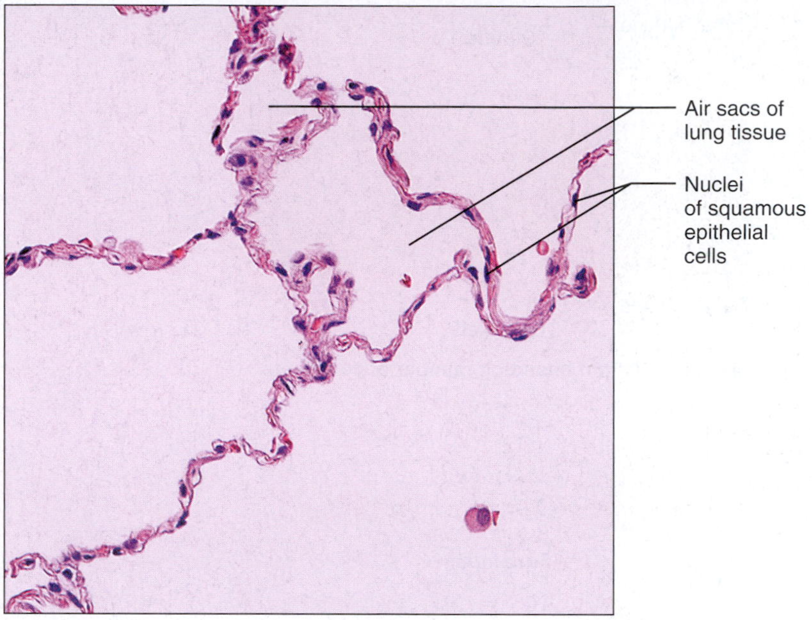

Air sacs of lung tissue

Nuclei of squamous epithelial cells

Photomicrograph: Simple squamous epithelium forming part of the alveolar (air sac) walls (300×).

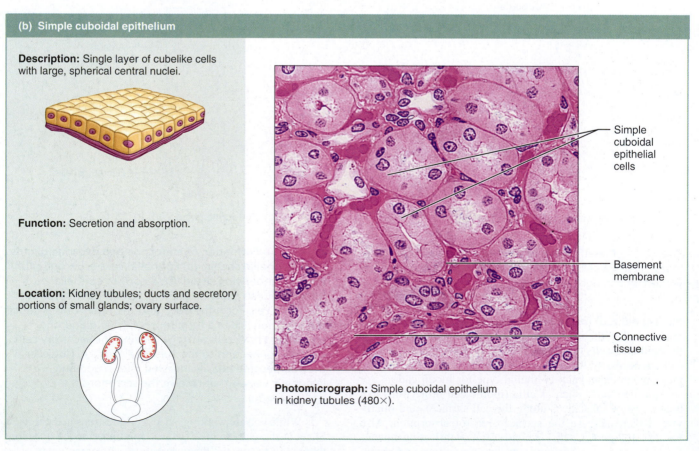

(b) Simple cuboidal epithelium

Description: Single layer of cubelike cells with large, spherical central nuclei.

Function: Secretion and absorption.

Location: Kidney tubules; ducts and secretory portions of small glands; ovary surface.

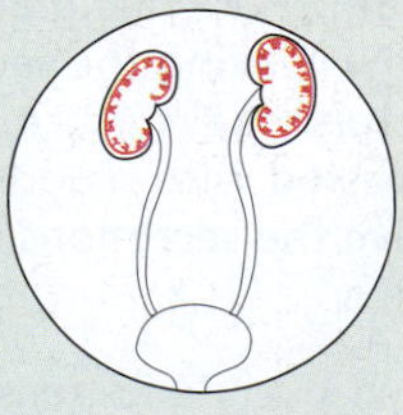

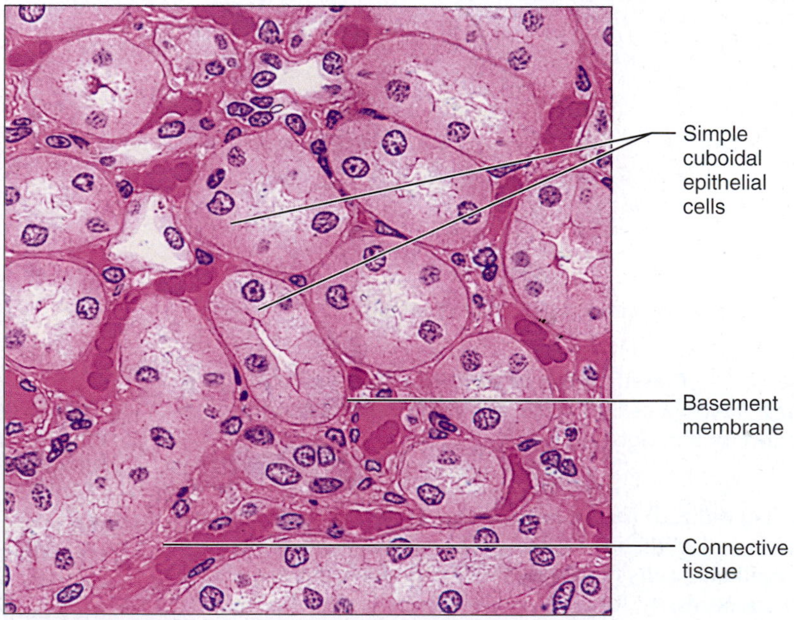

Simple cuboidal epithelial cells

Basement membrane

Connective tissue

Photomicrograph: Simple cuboidal epithelium in kidney tubules (480×).

Figure 5.3 Epithelial tissues. Simple epithelia **(a** and **b)**.

(c) Simple columnar epithelium

Description: Single layer of tall cells with *round* to *oval* nuclei; some cells bear cilia; layer may contain mucus-secreting unicellular glands (goblet cells).

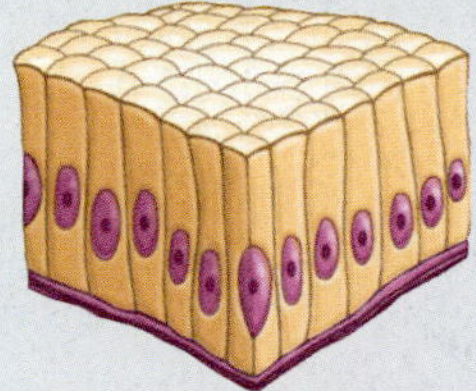

Function: Absorption; secretion of mucus, enzymes, and other substances; ciliated type propels mucus (or reproductive cells) by ciliary action.

Location: Nonciliated type lines most of the digestive tract (stomach to anal canal), gallbladder, and excretory ducts of some glands; ciliated variety lines small bronchi, uterine tubes, and some regions of the uterus.

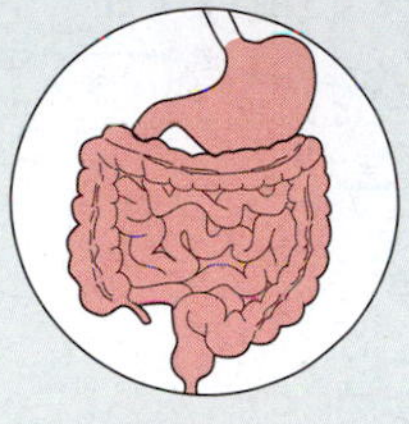

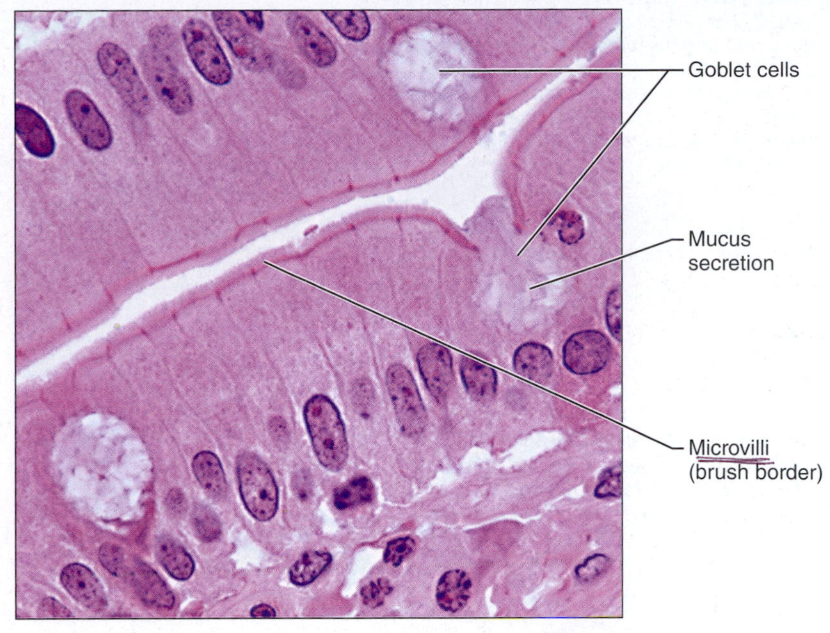

Photomicrograph: Simple columnar epithelium containing goblet cells from the small intestine (675×).

(d) Pseudostratified columnar epithelium

Description: Single layer of cells of differing heights, some not reaching the free surface; nuclei seen at different levels; may contain mucus-secreting goblet cells and bear cilia.

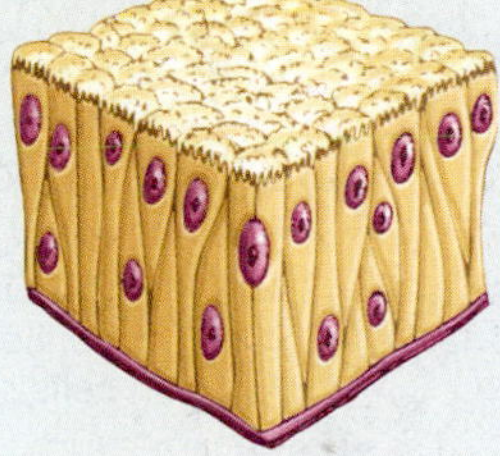

Function: Secretes substances, particularly mucus; propulsion of mucus by ciliary action.

Location: Nonciliated type in male's sperm-carrying ducts and ducts of large glands; ciliated variety lines the trachea, most of the upper respiratory tract.

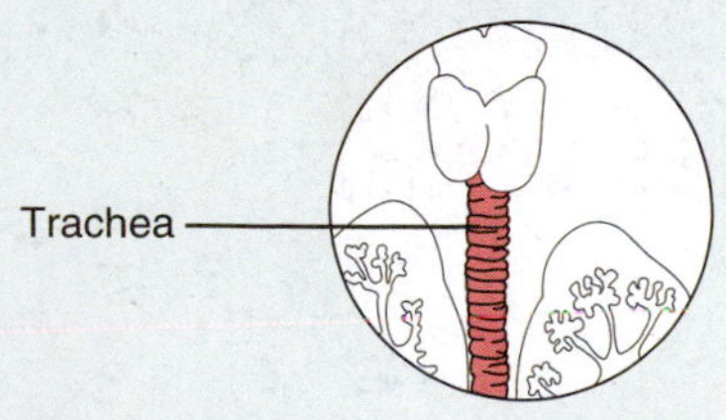

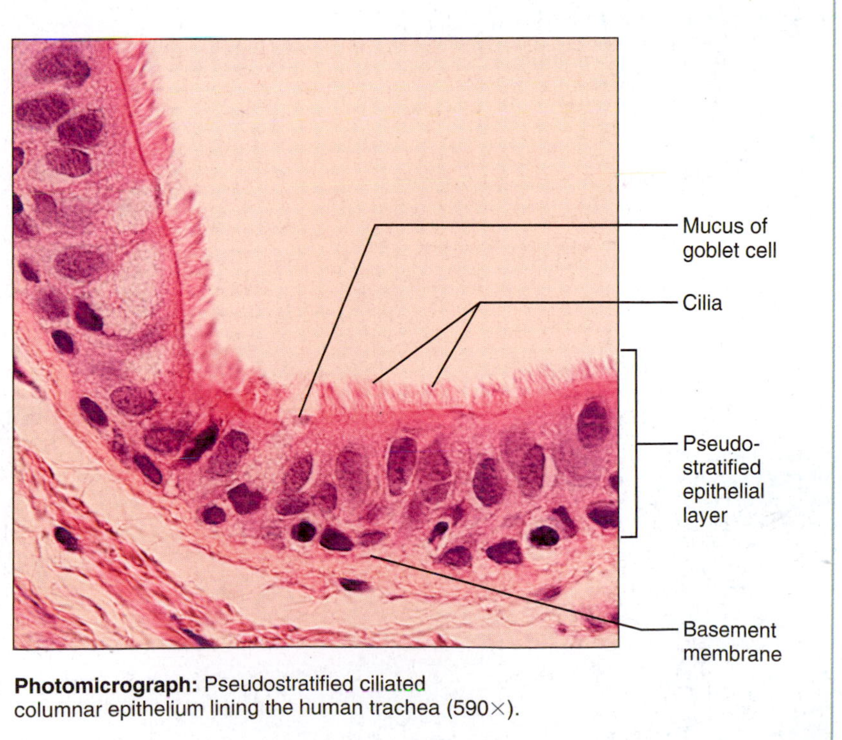

Photomicrograph: Pseudostratified ciliated columnar epithelium lining the human trachea (590×).

Figure 5.3 *(continued)* **Epithelial tissues.** Simple epithelia **(c** and **d)**.

(e) Stratified squamous epithelium

Description: Thick membrane composed of several cell layers; basal cells are cuboidal or columnar and metabolically active; surface cells are flattened (squamous); in the keratinized type, the surface cells are full of keratin and dead; basal cells are active in mitosis and produce the cells of the more superficial layers.

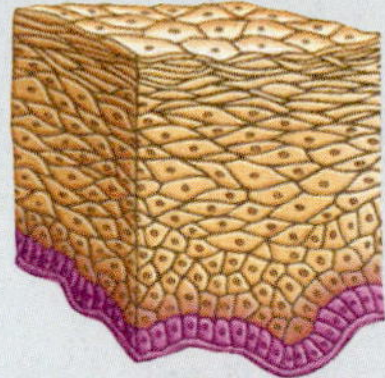

Function: Protects underlying tissues in areas subjected to abrasion.

Location: Nonkeratinized type forms the moist linings of the esophagus, mouth, and vagina; keratinized variety forms the epidermis of the skin, a dry membrane.

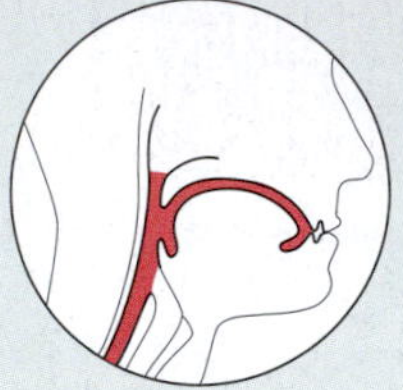

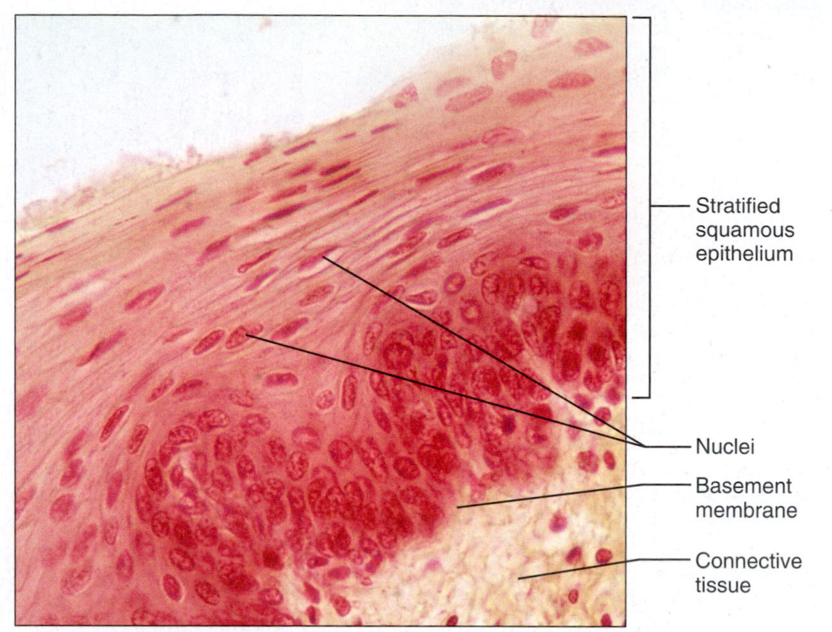

Photomicrograph: Stratified squamous epithelium lining the esophagus (590×).

(f) Stratified cuboidal epithelium

Description: Generally two layers of cubelike cells.

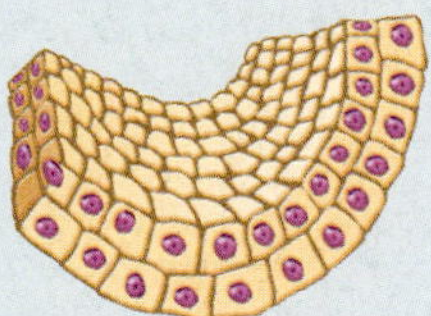

Function: Protection

Location: Largest ducts of sweat glands, mammary glands, and salivary glands.

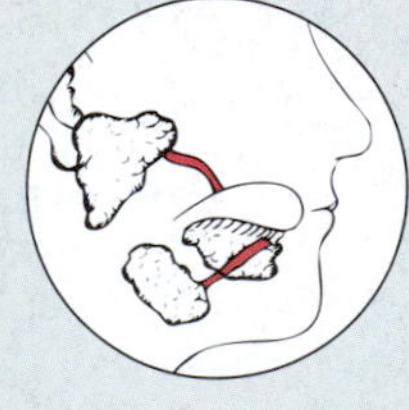

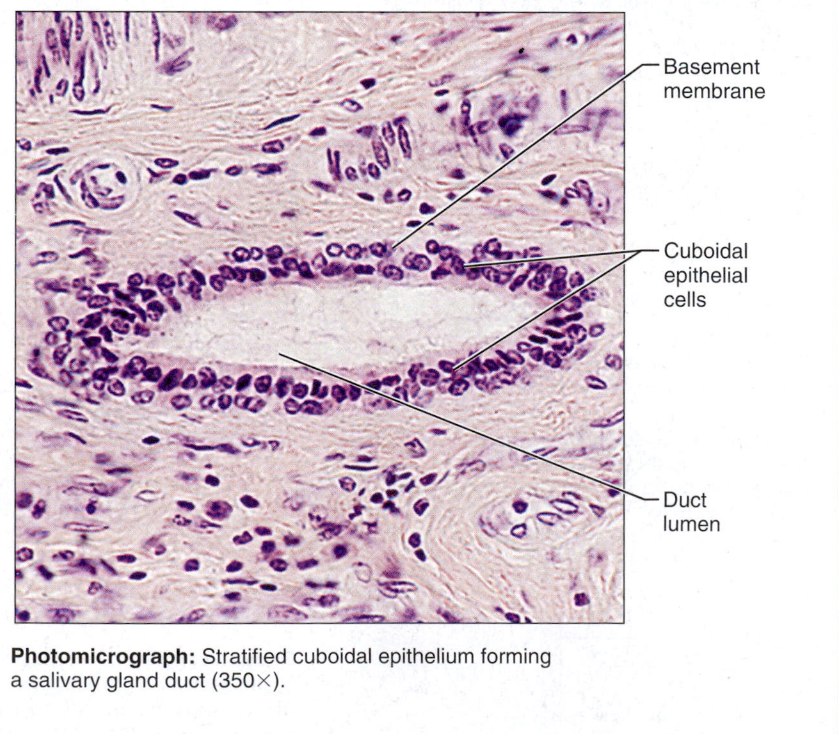

Photomicrograph: Stratified cuboidal epithelium forming a salivary gland duct (350×).

Figure 5.3 *(continued)* **Epithelial tissues.** Stratified epithelia **(e and f)**.

(g) Stratified columnar epithelium

Description: Several cell layers; basal cells usually cuboidal; superficial cells elongated and columnar.

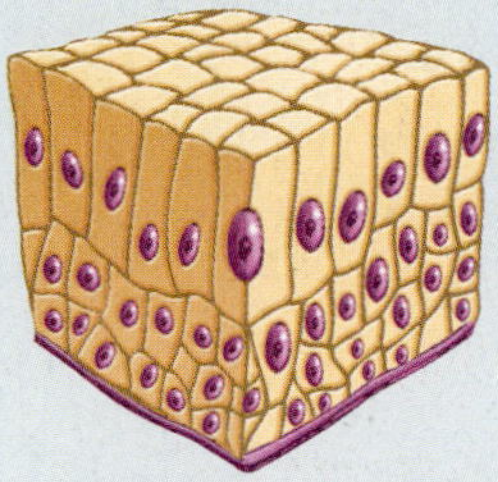

Function: Protection; secretion.

Location: Rare in the body; small amounts in male urethra and in large ducts of some glands.

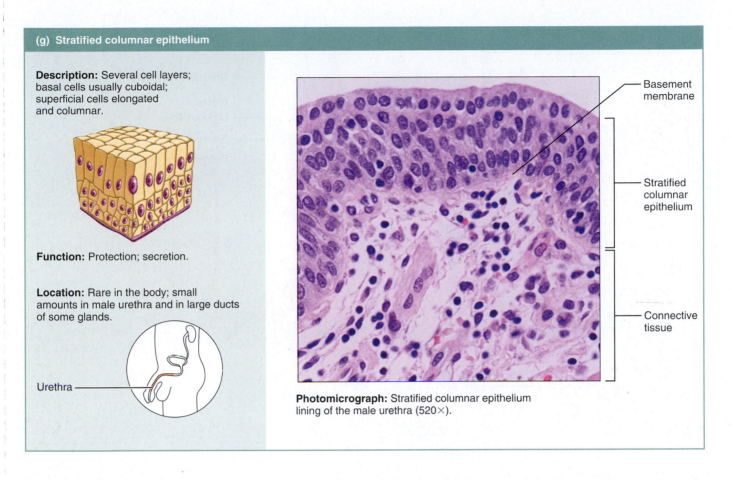

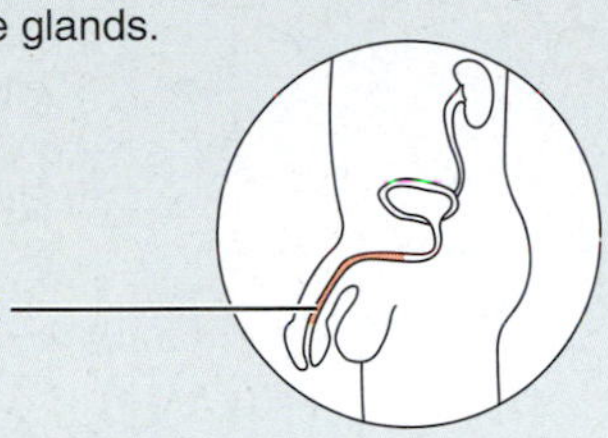

Urethra

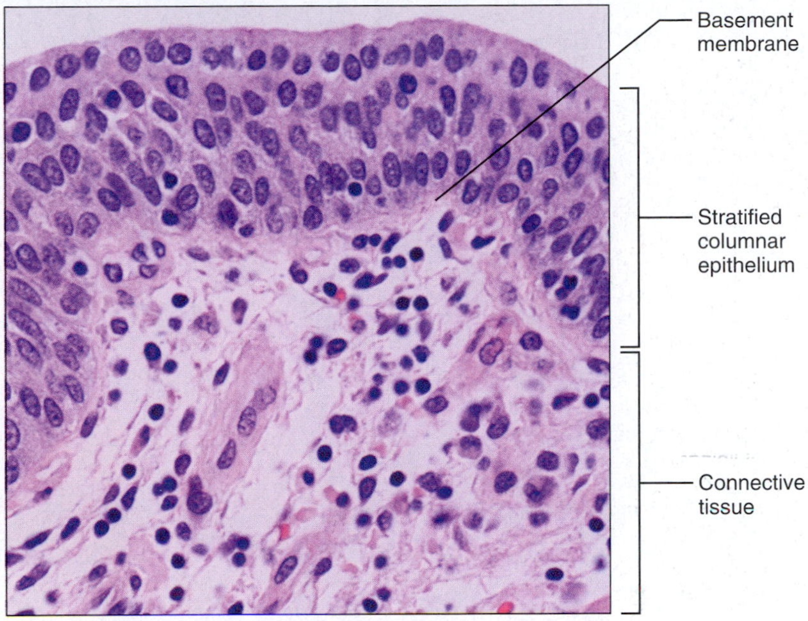

Basement membrane

Stratified columnar epithelium

Connective tissue

Photomicrograph: Stratified columnar epithelium lining of the male urethra (520×).

(h) Transitional epithelium

Description: Resembles both stratified squamous and stratified cuboidal; basal cells cuboidal or columnar; surface cells dome shaped or squamous-like, depending on degree of organ stretch.

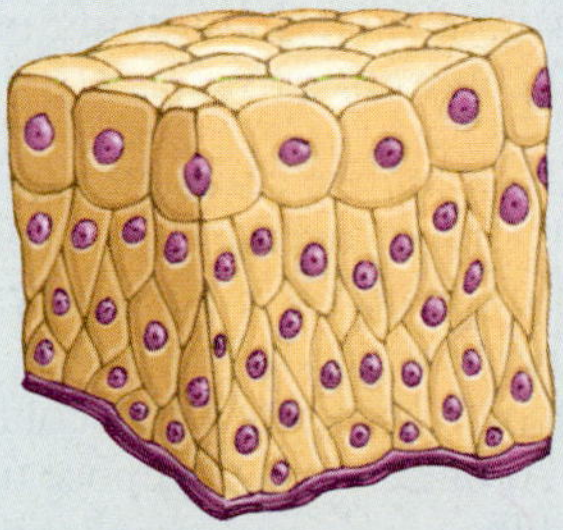

Function: Stretches readily and permits distension of urinary organ by contained urine.

Location: Lines the ureters, bladder, and part of the urethra.

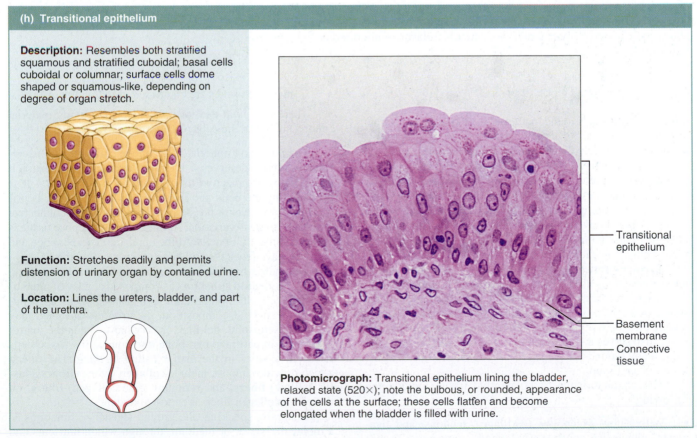

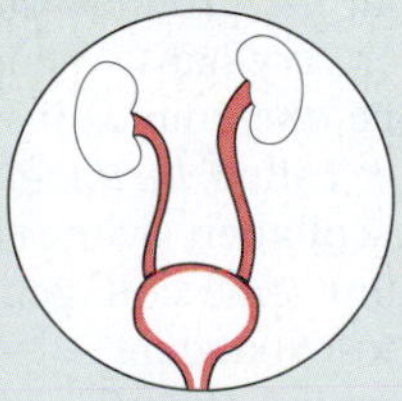

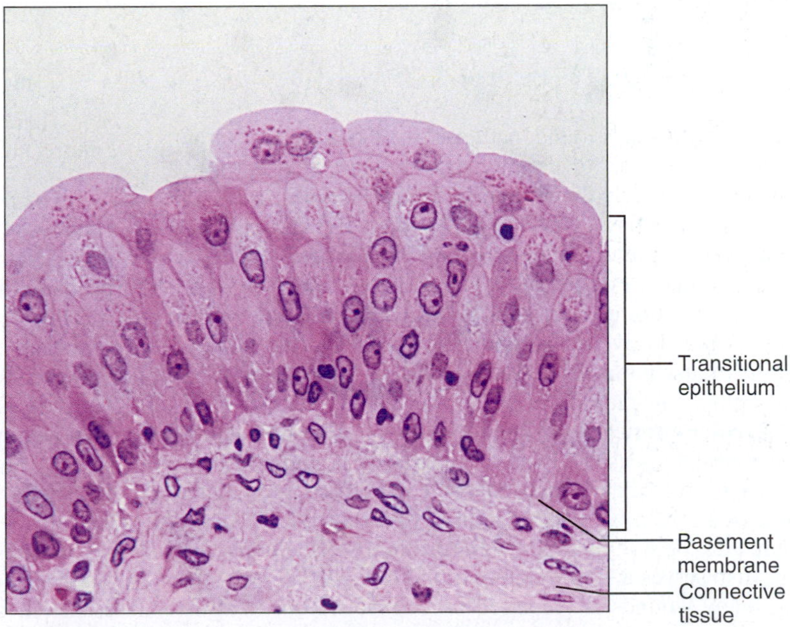

Transitional epithelium

Basement membrane

Connective tissue

Photomicrograph: Transitional epithelium lining the bladder, relaxed state (520×); note the bulbous, or rounded, appearance of the cells at the surface; these cells flatten and become elongated when the bladder is filled with urine.

Figure 5.3 (*continued*) **Epithelial tissues.** Stratified epithelia (**g** and **h**).

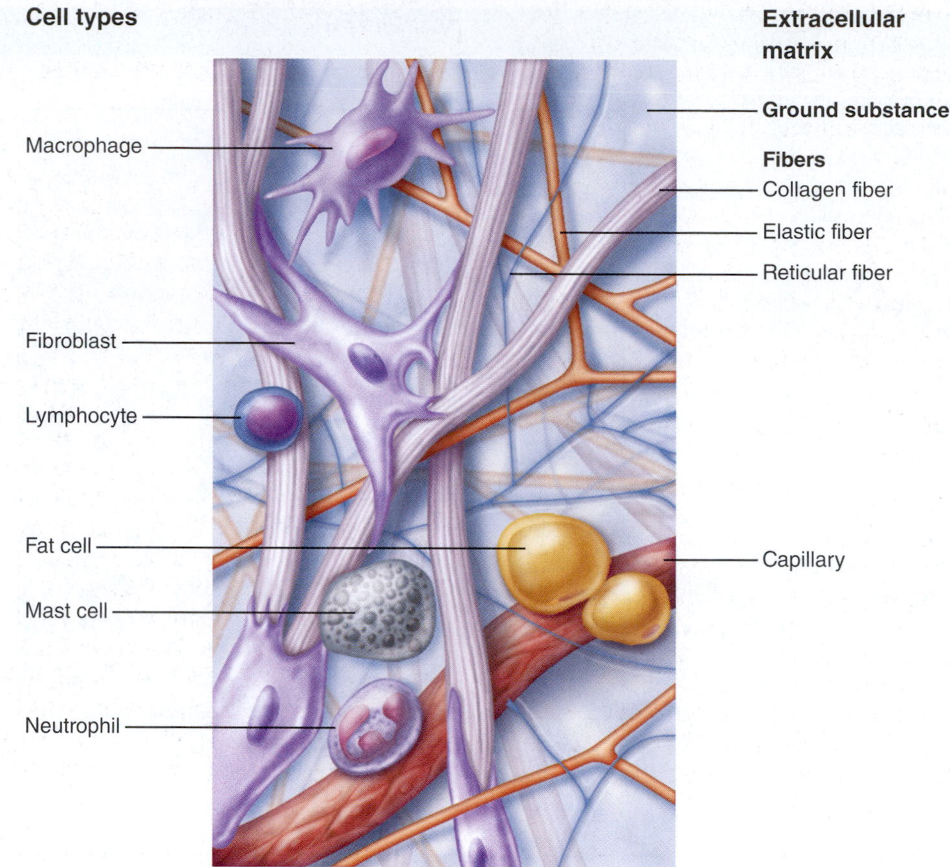

Figure 5.4 Areolar connective tissue: A prototype (model) connective tissue. This tissue underlies epithelia and surrounds capillaries. Note the various cell types and the three classes of fibers (collagen, reticular, elastic) embedded in the ground substance.

Connective Tissue

Connective tissue is found in all parts of the body as discrete structures or as part of various body organs. It is the most abundant and widely distributed of the tissue types.

Connective tissues perform a variety of functions, but primarily they protect, support, and bind together other tissues of the body. For example, bones are composed of connective tissue **(bone,** or **osseous tissue)**, and they protect and support other body tissues and organs. The ligaments and tendons **(dense connective tissue)** bind the bones together or bind skeletal muscles to bones.

Areolar connective tissue (Figure 5.4) is a soft packaging material that cushions and protects body organs. **Adipose** (fat) tissue provides insulation for the body tissues and a source of stored food. Blood-forming **(hematopoietic)** tissue replenishes the body's supply of red blood cells. Connective tissue also serves a vital function in the repair of all body tissues; many wounds are repaired by connective tissue in the form of scar tissue.

The characteristics of connective tissue include the following:

- With a few exceptions (cartilages, tendons, and ligaments, which are poorly vascularized), connective tissues have a rich supply of blood vessels.
- Connective tissues are composed of many types of cells.
- There is a great deal of noncellular, nonliving material (matrix) between the cells of connective tissue.

The nonliving material between the cells—the **extracellular matrix**—deserves a bit more explanation because it distinguishes connective tissue from all other tissues. It is produced by the cells and then extruded. The matrix is primarily responsible for the strength associated with connective tissue, but there is variation. At one extreme, adipose tissue is composed mostly of cells. At the opposite extreme, bone and cartilage have few cells and large amounts of matrix.

The matrix has two components—ground substance and fibers. The **ground substance** is composed chiefly of interstitial fluid, cell adhesion proteins, and proteoglycans. Depending on its specific composition, the ground substance may be liquid, semisolid, gel-like, or very hard. When the matrix is firm, as in cartilage and bone, the connective tissue cells reside in cavities in the matrix called *lacunae.* The fibers, which provide support, include **collagen** (white) **fibers, elastic** (yellow) **fibers,** and **reticular** (fine collagen) **fibers.** Of these, the collagen fibers are most abundant.

Generally speaking, the ground substance functions as a molecular sieve, or medium, through which nutrients and other dissolved substances can diffuse between the blood capillaries and the cells. The fibers in the matrix hinder diffusion somewhat and make the ground substance less pliable. The properties of the connective tissue cells and the makeup and arrangement of their matrix elements vary tremendously, accounting for the amazing diversity of this tissue type. Nonetheless, the connective tissues have a common structural plan seen best in *areolar connective tissue* (Figure 5.4), a soft packing tissue that occurs throughout the body. Because all other connective tissues are variations of areolar, it is considered the model or prototype of the connective tissues. Notice that areolar tissue has all three varieties of fibers, but they are sparsely arranged in its transparent gel-like ground substance (refer to Figure 5.4). The cell type that secretes its matrix is the *fibroblast,* but a wide variety of other cells, including phagocytic cells such as macrophages and certain white blood cells and mast cells that act in the inflammatory response, are present as well. The more durable connective tissues, such as bone, cartilage, and the dense fibrous varieties, characteristically have a firm ground substance and many more fibers.

There are four main types of adult connective tissue, all of which typically have large amounts of matrix. These are **connective tissue proper** (which includes areolar, adipose, reticular, and dense [fibrous] connective tissues), **cartilage, bone,** and **blood.** All of these derive from an embryonic tissue called *mesenchyme.* The general characteristics, location, and function of some of the connective tissues found in the body are listed in the illustrations on the next several pages **(Figure 5.5)**.

ACTIVITY 2

Examining Connective Tissue Under the Microscope

Obtain prepared slides of mesenchyme; of areolar, adipose, reticular, and dense irregular, regular, and elastic connective tissue; of hyaline and elastic cartilage and fibrocartilage; of osseous connective tissue (bone); and of blood. Compare your observations with the views illustrated (Figure 5.5).

Distinguish the living cells from the matrix, and pay particular attention to the denseness and arrangement of the matrix. For example, notice how the matrix of the dense fibrous connective tissues, which make up tendons and the dermis of the skin, is packed with collagen fibers. Note also that in the *regular* variety (tendon), the fibers are all running in the same direction, whereas in the *irregular* variety (dermis), they appear to be running in many directions.

While examining the areolar connective tissue, notice how much empty space there appears to be (*areol* = small empty space), and distinguish the collagen fibers from the coiled elastic fibers. Identify the starlike fibroblasts. Also, try to locate a **mast cell,** which has large, darkly staining granules in its cytoplasm (*mast* = stuffed full of granules). This cell type releases histamine which makes capillaries more permeable during inflammatory reactions and allergies and thus is partially responsible for that "runny nose" of some allergies.

(Text continues on page 67.)

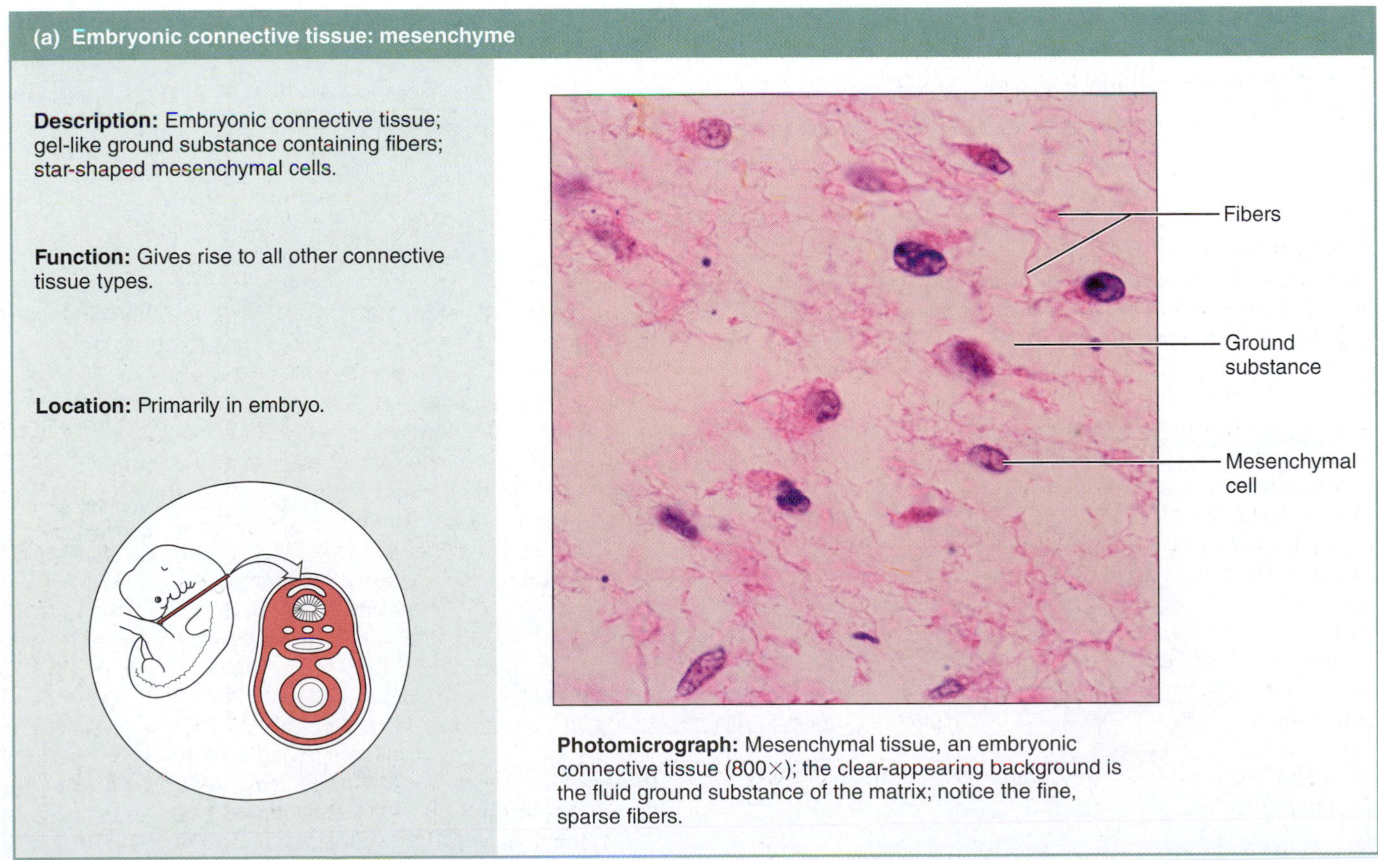
(a) Embryonic connective tissue: mesenchyme
Description: Embryonic connective tissue; gel-like ground substance containing fibers; star-shaped mesenchymal cells.
Function: Gives rise to all other connective tissue types.
Location: Primarily in embryo.
Fibers
Ground substance
Mesenchymal cell
Photomicrograph: Mesenchymal tissue, an embryonic connective tissue (800×); the clear-appearing background is the fluid ground substance of the matrix; notice the fine, sparse fibers.

Figure 5.5 Connective tissues. Embryonic connective tissue **(a)**.

(b) Connective tissue proper: loose connective tissue, areolar

Description: Gel-like matrix with all three fiber types; cells: fibroblasts, macrophages, mast cells, and some white blood cells.

Function: Wraps and cushions organs; its macrophages phagocytize bacteria; plays important role in inflammation; holds and conveys tissue fluid.

Location: Widely distributed under epithelia of body, e.g., forms lamina propria of mucous membranes; packages organs; surrounds capillaries.

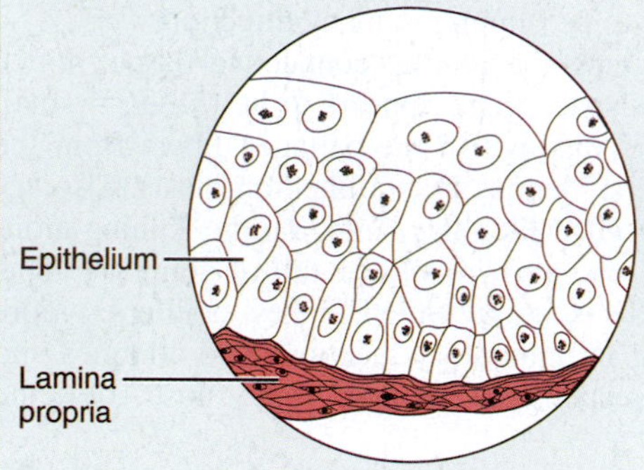

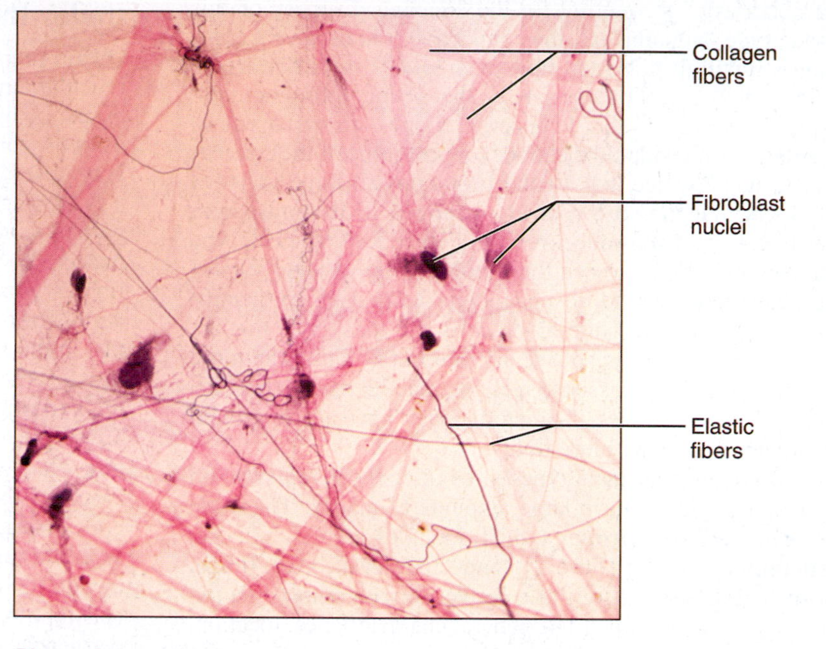

Photomicrograph: Areolar connective tissue, a soft packaging tissue of the body (350×).

(c) Connective tissue proper: loose connective tissue, adipose

Description: Matrix as in areolar, but very sparse; closely packed adipocytes, or fat cells, have nucleus pushed to the side by large fat droplet.

Function: Provides reserve food fuel; insulates against heat loss; supports and protects organs.

Location: Under skin in the hypodermis; around kidneys and eyeballs; within abdomen; in breasts.

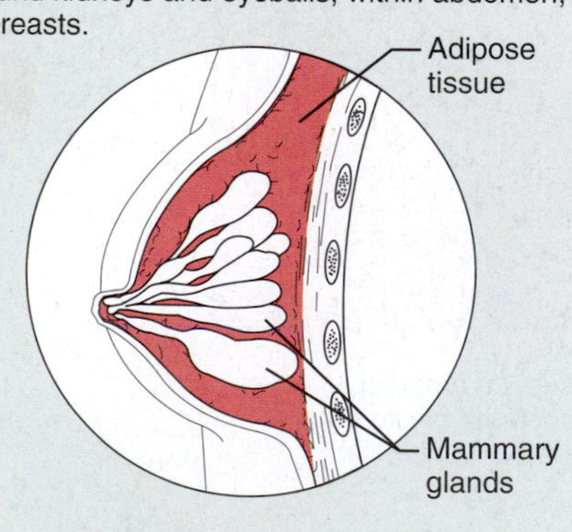

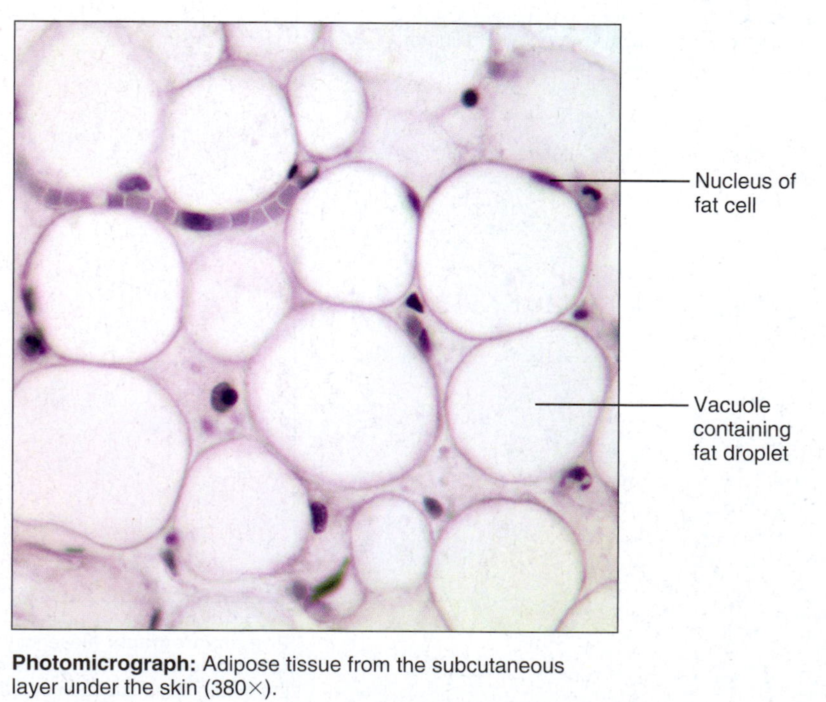

Photomicrograph: Adipose tissue from the subcutaneous layer under the skin (380×).

Figure 5.5 *(continued)* **Connective tissues.** Connective tissue proper **(b and c)**.

(d) Connective tissue proper: loose connective tissue, reticular

Description: Network of reticular fibers in a typical loose ground substance; reticular cells lie on the network.

Function: Fibers form a soft internal skeleton (stroma) that supports other cell types including white blood cells, mast cells, and macrophages.

Location: Lymphoid organs (lymph nodes, bone marrow, and spleen).

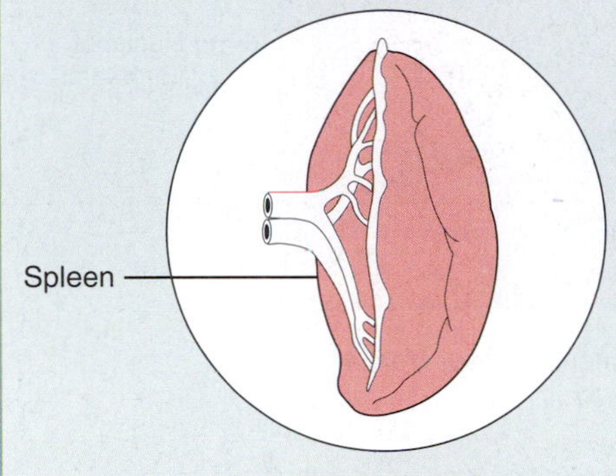

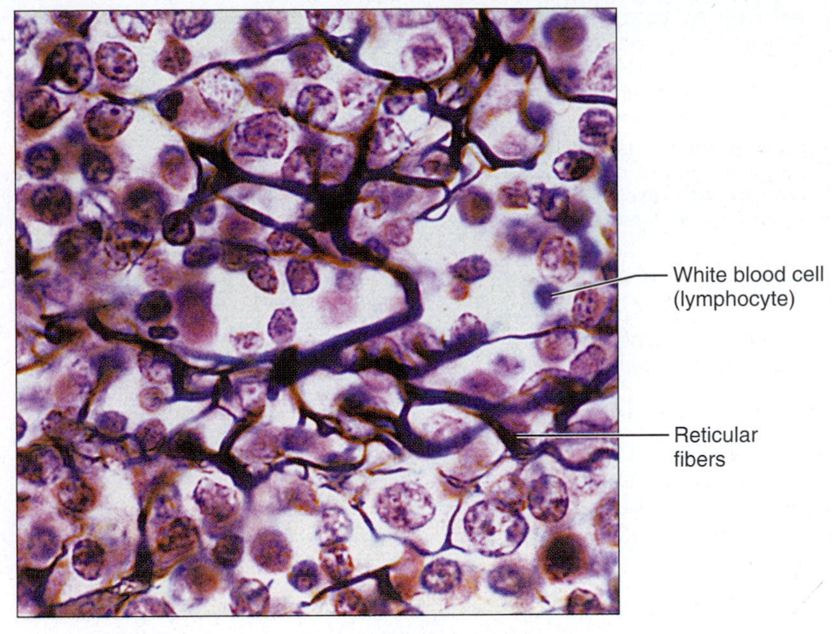

Photomicrograph: Dark-staining network of reticular connective tissue fibers forming the internal skeleton of the spleen (630×).

(e) Connective tissue proper: dense connective tissue, dense irregular

Description: Primarily irregularly arranged collagen fibers; some elastic fibers; major cell type is the fibroblast.

Function: Able to withstand tension exerted in many directions; provides structural strength.

Location: Fibrous capsules of organs and of joints; dermis of the skin; submucosa of digestive tract.

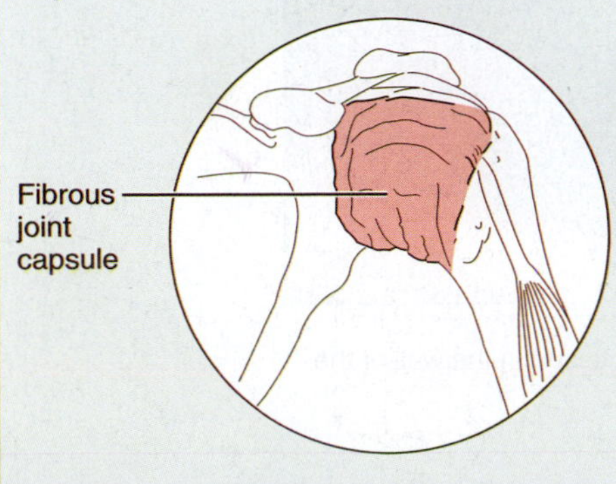

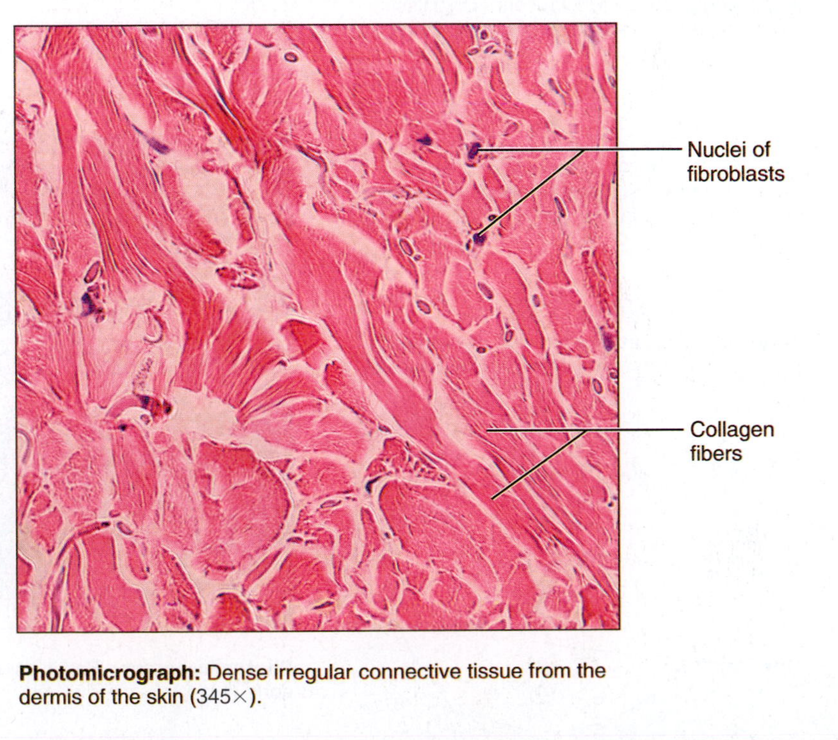

Photomicrograph: Dense irregular connective tissue from the dermis of the skin (345×).

Figure 5.5 *(continued)* **Connective tissues.** Connective tissue proper **(d and e)**.

(f) Connective tissue proper: dense connective tissue, dense regular

Description: Primarily parallel collagen fibers; a few elastic fibers; major cell type is the fibroblast.

Function: Attaches muscles to bones or to muscles; attaches bones to bones; withstands great tensile stress when pulling force is applied in one direction.

Location: Tendons, most ligaments, aponeuroses.

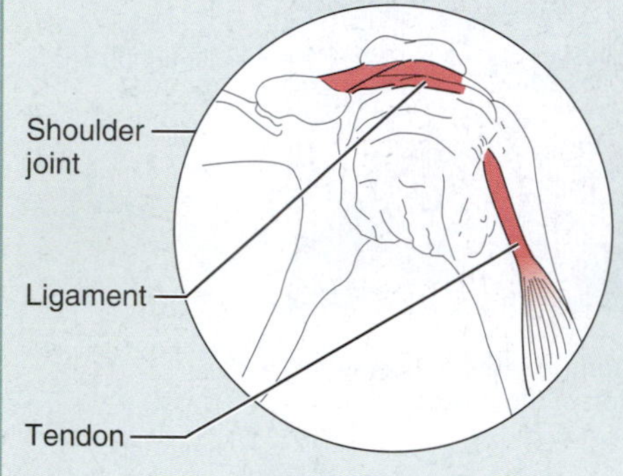

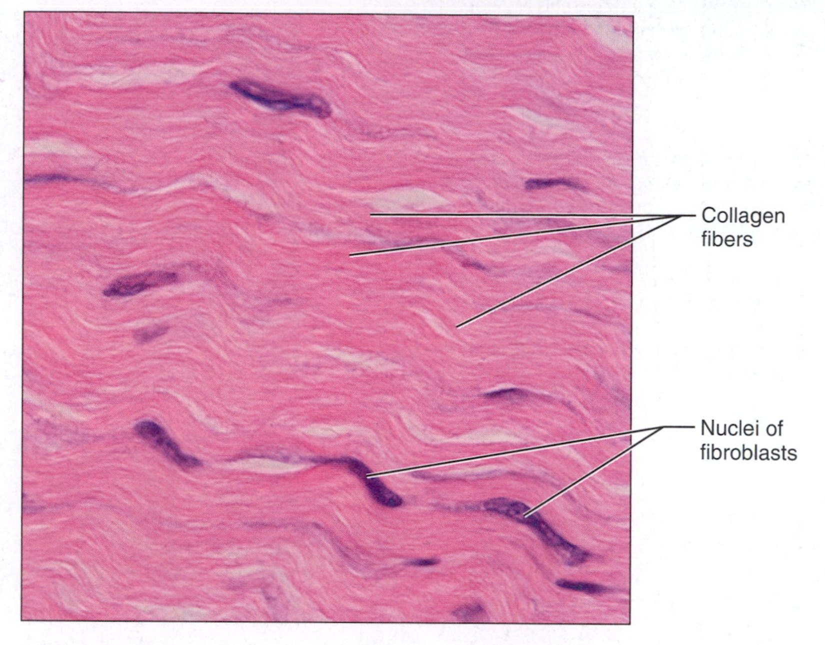

Photomicrograph: Dense regular connective tissue from a tendon (625×).

(g) Connective tissue proper: dense connective tissue, elastic

Description: Dense regular connective tissue containing a high proportion of elastic fibers.

Function: Allows recoil of tissue following stretching; maintains pulsatile flow of blood through arteries; aids passive recoil of lungs following inspiration.

Location: Walls of large arteries; within certain ligaments associated with the vertebral column; within the walls of the bronchial tubes.

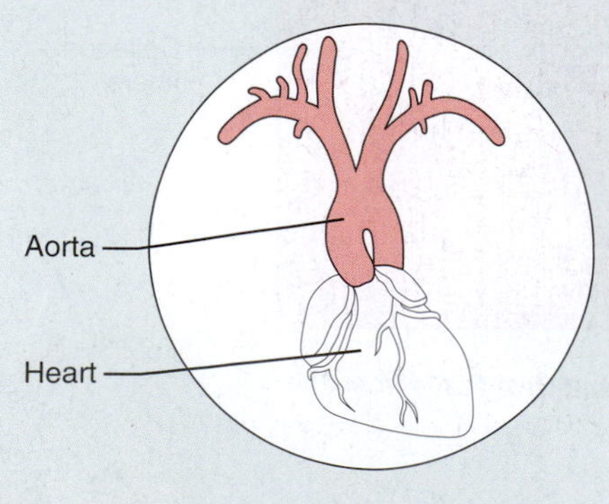

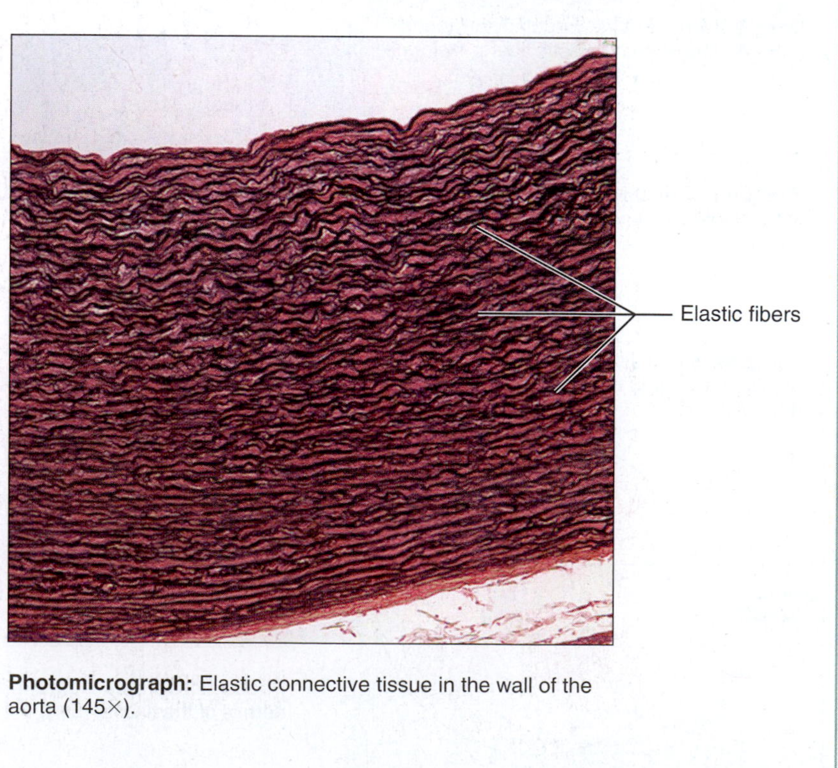

Photomicrograph: Elastic connective tissue in the wall of the aorta (145×).

Figure 5.5 *(continued)* **Connective tissues.** Connective tissue proper **(f and g)**.

(h) Cartilage: hyaline

Description: Amorphous but firm matrix; collagen fibers form an imperceptible network; chondroblasts produce the matrix and when mature (chondrocytes) lie in lacunae.

Function: Supports and reinforces; serves as resilient cushion; resists compressive stress.

Location: Forms most of the embryonic skeleton; covers the ends of long bones in joint cavities; forms costal cartilages of the ribs; cartilages of the nose, trachea, and larynx.

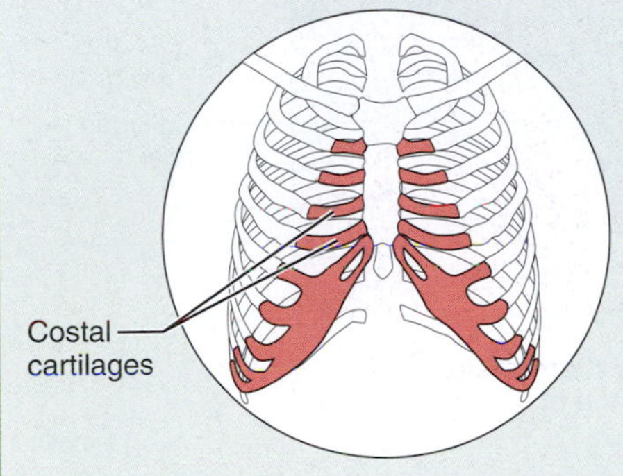

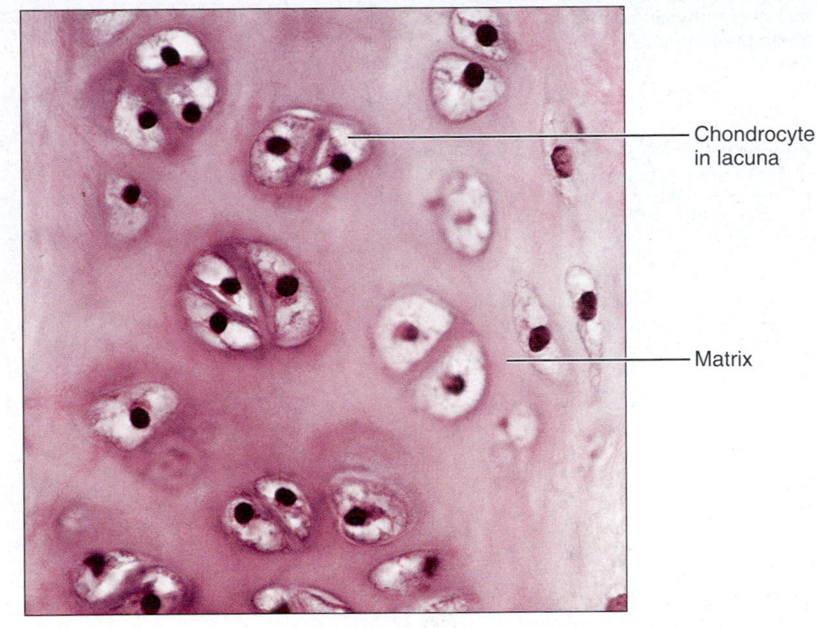

Photomicrograph: Hyaline cartilage from a costal cartilage of a rib (380×).

(i) Cartilage: elastic

Description: Similar to hyaline cartilage, but more elastic fibers in matrix.

Function: Maintains the shape of a structure while allowing great flexibility.

Location: Supports the external ear (auricle); epiglottis.

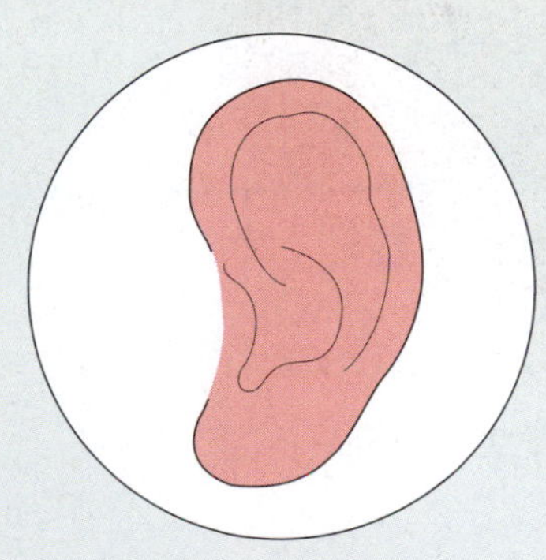

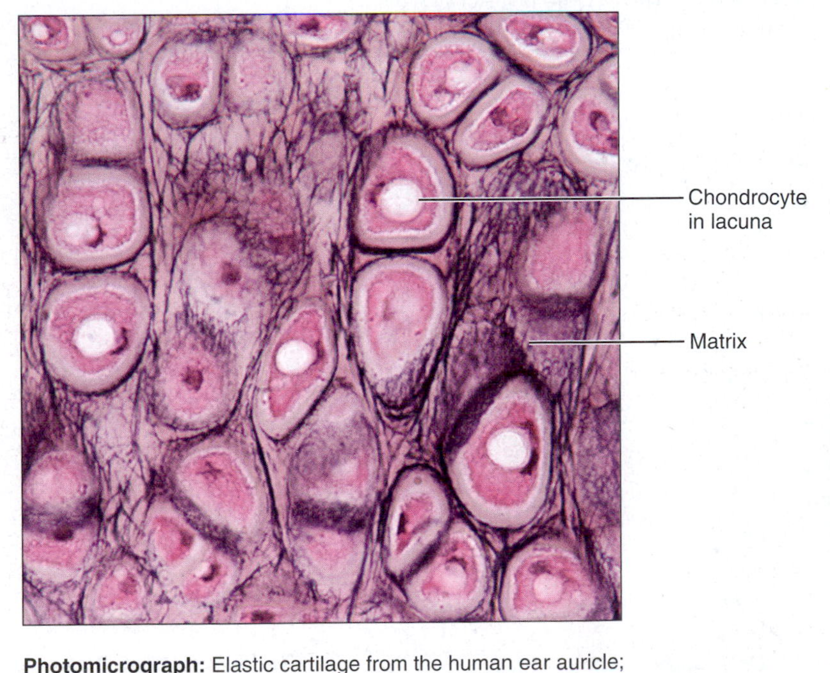

Photomicrograph: Elastic cartilage from the human ear auricle; forms the flexible skeleton of the ear (715×).

Figure 5.5 *(continued)* **Connective tissues.** Cartilage **(h and i)**.

(j) Cartilage: fibrocartilage

Description: Matrix similar to but less firm than that in hyaline cartilage; thick collagen fibers predominate.

Function: Tensile strength with the ability to absorb compressive shock.

Location: Intervertebral discs; pubic symphysis; discs of knee joint.

Intervertebral discs

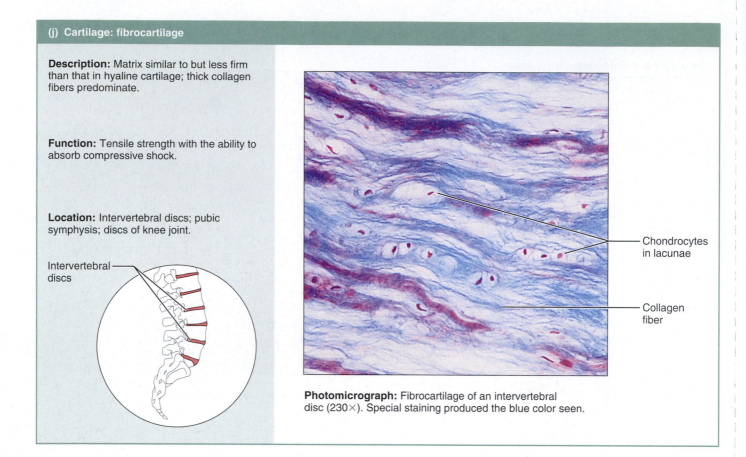

Chondrocytes in lacunae

Collagen fiber

Photomicrograph: Fibrocartilage of an intervertebral disc (230×). Special staining produced the blue color seen.

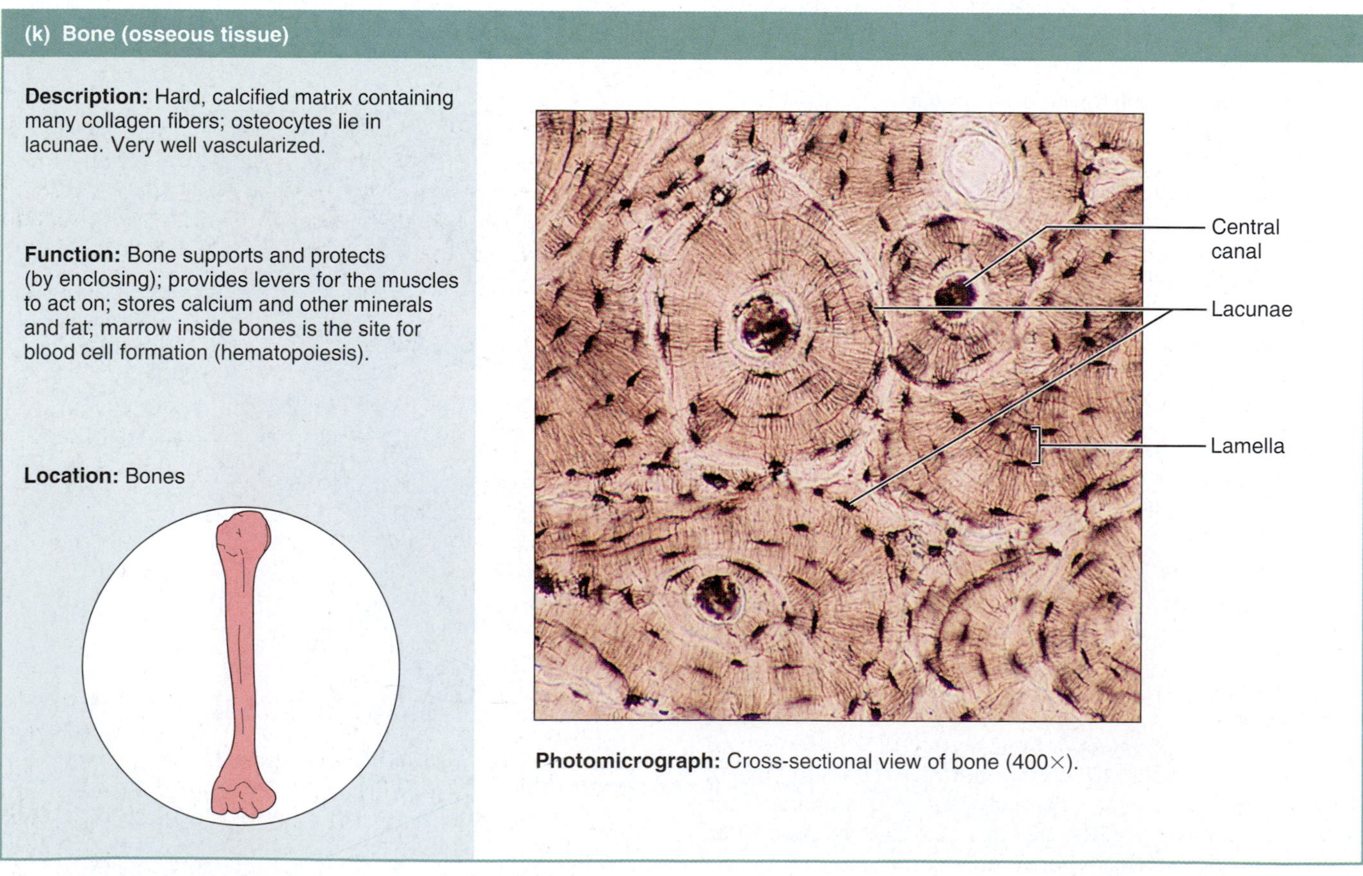

(k) Bone (osseous tissue)

Description: Hard, calcified matrix containing many collagen fibers; osteocytes lie in lacunae. Very well vascularized.

Function: Bone supports and protects (by enclosing); provides levers for the muscles to act on; stores calcium and other minerals and fat; marrow inside bones is the site for blood cell formation (hematopoiesis).

Location: Bones

Photomicrograph: Cross-sectional view of bone (400×).

Figure 5.5 *(continued)* **Connective tissues.** Cartilage **(j)** and bone **(k)**.

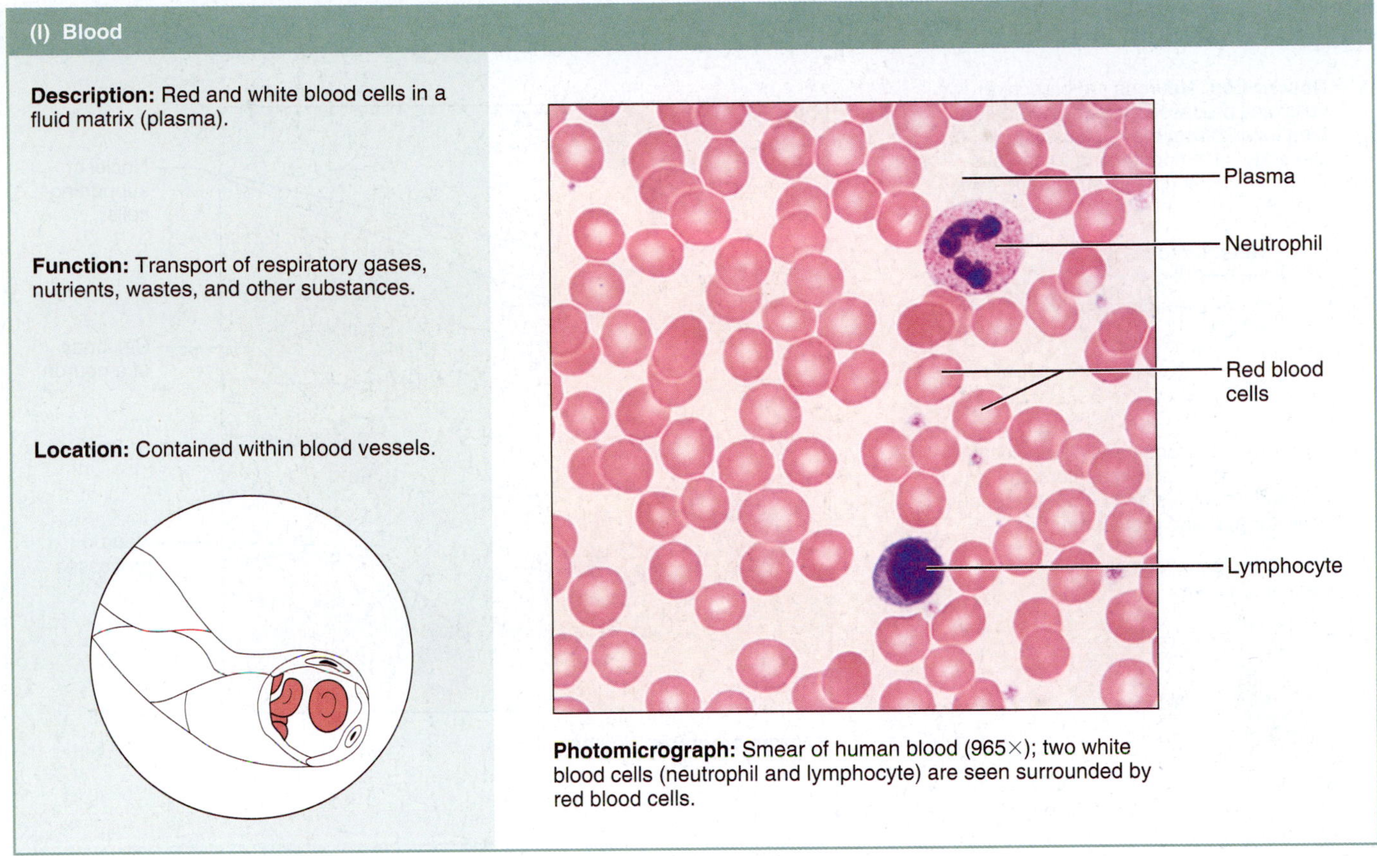
(l) Blood
Description: Red and white blood cells in a fluid matrix (plasma).
Function: Transport of respiratory gases, nutrients, wastes, and other substances.
Location: Contained within blood vessels.
Plasma
Neutrophil
Red blood cells
Lymphocyte
Photomicrograph: Smear of human blood (965×); two white blood cells (neutrophil and lymphocyte) are seen surrounded by red blood cells.

Figure 5.5 (*continued*) **Connective tissues.** Blood **(l)**.

In adipose tissue, locate a "signet ring" cell, a fat cell in which the nucleus can be seen pushed to one side by the large, fat-filled vacuole that appears to be a large empty space. Also notice how little matrix there is in adipose (fat) tissue. Distinguish the living cells from the matrix in the dense fibrous, bone, and hyaline cartilage preparations.

Scan the blood slide at low and then high power to examine the general shape of the red blood cells. Then, switch to the oil immersion lens for a closer look at the various types of white blood cells. How does blood differ from all other connective tissues?

____

____ ■

Nervous Tissue

Nervous tissue is made up of two major cell populations. The **neuroglia** are special supporting cells that protect, support, and insulate the more delicate neurons. The **neurons** are highly specialized to receive stimuli (excitability) and to generate electrical signals that may be sent to all parts of the body (conductivity). They are the cells that are most often associated with nervous system functioning.

The structure of neurons is markedly different from that of all other body cells. They have a nucleus-containing cell body, and their cytoplasm is drawn out into long extensions (cell processes)—sometimes as long as 1 m (about 3 feet), which allows a single neuron to conduct an electrical signal over relatively long distances. (More detail about the anatomy of the different classes of neurons and neuroglia appears in Exercise 13).

ACTIVITY 3

Examining Nervous Tissue Under the Microscope

Obtain a prepared slide of a spinal cord smear. Locate a neuron, and compare it to the photomicrograph **(Figure 5.6)**. Keep the light dim—this will help you see the cellular extensions of the neurons. (See also Figure 13.2 in Exercise 13.) ■

Muscle Tissue

Muscle tissue (Figure 5.7) is highly specialized to contract and produces most types of body movement. As you might expect, muscle cells tend to be elongated, providing a long axis for contraction. The three basic types of muscle tissue are described briefly here. (Cardiac and skeletal muscles are treated more completely in later exercises.)

Skeletal muscle, the "meat," or flesh, of the body, is attached to the skeleton. It is under conscious voluntary

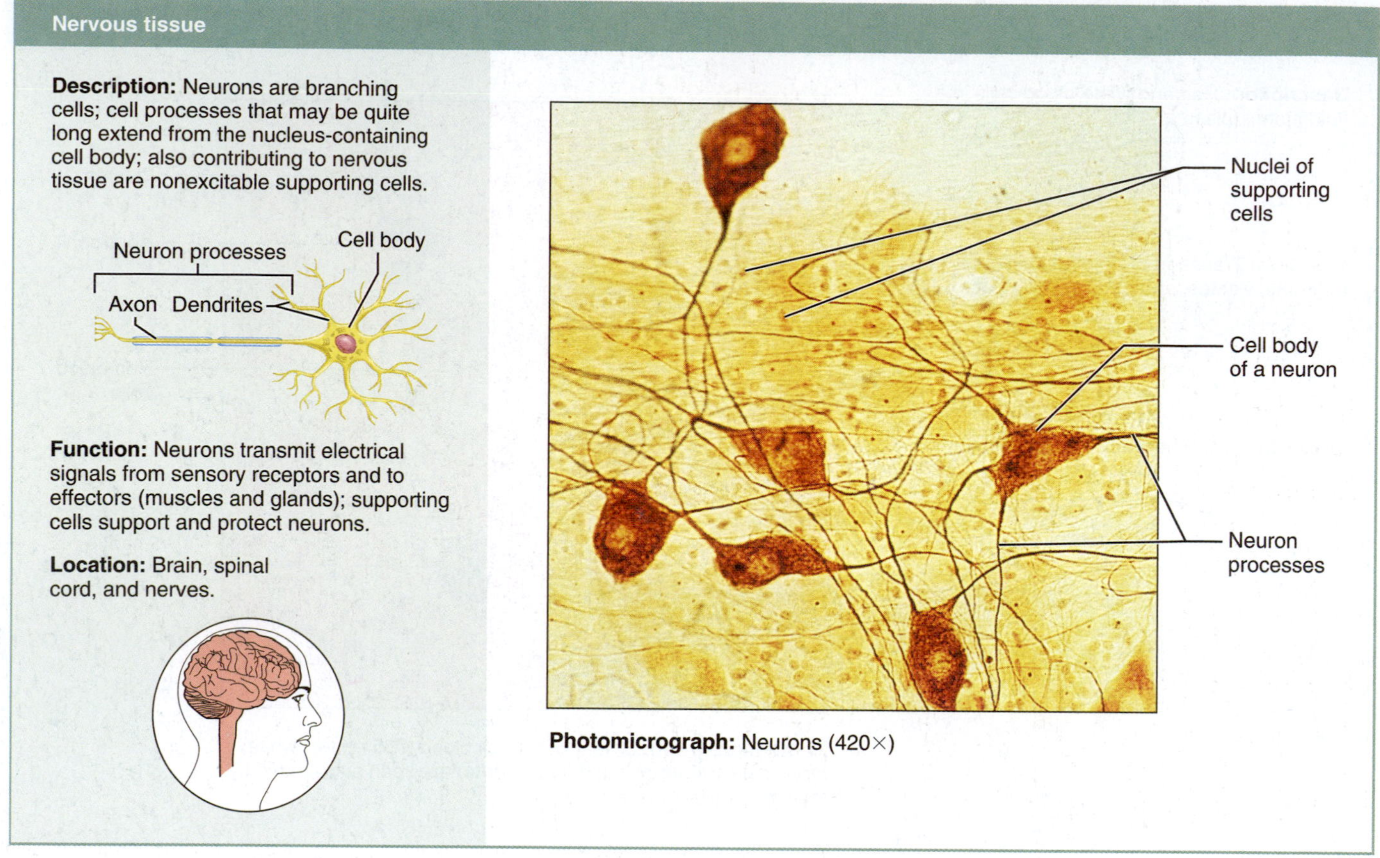

Nervous tissue
Description: Neurons are branching cells; cell processes that may be quite long extend from the nucleus-containing cell body; also contributing to nervous tissue are nonexcitable supporting cells.
Neuron processes
Cell body
Axon
Dendrites
Function: Neurons transmit electrical signals from sensory receptors and to effectors (muscles and glands); supporting cells support and protect neurons.
Location: Brain, spinal cord, and nerves.
Nuclei of supporting cells
Cell body of a neuron
Neuron processes
Photomicrograph: Neurons (420×)

Figure 5.6 **Nervous tissue.**

control, and its contraction moves the limbs and other body parts. The cells of skeletal muscles are long, cylindrical, and multinucleate (several nuclei per cell), with the nuclei pushed to the periphery of the cells; they have obvious *striations* (stripes).

Cardiac muscle is found only in the heart. As it contracts, the heart acts as a pump, propelling the blood into the blood vessels. Cardiac muscle, like skeletal muscle, has striations, but cardiac cells are branching uninucleate cells that interdigitate (fit together) at junctions called **intercalated discs.** These structural modifications allow the cardiac muscle to act as a unit. Cardiac muscle is under involuntary control, which means that we cannot voluntarily or consciously control the operation of the heart.

Smooth muscle, or *visceral muscle,* is found mainly in the walls of hollow organs (digestive and urinary tract organs, uterus, blood vessels). Typically it has two layers that run at right angles to each other; consequently its contraction can constrict or dilate the lumen (cavity) of an organ and propel substances along predetermined pathways. Smooth muscle cells are quite different in appearance from those of skeletal or cardiac muscle. No striations are visible, and the uninucleate smooth muscle cells are spindle-shaped. Like cardiac muscle, it is under involuntary control.

ACTIVITY 4

Examining Muscle Tissue Under the Microscope

Obtain and examine prepared slides of skeletal, cardiac, and smooth muscle. Notice their similarities and dissimilarities in your observations and in the illustrations and photomicrographs (Figure 5.7).

(a) Skeletal muscle

Description: Long, cylindrical, multinucleate cells; obvious striations.

Function: Voluntary movement; locomotion; manipulation of the environment; facial expression; voluntary control.

Location: In skeletal muscles attached to bones or occasionally to skin.

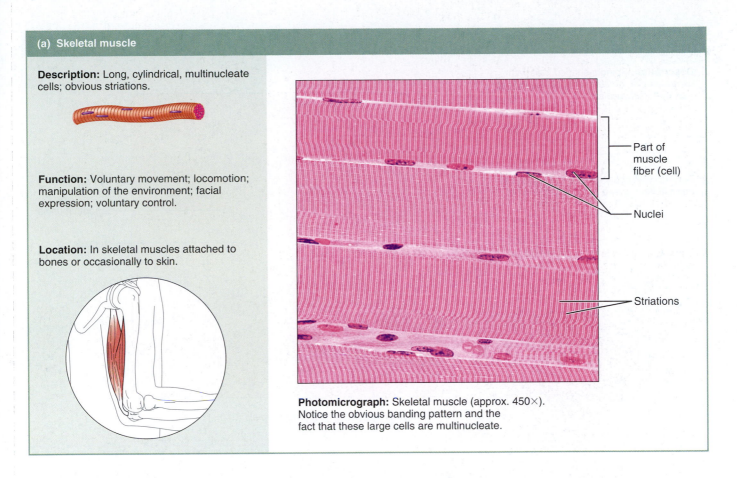

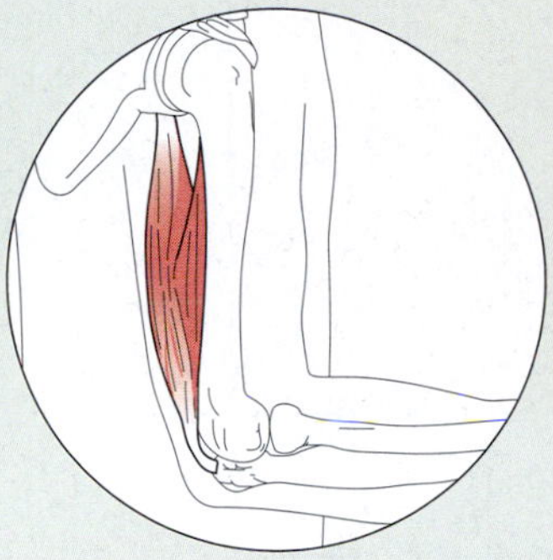

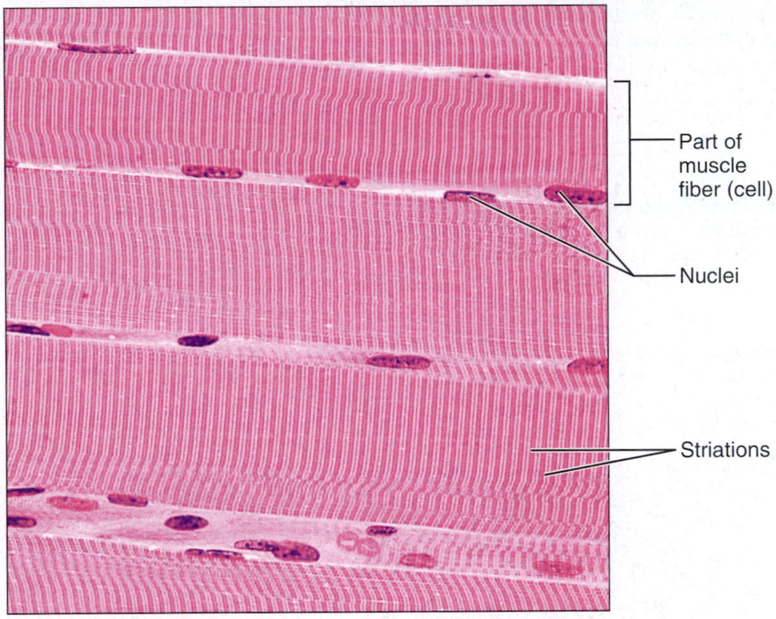

Part of muscle fiber (cell)

Nuclei

Striations

Photomicrograph: Skeletal muscle (approx. 450×). Notice the obvious banding pattern and the fact that these large cells are multinucleate.

(b) Cardiac muscle

Description: Branching, striated, generally uninucleate cells that interdigitate at specialized junctions called intercalated discs.

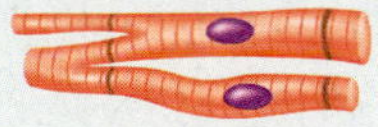

Function: As it contracts, it propels blood into the circulation; involuntary control.

Location: The walls of the heart.

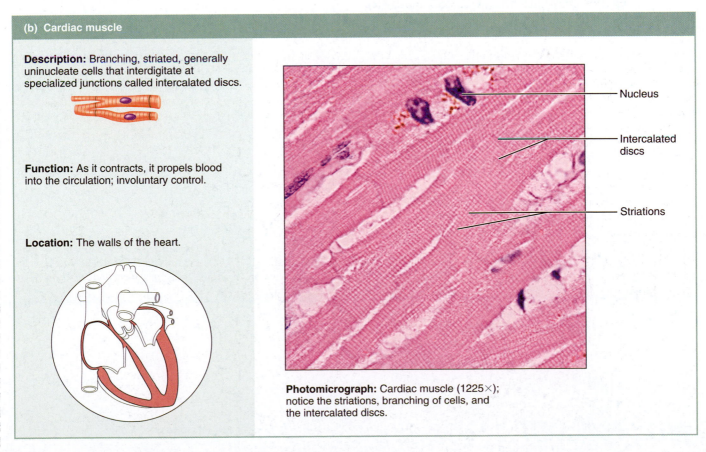

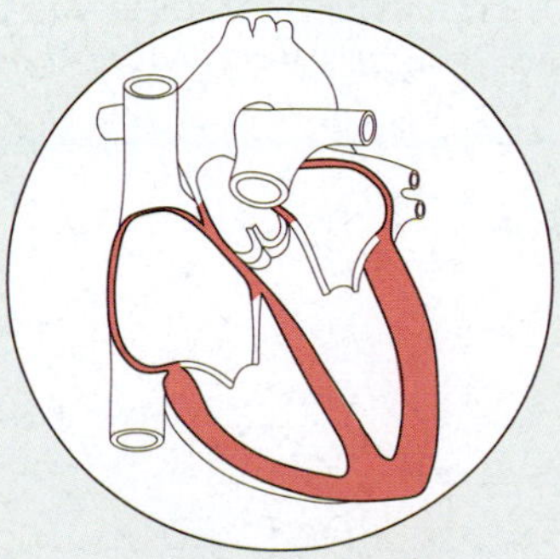

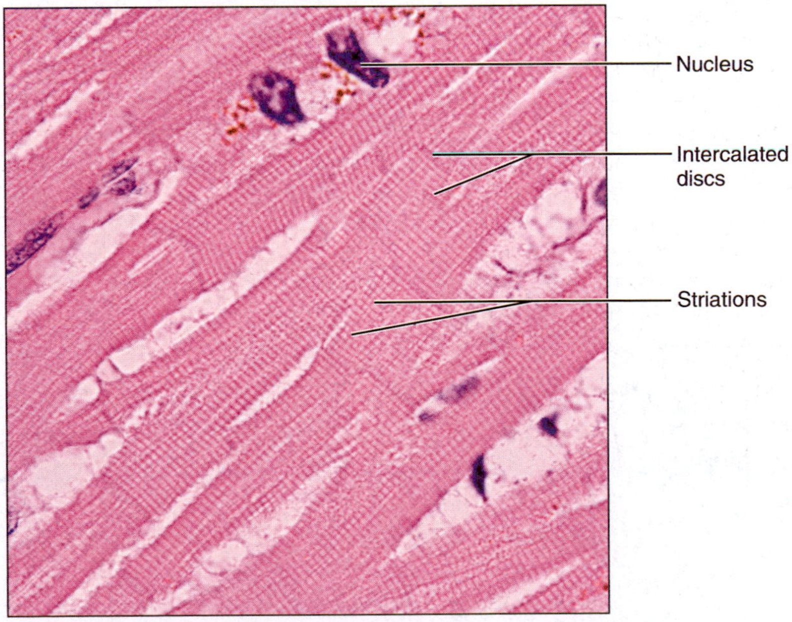

Nucleus

Intercalated discs

Striations

Photomicrograph: Cardiac muscle (1225×); notice the striations, branching of cells, and the intercalated discs.

Figure 5.7 Muscle tissues. Skeletal **(a)** and cardiac **(b)** muscles.

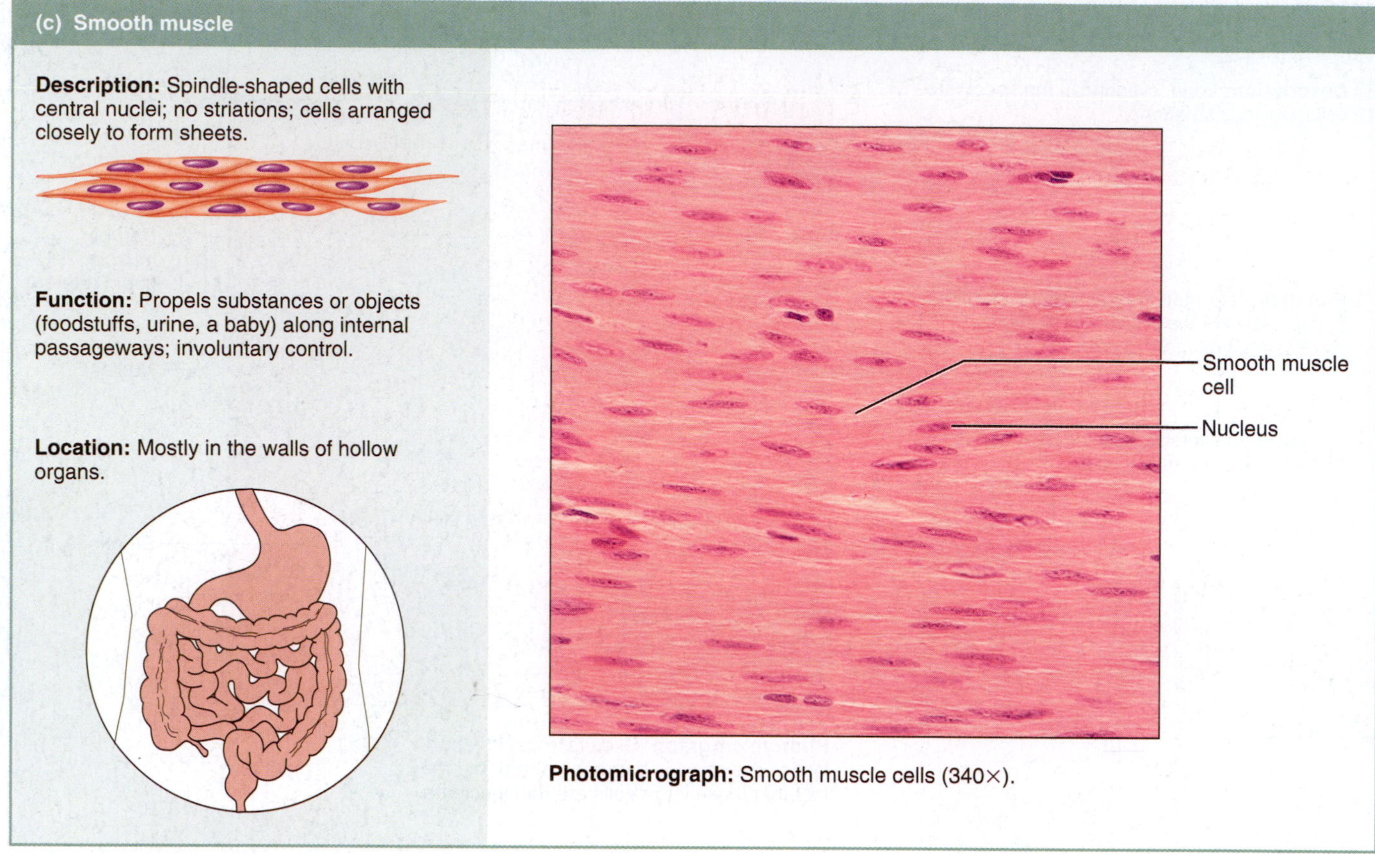
(c) Smooth muscle
Description: Spindle-shaped cells with central nuclei; no striations; cells arranged closely to form sheets.
Function: Propels substances or objects (foodstuffs, urine, a baby) along internal passageways; involuntary control.
Location: Mostly in the walls of hollow organs.
Smooth muscle cell
Nucleus
Photomicrograph: Smooth muscle cells (340×).

Figure 5.7 *(continued)* **Muscle tissues.** Smooth muscle **(c)**.

Name ______________________________

Lab Time/Date ______________________

EXERCISE 5

REVIEW SHEET

Classification of Tissues

Tissue Structure and Function—General Review

1. Define *tissue.* ____

____

2. Use the key to identify the major tissue types described below. Responses may be used more than once.

Key: a. connective tissue b. epithelium c. muscle d. nervous tissue

______________________ 1. lines body cavities and covers the body's external surface

______________________ 2. pumps blood, flushes urine out of the body, allows one to swing a bat

______________________ 3. transmits electrical signals

______________________ 4. anchors, packages, and supports body organs

______________________ 5. cells may absorb, secrete, and filter

______________________ 6. most involved in regulating and controlling body functions

______________________ 7. major function is to contract

______________________ 8. synthesizes hormones

______________________ 9. the most durable tissue type

______________________ 10. abundant nonliving extracellular matrix

______________________ 11. most widespread tissue in the body

______________________ 12. forms nerves and the brain

Epithelial Tissue

3. Describe five general characteristics of epithelial tissue. ______________________________

____

____

____

4. On what basis are epithelial tissues classified? ______________________________

____

____

5. List five major functions of epithelium in the body, and give examples of each.

Function 1: ______________________ Example: ______________________

Function 2: ______________________ Example: ______________________

Function 3: ______________________ Example: ______________________

Function 4: ______________________ Example: ______________________

Function 5: ______________________ Example: ______________________

6. How does the function of stratified epithelium differ from the function of simple epithelium? ______________________

7. Where is ciliated epithelium found? ______________________

What role does it play? ______________________

8. Transitional epithelium is actually stratified squamous epithelium, but there is something special about it.

How does it differ structurally from other stratified squamous epithelia? ______________________

How does the structural difference support its function in the body? ______________________

9. How do the endocrine and exocrine glands differ in structure and function? ______________________

10. Using the key, write the letter indicating the type of epithelial tissue that fits the description. Some responses are used more than once.

Key: a. simple squamous
b. simple cubodial
c. simple columnar
d. pseudostratified ciliated columnar
e. stratified squamous
f. transitional

______________ 1. lining of the esophagus

______________ 2. lining of the stomach

______________ 3. alveolar sacs of lungs

______________________ 4. tubules of the kidney

______________________ 5. epidermis of the skin

______________________ 6. lining of bladder; peculiar cells that have the ability to slide over each other

______________________ 7. forms the thin serous membranes; a single layer of flattened cells

Connective Tissue

11. What are three general characteristics of connective tissues? ______________________

____

____

12. What functions are performed by connective tissue? ______________________

____

13. How are the functions of connective tissue reflected in its structure? ______________________

____

____

14. Using the key, choose the best response to identify the connective tissues described below. Some responses are used more than once.

______________________ 1. attaches bones to bones and muscles to bones

______________________ 2. acts as a storage depot for fat

______________________ 3. the dermis of the skin

______________________ 4. makes up the intervertebral discs

______________________ 5. forms the hip bone

______________________ 6. composes basement membranes; a soft packaging tissue with a jellylike matrix

______________________ 7. forms the larynx, the costal cartilages of the ribs, and the embryonic skeleton

______________________ 8. provides a flexible framework for the external ear

______________________ 9. firm, structurally amorphous matrix heavily invaded with fibers; appears glassy and smooth

______________________ 10. matrix hard owing to calcium salts; provides levers for muscles to act on

______________________ 11. insulates against heat loss

______________________ 12. walls of large arteries

Key:
a. adipose connective tissue
b. areolar connective tissue
c. dense fibrous connective tissue
d. elastic cartilage
e. elastic connective tissue
f. fibrocartilage
g. hematopoietic tissue
h. hyaline cartilage
i. osseous tissue

15. Why do adipose cells remind people of a ring with a single jewel? ______________________

____

Nervous Tissue

16. What two physiological characteristics are highly developed in neurons (nerve cells)? ______

______

17. In what ways are neurons similar to other cells? ______

______

How are they different? ______

18. Describe how the unique structure of a neuron relates to its function in the body. ______

______

Muscle Tissue

19. The three types of muscle tissue exhibit similarities as well as differences. Check the appropriate space in the chart to indicate which muscle types exhibit each characteristic.

Characteristic	Skeletal	Cardiac	Smooth
Voluntarily controlled			
Involuntarily controlled			
Striated			
Has a single nucleus in each cell			
Has several nuclei per cell			
Found attached to bones			
Allows you to direct your eyeballs			
Found in the walls of the stomach, uterus, and arteries			
Contains spindle-shaped cells			
Contains branching cylindrical cells			
Contains long, nonbranching cylindrical cells			
Has intercalated discs			
Concerned with locomotion of the body as a whole			
Changes the internal volume of an organ as it contracts			
Tissue of the heart			

For Review

20. Label the tissue types illustrated here and on the next page, and identify all structures with leader lines.

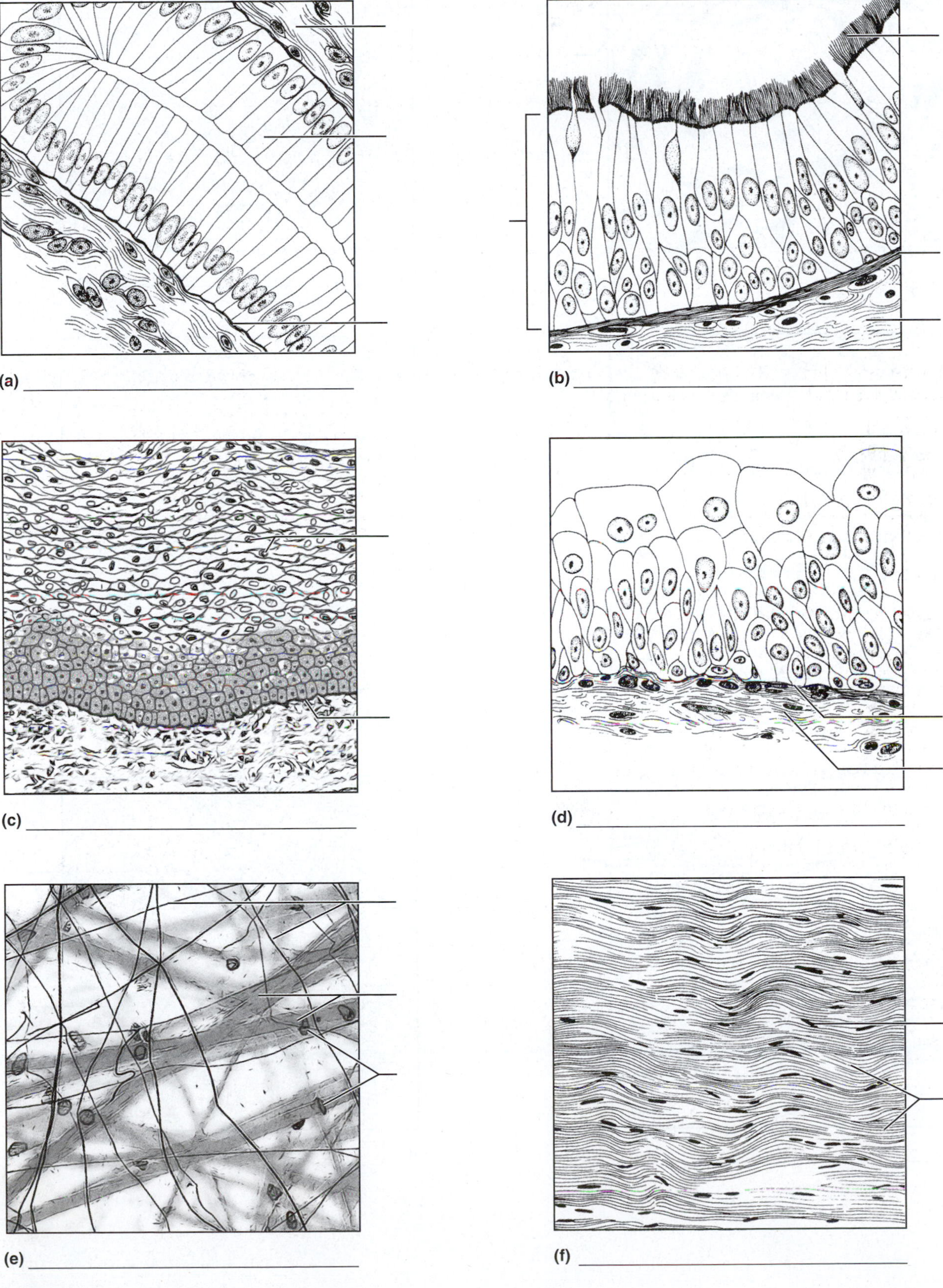

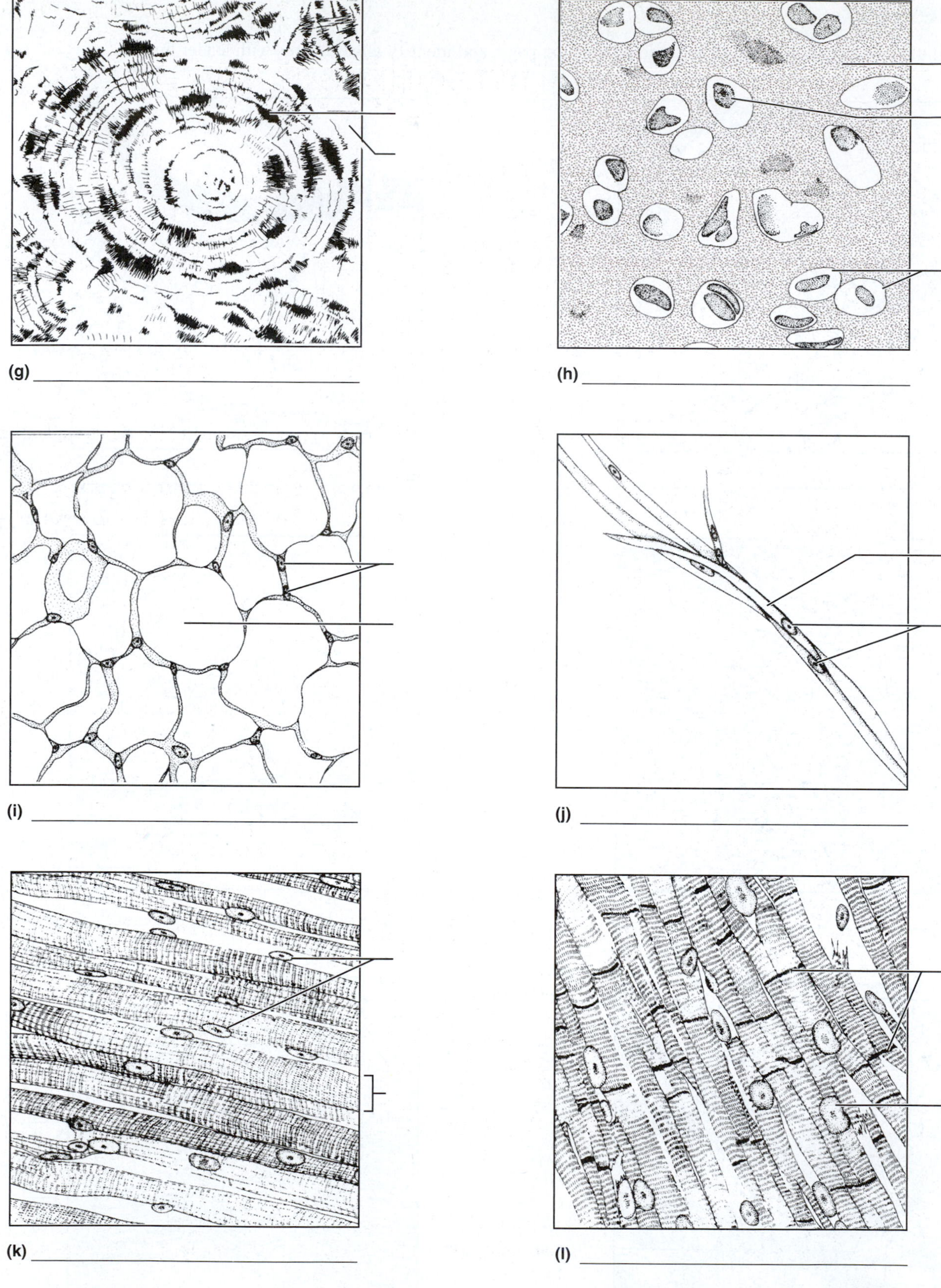
(g)
(h)
(i)
(j)
(k)
(l)

6

EXERCISE

The Integumentary System

MATERIALS

- ☐ Skin model (three-dimensional, if available)
- ☐ Compound microscope
- ☐ Prepared slide of human scalp
- ☐ Prepared slide of skin of palm or sole
- ☐ Sheet of 20# bond paper ruled to mark off 1-cm^2 areas
- ☐ Scissors
- ☐ Betadine® swabs, or Lugol's iodine and cotton swabs
- ☐ Adhesive tape
- ☐ Disposable gloves
- ☐ Data collection sheet for plotting distribution of sweat glands
- ☐ Porelon fingerprint pad or portable inking foils
- ☐ Ink cleaner towelettes
- ☐ Index cards (4 in. × 6 in.)
- ☐ Magnifying glasses

OBJECTIVES

1. List several important functions of the skin, or integumentary system.
2. Identify the following skin structures on a model, image, or microscope slide: epidermis, dermis (papillary and reticular layers), hair follicles and hair, sebaceous glands, and sweat glands.
3. Name and describe the layers of the epidermis.
4. List the factors that determine skin color, and describe the function of melanin.
5. Identify the major regions of nails.
6. Describe the distribution and function of hairs, sebaceous glands, and sweat glands.
7. Discuss the difference between eccrine and apocrine sweat glands.
8 Compare and contrast the structure and functions of the epidermis and the dermis.

PRE-LAB QUIZ

1. All the following are functions of the skin *except:*
   a. excretion of body wastes
   b. insulation
   c. protection from mechanical damage
   d. site of vitamin A synthesis
2. The skin has two distinct regions. The superficial layer is the ______________________________, and the underlying connective tissue, the ______________________________.
3. The most superficial layer of the epidermis is the:
   a. stratum basale
   b. stratum spinosum
   c. stratum granulosum
   d. stratum corneum
4. Thick skin of the epidermis contains ______________________________ layers.
5. ______________________________ is a yellow-orange pigment found in the stratum corneum and the hypodermis.
   a. Keratin
   b. Carotene
   c. Melanin
   d. Hemoglobin
6. These cells produce a brown-to-black pigment that colors the skin and protects DNA from ultraviolet radiation damage. The cells are:
   a. dendritic cells
   b. keratinocytes
   c. melanocytes
   d. tactile cells
7. Circle True or False. Nails are hornlike derivatives of the epidermis.
8. The portion of a hair that you see that projects from the scalp surface is known as the:
   a. bulb
   b. matrix
   c. root
   d. shaft
9. Circle the correct underlined term. The ducts of sebaceous / sweat glands usually empty into a hair follicle but may also open directly onto the skin surface.
10. Circle the correct underlined term. Eccrine / Apocrine glands are found primarily in the genital and axillary areas.

The **skin,** or **integument,** is considered an organ system because of its extent and complexity. It is much more than an external body covering; architecturally the skin is a marvel. It is tough yet pliable, a characteristic that enables it to withstand constant insult from outside agents.

The skin has many functions, most concerned with protection. It insulates and cushions the underlying body tissues and protects the entire body from abrasion, exposure to harmful chemicals, temperature extremes, and bacterial invasion. The hardened uppermost layer of the skin helps prevent water loss from the body surface. The skin's abundant capillary network (under the control of the nervous system) plays an important role in temperature regulation.

The skin has other functions as well. For example, it acts as a mini excretory system; urea, salts, and water are lost through the skin pores in sweat. The skin also has important metabolic duties. For example, like liver cells, it carries out some chemical conversions that activate or inactivate certain drugs and hormones, and it is the site of vitamin D synthesis for the body. Finally, the sense organs for touch, pressure, pain, and temperature are located here.

Basic Structure of the Skin

The skin has two distinct regions—the superficial *epidermis* composed of epithelium and an underlying connective tissue *dermis* **(Figure 6.1)**. These layers are firmly "cemented" together along a wavy border. But friction, such as the rubbing of a poorly fitting shoe, may cause them to separate, resulting in a blister. Immediately deep to the dermis is the **hypodermis,** or **superficial fascia,** which is not considered part of the skin. It consists primarily of adipose tissue. The main skin areas and structures are described next.

ACTIVITY 1

Locating Structures on a Skin Model

As you read, locate the following structures in the figure (Figure 6.1) and on a skin model. ■

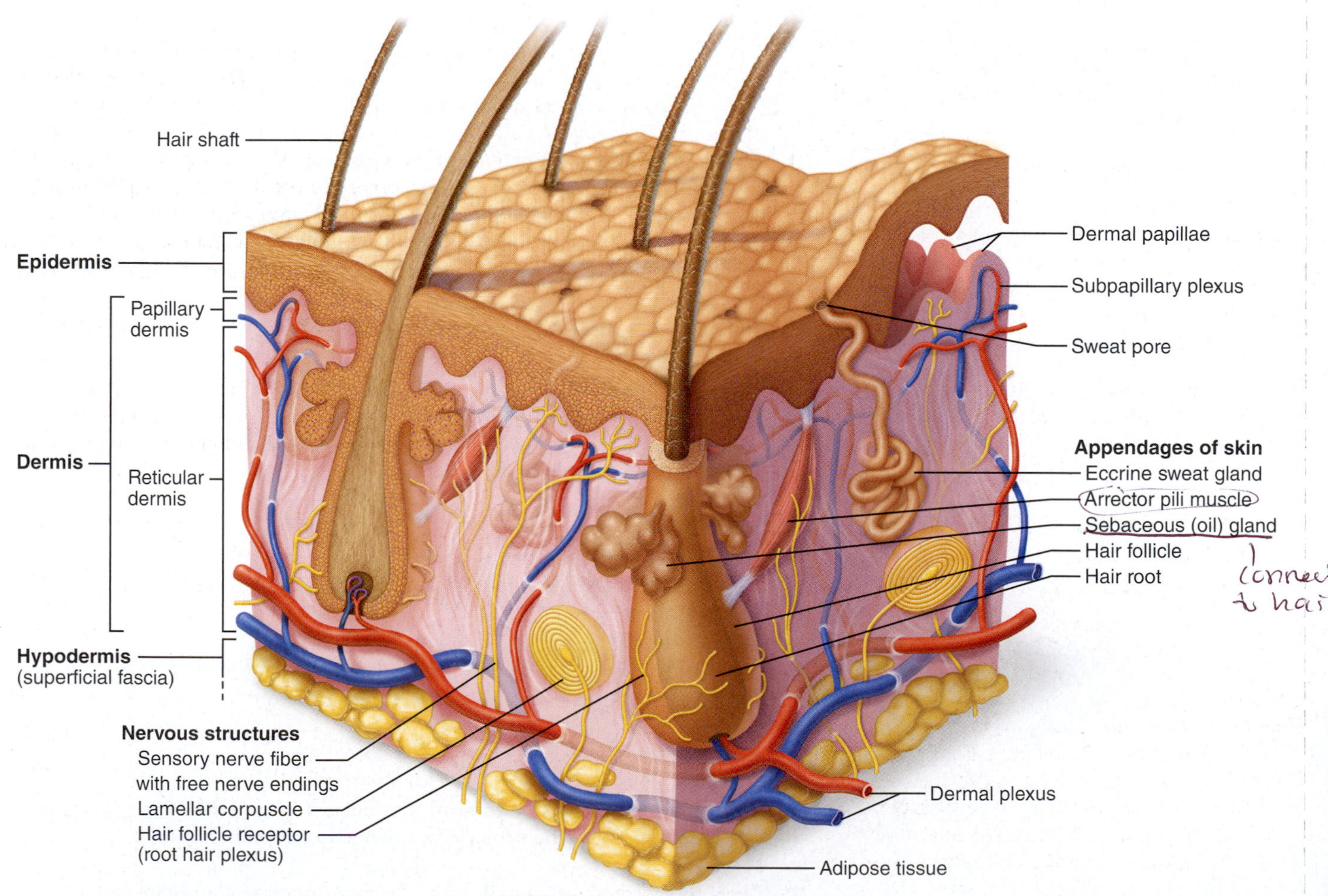

Figure 6.1 Skin structure. Three-dimensional view of the skin and the underlying hypodermis. The epidermis and dermis have been pulled apart at the right corner to reveal the dermal papillae.

Epidermis

Structurally, the avascular epidermis is a keratinized stratified squamous epithelium consisting of four distinct cell types and four or five distinct layers.

Cells of the Epidermis

- **Keratinocytes** (literally, keratin cells): The most abundant epidermal cells, their main function is to produce keratin fibrils. **Keratin** is a fibrous protein that gives the epidermis its durability and protective capabilities. Keratinocytes are tightly connected to each other by desmosomes.

Far less numerous are the following types of epidermal cells **(Figure 6.2)**:

- **Melanocytes:** Spidery black cells that produce the brown-to-black pigment called **melanin.** The skin tans because melanin production increases when the skin is exposed to sunlight. The melanin provides a protective pigment umbrella over the nuclei of the cells in the deeper epidermal layers, thus shielding their genetic material (DNA) from the damaging effects of ultraviolet radiation. A concentration of melanin in one spot is called a *freckle.*
- **Dendritic cells:** Also called *Langerhans cells,* these phagocytic cells (macrophages) play a role in immunity.
- **Tactile epithelial cells:** Occasional spiky hemispheres that, in conjunction with disklike sensory nerve endings, form sensitive touch receptors located at the epidermal-dermal junction.

Layers of the Epidermis

The epidermis consists of four layers in thin skin, which covers most of the body. Thick skin, found on the palms of the hands and soles of the feet, contains a fifth layer, the stratum lucidum. From deep to superficial, the layers of the epidermis are the stratum basale, stratum spinosum, stratum granulosum, stratum lucidum, and stratum corneum (Figure 6.2).

- **Stratum basale** (basal layer): A single row of cells immediately adjacent to the dermis. Its cells are constantly undergoing mitotic cell division to produce millions of new cells daily, hence its alternate name *stratum germinativum.* From 10% to 25% of the cells in this layer are melanocytes, which thread their processes through this and the adjacent layers of keratinocytes (Figure 6.2). Note also the tactile epithelial cell of this layer.
- **Stratum spinosum** (spiny layer): A stratum consisting of several cell layers immediately superficial to the basal layer. Its cells contain thick weblike bundles of intermediate filaments made of a pre-keratin protein. The stratum spinosum cells appear spiky (hence their name) because as the skin tissue is prepared for histological examination, they shrink but their desmosomes hold tight. Cells divide fairly rapidly in this layer, but less so than in the stratum basale. Cells in the basal and spiny layers are the only ones to receive adequate nourishment via diffusion of nutrients from the dermis. So as their daughter cells are pushed upward and away from the source of nutrition, they gradually die. Dendritic cells may occur in the spiny layer (Figure 6.2b).
- **Stratum granulosum** (granular layer): A thin layer named for the abundant granules its cells contain. These granules are of two types: (1) *lamellar granules,* which contain a waterproofing glycolipid that is secreted into the extracellular space; and (2) *keratohyalin granules,* which combine with the intermediate filaments in the more superficial layers to form the keratin fibrils. At the upper border of this layer, the cells are beginning to die.
- **Stratum lucidum** (clear layer): A very thin translucent band of flattened dead keratinocytes with indistinct boundaries. It is not present in regions of thin skin.
- **Stratum corneum** (horny layer): This outermost epidermal layer consists of some 20 to 30 cell layers and accounts for the bulk of the epidermal thickness. Cells in this layer, like those in the stratum lucidum (where it exists), are dead and their flattened scalelike remnants are fully keratinized. They are constantly being rubbed off and replaced by division of the deeper cells.

Dermis

The dense irregular connective tissue making up the dermis consists of two principal regions—the papillary and reticular areas (Figure 6.1). Like the epidermis, the dermis varies in thickness. For example, it is particularly thick on the palms of the hands and soles of the feet and is quite thin on the eyelids.

- **Papillary dermis:** The more superficial dermal region composed of areolar connective tissue. It is very uneven and has fingerlike projections from its superior surface, the **dermal papillae,** which attach it to the epidermis above. These projections lie on top of the larger dermal ridges. In the palms of the hands and soles of the feet, they produce the *fingerprints,* unique patterns of *epidermal ridges* that remain unchanged throughout life. The abundant capillary networks in the papillary layer furnish nutrients for the epidermal layers and allow heat to radiate to the skin surface. The pain (free nerve endings) and touch receptors (**tactile corpuscles** in hairless skin) are also found here.
- **Reticular dermis:** The deepest skin layer. It is composed of dense irregular connective tissue and contains many arteries and veins, sweat and sebaceous glands, and pressure receptors (**lamellar corpuscles**).

Both the papillary dermis and reticular dermis are heavily invested with collagenic and elastic fibers. The elastic fibers give skin its exceptional elasticity in youth. In old age, the number of elastic fibers decreases and the subcutaneous layer loses fat, which leads to wrinkling and inelasticity of the skin. Fibroblasts, adipose cells, various types of macrophages (which are important in the body's defense), and other cell types are found throughout the dermis.

The abundant dermal blood supply, consisting mainly of the deep *dermal plexus* (between the dermis and hypodermis) and the *subpapillary plexus* (located just deep to the dermal papillae), allows the skin to play a role in regulating body temperature. When body temperature is high, the arterioles serving the skin dilate, and the capillary network of the dermis becomes engorged with the heated blood. Thus body heat is allowed to radiate from the skin surface. If the environment is cool and body heat must be conserved, the arterioles constrict so that blood bypasses the dermal capillary networks temporarily.

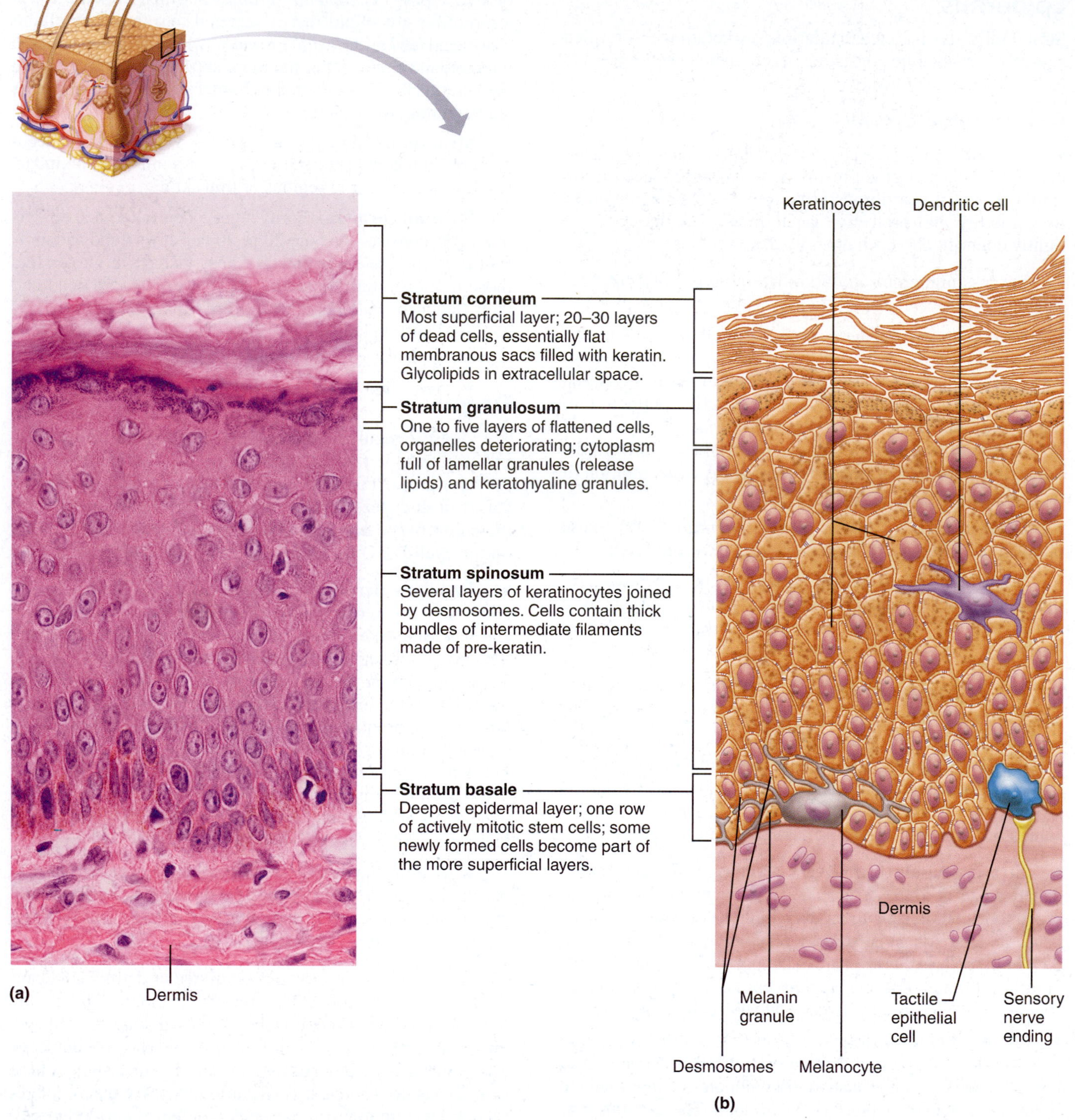

Stratum corneum
Most superficial layer; 20–30 layers of dead cells, essentially flat membranous sacs filled with keratin. Glycolipids in extracellular space.
Stratum granulosum
One to five layers of flattened cells, organelles deteriorating; cytoplasm full of lamellar granules (release lipids) and keratohyaline granules.
Stratum spinosum
Several layers of keratinocytes joined by desmosomes. Cells contain thick bundles of intermediate filaments made of pre-keratin.
Stratum basale
Deepest epidermal layer; one row of actively mitotic stem cells; some newly formed cells become part of the more superficial layers.
(a)
Dermis
Keratinocytes
Dendritic cell
Dermis
Melanin granule
Tactile epithelial cell
Sensory nerve ending
Desmosomes
Melanocyte
(b)

Figure 6.2 The main structural features in epidermis of thin skin. (a) Photomicrograph depicting the four major epidermal layers (435×). **(b)** Diagram showing the layers and relative distribution of the different cell types. The four cell types are keratinocytes (orange), melanocytes (gray), dendritic cells (purple), and tactile epithelial cells (blue). A sensory nerve ending (yellow), extending from the dermis (pink), is shown associated with the tactile cell forming the tactile disc (touch receptor). Notice that the keratinocytes are joined by numerous desmosomes. The stratum lucidum, present in thick skin, is not illustrated here.

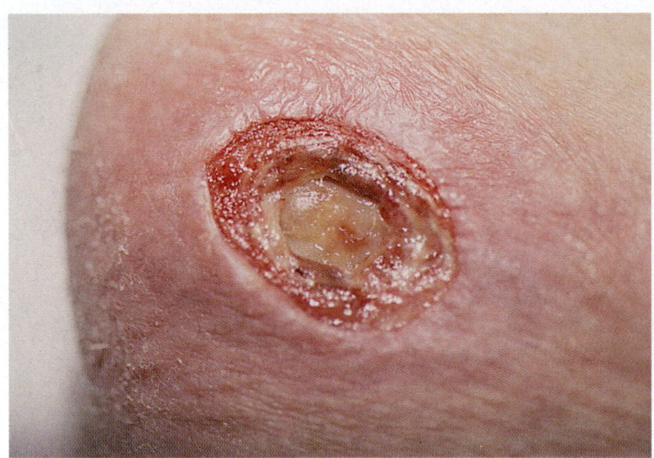

Figure 6.3 **Photograph of a deep (stage III) decubitus ulcer.**

Any restriction of the normal blood supply to the skin results in cell death and, if severe enough, skin ulcers **(Figure 6.3)**. **Bedsores (decubitus ulcers)** occur in bedridden patients who are not turned regularly enough. The weight of the body exerts pressure on the skin, especially over bony projections (hips, heels, etc.), which leads to restriction of the blood supply and tissue death. +

The dermis is also richly provided with lymphatic vessels and nerve fibers. Many of the nerve endings bear highly specialized receptor organs that, when stimulated by environmental changes, transmit messages to the central nervous system for interpretation. Some of these receptors include free nerve endings, lamellar corpuscles, and a hair follicle receptor (also called a *root hair plexus*) (Figure 6.1).

Skin Color

Skin color is a result of the relative amount of melanin in skin, the relative amount of carotene in skin, and the degree of oxygenation of the blood. People who produce large amounts of melanin have brown-toned skin. In light-skinned people, who have less melanin pigment, the dermal blood supply flushes through the rather transparent cell layers above, giving the skin a rosy glow. *Carotene* is a yellow-orange pigment present primarily in the stratum corneum and in the adipose tissue of the hypodermis. Its presence is most noticeable when large amounts of carotene-rich foods (carrots, for instance) are eaten.

Skin color can be an important diagnostic tool. For example, flushed skin may indicate hypertension, fever, or embarrassment, whereas pale skin is typically seen in anemic individuals. When the blood is inadequately oxygenated, as during asphyxiation and serious lung disease, both the blood and the skin take on a bluish, or cyanotic, cast. **Jaundice,** in which the tissues become yellowed, is almost always diagnostic for liver disease, whereas a bronzing of the skin hints that a person's adrenal cortex is hypoactive **(Addison's disease)**. +

Accessory Organs of the Skin

The accessory organs of the skin—cutaneous glands, hair, and nails—are all derivatives of the epidermis, but they reside in the dermis. They originate from the stratum basale and grow downward into the deeper skin regions.

Cutaneous Glands

The cutaneous glands fall primarily into two categories: the sebaceous glands and the sweat glands (Figure 6.1 and **Figure 6.4**).

Sebaceous (Oil) Glands

The sebaceous glands are found nearly all over the skin, except for the palms of the hands and the soles of the feet. Their ducts usually empty into a hair follicle, but some open directly on the skin surface.

Sebum is the product of sebaceous glands. It is a mixture of oily substances and fragmented cells that acts as a lubricant to keep the skin soft and moist (a natural skin cream) and keeps the hair from becoming brittle. The sebaceous glands become particularly active during puberty, when more male hormones (androgens) begin to be produced; thus the skin tends to become oilier during this period of life.

Blackheads are accumulations of dried sebum, bacteria, and melanin from epithelial cells in the oil duct. **Acne** is an active infection of the sebaceous glands. +

Sweat (Sudoriferous) Glands

Sweat, or sudoriferous, glands are exocrine glands and are widely distributed all over the skin. Outlets for the glands are epithelial openings called *pores.* Sweat glands are categorized by the composition of their secretions.

- **Eccrine glands:** Also called **merocrine sweat glands,** these glands are distributed all over the body. They produce clear perspiration consisting primarily of water, salts (mostly NaCl), and urea. Eccrine sweat glands, under the control of the nervous system, are an important part of the body's heat-regulating apparatus. They secrete perspiration when the external temperature or body temperature is high. When this water-based substance evaporates, it carries excess body heat with it. Thus evaporation of greater amounts of perspiration provides an efficient means of dissipating body heat when the capillary cooling system is not sufficient or is unable to maintain body temperature homeostasis.
- **Apocrine glands:** Found predominantly in the axillary and genital areas, these glands secrete the basic components of eccrine sweat plus proteins and fat-rich substances. Apocrine sweat is an excellent nutrient medium for the microorganisms typically found on the skin. This sweat is odorless, but when bacteria break down its organic components, it begins to smell unpleasant. The function of apocrine glands is not known. However, because their activity increases during sexual foreplay and the glands enlarge and recede with the phases of a women's menstrual cycle, they may be the human equivalent of the sexual scent glands of other animals.

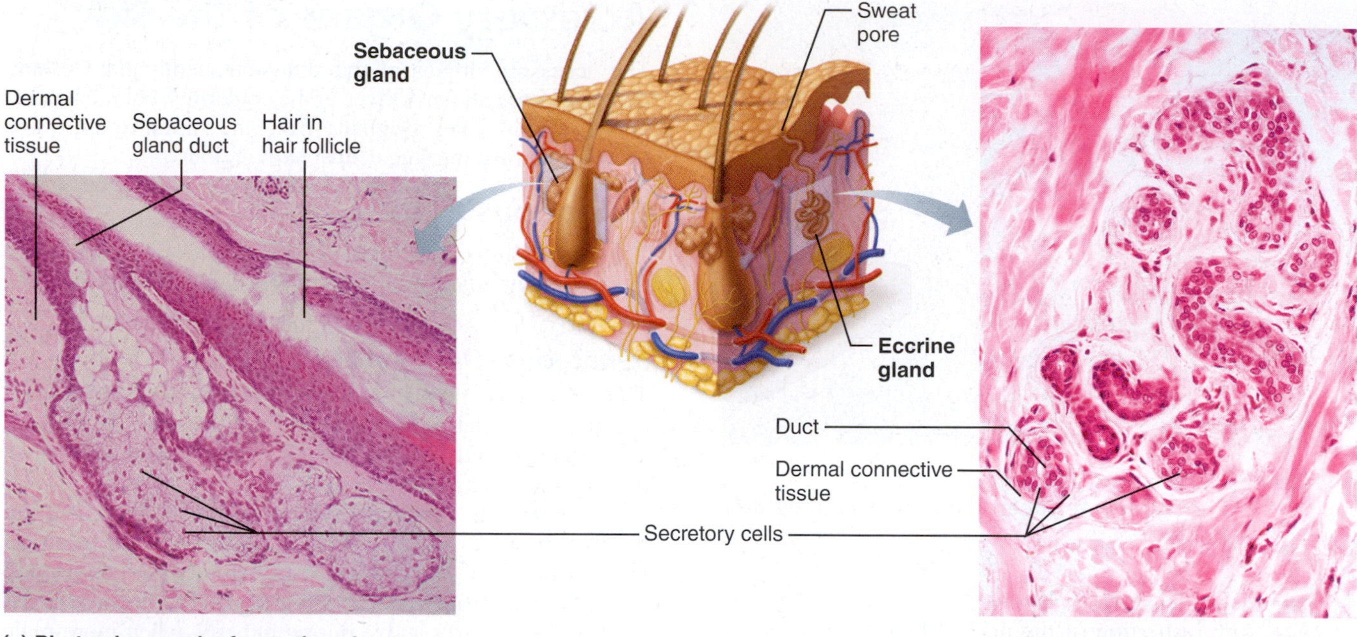

(a) Photomicrograph of a sectioned sebaceous gland (125×)

(b) Photomicrograph of a sectioned eccrine gland (220×)

Figure 6.4 Cutaneous glands.

ACTIVITY 2

Differentiating Sebaceous and Sweat Glands Microscopically

Using the slide *thin skin with hairs*, identify sebaceous and eccrine sweat glands (refer to Figure 6.4). What characteristics relating to location or gland structure allow you to differentiate these glands?

___

___

___

ACTIVITY 3

Plotting the Distribution of Sweat Glands

1. Form a hypothesis about the relative distribution of sweat glands on the palm and forearm. Justify your hypothesis.

2. The bond paper for this simple experiment has been preruled in cm^2—put on disposable gloves, and cut along the lines to obtain the required squares. You will need two squares of bond paper (each 1 cm × 1 cm), adhesive tape, and a Betadine® (iodine) swab *or* Lugol's iodine and a cotton-tipped swab.

3. Paint an area of the medial aspect of your left palm (avoid the crease lines) and a region of your left forearm with the iodine solution, and allow it to dry thoroughly. The painted area in each case should be slightly larger than the paper squares to be used.

4. Have your lab partner *securely* tape a square of bond paper over each iodine-painted area, and leave them in place for 20 minutes. (If it is very warm in the laboratory while this test is being conducted, good results may be obtained within 10 to 15 minutes.)

5. After 20 minutes, remove the paper squares, and count the number of blue-black dots on each square. The presence of a blue-black dot on the paper indicates an active sweat gland. The iodine in the pore is dissolved in the sweat and reacts chemically with the starch in the bond paper to produce the blue-black color. Thus "sweat maps" have been produced for the two skin areas.

6. Which skin area tested has the greater density of sweat glands?

___

___

7. Tape your results (bond paper squares) to a data collection sheet labeled "palm" and "forearm" at the front of the lab. Be sure to put your paper squares in the correct columns on the data sheet.

8. Once all the data has been collected, review the class results.

Nails

Nails are hornlike derivatives of the epidermis (Figure 6.5). They consist of the following parts:

- **Nail plate:** The visible attached portion.
- **Free edge:** The portion of the nail that grows out away from the body.
- **Hyponychium:** The region beneath the free edge of the nail.
- **Root:** The part that is embedded in the skin and adheres to an epithelial nail bed.
- **Nail folds:** Skin folds that overlap the borders of the nail.
- **Eponychium:** The thick proximal nail fold, commonly called the cuticle.

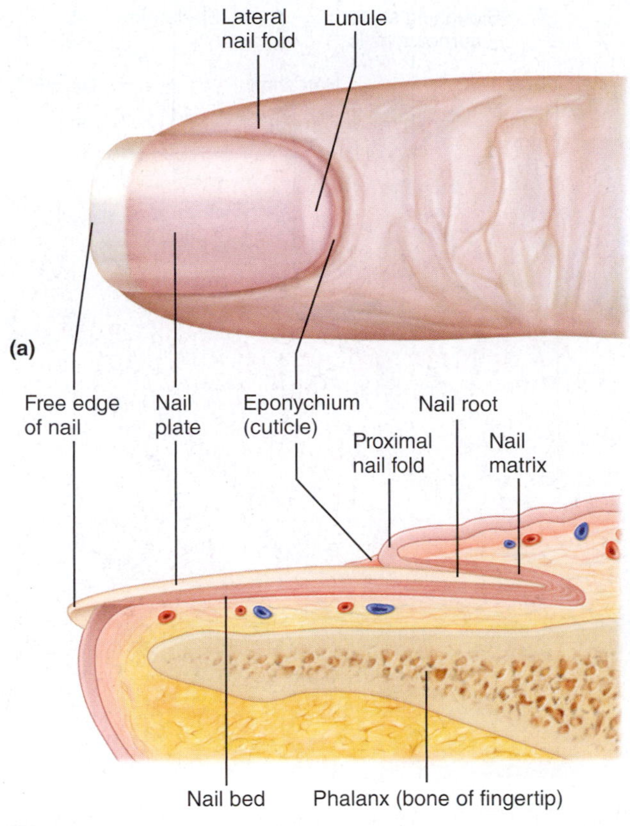

Figure 6.5 Structure of a nail. (a) Surface view of the distal part of a finger showing nail parts. The nail matrix that forms the nail lies beneath the lunule; the epidermis of the nail bed underlies the nail. **(b)** Sagittal section of the fingertip.

- **Nail bed:** Extension of the stratum basale beneath the nail.
- **Nail matrix:** The thickened proximal part of the nail bed containing germinal cells responsible for nail growth. As the matrix produces the nail cells, they become heavily keratinized and die. Thus nails, like hairs, are mostly nonliving material.
- **Lunule:** The proximal region of the thickened nail matrix, which appears as a white crescent. Everywhere else, nails are transparent and nearly colorless, but they appear pink because of the blood supply in the underlying dermis. When someone is cyanotic because of a lack of oxygen in the blood, the nail beds take on a blue cast.

ACTIVITY 4

Identifying Nail Structures

Identify the nail structures shown in the figure (Figure 6.5) on yourself or your lab partner.

Hairs and Associated Structures

Hairs, enclosed in hair follicles, are found all over the entire body surface, except for thick-skinned areas (the palms of the hands and the soles of the feet), parts of the external genitalia, the nipples, and the lips.

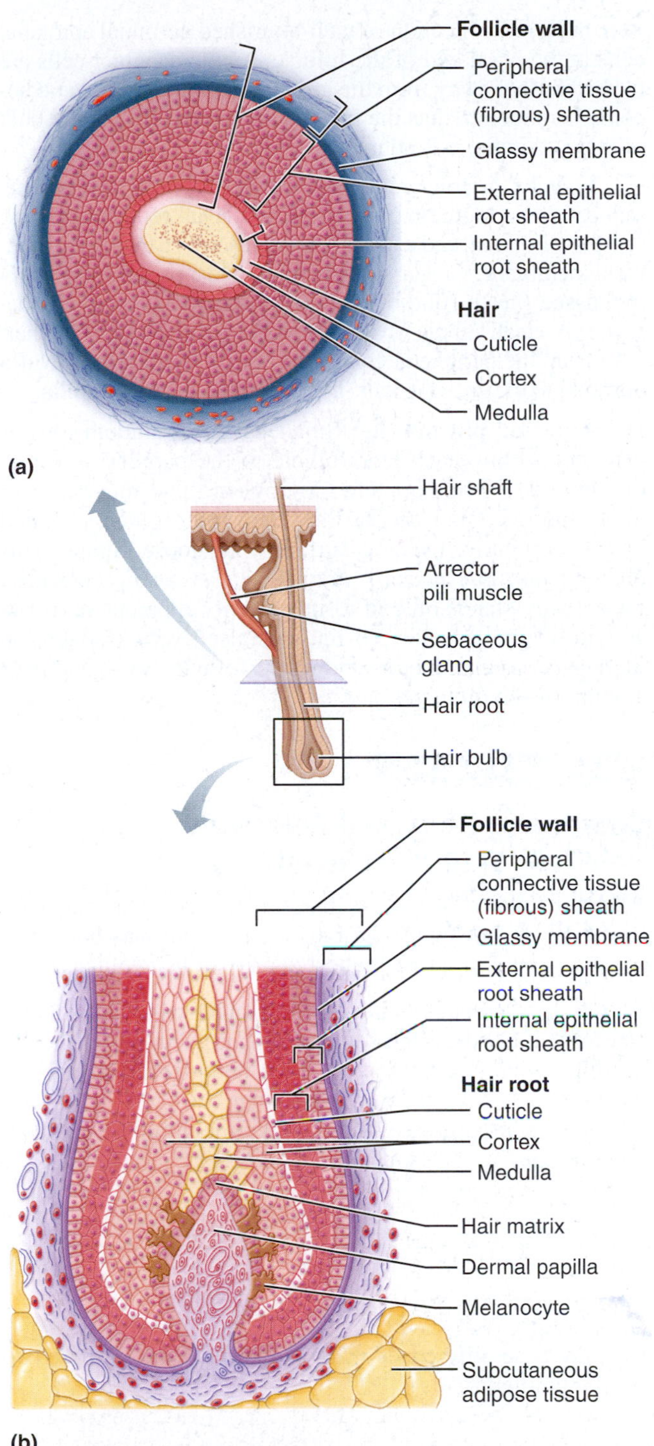

Figure 6.6 Structure of a hair and hair follicle. (a) Diagram of a cross section of a hair within its follicle. **(b)** Diagram of a longitudinal view of the expanded hair bulb of the follicle, which encloses the matrix.

- **Hair:** A structure consisting of three parts: a *medulla,* which is the central region, surrounded first by the *cortex* and then by a protective *cuticle* **(Figure 6.6)**. Abrasion of the cuticle results in split ends. Hair color depends on the amount and kind of melanin pigment within the hair cortex. The portion of the hair enclosed within the follicle is called the **root;** that portion projecting from the scalp surface is called the **shaft.** The

hair bulb is a collection of well-nourished germinal epithelial cells at the basal end of the follicle. As the daughter cells are pushed farther away from the growing region, they die and become keratinized; thus the bulk of the hair shaft, like the bulk of the epidermis, is dead material.

- **Follicle:** A structure formed from both epidermal and dermal cells (see Figure 6.6). Its inner epithelial root sheath, with two parts (internal and external), is enclosed by a thickened basement membrane, the glassy membrane, and a peripheral connective tissue sheath (fibrous sheath), which is essentially dermal tissue. A small nipple of dermal tissue protrudes into the hair bulb from the peripheral connective tissue sheath and provides nutrition to the growing hair. It is called the **dermal papilla.**
- **Arrector pili muscle:** Small bands of smooth muscle cells connecting each hair follicle to the papillary layer of the dermis (Figures 6.1 and 6.6). When these muscles contract (during cold or fright), the slanted hair follicle is pulled upright, dimpling the skin surface with goose bumps. This phenomenon is especially dramatic in a scared cat, whose fur actually stands on end to increase its apparent size. The activity of the arrector pili muscles also exerts pressure on the sebaceous glands surrounding the follicle, causing a small amount of sebum to be released.

ACTIVITY 5

Comparing Hairy and Relatively Hair-Free Skin Microscopically

Whereas thick skin has no hair follicles or sebaceous (oil) glands, thin skin typical of most of the body has both. The scalp, of course, has the highest density of hair follicles.

1. Obtain a prepared slide of the human scalp, and study it carefully under the microscope. Compare your tissue slide to the photomicrograph **(Figure 6.7a)**, and identify as many of the structures as possible (refer to Figure 6.1).

How is this stratified squamous epithelium different from that observed in the esophagus (Exercise 5)?

____

____

How do these differences relate to the functions of these two similar epithelia?

____

____

____

2. Obtain a prepared slide of hairless skin of the palm or sole (Figure 6.7b). Compare the slide to the photomicrograph (Figure 6.7a). In what ways does the thick skin of the palm or sole differ from the thin skin of the scalp?

____

____

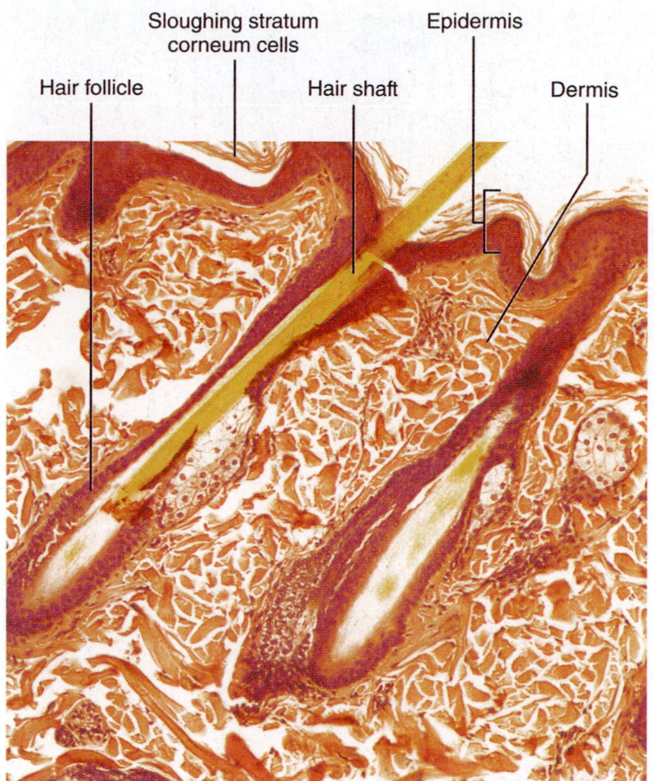

(a)

Epidermis
- Stratum corneum
- Stratum lucidum
- Stratum granulosum
- Stratum spinosum
- Stratum basale

Dermis

(b)

Figure 6.7 Photomicrographs of skin. (a) Thin skin with hairs (190×). **(b)** Thick hairless skin (155×).

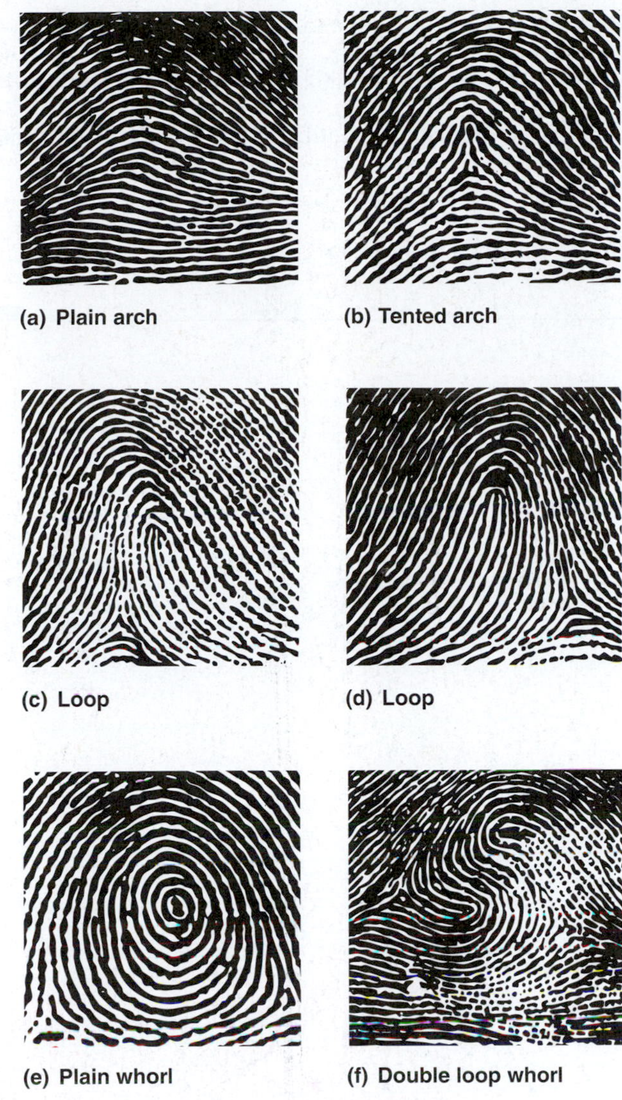

Figure 6.8 Main types of fingerprint patterns. (a–b) Arches. **(c–d)** Loops. **(e–f)** Whorls.

Dermography: Fingerprinting

Each of us has a unique, genetically determined set of fingerprints. Because of the usefulness of fingerprinting for identifying and apprehending criminals, most people associate this craft solely with criminal investigations. However, civil fingerprints are invaluable in quickly identifying amnesia victims, missing persons, and unknown deceased, such as those killed in major disasters.

The friction ridges responsible for fingerprints appear in several patterns, which are clearest when the fingertips are inked and then pressed against white paper. Impressions are also made when perspiration or any foreign material such as blood, dirt, or grease adheres to the ridges and the fingers are then pressed against a smooth, nonabsorbent surface. The three most common patterns are *arches, loops,* and *whorls* **(Figure 6.8)**. The *pattern area* in loops and whorls is the only area of the print used in identification, and it is delineated by the *type lines*—specifically the two innermost ridges that start parallel, diverge, and/or surround or tend to surround the pattern area.

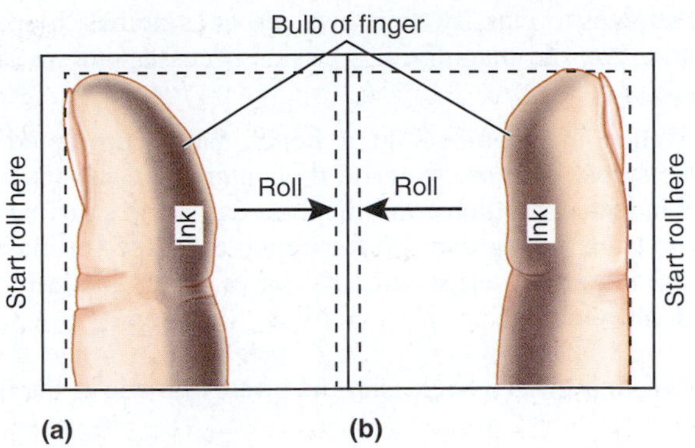

Figure 6.9 Fingerprinting. Method of inking **(a)** the thumb and **(b)** the index finger.

ACTIVITY 6

Taking and Identifying Inked Fingerprints

For this activity, you will be working as a group with your lab partners. Though the equipment for professional fingerprinting is fairly basic, consisting of a glass or metal inking plate, printer's ink (a heavy black paste), ink roller, and standard 8 in. × 8 in. cards, you will be using supplies that are even easier to handle. Each student will prepare two index cards, each bearing the thumbprint and index fingerprint of the right hand.

1. Obtain the following supplies and bring them to your bench: two 4 in. × 6 in. index cards per student, Porelon fingerprint pad or portable inking foils, ink cleaner towelettes, and a magnifying glass.

2. The subject should wash and dry the hands. Open the ink pad or peel back the covering over the ink foil, and position it close to the edge of the laboratory bench. The subject should position himself or herself at arm's length from the bench edge and inking object.

3. A second student, called the *operator,* will stand to the left of the subject and with two hands will hold and direct the movement of the subject's fingertip. During this process, the subject should look away, try to relax, and refrain from trying to help the operator.

4. The thumbprint is to be placed on the left side of the index card, the index fingerprint on the right. The operator should position the subject's right thumb or index finger on the side of the bulb of the finger in such a way that the area to be inked spans the distance from the fingertip to just beyond the first joint, and then roll the finger lightly across the inked surface until its bulb faces in the opposite direction. To prevent smearing, the thumb is rolled away from the body midline (from left to right as the subject sees it; see **Figure 6.9**) and the index finger is rolled toward the body midline (from right to left). The same ink foil can be reused for all the students at the bench; the ink pad is good for thousands of prints. Repeat the procedure (still using the subject's right hand) on the second index card.

5. If the prints are too light, too dark, or smeary, repeat the procedure.

6. While subsequent members are making clear prints of their thumb and index finger, those who have completed that activity should clean their inked fingers with a towelette

and attempt to classify their own prints as arches, loops, or whorls. Use the magnifying glass as necessary to see ridge details.

7. When all members at a bench have completed the above steps, they are to write their names on the backs of their index cards, then combine their cards and shuffle them before transferring them to the bench opposite for classification of pattern and identification of prints made by the same individuals.

How difficult was it to classify the prints into one of the three categories given? ________________________

Why do you think this was so? ________________________

Was it easy or difficult to identify the prints made by the same individual? ________________________

Why do you think this was so? ________________________

____

Name ______________________

Lab Time/Date ______________________

EXERCISE 6

REVIEW SHEET

The Integumentary System

Basic Structure of the Skin

1. Complete the following statements by writing the appropriate word or phrase on the correspondingly numbered blank:

The two basic tissues of which the skin is composed are dense irregular connective tissue, which makes up the dermis, and _1_, which forms the epidermis. The tough protective protein found in the epidermal cells is called _2_. The pigments melanin and _3_ contribute to skin color. A localized concentration of melanin is referred to as a _4_.

1. ______________________
2. ______________________
3. ______________________
4. ______________________

2. Name four protective functions of the skin:

a. ______________________ c. ______________________

b. ______________________ d. ______________________

3. Using the key choices, choose all responses that apply to the following descriptions. Some terms are used more than once.

Key:

a. stratum basale	d. stratum lucidum	g. reticular dermis
b. stratum corneum	e. stratum spinosum	h. epidermis as a whole
c. stratum granulosum	f. papillary dermis	i. dermis as a whole

______________ 1. layer of translucent cells containing dead keratinocytes

______________ 2. two layers of dead cells

______________ 3. dermal layer responsible for fingerprints

______________ 4. vascular region of the skin

______________ 5. major skin area that produces derivatives (nails and hair)

______________ 6. epidermal layer exhibiting the most rapid cell division

______________ 7. layer including scalelike dead cells, full of keratin, that constantly slough off

______________ 8. layer of mitotic cells filled with intermediate filaments

______________ 9. has abundant elastic and collagenic fibers

______________ 10. location of melanocytes and tactile epithelial cells

______________ 11. area where weblike pre-keratin filaments first appear

______________ 12. layer of areolar connective tissue

4. Label the skin structures and areas indicated in the accompanying diagram of thin skin. Then, complete the statements that follow.

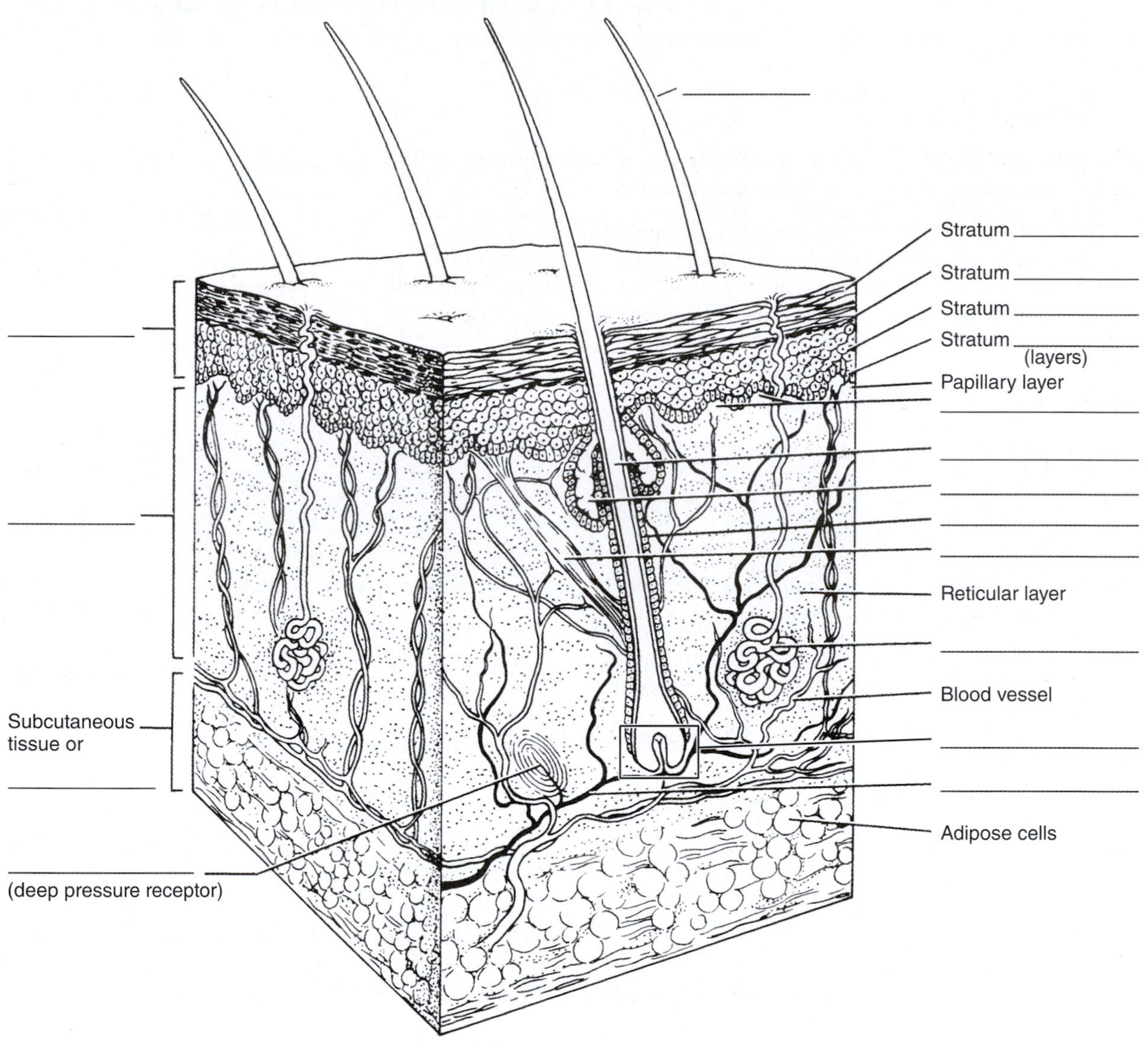

a. ______________________ granules contain glycolipids that prevent water loss from the skin.

b. Fibers in the dermis are produced by ______________________.

c. Glands that respond to rising androgen levels are the ______________________ glands.

d. Phagocytic cells that occupy the epidermis are called ______________________.

e. A unique touch receptor formed from a stratum basale cell and a nerve fiber is a ______________________.

f. Which layer is present in thick skin but not in thin skin? ______________________

g. What cell-to-cell structures hold the cells of the stratum spinosum tightly together? ______________________

5. What substance is manufactured in the skin (but is not a secretion) to play a role elsewhere in the body?

____

6. List the sensory receptors found in the dermis of the skin. ______________________________

____

____

7. A nurse tells a doctor that a patient is cyanotic. Define *cyanosis.* ______________________________

What does its presence imply? ______________________________

8. What is a bedsore (decubitus ulcer)? ______________________________

Why does it occur? ______________________________

____

Accessory Organs of the Skin

9. Using the key, match the terms with the appropriate descriptions. Some terms are used more than once.

Key:
a. arrector pili
b. cutaneous receptors
c. hair
d. hair follicle
e. nail
f. sebaceous gland
g. sweat gland—apocrine
h. sweat gland—eccrine

______________ 1. produces an accumulation of oily material that is known as a blackhead

______________ 2. tiny muscles, attached to hair follicles, that pull the hair upright during fright or cold

______________ 3. sweat gland with a role in temperature control

______________ 4. sheath formed of both epithelial and connective tissues

______________ 5. less numerous type of sweat-producing gland; found mainly in the pubic and axillary regions

______________ 6. found everywhere on the body except the palms of hands and soles of feet

______________ 7. primarily dead/keratinized cells

______________ 8. specialized nerve endings that respond to temperature, touch, etc.

______________ 9. its secretion is a lubricant for hair and skin

______________ 10. "sports" a lunule and a cuticle

10. Describe two integumentary system mechanisms that help regulate body temperature. ______________________________

____

____

____

11. Several structures or skin regions are listed below. Identify each by matching its letter with the appropriate area on the figure.

a. adipose cells

b. dermis

c. epidermis

d. hair follicle

e. hair shaft

f. sloughing stratum corneum cells

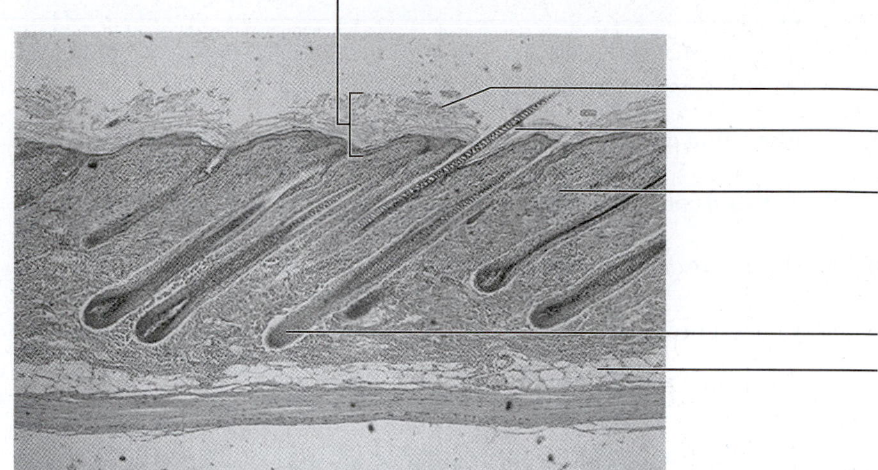

Plotting the Distribution of Sweat Glands

12. With what substance in the bond paper does the iodine painted on the skin react? ____________________

13. On the basis of class data, which skin area—the forearm or palm of hand—has more sweat glands?

____

Was this an expected result? ____________________ Explain. ____________________

____

Which other body areas would, if tested, prove to have a high density of sweat glands? ____________________

____

14. What organ system controls the activity of the eccrine sweat glands? ____________________

Dermography: Fingerprinting

15. Why can fingerprints be used to identify individuals? ____________________

____

16. Name the three common fingerprint patterns.

____________________, ____________________, and ____________________

7

EXERCISE

Overview of the Skeleton: Classification and Structure of Bones and Cartilages

MATERIALS

- ☐ Disarticulated bones (identified by number) that demonstrate classic examples of the four bone classifications (long, short, flat, and irregular)
- ☐ Long bone sawed longitudinally (beef bone from a slaughterhouse, if possible, or prepared laboratory specimen)
- ☐ Disposable gloves
- ☐ Long bone soaked in 10% hydrochloric acid (HCl) or vinegar until flexible
- ☐ Long bone baked at 250°F for more than 2 hours
- ☐ Compound microscope
- ☐ Prepared slide of ground bone (x.s.)
- ☐ Three-dimensional model of microscopic structure of compact bone
- ☐ Prepared slide of a developing long bone undergoing endochondral ossification
- ☐ Articulated skeleton

OBJECTIVES

1. Name the two tissue types that form the skeleton.
2. List the functions of the skeletal system.
3. Name the four main groups of bones based on shape.
4. Identify surface bone markings, and list their functions.
5. Identify the major anatomical areas on a longitudinally cut long bone or on an appropriate image.
6. Explain the role of inorganic salts and organic matrix in providing flexibility and hardness to bone.
7. Locate and identify the major parts of an osteon microscopically or on a histological model or appropriate image of compact bone.
8. Locate and identify the three major types of skeletal cartilages.

PRE-LAB QUIZ

1. All the following are functions of the skeleton *except:*
   a. attachment for muscles
   b. production of melanin
   c. site of red blood cell formation
   d. storage of lipids
2. Circle the correct underlined term. The axial / appendicular skeleton consists of bones that surround the body's center of gravity.
3. Circle the correct underlined term. Compact / Spongy bone looks smooth and homogeneous.
4. ________ bones are generally thin and have a layer of spongy bone between two layers of compact bone.
   a. Flat b. Irregular c. Long d. Short
5. The femur is an example of a(n) ________ bone.
   a. flat c. long
   b. irregular d. short
6. Circle the correct underlined term. The shaft of a long bone is known as the epiphysis / diaphysis.
7. The structural unit of compact bone is the:
   a. osteon b. canaliculus c. lacuna
8. Circle True or False. Embryonic skeletons consist primarily of elastic cartilage, which is gradually replaced by bone during development and growth.
9. The type of cartilage that has the greatest strength and is found in the knee joint and intervertebral discs is:
   a. elastic b. fibrocartilage c. hyaline
10. Circle True or False. Cartilage has a covering made of dense connective tissue called a periosteum.

The **skeleton,** the body's framework, is constructed of two of the most supportive tissues found in the human body—cartilage and bone. In embryos, the skeleton is predominantly made up of hyaline cartilage, but in the adult, most of the cartilage is replaced by more rigid bone. Cartilage persists only in such isolated areas as the external ear, bridge of the nose, larynx, trachea, joints, and parts of the rib cage (see Figure 7.5, page 99).

Besides supporting and protecting the body as an internal framework, the skeleton provides a system of levers with which the skeletal muscles work to move the body. In addition, the bones store lipids and many minerals

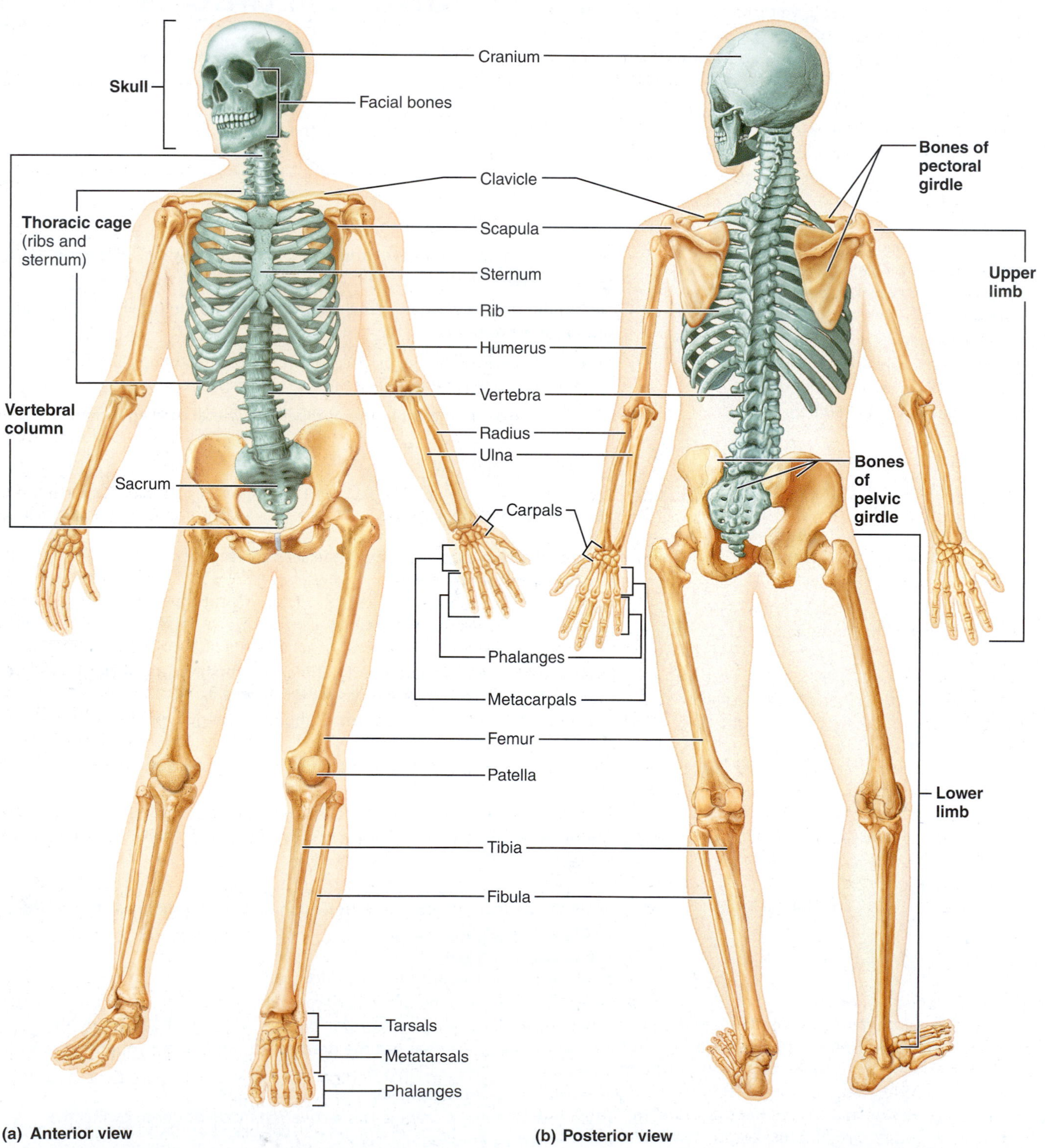

Figure 7.1 The human skeleton. The bones of the axial skeleton are colored green to distinguish them from the bones of the appendicular skeleton.

(the most important of which is calcium). Finally, the red marrow cavities of bones provide a site for hematopoiesis (blood cell formation).

The skeleton is made up of bones that are connected at *joints,* or *articulations.* The skeleton is subdivided into two divisions: the **axial skeleton** (those bones that lie around the body's center of gravity) and the **appendicular skeleton** (bones of the limbs, or appendages) **(Figure 7.1)**.

Before beginning your study of the skeleton, imagine for a moment that your bones have turned to putty. What if you were running when this change took place? Now imagine your bones forming a continuous metal framework within your body, somewhat like a network of plumbing pipes. What problems could you envision with this arrangement? These images should help you understand how well the skeletal system provides support and protection and facilitates movement.

Classification of Bones

The 206 bones of the adult skeleton are composed of two basic kinds of osseous tissue that differ in their texture. **Compact bone** looks smooth and homogeneous; **spongy** (or *cancellous*) **bone** is composed of small *trabeculae* (bars) of bone and lots of open space.

Bones may be classified further on the basis of their relative gross anatomy into four groups: long, short, flat, and irregular bones.

Long bones, such as the femur and bones of the fingers (phalanges) (Figure 7.1), are much longer than they are wide, generally consisting of a shaft with heads at either end. Long bones are composed predominantly of compact bone. **Short bones** are typically cube-shaped, and they contain more spongy bone than compact bone. The tarsals and carpals (Figure 7.1) are examples.

Flat bones are generally thin, with two waferlike layers of compact bone sandwiching a layer of spongy bone between them. Although the name "flat bone" implies a structure that is level or horizontal, many flat bones are curved (for example, the bones of the skull). Bones that do not fall into one of the preceding categories are classified as **irregular bones.** The vertebrae are irregular bones (see Figure 7.1).

Some anatomists also recognize two other subcategories of bones. **Sesamoid bones** are special types of short bones formed in tendons. The patellas (kneecaps) are sesamoid bones. **Sutural bones** are tiny bones between cranial bones. Except for the patellas, the sesamoid and sutural bones are not included in the bone count of 206 because they vary in number and location in different individuals.

Bone Markings

Even a casual observation of the bones will reveal that bone surfaces are not featureless smooth areas but are scarred with an array of bumps, holes, and ridges. These **bone markings** reveal where bones form joints with other bones, where muscles, tendons, and ligaments were attached, and where blood vessels and nerves passed. Bone markings fall into two categories: projections, or processes that grow out from the bone and serve as sites of muscle attachment or help form joints; and depressions or cavities, indentations or openings in the bone that often serve as conduits for nerves and blood vessels. Refer to the table of summarized bone markings **(Table 7.1)**.

ACTIVITY 1

Examining and Classifying Bones

Examine the isolated (disarticulated) bones on display to find specific examples of the bone markings described in the table (Table 7.1). Then classify each of the bones into one of the four anatomical groups by recording its number in the accompanying chart.

Long	Short	Flat	Irregular

Gross Anatomy of the Typical Long Bone

ACTIVITY 2

Examining a Long Bone

1. Obtain a long bone that has been sawed along its longitudinal axis. If a cleaned dry bone is provided, no special preparations need be made.

Note: If the bone supplied is a fresh beef bone, don disposable gloves before beginning your observations.

Identify the **diaphysis,** or shaft **(Figure 7.2)**. Observe its smooth surface, which is composed of compact bone. If you are using a fresh specimen, carefully pull away the **periosteum,** or fibrous membrane covering, to view the bone surface. Notice that many fibers of the periosteum penetrate into the bone. These fibers are called **perforating collagen fiber bundles** *(Sharpey's fibers).* Blood vessels and nerves travel through the periosteum and invade the bone. *Osteoblasts* (bone-forming cells) and *osteoclasts* (bone-destroying cells) are found on the inner, or osteogenic, layer of the periosteum.

2. Now inspect the **epiphysis,** the end of the long bone. Notice that it is composed of a thin layer of compact bone that encloses spongy bone.

3. Identify the **articular cartilage,** which covers the epiphyseal surface in place of the periosteum. The glassy hyaline cartilage provides a smooth surface to minimize friction at joints.

4. If the animal was still young and growing, you will be able to see the **epiphyseal plate,** a thin area of hyaline cartilage that provides for longitudinal growth of the bone during youth. Once the long bone has stopped growing, these areas are replaced with bone and appear as thin, barely discernible remnants—the **epiphyseal lines.**

5. In an adult animal, the central cavity of the shaft (*medullary cavity*) is essentially a storage region for adipose tissue, or **yellow marrow.** In the infant, this area is involved in forming blood cells, and so **red marrow** is found in the

Table 7.1 Bone Markings

Name of bone marking	Description
Projections That Are Sites of Muscle and Ligament Attachment	
Tuberosity (too″be-ros′ĭ-te)	Large rounded projection; may be roughened
Crest	Narrow ridge of bone; usually prominent
Trochanter (tro-kan′ter)	Very large, blunt, irregularly shaped process (the only examples are on the femur)
Line	Narrow ridge of bone; less prominent than a crest
Tubercle (too′ber-kl)	Small rounded projection or process
Epicondyle (ep″ĭ-kon′dīl)	Raised area on or above a condyle
Spine	Sharp, slender, often pointed projection
Process	Any bony prominence

Illustration:

Iliac crest
Ischial spine
Ischial tuberosity
Coxal bone
Trochanters
Intertrochanteric line
Adductor tubercle
Medial epicondyle
Condyle
Femur of thigh
Vertebra
Spinous process

Name of bone marking	Description
Surfaces That Form Joints	
Head	Bony expansion carried on a narrow neck
Facet	Smooth, nearly flat articular surface
Condyle (kon′dīl)	Rounded articular projection, often articulates with a corresponding fossa
Ramus (ra′mus)	Armlike bar of bone

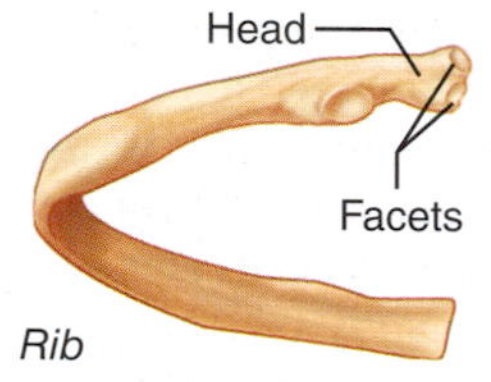

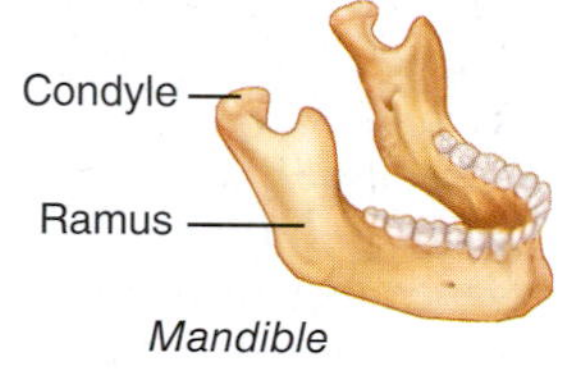

Name of bone marking	Description
Depressions and Openings	
For passage of vessels and nerves	
Foramen (fo-ra′men)	Round or oval opening through a bone
Groove	Furrow
Fissure	Narrow, slitlike opening
Notch	Indentation at the edge of a structure
Others	
Fossa (fos′ah)	Shallow basinlike depression in a bone, often serving as an articular surface
Meatus (me-a′tus)	Canal-like passageway
Sinus	Bone cavity, filled with air and lined with mucous membrane

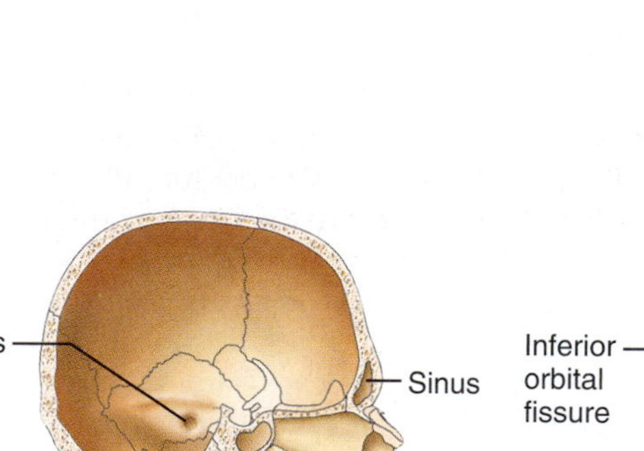

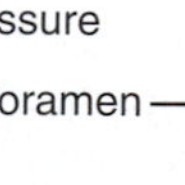

Inferior orbital fissure
Foramen
Skull

marrow cavities. In adult bones, the red marrow is confined to the interior of the epiphyses, where it occupies the spaces between the trabeculae of spongy bone.

6. If you are examining a fresh bone, look carefully to see if you can distinguish the delicate **endosteum** lining the shaft. The endosteum also covers the trabeculae of spongy bone and lines the canals of compact bone. Like the periosteum, the endosteum contains both osteoblasts and osteoclasts. As the bone grows in diameter on its external surface, it is constantly being broken down on its inner surface. Thus the thickness

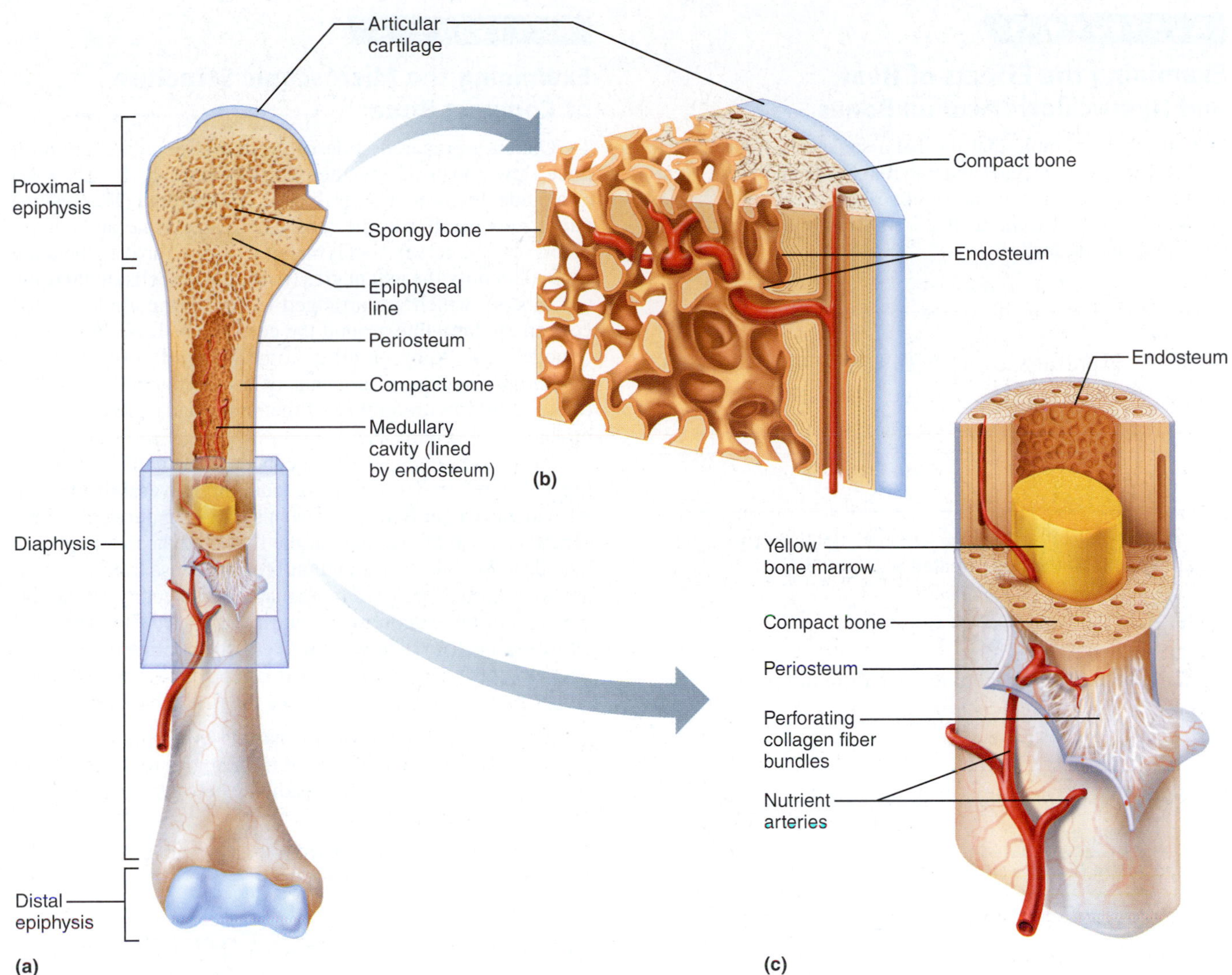

Figure 7.2 The structure of a long bone (humerus of the arm). (a) Anterior view with longitudinal section cut away at the proximal end. **(b)** Pie-shaped, three-dimensional view of spongy bone and compact bone of the epiphysis. **(c)** Cross section of diaphysis (shaft). Note that the external surface of the diaphysis is covered by a periosteum, but the articular surface of the epiphysis is covered with hyaline cartilage.

of the compact bone layer composing the shaft remains relatively constant.

7. If you have been working with a fresh bone specimen, return it to the appropriate area, and properly dispose of your gloves. Wash your hands before continuing on to the microscope study. ■

Longitudinal bone growth at epiphyseal plates (growth plates) follows a predictable sequence and provides a reliable indicator of the age of children exhibiting normal growth. In cases in which problems of long-bone growth are suspected (for example, pituitary dwarfism), X-ray films are taken to view the width of the growth plates. An abnormally thin epiphyseal plate indicates growth retardation. +

Chemical Composition of Bone

Bone is one of the hardest materials in the body. Although relatively light, bone has a remarkable ability to resist tension and shear forces that continually act on it. An engineer would tell you that a cylinder (like a long bone) is one of the strongest structures for its mass. Thus nature has given us an extremely strong, exceptionally simple (almost crude), and flexible supporting system without sacrificing mobility.

The hardness of bone is due to the inorganic calcium salts deposited in its ground substance. Its flexibility comes from the organic elements of the matrix, particularly the collagen fibers.

ACTIVITY 3

Examining the Effects of Heat and Hydrochloric Acid on Bones

Obtain a bone sample that has been soaked in hydrochloric acid (HCl) (or in vinegar) and one that has been baked. Heating removes the organic part of bone, whereas acid dissolves out the minerals. Do the treated bones retain the structure of untreated specimens?

____

Gently apply pressure to each bone sample. What happens to the heated bone?

____

____

What happens to the bone treated with acid?

____

What does the acid appear to remove from the bone?

____

What does baking appear to do to the bone?

____

____

In rickets, the bones are not properly calcified. Which of the demonstration specimens would more closely resemble the bones of a child with rickets?

____

Microscopic Structure of Compact Bone

As you have seen, spongy bone has a spiky, open-work appearance, resulting from the arrangement of the **trabeculae** that compose it, whereas compact bone appears to be dense and homogeneous. However, microscopic examination of compact bone reveals that it is riddled with passageways carrying blood vessels, nerves, and lymphatic vessels that provide the living bone cells with needed substances and a way to eliminate wastes **(Figure 7.3)**. Indeed, bone histology is much easier to understand when you recognize that bone tissue is organized around its blood supply.

ACTIVITY 4

Examining the Microscopic Structure of Compact Bone

1. Obtain a prepared slide of ground bone and examine it under low power. Using the photomicrograph (Figure 7.3c) as a guide, focus on a central canal. The **central (Haversian) canal** runs parallel to the long axis of the bone and carries blood vessels, nerves, and lymphatic vessels through the bony matrix. Identify the **osteocytes** (mature bone cells) in **lacunae** (chambers), which are arranged in concentric circles called **concentric lamellae** around the central canal. Because bone remodeling is going on all the time, you will also see some *interstitial lamellae,* remnants of *circumferential lamellae* (those at the circumference of the bony shaft) that have been broken down (Figure 7.3c).

A central canal and all the concentric lamellae surrounding it are referred to as an **osteon,** or **Haversian system.** Also identify **canaliculi,** tiny canals radiating outward from a central canal to the lacunae of the first lamella and then from lamella to lamella. The canaliculi form a dense transportation network through the hard bone matrix, connecting all the living cells of the osteon to the nutrient supply. The canaliculi allow each cell to take what it needs for nourishment and to pass along the excess to the next osteocyte. You may need a higher-power magnification to see the fine canaliculi.

2. Also note the **perforating canals** *(Volkmann's canals)* (Figure 7.3). These canals run at right angles to the shaft and complete the communication pathway between the bone interior and its external surface.

3. If a model of bone histology is available, identify the same structures on the model.

Ossification: Bone Formation and Growth in Length

Except for the collarbones (clavicles), all bones of the body inferior to the skull form in the embryo by the process of **endochondral ossification,** which uses hyaline cartilage "bones" as patterns for bone formation. The major events of this process, which begins in the (primary ossification) center of the shaft of a developing long bone, are as follows:

- Blood vessels invade the perichondrium covering the hyaline cartilage model and convert it to a periosteum.
- Osteoblasts at the inner surface of the periosteum secrete bone matrix around the hyaline cartilage model, forming a bone collar.
- Cartilage in the shaft center calcifies and then hollows out, forming an internal cavity.
- A *periosteal bud* (blood vessels, nerves, red marrow elements, osteoblasts, and osteoclasts) invades the cavity and forms spongy bone, which is removed by osteoclasts, thus

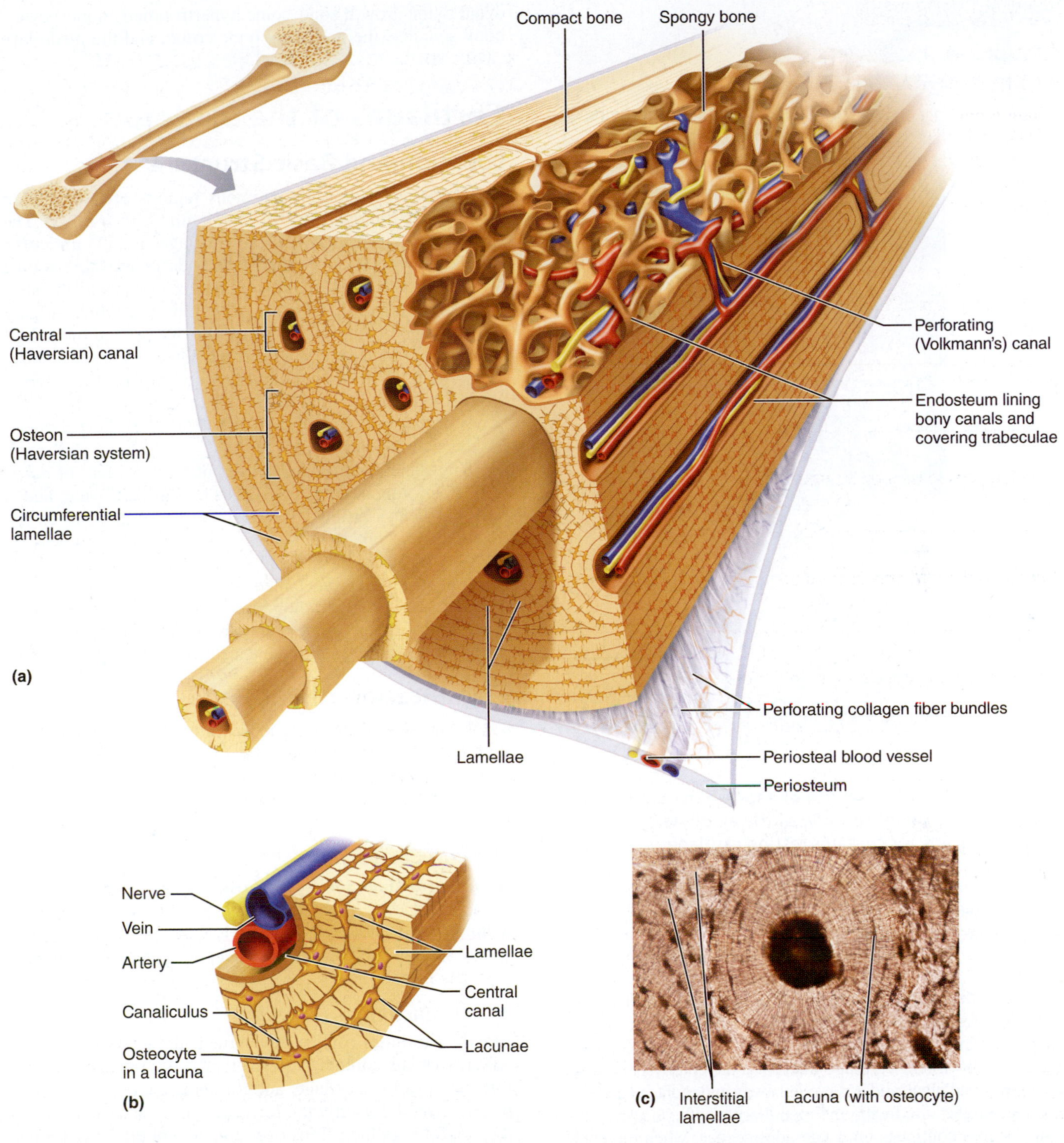

Figure 7.3 Microscopic structure of compact bone. (a) Diagram view of a pie-shaped segment of compact bone, illustrating its structural units (osteons). **(b)** Higher magnification view of a portion of one osteon. Note the position of osteocytes in lacunae. **(c)** Photomicrograph of a cross-sectional view of an osteon (405×).

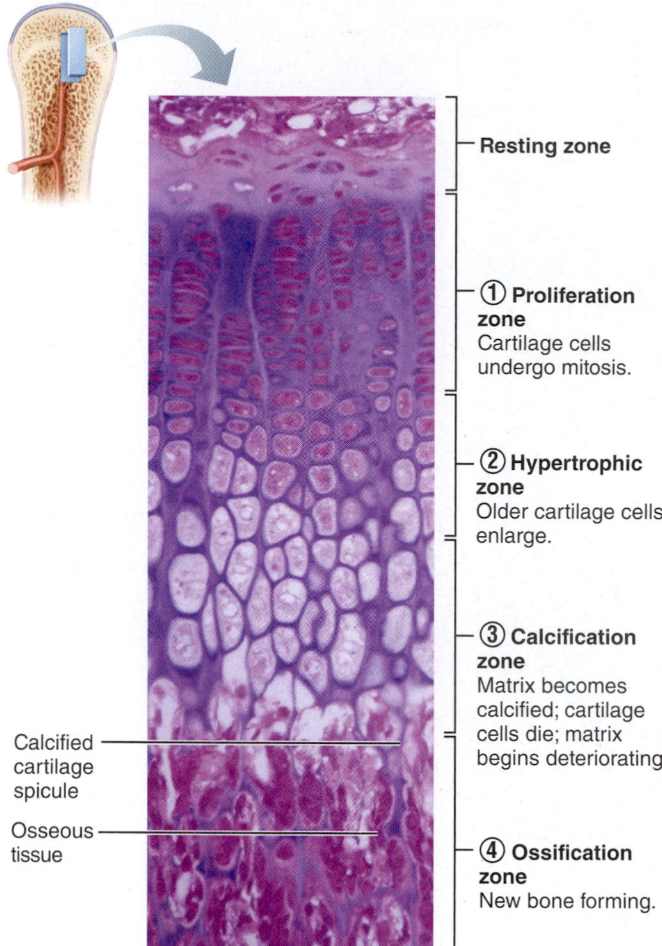
Resting zone
① Proliferation zone
Cartilage cells undergo mitosis.
② Hypertrophic zone
Older cartilage cells enlarge.
③ Calcification zone
Matrix becomes calcified; cartilage cells die; matrix begins deteriorating.
④ Ossification zone
New bone forming.
Calcified cartilage spicule
Osseous tissue

Figure 7.4 Growth in length of a long bone occurs at the epiphyseal plate. The side of the epiphyseal plate facing the epiphysis contains resting cartilage cells. Cells proximal to the resting zone are arranged in four zones: proliferation, hypertrophic, calcification, and ossification from the earliest stage of growth ① to the region where bone replaces cartilage ④ (300×).

producing the medullary cavity. This process proceeds in both directions from the *primary ossification center*. As bones grow longer, the medullary cavity gets larger and larger. Chondroblasts lay down new cartilage matrix on the epiphyseal face of the epiphyseal plate, and it is eroded away and replaced by bony spicules on the side facing the medullary cavity **(Figure 7.4)**. This process continues until late adolescence, when the entire epiphyseal plate is replaced by bone.

ACTIVITY 5

Examining the Osteogenic Epiphyseal Plate

Obtain a slide depicting endochondral ossification (cartilage bone formation) and bring it to your bench to examine under the microscope. Identify the proliferation, hypertrophic, calcification, and ossification zones of the epiphyseal plate (refer to Figure 7.4). Then, identify the area of resting cartilage cells distal to the growth zone, some hypertrophied chondrocytes, bony spicules, the periosteal bone collar, and the medullary cavity. ■

Cartilages of the Skeleton

Location and Basic Structure

As mentioned earlier, cartilaginous regions of the skeleton have a fairly limited distribution in adults **(Figure 7.5)**. The most important of these skeletal cartilages are (1) **articular cartilages,** which cover the bone ends at movable joints; (2) **costal cartilages,** found connecting the ribs to the sternum (breastbone); (3) **laryngeal cartilages,** which largely construct the larynx (voice box); (4) **tracheal** and **bronchial cartilages,** which reinforce other passageways of the respiratory system; (5) **nasal cartilages,** which support the external nose; (6) **intervertebral discs,** which separate and cushion bones of the spine (vertebrae); and (7) the cartilage supporting the external ear.

The skeletal cartilages consist of some variety of *cartilage tissue,* which typically consists primarily of water and is fairly resilient. Cartilage tissues are also distinguished by the fact that they contain no nerves and very few blood vessels. Like bones, each cartilage is surrounded by a covering of dense connective tissue, called a *perichondrium* (rather than a periosteum). The perichondrium acts like a girdle to resist distortion of the cartilage when the cartilage is subjected to pressure. It also plays a role in cartilage growth and repair.

Classification of Cartilage

The skeletal cartilages have representatives from each of the three cartilage tissue types—hyaline, elastic, and fibrocartilage. Because cartilage tissues are covered in another lab (Exercise 5), we will only briefly discuss that information here.

Hyaline Cartilage

Hyaline cartilage looks like frosted glass when viewed by the unaided eye. Most skeletal cartilages are composed of hyaline cartilage (Figure 7.5). Hyaline cartilage provides sturdy support with some resilience, or "give." (Review Figure 5.5h, page 65.)

Elastic Cartilage

Elastic cartilage can be envisioned as "hyaline cartilage with more elastic fibers." Consequently, it is much more flexible than hyaline cartilage, and it tolerates repeated bending better. Essentially, only the cartilages of the external ear and the epiglottis (which flops over and covers the larynx when we swallow) are made of elastic cartilage. (Review Figure 5.5i, page 65.)

Fibrocartilage

Fibrocartilage consists of rows of chondrocytes alternating with rows of thick collagen fibers. This tissue looks like a cartilage-dense regular connective tissue hybrid. Fibrocartilage has great tensile strength and can withstand heavy compression. Hence, its use to construct the intervertebral discs and the cartilages within the knee joint makes a lot of sense. (See Figure 7.5 and Figure 5.5j, page 66.)

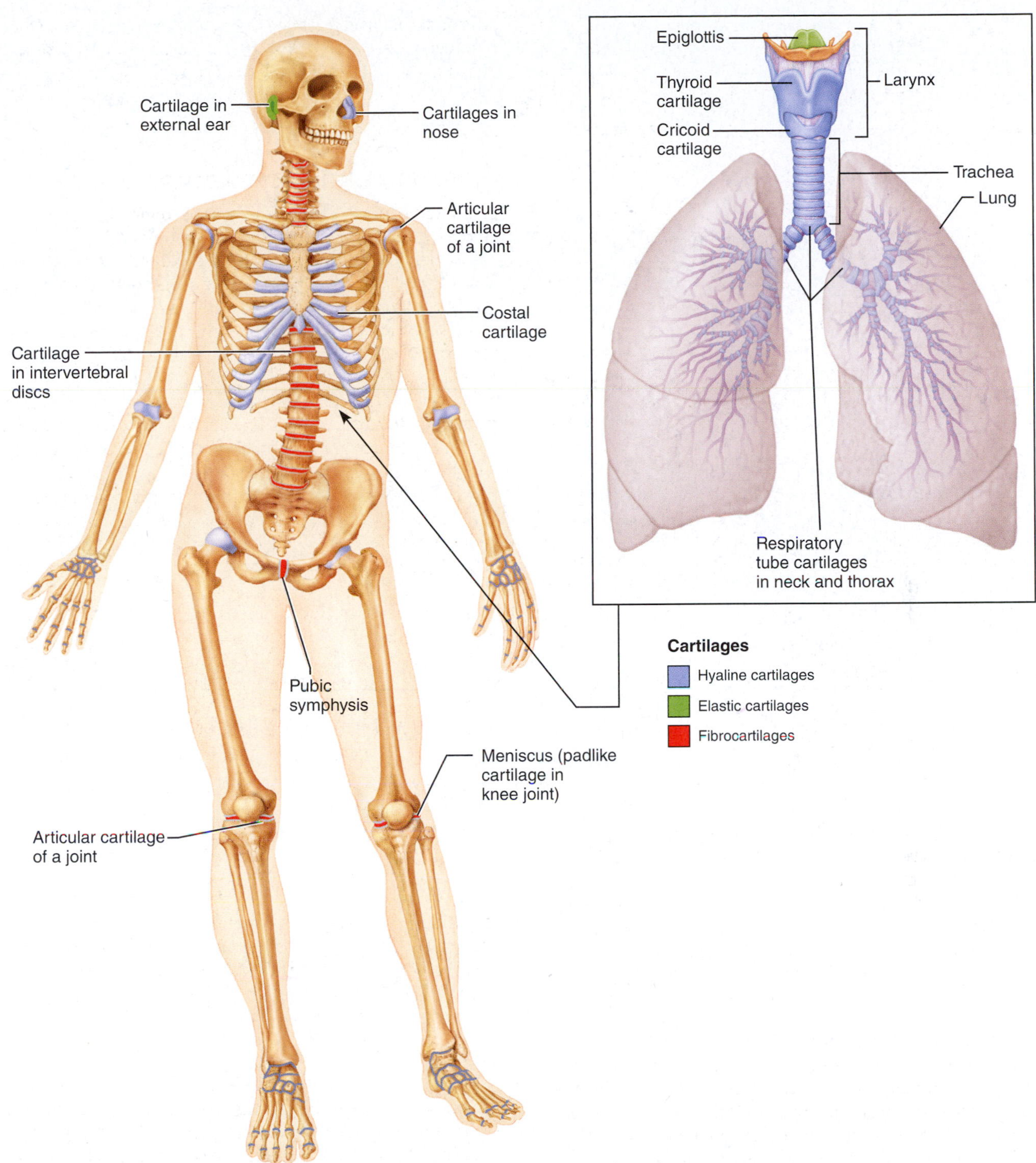

Figure 7.5 Cartilages in the adult skeleton and body. The cartilages that support the respiratory tubes and larynx are shown separately at the upper right.

Name ______________________

Lab Time/Date ______________________

EXERCISE 7

REVIEW SHEET

Overview of the Skeleton: Classification and Structure of Bones and Cartilages

Classification of Bones

1. Place the name of each labeled bone in the figure of the human skeleton (Figure 7.1, page 92), into the appropriate column of the chart below. Use appropriate references as necessary.

Long	Short	Flat	Irregular

2. The four major anatomical classifications of bones are long, short, flat, and irregular bones. Which category has the *least* amount of spongy bone relative to its total volume? ______________________

Bone Markings

3. Match the terms in column B with the appropriate description in column A.

	Column A	Column B
________	1. sharp, slender process*	a. condyle
________	2. small rounded projection*	b. crest
________	3. narrow ridge of bone*	c. epicondyle
________	4. large rounded projection*	d. facet
________	5. structure supported on neck†	e. fissure
________	6. armlike projection†	f. foramen
________	7. rounded, convex projection†	g. fossa
________	8. narrow opening‡	h. head
________	9. canal-like structure	i. meatus
________	10. round or oval opening through a bone‡	j. process
________	11. shallow depression	k. ramus

________________	12. air-filled cavity	l.	sinus
________________	13. large, irregularly shaped projection*	m.	spine
________________	14. raised area on or above a condyle*	n.	trochanter
________________	15. projection or prominence	o.	tubercle
________________	16. smooth, nearly flat articular surface†	p.	tuberosity

*a site of muscle and ligament attachment
†takes part in joint formation
‡a passageway for nerves or blood vessels

Gross Anatomy of the Typical Long Bone

4. Use the terms in the key below to identify the structures marked by leader lines and braces in the diagrams. Some terms are used more than once.

Key:
a. articular cartilage
b. compact bone
c. diaphysis
d. endosteum
e. epiphyseal line
f. epiphysis
g. medullary cavity
h. nutrient artery
i. periosteum
j. red marrow cavity
k. trabeculae of spongy bone
l. yellow marrow

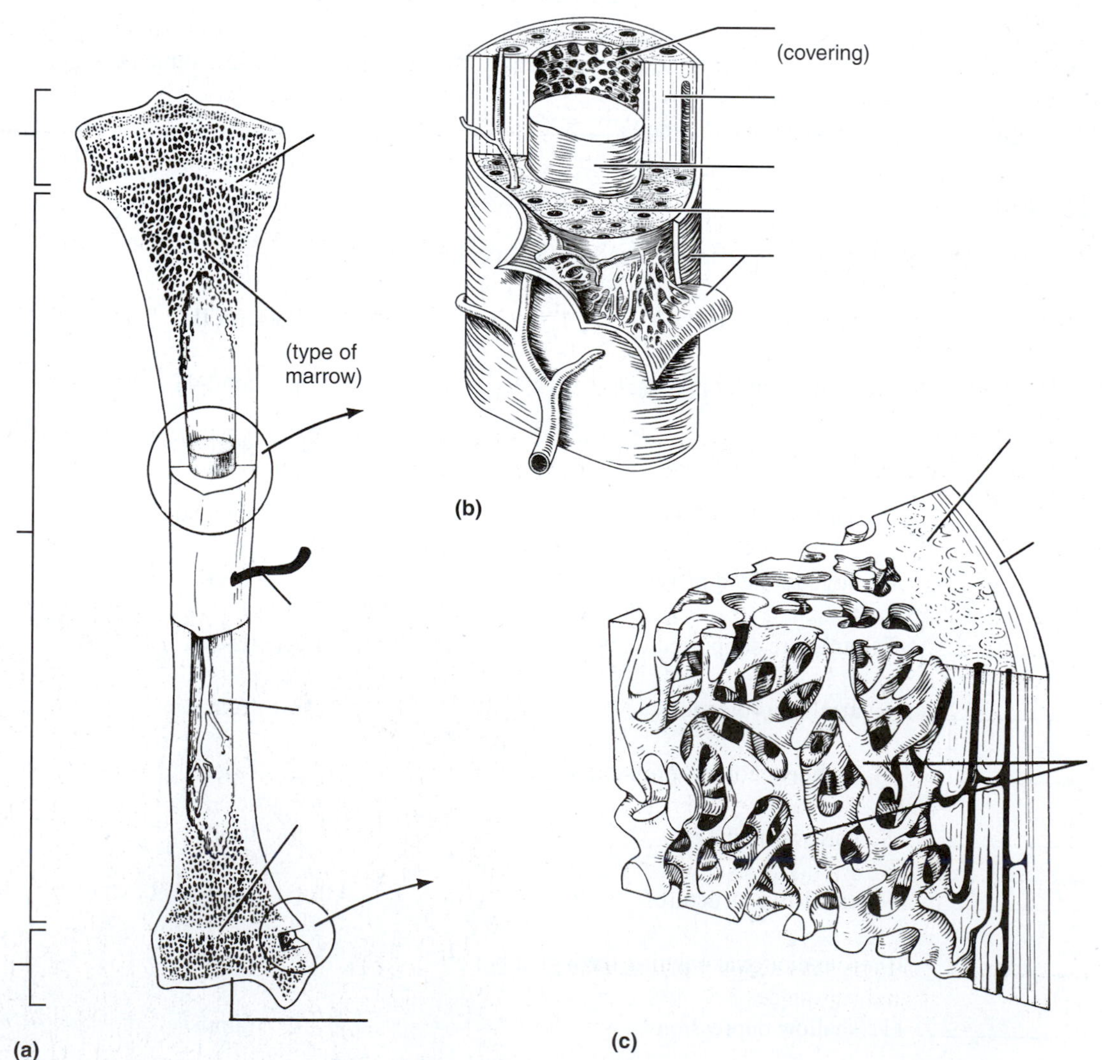

5. Match the letters of terms in question 4 with the information below.

_______ 1. contains spongy bone in adults

_______ 2. made of compact bone

_______ 3. site of blood cell formation

___, ___ 4. major submembranous sites of osteoclasts

_______ 5. scientific term for bone shaft

_______ 6. contains fat in adult bones

_______ 7. growth plate remnant

_______ 8. major submembranous site of osteoblasts

6. What differences between compact and spongy bone can be seen with the naked eye? _______________

____

7. What is the function of the periosteum? _______________

Chemical Composition of Bone

8. What is the function of the organic matrix in bone? _______________

9. Name the important organic bone components. _______________

10. Calcium salts form the bulk of the inorganic material in bone. What is the function of the calcium salts?

____

11. Baking removes _______ from bone. Soaking bone in acid removes _______.

Microscopic Structure of Compact Bone

12. Trace the route taken by nutrients through a bone, starting with the periosteum and ending with an osteocyte in a lacuna.

Periosteum → _______________ → _______________

_______________ → _______________ → osteocyte

13. Several descriptions of bone structure are given below. Identify the structure involved by choosing the appropriate term from the key and placing its letter in the blank. Then, on the photomicrograph of bone on the right, identify all structures that are named in the key, and bracket an osteon.

Key: a. canaliculi b. central canal c. concentric lamellae d. lacunae e. matrix

_______ 1. layers of bony matrix around a central canal

_______ 2. site of osteocytes

_______ 3. longitudinal canal carrying blood vessels, lymphatics, and nerves

_______ 4. minute canals connecting osteocytes of an osteon

_______ 5. inorganic salts deposited in organic ground substance

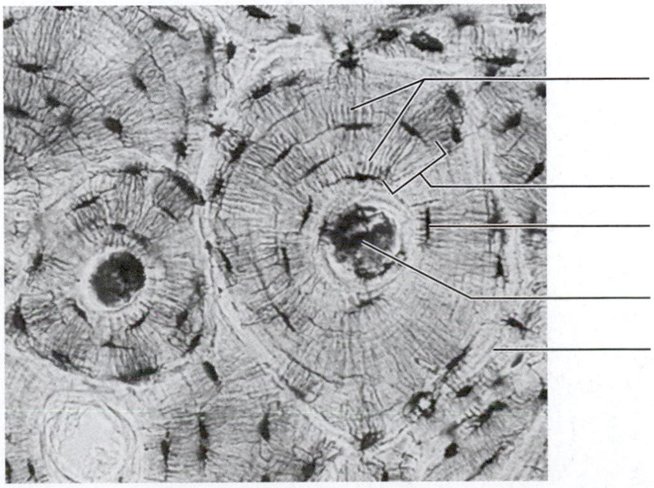

Ossification: Bone Formation and Growth in Length

14. Compare and contrast events occurring on the epiphyseal and diaphyseal faces of the epiphyseal plate.

epiphyseal face: ______________________________

diaphyseal face: ______________________________

____

Cartilages of the Skeleton

15. Using the key choices, identify each type of cartilage described (in terms of its body location or function) below. Terms may be used more than once.

Key: a. elastic b. fibrocartilage c. hyaline

______________ 1. supports the external ear

______________ 2. between the vertebrae

______________ 3. forms the walls of the voice box (larynx)

______________ 4. the epiglottis

______________ 5. articular cartilages

______________ 6. meniscus in a knee joint

______________ 7. connects the ribs to the sternum

______________ 8. most effective at resisting compression

______________ 9. most springy and flexible

______________ 10. most abundant

16. Identify the two types of cartilage diagrammed below. On each, label the *chondrocytes in lacunae* and the *matrix.*

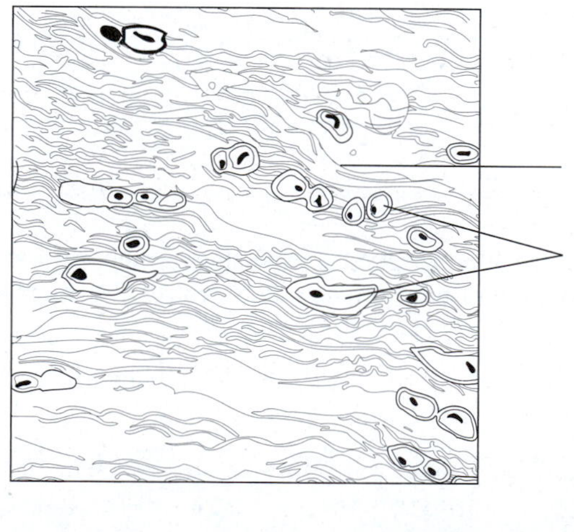

(a) ______________________________

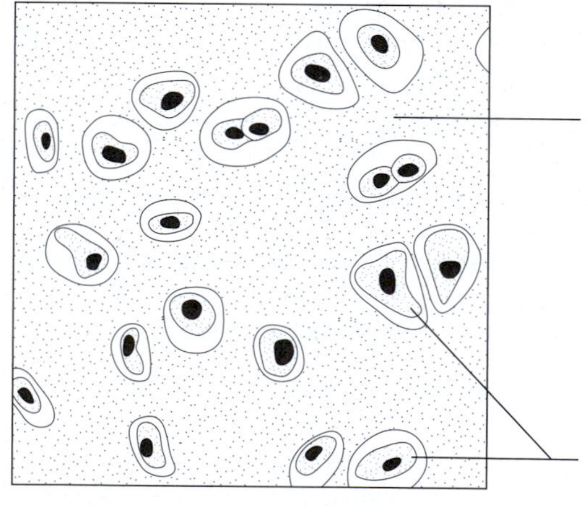

(b) ______________________________

8

EXERCISE

The Axial Skeleton

MATERIALS

- ☐ Intact skull and Beauchene skull
- ☐ X-ray images of individuals with scoliosis, lordosis, and kyphosis (if available)
- ☐ Articulated skeleton, articulated vertebral column, removable intervertebral discs
- ☐ Isolated cervical, thoracic, and lumbar vertebrae, sacrum, and coccyx
- ☐ Isolated fetal skull

OBJECTIVES

1. Name the three parts of the axial skeleton.
2. Identify the bones of the axial skeleton, either by examining isolated bones or by pointing them out on an articulated skeleton or skull, and name the important bone markings on each bone.
3. Name and describe the different types of vertebrae.
4. Discuss the importance of intervertebral discs and spinal curvatures.
5. Identify three abnormal spinal curvatures.
6. List the components of the thoracic cage.
7. Identify the bones of the fetal skull by examining an articulated skull or image.
8. Define *fontanelle*, and discuss the function and fate of fontanelles in the fetus.
9. Discuss important differences between the fetal and adult skulls.

PRE-LAB QUIZ

1. The axial skeleton can be divided into the skull, the vertebral column, and the:
   a. thoracic cage c. hip bones
   b. femur d. humerus
2. Eight bones make up the ________, which encloses and protects the brain.
   a. cranium b. face c. skull
3. How many bones of the skull are considered facial bones? ________
4. Circle the correct underlined term. The lower jawbone, or <u>maxilla</u> / <u>mandible</u>, articulates with the temporal bones in the only freely movable joints in the skull.
5. Circle the correct underlined term. The <u>body</u> / <u>spinous process</u> of a typical vertebra forms the rounded, central portion that faces anteriorly in the human vertebral column.
6. The seven bones of the neck are called ________ vertebrae.
   a. cervical c. spinal
   b. lumbar d. thoracic
7. The ________ vertebrae articulate with the corresponding ribs.
   a. cervical c. spinal
   b. lumbar d. thoracic
8. The ________, commonly referred to as the breastbone, is a flat bone formed by the fusion of three bones: the manubrium, the body, and the xiphoid process.
   a. coccyx b. sacrum c. sternum
9. Circle True or False. The first seven pairs of ribs are called floating ribs because they have only indirect cartilage attachments to the sternum.
10. A fontanelle:
    a. is found only in the fetal skull
    b. is a fibrous membrane
    c. allows for compression of the skull during birth
    d. all of the above

The **axial skeleton** (the green portion of Figure 7.1 on page 92) can be divided into three parts: the skull, the vertebral column, and the thoracic cage.

The Skull

The **skull** is composed of two sets of bones. The bones of the **cranium** enclose and protect the fragile brain tissue. The **facial bones** support the eyes and position them anteriorly. They also provide attachment sites for facial muscles, which make it possible for us to present our feelings to the world. All but one of the bones of the skull are joined by interlocking joints called *sutures.* The mandible, or lower jawbone, is attached to the rest of the skull by a freely movable joint.

ACTIVITY 1

Identifying the Bones of the Skull

The bones of the skull (**Figure 8.1** through **Figure 8.7**) are described below. As you read through this material, identify each bone on an intact and/or Beauchene skull (see Figure 8.6c).

Note: Important bone markings are listed beneath the bones on which they appear. The color-coded dot before each bone name corresponds to the bone color in the figures. ■

The Cranium

The cranium may be divided into two major areas for study—the **calvaria,** forming the superior, lateral, and posterior walls of the skull, and the **cranial base,** forming the skull bottom. Internally, the cranial base has three distinct concavities, the **anterior, middle,** and **posterior cranial fossae** (Figure 8.3). The brain sits in these fossae, completely enclosed by the calvaria.

Eight large bones construct the cranium. *With the exception of two paired bones (the parietals and the temporals), all are single bones.* Sometimes the six ossicles of the middle ear are also considered part of the cranium. Because the ossicles are functionally part of the hearing apparatus, we consider them in our study of hearing (Exercise 19, Special Senses: Hearing and Equilibrium).

● ***Frontal Bone*** (Figures 8.1, 8.3, 8.6, and **Figure 8.8**). Anterior portion of cranium; forms the forehead, superior part of the orbit, and floor of anterior cranial fossa.

Supraorbital foramen (notch): Opening above each orbit allowing blood vessels and nerves to pass.

Glabella: Smooth area between the eyes.

● ***Parietal Bone*** (Figures 8.1 and 8.6). Posterolateral to the frontal bone, forming sides of cranium.

Sagittal suture: Midline articulation point of the two parietal bones.

Coronal suture: Point of articulation of parietals with frontal bone.

● ***Temporal Bone*** (Figures 8.1 through 8.3 and 8.6) Inferior to parietal bone on lateral skull. The temporals can be divided into three major parts: the **squamous part** borders the parietals; the **tympanic part** surrounds the external ear opening; and the **petrous part** forms the lateral portion of the skull base and contains the mastoid process.

Important markings associated with the flaring squamous part (Figures 8.1 and 8.2) include the following:

Squamous suture: Point of articulation of the temporal bone with the parietal bone.

Zygomatic process: A bridgelike projection joining the zygomatic bone (cheekbone) anteriorly. Together these two bones form the *zygomatic arch.*

Mandibular fossa: Rounded depression on the inferior surface of the zygomatic process (anterior to the ear); forms the condylar process of the mandible, the point where the mandible (lower jaw) joins the cranium.

Tympanic part markings (Figures 8.1 and 8.2) include the following:

External acoustic meatus: Canal leading to eardrum and middle ear.

Styloid process: Needle-like (*stylo* = stake, pointed object) projection inferior to external acoustic meatus; attachment point for muscles and ligaments of the neck. This process is often broken off demonstration skulls.

The petrous part (Figures 8.2 and 8.3), which helps form the middle and posterior cranial fossae, contains the labyrinth (holding the organs of hearing and balance) and exhibits several obvious foramina with important functions and includes:

Jugular foramen: Opening medial to the styloid process through which the internal jugular vein and cranial nerves IX, X, and XI pass.

Carotid canal: Opening medial to the styloid process through which the internal carotid artery passes into the cranial cavity.

Internal acoustic meatus: Opening on posterior aspect (petrous part) of temporal bone allowing passage of cranial nerves VII and VIII (Figure 8.3).

Foramen lacerum: A jagged opening between the petrous temporal bone and the sphenoid providing passage for a number of small nerves and for the internal carotid artery to enter the middle cranial fossa (after it passes through part of the temporal bone).

Stylomastoid foramen: Tiny opening between the mastoid and styloid processes through which cranial nerve VII leaves the cranium.

Mastoid process: Rough projection inferior and posterior to external acoustic meatus; attachment site for muscles.

The mastoid process, full of air cavities and so close to the middle ear—a trouble spot for infections—often becomes infected too, a condition referred to as **mastoiditis.** Because the mastoid area is separated from the brain by only a thin layer of bone, an ear infection that has spread to the mastoid process can inflame the brain coverings, or the meninges. The latter condition is known as **meningitis.** +

● ***Occipital Bone*** (Figures 8.1 through 8.3 and 8.6) Most posterior bone of cranium—forms floor and back wall. Joins sphenoid bone anteriorly via its narrow basilar part.

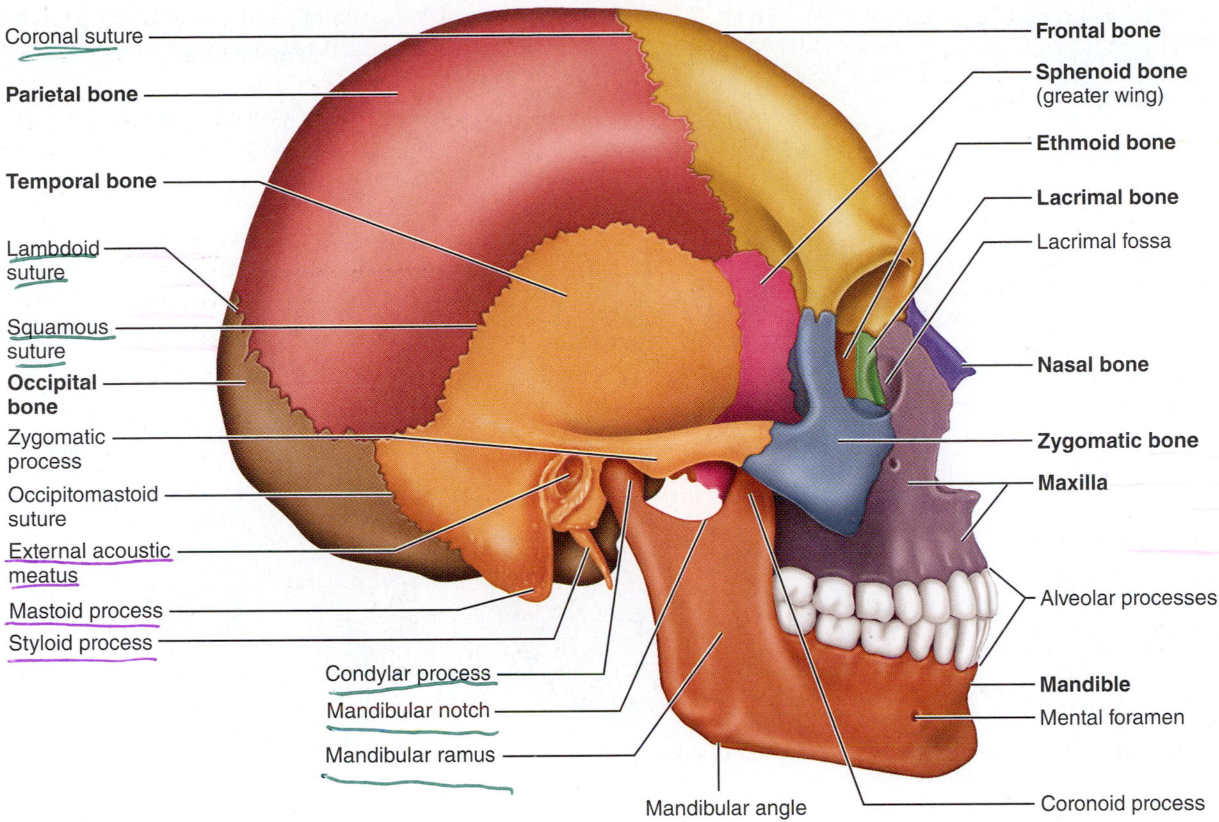

Figure 8.1 External anatomy of the right lateral aspect of the skull.

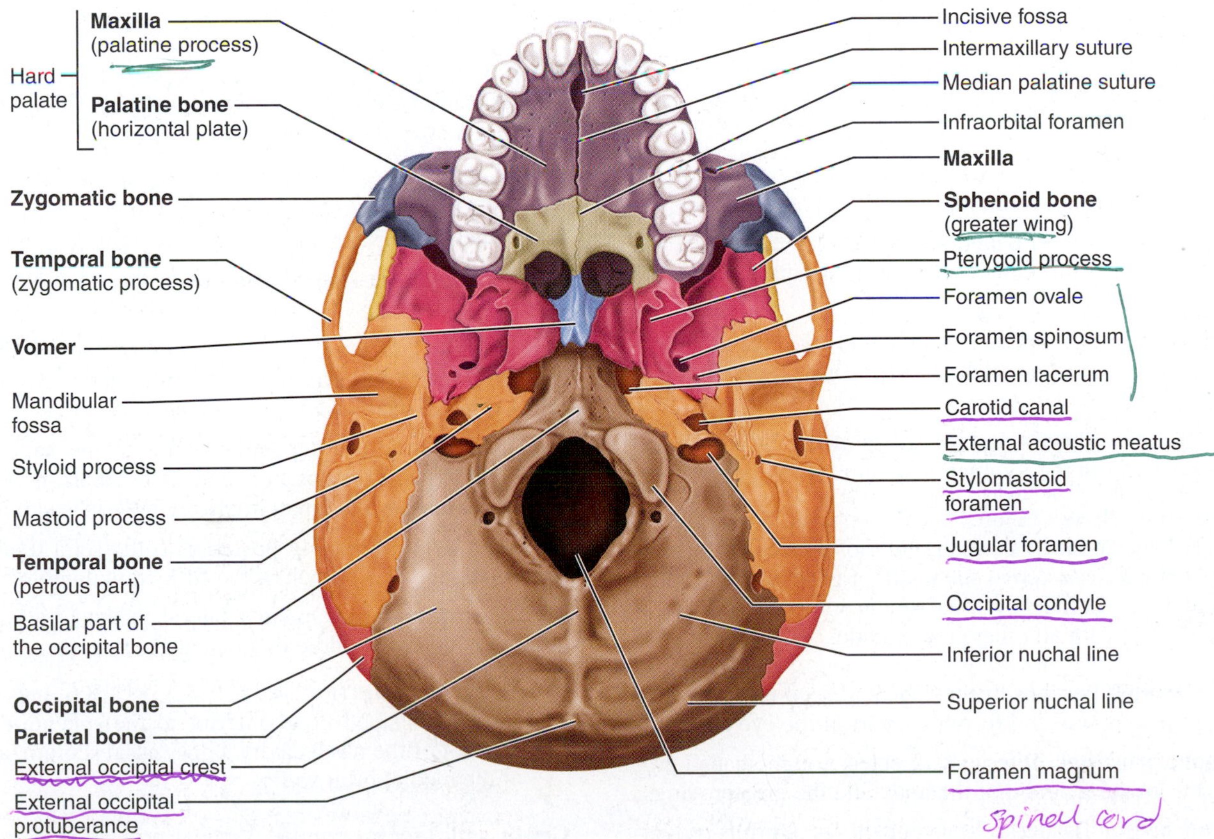

Figure 8.2 Inferior superficial view of the skull, mandible removed.

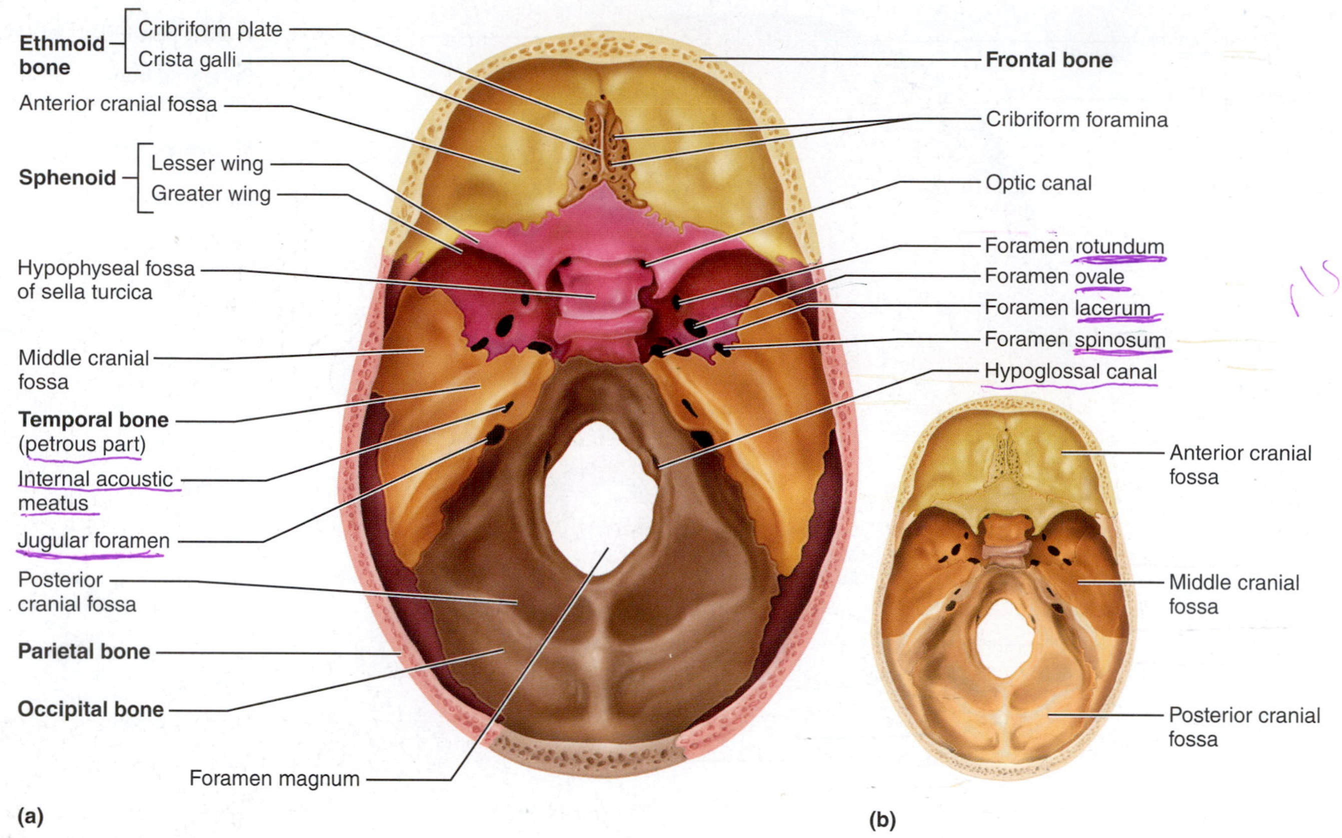

Figure 8.3 Internal anatomy of the inferior portion of the skull. (a) Superior view of the cranial cavity, calvaria removed. **(b)** Schematic view of the cranial base showing the extent of its major fossae.

Lambdoid suture: Site of articulation of occipital bone and parietal bones.

Foramen magnum: Large opening in base of occipital bone, which allows the spinal cord to join with the brain.

Occipital condyles: Rounded projections lateral to the foramen magnum that articulate with the first cervical vertebra (atlas).

Hypoglossal canal: Opening medial and superior to the occipital condyle through which the hypoglossal nerve (cranial nerve XII) passes.

External occipital crest and protuberance: Midline prominences posterior to the foramen magnum.

● ***Sphenoid Bone*** (Figures 8.1 through 8.4, 8.6, and 8.8). Bat-shaped bone forming the anterior plateau of the middle cranial fossa across the width of the skull. The sphenoid bone is the keystone of the cranium because it articulates with all other cranial bones.

Greater wings: Portions of the sphenoid seen exteriorly anterior to the temporal and forming a part of the eye orbits.

Pterygoid processes: Inferiorly directed trough-shaped projections from the junction of the body and the greater wings.

Superior orbital fissures: Jagged openings in orbits providing passage for cranial nerves III, IV, V, and VI to enter the orbit where they serve the eye.

The sphenoid bone can be seen in its entire width if the top of the cranium (calvaria) is removed (Figure 8.3).

Sella turcica (Turk's saddle): A saddle-shaped region in the sphenoid midline. The seat of this saddle, called the **hypophyseal fossa,** surrounds the pituitary gland (hypophysis).

Lesser wings: Bat-shaped portions of the sphenoid anterior to the sella turcica.

Optic canals: Openings in the bases of the lesser wings through which the optic nerves (cranial nerve II) enter the orbits to serve the eyes.

Foramen rotundum: Opening lateral to the sella turcica providing passage for a branch of the fifth cranial nerve. (This foramen is not visible on an inferior view of the skull.)

Foramen ovale: Opening posterior to the sella turcica that allows passage of a branch of the fifth cranial nerve.

Foramen spinosum: Opening lateral to the foramen ovale through which the middle meningeal artery passes.

● ***Ethmoid Bone*** (Figures 8.1, 8.3, 8.5, 8.6, and 8.8) Irregularly shaped bone anterior to the sphenoid. Forms the roof of the nasal cavity, upper nasal septum, and part of the medial orbit walls.

Crista galli (cock's comb): Vertical projection providing a point of attachment for the dura mater, helping to secure the brain within the skull.

Figure 8.4 The sphenoid bone.

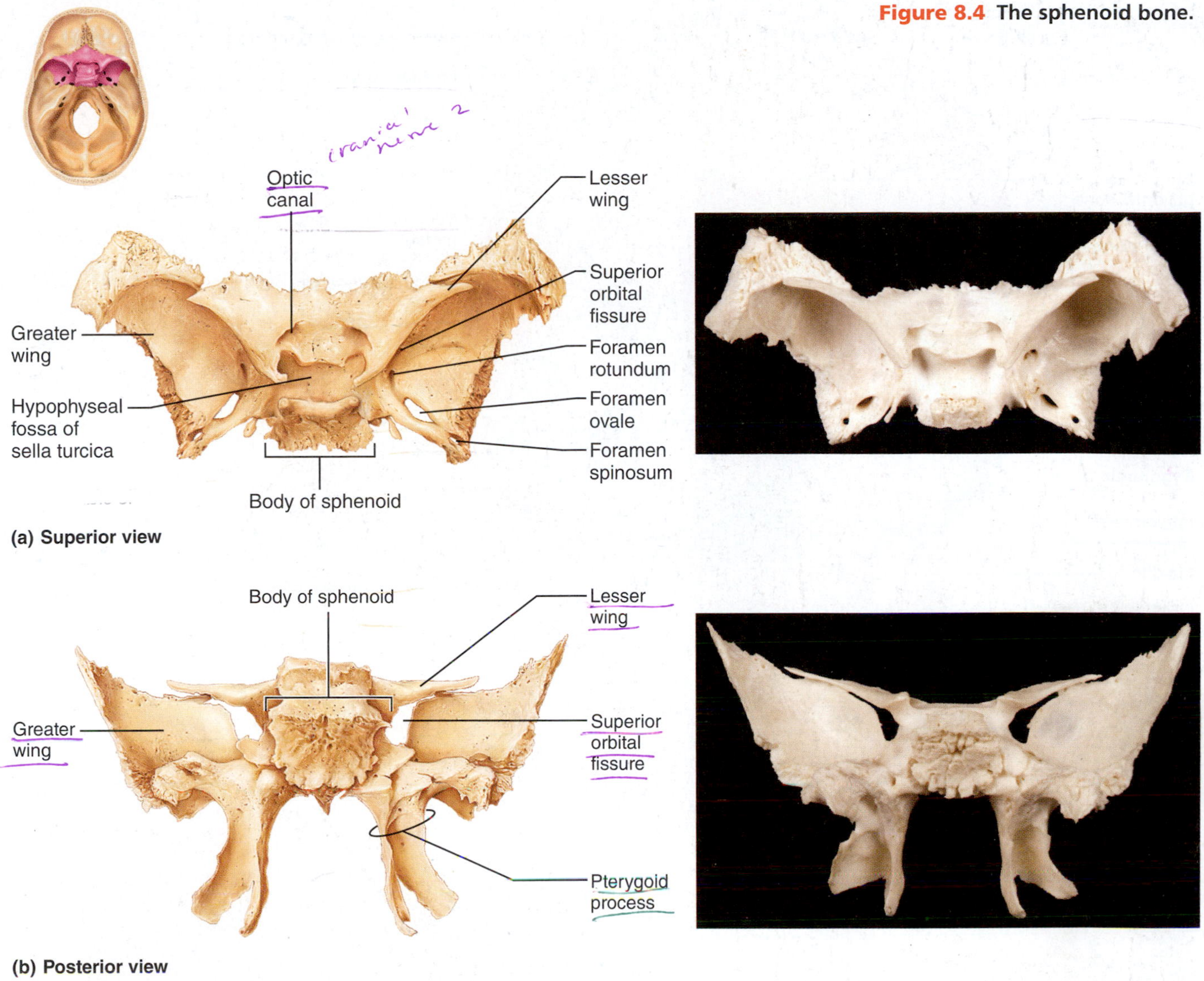

Optic canal
Lesser wing
Superior orbital fissure
Greater wing
Foramen rotundum
Hypophyseal fossa of sella turcica
Foramen ovale
Foramen spinosum
Body of sphenoid
(a) Superior view
Body of sphenoid
Lesser wing
Greater wing
Superior orbital fissure
Pterygoid process
(b) Posterior view

Crista galli
Orbital plate
Left ethmoidal labyrinth
Ethmoidal air cells
Perpendicular plate
Middle nasal concha

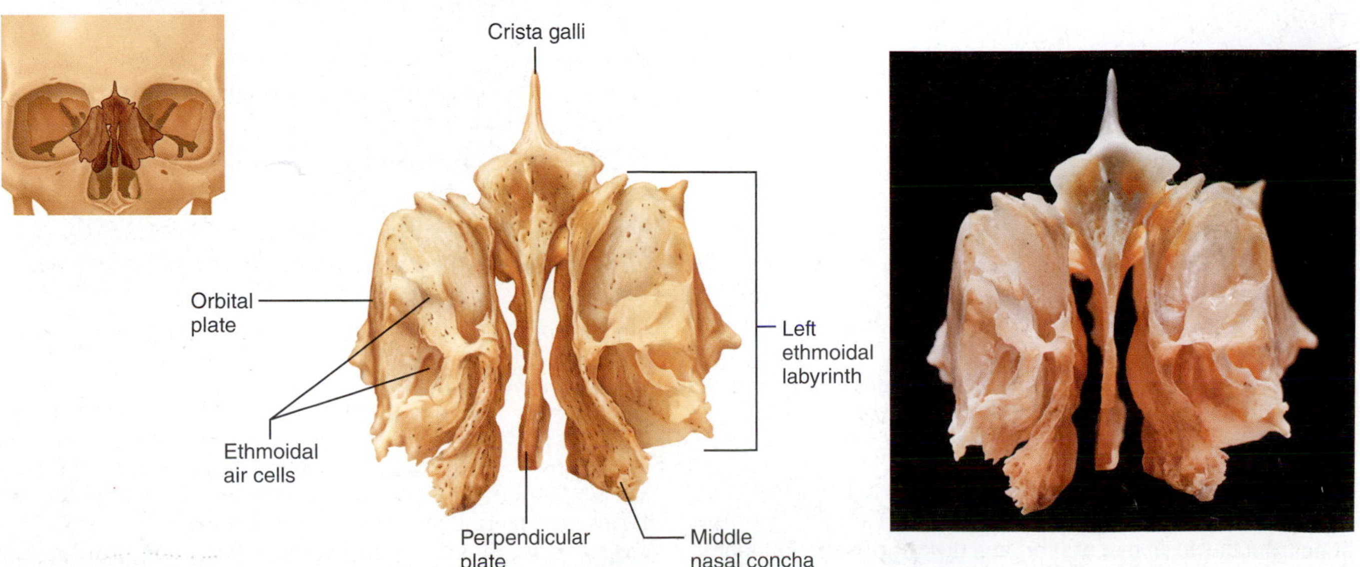

Figure 8.5 The ethmoid bone. Anterior view.

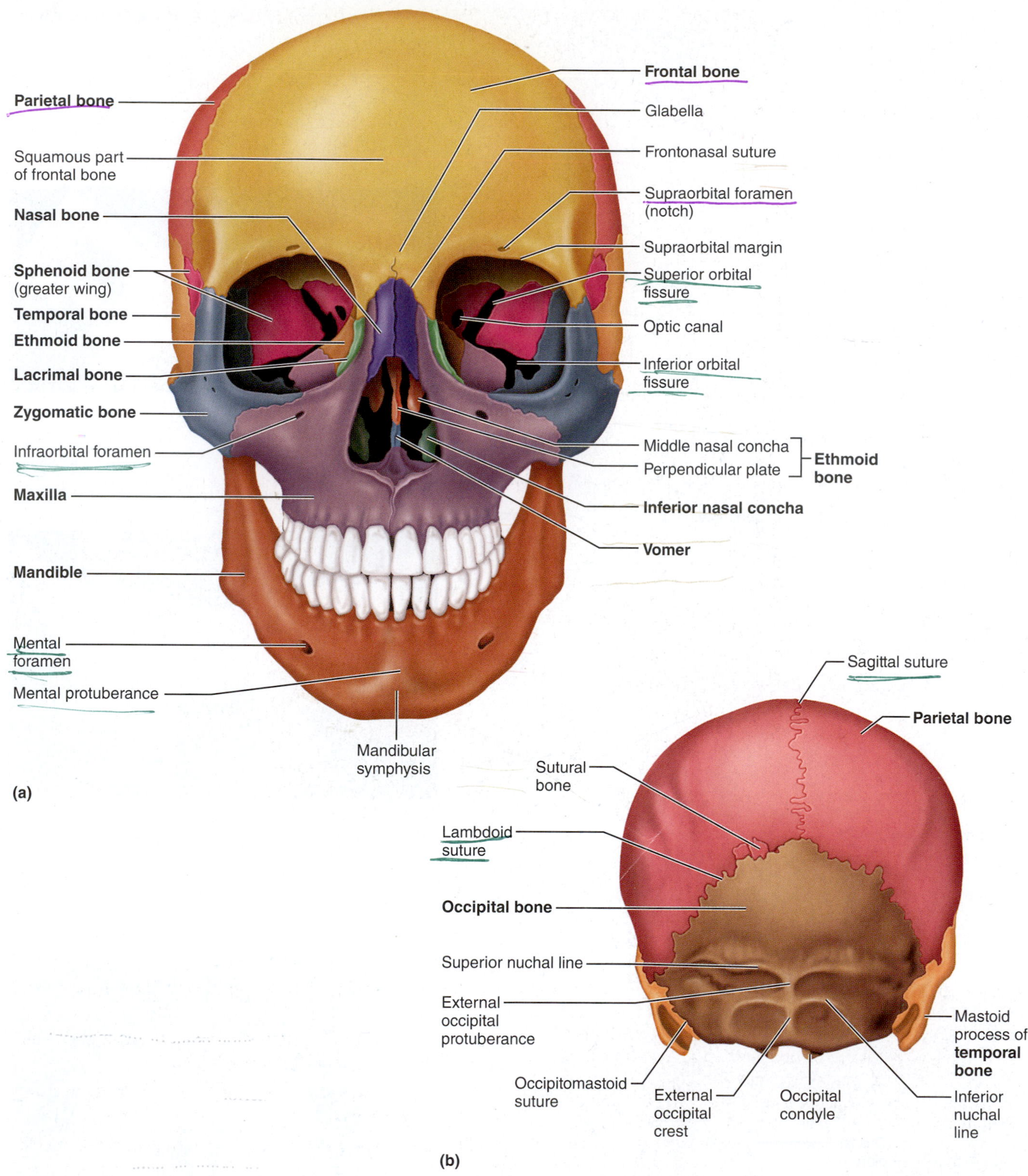

Figure 8.6 Anatomy of the anterior and posterior aspects of the skull. (a) Anterior aspect. **(b)** Posterior aspect.

Cribriform plates: Bony plates lateral to the crista galli through which olfactory fibers (cranial nerve I) pass to the brain from the nasal mucosa through the cribriform foramina. Together the cribriform plates and the midline crista galli form the *horizontal plate* of the ethmoid bone.

Perpendicular plate: Inferior projection of the ethmoid that forms the superior part of the nasal septum.

Ethmoidal labyrinths: Irregularly shaped thin-walled bony regions flanking the perpendicular plate laterally. Their lateral surfaces *(orbital plates)* shape part of the medial orbit wall.

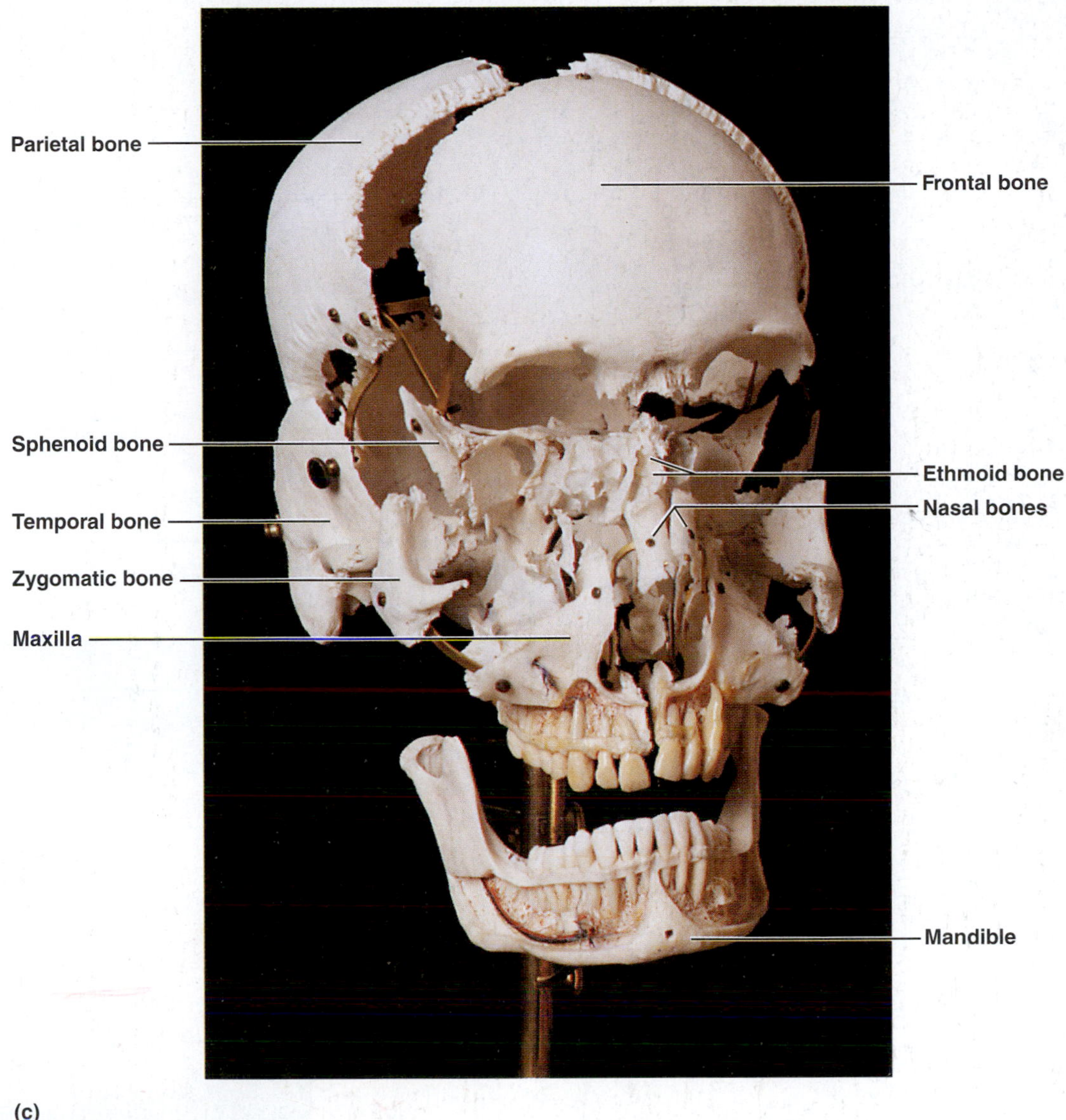

Figure 8.6 (*continued*) **(c)** Frontal view of the Beauchene skull.

Superior and middle nasal conchae (turbinates): Thin, delicately coiled plates of bone extending medially from the ethmoidal labyrinths into the nasal cavity. The conchae make airflow through the nasal cavity more efficient and greatly increase the surface area of the mucosa that covers them, thus increasing the mucosa's ability to warm and humidify incoming air.

Facial Bones

Of the 14 bones composing the face, 12 are paired. *Only the mandible and vomer are single bones.* An additional bone, the hyoid bone, although not a facial bone, is considered here because of its location.

- ***Mandible*** (Figures 8.1, 8.6, and 8.7) The lower jawbone, which articulates with the temporal bones in the only freely movable joints of the skull.

Mandibular body: Horizontal portion; forms the chin.

Mandibular ramus: Vertical extension of the body on either side.

Condylar process: Articulation point of the mandible with the mandibular fossa of the temporal bone.

Coronoid process: Jutting anterior portion of the ramus; site of muscle attachment.

Mandibular angle: Posterior point at which ramus meets the body.

Mental foramen: Prominent opening on the body (lateral to the midline) that transmits the mental blood vessels and nerve to the lower jaw.

Mandibular foramen: Open the lower jaw of the skull to identify this prominent foramen on the medial aspect of the mandibular ramus. This foramen permits passage of the nerve involved with tooth sensation (mandibular branch of cranial nerve V) and is the site where the dentist injects Novocain to prevent pain while working on the lower teeth.

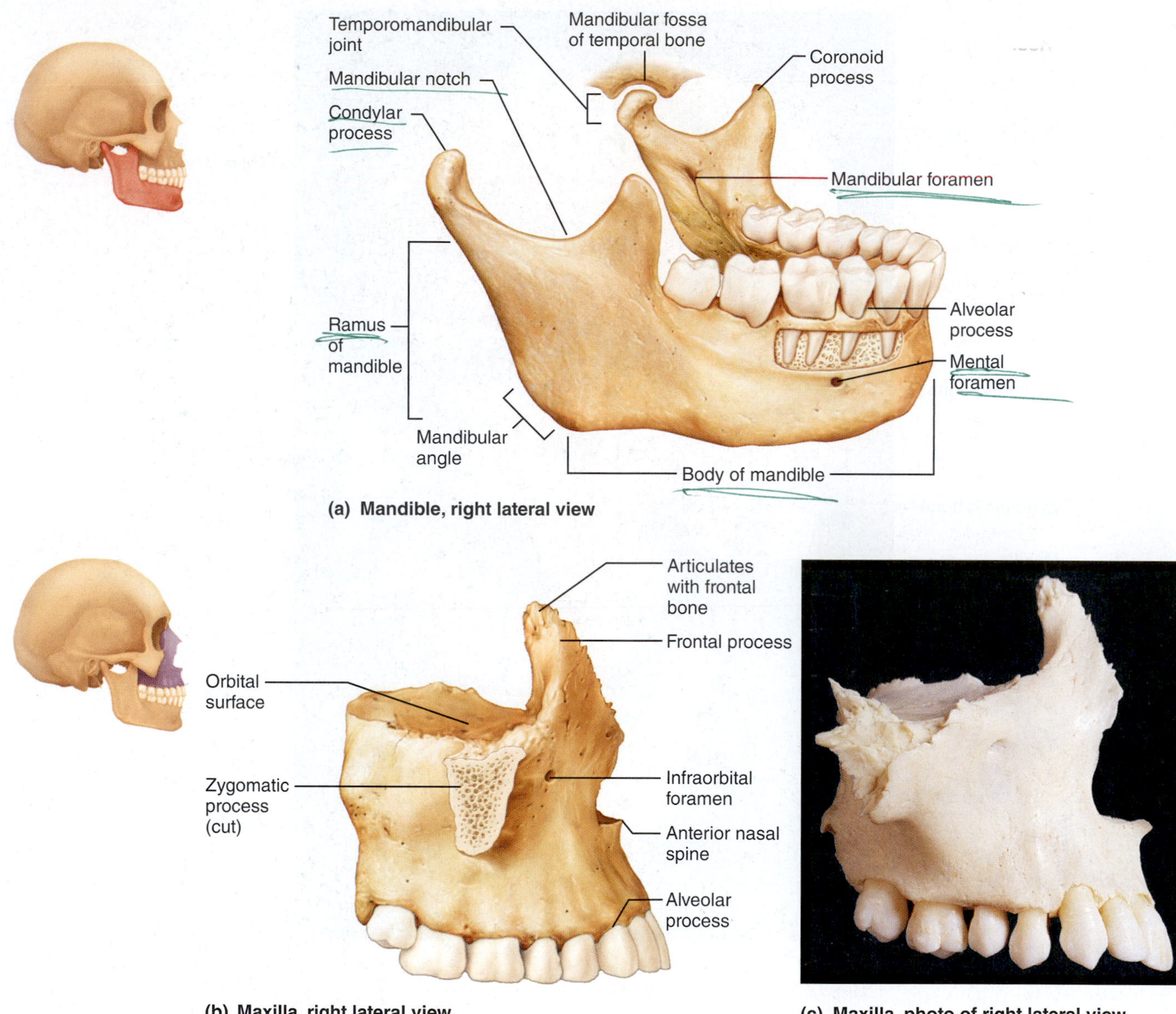

Figure 8.7 Right lateral view of some isolated facial bones. (Note that the mandible and maxilla are not drawn in proportion to each other.)

Alveolar process: Superior margin of mandible; contains sockets in which the teeth lie.

Mandibular symphysis: Anterior median depression indicating point of mandibular fusion.

- ***Maxillae*** (Figures 8.1, 8.2, 8.6, and 8.7) Two bones fused in a median suture; form the upper jawbone and part of the orbits. All facial bones, except the mandible, join the maxillae. Thus they are the main, or keystone, bones of the face.

Alveolar process: Inferior margin containing the sockets in which teeth lie.

Palatine processes: Form the anterior hard palate and meet medially in the intermaxillary suture.

Infraorbital foramen: Opening under the orbit carrying the infraorbital nerves and blood vessels to the nasal region.

Incisive fossa: Large bilateral opening located posterior to the central incisor tooth of the maxilla and piercing the hard palate; transmits the nasopalatine arteries and blood vessels.

- ***Palatine Bone*** (Figures 8.2 and 8.8) Paired bones posterior to the palatine processes; form posterior hard palate and part of the orbit; meet medially at the median palatine suture.

- ***Zygomatic Bone*** (Figures 8.1, 8.2, 8.6a, and 8.8) Lateral to the maxilla; forms the portion of the face commonly called the cheekbone, and forms part of the lateral orbit.

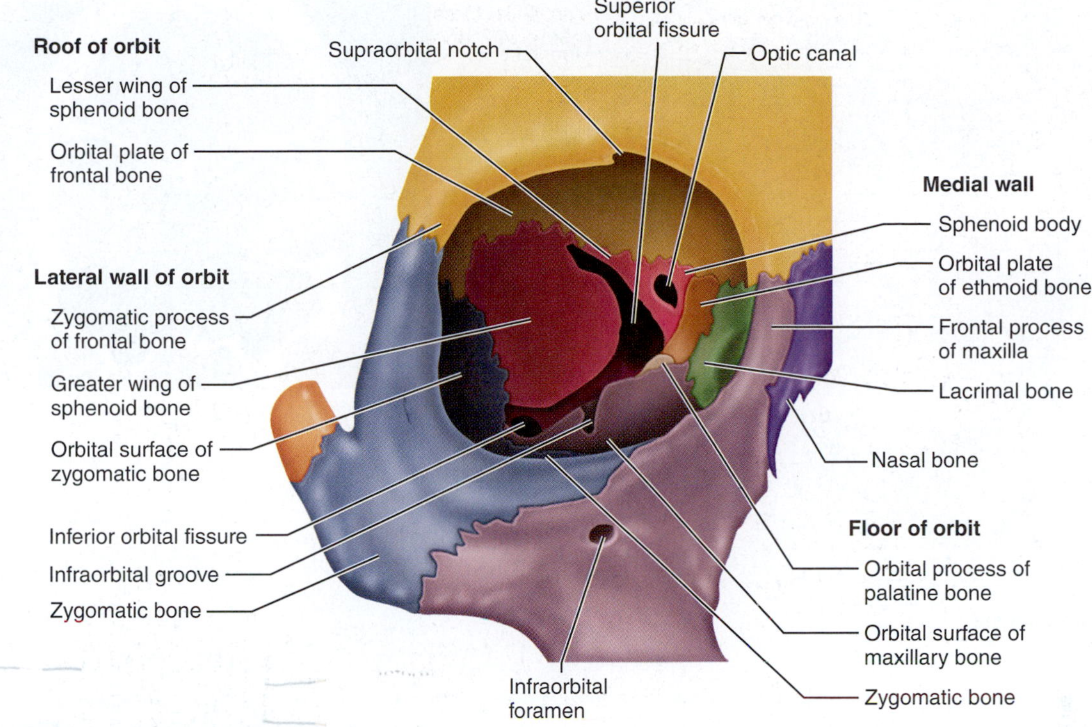

Figure 8.8 **Bones that form the orbit.**

Its three processes are named for the bones with which they articulate.

- ***Lacrimal Bone*** (Figures 8.1 and 8.6) Fingernail-sized bones forming a part of the medial orbit walls between the maxilla and the ethmoid. Each lacrimal bone is pierced by an opening, the **lacrimal fossa,** which serves as a passageway for tears (*lacrima* = tear).
- ***Nasal Bone*** (Figures 8.1 and 8.6a) Small rectangular bones forming the bridge of the nose.
- ***Vomer*** (*vomer* = plow) (Figures 8.2 and 8.6) Blade-shaped bone in median plane of nasal cavity that forms the posterior and inferior nasal septum.
- ***Inferior Nasal Conchae (Turbinates)*** (Figure 8.6) Thin curved bones protruding medially from the lateral walls of the nasal cavity; serve the same purpose as the turbinate portions of the ethmoid bone.

The Orbit

Seven bones of the skull form the orbit, the bony cavity that surrounds the eye: the frontal, sphenoid, ethmoid, lacrimal, maxilla, palatine, and zygomatic (Figure 8.8).

Hyoid Bone

Not really considered or counted as a skull bone, the hyoid bone is located in the throat above the larynx **(Figure 8.9)** where it serves as a point of attachment for many tongue and neck muscles. It does not articulate with any other bone and is thus unique. It is horseshoe-shaped with a body and two pairs of **horns,** or **cornua.**

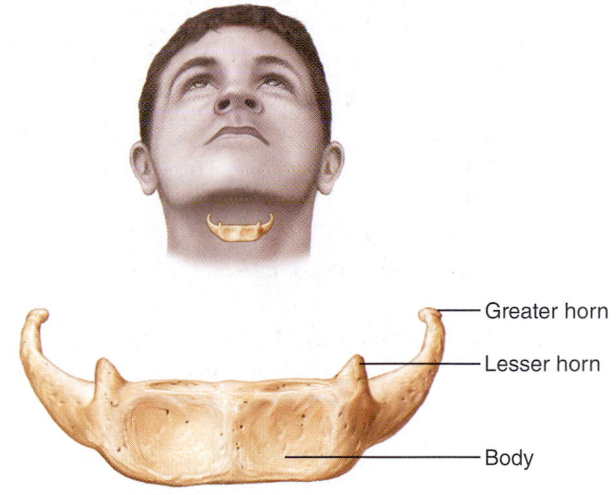

Figure 8.9 **Hyoid bone.**

Paranasal Sinuses

Four skull bones—maxillary, sphenoid, ethmoid, and frontal—contain sinuses (mucosa-lined air cavities), which lead into the nasal passages (Figure 8.5 and **Figure 8.10**). These paranasal sinuses lighten the facial bones and may act as resonance chambers for speech. The maxillary sinus is the largest of the sinuses found in the skull.

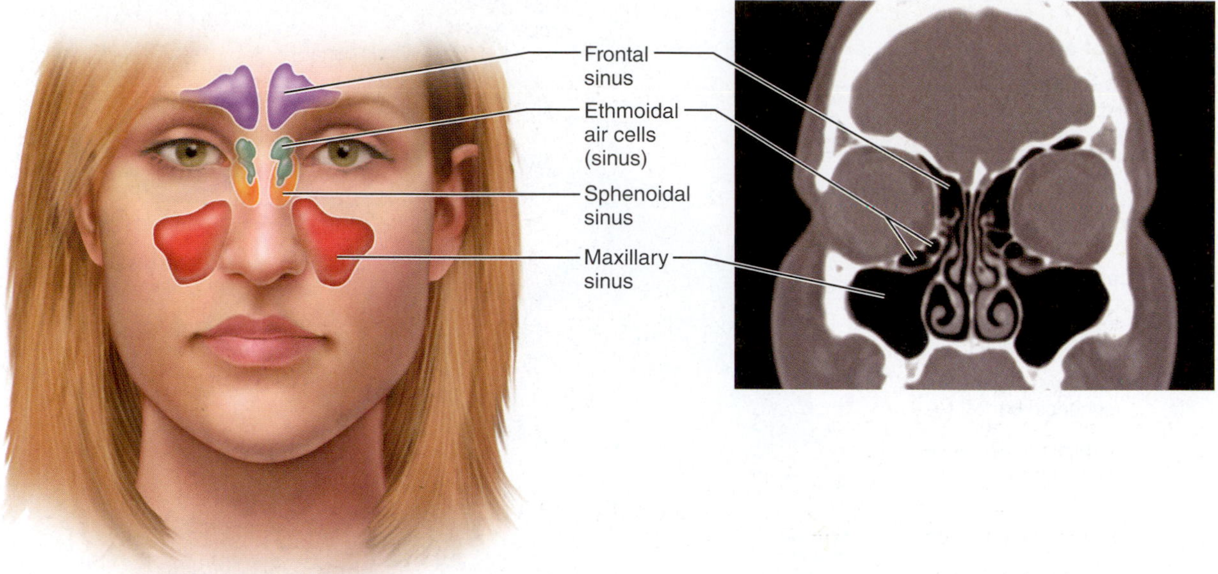

(a) Anterior aspect

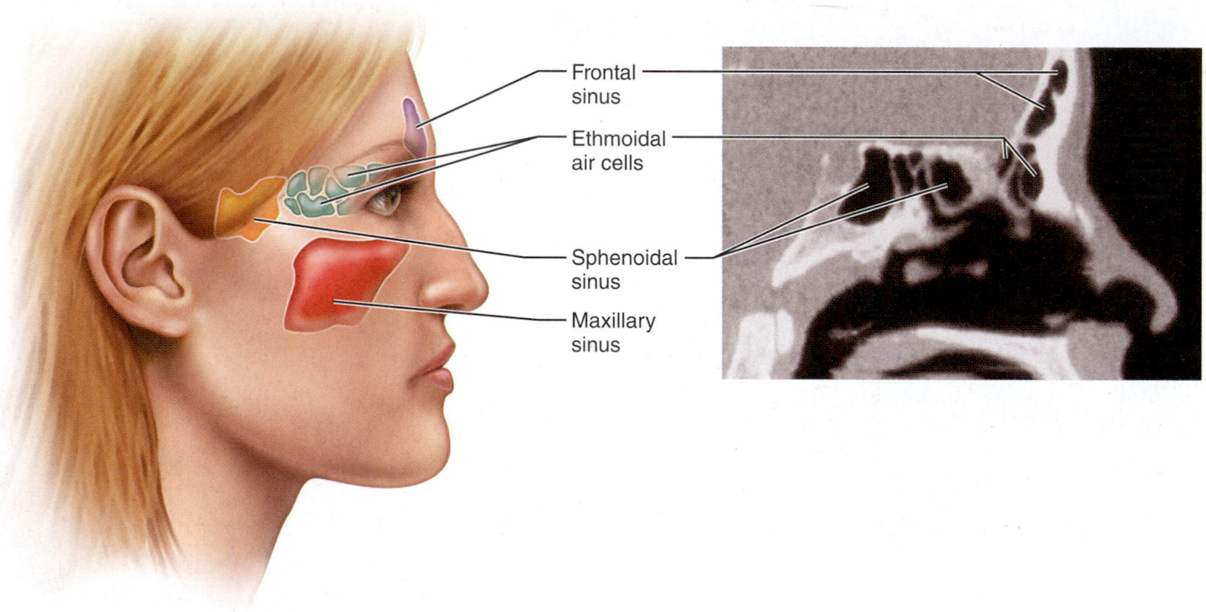

(b) Medial aspect

Figure 8.10 Paranasal sinuses. Illustrations on left, CT scans on right.

Sinusitis, or inflammation of the sinuses, sometimes occurs as a result of an allergy or bacterial invasion of the sinus cavities. In such cases, some of the connecting passageways between the sinuses and nasal passages may become blocked with thick mucus or infectious material. Then, as the air in the sinus cavities is absorbed, a partial vacuum forms. The result is a sinus headache localized over the inflamed sinus area. Severe sinus infections may require surgical drainage to relieve this painful condition. +

ACTIVITY 2

Palpating Skull Markings

Palpate the following areas on yourself. Place a check mark in the boxes as you locate the skull markings. Ask your instructor for help with any markings that you are unable to locate.

GROUP CHALLENGE

Odd Bone Out

Each box below contains four bones. One of the listed bones does *not* share a characteristic that the other three do. Circle the bone that doesn't belong with the others, and explain why it is singled out. What characteristic is it missing? Sometimes there may be multiple reasons why the bone doesn't belong with the others. Include as many as you can think of, but make sure the bone does not have the key characteristic(s). Use an articulated skull, disarticulated skull bones, and the pictures in your lab manual to help you select and justify your answer.

1. Which is the "odd bone"?	Why is it the odd one out?
Zygomatic bone Maxilla Vomer Nasal bone	
2. Which is the "odd bone"?	**Why is it the odd one out?**
Parietal bone Sphenoid bone Frontal bone Occipital bone	
3. Which is the "odd bone"?	**Why is it the odd one out?**
Lacrimal bone Nasal bone Zygomatic bone Maxilla	

☐ Zygomatic bone and arch. (The most prominent part of your cheek is your zygomatic bone. Follow the posterior course of the zygomatic arch to its junction with your temporal bone.)

☐ Mastoid process (the rough area behind your ear).

☐ Temporomandibular joints. (Open and close your jaws to locate these.)

☐ Greater wing of sphenoid. (Find the indentation posterior to the orbit and superior to the zygomatic arch on your lateral skull.)

☐ Supraorbital foramen. (Apply firm pressure along the superior orbital margin to find the indentation resulting from this foramen.)

☐ Infraorbital foramen. (Apply firm pressure just inferior to the inferomedial border of the orbit to locate this large foramen.)

☐ Mandibular angle (most inferior and posterior aspect of the mandible).

☐ Mandibular symphysis (midline of chin).

☐ Nasal bones. (Run your index finger and thumb along opposite sides of the bridge of your nose until they "slip" medially at the inferior end of the nasal bones.)

☐ External occipital protuberance. (This midline projection is easily felt by running your fingers up the furrow at the back of your neck to the skull.)

☐ Hyoid bone. (Place a thumb and index finger beneath the chin just anterior to the mandibular angles, and squeeze gently. Exert pressure with the thumb, and feel the horn of the hyoid with the index finger.)

The Vertebral Column

The **vertebral column,** extending from the skull to the pelvis, forms the body's major axial support. Additionally, it surrounds and protects the delicate spinal cord while allowing the spinal nerves to issue from the cord via openings between adjacent vertebrae. The term *vertebral column* might suggest a rather rigid supporting rod, but this is far from the truth. The vertebral column consists of 24 single bones called **vertebrae** and two composite, or fused, bones (the sacrum and coccyx) that are connected in such a way as to provide a flexible curved structure **(Figure 8.11)**. Of the 24 single vertebrae, the 7 bones of the neck are called *cervical vertebrae;* the next 12 are *thoracic vertebrae;* and the 5 supporting the lower back are *lumbar vertebrae.* Remembering common mealtimes for breakfast, lunch, and dinner (7 A.M., 12 noon, and 5 P.M.) may help you to remember the number of bones in each region.

The vertebrae are separated by pads of fibrocartilage, **intervertebral discs,** that cushion the vertebrae and absorb shocks. Each disc is composed of two major regions, a central gelatinous *nucleus pulposus* that behaves like a fluid, and an outer ring of encircling collagen fibers called the *anulus fibrosus* that stabilizes the disc and contains the pulposus.

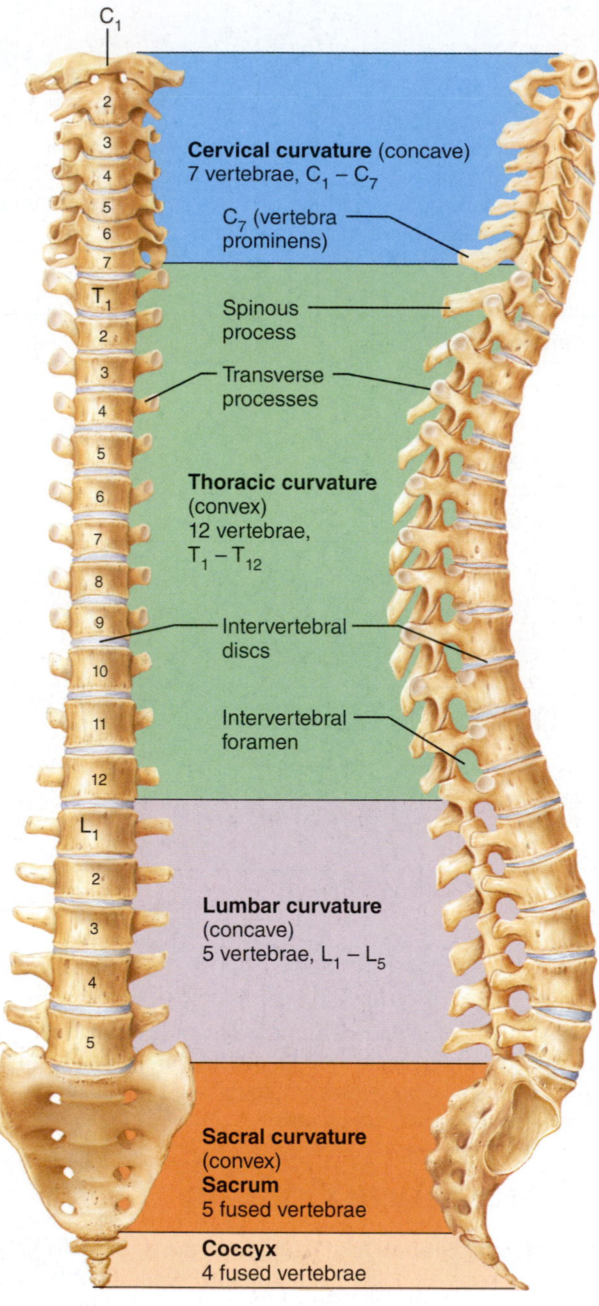

Figure 8.11 The vertebral column. Notice the curvatures in the lateral view. (The terms *convex* and *concave* refer to the curvature of the posterior aspect of the vertebral column.)

As a person ages, the water content of the discs decreases (as it does in other tissues throughout the body), and the discs become thinner and less compressible. This situation, along with other degenerative changes such as weakening of the ligaments and muscle tendons associated with the vertebral column, predisposes older people to ruptured discs. In a ruptured, or **herniated, disc,** the anulus fibrosus commonly ruptures, and the nucleus pulposus protudes (herniates) through it. This event typically compresses adjacent nerves causing pain. +

The presence of the discs and the curvatures create a springlike construction of the vertebral column that prevents shock to the head in walking and running and provide flexibility to the body trunk. The thoracic and sacral curvatures of the spine are referred to as *primary curvatures* because they are present and well developed at birth. Later the *secondary curvatures* are formed. The cervical curvature becomes prominent when the baby begins to hold its head up independently, and the lumbar curvature develops when the baby begins to walk.

ACTIVITY 3

Examining Spinal Curvatures

1. Observe the normal curvature of the vertebral column in the articulated vertebral column or laboratory skeleton, and compare it to the figure (Figure 8.11). Abnormal spinal curvatures, including *scoliosis, kyphosis,* and *lordosis,* may result from disease or poor posture **(Figure 8.12)**. Also examine X-ray films, if they are available, showing these same conditions in a living patient.

2. Then, using the articulated vertebral column (or an articulated skeleton), examine the freedom of movement between two lumbar vertebrae separated by an intervertebral disc.

When the fibrous disc is properly positioned, are the spinal cord or peripheral nerves impaired in any way?

____

Remove the disc, and put the two vertebrae back together. What happens to the nerve?

____

What would happen to the spinal nerves in areas of malpositioned, or "slipped," discs?

____

____

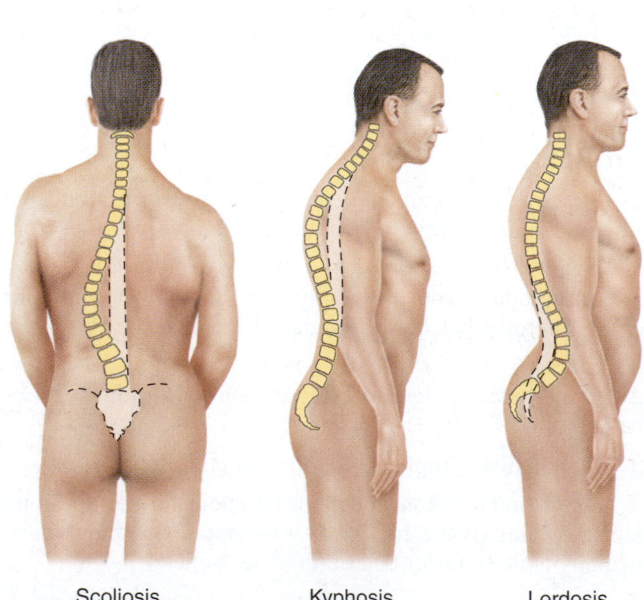

Figure 8.12 Abnormal spinal curvatures.

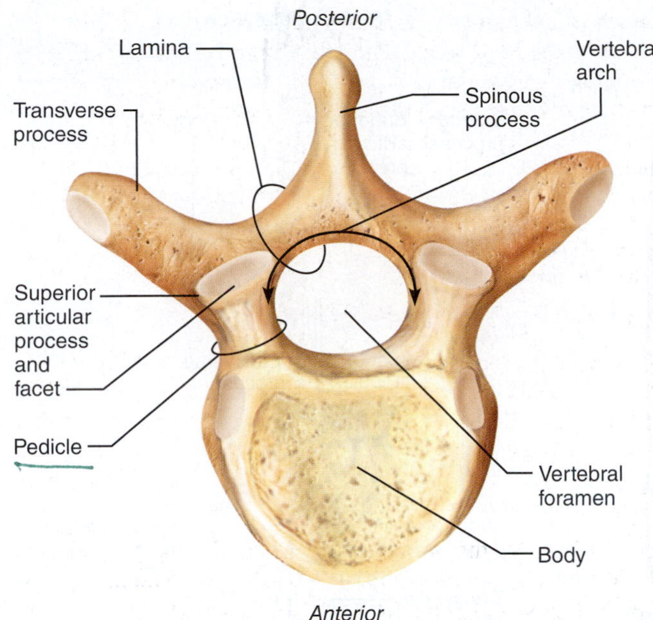

Figure 8.13 A typical vertebra, superior view. Inferior articulating surfaces not shown.

Structure of a Typical Vertebra

Although they differ in size and specific features, all vertebrae have some features in common **(Figure 8.13)**.

Body: Rounded central portion of the vertebra, which faces anteriorly in the human vertebral column.

Vertebral arch: Composed of pedicles and laminae, it represents the junction of all posterior extensions from the vertebral body.

Vertebral foramen: Opening enclosed by the body and vertebral arch; a conduit for the spinal cord.

Transverse processes: Two lateral projections from the vertebral arch.

Spinous process: Single medial and posterior projection from the vertebral arch.

Superior and inferior articular processes: Paired projections lateral to the vertebral foramen that enable articulation with adjacent vertebrae. The superior articular processes typically face toward the spinous process (posteriorly), whereas the inferior articular processes face (anteriorly) away from the spinous process.

Intervertebral foramina: The right and left pedicles have notches **(Figure 8.15)** on their inferior and superior surfaces that create openings, the intervertebral foramina (see Figure 8.10), for spinal nerves to leave the spinal cord between adjacent vertebrae.

The following sections describe how specific vertebrae differ with respect to structure and function (refer to **Figure 8.14** through **8.16** and **Table 8.1**).

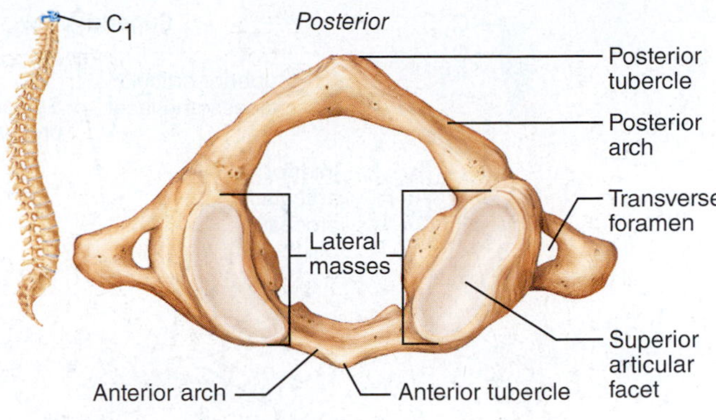

(a) Superior view of atlas (C_1)

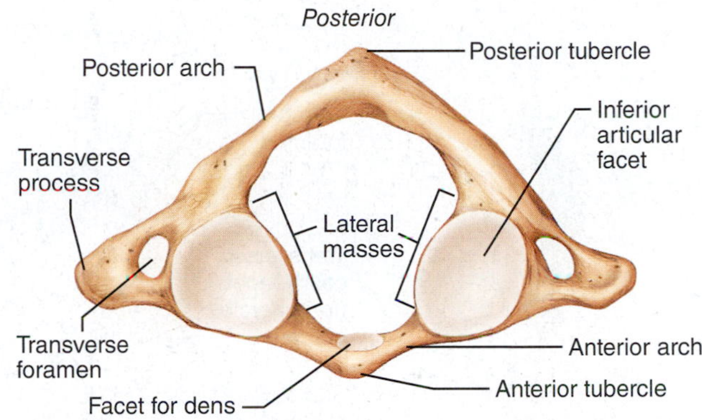

(b) Inferior view of atlas (C_1)

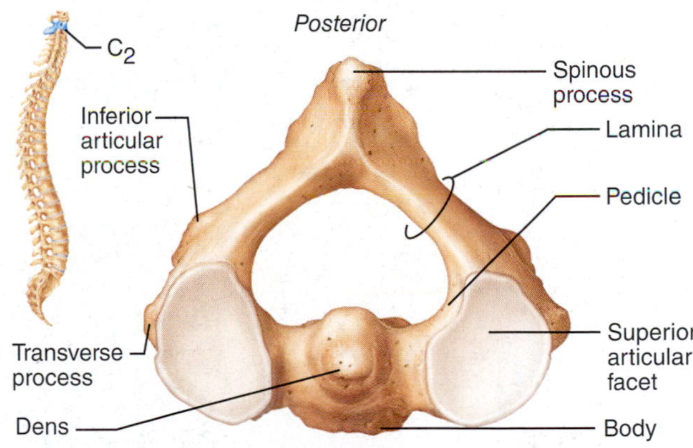

(c) Superoposterior view of axis (C_2)

Figure 8.14 The first and second cervical vertebrae.

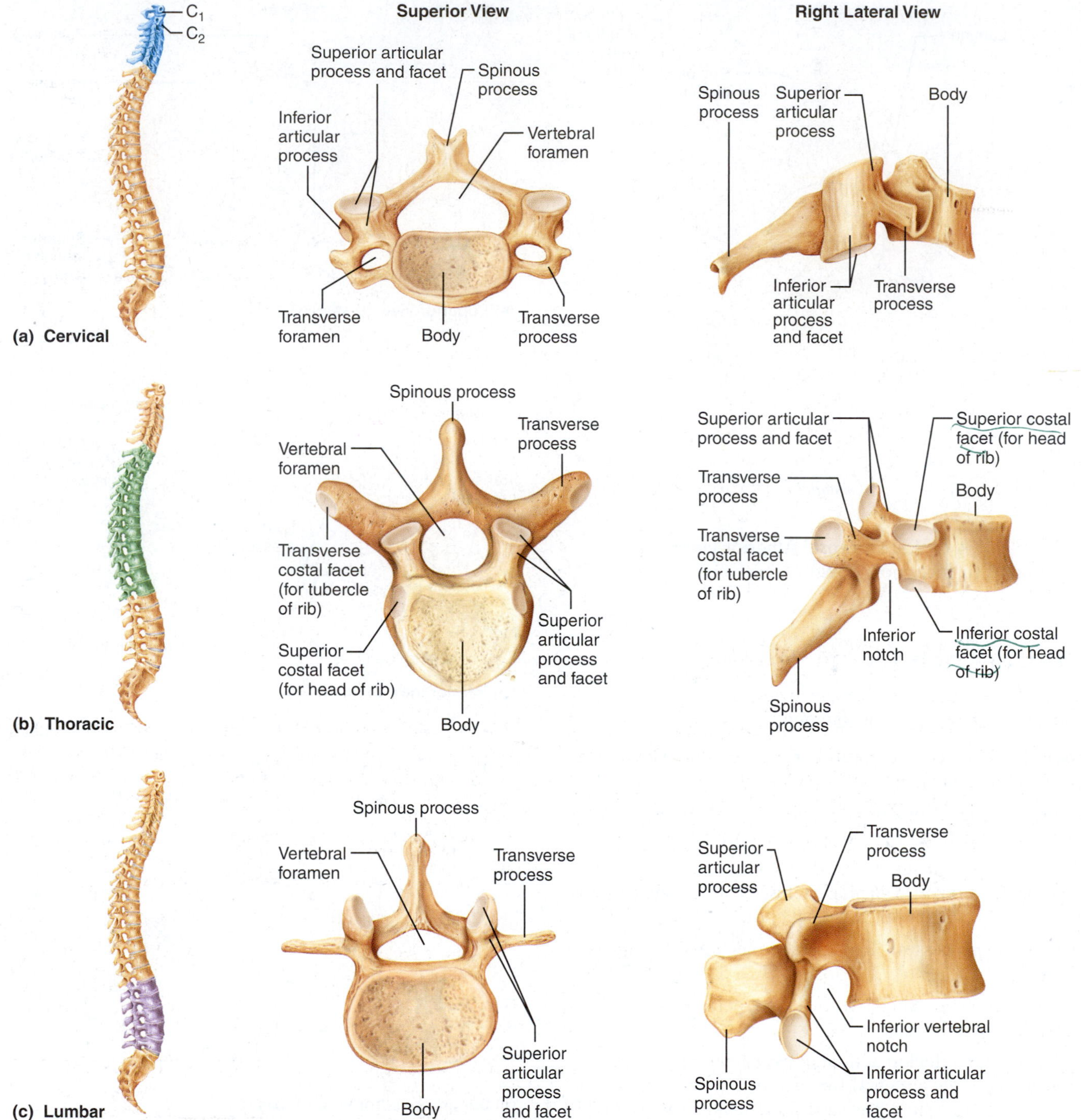

Figure 8.15 Superior and right lateral views of typical vertebrae. (a) Cervical. **(b)** Thoracic. **(c)** Lumbar.

Cervical Vertebrae

The seven cervical vertebrae (referred to as C_1 through C_7) form the neck portion of the vertebral column. The first two cervical vertebrae (atlas and axis) are highly modified to perform special functions (see Figure 8.14). The **atlas** (C_1) lacks a body, and its lateral processes contain large concave depressions on their superior surfaces that receive the occipital condyles of the skull. This joint enables you to nod "yes." The **axis** (C_2) acts as a pivot for the rotation of the atlas (and skull) above. It bears a large vertical process, the **dens,** that serves as the pivot point. The articulation between C_1 and C_2 allows you to rotate your head from side to side to indicate "no."

The more typical cervical vertebrae (C_3 through C_7) are distinguished from the thoracic and lumbar vertebrae by several features (see Table 8.1 and Figure 8.15). They are the

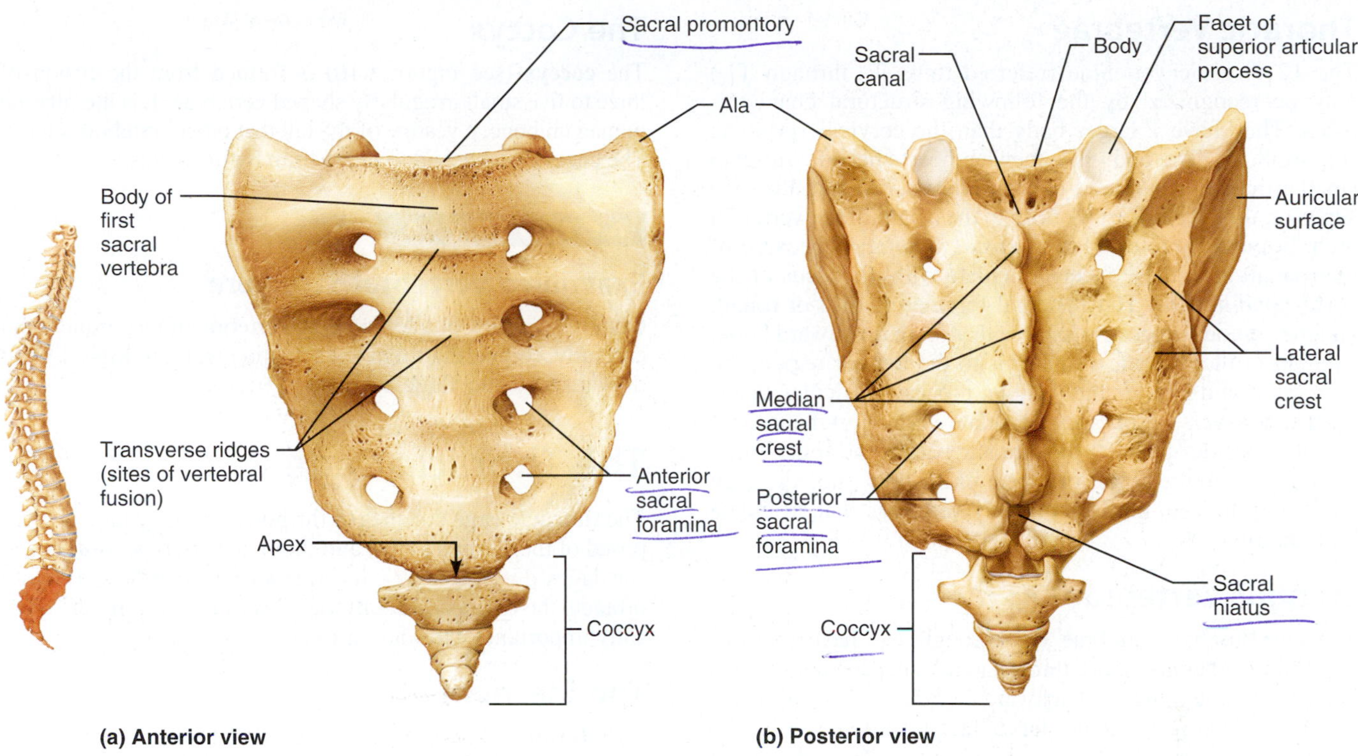

Figure 8.16 **Sacrum and coccyx.**

smallest, lightest vertebrae, and the vertebral foramen is triangular. The spinous process is short and often bifurcated (divided into two branches). The spinous process of C_7 is not branched, however, and is substantially longer than that of the other cervical vertebrae. Because the spinous process of C_7 is visible through the skin, it is called the *vertebra prominens* (Figure 8.11) and is used as a landmark for counting the vertebrae. Transverse processes of the cervical vertebrae are wide, and they contain foramina through which the vertebral arteries pass superiorly on their way to the brain. Any time you see these foramina in a vertebra, you can be sure that it is a cervical vertebra.

☐ Palpate your vertebra prominens.

Place a check mark in the box when you locate the structure.

Table 8.1 **Regional Characteristics of Cervical, Thoracic, and Lumbar Vertebrae**

Characteristic	(a) Cervical (C_3–C_7)	(b) Thoracic	(c) Lumbar
Body	Small, wide side to side	Larger than cervical; heart-shaped; bears costal facets	Massive; kidney-shaped
Spinous process	Short; bifid; projects directly posteriorly	Long; sharp; projects inferiorly	Short; blunt; projects directly posteriorly
Vertebral foramen	Triangular	Circular	Triangular
Transverse processes	Contain foramina	Bear facets for ribs (except T_{11} and T_{12})	Thin and tapered
Superior and inferior articulating processes	Superior facets directed superoposteriorly	Superior facets directed posteriorly	Superior facets directed posteromedially (or medially)
	Inferior facets directed inferoanteriorly	Inferior facets directed anteriorly	Inferior facets directed anterolaterally (or laterally)
Movements allowed	Flexion and extension; lateral flexion; rotation; the spine region with the greatest range of movement	Rotation; lateral flexion possible but limited by ribs; flexion and extension prevented	Flexion and extension; some lateral flexion; rotation prevented

Thoracic Vertebrae

The 12 thoracic vertebrae (referred to as T_1 through T_{12}) may be recognized by the following structural characteristics. They have a larger body than the cervical vertebrae (Figure 8.15). The body is somewhat heart-shaped, with two small articulating surfaces, or **costal facets,** on each side (one superior, the other inferior) close to the origin of the vertebral arch. Sometimes referred to as *costal demifacets* because of their small size, these facets articulate with the heads of the corresponding ribs. The vertebral foramen is oval or round, and the spinous process is long, with a sharp downward hook. The closer the thoracic vertebra is to the lumbar region, the less sharp and shorter the spinous process. Articular facets on the transverse processes articulate with the tubercles of the ribs. Besides forming the thoracic part of the spine, these vertebrae form the posterior aspect of the bony thoracic cage (rib cage). Indeed, they are the only vertebrae that articulate with the ribs.

Lumbar Vertebrae

The five lumbar vertebrae (L_1 through L_5) have massive blocklike bodies and short, thick, hatchet-shaped spinous processes extending directly backward (see Table 8.1 and Figure 8.15). The superior articular facets face posteromedially; the inferior ones are directed anterolaterally. These structural features reduce the mobility of the lumbar region of the spine. Because most stress on the vertebral column occurs in the lumbar region, these are also the sturdiest of the vertebrae.

The spinal cord ends at the superior edge of L_2, but the outer covering of the cord, filled with cerebrospinal fluid, extends an appreciable distance beyond. Thus a *lumbar puncture* (for examination of the cerebrospinal fluid) or the administration of "saddle block" anesthesia for childbirth is normally done between L_3 and L_4 or L_4 and L_5, where there is little or no chance of injuring the delicate spinal cord.

The Sacrum

The **sacrum** (Figure 8.16) is a composite bone formed from the fusion of five vertebrae. Superiorly it articulates with L_5, and inferiorly it connects with the coccyx. The **median sacral crest** is a remnant of the spinous processes of the fused vertebrae. The winglike **alae,** formed by fusion of the transverse processes, articulate laterally with the hip bones. The sacrum is concave anteriorly and forms the posterior border of the pelvis. Four ridges (lines of fusion) cross the anterior part of the sacrum, and **sacral foramina** are located at either end of these ridges. These foramina allow blood vessels and nerves to pass. The vertebral canal continues inside the sacrum as the **sacral canal** and terminates near the coccyx via an enlarged opening called the **sacral hiatus.** The **sacral promontory** (anterior border of the body of S_1) is an important anatomical landmark for obstetricians.

☐ Attempt to palpate the median sacral crest of your sacrum. (This is more easily done by thin people and obviously in privacy.)

Place a check mark in the box when you locate the structure.

The Coccyx

The **coccyx** (see Figure 8.16) is formed from the fusion of three to five small irregularly shaped vertebrae. It is literally the human tailbone, a vestige of the tail that other vertebrates have. The coccyx is attached to the sacrum by ligaments.

ACTIVITY 4

Examining Vertebral Structure

Obtain examples of each type of vertebra and examine them carefully, comparing them to each other (refer to Figures 8.14 through 8.16 and Table 8.1). ■

The Thoracic Cage

The **thoracic cage** consists of the bony thorax, which is composed of the sternum, ribs, and thoracic vertebrae, plus costal cartilages **(Figure 8.17)**. Its cone-shaped, cagelike structure protects the organs of the thoracic cavity, including the critically important heart and lungs.

The Sternum

The **sternum** (breastbone), a typical flat bone, is a result of the fusion of three bones—the manubrium, body, and xiphoid process. It is attached to the first seven pairs of ribs. The superiormost **manubrium** looks like the knot of a tie; it articulates with the clavicle (collarbone) laterally. The **body** forms the bulk of the sternum. The **xiphoid process** constructs the inferior end of the sternum and lies at the level of the fifth intercostal space. Although it is made of hyaline cartilage in children, it is usually ossified in adults.

In some people, the xiphoid process projects dorsally. This may present a problem because underlying physical trauma to the chest can push such a xiphoid into the underlying heart or liver, causing massive hemorrhage. ✚

The sternum has three important bony landmarks—the jugular notch, the sternal angle, and the xiphisternal joint. The **jugular notch** (concave upper border of the manubrium) can be palpated easily; generally it is at the level of the third thoracic vertebra. The **sternal angle** is a result of the manubrium and body meeting at a slight angle to each other, so that a transverse ridge is formed at the level of the second ribs. It provides a handy reference point for counting ribs to locate the second intercostal space for listening to certain heart valves, and is an important anatomical landmark for thoracic surgery. The **xiphisternal joint,** the point where the sternal body and xiphoid process fuse, lies at the level of the ninth thoracic vertebra.

☐ Palpate your sternal angle and jugular notch.

Place a check mark in the box when you locate the structure.

Because of its accessibility, the sternum is a favored site for obtaining samples of blood-forming (hematopoietic) tissue for the diagnosis of suspected blood diseases. A needle is inserted into the marrow of the sternum, and the sample is withdrawn (sternal puncture).

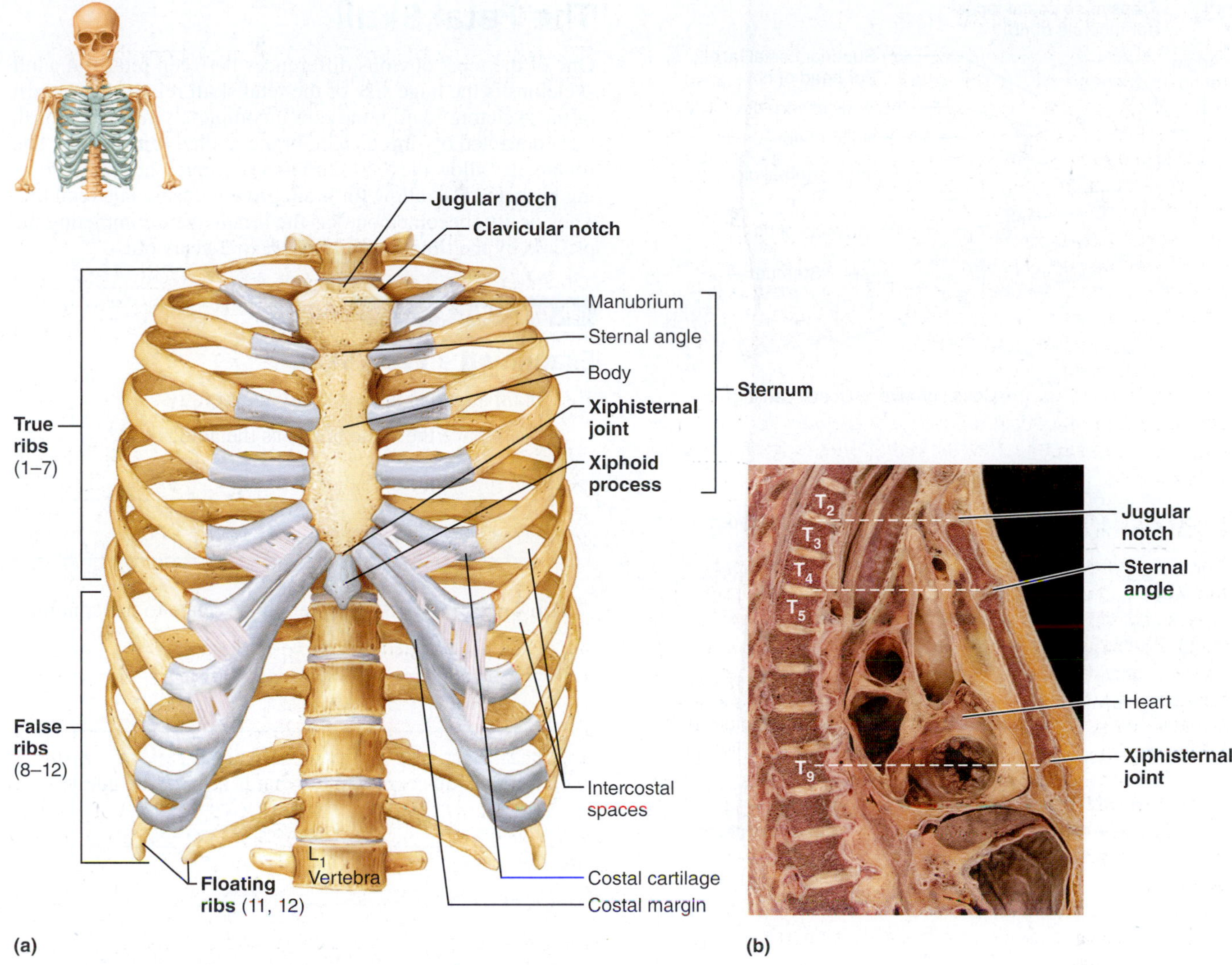

Figure 8.17 **The thoracic cage. (a)** Anterior view with costal cartilages shown in blue. **(b)** Midsagittal section of the thorax, illustrating the relationship of the surface anatomical landmarks of the thorax to the vertebral column (thoracic portion).

The Ribs

The 12 pairs of **ribs** form the walls of the thoracic cage (see Figure 8.17 and **Figure 8.18**). All of the ribs articulate posteriorly with the vertebral column via their heads and tubercles and then curve downward and toward the anterior body surface. The first seven pairs, called the *true,* or *vertebrosternal, ribs,* attach directly to the sternum by their "own" costal cartilages. The next five pairs are called *false ribs;* they attach indirectly to the sternum or entirely lack a sternal attachment. Of these, rib pairs 8–10, which are also called *vertebrochondral ribs,* have indirect cartilage attachments to the sternum via the costal cartilage of rib 7. The last two pairs, called *floating,* or *vertebral, ribs,* have no sternal attachment.

ACTIVITY 5

Examining the Relationship Between Ribs and Vertebrae

First take a deep breath to expand your chest. Notice how your ribs seem to move outward and how your sternum rises. Then examine an articulated skeleton to observe the relationship between the ribs and the vertebrae. (Refer to Activity 3, "Palpating Landmarks of the Trunk," and Activity 4, "Palpating Landmarks of the Abdomen," in Exercise 30, Surface Anatomy Roundup.) ■

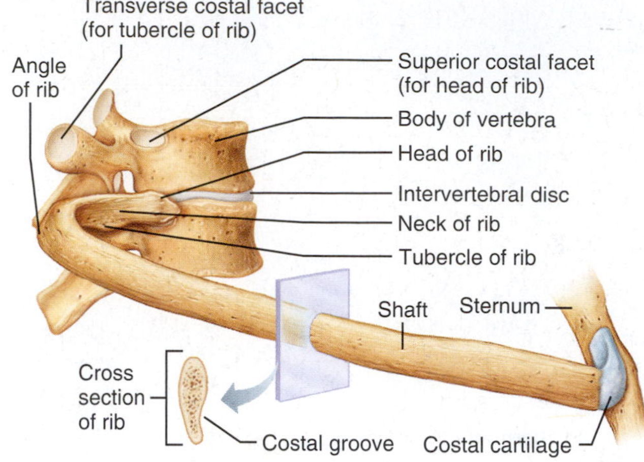

(a)

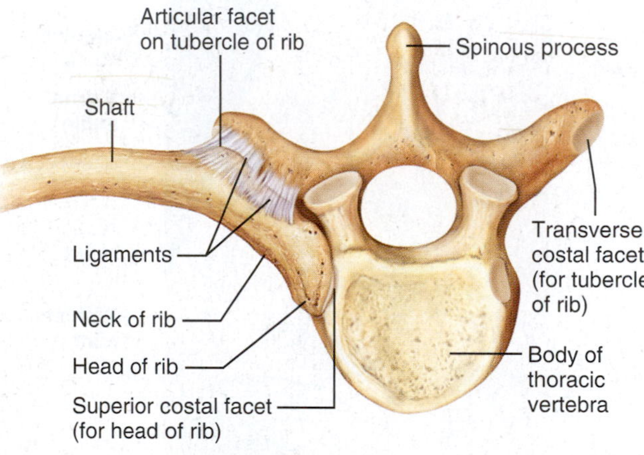

(b)

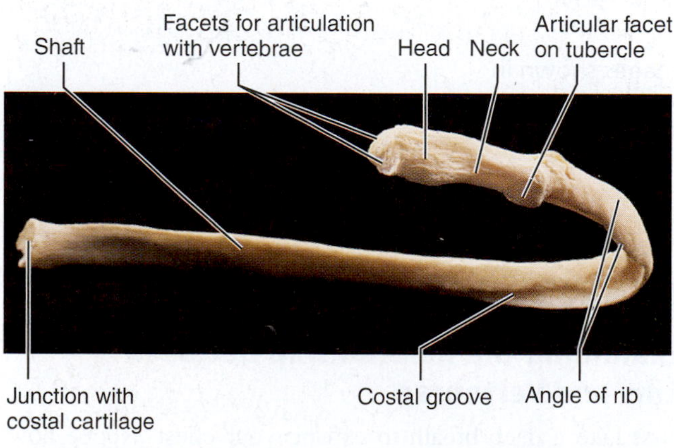

(c)

Figure 8.18 Structure of a "typical" true rib and its articulations. (a) Vertebral and sternal articulations of a typical true rib. **(b)** Superior view of the articulation between a rib and a thoracic vertebra, with costovertebral ligaments shown on left side only. **(c)** Right rib 6, posterior view.

The Fetal Skull

One of the most obvious differences between fetal and adult skeletons is the huge size of the fetal skull relative to the rest of the skeleton. Skull bones are incompletely formed at birth and connected by fibrous membranes called **fontanelles.** The fontanelles allow the fetal skull to be compressed slightly during birth and also allow for brain growth during late fetal life. They ossify (become bone) as the infant ages, completing the process by the time the child is 1½ to 2 years old.

ACTIVITY 6

Examining a Fetal Skull

1. Obtain a fetal skull and study it carefully.

   Does it have the same bones as the adult skull?

   ____

   ____

   How does the size of the fetal face relate to the cranium?

   ____

   ____

   How does this compare to what is seen in the adult?

   ____

   ____

2. Locate the following fontanelles on the fetal skull **(Figure 8.19)**: *anterior* (or *frontal*) *fontanelle*, *mastoid fontanelle*, *sphenoidal fontanelle*, and *posterior* (or *occipital*) *fontanelle*.

3. Notice that some of the cranial bones have conical protrusions. These are **ossification (growth) centers.** Notice also that the frontal bone is still bipartite and that the temporal bone is incompletely ossified—it is little more than a ring of bone.

4. Before completing this study, check the questions on the Review Sheet at the end of this exercise to ensure that you have made all of the necessary observations. ■

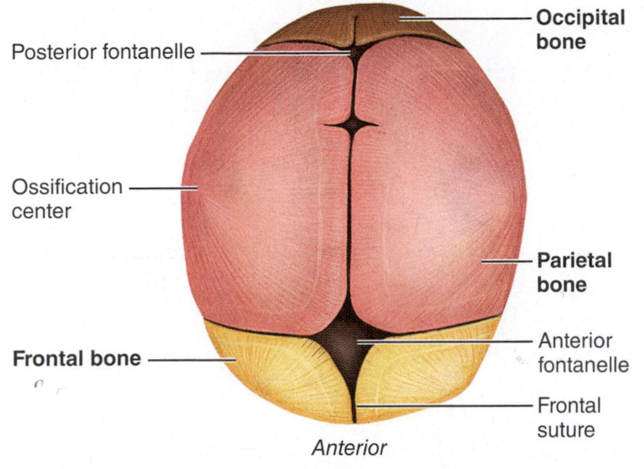

(a) Superior view

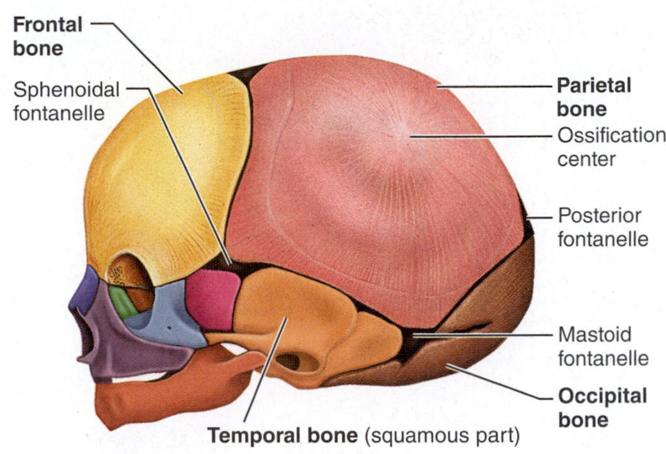

(b) Left lateral view

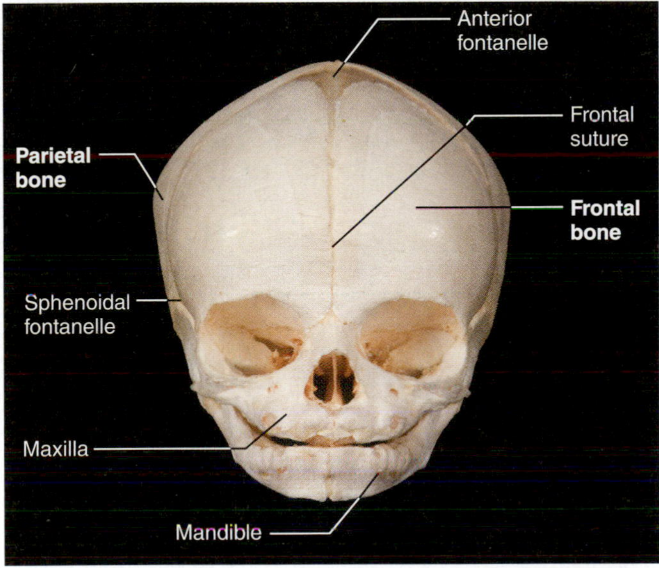

(c) Anterior view

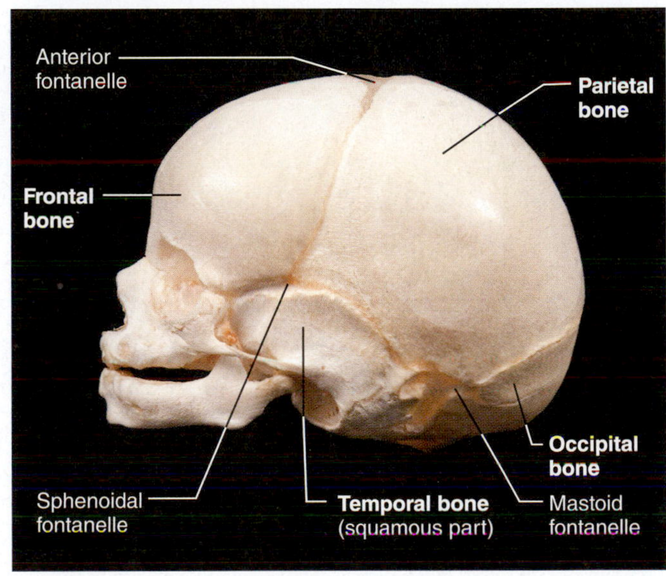

(d) Left lateral view

Figure 8.19 Skull of a newborn.

Name ______________________

Lab Time/Date ______________________

EXERCISE 8

REVIEW SHEET

The Axial Skeleton

The Skull

1. Match the bone names in column B with the descriptions in column A. Some responses in column B may be used more than once.

Column A

______ 1. forehead bone

______ 2. cheekbone

______ 3. lower jaw

______ 4. bridge of nose

______ 5. posterior bones of the hard palate

______ 6. much of the lateral and superior cranium

______ 7. most posterior part of cranium

______ 8. single, irregular, bat-shaped bone forming part of the cranial base

______ 9. tiny bones bearing tear ducts

______ 10. anterior part of hard palate

______ 11. superior and middle nasal conchae formed from its projections

______ 12. site of mastoid process

______ 13. site of sella turcica

______ 14. site of cribriform plate

______ 15. site of mental foramen

______ 16. site of styloid processes

______, ______, ______, and ______ 17. four bones containing paranasal sinuses

______ 18. condyles here articulate with the atlas

______ 19. foramen magnum contained here

______ 20. small U-shaped bone in neck, where many tongue muscles attach

______ 21. organ of hearing found here

______, ______ 22. two bones that form the nasal septum

______ 23. bears an upward protrusion, the "cock's comb," or crista galli

______, ______ 24. contain sockets that bear teeth

______ 25. forms the most inferior turbinate

Column B

a. ethmoid

b. frontal

c. hyoid

d. inferior nasal concha

e. lacrimal

f. mandible

g. maxilla

h. nasal

i. occipital

j. palatine

k. parietal

l. sphenoid

m. temporal

n. vomer

o. zygomatic

2. Using choices from the numbered key to the right, identify all bones (———●), sutures (—➤), and bone markings (———), provided with various leader lines in the two diagrams below. Some responses may be used more than once.

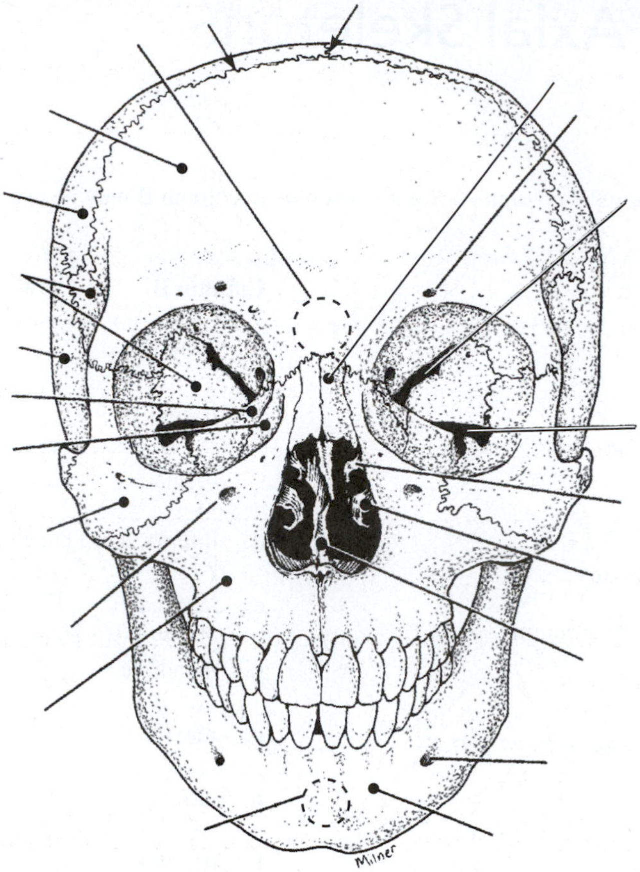

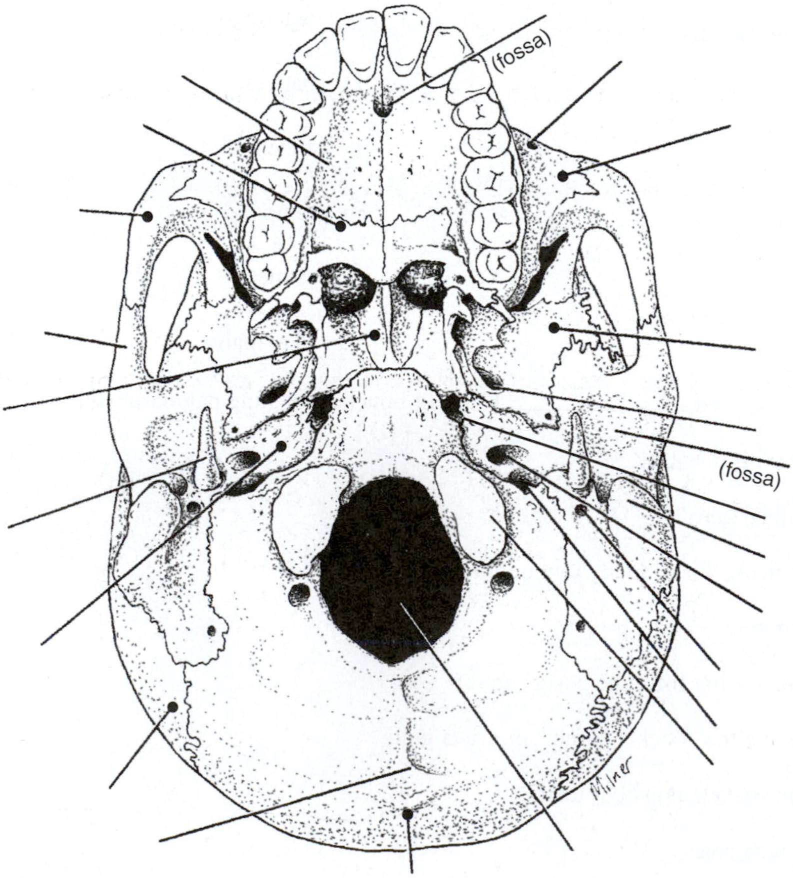

Key:
1. carotid canal
2. coronal suture
3. ethmoid bone
4. external occipital protuberance
5. foramen lacerum
6. foramen magnum
7. foramen ovale
8. frontal bone
9. glabella
10. incisive fossa
11. inferior nasal concha
12. inferior orbital fissure
13. infraorbital foramen
14. jugular foramen
15. lacrimal bone
16. mandible
17. mandibular fossa
18. mandibular symphysis
19. mastoid process
20. maxilla
21. mental foramen
22. middle nasal concha of ethmoid
23. nasal bone
24. occipital bone
25. occipital condyle
26. palatine bone
27. palatine process of maxilla
28. parietal bone
29. sagittal suture
30. sphenoid bone
31. styloid process
32. stylomastoid foramen
33. superior orbital fissure
34. supraorbital foramen
35. temporal bone
36. vomer bone
37. zygomatic bone
38. zygomatic process of temporal bone

3. Name the eight bones of the cranium. (Remember to include left and right).

________________ ________________ ________________ ________________

________________ ________________ ________________ ________________

4. Give two possible functions of the sinuses. ________________

____

5. What is the orbit? ________________

What bones contribute to the formation of the orbit? ________________

____

6. Why can the sphenoid bone be called the keystone of the cranial base? ________________

____

The Vertebral Column

7. The distinguishing characteristics of the vertebrae that make up the vertebral column are noted below. Correctly identify each described structure/region by choosing a response from the key.

Key:
a. atlas
b. axis
c. cervical vertebra—typical
d. coccyx
e. lumbar vertebra
f. sacrum
g. thoracic vertebra

________________ 1. vertebral type containing foramina in the transverse processes, through which the vertebral arteries ascend to reach the brain

________________ 2. dens here provides a pivot for rotation of the first cervical vertebra (C_1)

________________ 3. transverse processes faceted for articulation with ribs; spinous process pointing sharply downward

________________ 4. composite bone; articulates with the hip bone laterally

________________ 5. massive vertebrae; weight-sustaining

________________ 6. “tailbone”; vestigial fused vertebrae

________________ 7. supports the head; allows a rocking motion in conjunction with the occipital condyles

8. Using the key, correctly identify the vertebral parts/areas described below. (More than one choice may apply in some cases.) Also use the key letters to correctly identify the vertebral areas in the diagram.

Key:
a. body
b. intervertebral foramina
c. lamina
d. pedicle
e. spinous process
f. superior articular facet
g. transverse process
h. vertebral arch
i. vertebral foramen

________ 1. cavity enclosing the spinal cord

________ 2. weight-bearing portion of the vertebra

________ 3. provide levers against which muscles pull

________ 4. provide an articulation point for the ribs

________ 5. openings providing for exit of spinal nerves

________ 6. structures that form an enclosure for the spinal cord

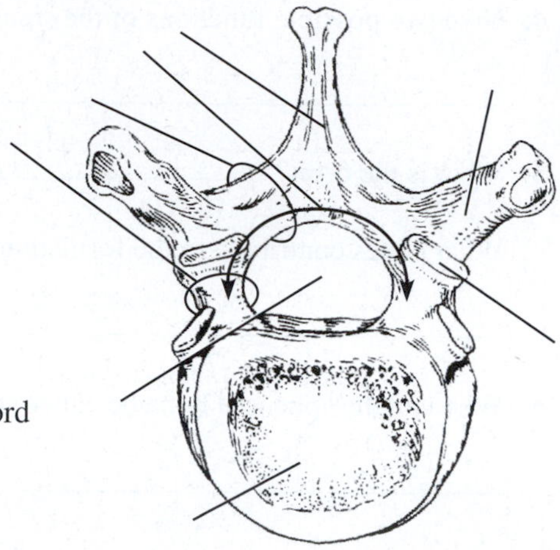

9. Describe how a spinal nerve exits from the vertebral column. ________

____

10. Name two factors/structures that permit flexibility of the vertebral column.

________ and ________

11. What kind of tissue makes up the intervertebral discs? ________

12. What is a herniated disc? ________

What problems might it cause? ________

____

13. Which two spinal curvatures are obvious at birth? ________ and ________

Under what conditions do the secondary curvatures develop? ________

____

____

14. On this illustration of an articulated vertebral column, identify each curvature indicated, and label it as a primary or a secondary curvature. Also identify the structures provided with leader lines, using the letters of the terms listed in the key below.

Key: a. atlas

b. axis

c. intervertebral disc

d. sacrum

e. two thoracic vertebrae

f. two lumbar vertebrae

g. vertebra prominens

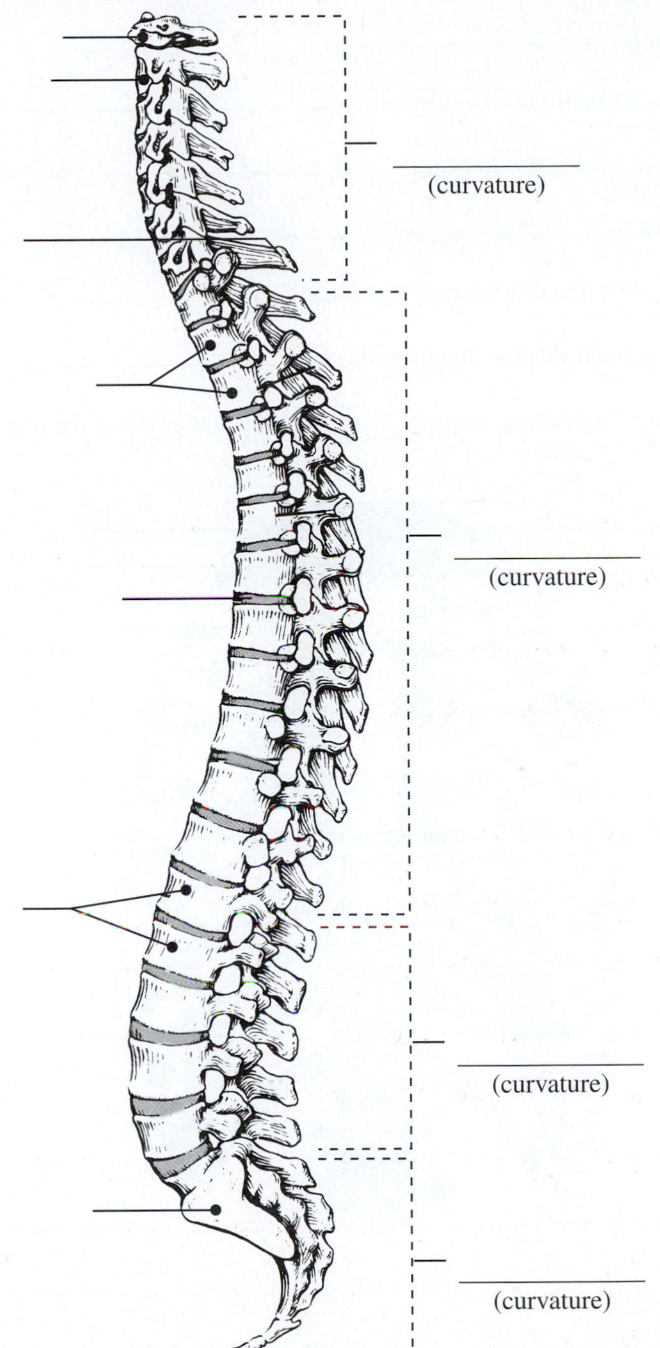

The Thoracic Cage

15. The major components of the thorax (excluding the vertebral column) are the ______________________________

and the ______________________________.

16. Differentiate a true rib from a false rib. ______________________________

____

____

Is a floating rib a true or a false rib? ______________________________

17. What is the general shape of the thoracic cage? ______________________________

18. Using the terms in the key, identify the regions and landmarks of the thoracic cage.

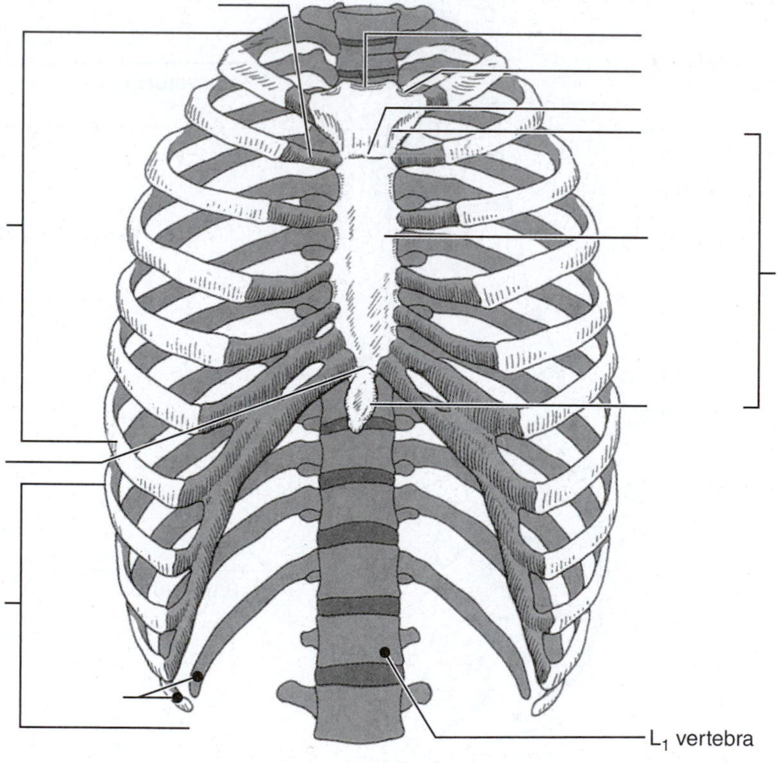

Key:
a. body
b. clavicular notch
c. costal cartilage
d. false ribs
e. floating ribs
f. jugular notch
g. manubrium
h. sternal angle
i. sternum
j. true ribs
k. xiphisternal joint
l. xiphoid process

The Fetal Skull

19. Are the same skull bones seen in the adult also found in the fetal skull? ______

20. How does the size of the fetal face compare to the fetal cranium? ______

______

How does this compare to the adult skull? ______

______

______

21. What are the outward conical projections on some of the fetal cranial bones? ______

22. What is a fontanelle? ______

What is its fate? ______

What is the function of the fontanelles in the fetal skull? ______

______

23. Using the terms listed, identify each of the fontanelles shown on the fetal skull below.

Key: a. anterior fontanelle b. mastoid fontanelle c. posterior fontanelle d. sphenoidal fontanelle

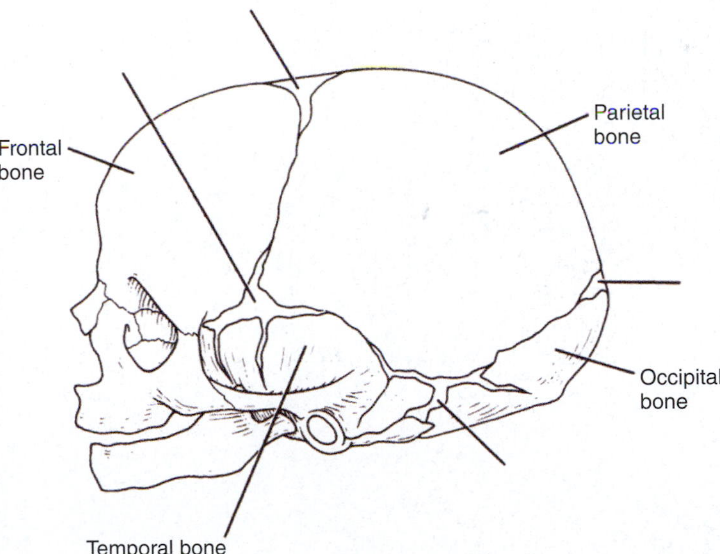

EXERCISE 9

The Appendicular Skeleton

MATERIALS

- ☐ Articulated skeletons
- ☐ Disarticulated skeletons (complete)
- ☐ Articulated pelves (male and female for comparative study)
- ☐ X-ray images of bones of the appendicular skeleton

OBJECTIVES

1. Identify the bones of the pectoral and pelvic girdles and their attached limbs by examining isolated bones or an articulated skeleton, and name the important bone markings on each.
2. Describe the differences between a male and a female pelvis, and explain the importance of these differences.
3. Compare the features of the human pectoral and pelvic girdles, and discuss how their structures relate to their specialized functions.
4. Arrange unmarked, disarticulated bones in their proper places to form an entire skeleton.

PRE-LAB QUIZ

1. The ______________________ skeleton is made up of 126 bones of the limbs and girdles.
2. Circle the correct underlined term. The <u>pectoral</u> / <u>pelvic</u> girdle attaches the upper limb to the axial skeleton.
3. The ______________________, on the posterior thorax, are roughly triangular in shape. They have no direct attachment to the axial skeleton but are held in place by trunk muscles.
4. The arm consists of one long bone, the:
   a. femur
   b. humerus
   c. tibia
   d. ulna
5. The hand consists of three groups of bones. The carpals make up the wrist. The ______________________ make up the palm, and the phalanges make up the fingers.
6. You are studying a pelvis that is wide and shallow. The acetabula are small and far apart. The pubic arch/angle is rounded and greater than 90°. It appears to be tilted forward, with a wide, short sacrum. Is this a male or a female pelvis? ________________
7. The strongest, heaviest bone of the body is in the thigh. It is the:
   a. femur
   b. fibula
   c. tibia
8. The ________________, or "kneecap," is a sesamoid bone that is found within the quadriceps tendon. It guards the knee joint anteriorly and improves leverage of the thigh muscles acting across the knee joint.
9. Circle True or False. The fingers of the hand and the toes of the foot—with the exception of the great toe and the thumb—each have three phalanges.
10. Each foot has a total of ________________ bones.

The **appendicular skeleton** (the gold-colored portion of Figure 7.1 on page 92) is composed of the 126 bones of the appendages and the pectoral and pelvic girdles, which attach the limbs to the axial skeleton. Although the upper and lower limbs differ in their functions and mobility, they have the same fundamental plan, with each limb made up of three major segments connected together by freely movable joints.

ACTIVITY 1

Examining and Identifying Bones of the Appendicular Skeleton

Carefully examine each of the bones described in this exercise, and identify the characteristic bone markings of each. The markings help you determine whether a bone is the right or left member of its pair; for example, the glenoid cavity is on the lateral aspect of the scapula, and the spine is on its posterior aspect. *This is a very important instruction because you will be constructing your own skeleton to finish this laboratory exercise.* Additionally, when corresponding X-ray films are available, compare the actual bone specimen to its X-ray image. ■

Bones of the Pectoral Girdle and Upper Limb

The Pectoral (Shoulder) Girdle

The paired **pectoral,** or **shoulder, girdles** (Figure 9.1) each consist of two bones—the anterior clavicle and the posterior scapula. The shoulder girdles attach the upper limbs to the axial skeleton, and they serve as attachment points for many trunk and neck muscles.

The **clavicle,** or collarbone, is a slender, doubly curved bone—convex forward on its medial two-thirds and concave laterally. Its *sternal* (medial) *end,* which attaches to the sternal manubrium, is rounded or triangular in cross section. The sternal end projects above the manubrium and can be felt and (usually) seen forming the lateral walls of the *jugular notch* (Figure 8.17, page 121). The *acromial* (lateral) *end* of the clavicle is flattened where it articulates with the scapula to form part of the shoulder joint. On its posteroinferior surface is the prominent **conoid tubercle** (Figure 9.2b). This projection anchors a ligament and provides a handy landmark for determining whether a given clavicle is from the right or left side of the body. The clavicle serves as an anterior brace, or strut, to hold the arm away from the top of the thorax.

The **scapulae** (Figure 9.2c–e), or shoulder blades, are generally triangular and are commonly called the "wings" of humans. Each scapula has a flattened body and two important processes—the **acromion** (the enlarged end of the spine of the scapula) and the beaklike **coracoid process** (*corac* = crow, raven). The acromion connects with the clavicle. The coracoid process points anteriorly over the tip of the shoulder joint and serves as an attachment point for some of the upper limb muscles. The **suprascapular notch** at the base of the coracoid process allows nerves to pass. The scapula has no direct attachment to the axial skeleton but is loosely held in place by trunk muscles.

The scapula has three angles: superior, inferior, and lateral. The inferior angle provides a landmark for auscultating (listening to) lung sounds. The **glenoid cavity,** a shallow socket that receives the head of the arm bone (humerus), is located in the lateral angle. The scapula also has three named borders: superior, medial (vertebral), and lateral (axillary). Several shallow depressions (fossae) appear on both sides of the scapula and are named according to location; there are the anterior *subscapular fossa* and the posterior *infraspinous* and *supraspinous fossae.*

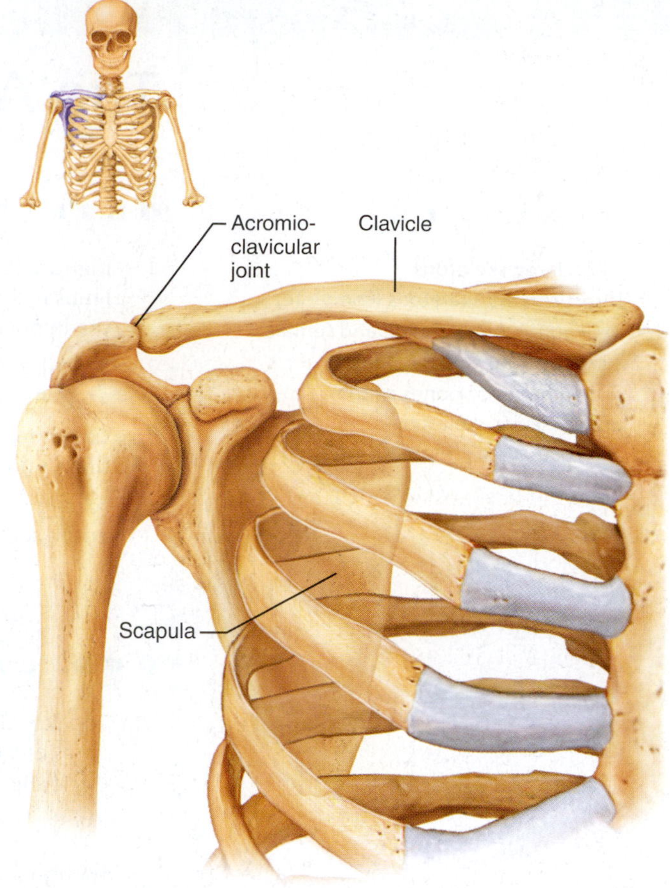

Figure 9.1 Articulated bones of the pectoral (shoulder) girdle. The right pectoral girdle is articulated to show the relationship of the girdle to the bones of the thorax and arm.

The shoulder girdle is exceptionally light and allows the upper limb a degree of mobility not seen anywhere else in the body. This is due to the following factors:

- The sternoclavicular joints are the *only* site where the shoulder girdles attach to the axial skeleton.
- The relative looseness of the scapular attachment allows it to slide back and forth against the thorax with muscular activity.
- The glenoid cavity is shallow and does little to stabilize the shoulder joint.

However, this exceptional flexibility exacts a price: The arm bone (humerus) is very susceptible to dislocation, and fracture of the clavicle disables the entire upper limb.

The Arm

The arm, or brachium (Figure 9.3), contains a single bone—the **humerus,** a typical long bone. Proximally, its rounded *head* fits into the shallow glenoid cavity of the scapula. The head is separated from the shaft by the **anatomical neck** and by the more constricted **surgical neck,** which is a common site of fracture. Opposite the head are two prominences, the **greater** and **lesser tubercles** (from lateral to medial aspect),

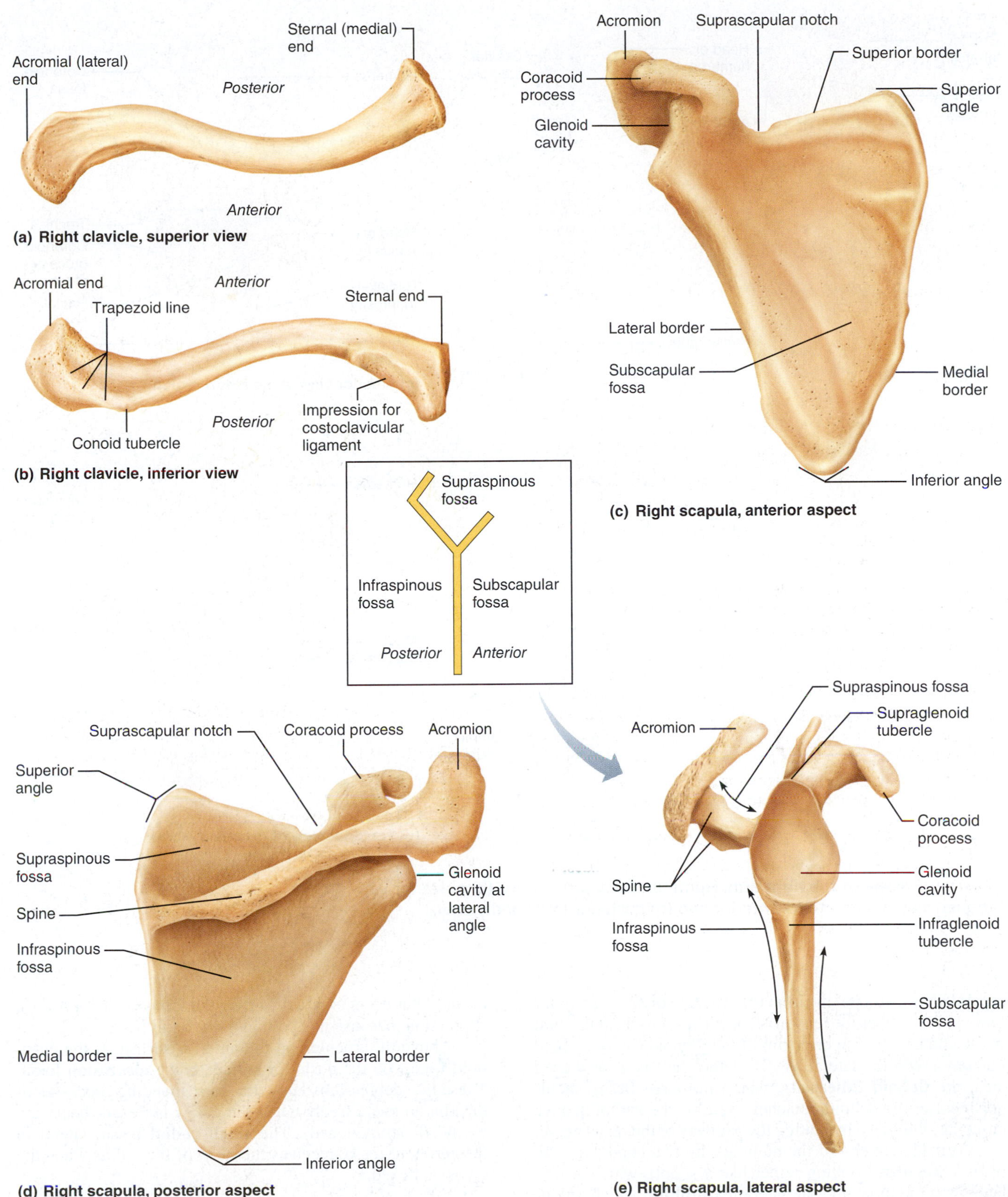

Figure 9.2 Individual bones of the pectoral (shoulder) girdle. View **(e)** is accompanied by a schematic representation of its orientation.

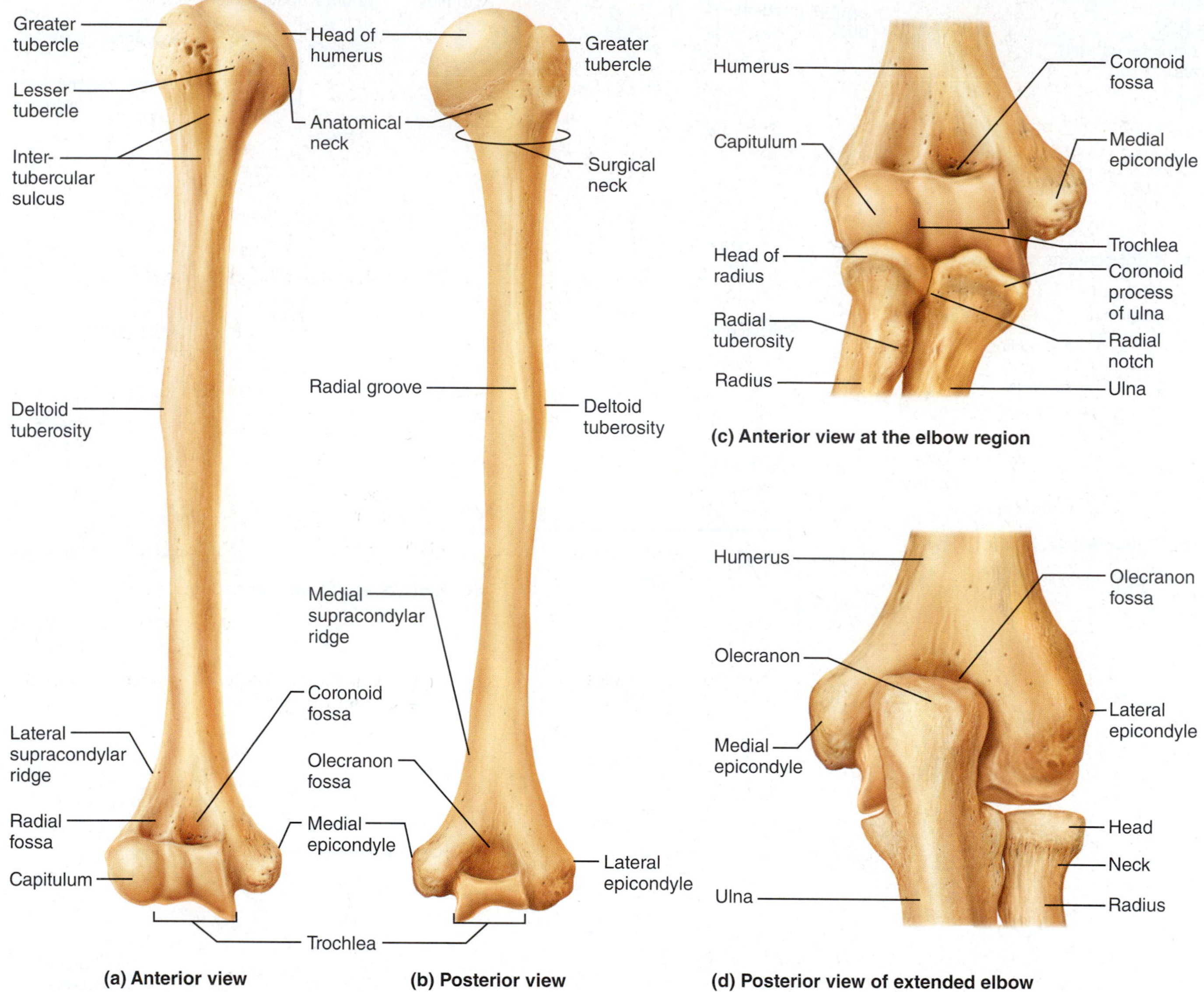

Figure 9.3 Bone of the right arm. Humerus, **(a)** anterior view, **(b)** posterior view. Detailed views illustrate **(c)** anterior and **(d)** posterior extended elbow.

separated by a groove (the **intertubercular sulcus,** or *bicipital groove*) that guides the tendon of the biceps muscle to its point of attachment (the superior rim of the glenoid cavity). About midway down the lateral aspect of the shaft is a roughened area, the **deltoid tuberosity,** where the large fleshy shoulder muscle, the deltoid, attaches. Nearby, the **radial groove** descends obliquely, indicating the pathway of the radial nerve.

At the distal end of the humerus are two condyles—the medial **trochlea** (looking rather like a spool), which articulates with the ulna, and the lateral **capitulum,** which articulates with the radius of the forearm. This condyle pair is flanked medially by the **medial epicondyle** and laterally by the **lateral epicondyle.**

The medial epicondyle is commonly called the "funny bone." The ulnar nerve runs in a groove beneath the medial epicondyle, and when this region is sharply bumped, a temporary, but excruciatingly painful, tingling sensation often occurs. This event is called "hitting the funny bone," a strange expression, because it is certainly *not* funny!

Above the trochlea on the anterior surface is the **coronoid fossa;** on the posterior surface is the **olecranon fossa.** These two depressions allow the corresponding processes of the ulna to move freely when the elbow is flexed (bent) and extended (straightened). The small **radial fossa,** lateral to the coronoid fossa, receives the head of the radius when the elbow is flexed.

The Forearm

Two bones, the radius and the ulna, compose the skeleton of the forearm, or antebrachium **(Figure 9.4)**. When the body is in the anatomical position, the **radius** is in the lateral position in the forearm, and the radius and ulna are parallel. Proximally, the disc-shaped head of the radius articulates

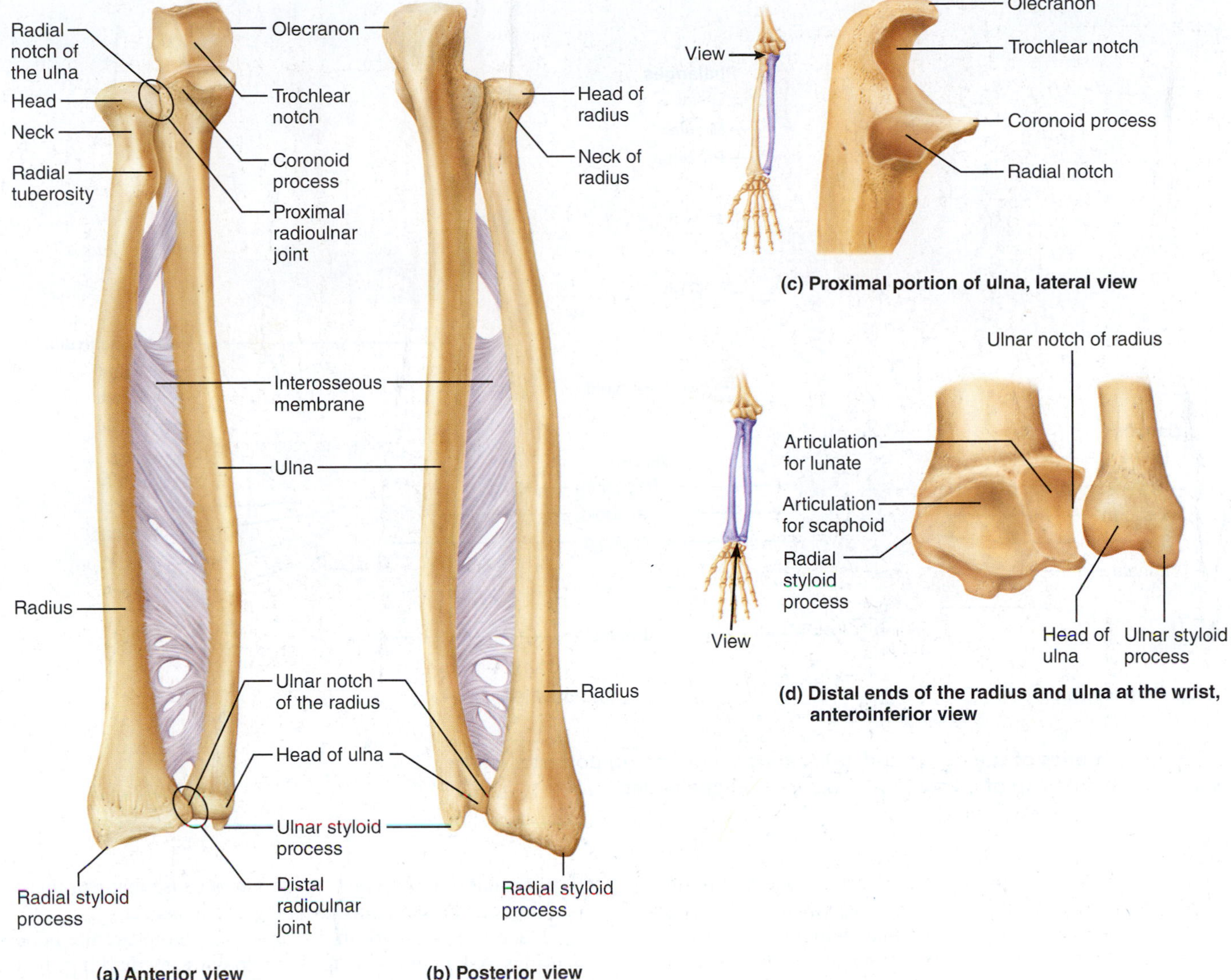

Figure 9.4 Bones of the right forearm. (a, b) Radius and ulna in anterior and posterior views. **(c, d)** Structural details of the articular surfaces between the radius and ulna and between the radius and bones of the wrist.

with the capitulum of the humerus. Just below the head, on the medial aspect of the shaft, is a prominence called the **radial tuberosity,** the point of attachment for the tendon of the biceps muscle of the arm. Distally, the small **ulnar notch** reveals where the radius articulates with the end of the ulna.

The **ulna** is the medial bone of the forearm. Its proximal end bears the anterior **coronoid process** and the posterior **olecranon,** which are separated by the **trochlear notch.** Together these processes grip the trochlea of the humerus in a plierslike joint. The small **radial notch** on the lateral side of the coronoid process articulates with the head of the radius. The slimmer distal end, the ulnar **head,** bears the small medial **ulnar styloid process,** which serves as a point of attachment for the ligaments of the wrist.

The Hand

The skeleton of the hand, or manus **(Figure 9.5)**, includes three groups of bones, those of the carpus (wrist), the metacarpals (bones of the palm), and the phalanges (bones of the fingers).

The wrist is the proximal portion of the hand. It is referred to anatomically as the **carpus;** the eight marble-sized bones composing it are the **carpals.** (So you actually wear your wristwatch over the distal part of your forearm.) The carpals are arranged in two irregular rows of four bones each (illustrated in Figure 9.5). In the proximal row (lateral to medial) are the *scaphoid, lunate, triquetrum,* and *pisiform bones;* the scaphoid and lunate articulate with the distal end of the radius. In the distal row are the *trapezium, trapezoid, capitate,*

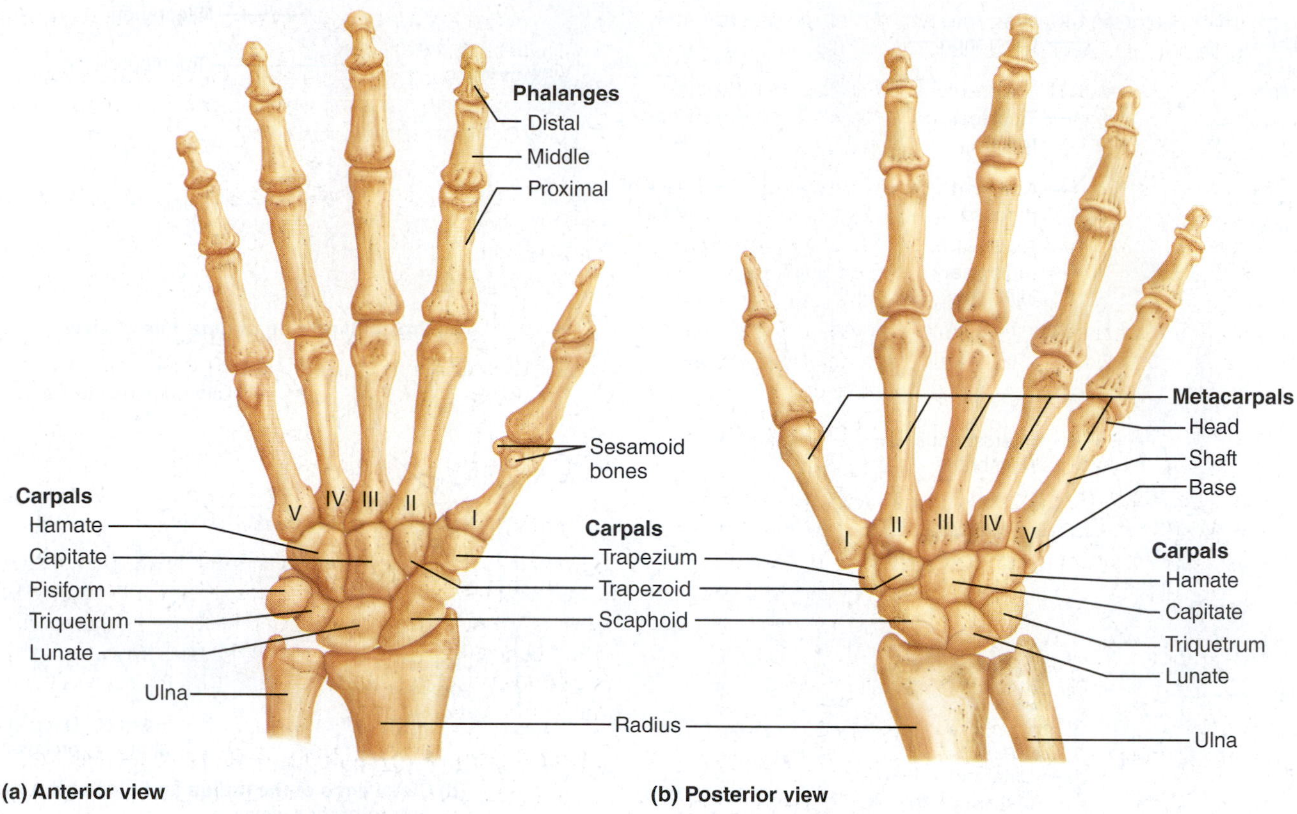

Figure 9.5 Bones of the right hand. (a) Anterior view and **(b)** posterior view, showing relationship of carpals, metacarpals, and phalanges.

and *hamate.* The carpals are bound closely together by ligaments, which restrict movements between them to allow only gliding movements. A mnemonic may help you to remember the names of the carpals in order from the lateral aspect of the proximal row and continuing with the distal row, lateral to medial: "**S**ally **L**eft **T**he **P**arty **T**o **T**ake **C**armen **H**ome."

The **metacarpals,** numbered I to V from the thumb side of the hand toward the little finger, radiate out from the wrist like spokes to form the palm of the hand. The *bases* of the metacarpals articulate with the carpals of the wrist; their more bulbous *heads* articulate with the phalanges of the fingers distally. When the fist is clenched, the heads of the metacarpals become prominent as the knuckles.

Like the bones of the palm, the fingers are numbered from I to V, beginning from the thumb *(pollex)* side of the hand. The 14 bones of the fingers, or digits, are miniature long bones, called **phalanges** (singular, *phalanx*) as noted above. Each finger contains three phalanges (proximal, middle, and distal) except the thumb, which has only two (proximal and distal).

ACTIVITY 2

Palpating the Surface Anatomy of the Pectoral Girdle and Upper Limb

Before continuing on to study the bones of the pelvic girdle, take the time to identify the following bone markings on the skin surface of the upper limb. It is usually preferable to palpate the bone markings on your lab partner because many of these markings can be seen only from the dorsal aspect.

Place a check mark in the boxes as you locate the bone markings. Ask your instructor for help with any markings that you are unable to locate.

☐ Clavicle: Palpate the clavicle along its entire length from sternum to shoulder.

☐ Acromioclavicular joint: The high point of the shoulder, which represents the junction point between the clavicle and the acromion of the scapular spine.

☐ Spine of the scapula: Extend your arm at the shoulder so that your scapula moves posteriorly. As you do this, your scapular spine will be seen as a winglike protrusion on your dorsal thorax and can be easily palpated by your lab partner.

☐ Lateral epicondyle of the humerus: The inferiormost projection at the lateral aspect of the distal humerus. After you have located the epicondyle, run your finger posteriorly into the hollow immediately dorsal to the epicondyle. This is the site where the extensor muscles of the hand are attached and is a common site of the excruciating pain of tennis elbow, a condition in which those muscles and their tendons are abused physically.

☐ Medial epicondyle of the humerus: Feel this medial projection at the distal end of the humerus.

☐ Olecranon of the ulna: Work your elbow—flexing and extending—as you palpate its dorsal aspect to feel the olecranon process of the ulna moving into and out of the olecranon fossa on the dorsal aspect of the humerus.

☐ Ulnar styloid process: With the hand in the anatomical position, feel out this small inferior projection on the medial aspect of the distal end of the ulna.

☐ Radial styloid process: Find this projection at the distal end of the radius (lateral aspect). It is most easily located by moving the hand medially at the wrist. Once you have palpated the radial styloid process, move your fingers just medially onto the anterior wrist. Press firmly, and then let up slightly on the pressure. You should be able to feel your pulse at this pressure point, which lies over the radial artery (radial pulse).

☐ Pisiform: Just distal to the ulnar styloid process, feel the rounded pealike pisiform bone.

☐ Metacarpophalangeal joints (knuckles): Clench your fist, and find the first set of flexed-joint protrusions beyond the wrist—these are your metacarpophalangeal joints. ■

Bones of the Pelvic Girdle and Lower Limb

The Pelvic (Hip) Girdle

As with the bones of the pectoral girdle and upper limb, pay particular attention to bone markings needed to identify right and left bones.

The **pelvic girdle,** or **hip girdle** (Figure 9.6), is formed by the two **coxal** (*coxa* = hip) **bones** (also called the **ossa coxae,** or hip bones) and the sacrum. The deep structure formed by the hip bones, sacrum, and coccyx is called the **pelvis** or *bony pelvis.* In contrast to the bones of the shoulder girdle, those of the pelvic girdle are heavy and massive, and they attach securely to the axial skeleton. The sockets for the heads of the femurs (thigh bones) are deep and heavily reinforced by ligaments to ensure a stable, strong limb attachment. The ability to bear weight is more important here than mobility and flexibility. The combined weight of the upper body rests on the pelvic girdle (specifically, where the hip bones meet the sacrum).

Each coxal bone is a result of the fusion of three bones—the ilium, ischium, and pubis—which are distinguishable in the young child. The **ilium,** a large flaring bone, forms the major portion of the coxal bone. It connects posteriorly, via its **auricular surface,** with the sacrum at the **sacroiliac joint.** The superior margin of the iliac bone, the **iliac crest,** is rough; when you rest your hands on your hips, you are palpating your iliac crests. The iliac crest terminates anteriorly in the **anterior superior spine** and posteriorly in the **posterior superior spine.** Two inferior spines are located below these. The shallow **iliac fossa** marks its internal surface, and a prominent ridge, the **arcuate line,** outlines the pelvic inlet, or pelvic brim.

The **ischium** is the "sit-down" bone, forming the most inferior and posterior portion of the coxal bone. The most outstanding marking on the ischium is the rough **ischial tuberosity,** which receives the weight of the body when sitting. The **ischial spine,** superior to the ischial tuberosity, is an important anatomical landmark of the pelvic cavity. (See **Table 9.1** for a comparison of male and female pelves.) The obvious **lesser** and **greater sciatic notches** allow nerves and blood vessels to pass to and from the thigh. The sciatic nerve passes through the latter.

The **pubis** is the most anterior portion of the coxal bone. Fusion of the pubic **rami** anteriorly and the ischium posteriorly forms a bar of bone enclosing the **obturator foramen,** through which blood vessels and nerves run from the pelvic cavity into the thigh. The pubis of each hip bone meets anteriorly at the **pubic crest** to form a cartilaginous joint called the **pubic symphysis.** At the lateral end of the pubic crest is the *pubic tubercle* (see Figure 9.6c) to which the important *inguinal ligament* attaches.

The ilium, ischium, and pubis fuse at the deep hemispherical socket called the **acetabulum** (literally, "wine cup"), which receives the head of the thigh bone.

ACTIVITY 3

Observing Pelvic Articulations

Before continuing with the bones of the lower limbs, take the time to examine an articulated pelvis. Notice how each coxal bone articulates with the sacrum posteriorly and how the two coxal bones join at the pubic symphysis. The sacroiliac joint is a common site of lower back problems because of the pressure it must bear. ■

Comparison of the Male and Female Pelves

Although bones of males are usually larger, heavier, and have more prominent bone markings, the male and female skeletons are very similar. The exception to this generalization is pelvic structure.

The female pelvis reflects modifications for childbearing—it is wider, shallower, lighter, and rounder than that of the male. Her pelvis must not only support the increasing size of a fetus, but also be large enough to allow the infant's head (its largest dimension) to descend through the birth canal at birth.

To describe pelvic sex differences, we need to introduce a few more terms. The **false pelvis** is that portion superior to the arcuate line; it is bounded by the alae of the ilia laterally and the sacral promontory and lumbar vertebrae posteriorly. Although the false pelvis supports the abdominal viscera, it does not restrict childbirth in any way. The **true pelvis** is the region inferior to the arcuate line that is almost entirely surrounded by bone. Its posterior boundary is formed by the sacrum. The ilia, ischia, and pubic bones define its limits laterally and anteriorly.

The dimensions of the true pelvis, particularly its inlet and outlet, are critical if delivery of a baby is to be uncomplicated; and they are carefully measured by the obstetrician. The **pelvic inlet,** or **pelvic brim,** is the opening delineated by the sacral promontory posteriorly and the arcuate lines of the ilia anterolaterally. It is the superiormost margin of the true pelvis. Its widest dimension is from left to right, that is, along the frontal plane. The **pelvic outlet** is the inferior margin of the true pelvis. It is bounded anteriorly by the pubic arch, laterally by the ischia, and posteriorly by the sacrum and coccyx. Because both the coccyx and the ischial spines protrude into the outlet opening, a sharply angled coccyx or large, sharp ischial spines can dramatically narrow the outlet. The largest dimension of the outlet is the anterior-posterior diameter.

The major differences between the male and female pelves are summarized in the table (Table 9.1).

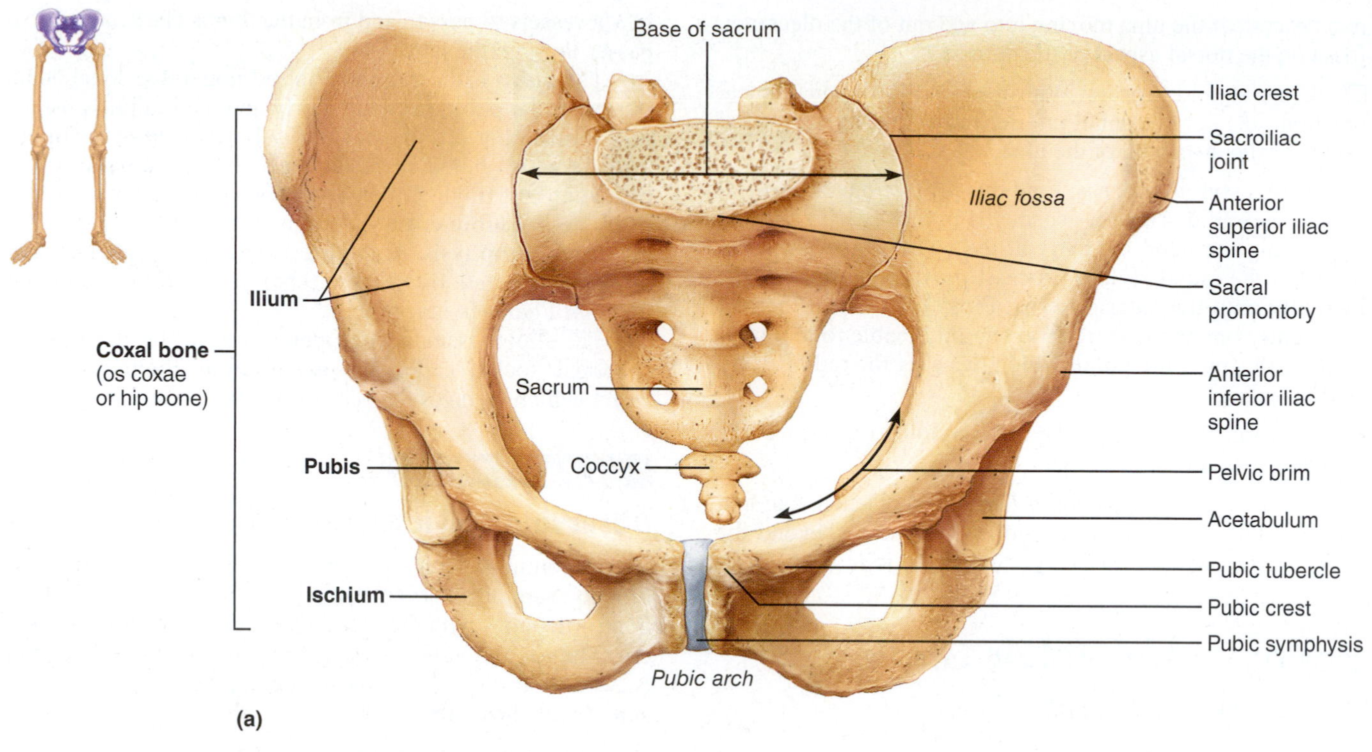

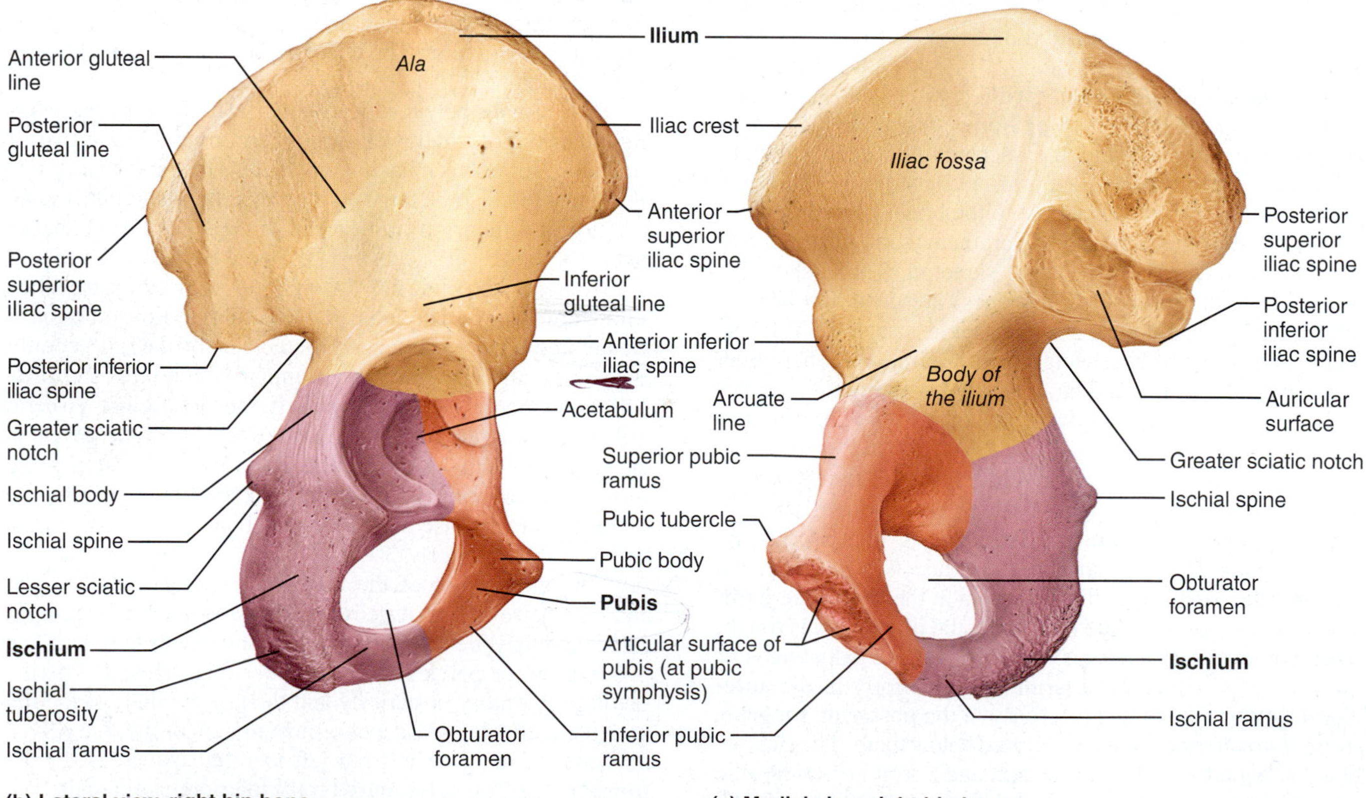

Figure 9.6 Bones of the pelvic girdle. (a) Articulated bony pelvis, showing the two coxal bones, which together with the sacrum comprise the pelvic girdle, and the coccyx. **(b)** Right hip bone, lateral view, showing the point of fusion of the ilium, ischium, and pubis. **(c)** Right hip bone, medial view.

Table 9.1 Comparison of the Male and Female Pelves

Characteristic	Female	Male
General structure and functional modifications	Tilted forward; adapted for childbearing; true pelvis defines the birth canal; cavity of the true pelvis is broad, shallow, and has a greater capacity	Tilted less far forward; adapted for support of a male's heavier build and stronger muscles; cavity of the true pelvis is narrow and deep
Bone thickness	Less; bones lighter, thinner, and smoother	Greater; bones heavier and thicker, and markings are more prominent
Acetabula	Smaller; farther apart	Larger; closer
Pubic angle/arch	Broader angle (80°–90°); more rounded	Angle is more acute (50°–60°)
Anterior view	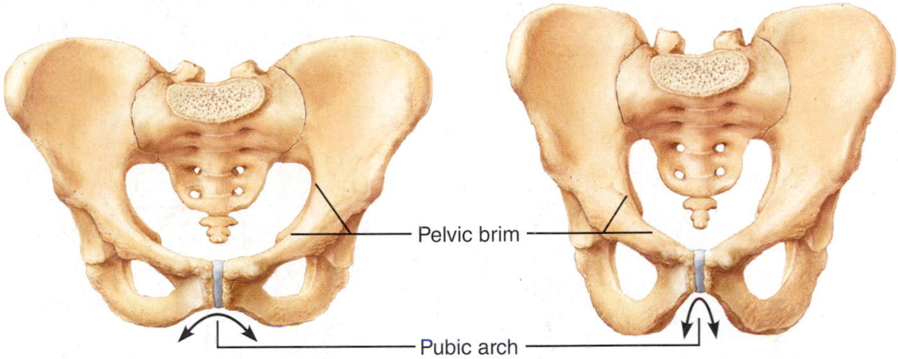 Pelvic brim Pubic arch	Pelvic brim Pubic arch
Sacrum	Wider; shorter; sacrum is less curved	Narrow; longer; sacral promontory more ventral
Coccyx	More movable; straighter, projects inferiorly	Less movable; curves and projects anteriorly
Left lateral view	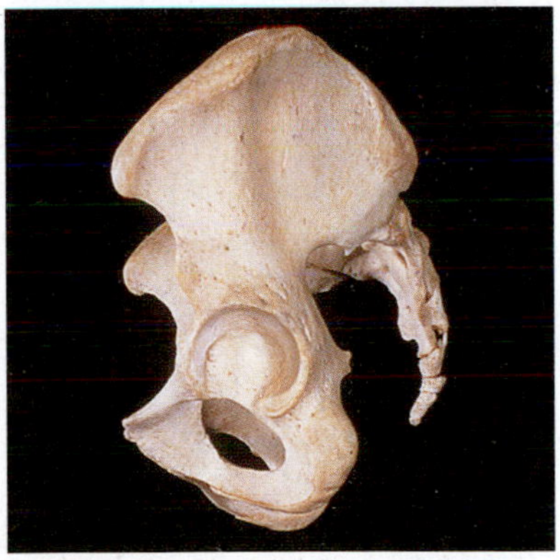	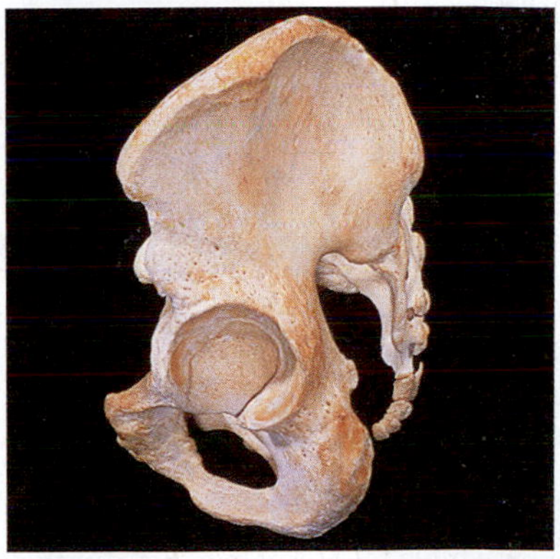
Pelvic inlet (brim)	Wider; oval from side to side	Narrow; basically heart-shaped
Pelvic outlet	Wider; ischial spines shorter, farther apart, and everted	Narrower; ischial spines longer, sharper, and point more medially
Posteroinferior view	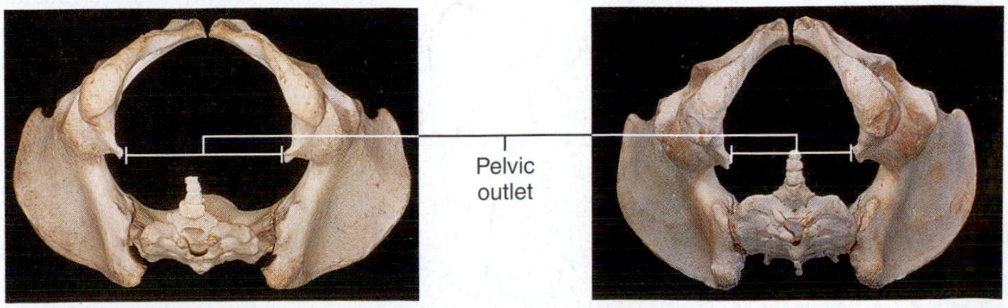 Pelvic outlet	Pelvic outlet

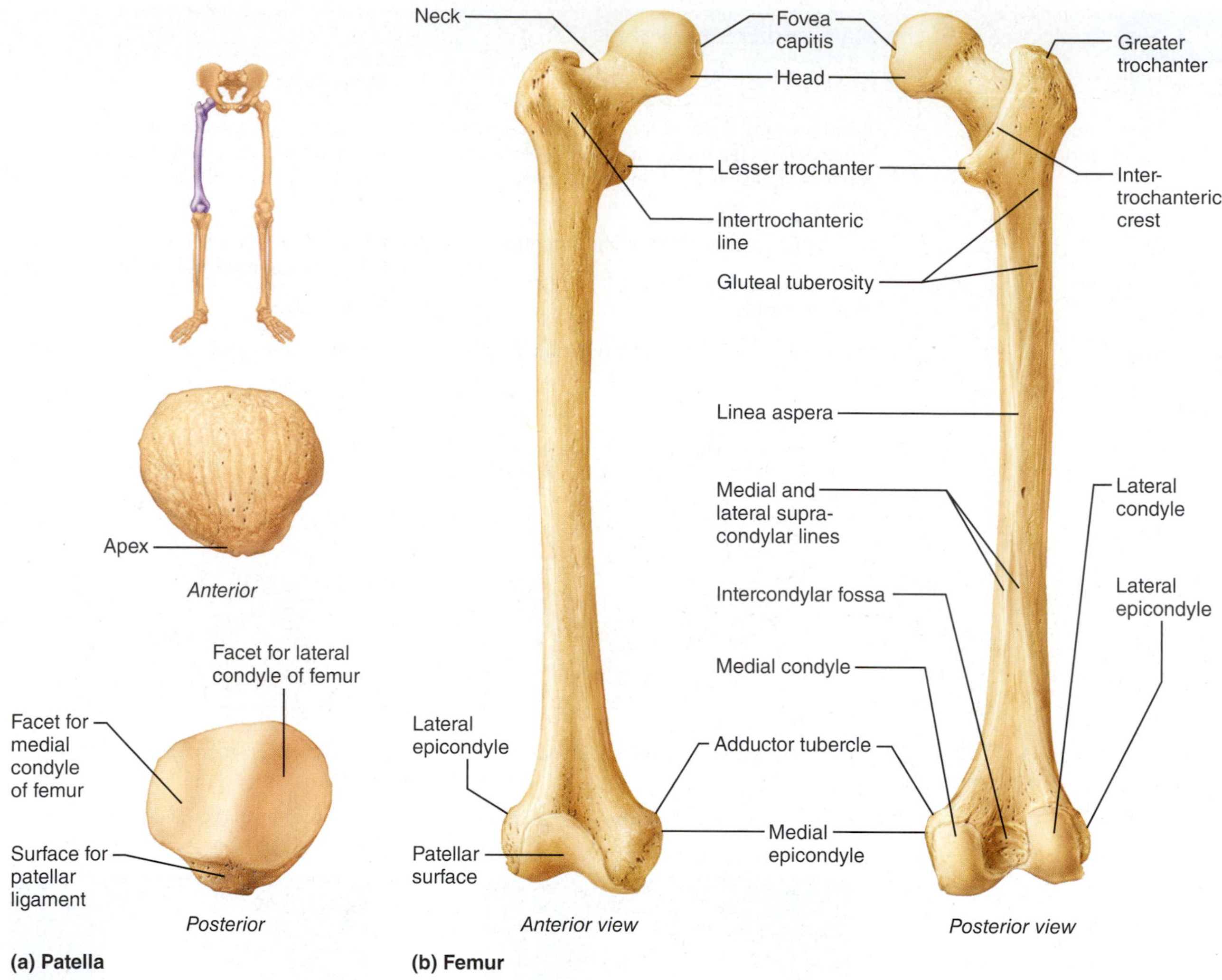

Figure 9.7 Bones of the right knee and thigh. (a) The patella (kneecap). **(b)** The femur (thigh bone).

ACTIVITY 4

Comparing Male and Female Pelves

Examine male and female pelves for the following differences:

- The female inlet is larger and more circular.
- The female pelvis as a whole is shallower, and the bones are lighter and thinner.
- The female sacrum is broader and less curved, and the pubic arch is more rounded.
- The female acetabula are smaller and farther apart, and the ilia flare more laterally.
- The female ischial spines are shorter, farther apart, and everted, thus enlarging the pelvic outlet. ■

The Thigh

The **femur,** or thigh bone **(Figure 9.7a)**, is the only bone of the thigh. It is the heaviest, strongest bone in the body. The ball-like head of the femur articulates with the hip bone via the deep, secure socket of the acetabulum. Obvious in the femur's head is a small central pit called the **fovea capitis** ("pit of the head") from which a small ligament runs to the acetabulum. The head of the femur is carried on a short, constricted *neck,* which angles laterally to join the shaft. The neck is the weakest part of the femur and is a common fracture site (an injury called a broken hip), particularly in the elderly. At the junction of the shaft and neck are the **greater** and **lesser trochanters,** separated posteriorly by the **intertrochanteric crest** and anteriorly by the **intertrochanteric line.** The trochanters and trochanteric crest, as well as the **gluteal tuberosity** and the **linea aspera** located on the shaft, are sites of muscle attachment.

The femur inclines medially as it runs downward to the leg bones; this brings the knees in line with the body's center of gravity, or maximum weight. The medial course of the femur is more noticeable in females because of the wider female pelvis.

Distally, the femur terminates in the **lateral** and **medial condyles,** which articulate with the tibia below, and the **patellar surface,** which forms a joint with the patella (kneecap) anteriorly. The **lateral** and **medial epicondyles,** just superior to the condyles, are separated by the **intercondylar fossa.**

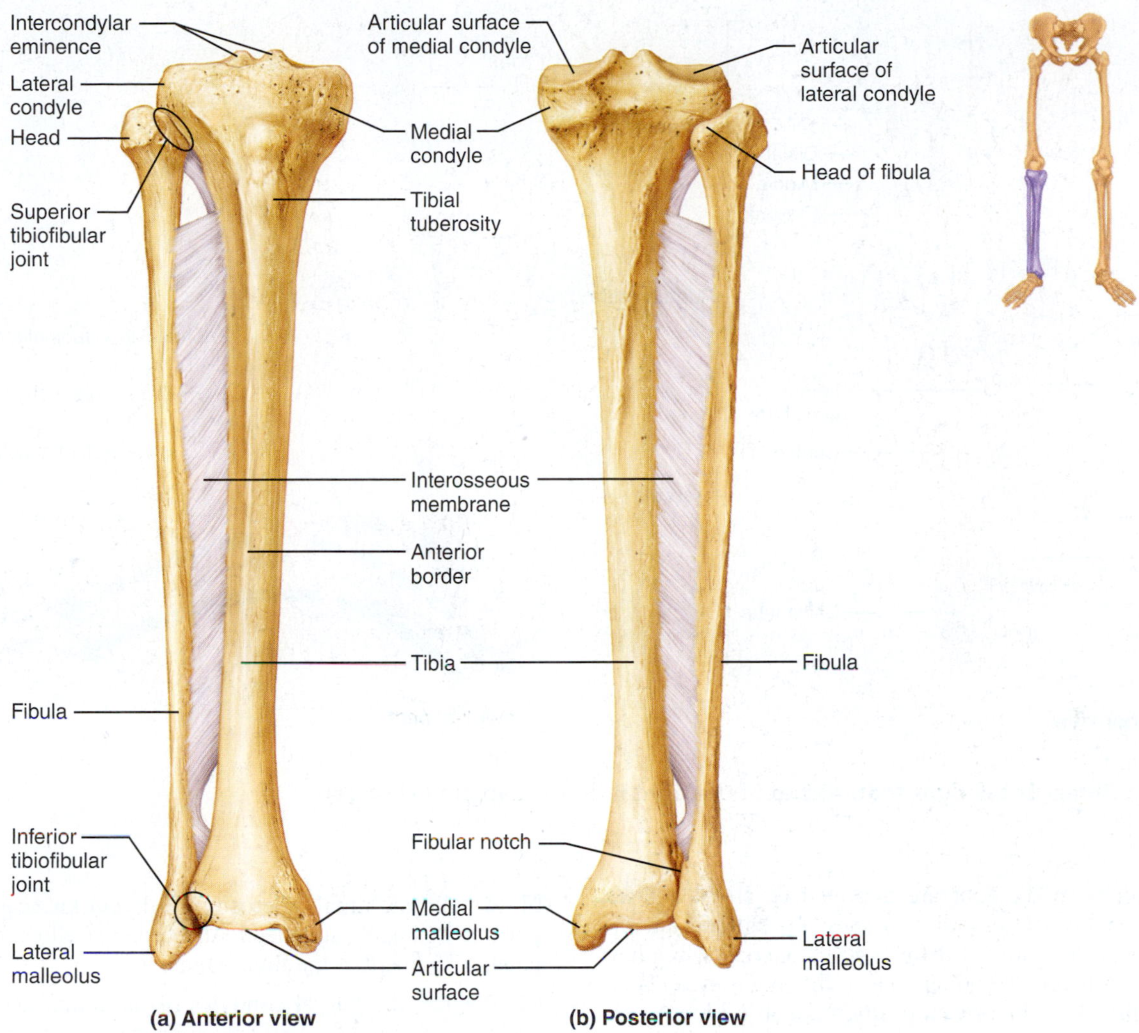

Figure 9.8 **Bones of the right leg.** Tibia and fibula, **(a)** anterior view, **(b)** posterior view.

On the superior part of the medial epicondyle is a bump, the **adductor tubercle,** to which the large adductor magnus muscle attaches.

The **patella** (Figure 9.7a) is a triangular sesamoid bone enclosed in the (quadriceps) tendon that secures the anterior thigh muscles to the tibia. It guards the knee joint anteriorly and improves the leverage of the thigh muscles acting across the knee joint.

The Leg

Two bones, the tibia and the fibula, form the skeleton of the leg (see **Figure 9.8**). The **tibia,** or *shinbone,* is the larger and more medial of the two leg bones. At the proximal end, the **medial** and **lateral condyles** (separated by the **intercondylar eminence**) receive the distal end of the femur to form the knee joint. The **tibial tuberosity,** a roughened protrusion on the anterior tibial surface (just below the condyles), is the site of attachment of the patellar (kneecap) ligament. Small facets on the superior and inferior surface of the lateral condyle of the tibia articulate with the fibula. Distally, a process called the **medial malleolus** forms the inner (medial) bulge of the ankle. The distal articular surface of the tibia articulates with the talus bone of the foot. The anterior surface of the tibia bears a sharpened ridge that is relatively unprotected by muscles. This so-called **anterior border** is easily felt beneath the skin.

The **fibula,** which lies parallel to the tibia, takes no part in forming the knee joint. Its proximal head articulates with the lateral condyle of the tibia. The fibula is thin and sticklike with a sharp anterior crest. It terminates distally in the **lateral malleolus,** which forms the outer part, or lateral bulge, of the ankle.

The Foot

The bones of the foot include the 7 **tarsal** bones, 5 **metatarsals,** which form the instep, and 14 **phalanges,** which form the toes **(Figure 9.9)**. Body weight is concentrated on the two largest tarsals, which form the posterior aspect of the foot, the *calcaneus* (heel bone) and the *talus,* which lies between the tibia and the calcaneus. The other tarsals are named and identified in the figure (Figure 9.9). The metatarsals are numbered I through V, medial to lateral. Like the fingers of the hand, each toe has three phalanges except the great toe, which has two.

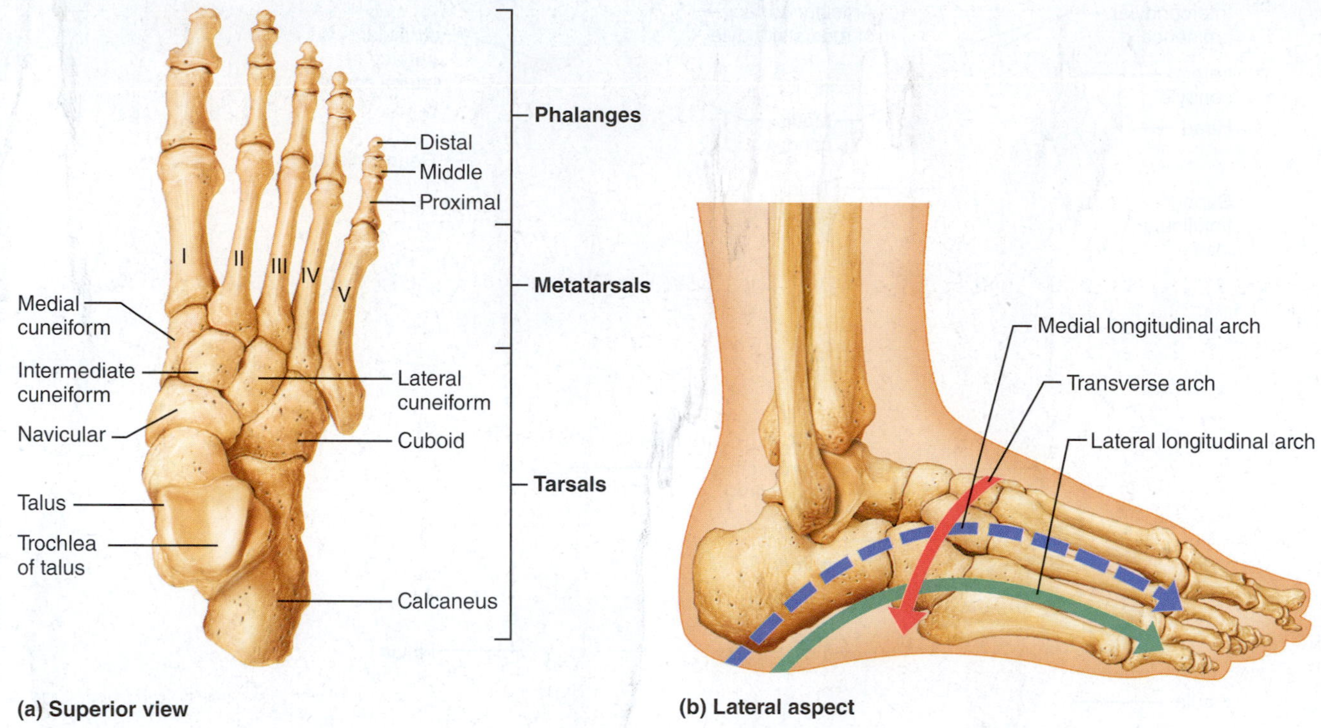

Figure 9.9 Bones of the right foot. Arches of the right foot are diagrammed in (b).

The bones in the foot are arranged to produce three strong arches—two longitudinal arches (medial and lateral) and one transverse arch (Figure 9.9b). Ligaments, binding the foot bones together, and tendons of the foot muscles hold the bones firmly in the arched position but still allow a certain degree of give. Weakened arches are referred to as *fallen arches* or *flat feet.*

ACTIVITY 5

Palpating the Surface Anatomy of the Pelvic Girdle and Lower Limb

Locate and palpate the following bone markings on yourself and/or your lab partner.

Place a check mark in the boxes as you locate the bone markings. Ask your instructor for help with any markings that you are unable to locate.

☐ Iliac crest and anterior superior iliac spine: Rest your hands on your hips—they will be overlying the iliac crests. Trace the crest as far posteriorly as you can, and then follow it anteriorly to the anterior superior iliac spine. This latter bone marking is easily felt in almost everyone, and is clearly visible through the skin (and perhaps the clothing) of very slim people. (The posterior superior iliac spine is much less obvious and is usually indicated only by a dimple in the overlying skin. Check it out in the mirror tonight.)

☐ Greater trochanter of the femur: This is easier to locate in females than in males because of the wider female pelvis; also it is more likely to be clothed by bulky muscles in males. Try to locate it on yourself as the most lateral point of the proximal femur. It typically lies about 6 to 8 inches below the iliac crest.

☐ Patella and tibial tuberosity: Feel your kneecap, and palpate the ligaments attached to its borders. Follow the inferior patellar ligament to the tibial tuberosity.

☐ Medial and lateral condyles of the femur and tibia: As you move from the patella inferiorly on the medial (and then the lateral) knee surface, you will feel first the femoral and then the tibial condyle.

☐ Medial malleolus: Feel the medial protrusion of your ankle, the medial malleolus of the distal tibia.

☐ Lateral malleolus: Feel the bulge of the lateral aspect of your ankle, the lateral malleolus of the fibula.

☐ Calcaneus: Attempt to follow the extent of your calcaneus, or heel bone. ■

ACTIVITY 6

Constructing a Skeleton

1. When you finish examining yourself and the disarticulated bones of the appendicular skeleton, work with your lab partner to arrange the disarticulated bones on the laboratory bench in their proper relative positions to form an entire skeleton. Careful observation of bone markings should help you distinguish between right and left members of bone pairs.

2. When you believe that you have accomplished this task correctly, ask the instructor to check your arrangement to ensure that it is correct. If it is not, go to the articulated skeleton and check your bone arrangements. Also review the descriptions of the bone markings as necessary to correct your bone arrangement. ■

Name ______________________________

Lab Time/Date ______________________

EXERCISE 9

REVIEW SHEET

The Appendicular Skeleton

Bones of the Pectoral Girdle and Upper Limb

1. Match the bone names or markings in column B with the descriptions in column A. The items in column B may be used more than once.

Column A

_______________ 1. raised area on lateral surface of humerus to which deltoid muscle attaches

_______________ 2. arm bone

_______________, _______________ 3. bones of the shoulder girdle

_______________, _______________ 4. forearm bones

_______________ 5. scapular region to which the clavicle connects

_______________ 6. shoulder girdle bone that does not attach to the axial skeleton

_______________ 7. shoulder girdle bone that transmits forces from the upper limb to the bony thorax

_______________ 8. depression in the scapula that articulates with the humerus

_______________ 9. process above the glenoid cavity that permits muscle attachment

_______________ 10. the "collarbone"

_______________ 11. distal condyle of the humerus that articulates with the ulna

_______________ 12. medial bone of forearm in anatomical position

_______________ 13. rounded knob on the humerus; adjoins the radius

_______________ 14. anterior depression, superior to the trochlea, which receives part of the ulna when the forearm is flexed

_______________ 15. forearm bone involved in formation of the elbow joint

_______________ 16. wrist bones

_______________ 17. finger bones

_______________ 18. heads of these bones form the knuckles

_______________, _______________ 19. bones that articulate with the clavicle

Column B

a. acromion
b. capitulum
c. carpals
d. clavicle
e. coracoid process
f. coronoid fossa
g. deltoid tuberosity
h. glenoid cavity
i. humerus
j. metacarpals
k. olecranon
l. olecranon fossa
m. phalanges
n. radial styloid process
o. radial tuberosity
p. radius
q. scapula
r. sternum
s. trochlea
t. ulna

2. How is the arm held clear of the widest dimension of the thoracic cage? ______________________________

____

____

3. What is the total number of phalanges in the hand? ______________

4. What is the total number of carpals in the wrist? ______________

Name the carpals (medial to lateral) in the proximal row. ______________________________

In the distal row, they are (medial to lateral): ______________________________

5. Use letters from the key to identify the anatomical landmarks and regions of the scapula.

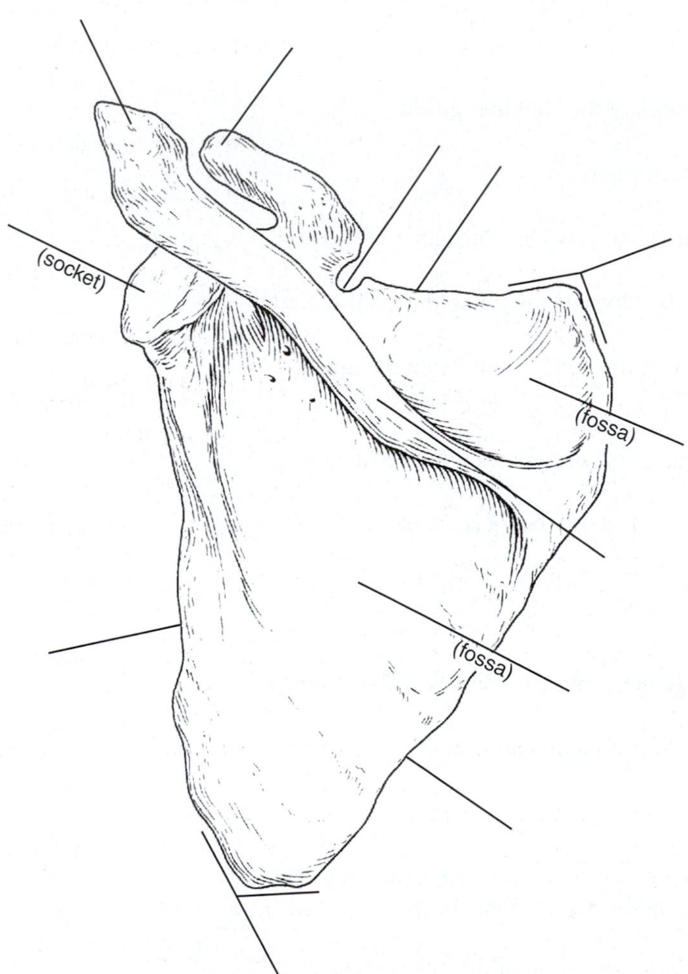

Key:

a. acromion

b. coracoid process

c. glenoid cavity

d. inferior angle

e. infraspinous fossa

f. lateral border

g. medial border

h. spine

i. superior angle

j. superior border

k. suprascapular notch

l. supraspinous fossa

6. Match the terms in the key with the appropriate leader lines on the drawings of the humerus and the radius and ulna. Also decide whether the bones shown are right or left bones and whether the view shown is an anterior or a posterior view.

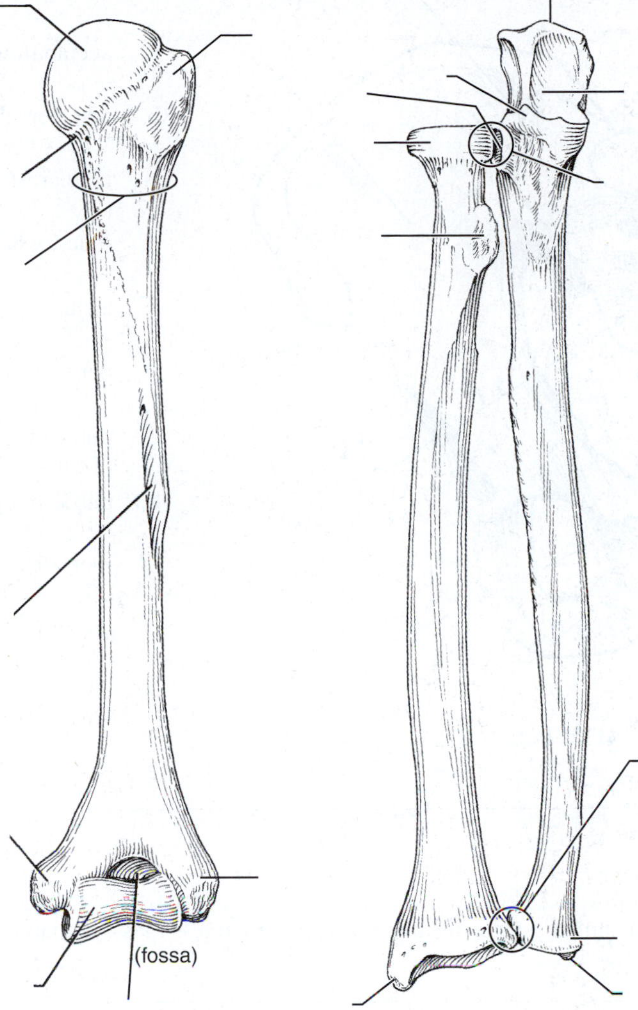

Key:

a. anatomical neck
b. coronoid process
c. distal radioulnar joint
d. greater tubercle
e. head of humerus
f. head of radius
g. head of ulna
h. lateral epicondyle
i. medial epicondyle
j. olecranon
k. olecranon fossa
l. proximal radioulnar joint
m. radial groove
n. radial notch
o. radial styloid process
p. radial tuberosity
q. surgical neck
r. trochlea
s. trochlear notch
t. ulnar styloid process

Circle the correct term for each pair in parentheses:

The humerus is the (right/left) bone in (an anterior/a posterior) view. The radius and ulna are (right/left) bones in (an anterior/a posterior) view.

Bones of the Pelvic Girdle and Lower Limb

7. Compare the pectoral and pelvic girdles by choosing appropriate descriptive terms from the key.

Key:
a. flexibility most important
b. massive
c. lightweight
d. insecure axial and limb attachments
e. secure axial and limb attachments
f. weight-bearing most important

Pectoral: ____________, ____________, ____________ Pelvic: ____________, ____________, ____________

8. What organs are protected, at least in part, by the pelvic girdle? ______________________________

____

9. Distinguish the true pelvis from the false pelvis.

____

10. Use letters from the key to identify the bone markings on this illustration of an articulated pelvis. Make an educated guess as to whether the illustration shows a male or female pelvis, and provide two reasons for your decision.

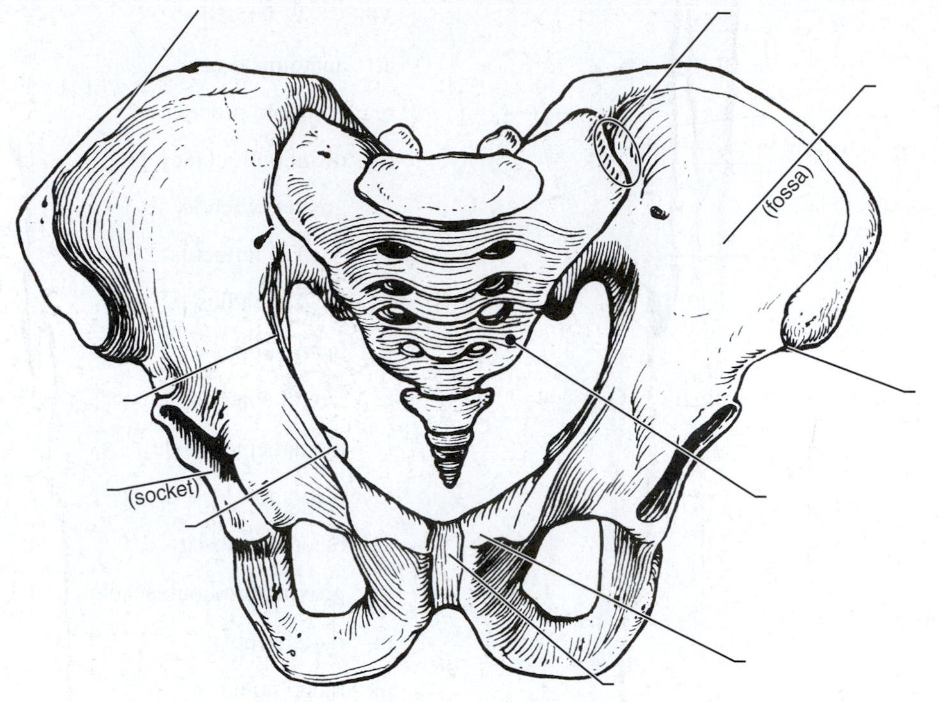

Key:

a. acetabulum

b. anterior superior iliac spine

c. iliac crest

d. iliac fossa

e. ischial spine

f. pelvic brim

g. pubic crest

h. pubic symphysis

i. sacroiliac joint

j. sacrum

This is a ________________ (female/male) pelvis because:

____

____

11. Deduce why the pelvic bones of a four-legged animal such as the cat or pig are much less massive than those of the human.

____

____

12. A person instinctively curls over his abdominal area in times of danger. Why? ________________

____

13. For what anatomical reason do many women appear to be slightly knock-kneed? ________________

____

How might this anatomical arrangement contribute to knee injuries? ________________

____

14. What structural changes result in *fallen arches*? ________________

____

15. Match the bone names and markings in column B with the descriptions in column A. The items in column B may be used more than once.

Column A	Column B
______________, ______________, and	a. acetabulum
______________ 1. fuse to form the coxal bone	b. calcaneus
______________ 2. inferoposterior "bone" of the coxal bone	c. femur
______________ 3. point where the coxal bones join anteriorly	d. fibula
______________ 4. superiormost margin of the coxal bone	e. gluteal tuberosity
______________ 5. deep socket in the coxal bone that receives the head of the thigh bone	f. greater and lesser trochanters
	g. greater sciatic notch
______________ 6. joint between axial skeleton and pelvic girdle	h. iliac crest
______________ 7. longest, strongest bone in body	i. ilium
______________ 8. thin lateral leg bone	j. ischial tuberosity
______________ 9. heavy medial leg bone	k. ischium
______________, ______________ 10. bones forming knee joint	l. lateral malleolus
______________ 11. point where the patellar ligament attaches	m. lesser sciatic notch
______________ 12. kneecap	n. linea aspera
______________ 13. shinbone	o. medial malleolus
______________ 14. medial ankle projection	p. metatarsals
______________ 15. lateral ankle projection	q. obturator foramen
______________ 16. largest tarsal bone	r. patella
______________ 17. ankle bones	s. pubic symphysis
______________ 18. bones forming the instep of the foot	t. pubis
______________ 19. opening in hip bone formed by the pubic and ischial rami	u. sacroiliac joint
______________ and ______________ 20. sites of muscle attachment on the proximal femur	v. talus
______________ 21. tarsal bone that "sits" on the calcaneus	w. tarsals
______________ 22. weight-bearing bone of the leg	x. tibia
______________ 23. tarsal bone that articulates with the tibia	y. tibial tuberosity

16. Match the terms in the key with the appropriate leader lines on the drawings of the femur and the tibia and fibula (some terms may be used more than once). Also decide whether these bones are right or left bones and whether it is their anterior or posterior view that is illustrated.

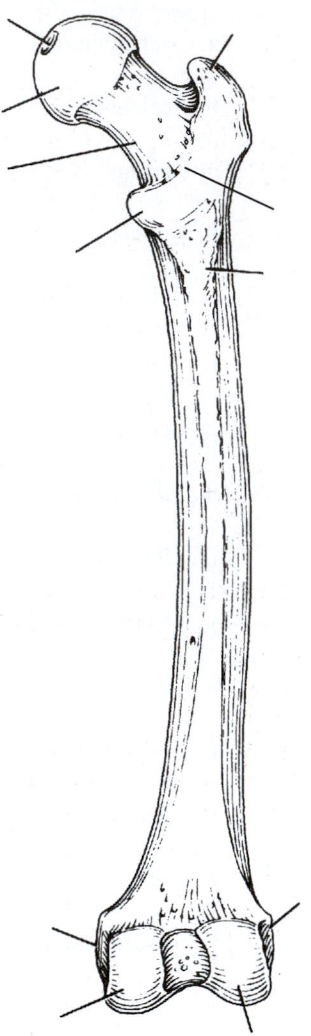

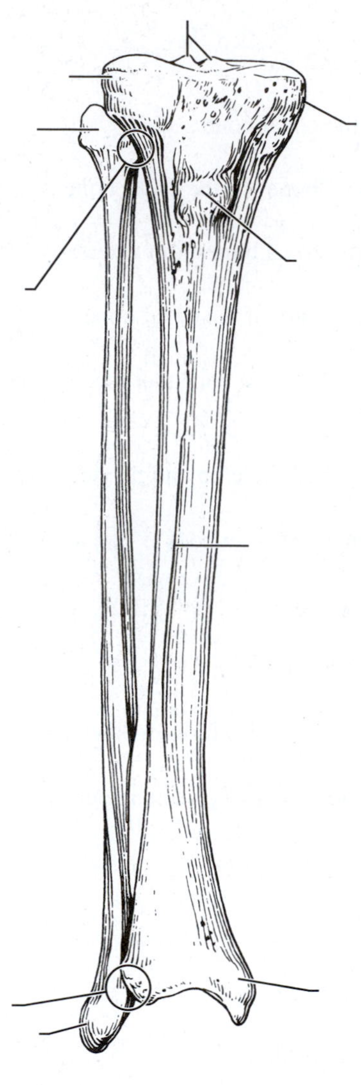

Key:

a. fovea capitis

b. gluteal tuberosity

c. greater trochanter

d. head of femur

e. head of fibula

f. inferior tibiofibular joint

g. intercondylar eminence

h. intertrochanteric crest

i. lateral condyle

j. lateral epicondyle

k. lateral malleolus

l. lesser trochanter

m. medial condyle

n. medial epicondyle

o. medial malleolus

p. neck of femur

q. superior tibiofibular joint

r. tibial anterior border

s. tibial tuberosity

Circle the corrrect term for each pair in parentheses:

The femur is a (right/left) bone in (an anterior/a posterior) view. The tibia and fibula are (right/left) bones in (an anterior/a posterior) view.

Summary of the Skeleton

17. Identify all indicated bones (or groups of bones) in the diagram of the articulated skeleton (page 151).

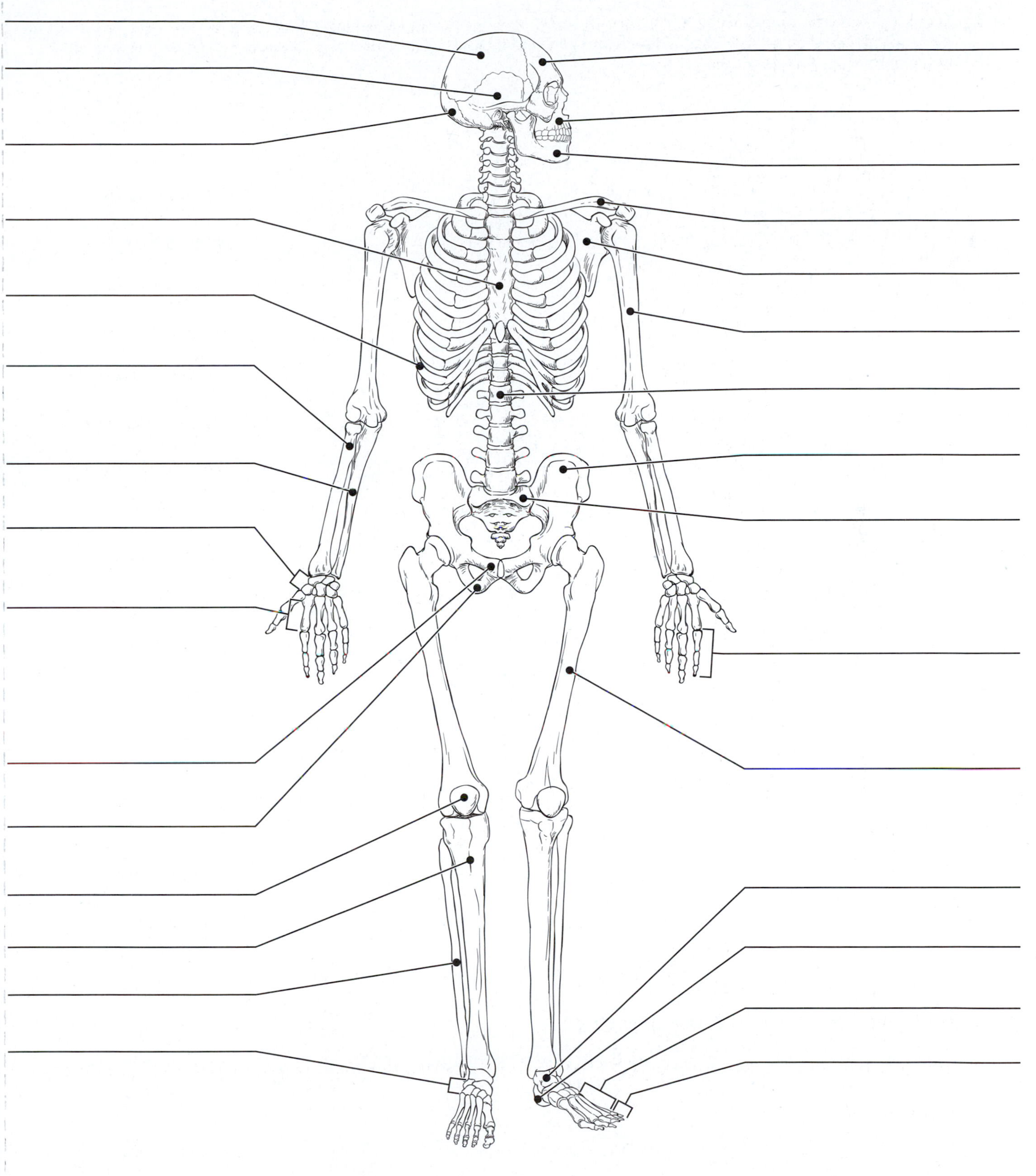

10

EXERCISE

Articulations and Body Movements

MATERIALS

- ☐ Skull
- ☐ Articulated skeleton
- ☐ Anatomical chart of joint types (if available)
- ☐ Diarthrotic joint (fresh or preserved), preferably a beef knee joint sectioned sagittally; alternatively, pig's feet with phalanges sectioned frontally
- ☐ Disposable gloves
- ☐ Water balloons and clamps
- ☐ Functional models of hip, knee, and shoulder joints (if available)
- ☐ X-ray images of normal and arthritic joints (if available)

OBJECTIVES

1. Name and describe the three functional categories of joints.
2. Name and describe the three structural categories of joints, and discuss how their structure is related to mobility.
3. Identify the types of synovial joints; indicate whether they are nonaxial, uniaxial, biaxial, or multiaxial; and describe the movements each type makes.
4. Define *origin* and *insertion* of muscles.
5. Demonstrate or describe the various body movements.
6. Compare and contrast the structure and function of the shoulder and hip joints.
7. Describe the structure and function of the knee and temporomandibular joints.

PRE-LAB QUIZ

1. Name one of the two functions of an articulation, or joint. ______________________________
2. The functional classification of joints is based on:
   a. a joint cavity
   b. amount of connective tissue
   c. amount of movement allowed by the joint
3. Structural classification of joints includes fibrous, cartilaginous, and ______________, which have a fluid-filled cavity between articulating bones.
4. Circle the correct underlined term. Sutures, whose irregular edges of bone are joined by short fibers of connective tissue, are an example of fibrous / cartilaginous joints.
5. Circle True or False. All synovial joints are diarthroses, or freely movable joints.
6. Circle the correct underlined term. Every muscle of the body is attached to a bone or other connective tissue structure at two points. The origin / insertion is the more movable attachment.
7. The hip joint is an example of a ______________ synovial joint.
   a. ball-and-socket c. pivot
   b. hinge d. plane
8. Movement of a limb away from the midline or median plane of the body in the frontal plane is known as:
   a. abduction c. extension
   b. eversion d. rotation
9. Circle the correct underlined term. This type of movement is common in ball-and-socket joints and can be described as the movement of a bone around its longitudinal axis. It is rotation / flexion.
10. Circle True or False. The knee joint is the most freely movable joint in the body.

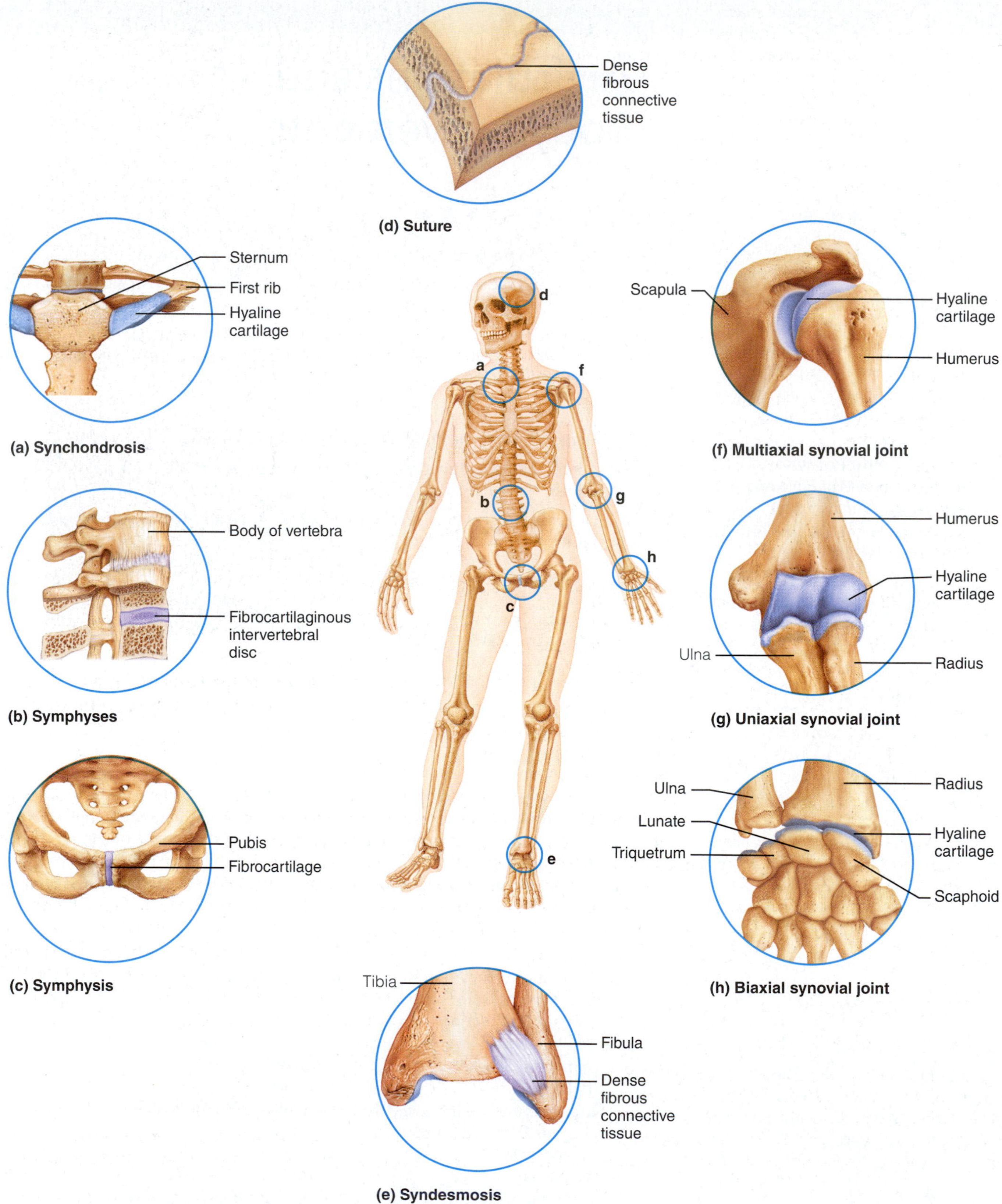

Figure 10.1 Types of joints. Joints to the left of the skeleton are cartilaginous joints; joints above and below the skeleton are fibrous joints; joints to the right of the skeleton are synovial joints. **(a)** Joint between costal cartilage of rib 1 and the sternum. **(b)** Intervertebral discs of fibrocartilage connecting adjacent vertebrae. **(c)** Fibrocartilaginous pubic symphysis connecting the pubic bones anteriorly. **(d)** Dense fibrous connective tissue connecting interlocking skull bones. **(e)** Ligament connecting the inferior ends of the tibia and fibula. **(f)** Shoulder joint. **(g)** Elbow joint. **(h)** Wrist joint.

With rare exceptions, every bone in the body is connected to, or forms a joint with, at least one other bone. **Articulations,** or joints, perform two functions for the body. They (1) hold the bones together and (2) allow the rigid skeletal system some flexibility so that gross body movements can occur.

Joints may be classified structurally or functionally. The structural classification is based on the presence of connective tissue fiber, cartilage, or a joint cavity between the articulating bones. Structurally, there are *fibrous, cartilaginous,* and *synovial joints.*

The functional classification focuses on the amount of movement allowed at the joint. On this basis, there are **synarthroses,** or immovable joints; **amphiarthroses,** or slightly movable joints; and **diarthroses,** or freely movable joints. Freely movable joints predominate in the limbs, whereas immovable and slightly movable joints are largely restricted to the axial skeleton, where firm bony attachments and protection of enclosed organs are a priority.

As a general rule, fibrous joints are immovable, and synovial joints are freely movable. Cartilaginous joints offer both rigid and slightly movable examples. Because the structural categories are more clear-cut, we will use the structural classification here and indicate functional properties as appropriate.

Fibrous Joints

In **fibrous joints,** the bones are joined by fibrous tissue. No joint cavity is present. The amount of movement allowed depends on the length of the fibers uniting the bones. Although some fibrous joints are slightly movable, most are synarthrotic and permit virtually no movement.

The two major types of fibrous joints are sutures and syndesmoses. In **sutures** (**Figure 10.1d**), the irregular edges of the bones interlock and are united by very short connective tissue fibers. Most joints of the skull are sutures. In **syndesmoses,** the articulating bones are connected by short ligaments of dense fibrous tissue; the bones do not interlock. The joint at the distal end of the tibia and fibula is an example of a syndesmosis (Figure 10.1e). Although this syndesmosis allows some give, it is classified functionally as a synarthrosis. Not illustrated here is a **gomphosis,** in which a tooth is secured in a bony socket by the periodontal ligament (see Figure 27.12).

ACTIVITY 1

Identifying Fibrous Joints

Examine a human skull again. Notice that adjacent bone surfaces do not actually touch but are separated by fibrous connective tissue. Also examine a skeleton and anatomical chart of joint types and the table of joints (**Table 10.1**, pages 164–165) for examples of fibrous joints. ■

Cartilaginous Joints

In **cartilaginous joints,** the articulating bone ends are connected by a plate or pad of cartilage. No joint cavity is present. The two major types of cartilaginous joints are synchondroses and symphyses. Although there is variation, most cartilaginous joints are *slightly movable* (amphiarthrotic) functionally. In **symphyses** (*symphysis* = a growing together), the bones are connected by a broad, flat disc of fibrocartilage. The intervertebral joints between adjacent vertebral bodies and the pubic symphysis of the pelvis are symphyses (see Figure 10.1b and c). In **synchondroses,** the bony portions are united by hyaline cartilage. The articulation of the costal cartilage of the first rib with the sternum (Figure 10.1a) is a synchondrosis, but perhaps the best examples of synchondroses are the epiphyseal plates in the long bones of growing children (**Figure 10.2**). The epiphyseal plates are flexible during childhood, but eventually they are totally ossified.

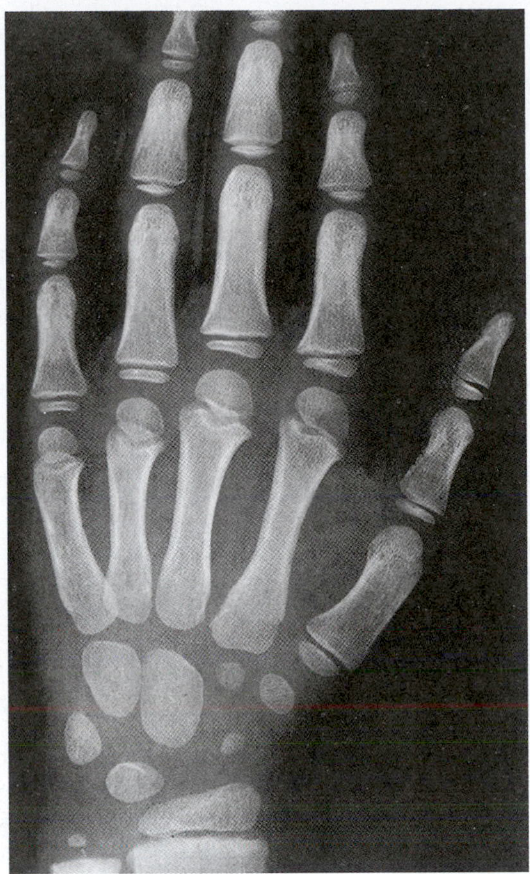

Figure 10.2 X-ray image of the hand of a child. Notice the cartilaginous epiphyseal plates, examples of temporary synchondroses.

ACTIVITY 2

Identifying Cartilaginous Joints

Identify the cartilaginous joints on a human skeleton, in the table of joints (Table 10.1), and on an anatomical chart of joint types. ■

Synovial Joints

Synovial joints are joints in which the articulating bone ends are separated by a joint cavity containing synovial fluid (see Figure 10.1f–h). All synovial joints are diarthroses, or freely movable joints. Their mobility varies, however; some synovial joints can move in only one plane, and others can move in several directions (multiaxial movement). Most joints in the body are synovial joints.

All synovial joints have the following structural characteristics **(Figure 10.3)**:

- The joint surfaces are enclosed by a two-layered *articular capsule* (a sleeve of connective tissue), creating a joint cavity.
- The inner layer is a smooth connective tissue membrane, called the *synovial membrane,* which produces a lubricating fluid (synovial fluid) that reduces friction. The outer layer, or *fibrous layer,* is dense irregular connective tissue.
- *Articular* (hyaline) *cartilage* covers the surfaces of the bones forming the joint.
- The articular capsule is typically reinforced with ligaments and may contain *bursae* (fluid-filled sacs that reduce friction where tendons cross bone).
- Fibrocartilage pads *(articular discs)* may be present within the capsule.

ACTIVITY 3

Examining Synovial Joint Structure

Examine a beef or pig joint to identify the general structural features of synovial joints in the previous list.

! If the joint is fresh and you will be handling it, don disposable gloves before beginning your observations.

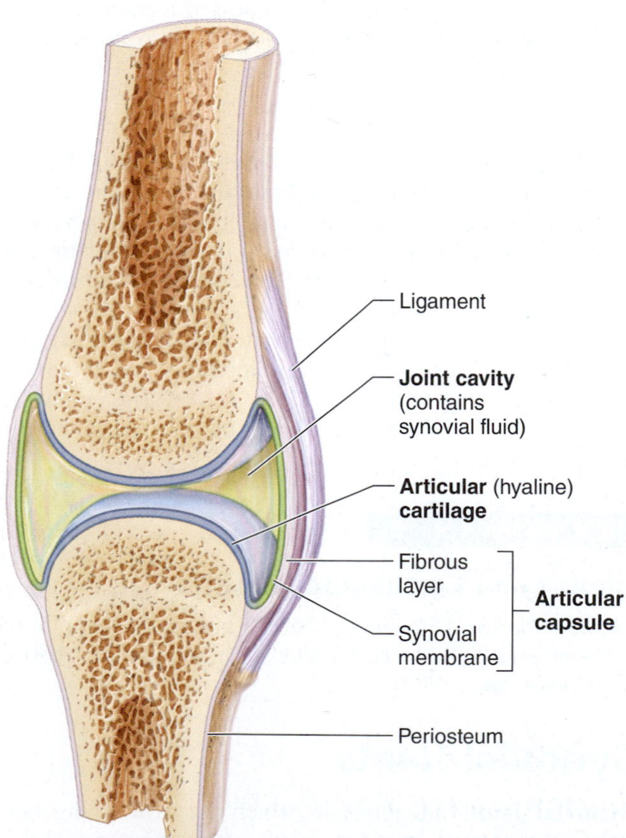

Figure 10.3 Major structural features of a synovial joint.

ACTIVITY 4

Demonstrating the Importance of Friction-Reducing Structures

1. Obtain a small water balloon and clamp. Partially fill the balloon with water (it should still be flaccid), and clamp it closed.

2. Position the balloon atop one of your fists, and press down on its top surface with the other fist. Push on the balloon until your two fists touch, and move your fists back and forth over one another. Assess the amount of friction generated.

3. Unclamp the balloon, and add more water. The goal is to get just enough water in the balloon so that your fists cannot come into contact with one another, but instead remain separated by a thin water layer when pressure is applied to the balloon.

4. Repeat the movements in step 2 to assess the amount of friction generated.

How does the presence of a sac containing fluid influence the amount of friction generated?

____

What anatomical structure(s) does the water-containing balloon mimic?

____

What anatomical structures might be represented by your fists?

____

Types of Synovial Joints

The many types of synovial joints can be subdivided according to their function and structure. The shape of the articular surfaces determines the types of movements that can occur at the joint, and it also determines the structural classification of the joints **(Figure 10.4)**:

- **Plane (nonaxial):** Articulating surfaces are flat or slightly curved. These surfaces allow only gliding movements as the surfaces slide past one another. Examples include intercarpal joints, intertarsal joints, and joints between vertebral articular surfaces.
- **Hinge (uniaxial):** The rounded or cylindrical process of one bone fits into the concave surface of another bone, allowing movement in one plane, usually flexion and extension. Examples include the elbow and interphalangeal joints.
- **Pivot (uniaxial):** The rounded surface of one bone articulates with a shallow depression or foramen in another bone, permitting rotational movement in one plane. Examples include the proximal radioulnar joint and the atlantoaxial joint (between the atlas and axis—C_1 and C_2).
- **Condylar (biaxial):** The oval condyle of one bone fits into an ellipsoidal depression in another bone to allow movement in two planes, usually flexion/extension and abduction/adduction. Examples include the wrist and metacarpophalangeal (knuckle) joints.

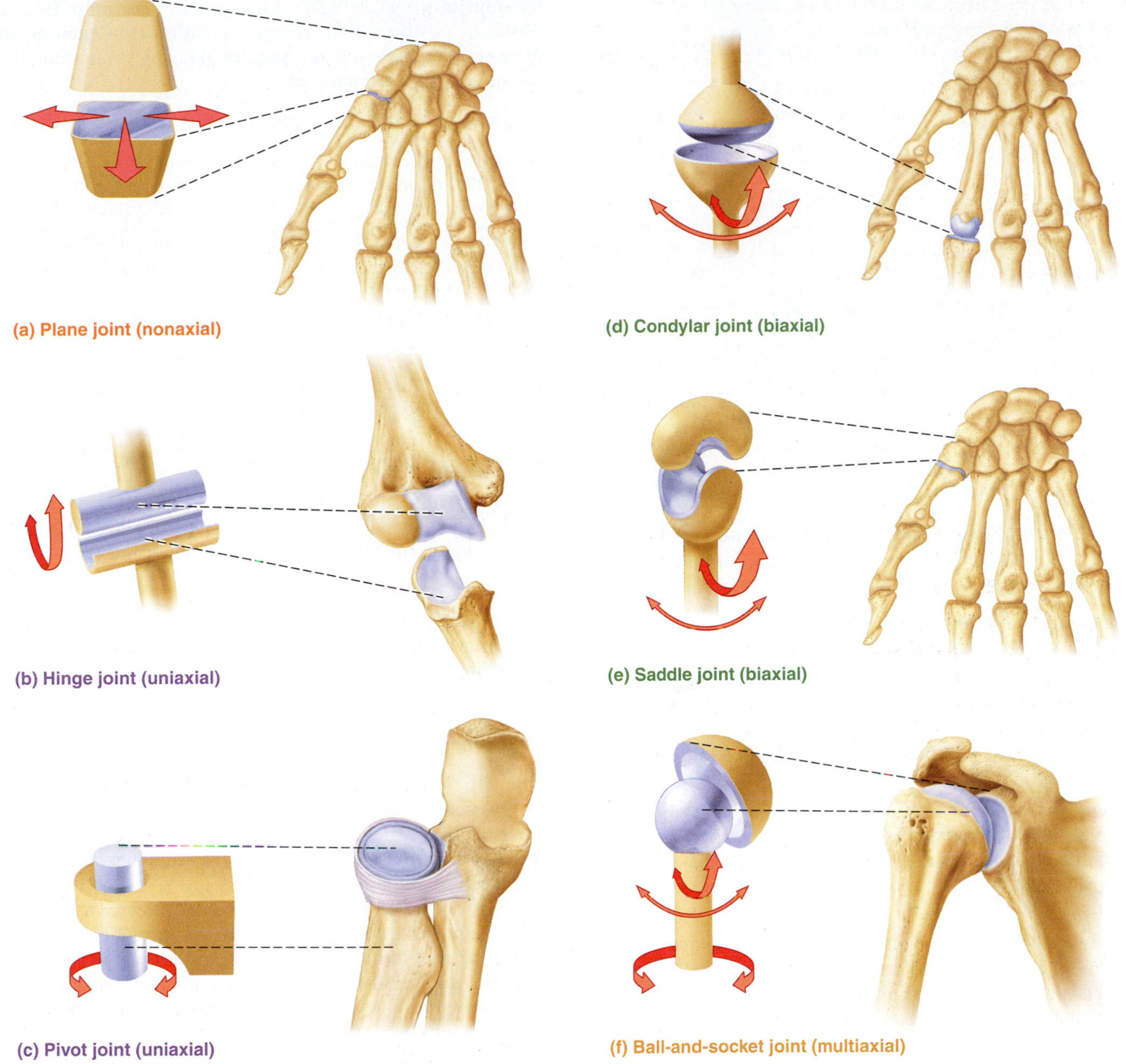

(a) Plane joint (nonaxial)

(b) Hinge joint (uniaxial)

(c) Pivot joint (uniaxial)

(d) Condylar joint (biaxial)

(e) Saddle joint (biaxial)

(f) Ball-and-socket joint (multiaxial)

Figure 10.4 Types of synovial joints. Dashed lines indicate the articulating bones.

- **Saddle (biaxial):** Articulating surfaces are saddle-shaped; one surface is convex, and the other is concave. This type of joint permits movement in two planes, flexion/extension and abduction/adduction. Examples include the carpometacarpal joints of the thumbs.
- **Ball-and-socket (multiaxial):** The ball-shaped head of one bone fits into a cuplike depression of another bone. These joints permit flexion/extension, abduction/adduction, and rotation, which combine to allow movement in many planes. Examples include the shoulder and hip joints.

Movements Allowed by Synovial Joints

Every muscle of the body is attached to bone (or other connective tissue structures) at two points—the **origin** (the stationary, immovable, or less movable attachment) and the **insertion** (the movable attachment). Body movement occurs when muscles contract across diarthrotic synovial joints **(Figure 10.5)**. When the muscle contracts and its fibers shorten, the insertion moves toward the origin. The type of movement depends on the construction of the joint and on the placement of the muscle relative to the joint. The most common types of body movements are described below and illustrated in the figure **(Figure 10.6)**.

ACTIVITY 5

Demonstrating Movements of Synovial Joints

Attempt to demonstrate each movement as you read through the following material:

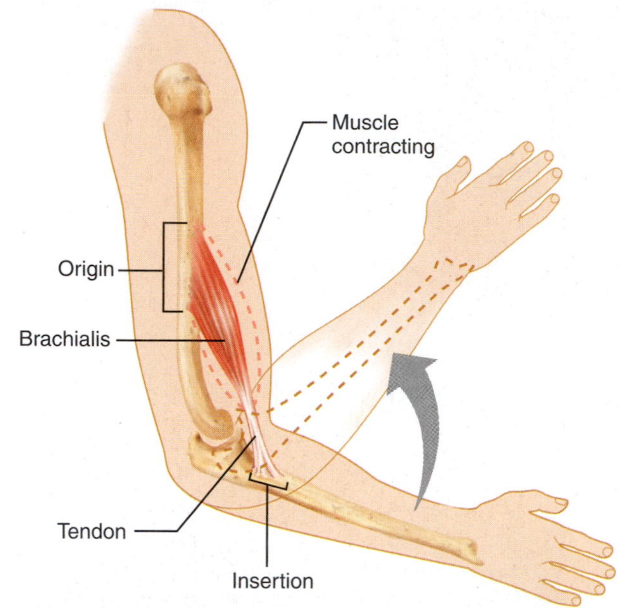

Figure 10.5 Muscle attachments (origin and insertion). When a skeletal muscle contracts, its insertion moves toward its origin.

Flexion (Figure 10.6a, b, and c): A movement, generally in the sagittal plane, that decreases the angle of the joint and reduces the distance between the two bones. Flexion is typical of hinge joints (bending the knee or elbow) but is also common at ball-and-socket joints (bending forward at the hip).

Extension (Figure 10.6a, b, and c): A movement that increases the angle of a joint and the distance between two bones or parts of the body (straightening the knee or elbow); the opposite of flexion. If extension proceeds beyond anatomical position (bends the trunk backward), it is called *hyperextension.*

Abduction (Figure 10.6d): Movement of a limb away from the midline or median plane of the body, generally on the frontal plane, or the fanning movement of fingers or toes when they are spread apart.

Adduction (Figure 10.6d): Movement of a limb toward the midline of the body or drawing the fingers or toes together; the opposite of abduction.

Rotation (Figure 10.6e): Movement of a bone around its longitudinal axis without lateral or medial displacement. Rotation, a common movement of ball-and-socket joints, also describes the movement of the atlas around the dens of the axis.

Circumduction (Figure 10.6d): A combination of flexion, extension, abduction, and adduction, commonly observed in ball-and-socket joints such as the shoulder. The proximal end of the limb remains stationary, and the distal end moves in a circle. The limb as a whole outlines a cone. Condylar and saddle joints also allow circumduction.

Pronation (Figure 10.6f): Movement of the palm of the hand from an anterior or upward-facing position to a posterior or downward-facing position. The distal end of the radius moves across the ulna so that the bones form an X.

Supination (Figure 10.6f): Movement of the palm from a posterior position to an anterior position (the anatomical position); the opposite of pronation. During supination, the radius and ulna are parallel.

The last four terms refer to movements of the foot:

Dorsiflexion (Figure 10.6g): A movement of the ankle joint that lifts the foot so that its superior surface approaches the shin.

Plantar flexion (Figure 10.6g): A movement of the ankle joint in which the foot is flexed downward as in standing on one's toes or pointing the toes.

Inversion (Figure 10.6h): A movement that turns the sole of the foot medially.

Eversion (Figure 10.6h): A movement that turns the sole of the foot laterally; the opposite of inversion. ■

Selected Synovial Joints

Now you will compare and contrast the structure of the hip and knee joints and investigate the structure and movements of the temporomandibular joint and shoulder joint.

The Hip and Knee Joints

Both the hip and knee joints are large weight-bearing joints of the lower limb, but they differ substantially in their security. Read through the brief descriptive material below, and look at the "Selected Synovial Joints" section of the Review Sheet for the questions that pertain to these joints before beginning your comparison.

The Hip Joint The hip joint is a ball-and-socket joint, so movements can occur in all possible planes. However, its movements are definitely limited by its deep socket and strong reinforcing ligaments, the two factors that account for its exceptional stability **(Figure 10.7)**.

The deeply cupped acetabulum that receives the head of the femur is enhanced by a circular rim of fibrocartilage called the **acetabular labrum.** Because the diameter of the labrum is smaller than that of the femur's head, dislocations of the hip are rare. A short ligament, the **ligament of the head of the femur** *(ligamentum teres),* runs from the pitlike **fovea capitis** on the femur head to the acetabulum, where it helps to secure the femur. Several strong ligaments, including the **iliofemoral** and **pubofemoral** anteriorly and the **ischiofemoral** that spirals posteriorly (not shown), are arranged so that they "screw" the femur head into the socket when a person stands upright.

ACTIVITY 6

Demonstrating Actions at the Hip Joint

If a functional hip joint model is available, identify the joint parts, and manipulate it to demonstrate the following movements that can occur at this joint: flexion, extension, abduction, and medial and lateral rotation.

Reread the information on what movements the associated ligaments restrict, and verify that information during your joint manipulations. ■

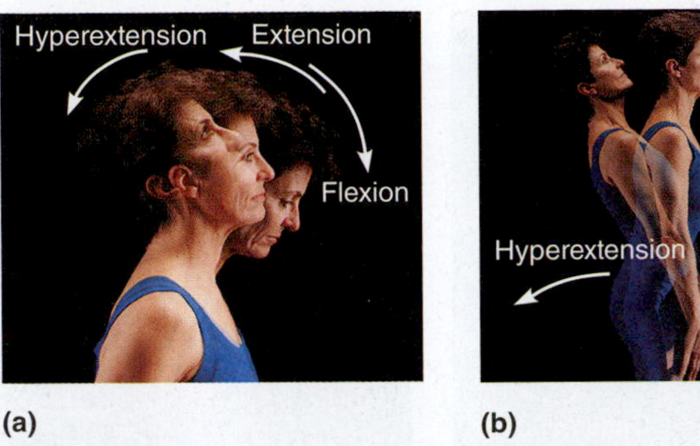

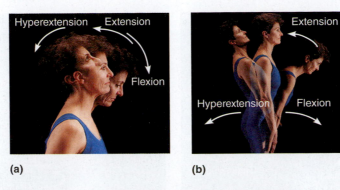

(a)

(b)

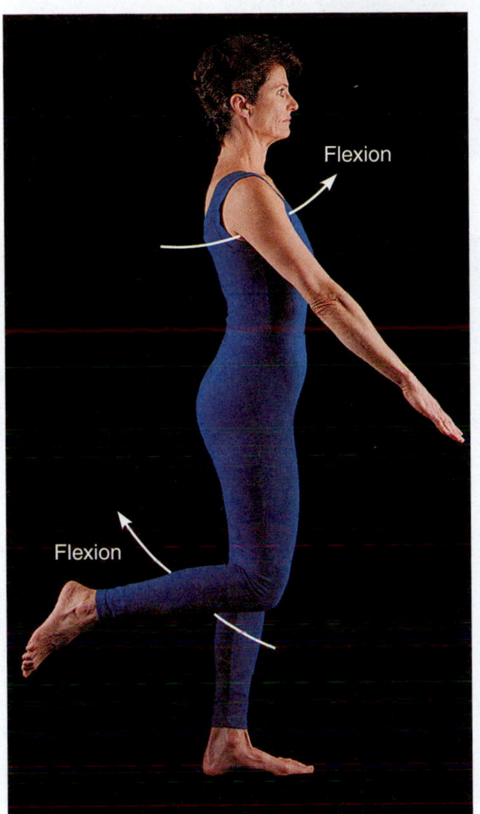

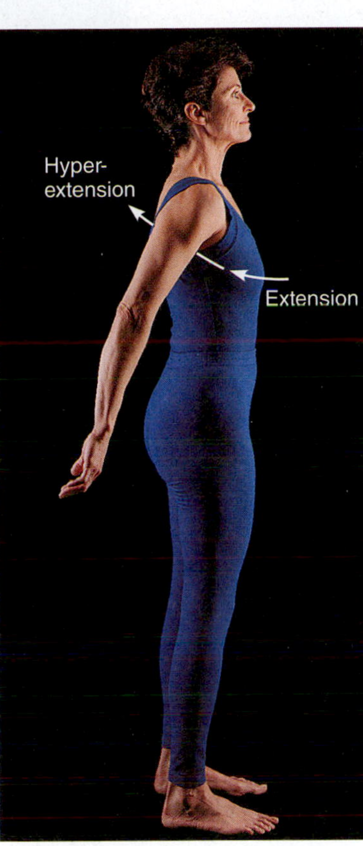

(c)

Figure 10.6 Movements occurring at synovial joints of the body. (a) Flexion, extension, and hyperextension of the neck. **(b)** Flexion, extension, and hyperextension of the vertebral column. **(c)** Flexion and extension of the knee and shoulder, and hyperextension of the shoulder.

The Knee Joint The knee is the largest and most complex joint in the body. Three joints in one **(Figure 10.8)**, it allows extension, flexion, and a little rotation. The **tibiofemoral joint,** actually a duplex joint between the femoral condyles above and the **menisci** (semilunar cartilages) of the tibia below, is functionally a hinge joint, a very unstable one made slightly more secure by the menisci (Figure 10.8b and d). Some rotation occurs when the knee is partly flexed, but during extension, the menisci and ligaments counteract rotation and side-to-side movements. The other joint is the **femoropatellar joint,** the intermediate joint anteriorly (Figure 10.8a and c).

The knee is unique in that it is only partly enclosed by an articular capsule. Anteriorly, where the capsule is absent, are three broad ligaments, the **patellar ligament** and the **medial** and **lateral patellar retinacula,** which run from the patella to the tibia below and merge with the capsule on either side.

Capsular ligaments, including the **fibular** and **tibial collateral ligaments** (which prevent rotation during extension) and the **oblique popliteal** and **arcuate popliteal ligaments,** are crucial in reinforcing the knee. The knees have a built-in locking device that must be "unlocked" by the popliteus muscles (Figure 10.8e) before the knees can be flexed again. The **cruciate ligaments** are intracapsular ligaments that cross (*cruci* = cross) in the notch between the femoral condyles. They prevent anterior-posterior displacement of the joint and overflexion and hyperextension of the joint.

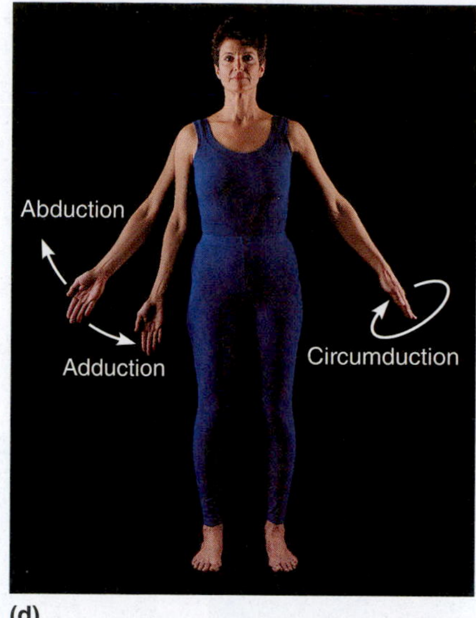

(d)

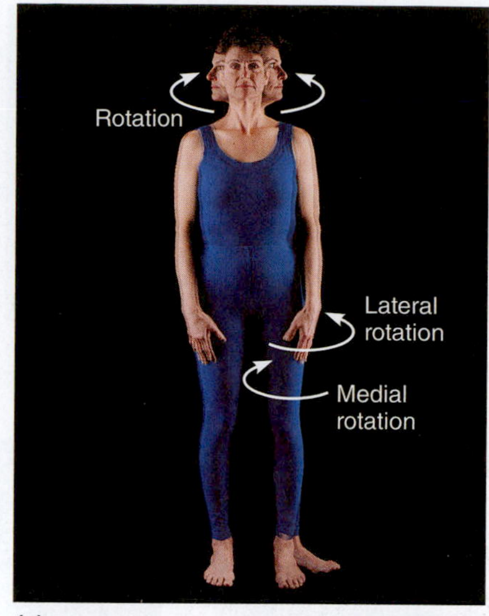

(e)

(f)

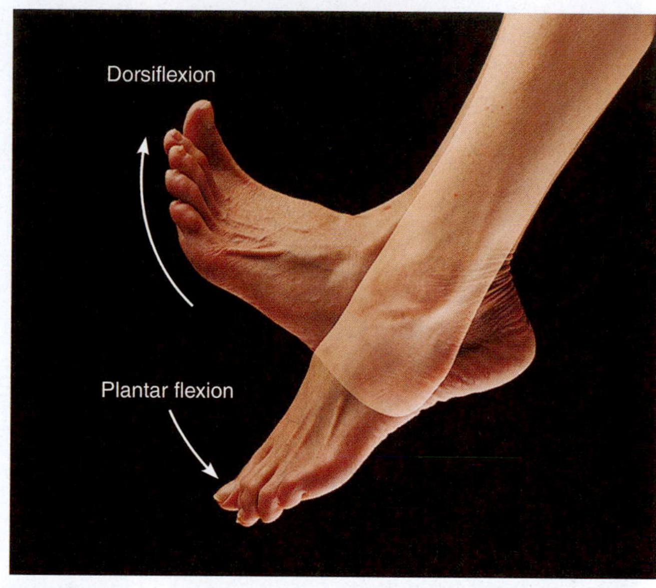

(g)

(h)

Figure 10.6 ***(continued)*** **(d)** Abduction, adduction, and circumduction of the upper limb. **(e)** Rotation of the head and lower limb. **(f)** Pronation and supination of the forearm. **(g)** Dorsiflexion and plantar flexion of the foot. **(h)** Inversion and eversion of the foot.

ACTIVITY 7

Demonstrating Actions at the Knee Joint

If a functional model of a knee joint is available, identify the joint parts, and manipulate it to illustrate the following movements: flexion, extension, and medial and lateral rotation.

Reread the information on what movements the various associated ligaments restrict, and verify that information during your joint manipulations. ■

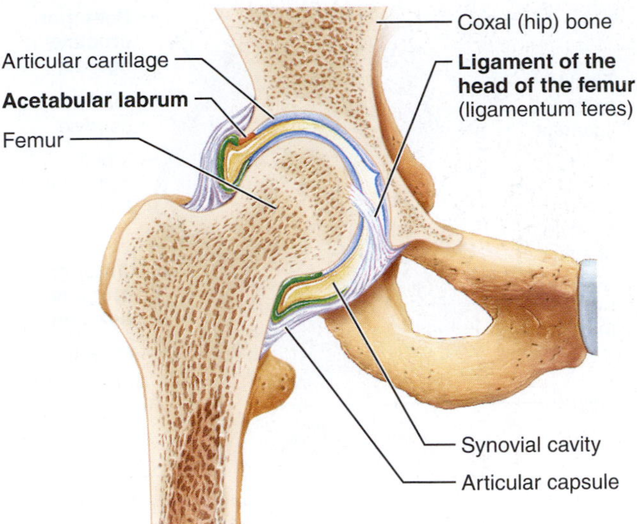

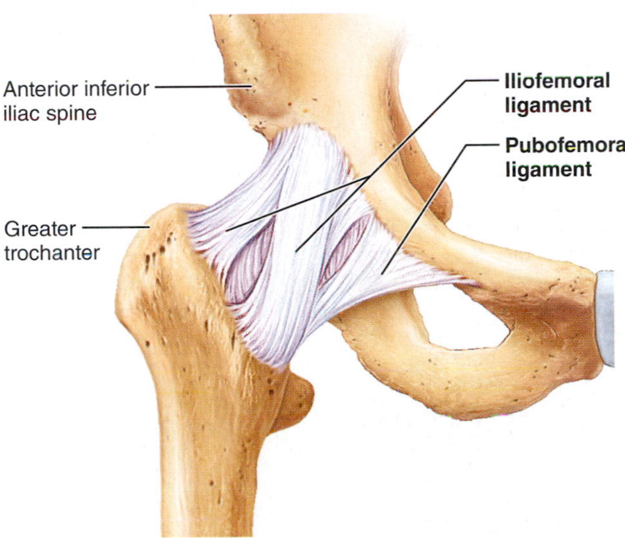

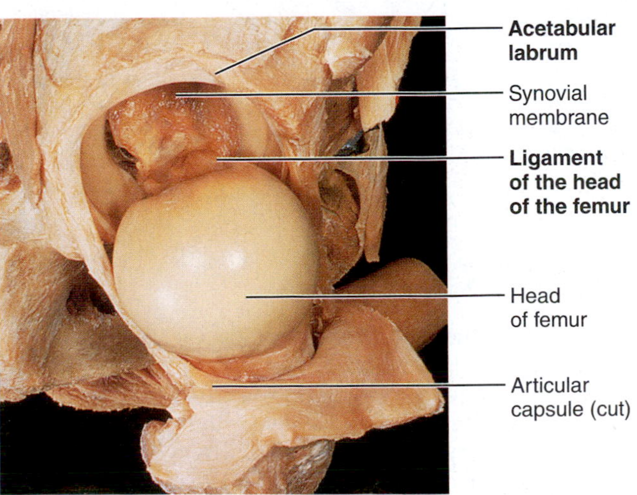

The Temporomandibular Joint

The **temporomandibular joint (TMJ)** lies just anterior to the ear **(Figure 10.9)**, where the egg-shaped condylar process of the mandible articulates with the inferior surface of the squamous region of the temporal bone. The temporal bone joint surface has a complicated shape: posteriorly is the **mandibular fossa,** and anteriorly is a bony knob called the **articular tubercle.** The joint's articular capsule, though strengthened by the **lateral ligament,** is slack; an articular disc divides the joint cavity into superior and inferior compartments. Typically, the condylar process–mandibular fossa connection allows the familiar hinge-like movements of elevating and depressing the mandible to open and close the mouth. However, when the mouth is opened wide, the condylar process glides anteriorly and is braced against the dense bone of the articular tubercle so that the mandible is not forced superiorly when we bite hard foods.

ACTIVITY 8

Examining the Action at the Temporomandibular Joint

While placing your fingers over the area just anterior to the ear, open and close your mouth to feel the hinge action at the TMJ. Then, keeping your fingers on the TMJ, yawn to demonstrate the anterior gliding of the condylar process of the mandible. ■

The Shoulder Joint

The shoulder joint, or **glenohumeral joint,** is the most freely moving joint of the body. The rounded head of the humerus fits the shallow glenoid cavity of the scapula **(Figure 10.10)**. A rim of fibrocartilage, the **glenoid labrum,** deepens the cavity slightly.

The articular capsule enclosing the joint is thin and loose, contributing to ease of movement. Few ligaments reinforce the shoulder; most are located anteriorly. The **coracohumeral ligament** helps support the weight of the upper limb; three weak **glenohumeral ligaments** strengthen the front of the capsule. Sometimes they are absent. Muscle tendons from the biceps brachii and **rotator cuff** muscles (subscapularis, supraspinatus, infraspinatus, and teres minor) contribute most to shoulder stability.

ACTIVITY 9

Demonstrating Actions at the Shoulder Joint

If a functional shoulder joint model is available, identify the joint parts and manipulate it to demonstrate the following movements: flexion, extension, abduction, adduction, circumduction, and medial and lateral rotation.

Note where the joint is weakest, and verify the most common direction of a dislocated humerus. ■

Figure 10.7 Hip joint relationships. (a) Frontal section through the right hip joint. **(b)** Anterior superficial view of the right hip joint. **(c)** Photograph of the interior of the hip joint, lateral view.

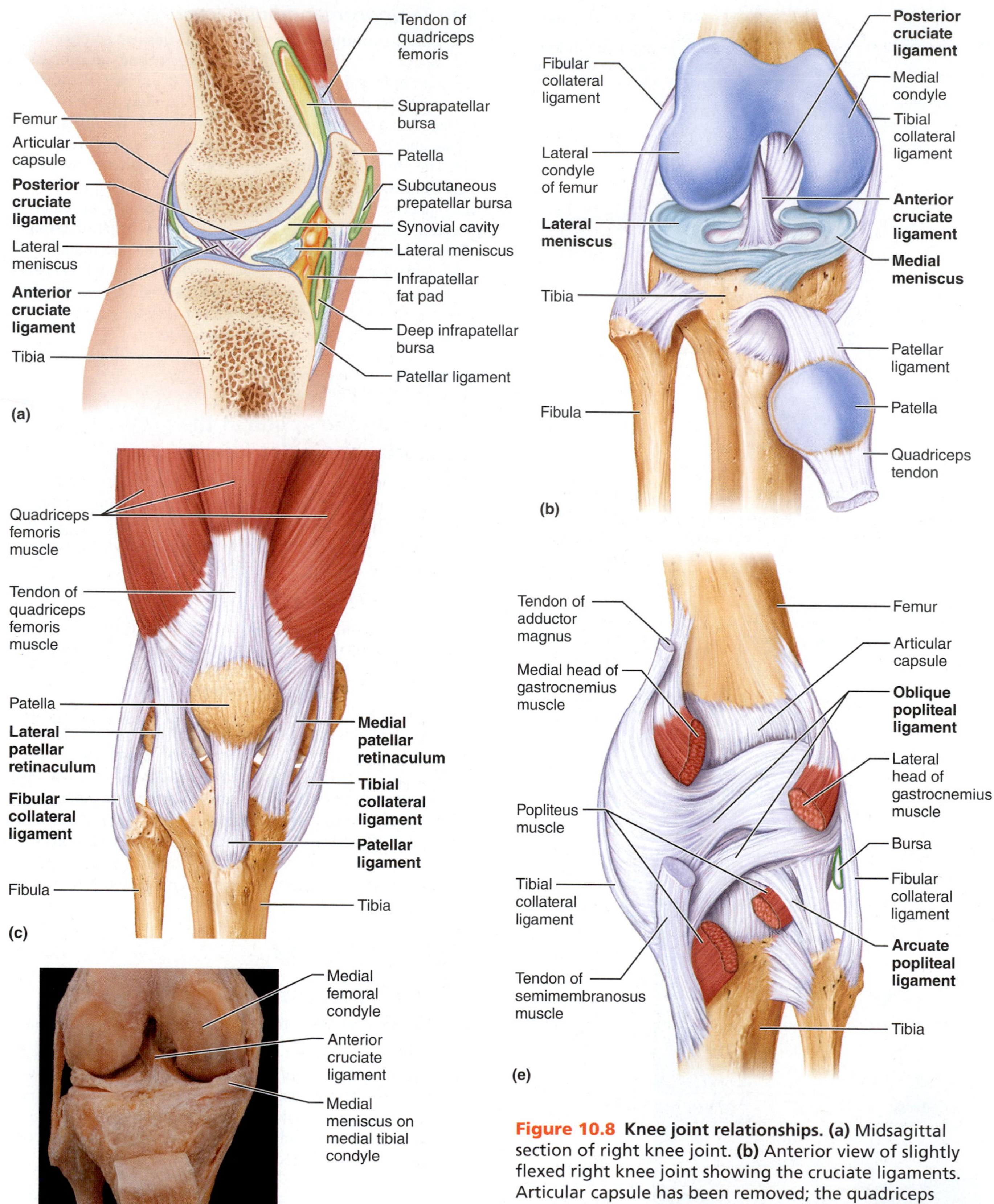

Figure 10.8 Knee joint relationships. (a) Midsagittal section of right knee joint. **(b)** Anterior view of slightly flexed right knee joint showing the cruciate ligaments. Articular capsule has been removed; the quadriceps tendon has been cut and reflected distally. **(c)** Anterior superficial view of the right knee. **(d)** Photograph of an opened knee joint corresponds to view in (b). **(e)** Posterior superficial view of the ligaments clothing the knee joint.

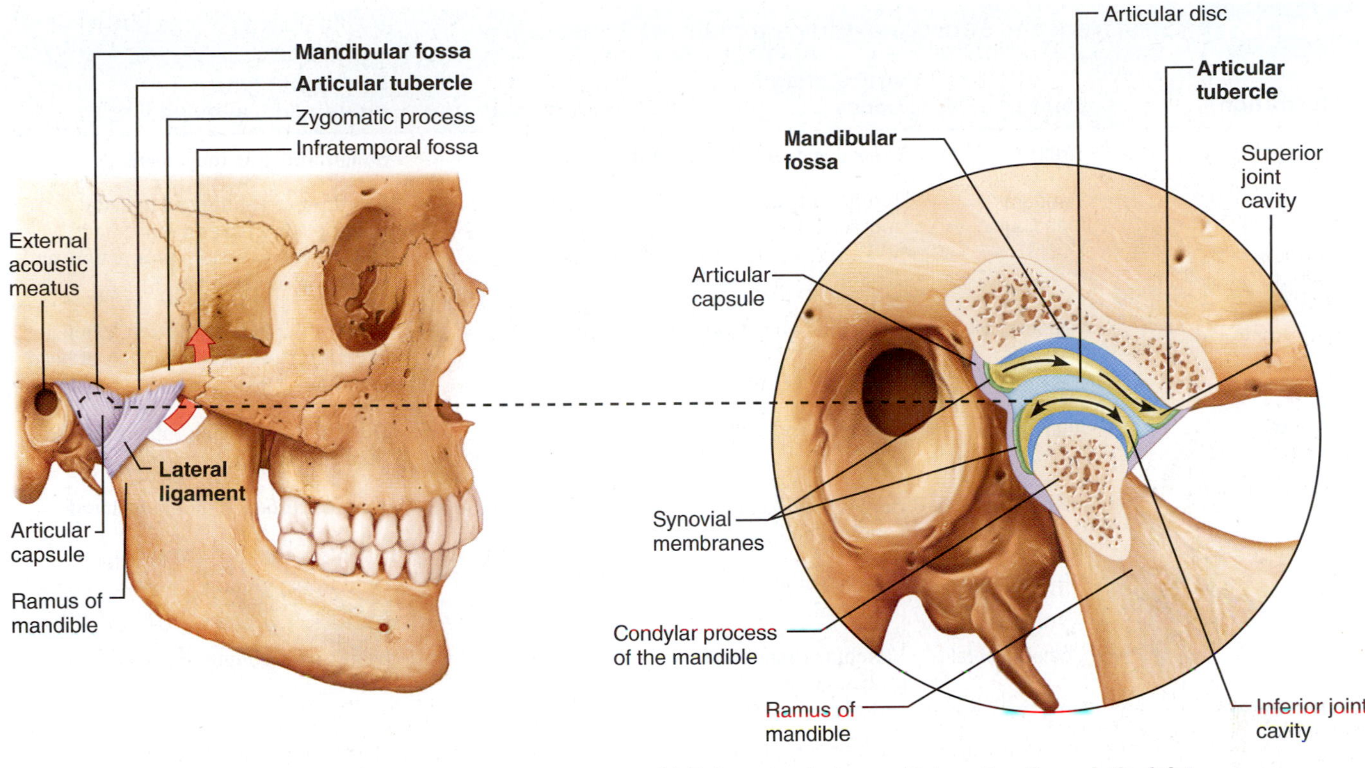

Figure 10.9 The temporomandibular (jaw) joint relationships. Note that the superior and inferior compartments of the joint cavity allow different movements indicated by arrows.

GROUP CHALLENGE

Articulations: "Simon Says"

Working in groups of three or four, play a game of "Simon Says" using the movements defined in Activity 5 (see pages 157–160). One student will play the role of "Simon" while the others perform the movement. For example, when "Simon" says, "Simon says, perform flexion at the elbow," the remaining students flex their arm. Take turns playing the role of Simon. As you perform the movements, consider and discuss whether the joint allows other movements and whether the joint is uniaxial, biaxial, or multiaxial. (Use Table 10.1 as a guide.) After playing for 15–20 minutes, complete the tables below.

1. List two uniaxial joints, and describe the movement at each.

Uniaxial Joints

Name of joint	Movements allowed

2. List two biaxial joints, and describe the movement at each.

Biaxial Joints

Name of joint	Movements allowed

3. List two multiaxial joints, and describe the movement at each.

Multiaxial Joints

Name of joint	Movements allowed

Table 10.1 Structural and Functional Characteristics of Body Joints

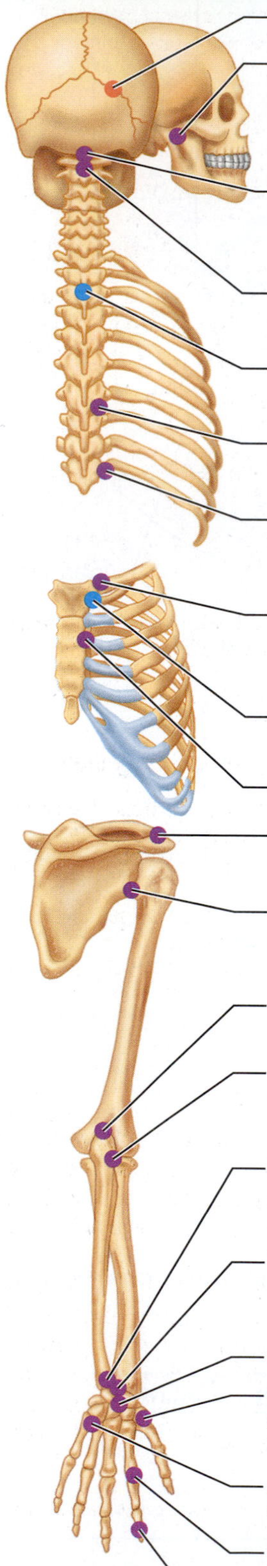

Illustration	Joint	Articulating bones	Structural type*	Functional type; movements allowed
	Skull	Cranial and facial bones	Fibrous; suture	Synarthrotic; no movement
	Temporo-mandibular	Temporal bone of skull and mandible	Synovial; modified hinge[†] (contains articular disc)	Diarthrotic; gliding and uniaxial rotation; slight lateral movement, elevation, depression, protraction, and retraction of mandible
	Atlanto-occipital	Occipital bone of skull and atlas	Synovial; condylar	Diarthrotic; biaxial; flexion, extension, lateral flexion, circumduction of head on neck
	Atlantoaxial	Atlas (C_1) and axis (C_2)	Synovial; pivot	Diarthrotic; uniaxial; rotation of the head
	Intervertebral	Between adjacent vertebral bodies	Cartilaginous; symphysis	Amphiarthrotic; slight movement
	Intervertebral	Between articular processes	Synovial; plane	Diarthrotic; gliding
	Costovertebral	Vertebrae (transverse process or bodies) and ribs	Synovial; plane	Diarthrotic; gliding of ribs
	Sternoclavicular	Sternum and clavicle	Synovial; shallow saddle (contains articular disc)	Diarthrotic; multiaxial (allows clavicle to move in all axes)
	Sternocostal (first)	Sternum and rib 1	Cartilaginous; synchondrosis	Synarthrotic; no movement
	Sternocostal	Sternum and ribs 2–7	Synovial; double plane	Diarthrotic; gliding
	Acromio-clavicular	Acromion of scapula and clavicle	Synovial; plane (contains articular disc)	Diarthrotic; gliding and rotation of scapula on clavicle
	Shoulder (glenohumeral)	Scapula and humerus	Synovial; ball and socket	Diarthrotic; multiaxial; flexion, extension, abduction, adduction, circumduction, rotation of humerus
	Elbow	Ulna (and radius) with humerus	Synovial; hinge	Diarthrotic; uniaxial; flexion, extension of forearm
	Proximal radioulnar	Radius and ulna	Synovial; pivot	Diarthrotic; uniaxial; pivot (head of radius rotates in radial notch of ulna)
	Distal radioulnar	Radius and ulna	Synovial; pivot (contains articular disc)	Diarthrotic; uniaxial; rotation of radius around long axis of forearm to allow pronation and supination
	Wrist	Radius and proximal carpals	Synovial; condylar	Diarthrotic; biaxial; flexion, extension, abduction, adduction, circumduction of hand
	Intercarpal	Adjacent carpals	Synovial; plane	Diarthrotic; gliding
	Carpometacarpal of digit 1 (thumb)	Carpal (trapezium) and metacarpal I	Synovial; saddle	Diarthrotic; biaxial; flexion, extension, abduction, adduction, circumduction, opposition of metacarpal I
	Carpometacarpal of digits 2–5	Carpal(s) and metacarpal(s)	Synovial; plane	Diarthrotic; gliding of metacarpals
	Metacarpo-phalangeal (knuckle)	Metacarpal and proximal phalanx	Synovial; condylar	Diarthrotic; biaxial; flexion, extension, abduction, adduction, circumduction of fingers
	Interphalangeal (finger)	Adjacent phalanges	Synovial; hinge	Diarthrotic; uniaxial; flexion, extension of fingers

Table 10.1 *(continued)*

Illustration	Joint	Articulating bones	Structural type	Functional type; movements allowed
	Sacroiliac	Sacrum and coxal bone	Synovial; plane	Diarthrotic; little movement, slight gliding possible (more during pregnancy)
	Pubic symphysis	Pubic bones	Cartilaginous; symphysis	Amphiarthrotic; slight movement (enhanced during pregnancy)
	Hip (coxal)	Hip bone and femur	Synovial; ball and socket	Diarthrotic; multiaxial; flexion, extension, abduction, adduction, rotation, circumduction of femur
	Knee (tibiofemoral)	Femur and tibia	Synovial; modified hinge[†] (contains articular discs)	Diarthrotic; biaxial; flexion, extension of leg, some rotation allowed
	Knee (femoropatellar)	Femur and patella	Synovial; plane	Diarthrotic; gliding of patella
	Superior tibiofibular	Tibia and fibula (proximally)	Synovial; plane	Diarthrotic; gliding of fibula
	Inferior tibiofibular	Tibia and fibula (distally)	Fibrous; syndesmosis	Synarthrotic; slight "give" during dorsiflexion
	Ankle	Tibia and fibula with talus	Synovial; hinge	Diarthrotic; uniaxial; dorsiflexion and plantar flexion of foot
	Intertarsal	Adjacent tarsals	Synovial; plane	Diarthrotic; gliding; inversion and eversion of foot
	Tarsometatarsal	Tarsal(s) and metatarsal(s)	Synovial; plane	Diarthrotic; gliding of metatarsals
	Metatarso-phalangeal	Metatarsal and proximal phalanx	Synovial; condylar	Diarthrotic; biaxial; flexion, extension, abduction, adduction, circumduction of great toe
	Interphalangeal (toe)	Adjacent phalanges	Synovial; hinge	Diarthrotic; uniaxial; flexion, extension of toes

[*]**Fibrous joints** are indicated by orange circles; **cartilaginous joints** by blue circles; **synovial joints** by purple circles.
[†] These modified hinge joints are structurally bicondylar.

Joint Disorders

Joint pains and malfunctions can be due to a number of causes. For example, a hard blow to the knee can cause a painful bursitis, known as "water on the knee," due to damage to, or inflammation of, the patellar bursa. Slippage of a fibrocartilage pad may result in a painful condition that heals slowly.

Sprains and dislocations are other types of joint problems. In a **sprain,** the ligaments reinforcing a joint are damaged by overstretching or are torn away from the bony attachment. Because both ligaments and tendons are cords of dense connective tissue with a poor blood supply, sprains heal slowly and are quite painful.

Dislocations occur when bones are forced out of their normal position in the joint cavity. They are normally accompanied by torn or stressed ligaments and considerable inflammation. The process of returning the bone to its proper position, called *reduction,* should only be done by a physician. Attempts by an untrained person to "snap the bone back into its socket" are often more harmful than helpful.

Advancing years also take their toll on joints. Weight-bearing joints in particular eventually begin to degenerate. *Adhesions* (fibrous bands) may form between the surfaces where bones join, and extraneous bone tissue *(spurs)* may grow along the joint edges. Such degenerative changes lead to the complaint so often heard from the elderly: "My joints are getting so stiff. . . ."

- If possible, compare an X-ray image of an arthritic joint to one of a normal joint. ✚

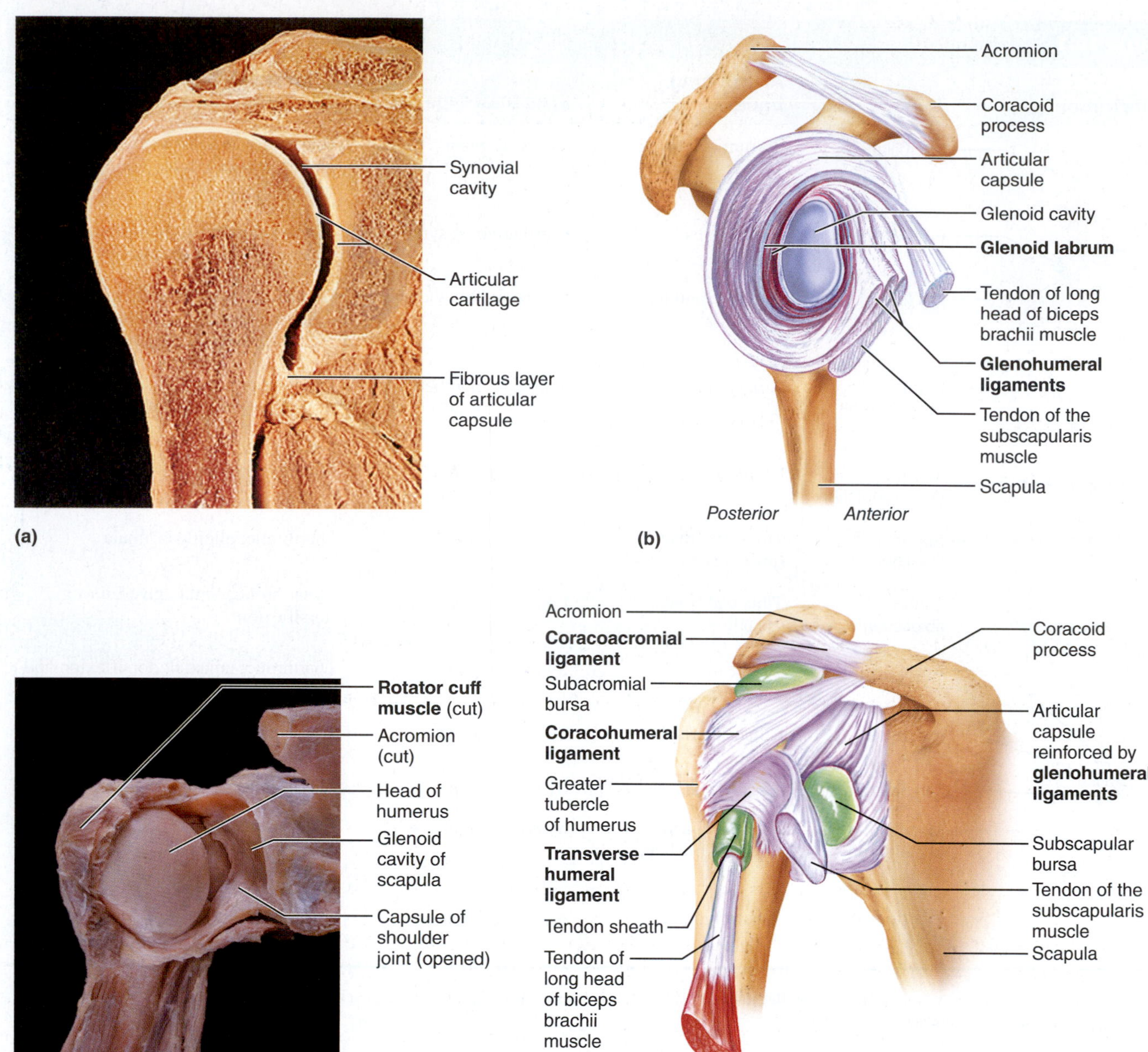

Figure 10.10 The shoulder joint. (a) Frontal section through the shoulder. **(b)** The right shoulder joint, cut open and viewed from the lateral aspect; the humerus has been removed. **(c)** Photo of an opened shoulder joint, anterior view. **(d)** Anterior superficial view of the right shoulder joint.

NAME ______________________

LAB TIME/DATE ______________________

EXERCISE 10

REVIEW SHEET

Articulations and Body Movements

Fibrous, Cartilaginous, and Synovial Joints

1. Use the key to identify the joint types described below. Some responses may be used more than once.

Key: a. cartilaginous b. fibrous c. synovial

______________ 1. typically allows a slight degree of movement

______________ 2. includes joints between the vertebral bodies and the pubic symphysis

______________ 3. essentially immovable joints

______________ 4. sutures are the most remembered examples

______________ 5. characterized by cartilage connecting the bony portions

______________ 6. all characterized by a fibrous articular capsule lined with a synovial membrane surrounding a joint cavity

______________ 7. all are freely movable or diarthrotic

______________ 8. bone regions are united by fibrous connective tissue

______________ 9. include the hip, knee, and elbow joints

2. Describe the tissue type and function of the following structures or tissues in relation to a synovial joint, and label the structures indicated by leader lines in the diagram. (Use an appropriate reference.)

ligament: ______________________

tendon: ______________________

articular cartilage: ______________________

synovial membrane: ______________________

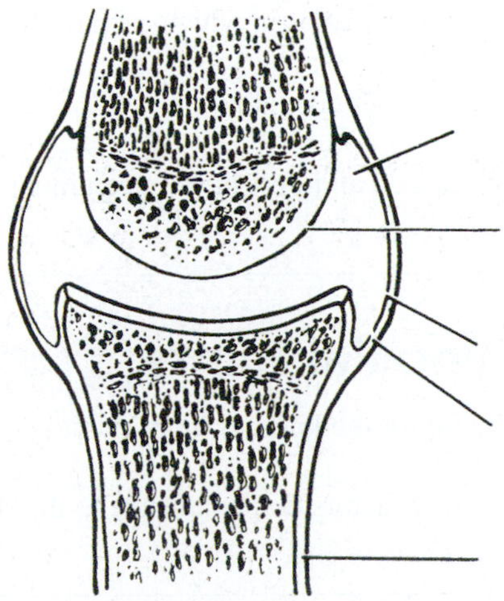

bursa: ______________________

3. Match the joint subcategories in column B with their descriptions in column A, and place an asterisk (*) beside all choices that are examples of synovial joints. Some responses may be used more than once.

	Column A	Column B
______	1. joint between skull bones	a. ball-and-socket
______	2. joint between the axis and atlas	b. condylar
______	3. hip joint	c. hinge
______	4. intervertebral joints (between articular processes)	d. pivot
______	5. joint between forearm bones and wrist	e. plane
______	6. elbow	f. saddle
______	7. interphalangeal joints	g. suture
______	8. intercarpal joints	h. symphysis
______	9. joint between talus and tibia/fibula	i. synchondrosis
______	10. joint between skull and vertebral column	j. syndesmosis
______	11. joint between jaw and skull	
______	12. joints between proximal phalanges and metacarpal bones	
______	13. epiphyseal plate of a child's long bone	
______	14. a multiaxial joint	
______, ______	15. biaxial joints	
______, ______	16. uniaxial joints	

4. Indicate the number of planes in which each of the following joint types can move.

______ in uniaxial joints, ______ in biaxial joints, ______ in multiaxial joints

5. What characteristics do all joints have in common? ______

______

Selected Synovial Joints

6. Which joint, the hip or the knee, is more stable? ______

Name two important factors that contribute to the stability of the hip joint.

______ and ______

Name two important factors that contribute to the stability of the knee.

______ and ______

7. The diagram shows a frontal section of the hip joint. Identify its major structural elements by using the letters in the key.

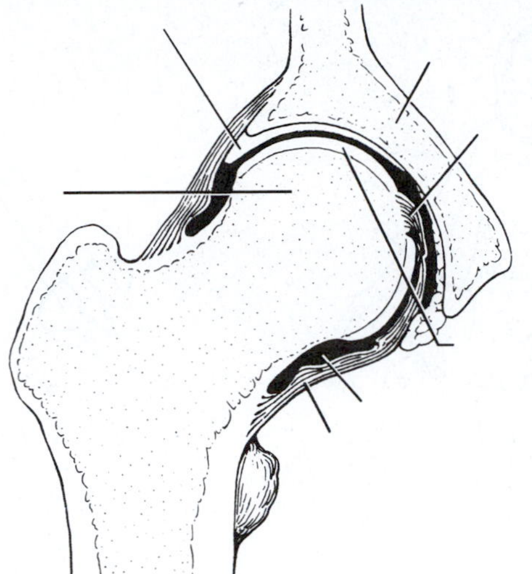

Key:

a. acetabular labrum

b. articular capsule

c. articular cartilage

d. coxal bone

e. head of femur

f. ligament of the head of the femur

g. synovial cavity

8. The shoulder joint is built for mobility. List four factors that contribute to the large range of motion at the shoulder:

1. ______________________________

2. ______________________________

3. ______________________________

4. ______________________________

9. In which direction does the shoulder usually dislocate? ______________________________

Movements Allowed by Synovial Joints

10. Which letter on the adjacent diagram marks the origin of the muscle? ______________

Which letter marks the insertion? ______________

Insert the words *origin* and *insertion* into the following sentence:

During muscle contraction, the ______________ moves toward the ______________.

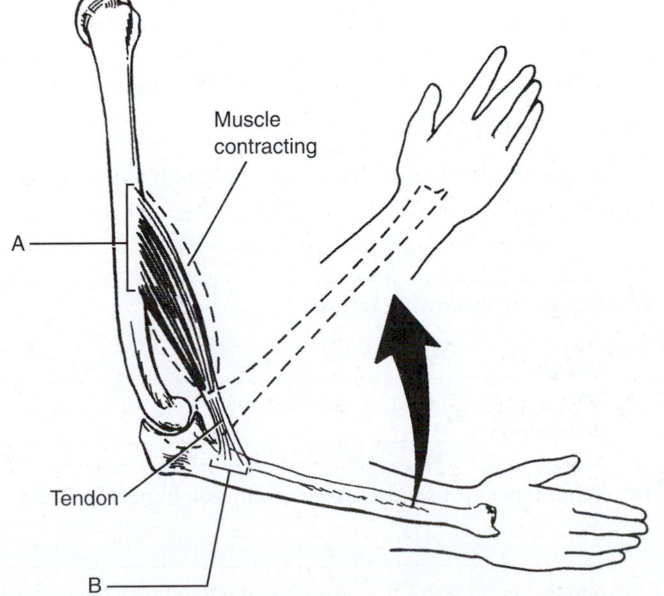

11. Complete the descriptions below the diagrams by inserting the type of movement in each answer blank.

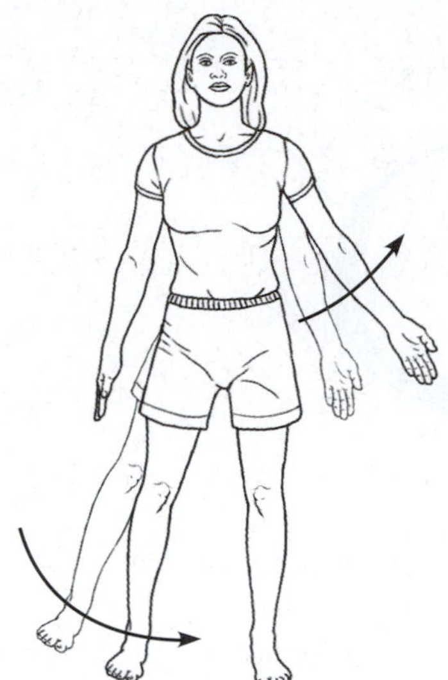

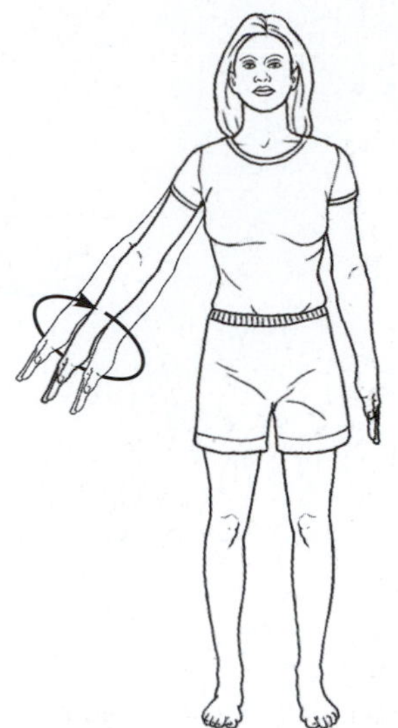

(a) ________________ **of the elbow**

(b) ________________ **of the knee**

(c) ________________ **of the shoulder**

(d) ________________ **of the hip**

(e) ________________ **of the shoulder**

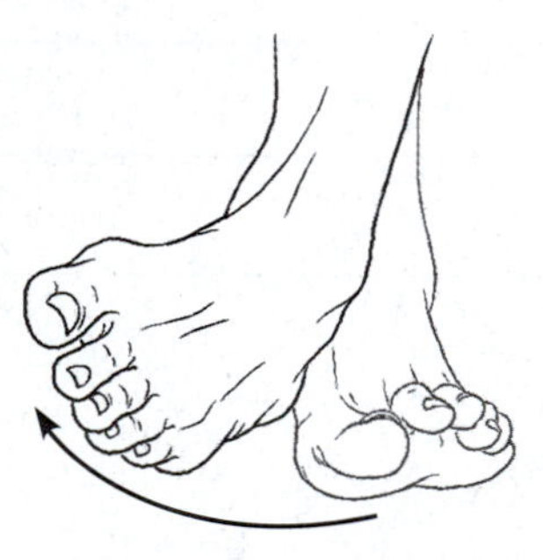

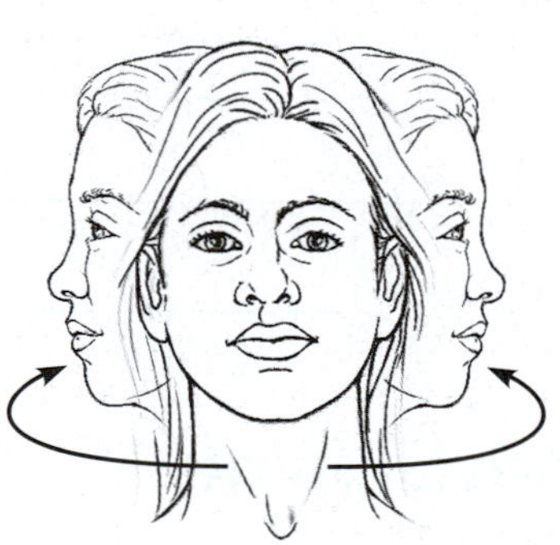

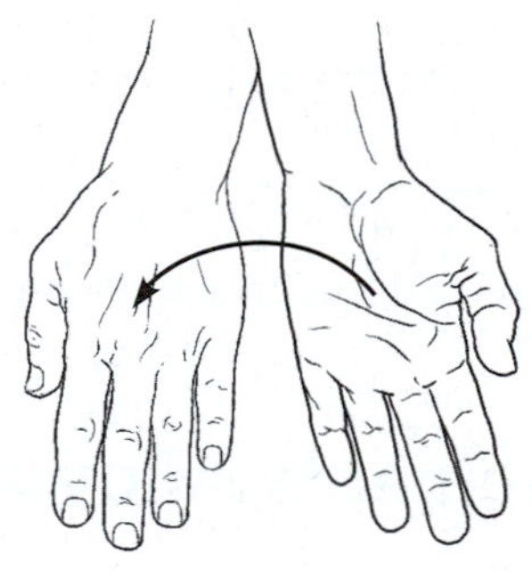

(f) ________________ **of the foot**

(g) ________________ **of the head**

(h) ________________ **of the hand**

Joint Disorders

12. What structural joint changes are common among elderly people? ____

____

13. Define the following terms.

sprain: ____

dislocation: ____

14. What types of tissue damage might you expect to find in a dislocated joint?

____

____

11

EXERCISE

Microscopic Anatomy and Organization of Skeletal Muscle

MATERIALS

- ☐ Three-dimensional model of skeletal muscle cells (if available)
- ☐ Forceps
- ☐ Dissecting needles
- ☐ Clean microscope slides and coverslips
- ☐ 0.9% saline solution in dropper bottles
- ☐ Chicken breast or thigh muscle (freshly obtained)
- ☐ Compound microscope
- ☐ Prepared slides of skeletal muscle (l.s. and x.s. views) and skeletal muscle showing neuromuscular junctions
- ☐ Three-dimensional model of skeletal muscle showing neuromuscular junction (if available)

OBJECTIVES

1. Define *fiber, myofibril,* and *myofilament*, and describe the structural relationships among them.
2. Describe thick (myosin) and thin (actin) filaments and their relation to the sarcomere.
3. Discuss the structure and location of T tubules and terminal cisterns.
4. Define *endomysium, perimysium,* and *epimysium*, and relate them to muscle fibers, fascicles, and entire muscles.
5. Define *tendon* and *aponeurosis,* and describe the difference between them.
6. Describe the structure of skeletal muscle from gross to microscopic levels.
7. Explain the connection between motor neurons and skeletal muscle, and discuss the structure and function of the neuromuscular junction.

PRE-LAB QUIZ

1. Which is *not* true of skeletal muscle?
   a. It enables you to manipulate your environment.
   b. It influences the body's contours and shape.
   c. It is one of the major components of hollow organs.
   d. It provides a means of locomotion.
2. Circle the correct underlined term. Because the cells of skeletal muscle are relatively large and cylindrical in shape, they are also known as fibers / tubules.
3. Circle True or False. Skeletal muscle cells have more than one nucleus.
4. The two contractile proteins that make up the myofilaments of skeletal muscle are ________________ and ________________.
5. Each muscle cell is surrounded by thin connective tissue called the:
   a. aponeurosis　　c. epimysium
   b. endomysium　　d. perimysium
6. A cordlike structure that connects a muscle to another muscle or bone is:
   a. a fascicle
   b. a tendon
   c. deep fascia
7. The junction between an axon and a muscle fiber is called a ________________________________.
8. Circle True or False. The neuron and muscle fiber membranes do not actually touch but are separated by a fluid-filled gap.
9. Circle the correct underlined term. The contractile unit of muscle is the sarcolemma / sarcomere.
10. Circle True or False. Larger, more powerful muscles have relatively less connective tissue than smaller muscles.

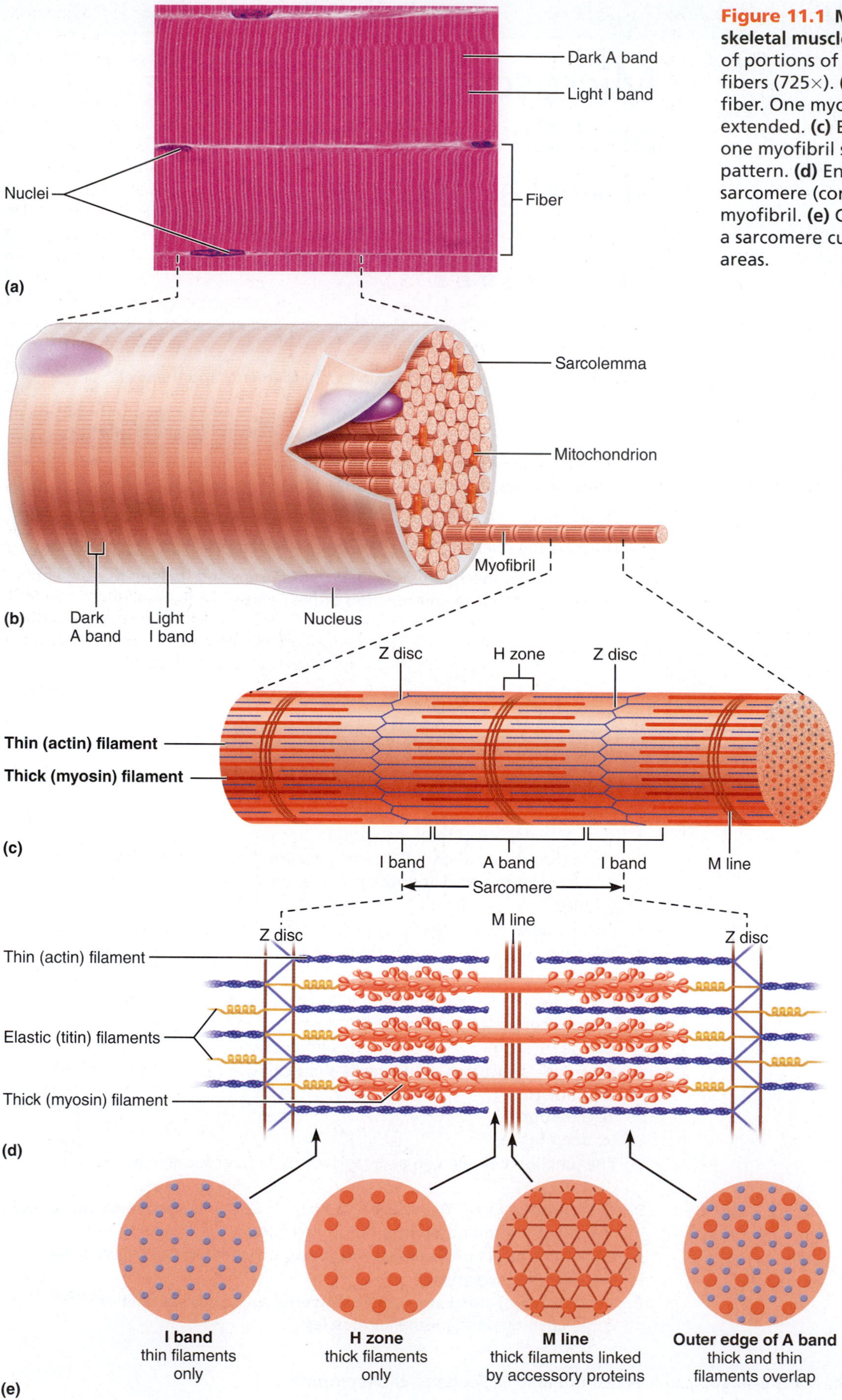

Figure 11.1 Microscopic anatomy of skeletal muscle. (a) Photomicrograph of portions of two isolated muscle fibers (725×). **(b)** Part of a muscle fiber. One myofibril has been extended. **(c)** Enlarged view of one myofibril showing its banding pattern. **(d)** Enlarged view of one sarcomere (contractile unit) of a myofibril. **(e)** Cross-sectional view of a sarcomere cut through in different areas.

Most of the muscle tissue in the body is **skeletal muscle,** which attaches to the skeleton or associated connective tissue. Skeletal muscle shapes the body and gives you the ability to move—to walk, run, jump, and dance; to draw, paint, and play a musical instrument; and to smile and frown. The remaining muscle tissue of the body consists of smooth muscle, which forms the walls of hollow organs, and cardiac muscle, which forms the walls of the heart. Smooth and cardiac muscle move materials within the body. For example, smooth muscle moves digesting food through the gastrointestinal system and moves urine from the kidneys to the exterior of the body. Cardiac muscle moves blood through the blood vessels.

Each of the three muscle types has a structure and function uniquely suited to its task in the body. However, because the term *muscular system* applies specifically to skeletal muscle, our primary objective in this exercise is to investigate the structure and function of skeletal muscle.

Skeletal muscle is also known as *voluntary muscle* (because it can be consciously controlled) and as *striated muscle* (because it has a striped appearance). As you might guess from both of these alternative names, skeletal muscle has some very special characteristics. Thus, we should begin our investigation of skeletal muscle at the cellular level.

The Cells of Skeletal Muscle

Skeletal muscle is made up of relatively large, long cylindrical cells, sometimes called **fibers.** These cells range from 10 to 100 μm in diameter and up to 6 cm in length.

Because hundreds of embryonic cells fuse to produce each muscle cell, the skeletal muscle cells **(Figure 11.1a** and **b)** are multinucleate; multiple oval nuclei can be seen just beneath the plasma membrane (called the *sarcolemma* in these cells). The nuclei are pushed peripherally by the longitudinally arranged **myofibrils,** which nearly fill the sarcoplasm. Alternating light (I) and dark (A) bands along the length of the perfectly aligned myofibrils give the muscle fiber as a whole its striped appearance.

Electron microscope studies have revealed that the myofibrils are made up of even smaller threadlike structures called **myofilaments** (Figure 11.1d). The myofilaments are composed largely of two varieties of contractile proteins—**actin** and **myosin**—which slide past each other during muscle activity to bring about shortening or contraction of the muscle cells. It is the highly specific arrangement of the myofilaments within the myofibrils that is responsible for the banding pattern in skeletal muscle. The actual contractile units of muscle, called **sarcomeres,** extend from the middle of one I band (its Z disc) to the middle of the next along the length of the myofibrils (Figure 11.1c and d). Cross sections of the sarcomere in areas where **thick filaments** and **thin filaments** overlap show that each thick filament is surrounded by six thin filaments; each thin filament is enclosed by three thick filaments (Figure 11.1e).

At each junction of the A and I bands, the sarcolemma indents into the muscle cell, forming a **transverse tubule (T tubule).** These tubules run deep into the muscle cell between cross channels, or **terminal cisterns,** of the elaborate smooth endoplasmic reticulum called the **sarcoplasmic reticulum (SR) (Figure 11.2)**. Regions where the SR terminal cisterns abut a T tubule on each side are called **triads.**

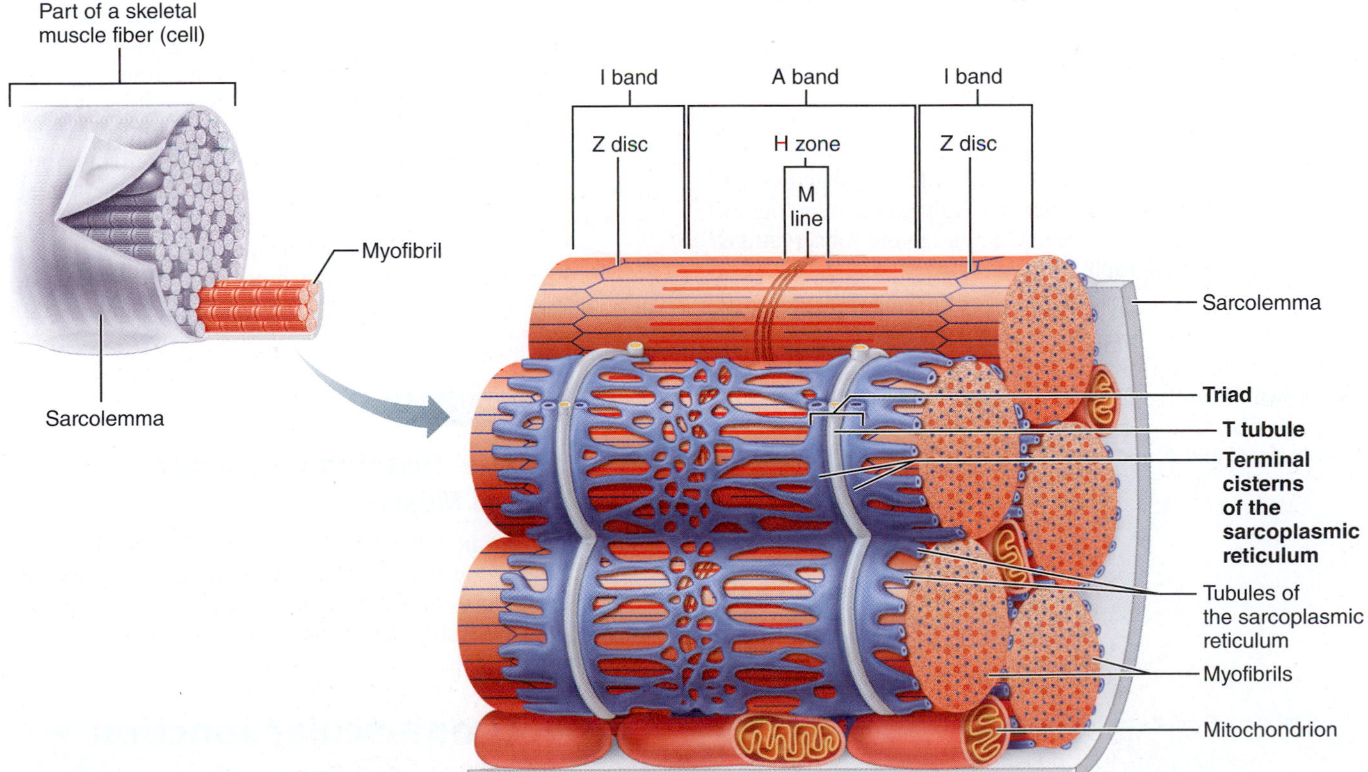

Figure 11.2 Relationship of the sarcoplasmic reticulum and T tubules to the myofibrils of skeletal muscle.

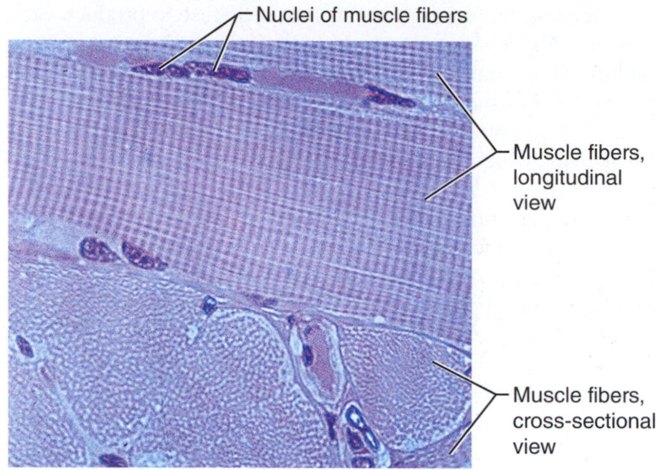

Figure 11.3 Photomicrograph of muscle fibers, longitudinal and cross sections (800×).

ACTIVITY 1

Examining Skeletal Muscle Cell Anatomy

1. Look at the three-dimensional model of skeletal muscle cells, noting the relative shape and size of the cells. Identify the nuclei, myofibrils, and light and dark bands.

2. Obtain forceps, two dissecting needles, a slide and coverslip, and a dropper bottle of saline solution. With forceps, remove a very small piece of muscle (about 1 mm diameter) from a fresh chicken breast or thigh. Place the tissue on a clean microscope slide, and add a drop of the saline solution.

3. Pull the muscle fibers apart (tease them) with the dissecting needles until you have a fluffy-looking mass of tissue. Cover the teased tissue with a coverslip, and observe under the high-power lens of a compound microscope. Look for the banding pattern by examining muscle fibers isolated at the edge of the tissue mass. Regulate the light carefully to obtain the highest possible contrast. Compare your observations with the photomicrograph (Figure 11.3).

4. Now compare your observations with what can be seen with professionally prepared muscle tissue. Obtain a slide of skeletal muscle (longitudinal section), and view it under high power. From your observations, draw a small section of a muscle fiber in the space provided below. Label the nuclei, sarcolemma, and A and I bands.

What structural details become apparent with the prepared slide?

____

____

Organization of Skeletal Muscle Cells into Muscles

Muscle fibers are soft and surprisingly fragile. Thousands of muscle fibers are bundled together with connective tissue to form the organs we refer to as skeletal muscles (Figure 11.4). Each muscle fiber is enclosed in a delicate, areolar connective tissue sheath called **endomysium.** Several sheathed muscle fibers are wrapped by a collagenic membrane called **perimysium,** forming a bundle of fibers called a **fascicle.** A large number of fascicles are bound together by a substantially coarser "overcoat" of dense connective tissue called an **epimysium,** which sheathes the entire muscle. These epimysia blend into the **deep fascia,** still coarser sheets of dense connective tissue that bind muscles into functional groups, and into strong cordlike **tendons** or sheetlike **aponeuroses,** which attach muscles to each other or indirectly to bones. A muscle's more movable attachment is called its *insertion*, whereas its fixed (or immovable) attachment is the *origin* (Exercise 10).

Tendons perform several functions. Two of the most important are to provide durability and to conserve space. Because tendons are tough collagenic connective tissue, they can span rough bony projections that would destroy the more delicate muscle tissues. Because of their relatively small size, more tendons than fleshy muscles can pass over a joint.

In addition to supporting and binding the muscle fibers, and providing strength to the muscle as a whole, the connective tissue wrappings provide a route for the entry and exit of nerves and blood vessels that serve the muscle fibers. The larger, more powerful muscles have relatively more connective tissue than muscles involved in fine or delicate movements.

As we age, the mass of the muscle fibers decreases, and the amount of connective tissue increases; thus the skeletal muscles gradually become more sinewy, or "stringier." +

ACTIVITY 2

Observing the Histologic Structure of a Skeletal Muscle

Obtain a slide showing a cross section of skeletal muscle tissue. Identify the muscle fibers, their peripherally located nuclei, and their connective tissue wrappings, the endomysium, perimysium, and epimysium if visible (use Figure 11.4 as a reference).

The Neuromuscular Junction

The voluntary skeletal muscle cells must be stimulated by motor neurons via nerve impulses. The junction between a nerve fiber (axon) and a muscle cell is called a **neuromuscular,** or **myoneural, junction** (Figure 11.5).

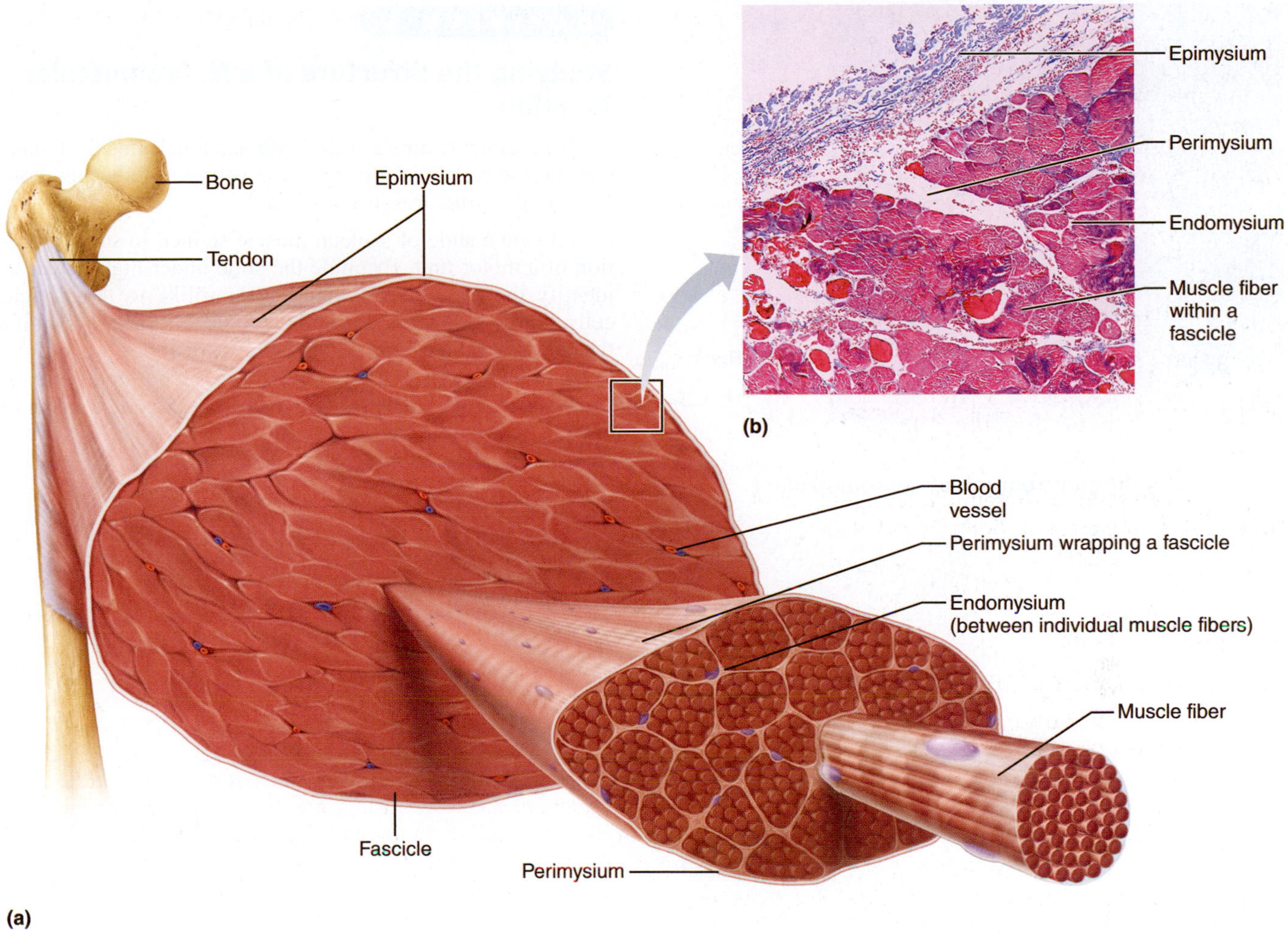

Figure 11.4 Connective tissue coverings of skeletal muscle. (a) Diagram view. **(b)** Photomicrograph of a cross section of skeletal muscle (55×).

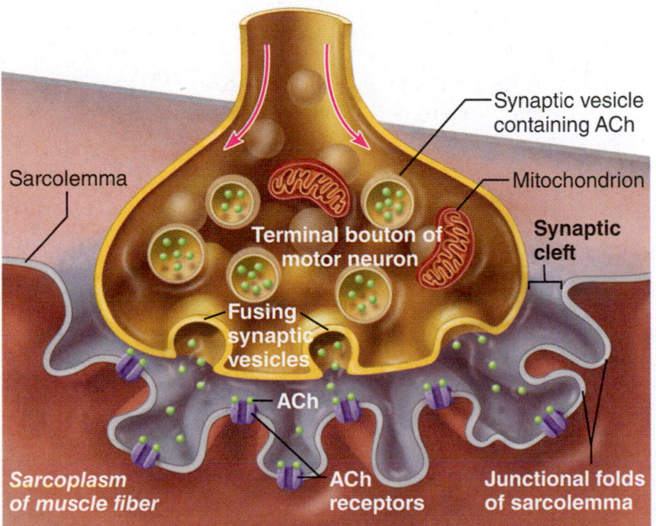

Figure 11.5 The neuromuscular junction. Red arrows indicate arrival of the nerve impulse (action potential), which ultimately causes vesicles to release acetylcholine (ACh). The ACh receptor is part of an ion channel that opens briefly, causing depolarization of the sarcolemma.

The axon of each motor neuron breaks up into many branches called **terminal boutons** or *axon terminals* as it approaches the muscle, and each of these branches participates in forming a neuromuscular junction with a single muscle cell. Thus, a single neuron may stimulate many muscle fibers. Together, a neuron and all the muscle fibers it stimulates make up the functional structure called the **motor unit.** (Part of a motor unit is shown in **Figure 11.6**) The neuron and muscle fiber membranes, close as they are, do not actually touch. They are separated by a small fluid-filled gap called the **synaptic cleft** (see Figure 11.5).

Within the terminal boutons are many mitochondria and vesicles containing a neurotransmitter chemical called acetylcholine (ACh). When a nerve impulse reaches the terminal boutons, some of these vesicles release their contents into the synaptic cleft. The ACh diffuses across the junction and combines with the receptors on the sarcolemma. When receptors bind ACh, the permeability of the sarcolemma changes. Channels that allow both sodium ions (Na^+) and potassium ions (K^+) to pass open briefly. Because more Na^+ diffuses into the muscle fiber than K^+ diffuses out, the sarcolemma undergoes depolarization, and the muscle fiber subsequently contracts.

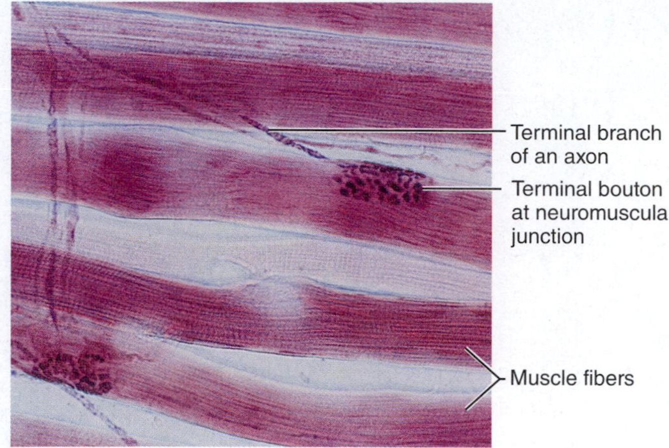

Figure 11.6 Photomicrograph of neuromuscular junction (750×).

ACTIVITY 3

Studying the Structure of a Neuromuscular Junction

1. If possible, examine a three-dimensional model of skeletal muscle cells that illustrates the neuromuscular junction. Identify the structures just described.

2. Obtain a slide of skeletal muscle stained to show a portion of a motor unit. Examine the slide under high power to identify the axon fibers extending leashlike to the muscle cells. Follow one of the axon fibers to its terminus to identify the oval-shaped terminal bouton. Compare your observations to the photomicrograph (Figure 11.6). Sketch a small section in the space provided below. Label the motor axon, a terminal branch, and muscle fibers. ■

NAME ____________________

LAB TIME/DATE ____________________

EXERCISE 11

REVIEW SHEET

Microscopic Anatomy and Organization of Skeletal Muscle

Skeletal Muscle Cells and Their Packaging into Muscles

1. Use the terms in the key to correctly identify the structures described below. Not all terms will be used.

____________ 1. connective tissue covering a bundle of muscle cells

____________ 2. bundle of muscle cells

____________ 3. contractile unit of muscle

____________ 4. a muscle cell

____________ 5. thin areolar connective tissue investing each muscle cell

____________ 6. plasma membrane of the muscle fiber

____________ 7. a long filamentous organelle with a banded appearance found within muscle cells

____________ 8. actin- or myosin-containing structure

____________ 9. cord of collagen fibers that attaches a muscle to a bone

Key:

a. endomysium
b. epimysium
c. fascicle
d. fiber
e. myofibril
f. myofilament
g. perimysium
h. sarcolemma
i. sarcomere
j. sarcoplasm
k. tendon

2. List three reasons the connective tissue wrappings of skeletal muscle are important.

____

____

____

3. Why are there more indirect—that is, tendinous—muscle attachments to bone than there are direct attachments?

____

____

____

4. How does an aponeurosis differ from a tendon structurally? ____

____

How is an aponeurosis functionally similar to a tendon? ____

____

5. The diagram illustrates a small portion of several myofibrils. Using letters from the key, correctly identify each structure indicated by a leader line or a bracket.

Key:					
a.	A band	d.	myosin filament	g.	triad
b.	actin filament	e.	T tubule	h.	sarcomere
c.	I band	f.	terminal cistern	i.	Z disc

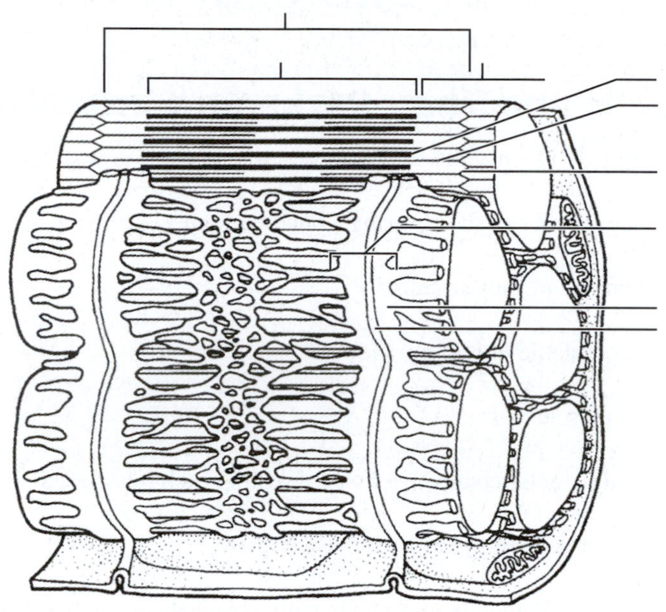

6. On the following figure, label a blood vessel, endomysium, epimysium, a fascicle, a muscle cell, perimysium, and the tendon.

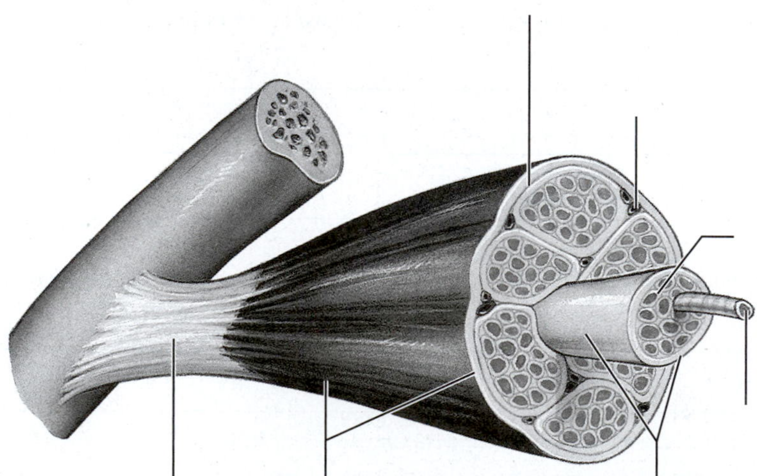

The Neuromuscular Junction

7. Complete the following statements:

The junction between a motor neuron's terminal bouton and the muscle cell membrane is called a __1__ junction. A motor neuron and all of the skeletal muscle cells it stimulates is called a __2__. The actual gap between the terminal bouton and the muscle cell is called a __3__. Within the terminal bouton are many small vesicles containing a neurotransmitter substance called __4__. When the __5__ reaches the ends of the axon, the neurotransmitter is released and diffuses to the muscle cell membrane to combine with receptors there. The combining of the neurotransmitter with the muscle membrane receptors causes the membrane to become permeable to both sodium and potassium. The greater influx of sodium ions results in __6__ of the membrane. Then contraction of the muscle cell occurs.

1. ____________________
2. ____________________
3. ____________________
4. ____________________
5. ____________________
6. ____________________

8. The events that occur at a neuromuscular junction are depicted below. Identify by labeling every structure provided with a leader line.

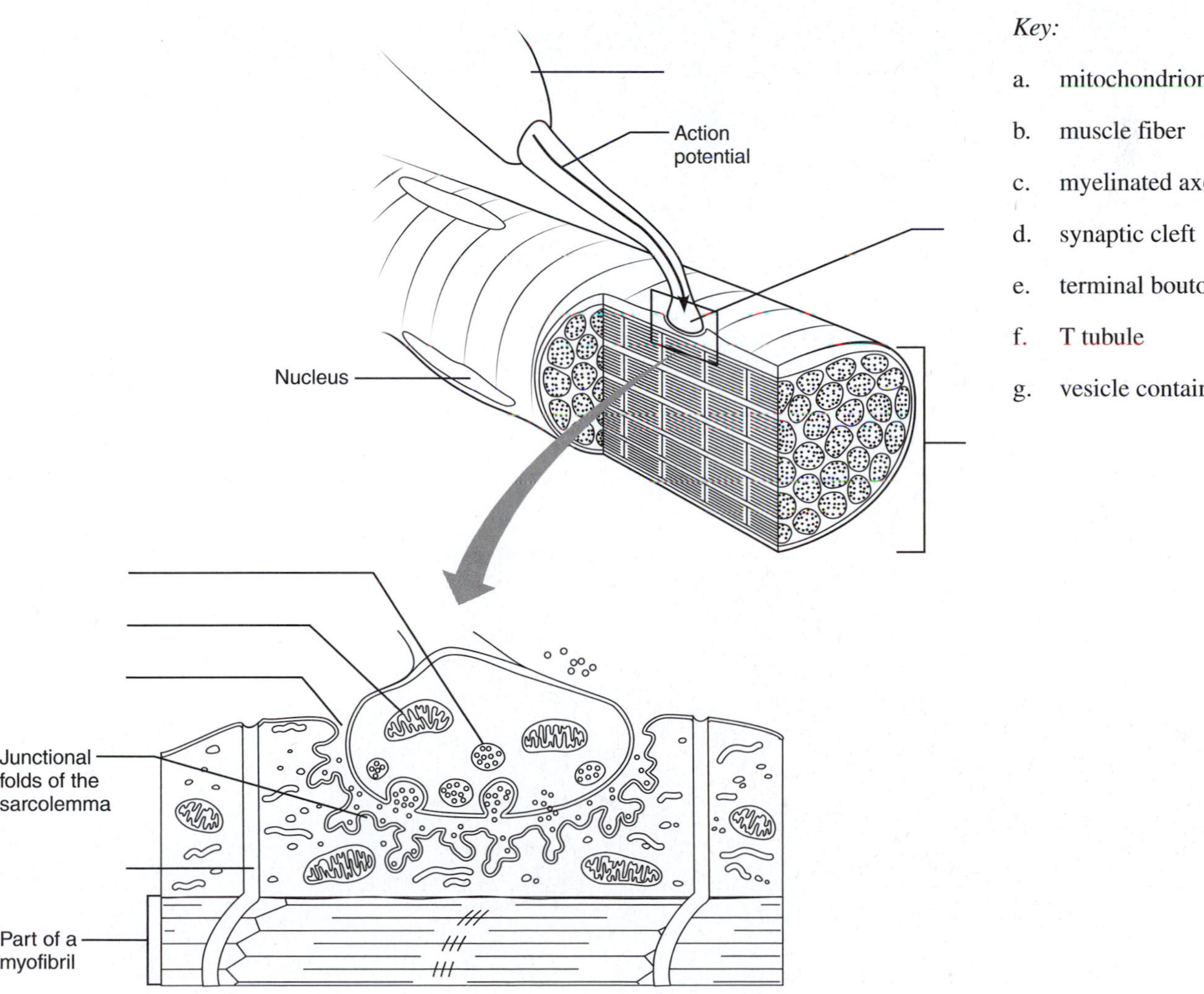

Key:

a. mitochondrion
b. muscle fiber
c. myelinated axon
d. synaptic cleft
e. terminal bouton
f. T tubule
g. vesicle containing ACh

EXERCISE 12

Gross Anatomy of the Muscular System

MATERIALS

- ☐ Human torso model or large anatomical chart showing human musculature
- ☐ Human cadaver for demonstration (if available)
- ☐ Disposable gloves or protective skin cream
- ☐ *Human Musculature* video
- ☐ Tubes of body (or face) paint
- ☐ 1-inch-wide artist's brushes

Cat dissection

- ☐ Lab coat or apron
- ☐ Disposable gloves
- ☐ Preserved and injected cat (one for every two to four students)
- ☐ Dissection tray and instruments
- ☐ Name tag and large plastic bag
- ☐ Paper towels
- ☐ Embalming fluid
- ☐ Organic debris container

OBJECTIVES

1. Define *prime mover (agonist), antagonist, synergist,* and *fixator.*
2. List the criteria used in naming skeletal muscles.
3. Identify the major muscles of the human body on a torso model, a human cadaver, lab chart, or image, and state the action of each.
4. Name muscle origins and insertions as required by the instructor.
5. Explain how muscle actions are related to their location.
6. List antagonists for the major prime movers.
7. Name and locate muscles on a dissected cat.
8. Recognize similarities and differences between human and cat musculature.

PRE-LAB QUIZ

1. A prime mover, or ___________, produces a particular type of movement.
   a. agonist c. fixator
   b. antagonist d. synergist
2. Skeletal muscles are named on the basis of many criteria. Name one.
   ____
3. Circle True or False. Muscles of facial expression differ from most skeletal muscles because they usually do not insert into a bone.
4. The ___________ musculature includes muscles that move the vertebral column and muscles that move the ribs.
   a. head and neck b. lower limb c. trunk
5. Muscles that act on the ___________ cause movement at the hip, knee, and foot joints.
   a. lower limb b. trunk c. upper limb
6. This two-headed muscle bulges when the forearm is flexed. It is the most familiar muscle of the anterior humerus. It is the:
   a. biceps brachii c. flexor carpii radialis
   b. extensor digitorum d. triceps brachii
7. These abdominal muscles are responsible for giving me my "six-pack." They also stabilize my pelvis when walking. They are the ___________ muscles.
   a. internal intercostal c. rectus abdominis
   b. quadriceps d. triceps femoris
8. Circle the correct underlined term. This lower limb muscle, which attaches to the calcaneus via the calcaneal tendon and plantar flexes the foot when the knee is extended, is the <u>tibialis anterior</u> / <u>gastrocnemius</u>.
9. The ___________ is the largest and most superficial of the gluteal muscles.
   a. gluteus internus c. gluteus medius
   b. gluteus maximus d. gluteus minimus
10. Circle True or False. The biceps femoris is located in the anterior compartment.

Classification of Skeletal Muscles

Types of Muscles

Most often, body movements do not result from the contraction of a single muscle but instead reflect the coordinated action of several muscles acting together. Muscles that are primarily responsible for producing a particular movement are called **prime movers,** or **agonists.**

Muscles that oppose or reverse a movement are called **antagonists.** When a prime mover is active, the fibers of the antagonist are stretched and in the relaxed state. The antagonist can also regulate the prime mover by providing some resistance, to prevent overshoot or to stop its action.

It should be noted that antagonists can be prime movers in their own right. For example, the biceps muscle of the arm (a prime mover of elbow flexion) is antagonized by the triceps (a prime mover of elbow extension).

Synergists help the action of agonists by reducing undesirable or unnecessary movement. Contraction of a muscle crossing two or more joints would cause movement at all joints spanned if the synergists were not there to stabilize them. For example, you can make a fist without bending your wrist only because synergist muscles stabilize the wrist joint and allow the prime mover to exert its force at the finger joints.

Fixators, or fixation muscles, are specialized synergists. They immobilize the origin of a prime mover so that all the tension is exerted at the insertion. Muscles that help maintain posture are fixators—so too are muscles of the back that stabilize, or "fix," the scapula during arm movements.

Naming Skeletal Muscles

Remembering the names of the skeletal muscles is a monumental task, but certain clues help. Muscles are named on the basis of the following criteria:

- **Direction of muscle fibers:** Some muscles are named in reference to some imaginary line, usually the midline of the body or the longitudinal axis of a limb bone. A muscle with fibers (and fascicles) running parallel to that imaginary line will have the term *rectus* (straight) in its name. For example, the rectus abdominis is the straight muscle of the abdomen. Likewise, the terms *transverse* and *oblique* indicate that the muscle fibers run at right angles and obliquely, respectively, to the imaginary line. Muscle structure is determined by fascicle arrangement **(Figure 12.1)**.
- **Relative size of the muscle:** Terms such as *maximus* (largest), *minimus* (smallest), *longus* (long), and *brevis* (short) are often used in naming muscles—as in gluteus maximus and gluteus minimus.
- **Location of the muscle:** Some muscles are named for the bone with which they are associated. For example, the temporalis muscle overlies the temporal bone.
- **Number of origins:** When the term *biceps, triceps,* or *quadriceps* forms part of a muscle name, you can generally assume that the muscle has two, three, or four origins (respectively). For example, the biceps muscle of the arm has two heads, or origins.
- **Location of the muscle's origin and insertion:** For example, the sternocleidomastoid muscle has its origin on the sternum *(sterno)* and clavicle *(cleido)* and inserts on the mastoid process of the temporal bone.
- **Shape of the muscle:** For example, the deltoid muscle is roughly triangular (*deltoid* = triangle), and the trapezius muscle resembles a trapezoid.
- **Action of the muscle:** For example, all the adductor muscles of the anterior thigh bring about its adduction, and all the extensor muscles of the wrist extend the wrist.

Identification of Human Muscles

While reading the tables and identifying the various human muscles in the figures, try to visualize what happens when the muscle contracts. Since muscles have many actions, we have indicated the primary action of each muscle in blue type in the tables. Then, use a torso model or an anatomical chart to again identify as many of these muscles as possible. If a human cadaver is available for observation, your instructor will provide specific instructions for muscle examination. Then carry out the instructions for demonstrating and palpating muscles.

Muscles of the Head and Neck

The muscles of the head serve many specific functions. For instance, the muscles of facial expression differ from most skeletal muscles because they insert into the skin or other muscles rather than into bone. As a result, they move the facial skin, allowing the face to show a wide range of emotions. Other muscles of the head are the muscles of mastication, which move the mandible during chewing, and the six extrinsic eye muscles located within the orbit, which aim the eye. (Orbital muscles are studied in Exercise 17.)

ACTIVITY 1

Identifying Muscles of the Head and Neck

Neck muscles are concerned primarily with the movement of the head and shoulder girdle. (**Figure 12.2** and **Figure 12.3** are summary figures illustrating the superficial musculature of the body). Read the descriptions of specific head and neck muscles, and identify the various muscles in the figures (**Table 12.1**, **Table 12.2**, Figure 12.3, **Figure 12.4**, and **Figure 12.5**), trying to visualize their action when they contract. Then identify them on a torso model or anatomical chart.

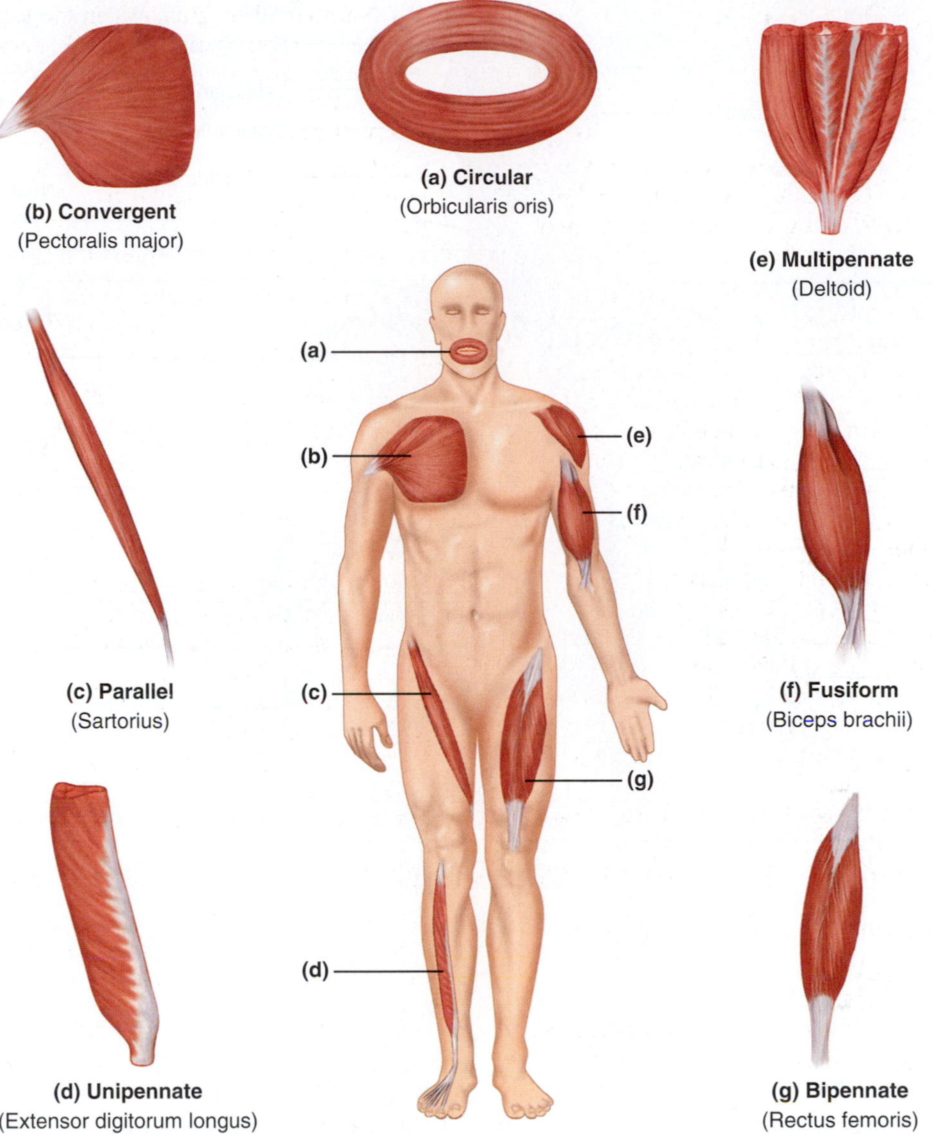

Figure 12.1 Patterns of fascicle arrangement in muscles.

Demonstrating Operations of Head Muscles

1. Raise your eyebrow to wrinkle your forehead. You are using the *frontal belly* of the *epicranius* muscle.

2. Blink your eyes; wink. You are contracting *orbicularis oculi.*

3. Close your lips and pucker up. This requires contraction of *orbicularis oris.*

4. Smile. You are using *zygomaticus.*

5. To demonstrate the temporalis, place your hands on your temples, and clench your teeth. Now you can also palpate the masseter at the angle of the jaw. ■

Muscles of the Trunk

The trunk musculature includes muscles that move the vertebral column; anterior thorax muscles that act to move ribs, head, and arms; and muscles of the abdominal wall that play a role in the movement of the vertebral column but, even more important, form the "natural girdle," or the major portion of the abdominal body wall.

(Text continues on page 191.)

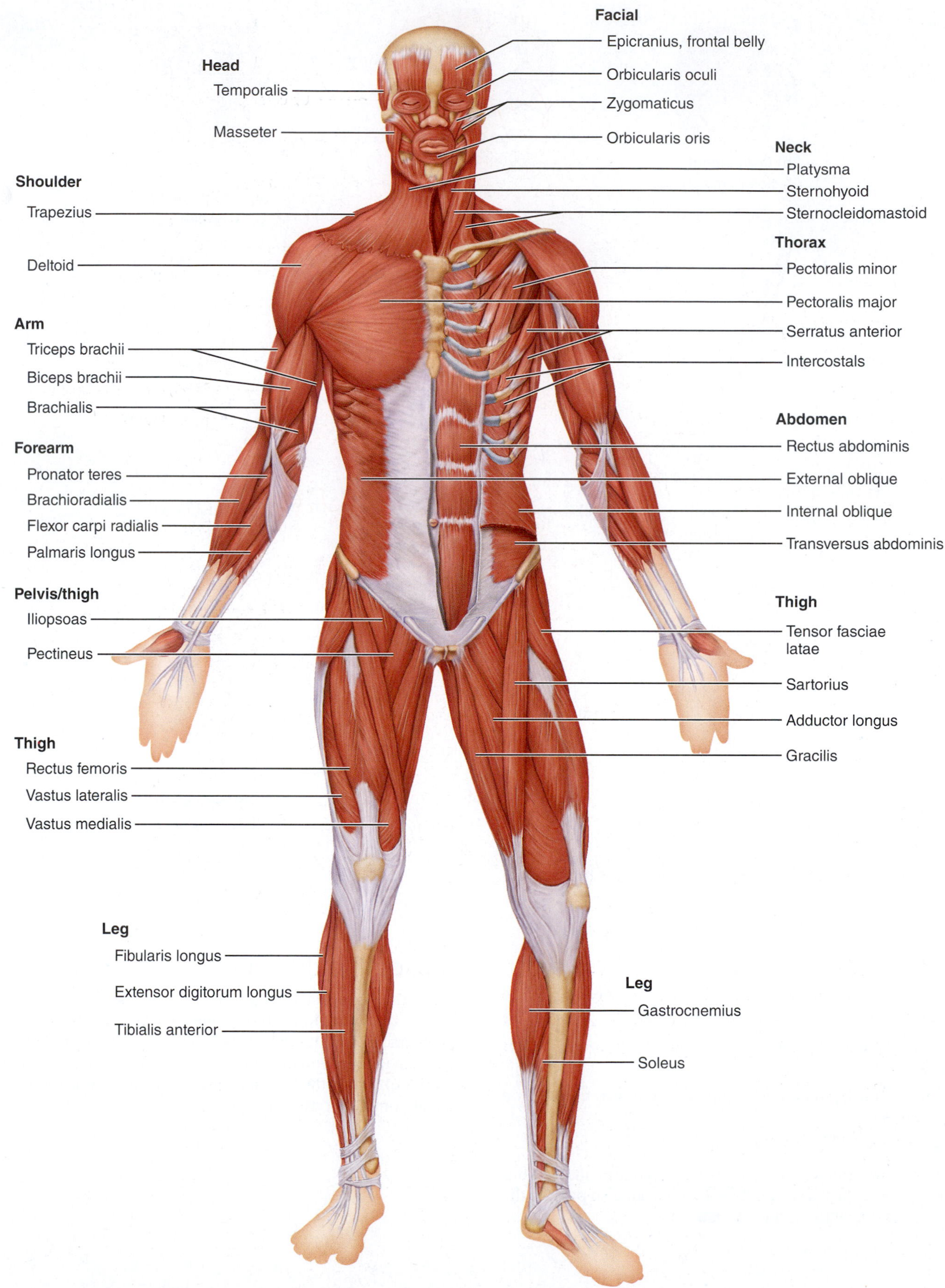

Figure 12.2 Anterior view of superficial muscles of the body. The abdominal surface has been partially dissected on the left side of the body to show somewhat deeper muscles.

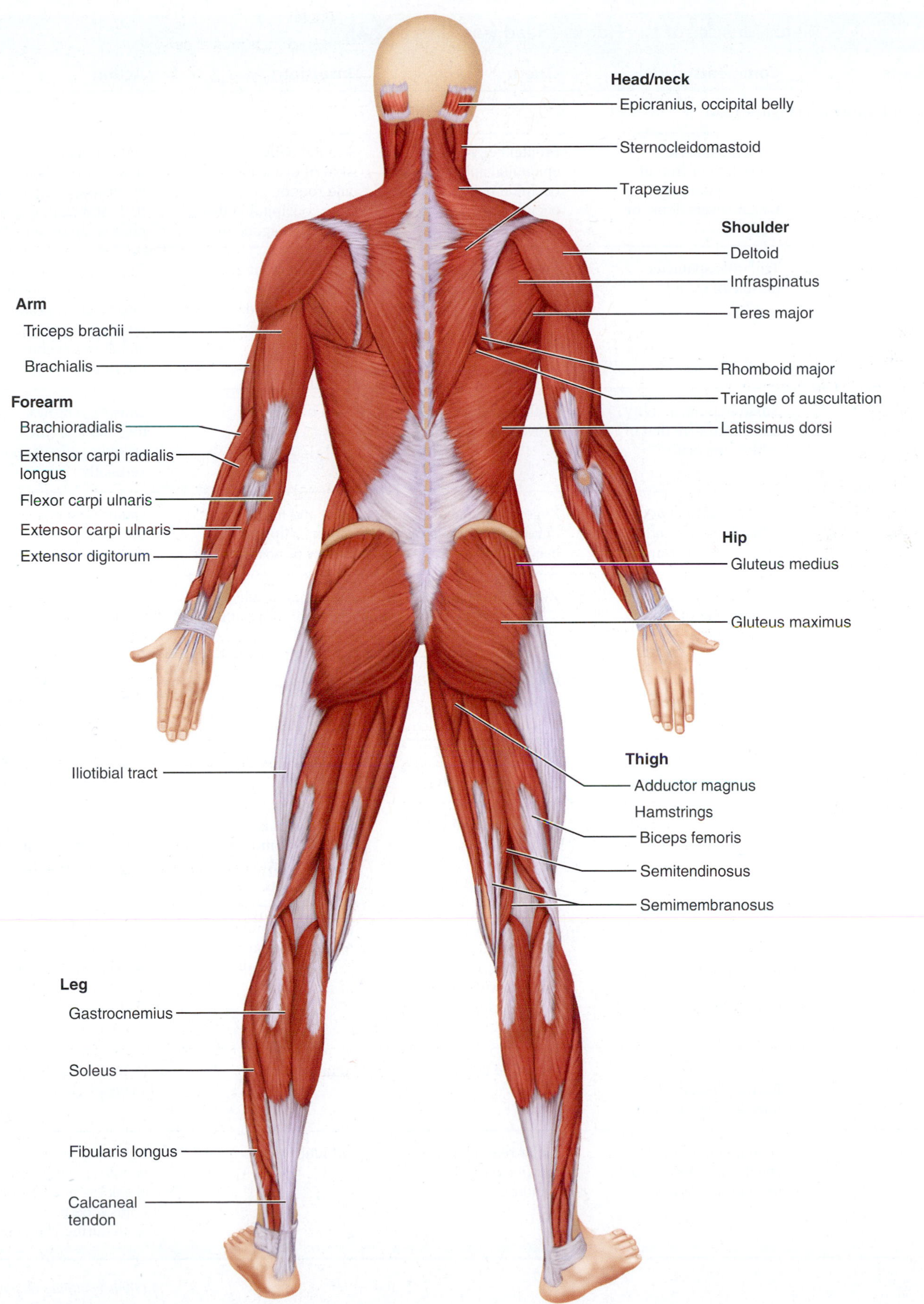

Figure 12.3 Posterior view of superficial muscles of the body.

Table 12.1 Major Muscles of the Human Head (see Figure 12.4)

Muscle	Comments	Origin	Insertion	Action
Facial Expression (Figure 12.4)				
Epicranius—frontal and occipital bellies	Bipartite muscle consisting of frontal and occipital parts, which covers dome of skull	Frontal belly—epicranial aponeurosis; occipital belly—occipital and temporal bones	Frontal belly—skin of eyebrows and root of nose; occipital belly—epicranial aponeurosis	With aponeurosis fixed, frontal belly raises eyebrows; occipital belly fixes aponeurosis and pulls scalp posteriorly
Orbicularis oculi	Tripartite sphincter muscle of eyelids	Frontal and maxillary bones and ligaments around orbit	Encircles orbit and inserts in tissue of eyelid	Closes eye; various parts can be activated individually; produces blinking, squinting, and draws eyebrows inferiorly
Corrugator supercilii	Small muscle; activity associated with that of orbicularis oculi	Arch of frontal bone above nasal bone	Skin of eyebrow	Draws eyebrows together and inferiorly; wrinkles skin of forehead vertically
Levator labii superioris	Thin muscle between orbicularis oris and inferior eye margin	Zygomatic bone and infraorbital margin of maxilla	Skin and muscle of upper lip and border of nostril	Opens lips; raises and furrows upper lip
Zygomaticus—major and minor	Extends diagonally from corner of mouth to cheekbone	Zygomatic bone	Skin and muscle at corner of mouth	Raises lateral corners of mouth upward (smiling muscle)
Risorius	Slender muscle; runs inferior and lateral to zygomaticus	Fascia of masseter muscle	Skin at angle of mouth	Draws corner of lip laterally; tenses lip; zygomaticus synergist
Depressor labii inferioris	Small muscle from lower lip to mandible	Body of mandible lateral to its midline	Skin and muscle of lower lip	Draws lower lip inferiorly
Depressor anguli oris	Small muscle lateral to depressor labii inferioris	Body of mandible below incisors	Skin and muscle at angle of mouth below insertion of zygomaticus	Draws corners of mouth downward and laterally; zygomaticus antagonist
Orbicularis oris	Multilayered muscle of lips with fibers that run in many different directions; most run circularly	Arises indirectly from maxilla and mandible; fibers blended with fibers of other muscles associated with lips	Encircles mouth; inserts into muscle and skin at angles of mouth	Closes lips; purses and protrudes lips (kissing and whistling muscle)
Mentalis	One of muscle pair forming V-shaped muscle mass on chin	Mandible below incisors	Skin of chin	Protrudes lower lip; wrinkles chin
Buccinator	Principal muscle of cheek; runs horizontally, deep to the masseter	Molar region of maxilla and mandible	Orbicularis oris	Draws corner of mouth laterally; compresses cheek (as in whistling); holds food between teeth during chewing

(Table continues on page 188.)

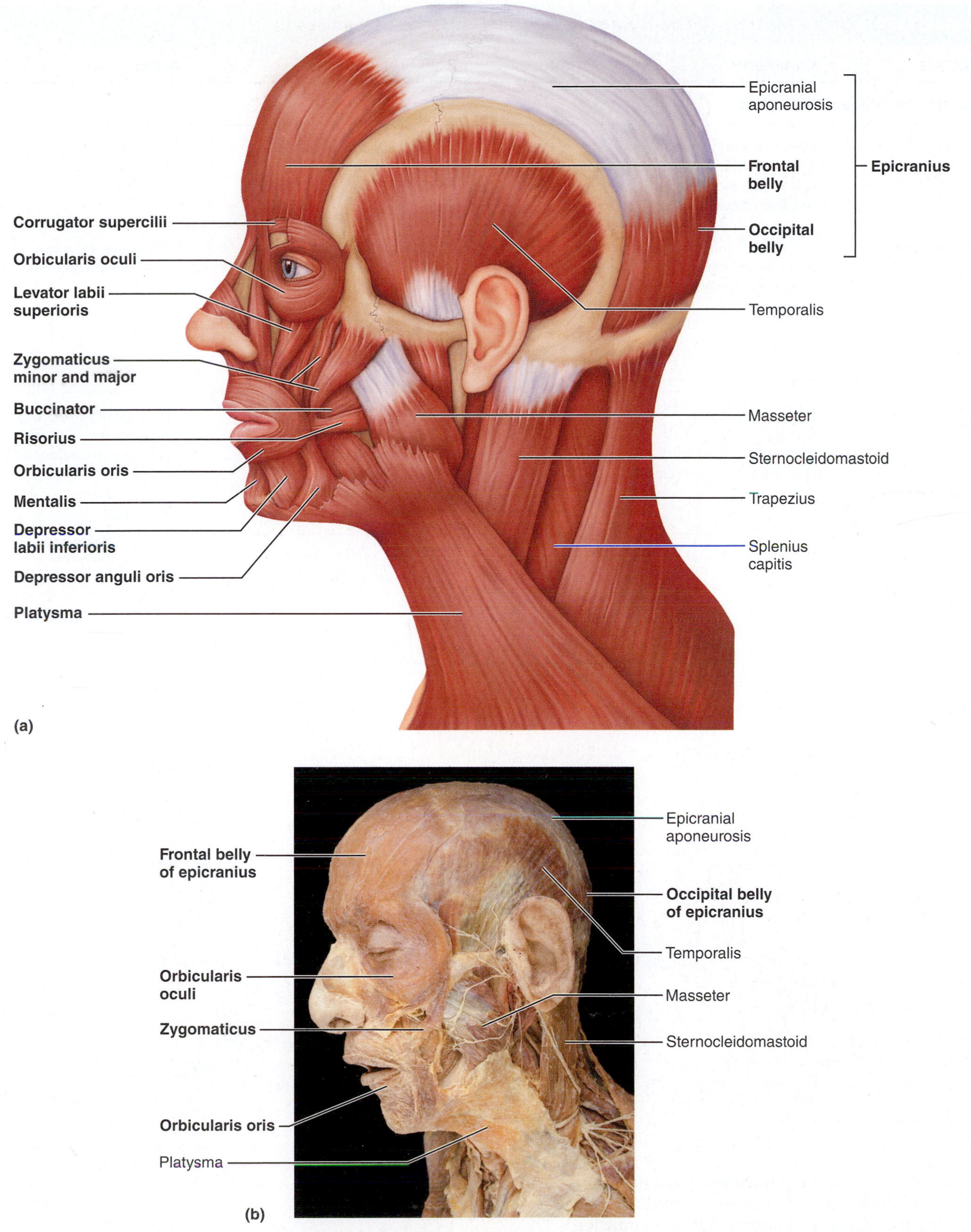

Figure 12.4 Muscles of the head (left lateral view). (a) Superficial muscles. **(b)** Cadaver photo of superficial structures of the head and neck.

Table 12.1 Major Muscles of the Human Head (see Figure 12.4) *(continued)*

Muscle	Comments	Origin	Insertion	Action
Mastication (Figure 12.4c and d)				
Masseter	Covers lateral aspect of mandibular ramus; can be palpated on forcible closure of jaws	Zygomatic arch and maxilla	Angle and ramus of mandible	Prime mover of jaw closure and elevates mandible
Temporalis	Fan-shaped muscle lying over parts of frontal, parietal, and temporal bones	Temporal fossa	Coronoid process of mandible	Closes jaw; elevates and retracts mandible
Buccinator	(See muscles of facial expression.)			
Medial pterygoid	Runs along internal (medial) surface of mandible (thus largely concealed by that bone)	Sphenoid, palatine, and maxillary bones	Medial surface of mandible, near its angle	Synergist of temporalis and masseter; elevates mandible; in conjunction with lateral pterygoid, aids in grinding movements
Lateral pterygoid	Superior to medial pterygoid	Greater wing of sphenoid bone	Condylar process of mandible	Protracts mandible (moves it anteriorly); in conjunction with medial pterygoid, aids in grinding movements of teeth

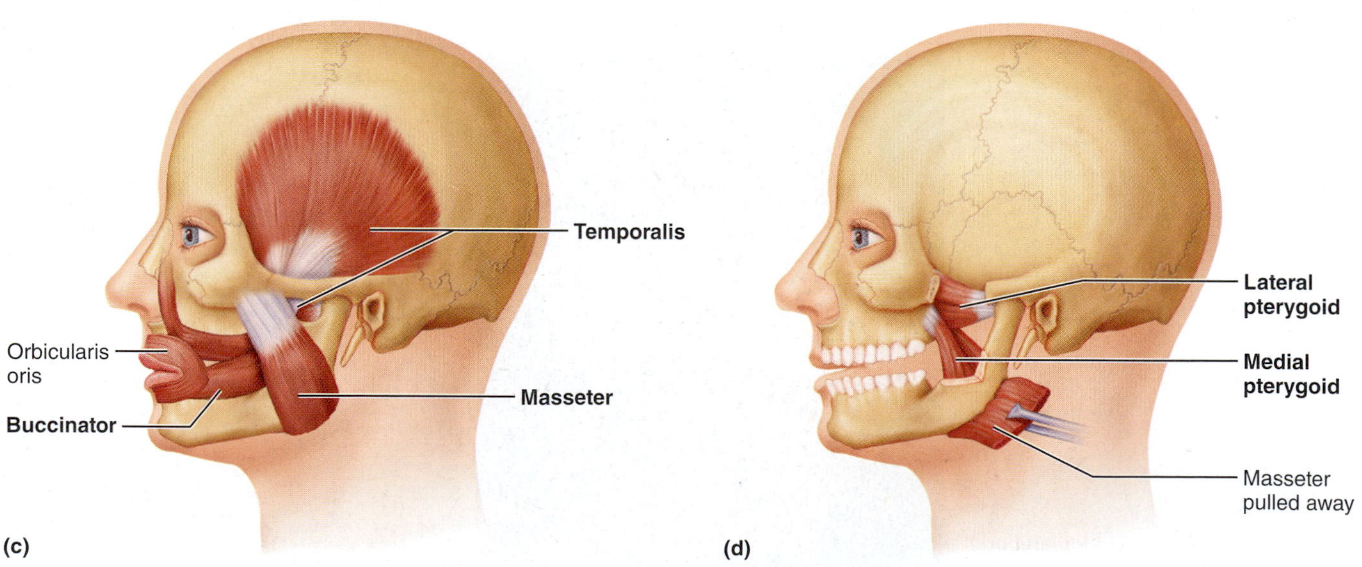

Figure 12.4 *(continued)* **Muscles of the head: (mastication). (c)** Lateral view of the temporalis, masseter, and buccinator muscles. **(d)** Lateral view of the deep chewing muscles, the medial and lateral pterygoid muscles.

Table 12.2 Anterolateral Muscles of the Human Neck (see Figure 12.5)

Muscle	Comments	Origin	Insertion	Action
Superficial				
Platysma	Unpaired muscle: thin, sheetlike superficial neck muscle, not strictly a head muscle but plays role in facial expression (see also Figure 12.4a)	Fascia of chest (over pectoral muscles) and deltoid	Lower margin of mandible, skin, and muscle at corner of mouth	Depresses mandible; pulls lower lip back and down (i.e., produces downward sag of the mouth); tenses skin of neck
Sternocleidomastoid	Two-headed muscle located deep to platysma on anterolateral surface of neck; fleshy parts on either side indicate limits of anterior and posterior triangles of neck	Manubrium of sternum and medial portion of clavicle	Mastoid process of temporal bone and superior nuchal line of occipital bone	Simultaneous contraction of both muscles of pair causes flexion of neck forward, generally against resistance (as when lying on the back); acting independently, rotate head toward shoulder on opposite side
Scalenes—anterior, middle, and posterior	Located more on lateral than anterior neck; deep to platysma and sternocleidomastoid (see Figure 12.5c)	Transverse processes of cervical vertebrae	Anterolaterally on ribs 1–2	Elevate ribs 1–2 (aid in inspiration); flex and slightly rotate neck

→

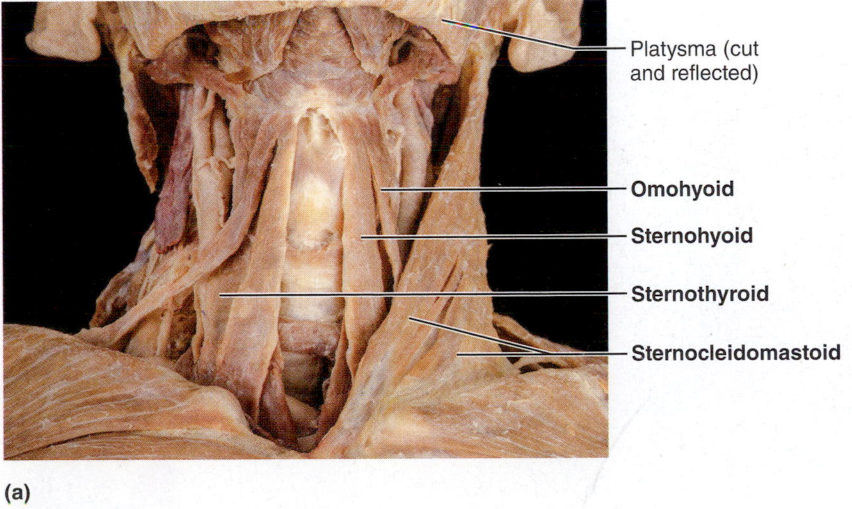

Figure 12.5 Muscles of the anterolateral neck and throat. (a) Cadaver photo of the anterior and lateral regions of the neck.

Table 12.2 Anterolateral Muscles of the Human Neck (see Figure 12.5) *(continued)*				
Muscle	**Comments**	**Origin**	**Insertion**	**Action**
Deep (Figure 12.5a and b)				
Digastric	Consists of two bellies united by an intermediate tendon; assumes a V-shaped configuration under chin	Lower margin of mandible (anterior belly) and mastoid process (posterior belly)	By a connective tissue loop to hyoid bone	Open mouth and depress mandible; acting in concert, elevate hyoid bone
Stylohyoid	Slender muscle parallels posterior border of digastric; below angle of jaw	Styloid process of temporal bone	Hyoid bone	Elevates and retracts hyoid bone
Mylohyoid	Just deep to digastric; forms floor of mouth	Medial surface of mandible	Hyoid bone and median raphe	Elevates hyoid bone and floor of mouth during swallowing
Sternohyoid	Runs most medially along neck; straplike	Manubrium and medial end of clavicle	Lower margin of body of hyoid bone	Depresses larynx and hyoid bone if mandible is fixed; may also flex skull
Sternothyroid	Lateral and deep to sternohyoid	Posterior surface of manubrium	Thyroid cartilage of larynx	Pulls larynx and hyoid bone inferiorly
Omohyoid	Straplike with two bellies; lateral to sternohyoid	Superior surface of scapula	Hyoid bone; inferior border	Depresses and retracts hyoid bone
Thyrohyoid	Appears as a superior continuation of sternothyroid muscle	Thyroid cartilage	Hyoid bone	Depresses hyoid bone; elevates larynx if hyoid is fixed

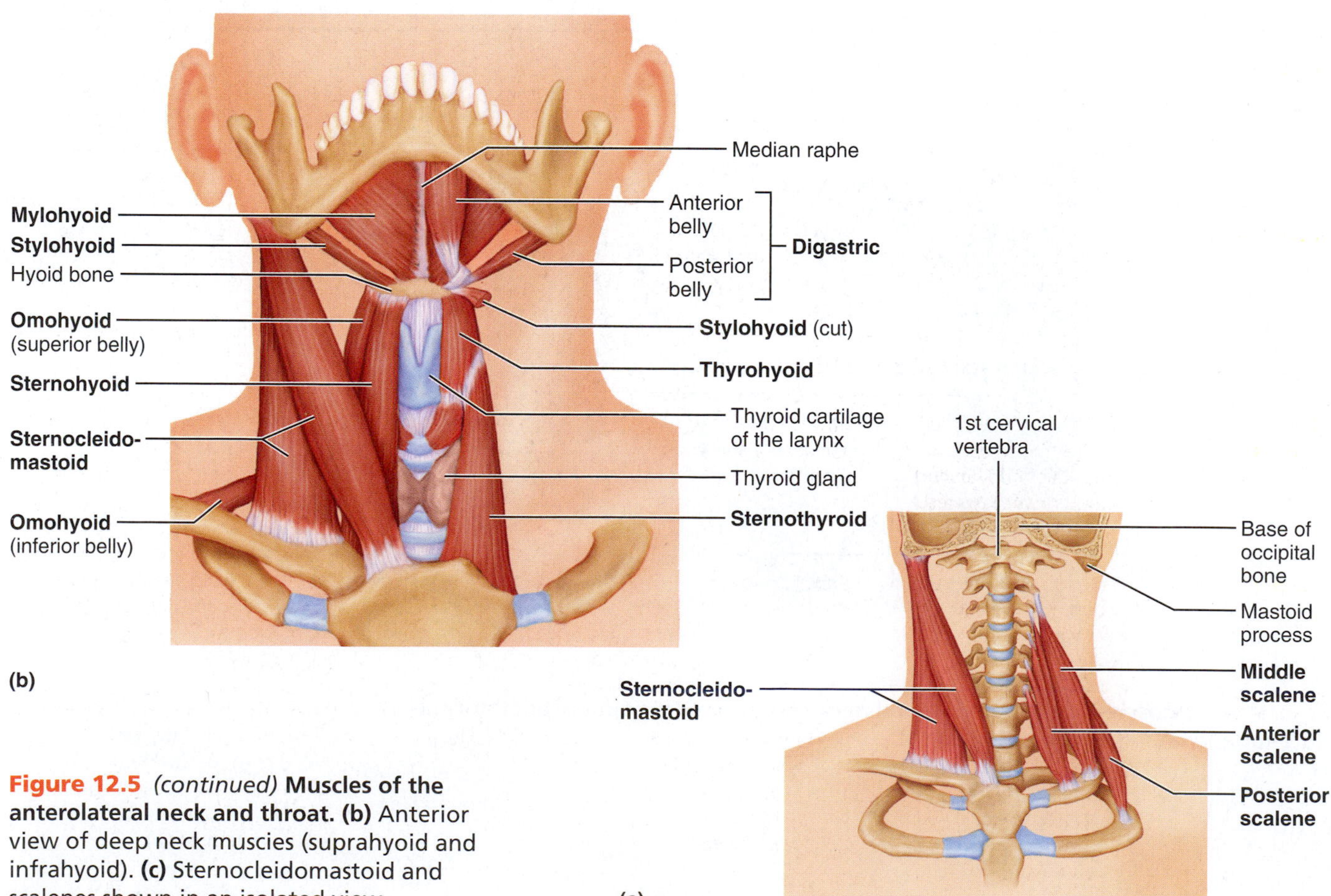

Figure 12.5 *(continued)* **Muscles of the anterolateral neck and throat. (b)** Anterior view of deep neck muscles (suprahyoid and infrahyoid). **(c)** Sternocleidomastoid and scalenes shown in an isolated view.

ACTIVITY 2

Identifying Muscles of the Trunk

Read the descriptions of specific trunk muscles, and identify them in the figures (**Table 12.3** and **Table 12.4** and **Figures 12.6**, **12.7**, **12.8**, and **12.9**), visualizing their action when they contract. Then identify them on a torso model or anatomical chart.

Demonstrating Operation of the Trunk Muscles

Now, work with a partner to demonstrate the operation of the following muscles. One of you can demonstrate the movement; the following steps are addressed to this partner. The other can supply resistance and palpate the muscle being tested.

1. Fully abduct the arm and extend the elbow. Now adduct the arm against resistance. You are using the *latissimus dorsi.*

2. To observe the *deltoid,* try to abduct your arm against resistance. Now attempt to elevate your shoulder against resistance; you are contracting the upper portion of the *trapezius.*

3. You use the *pectoralis major* when you press your hands together at chest level with your elbows widely abducted. ■

(Text continues on page 197.)

Table 12.3 Anterior Muscles of the Human Thorax, Shoulder, and Abdominal Wall (see Figures 12.6, 12.7, and 12.8)

Muscle	Comments	Origin	Insertion	Action
Thorax and Shoulder, Superficial (Figure 12.6)				
Pectoralis major	Large fan-shaped muscle covering upper portion of chest	Clavicle, sternum, cartilage of ribs 1–6 (or 7), and aponeurosis of external oblique muscle	Fibers converge to insert by short tendon into intertubercular sulcus of humerus	Prime mover of arm flexion; adducts, medially rotates arm; with arm fixed, pulls chest upward (thus also acts in forced inspiration)
Serratus anterior	Fan-shaped muscle deep to scapula; beneath and inferior to pectoral muscles on lateral rib cage	Lateral aspect of ribs 1–8 (or 9)	Vertebral border of anterior surface of scapula	Prime mover to protect and hold scapula against chest wall; rotates scapula, causing inferior angle to move laterally and upward; abduction and raising of arm (called "boxer's muscle")
Deltoid (see also Figure 12.9a)	Fleshy triangular muscle forming shoulder muscle mass; intramuscular injection site	Lateral third of clavicle; acromion and spine of scapula	Deltoid tuberosity of humerus	Acting as a whole, prime mover of arm abduction; when only specific fibers are active, can aid in flexion, extension, and rotation of humerus
Pectoralis minor	Flat, thin muscle directly beneath and obscured by pectoralis major	Anterior surface of ribs 3–5, near their costal cartilages	Coracoid process of scapula	With ribs fixed, draws scapula forward and inferiorly; with scapula fixed, draws rib cage superiorly
Thorax, Deep: Muscles of Respiration (Figure 12.7)				
External intercostals	11 pairs lie between ribs; fibers run obliquely downward and forward toward sternum	Inferior border of rib above (not shown in figure)	Superior border of rib below	Pull ribs toward one another to elevate rib cage; aid in inspiration
Internal intercostals	11 pairs lie between ribs; fibers run deep and at right angles to those of external intercostals	Superior border of rib below	Inferior border of rib above (not shown in figure)	Draw ribs together to depress rib cage; aid in forced expiration; antagonistic to external intercostals
Diaphragm	Broad muscle; forms floor of thoracic cavity; dome-shaped in relaxed state; fibers converge toward a central tendon	Inferior border of rib cage and sternum, costal cartilages of last six ribs and lumbar vertebrae	Central tendon	Prime mover of inspiration: flattens on contraction, increasing vertical dimensions of thorax; increases intra-abdominal pressure

→

Table 12.3 Anterior Muscles of the Human Thorax, Shoulder, and Abdominal Wall (see Figures 12.6, 12.7, and 12.8) *(continued)*

Muscle	Comments	Origin	Insertion	Action
Abdominal Wall (Figure 12.8)				
Rectus abdominis	Medial superficial muscle, extends from pubis to rib cage; ensheathed by aponeuroses of oblique muscles; segmented	Pubic crest and symphysis	Xiphoid process and costal cartilages of ribs 5–7	Flexes and rotates vertebral column; increases abdominal pressure; fixes and depresses ribs; stabilizes pelvis during walking; used in sit-ups and curls
External oblique	Most superficial lateral muscle; fibers run downward and medially; ensheathed by an aponeurosis	Anterior surface of last eight ribs	Linea alba,* pubic crest and tubercles, and iliac crest	Flex vertebral column and compress abdominal wall; also aids muscles of back in trunk rotation and lateral flexion; used in oblique curls
Internal oblique	Most fibers run at right angles to those of external oblique, which it underlies	Lumbar fascia, iliac crest, and inguinal ligament	Linea alba, pubic crest, and costal cartilages of last three ribs	As for external oblique
Transversus abdominis	Deepest muscle of abdominal wall; fibers run horizontally	Inguinal ligament, iliac crest, cartilages of last five or six ribs, and lumbar fascia	Linea alba and pubic crest	Compresses abdominal contents

*The linea alba (white line) is a narrow, tendinous sheath that runs along the middle of the abdomen from the sternum to the pubic symphysis. It is formed by the fusion of the aponeurosis of the external oblique and transversus muscles.

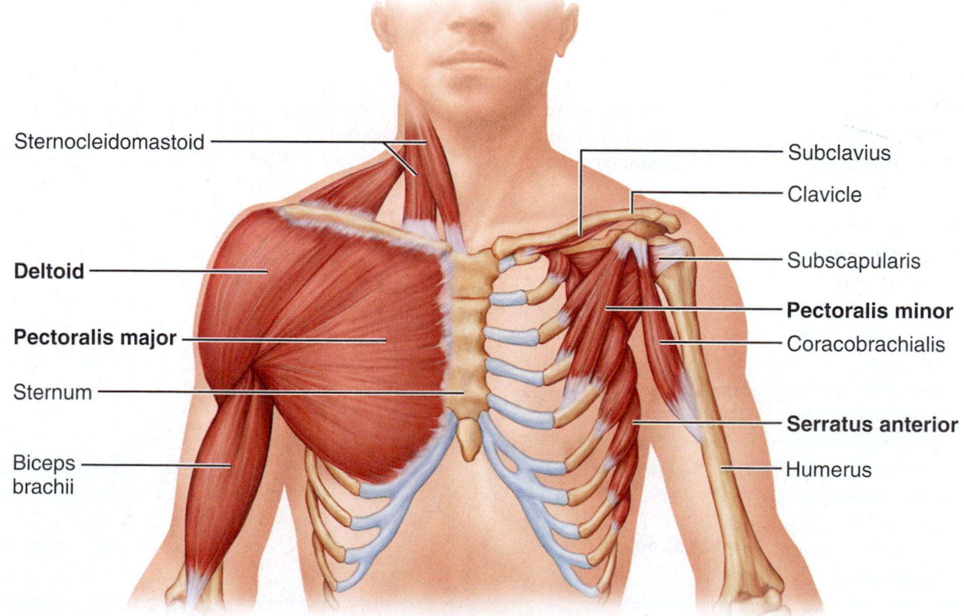

Figure 12.6 Superficial muscles of the thorax and shoulder acting on the scapula and arm (anterior view). The superficial muscles, which effect arm movements, are shown on the left. These muscles have been removed on the right side of the figure to show the muscles that stabilize or move the pectoral girdle.

Figure 12.7 Deep muscles of the thorax: muscles of respiration. (a) The external intercostals (inspiratory muscles) (left) and the internal intercostals (expiratory muscles) (right) run obliquely and at right angles to each other. **(b)** Inferior view of the diaphragm. Notice that its muscle fibers converge toward a central tendon, which causes the diaphragm to flatten and move inferiorly as it contracts.

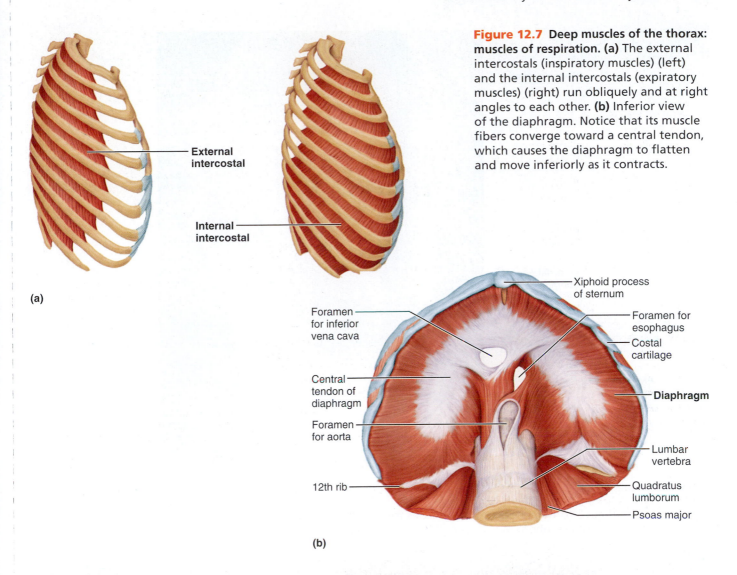

External intercostal
Internal intercostal
(a)

Xiphoid process of sternum
Foramen for inferior vena cava
Foramen for esophagus
Costal cartilage
Central tendon of diaphragm
Diaphragm
Foramen for aorta
Lumbar vertebra
12th rib
Quadratus lumborum
Psoas major
(b)

Figure 12.8 Anterior view of the muscles forming the anterolateral abdominal wall. (a) The superficial muscles have been partially cut away on the left side of the diagram to reveal the deeper internal oblique and transversus abdominis muscles.

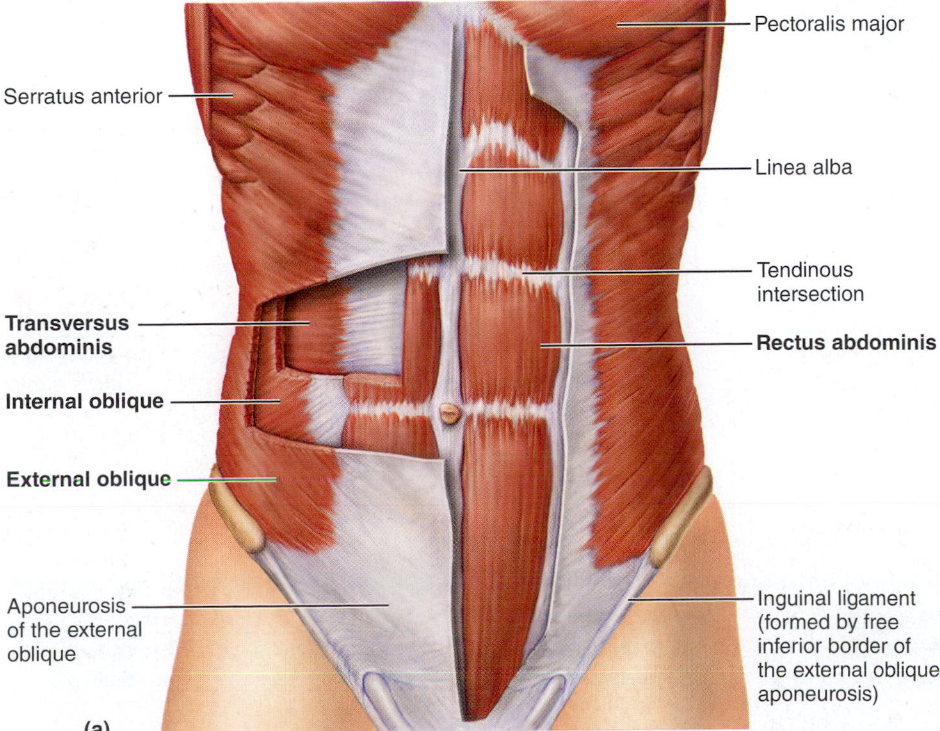

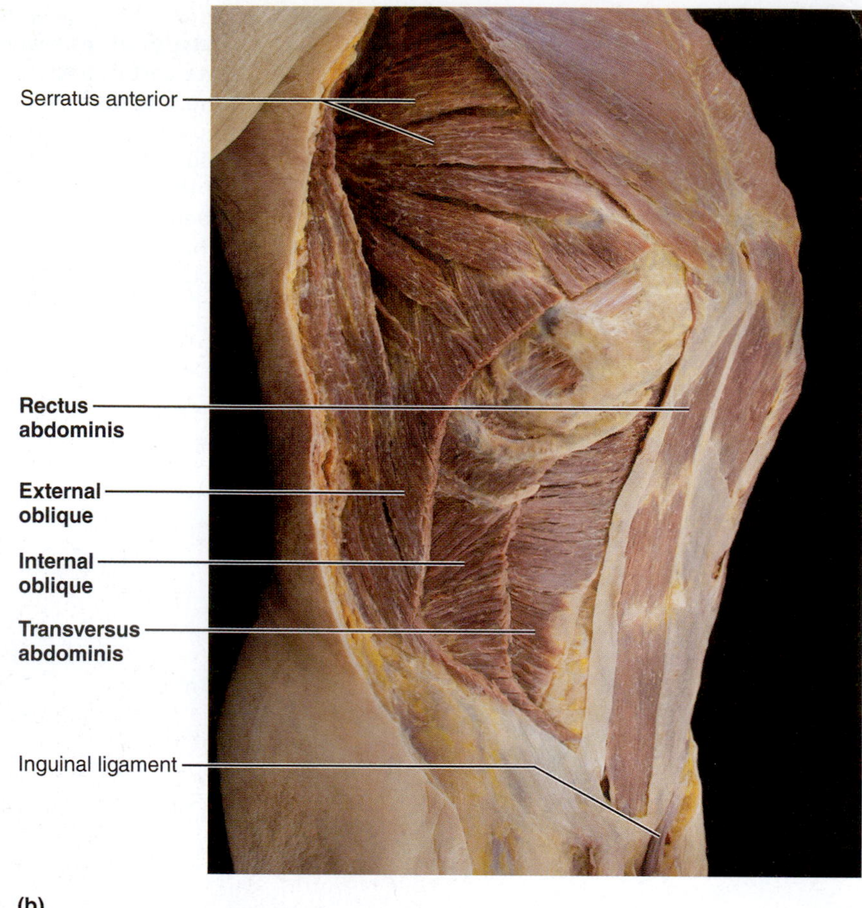

Figure 12.8 *(continued)* **Anterior view of the muscles forming the anterolateral abdominal wall. (b)** Cadaver photo of the anterolateral abdominal wall.

Table 12.4 Posterior Muscles of the Human Trunk (see Figure 12.9)

Muscle	Comments	Origin	Insertion	Action
Muscles of the Neck, Shoulder, and Thorax (Figure 12.9a)				
Trapezius	Most superficial muscle of posterior thorax; very broad origin and insertion	Occipital bone; ligamentum nuchae; spines of C_7 and all thoracic vertebrae	Acromion and spinous process of scapula; lateral third of clavicle	Stabilizes, raises, rotates, and retracts scapula; superior fibers elevate scapula (as in shrugging the shoulders) or can extend head; inferior fibers depress scapula
Latissimus dorsi	Broad flat muscle of lower back (lumbar region); extensive superficial origins	Indirect attachment to spinous processes of lower six thoracic vertebrae, lumbar vertebrae, last three to four ribs, and iliac crest	Floor of intertubercular sulcus of humerus	Prime mover of arm extension; adducts and medially rotates arm; brings arm down in power stroke, as in striking a blow
Infraspinatus	Partially covered by deltoid and trapezius; a rotator cuff muscle	Infraspinous fossa of scapula	Greater tubercle of humerus	Lateral rotation of humerus; helps hold head of humerus in glenoid cavity; stabilizes shoulder
Teres minor	Small muscle inferior to infraspinatus; a rotator cuff muscle	Lateral margin of scapula	Greater tubercle of humerus	As for infraspinatus

Table 12.4 *(continued)*

Muscle	Comments	Origin	Insertion	Action
Teres major	Located inferiorly to teres minor	Posterior surface at inferior angle of scapula	Intertubercular sulcus of humerus	Extends, medially rotates, and adducts humerus; synergist of latissimus dorsi
Supraspinatus	Obscured by trapezius; a rotator cuff muscle	Supraspinous fossa of scapula	Greater tubercle of humerus	Initiates abduction of humerus; stabilizes shoulder joint
Levator scapulae	Located at back and side of neck, deep to trapezius	Transverse processes of C_1–C_4	Medial border of scapula superior to spine	Elevates and adducts scapula; with fixed scapula, laterally flexes neck to the same side
Rhomboids—major and minor	Beneath trapezius and inferior to levator scapulae; rhomboid minor is the more superior muscle	Spinous processes of C_7 and T_1–T_5	Medial border of scapula	Pulls scapula medially (retraction); stabilizes scapula; rotates glenoid cavity downward
Muscles Associated with the Vertebral Column (Figure 12.9b)				
Semispinalis	Deep composite muscle of the back—thoracis, cervicis, and capitis portions	Transverse processes of C_7–T_{12}	Occipital bone and spinous processes of cervical vertebrae and T_1–T_4	Acting together, extend head and vertebral column; independently cause rotation toward opposite side

(Table continues on page 197.)

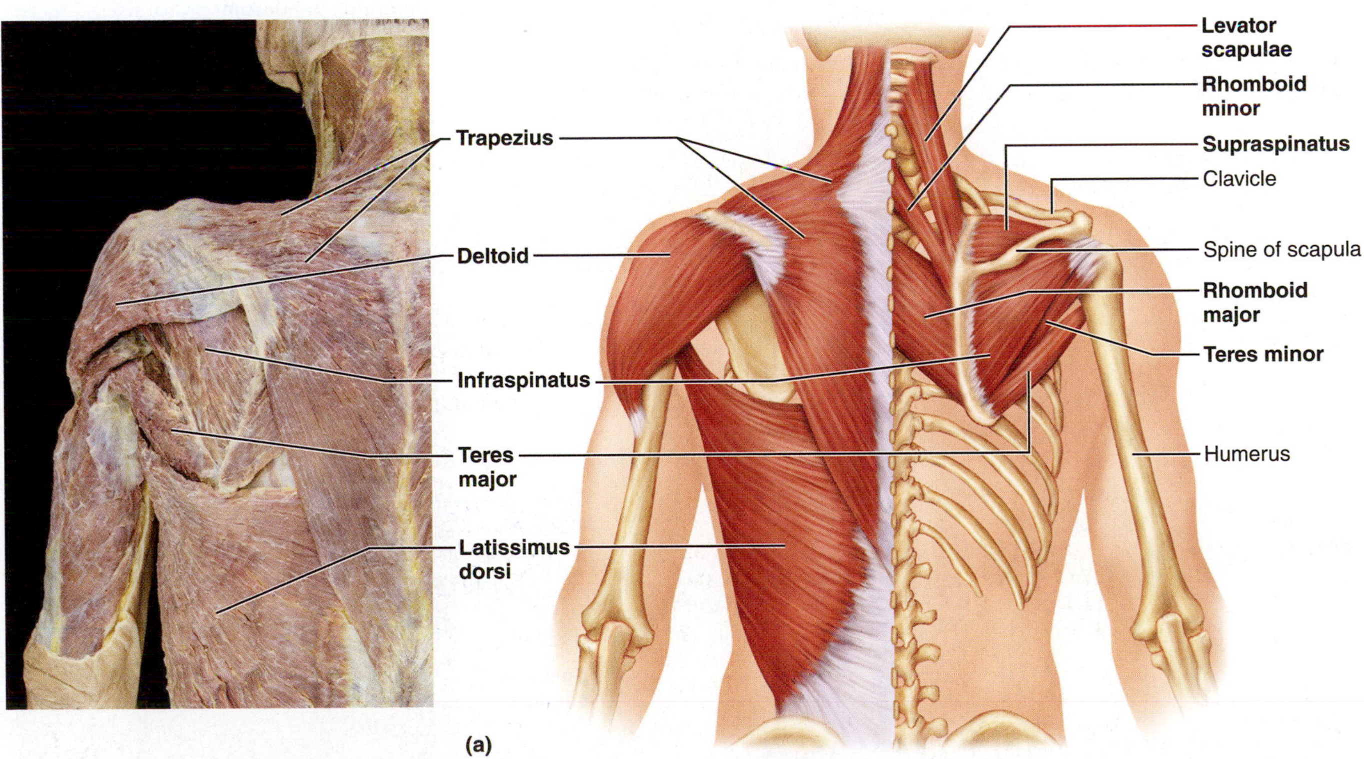

Figure 12.9 Muscles of the neck, shoulder, and thorax (posterior view). (a) The superficial muscles of the back are shown for the left side of the body, with a corresponding cadaver photograph. The superficial muscles are removed on the right side of the diagram to reveal the deeper muscles acting on the scapula and the rotator cuff muscles that help to stabilize the shoulder joint.

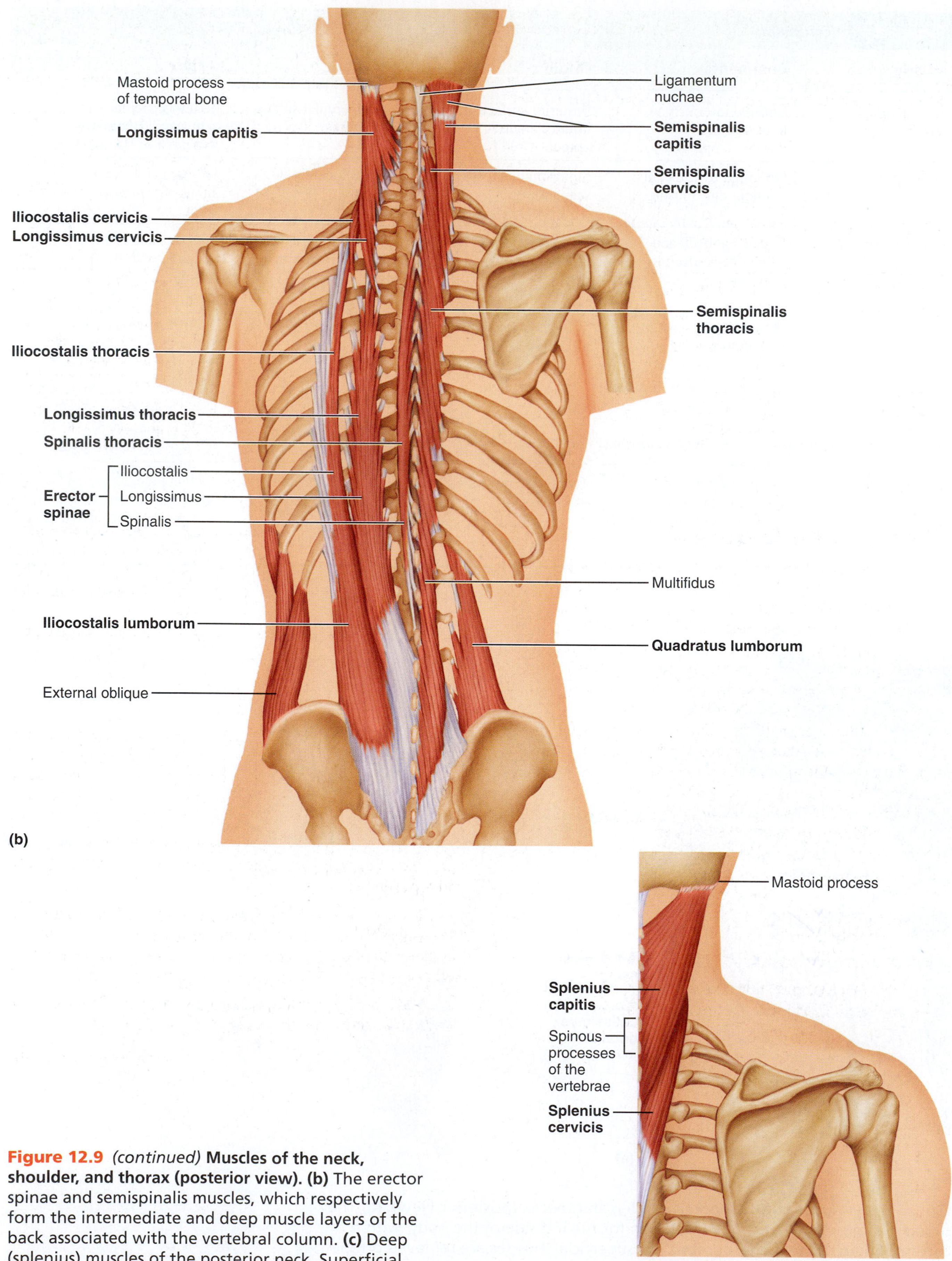

Figure 12.9 *(continued)* **Muscles of the neck, shoulder, and thorax (posterior view). (b)** The erector spinae and semispinalis muscles, which respectively form the intermediate and deep muscle layers of the back associated with the vertebral column. **(c)** Deep (splenius) muscles of the posterior neck. Superficial muscles have been removed.

Table 12.4 Posterior Muscles of the Human Trunk (see Figure 12.9) *(continued)*

Muscle	Comments	Origin	Insertion	Action
Muscles Associated with the Vertebral Column (Figure 12.9b), *(continued)*				
Erector spinae	A long tripartite muscle composed of iliocostalis (lateral), longissimus, and spinalis (medial) muscle columns; superficial to semispinalis muscles; extends from pelvis to head	Iliac crest, transverse processes of lumbar, thoracic, and cervical vertebrae, and/or ribs 3–6 depending on specific part	Ribs and transverse processes of vertebrae about six segments above origin; longissimus also inserts into mastoid process	Extend and laterally flex the vertebral column; fibers of the longissimus also extend head
Splenius (see Figure 12.9c)	Superficial muscle (capitis and cervicis parts) extending from upper thoracic region to skull	Ligamentum nuchae and spinous processes of C_7–T_6	Mastoid process, occipital bone, and transverse processes of C_2–C_4	As a group, extend or hyperextend head; when only one side is active, head is rotated and bent toward the same side
Quadratus lumborum	Forms greater portion of posterior abdominal wall	Iliac crest and lumbar fascia	Inferior border of rib 12; transverse processes of lumbar vertebrae	Each flexes vertebral column laterally; together extend the lumbar spine and fix rib 12; maintains upright posture

Muscles of the Upper Limb

The muscles that act on the upper limb fall into three groups: those that move the arm, those causing movement at the elbow, and those moving the wrist and hand.

The muscles that cross the shoulder joint to insert on the humerus and move the arm (subscapularis, supraspinatus and infraspinatus, deltoid, and so on) are primarily trunk muscles that originate on the axial skeleton or shoulder girdle. These muscles are included with the trunk muscles.

The second group of muscles, which cross the elbow joint and move the forearm, consists of muscles forming the musculature of the humerus. These muscles arise primarily from the humerus and insert in forearm bones. They are responsible for flexion, extension, pronation, and supination.

The third group forms the musculature of the forearm. For the most part, these muscles insert on the digits and produce movements at the wrist and fingers.

ACTIVITY 3

Identifying Muscles of the Upper Limb

Study the origins, insertions, and actions of muscles that move the forearm, and identify them in the figure (**Table 12.5** and **Figure 12.10**).

Do the same for muscles acting on the wrist and hand (**Table 12.6** and **Figure 12.11**). You can identify them more easily if you locate their insertion tendons first. (Muscles of the hand are described in **Table 12.7** and illustrated in **Figure 12.12**.)

As before, identify these muscles on a torso model, anatomical chart, or cadaver.

Demonstrating Operations of the Upper Limb Muscles

1. To observe the *biceps brachii,* attempt to flex your forearm (hand supinated) against resistance. You can also feel the insertion tendon of this biceps muscle in the lateral aspect of the cubital fossa (where it runs toward the radius to attach).

2. If you acutely flex your elbow and then try to extend it against resistance, you can demonstrate the action of your *triceps brachii.*

3. Strongly flex your wrist, and make a fist. Palpate your contracting wrist flexor muscles (which originate from the medial epicondyle of the humerus) and their insertion tendons, which you can easily feel at the anterior aspect of the wrist.

4. Flare your fingers to identify the tendons of the *extensor digitorum* muscle on the dorsum of your hand. ■

(Text continues on page 204.)

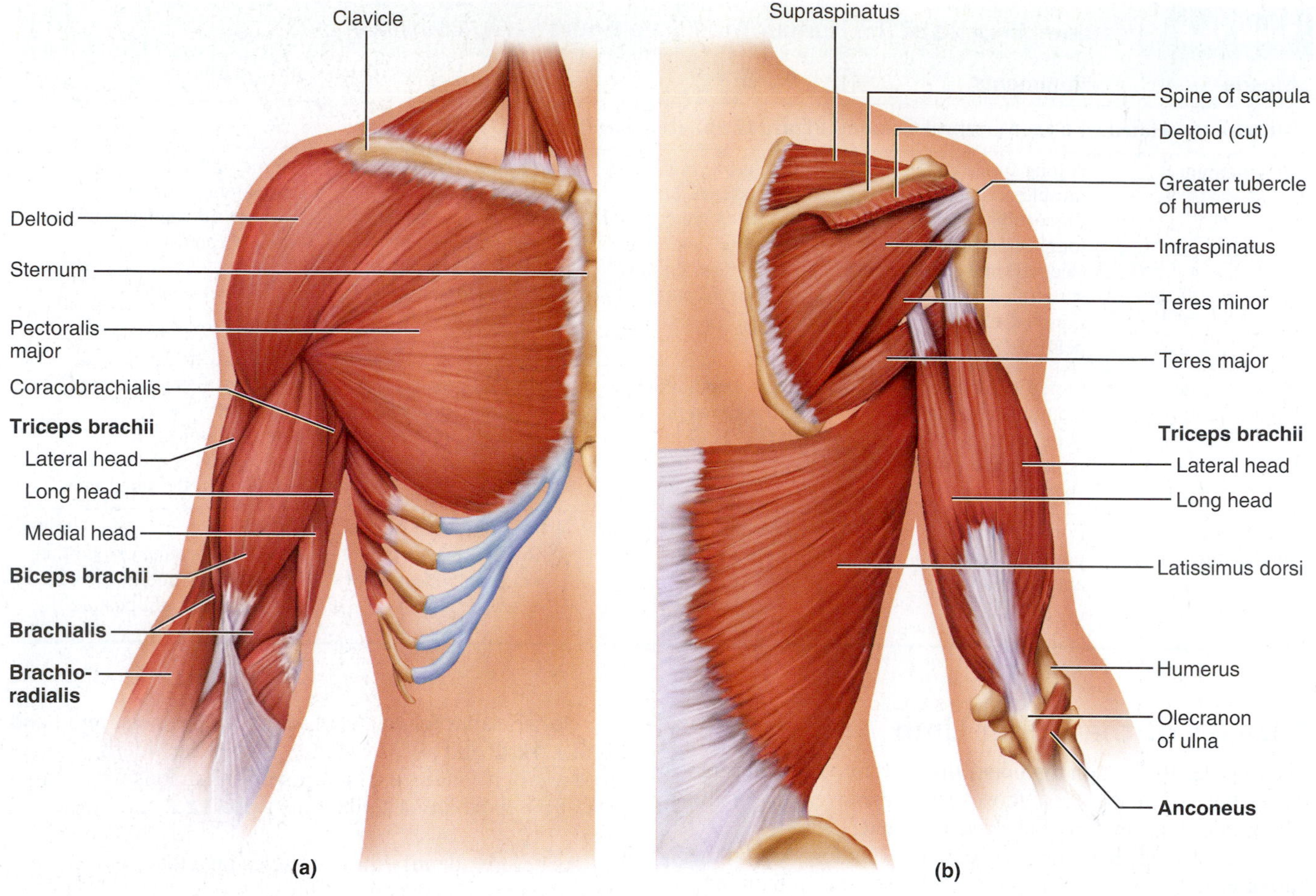

Figure 12.10 Muscles causing movements of the arm and forearm. (a) Superficial muscles of the anterior thorax, shoulder, and arm, anterior view. **(b)** Posterior aspect of the arm showing the lateral and long heads of the triceps brachii muscle. The supraspinatus, infraspinatus, teres minor, and subscapularis (Figure 12.6) are rotator cuff muscles.

Table 12.5 Muscles of the Human Humerus That Act on the Forearm (see Figure 12.10)

Muscle	Comments	Origin	Insertion	Action
Triceps brachii	Sole, large fleshy muscle of posterior humerus; three-headed origin	Long head—inferior margin of glenoid cavity; lateral head—posterior humerus; medial head—distal radial groove on posterior humerus	Olecranon of ulna	Powerful forearm extensor; antagonist of forearm flexors (brachialis and biceps brachii)
Anconeus (see also Figure 12.11d and e)	Short triangular muscle blended with triceps	Lateral epicondyle of humerus	Lateral aspect of olecranon of ulna	Abducts ulna during forearm pronation; extends elbow
Biceps brachii	Most familiar muscle of anterior humerus because this two-headed muscle bulges when forearm is flexed	Short head: coracoid process; long head; supraglenoid tubercle and tip of glenoid cavity; tendon of long head runs in intertubercular sulcus and within capsule of shoulder joint	Radial tuberosity	Flexion (powerful) of elbow and supination of forearm; "it turns the corkscrew and pulls the cork"; weak arm flexor
Brachioradialis (see also Figure 12.11a)	Superficial muscle of lateral forearm; forms lateral boundary of antecubital fossa	Lateral ridge at distal end of humerus	Base of radial styloid process	Synergist in forearm flexion
Brachialis	Immediately deep to biceps brachii	Distal portion of anterior humerus	Coronoid process of ulna	A major flexor of forearm

Table 12.6 Muscles of the Human Forearm That Act on the Hand and Fingers (see Figure 12.11)

Muscle	Comments	Origin	Insertion	Action
Anterior Compartment (Figure 12.11a, b, c)				
Superficial				
Pronator teres	Seen in a superficial view between proximal margins of brachioradialis and flexor carpi radialis	Medial epicondyle of humerus and coronoid process of ulna	Midshaft of radius	Acts synergistically with pronator quadratus to pronate forearm; weak elbow flexor
Flexor carpi radialis	Superficial; runs diagonally across forearm	Medial epicondyle of humerus	Base of metacarpals II and III	Powerful flexor of wrist; abducts hand
Palmaris longus	Small fleshy muscle with a long tendon; medial to flexor carpi radialis	Medial epicondyle of humerus	Palmar aponeurosis; skin and fascia of palm	Flexes wrist (weak); tenses skin and fascia of palm

→

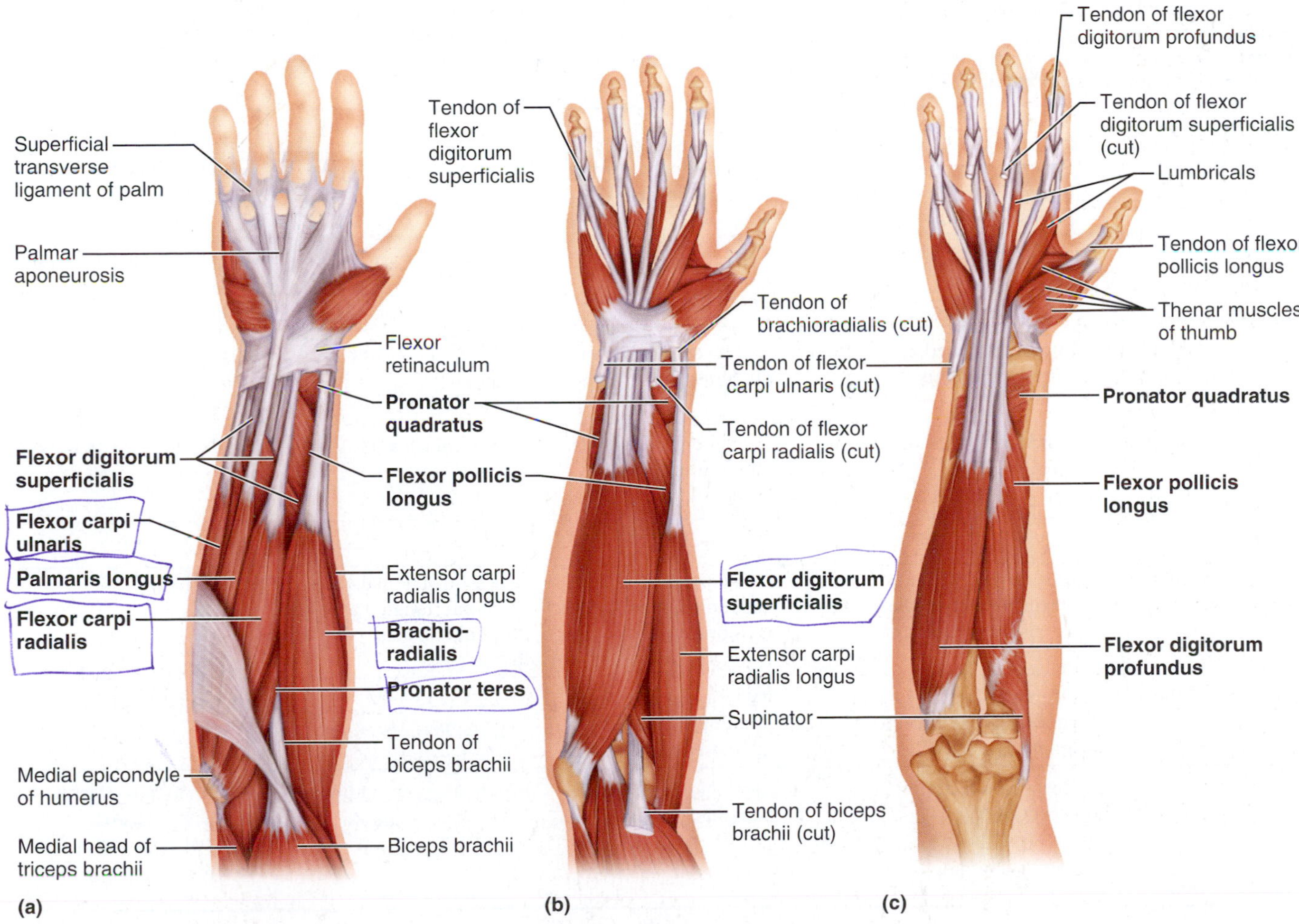

Figure 12.11 Muscles of the forearm and wrist. (a) Superficial anterior view of right forearm and hand. **(b)** The brachioradialis, flexors carpi radialis and ulnaris, and palmaris longus muscles have been removed to reveal the position of the somewhat deeper flexor digitorum superficialis. **(c)** Deep muscles of the anterior compartment. Superficial muscles have been removed. *Note:* The thenar muscles of the thumb and the lumbricals that help move the fingers are illustrated here but not described in the table (Table 12.6).

Table 12.6 Muscles of the Human Forearm That Act on the Hand and Fingers (see Figure 12.11) *(continued)*

Muscle	Comments	Origin	Insertion	Action
Flexor carpi ulnaris	Superficial; medial to palmaris longus	Medial epicondyle of humerus, olecranon process and posterior surface of ulna	Base of metacarpal V; pisiform and hamate bones	Powerful flexor of wrist; adducts hand
Flexor digitorum superficialis	Deeper muscle (deep to muscles named above); visible at distal end of forearm	Medial epicondyle of humerus, coronoid process of ulna, and shaft of radius	Middle phalanges of fingers 2–5	Flexes wrist and middle phalanges of fingers 2–5
Deep				
Flexor pollicis longus	Deep muscle of anterior forearm; distal to and paralleling lower margin of flexor digitorum superficialis	Anterior surface of radius, and interosseous membrane	Distal phalanx of thumb	Flexes distal phalanx of thumb
Flexor digitorum profundus	Deep muscle; overlain entirely by flexor digitorum superficialis	Anteromedial surface of ulna, interosseous membrane, and coronoid process	Distal phalanges of fingers 2–5	Sole muscle that flexes distal phalanges; assists in wrist flexion
Pronator quadratus	Deepest muscle of distal forearm	Distal portion of anterior ulnar surface	Anterior surface of radius, distal end	Pronates forearm
Posterior Compartment (Figure 12.11d, e, f)				
Superficial				
Extensor carpi radialis longus	Superficial; parallels brachioradialis on lateral forearm	Lateral supracondylar ridge of humerus	Base of metacarpal II	Extends and abducts wrist
Extensor carpi radialis brevis	Deep to extensor carpi radialis longus	Lateral epicondyle of humerus	Base of metacarpal III	Extends and abducts wrist; steadies wrist during finger flexion
Extensor digitorum	Superficial; medial to extensor carpi radialis brevis	Lateral epicondyle of humerus	By four tendons into distal phalanges of fingers 2–5	Prime mover of finger extension; extends wrist; can abduct (flare) fingers
Extensor carpi ulnaris	Superficial; medial posterior forearm	Lateral epicondyle of humerus; posterior border of ulna	Base of metacarpal V	Extends and adducts wrist
Deep				
Extensor pollicis longus and brevis	Muscle pair with a common origin and action; deep to extensor carpi ulnaris	Dorsal shaft of ulna and radius, interosseous membrane	Base of distal phalanx of thumb (longus) and proximal phalanx of thumb (brevis)	Extends thumb
Abductor pollicis longus	Deep muscle; lateral and parallel to extensor pollicis longus	Posterior surface of radius and ulna; interosseous membrane	Metacarpal I and trapezium	Abducts and extends thumb
Supinator	Deep muscle at posterior aspect of elbow	Lateral epicondyle of humerus; proximal ulna	Proximal end of radius	Acts with biceps brachii to supinate forearm; antagonist of pronator muscles

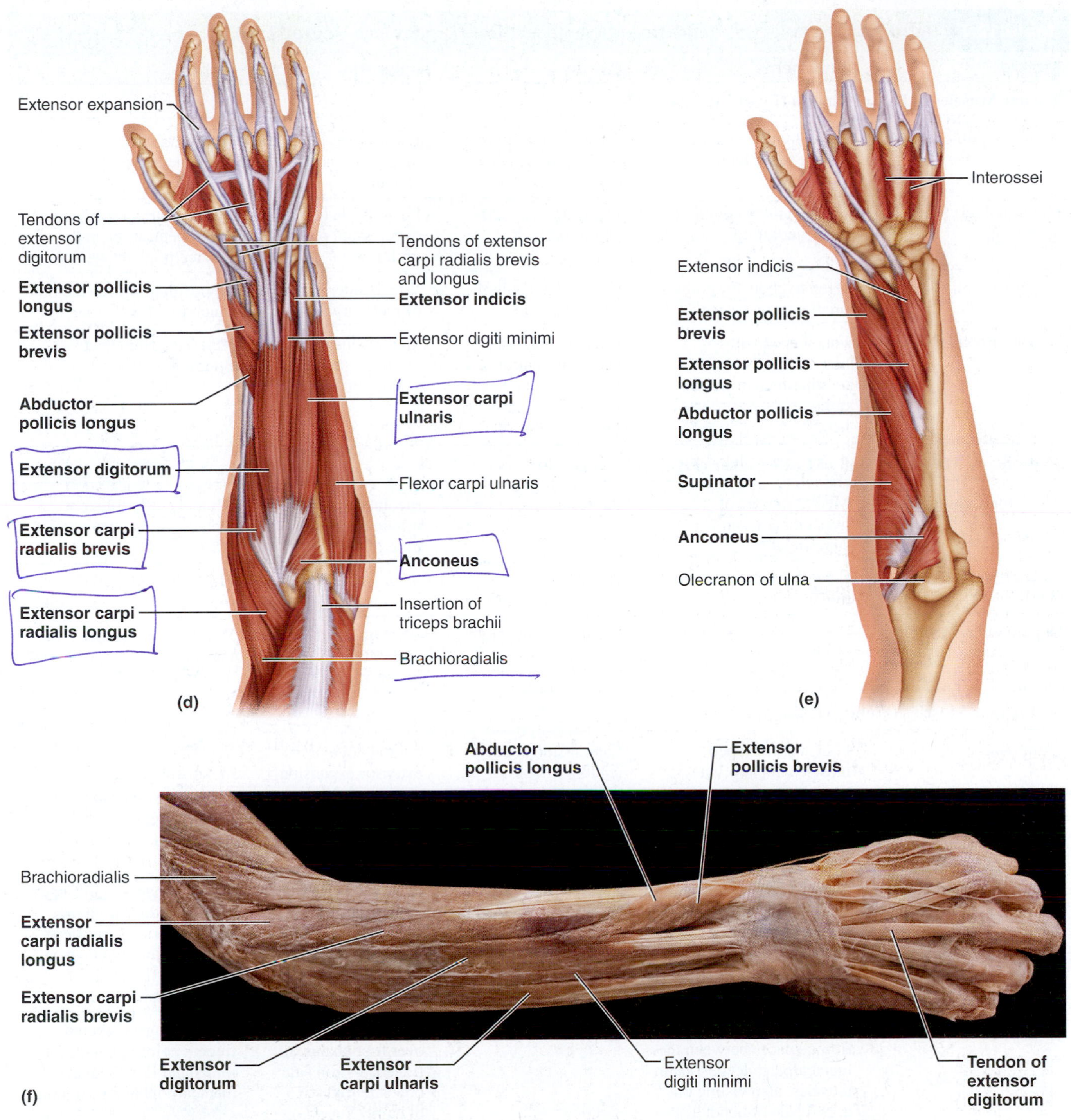

Figure 12.11 *(continued)* **Muscles of the forearm and wrist. (d)** Superficial muscles, posterior view. **(e)** Deep posterior muscles; superficial muscles have been removed. The interossei, the deepest layer of instrinsic hand muscles, are also illustrated. **(f)** Cadaver photo of posterior muscles of the right forearm.

Table 12.7 Intrinsic Muscles of the Hand: Fine Movements of the Fingers (see Figure 12.12)

Muscle	Comments	Origin	Insertion	Action
Thenar Muscles in Ball of Thumb (Figure 12.12a and b)				
Abductor pollicis brevis	Lateral muscle of thenar group; superficial	Flexor retinaculum and nearby carpals	Lateral base of thumb's proximal phalanx	Abducts thumb (at carpometacarpal joint)
Flexor pollicis brevis	Medial and deep muscle of thenar group	Flexor retinaculum and trapezium	Lateral side of base of proximal phalanx of thumb	Flexes thumb (at carpometacarpal and metacarpophalangeal joints)
Opponens pollicis	Deep to abductor pollicis brevis, on metacarpal I	Flexor retinaculum and trapezium	Whole anterior side of metacarpal I	Opposition: moves thumb to touch tip of little finger
Adductor pollicis	Fan-shaped with horizontal fibers; distal to other thenar muscles; oblique and transverse heads	Capitate bone and bases of metacarpals II–IV (oblique head); front of metacarpal III (transverse head)	Medial side of base of proximal phalanx of thumb	Adducts and helps to oppose thumb
Hypothenar Muscles in Ball of Little Finger (Figure 12.12a and b)				
Abductor digiti minimi	Medial muscle of hypothenar group; superficial	Pisiform bone	Medial side of proximal phalanx of little finger	Abducts little finger at metacarpophalangeal joint
Flexor digiti minimi brevis	Lateral deep muscle of hypothenar group	Hamate bone and flexor retinaculum	Same as abductor digiti minimi	Flexes little finger at metacarpophalangeal joint
Opponens digiti minimi	Deep to abductor digiti minimi	Same as flexor digiti minimi brevis	Most of length of medial side of metacarpal V	Helps in opposition: brings metacarpal V toward thumb to cup the hand
Midpalmar Muscles (Figure 12.12a, b, c, d)				
Lumbricals	Four worm-shaped muscles in palm, one to each finger (except thumb); odd because they originate from the tendons of another muscle	Lateral side of each tendon of flexor digitorum profundus in palm	Lateral edge of extensor expansion on first phalanx of fingers 2–5	Flex fingers at metacarpophalangeal joints but extend fingers at interphalangeal joints
Palmar interossei	Four long, cone-shaped muscles; lie ventral to the dorsal interossei and between the metacarpals	The side of each metacarpal that faces the midaxis of the hand (metacarpal III) where it's absent	Extensor expansion on first phalanx of each finger (except finger 3), on side facing midaxis of hand	Adduct fingers; pull fingers in toward third digit; act with lumbricals to extend fingers at interphalangeal joints and flex them at metacarpophalangeal joints
Dorsal interossei	Four bipennate muscles filling spaces between the metacarpals; deepest palm muscles, also visible on dorsal side of hand	Sides of metacarpals	Exterior expansion over first phalanx of fingers 2–4 on side opposite midaxis of hand (finger 3), but on *both* sides of finger 3	Abduct fingers; extend fingers at interphalangeal joints and flex them at metacarpophalangeal joints

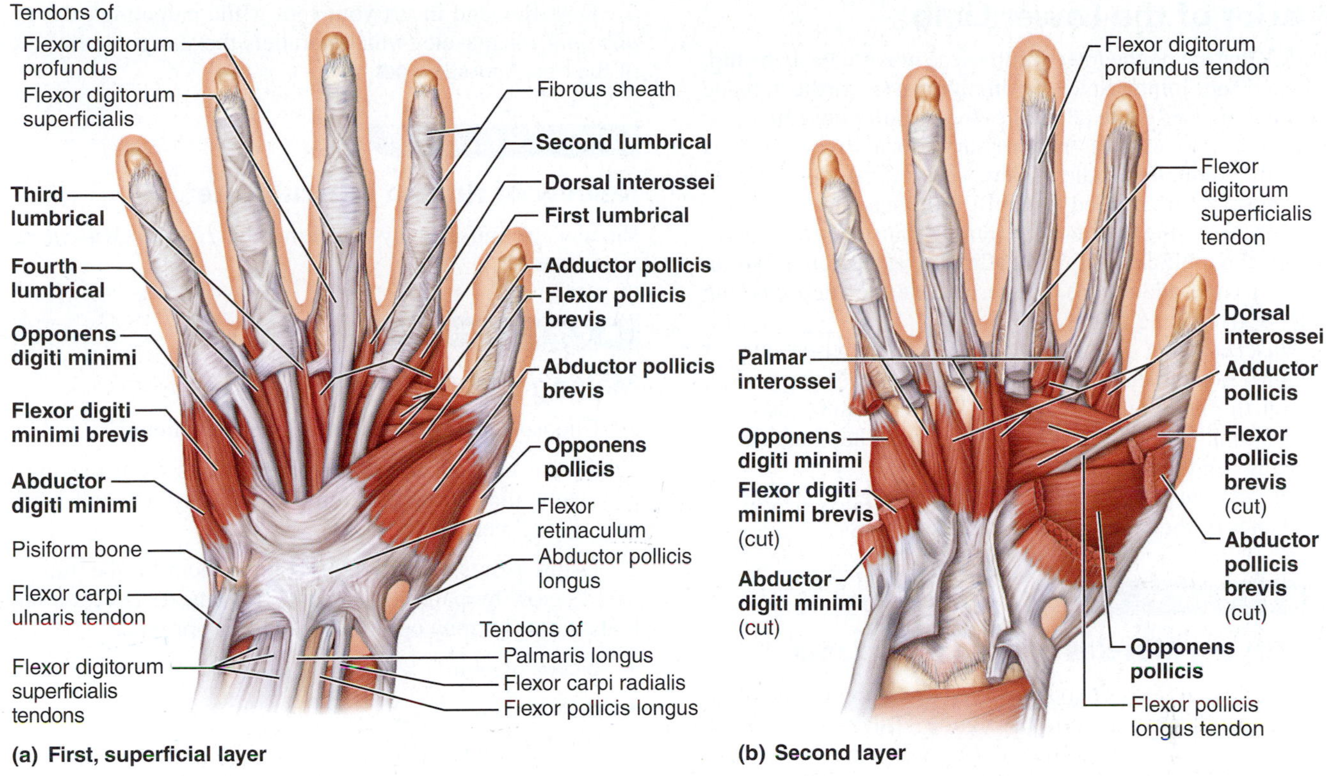

(a) First, superficial layer

(b) Second layer

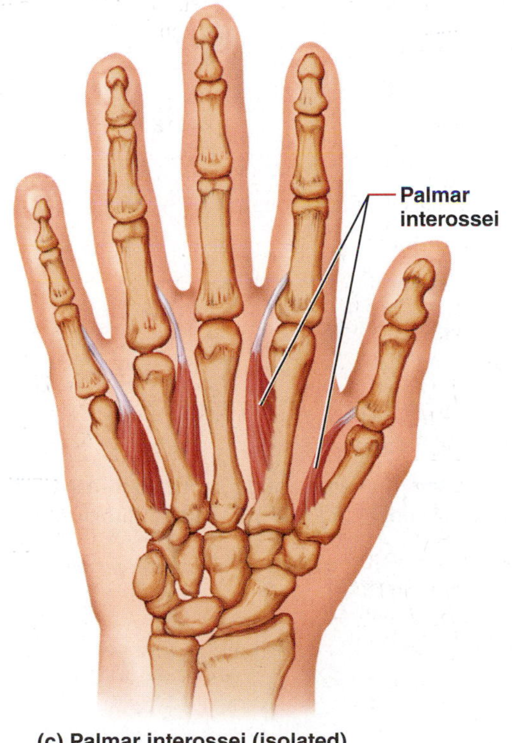

(c) Palmar interossei (isolated)

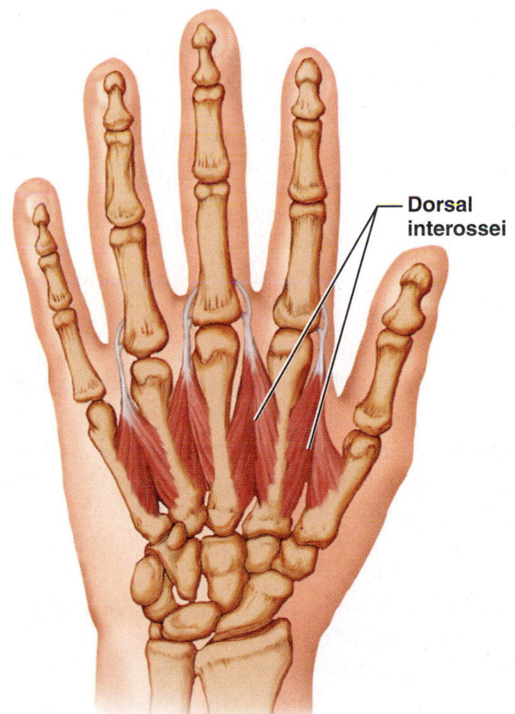

(d) Dorsal interossei (isolated)

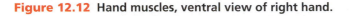

Figure 12.12 **Hand muscles, ventral view of right hand.**

Muscles of the Lower Limb

Muscles that act on the lower limb cause movement at the hip, knee, and foot joints. Since the human pelvic girdle is made up of heavy fused bones that allow very little movement, no special group of muscles is necessary to stabilize it. This is unlike the shoulder girdle, where several muscles (mainly trunk muscles) are needed to stabilize the scapulae.

Muscles acting on the thigh (femur) cause various movements at the multiaxial hip joint (flexion, extension, rotation, abduction, and adduction). These include the iliopsoas, the adductor group, and others.

Muscles acting on the leg form the major musculature of the thigh. (Anatomically, the term *leg* refers only to that portion between the knee and the ankle.) The thigh muscles cross the knee to allow its flexion and extension. They include the hamstrings and the quadriceps.

The muscles originating on the leg, as well as the intrinsic muscles of the foot, act on the foot and toes.

ACTIVITY 4

Identifying Muscles of the Lower Limb

Read the descriptions of specific muscles acting on the thigh and leg, and identify them in the figures (**Table 12.8**, **Table 12.9**, **Figure 12.13**, and **Figure 12.14**), trying to visualize their action when they contract. Because some of the muscles acting on the leg also have attachments on the pelvic girdle, they can cause movement at the hip joint.

Do the same for muscles acting on the foot and toes (**Table 12.10**, **Table 12.11**, **Figure 12.15**, **Figure 12.16**, and **Figure 12.17**).

Identify the muscles acting on the thigh, leg, foot, and toes as instructed previously for other muscle groups.

Demonstrating Operations of the Lower Limb Muscles

Complete this activity by performing the following palpation demonstrations with your lab partner.

1. Go into a deep knee bend, and palpate your own *gluteus maximus* muscle as you extend your hip to resume the upright posture.

2. Demonstrate the contraction of the anterior *quadriceps femoris* by trying to extend your knee against resistance. Do this while seated, and note how the patellar tendon reacts. The *biceps femoris* of the posterior thigh comes into play when you flex your knee against resistance.

3. Now stand on your toes. Have your partner palpate the lateral and medial heads of the *gastrocnemius* and follow it to its insertion in the calcaneal tendon.

4. Dorsiflex and invert your foot while palpating your *tibialis anterior* muscle (which parallels the sharp anterior crest of the tibia laterally). ■

ACTIVITY 5

Review of Human Musculature

Review the muscles by watching the *Human Musculature* video. ■

ACTIVITY 6

Making a Muscle Painting

1. Choose a male student to be "muscle painted."

2. Obtain brushes and water-based paints from the supply area while the "volunteer" removes his shirt and rolls up his pant legs (if necessary).

3. Using paints of different colors, identify the muscles listed below by painting his skin. If a muscle covers a large body area, you may opt to paint only its borders.

- biceps brachii
- deltoid
- erector spinae
- pectoralis major
- rectus femoris
- tibialis anterior
- triceps brachii
- vastus lateralis
- biceps femoris
- extensor carpi radialis longus
- latissimus dorsi
- rectus abdominis
- sternocleidomastoid
- trapezius
- triceps surae
- vastus medialis

4. Check your "human painting" with your instructor before cleaning your bench and leaving the laboratory. ■

(Text continues on page 215.)

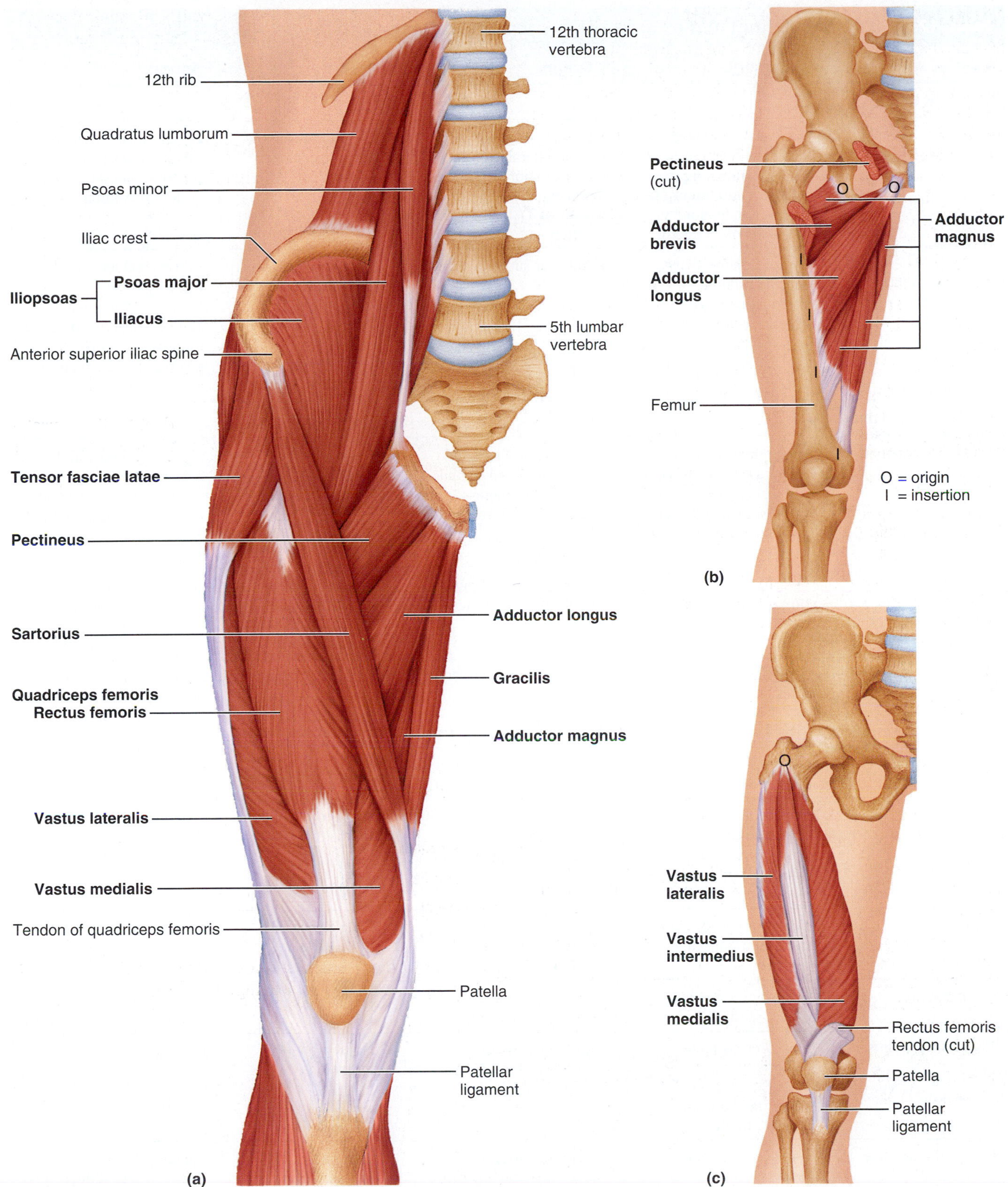

Figure 12.13 Anterior and medial muscles promoting movements of the thigh and leg. (a) Anterior view of the deep muscles of the pelvis and superficial muscles of the right thigh. **(b)** Adductor muscles of the medial compartment of the thigh. **(c)** The vastus muscles (isolated) of the quadriceps group.

Table 12.8 Muscles Acting on the Human Thigh and Leg, Anterior and Medial Aspects (see Figure 12.13)

Muscle	Comments	Origin	Insertion	Action
Origin on the Pelvis				
Iliopsoas—iliacus and psoas major	Two closely related muscles; fibers pass under inguinal ligament to insert into femur via a common tendon; iliacus is more lateral	Iliacus—iliac fossa and crest, lateral sacrum; psoas major—transverse processes, bodies, and discs of T_{12} and lumbar vertebrae	On and just below lesser trochanter of femur	Prime mover of thigh flexion and flexing trunk on thigh; lateral flexion of vertebral column (psoas)
Sartorius	Straplike superficial muscle running obliquely across anterior surface of thigh to knee	Anterior superior iliac spine	By an aponeurosis into medial aspect of proximal tibia	Flexes, abducts, and laterally rotates thigh; flexes knee; known as "tailor's muscle" because it helps effect crosslegged position in which tailors are often depicted
Medial Compartment				
Adductors—magnus, longus, and brevis	Large muscle mass forming medial aspect of thigh; arise from front of pelvis and insert at various levels on femur	Magnus—ischial and pubic rami and ischial tuberosity; longus—pubis near pubic symphysis; brevis—body and inferior pubic ramus	Magnus—linea aspera and adductor tubercle of femur; longus and brevis—linea aspera	Adduct and medially rotate and flex thigh; posterior part of magnus is also a synergist in thigh extension
Pectineus	Overlies adductor brevis on proximal thigh	Pectineal line of pubis (and superior pubic ramus)	Inferior from lesser trochanter to linea aspera of femur	Adducts, flexes, and medially rotates thigh
Gracilis	Straplike superficial muscle of medial thigh	Inferior ramus and body of pubis	Medial surface of tibia just inferior to medial condyle	Adducts thigh; flexes and medially rotates leg, especially during walking
Anterior Compartment				
*Quadriceps femoris**				
Rectus femoris	Superficial muscle of thigh; runs straight down thigh; only muscle of group to cross hip joint	Anterior inferior iliac spine and superior margin of acetabulum	Tibial tuberosity and patella	Extends knee and flexes thigh at hip
Vastus lateralis	Forms lateral aspect of thigh; intramuscular injection site	Greater trochanter, intertrochanteric line, and linea aspera	Tibial tuberosity and patella	Extends and stabilizes knee
Vastus medialis	Forms inferomedial aspect of thigh	Linea aspera and intertrochanteric line	Tibial tuberosity and patella	Extends knee; stabilizes patella
Vastus intermedius	Obscured by rectus femoris; lies between vastus lateralis and vastus medialis on anterior thigh	Anterior and lateral surface of femur	Tibial tuberosity and patella	Extends knee
Tensor fasciae latae	Enclosed between fascia layers of thigh	Anterior aspect of iliac crest and anterior superior iliac spine	Iliotibial tract (lateral portion of fascia lata)	Steadies the trunk on thigh; flexes, abducts, and medially rotates thigh

*The quadriceps form the flesh of the anterior thigh and have a common insertion in the tibial tuberosity via the patellar tendon. They are powerful leg extensors, enabling humans to kick a football, for example.

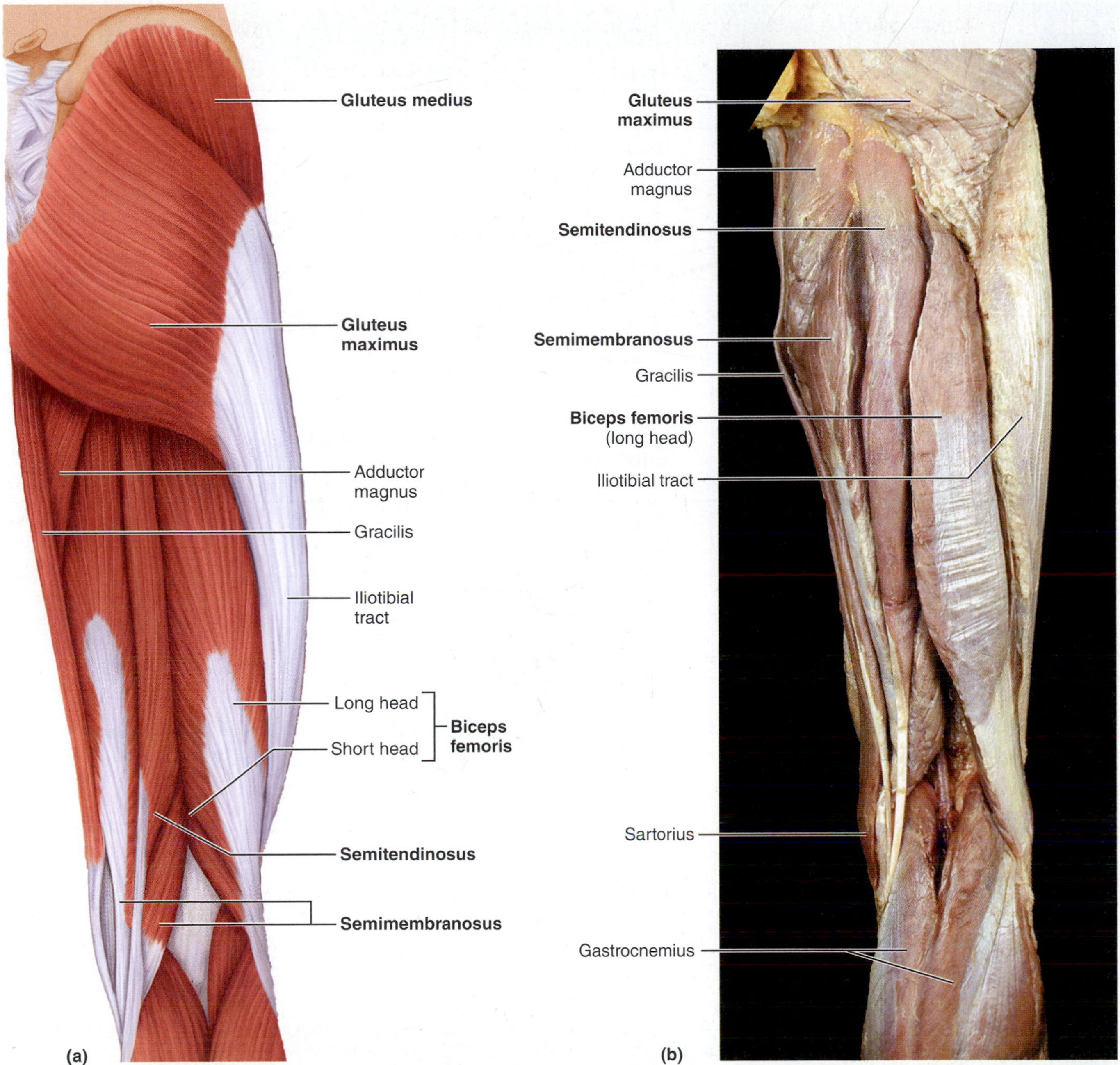

Figure 12.14 Muscles of the posterior aspect of the right hip and thigh. **(a)** Superficial view showing the gluteus muscles of the buttock and hamstring muscles of the thigh. **(b)** Cadaver photo of muscles of the posterior thigh.

Table 12.9 Muscles Acting on the Human Thigh and Leg, Posterior Aspect (see Figure 12.14)

Muscle	Comments	Origin	Insertion	Action
Origin on the Pelvis				
Gluteus maximus	Largest and most superficial of gluteal muscles (which form buttock mass); intramuscular injection site	Dorsal ilium, sacrum, and coccyx	Gluteal tuberosity of femur and iliotibial tract*	Major extensor of thigh; complex, powerful, and most effective when thigh is flexed, as in climbing stairs—but not as in walking; antagonist of iliopsoas; laterally rotates and abducts thigh
Gluteus medius	Partially covered by gluteus maximus; intramuscular injection site	Upper lateral surface of ilium	Greater trochanter of femur	Abducts and medially rotates thigh; steadies pelvis during walking
Gluteus minimus (not shown in figure)	Smallest and deepest gluteal muscle	External inferior surface of ilium	Greater trochanter of femur	Abducts and medially rotates thigh; steadies pelvis
Posterior Compartment *Hamstrings†*				
Biceps femoris	Most lateral muscle of group; arises from two heads	Ischial tuberosity (long head); linea aspera and distal femur (short head)	Tendon passes laterally to insert into head of fibula and lateral condyle of tibia	Extends thigh and flexes knee; laterally rotates leg
Semitendinosus	Medial to biceps femoris	Ischial tuberosity	Medial aspect of upper tibial shaft	Extends thigh; flexes knee; medially rotates leg
Semimembranosus	Deep to semitendinosus	Ischial tuberosity	Medial condyle of tibia; lateral condyle of femur	Extends thigh; flexes knee; medially rotates leg

*The iliotibial tract, a thickened lateral portion of the fascia lata, ensheathes all the muscles of the thigh. It extends as a tendinous band from the iliac crest to the knee.

†The hamstrings are the fleshy muscles of the posterior thigh. The name comes from the butchers' practice of using the tendons of these muscles to hang hams for smoking. As a group, they are strong extensors of the hip; they counteract the powerful quadriceps by stabilizing the knee joint when standing.

Table 12.10 Muscles Acting on the Human Foot and Ankle (see Figures 12.15 and 12.16)

Muscle	Comments	Origin	Insertion	Action
Lateral Compartment (Figure 12.15a, b and Figure 12.16b)				
Fibularis (peroneus) longus	Superficial lateral muscle; overlies fibula	Head and upper portion of fibula	By long tendon under foot to metatarsal I and medial cuneiform	Plantar flexes and everts foot; helps keep foot flat on ground
Fibularis (peroneus) brevis	Smaller muscle; deep to fibularis longus	Distal portion of fibula shaft	By tendon running behind lateral malleolus to insert on proximal end of metatarsal V	Plantar flexes and everts foot, as part of fibular group

(Table continues on page 210.)

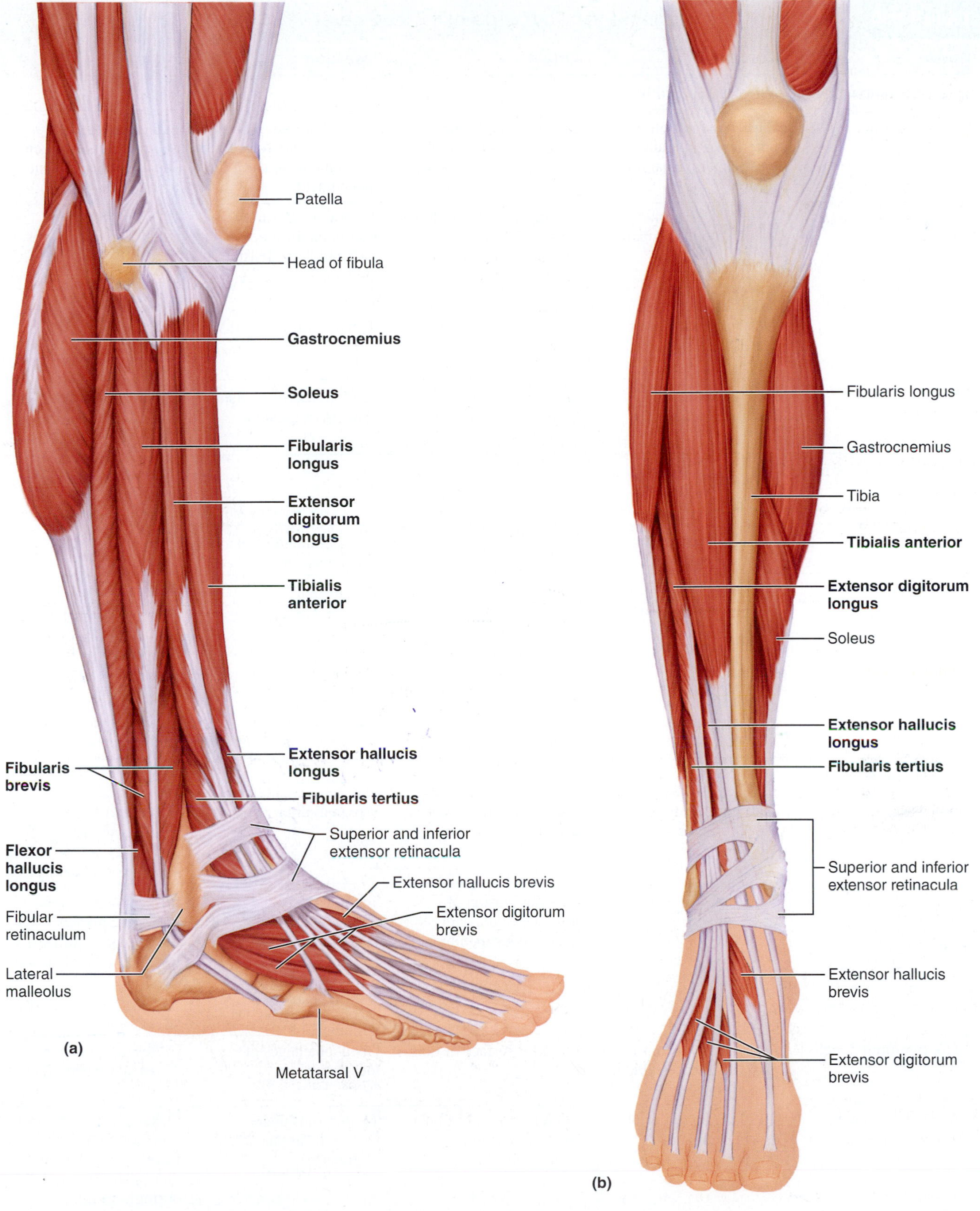

Figure 12.15 Muscles of the anterolateral aspect of the right leg. (a) Superficial view of lateral aspect of the leg, illustrating the positioning of the lateral compartment muscles (fibularis longus and brevis) relative to anterior and posterior leg muscles. **(b)** Superficial view of anterior leg muscles.

Table 12.10 Muscles Acting on the Human Foot and Ankle (see Figures 12.15 and 12.16) *(continued)*

Muscle	Comments	Origin	Insertion	Action
Anterior Compartment (Figure 12.15a and b)				
Tibialis anterior	Superficial muscle of anterior leg; parallels sharp anterior margin of tibia	Lateral condyle and upper two-thirds of tibia; interosseous membrane	By tendon into inferior surface of first cuneiform and metatarsal I	Prime mover of dorsiflexion; inverts foot; supports longitudinal arch of foot
Extensor digitorum longus	Anterolateral surface of leg; lateral to tibialis anterior	Lateral condyle of tibia; proximal third-fourths of fibula; interosseous membrane	Tendon divides into four parts; inserts into middle and distal phalanges of toes 2–5	Prime mover of toe extension; dorsiflexes foot
Fibularis (peroneus) tertius	Small muscle; often fused to distal part of extensor digitorum longus	Distal anterior surface of fibula and interosseous membrane	Tendon inserts on dorsum of metatarsal V	Dorsiflexes and everts foot
Extensor hallucis longus	Deep to extensor digitorum longus and tibialis	Anteromedial shaft of fibula and interosseous membrane	Tendon inserts on distal phalanx of great toe	Extends great toe; dorsiflexes foot
Posterior Compartment				
Superficial (Figures 12.15 and 12.16)				
Triceps surae	Muscle pair that shapes posterior calf		Via common tendon (calcaneal) into heel	Plantar flex foot
Gastrocnemius	Superficial muscle of pair; two prominent bellies	By two heads from medial and lateral condyles of femur	Calcaneus via calcaneal tendon	Plantar flexes foot when knee is extended; crosses knee joint; thus can flex knee (when foot is dorsiflexed)
Soleus	Deep to gastrocnemius	Proximal portion of tibia and fibula; interosseous membrane	Calcaneus via calcaneal tendon	Plantar flexes foot; is an important muscle for locomotion
Deep (Figure 12.16b)				
Popliteus	Thin muscle at posterior aspect of knee	Lateral condyle of femur and lateral meniscus	Proximal tibia	Flexes and rotates leg medially to "unlock" extended knee when knee flexion begins
Tibialis posterior	Thick muscle deep to soleus	Superior portion of tibia and fibula and interosseous membrane	Tendon passes obliquely behind medial malleolus and under arch of foot; inserts into several tarsals and metatarsals II–IV	Prime mover of foot inversion; plantar flexes foot; stabilizes longitudinal arch of foot
Flexor digitorum longus	Runs medial to and partially overlies tibialis posterior	Posterior surface of tibia	Distal phalanges of toes 2–5	Flexes toes; plantar flexes and inverts foot
Flexor hallucis longus (see also Figure 12.15a)	Lies lateral to inferior aspect of tibialis posterior	Middle portion of fibula shaft; interosseous membrane	Tendon runs under foot to distal phalanx of great toe	Flexes great toe (*hallux* = great toe); plantar flexes and inverts foot; the "push-off muscle" during walking

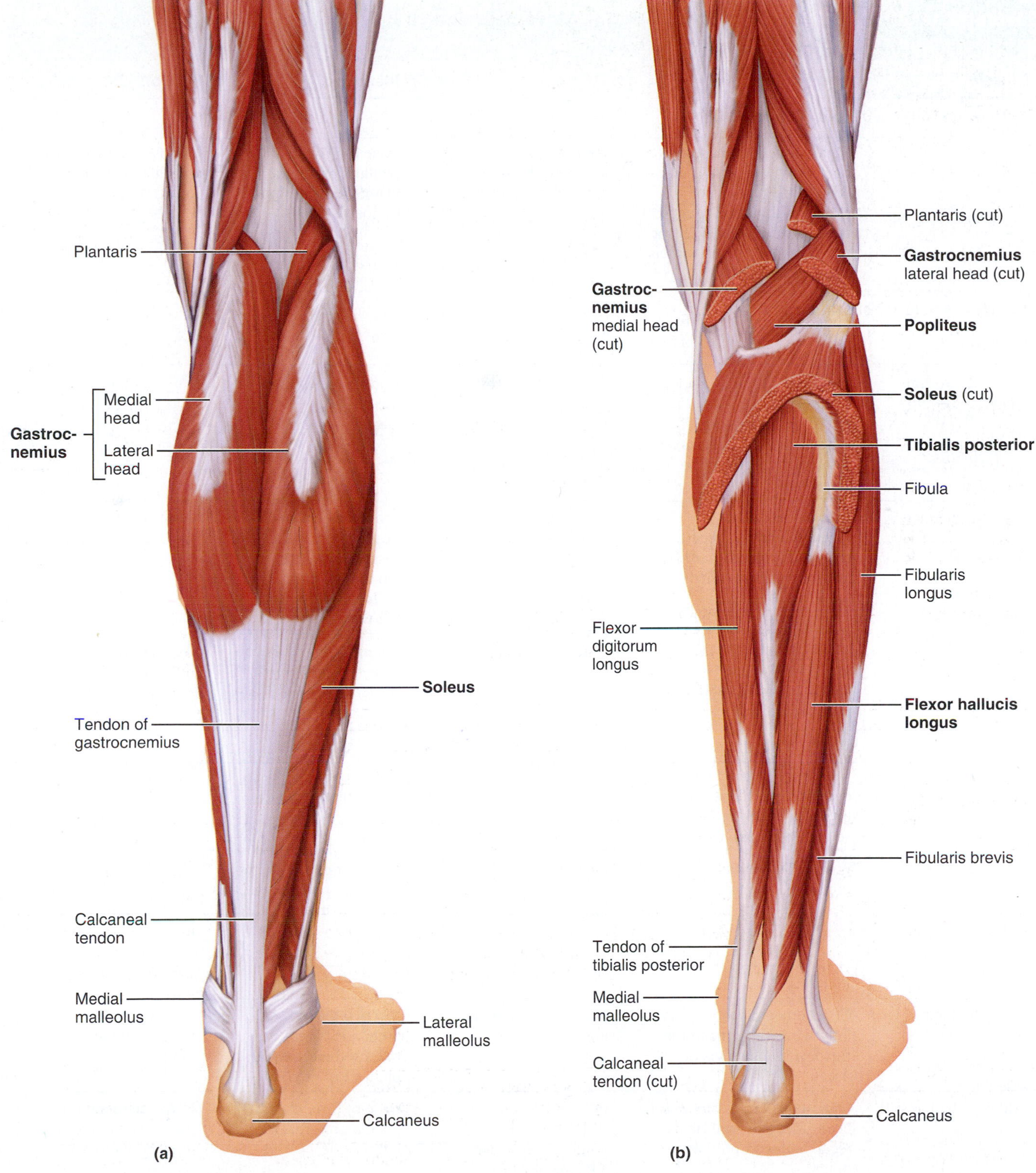

Figure 12.16 Muscles of the posterior aspect of the right leg. (a) Superficial view of the posterior leg. **(b)** The triceps surae has been removed to show the deep muscles of the posterior compartment.

Table 12.11 Intrinsic Muscles of the Foot: Toe Movement and Foot Support (see Figures 12.15 and 12.17)

Muscle	Comments	Origin	Insertion	Action
Muscles on Dorsum of Foot (Figure 12.15)				
Extensor digitorum brevis	Small, four-part muscle on dorsum of foot; deep to the tendons of extensor digitorum longus; corresponds to the extensor indicis and extensor pollicis muscles of forearm	Anterior part of calcaneus bone; extensor retinaculum	Base of proximal phalanx of big toe; extensor expansions on toes 2–4	Helps extend toes at metatarsophalangeal joints
Muscles on Sole of Foot (Figure 12.17)				
First layer				
Flexor digitorum brevis	Bandlike muscle in middle of sole; corresponds to flexor digitorum superficialis of forearm and inserts into digits in the same way	Calcaneal tuberosity	Middle phalanx of toes 2–4	Helps flex toes
Abductor hallucis	Lies medial to flexor digitorum brevis; recall the similar thumb muscle, abductor pollicis brevis	Calcaneal tuberosity and flexor retinaculum	Proximal phalanx of great toe, via a sesamoid bone in tendon of flexor hallucis brevis (see below)	Abducts great toe
Abductor digiti minimi	Most lateral of the three superficial sole muscles; recall the similar abductor muscle in palm	Calcaneal tuberosity	Lateral side of base of little toe's proximal phalanx	Abducts and flexes little toe
Second layer				
Flexor accessorius (quadratus plantae)	Rectangular muscle just deep to flexor digitorum brevis in posterior half of sole; two heads	Medial and lateral sides of calcaneus	Tendon of flexor digitorum longus in midsole	Straightens out the oblique pull of flexor digitorum longus
Lumbricals	Four little "worms," like lumbricals in hand	From each tendon of flexor digitorum longus	Extensor expansion on proximal phalanx of toes 2–5, medial side	By pulling on extensor expansion, flex toes at metatarsophalangeal joints and extend toes at interphalangeal joints
Third layer				
Flexor hallucis brevis	Covers metatarsal I; splits into two bellies; recall the flexor pollicis brevis of thumb	Mostly from cuboid bone	Via two tendons onto both sides of the base of the proximal phalanx of great toe	Flexes great toe's metatarsophalangeal joint
Adductor hallucis	Oblique and transverse heads; deep to lumbricals; recall adductor pollicis in thumb	From metatarsals II–IV, fibularis longus tendon, and a ligament across metatarsophalangeal joints	Base of proximal phalanx of great toe, lateral side	Helps maintain the transverse arch of foot; weak adductor of great toe

(Table continues on page 214.)

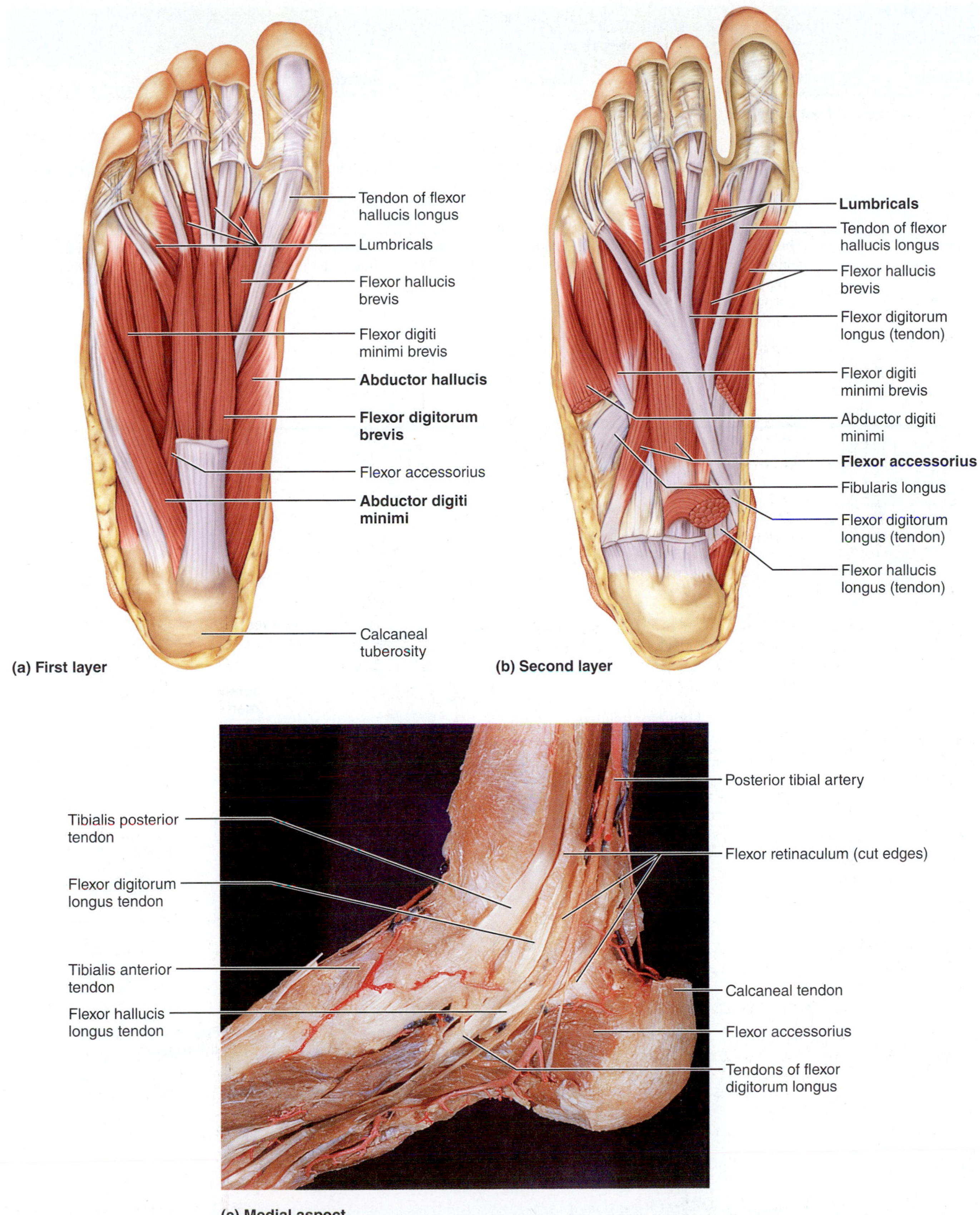

Tendon of flexor hallucis longus
Lumbricals
Flexor hallucis brevis
Flexor digiti minimi brevis
Abductor hallucis
Flexor digitorum brevis
Flexor accessorius
Abductor digiti minimi
Calcaneal tuberosity
(a) First layer
Lumbricals
Tendon of flexor hallucis longus
Flexor hallucis brevis
Flexor digitorum longus (tendon)
Flexor digiti minimi brevis
Abductor digiti minimi
Flexor accessorius
Fibularis longus
Flexor digitorum longus (tendon)
Flexor hallucis longus (tendon)
(b) Second layer
Posterior tibial artery
Tibialis posterior tendon
Flexor retinaculum (cut edges)
Flexor digitorum longus tendon
Tibialis anterior tendon
Calcaneal tendon
Flexor hallucis longus tendon
Flexor accessorius
Tendons of flexor digitorum longus
(c) Medial aspect

Figure 12.17 Muscles of the right foot, plantar and medial aspects.

Table 12.11 Intrinsic Muscles of the Foot: Toe Movement and Foot Support (see Figures 12.15 and 12.17) *(continued)*

Muscle	Comments	Origin	Insertion	Action
Muscles on Sole of Foot *(continued)*				
Flexor digiti minimi brevis	Covers metatarsal V; recall same-named muscle in hand	Base of metatarsal V and tendon of fibularis longus	Base of proximal phalanx of toe 5	Flexes little toe at metatarsophalangeal joint
Fourth layer				
Plantar and dorsal interossei	Three plantar and four dorsal interossei; similar to the palmar and dorsal interossei of hand in locations, attachments, and actions; however, the long axis of foot around which these muscles orient is the second digit, not the third	See palmar and dorsal interossei (Table 12.7)	See palmar and dorsal interossei (Table 12.7)	See palmar and dorsal interossei (Table 12.7)

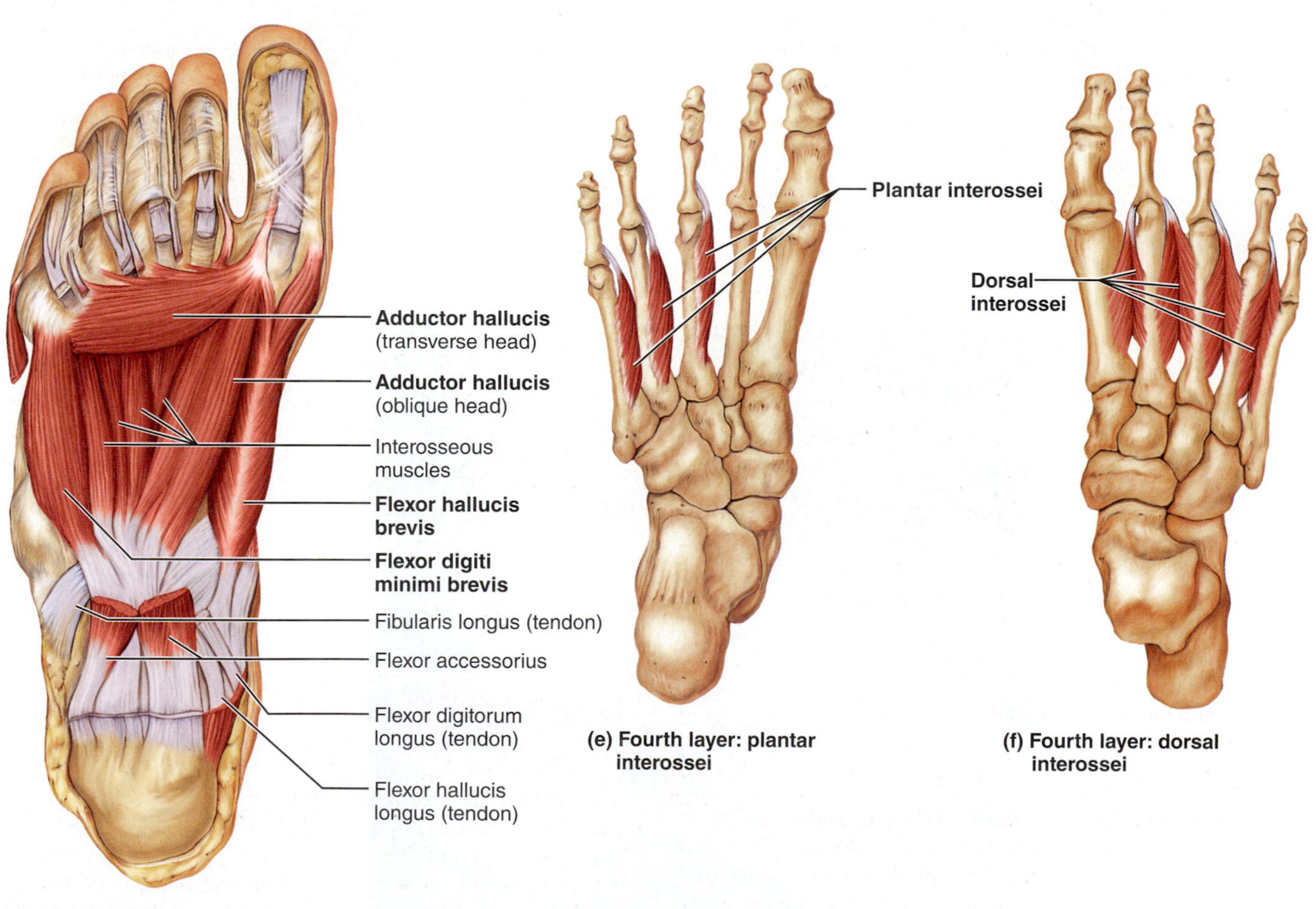

Figure 12.17 *(continued)* **Muscles of the right foot, plantar and medial aspects.**

GROUP CHALLENGE

Muscle IDs

Work in groups of three or four to fill out the chart below. Do not look back at the tables; use the "brain power" of your group and the appropriate muscle models. To help complete this task, recall that when a muscle contracts, the muscle's insertion moves toward the muscle's origin. Also, in the muscles of the limbs, the origin typically lies proximal to the insertion. Sometimes the origin and insertion are even part of the muscle's name!

Group Challenge: Muscle IDs

Origin	Insertion	Muscle	Primary action
Zygomatic arch and maxilla	Angle and ramus of the mandible		
Anterior surface of ribs 3–5	Coracoid process of the scapula		
Inferior border of rib above	Superior border of rib below		
Distal portion of anterior humerus	Coronoid process of the ulna		
Anterior inferior iliac spine and superior margin of acetabulum	Tibial tuberosity and patella		
By two heads from medial and lateral condyles of femur	Calcaneus via calcaneal tendon		

DISSECTION AND IDENTIFICATION

Cat Muscles

The skeletal muscles of all mammals are named in a similar fashion. However, some muscles that are separate in other animals are fused in humans, and some muscles present in other animals are absent in humans. This exercise involves dissecting the cat musculature to enhance your knowledge of the human muscular system. Because the aim is to become familiar with the muscles of the human body, you should pay particular attention to the similarities between cat and human muscles. However, pertinent differences will be pointed out as they are encountered.

Wear a lab coat or apron when dissecting to protect your clothes. ■

ACTIVITY 7

Preparing the Cat for Dissection

The preserved laboratory animals purchased for dissection have been embalmed with a solution that prevents the tissues from deteriorating. The animals are generally delivered in plastic bags that contain a small amount of the embalming fluid. *Do not dispose of this fluid* when you remove the cat; the fluid prevents the cat from drying out. It is very important to keep the cat's tissues moist because you will probably use the same cat from now until the end of the course. The embalming fluid may cause your eyes to smart and may dry your skin, but these small irritants are preferable to working with a cat that has become hard and odoriferous as a result of bacterial action.

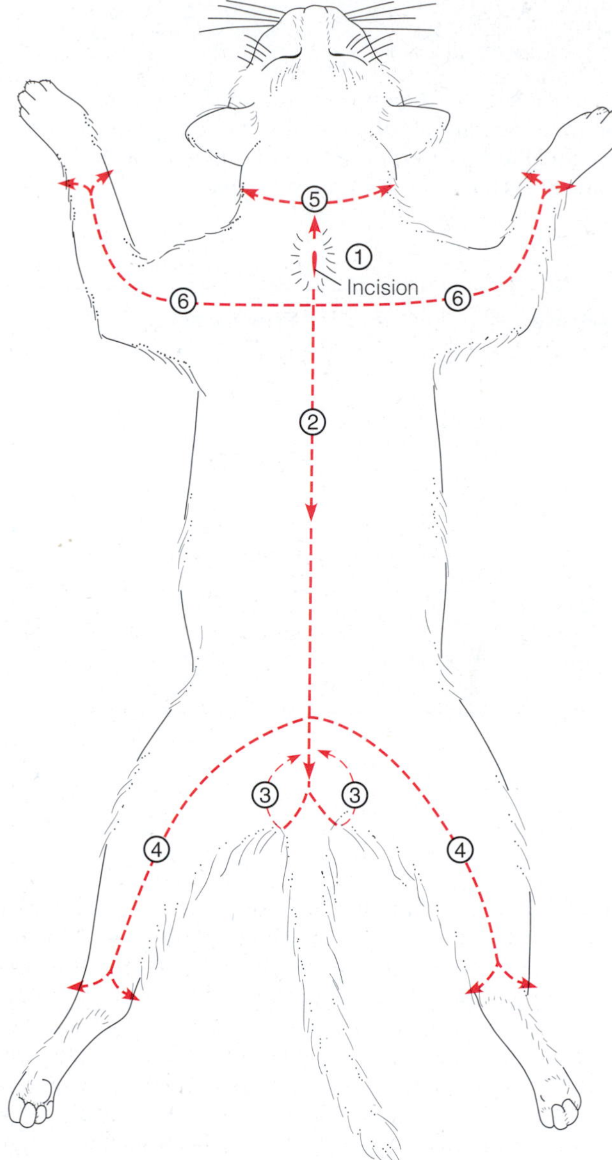

Figure 12.18 Incisions to be made in skinning a cat. Numbers indicate sequence.

1. Don disposable gloves and then obtain a cat, dissection tray, dissection instruments, and a name tag. Mark the name tag with the names of the members of your group, and set it aside. The name tag will be attached to the plastic bag at the end of the dissection so that you may identify your animal in subsequent laboratories.

2. To begin removing the skin, place the cat ventral side down on the dissecting tray. Cutting away from yourself with a scalpel, make a short, shallow incision in the midline of the neck, just to penetrate the skin. From this point on, use scissors. Continue the cut the length of the back to the sacrolumbar region, stopping at the tail **(Figure 12.18)**.

3. From the dorsal surface of the tail region, continue the incision around the tail, encircling the anus and genital organs. Do not remove the skin from this region.

4. Beginning again at the dorsal tail region, make an incision through the skin down each hind leg nearly to the ankle. Continue the cuts completely around the ankles.

5. Return to the neck. Cut the skin around the neck.

6. Cut down each foreleg to the wrist. Completely cut through the skin around the wrists.

7. Now free the skin from the loose connective tissue (superficial fascia) that binds it to the underlying structures. With one hand, grasp the skin on one side of the midline dorsal incision. Then, using your fingers or a blunt probe, break through the "cottony" connective tissue fibers to release the skin from the muscle beneath. Work toward the ventral surface and then toward the neck. As you pull the skin from the body, you should see small, white, cordlike structures extending from the skin to the muscles at fairly regular intervals. These are the cutaneous nerves that serve the skin. You will also see (particularly as you approach the ventral surface) that a thin layer of muscle fibers remains adhered to the skin. This is the **cutaneous maximus** muscle, which enables the cat to move its skin, rather like our facial muscles allow us to express emotion. Where the cutaneous maximus fibers cling to those of the deeper muscles, they should be carefully cut free. Along the ventral surface of the trunk notice the two lines of nipples associated with the mammary glands. These are more prominent in females.

8. You will notice as you start to free the skin in the neck that it is more difficult to remove. Take extra care and time in this area. The large flat **platysma** muscle in the ventral neck region (a skin muscle like the cutaneous maximus) will remain attached to the skin. You will not remove the skin from the head because the cat's muscles are not sufficiently similar to human head muscles to merit study.

9. Complete the skinning process by freeing the skin from the forelimbs, the lower torso, and the hindlimbs in the same manner. The skin is more difficult to remove as you approach the paws, so you may need to spend additional time on these areas. *Do not discard the skin.*

10. Inspect your skinned cat. Notice that it is difficult to see any cleavage lines between the muscles because of the overlying connective tissue, which is white or yellow. If time allows, carefully remove as much of the fat and fascia from the surface of the muscles as possible, using forceps or your fingers. The muscles, when exposed, look grainy or threadlike and are light brown. If you carry out this clearing process carefully and thoroughly, you will be ready to begin your identification of the superficial muscles.

11. If the muscle dissection exercises are to be done at a later laboratory session, follow the cleanup instructions noted in the following box. *Prepare your cat for storage in this way every time you use the cat.* ■

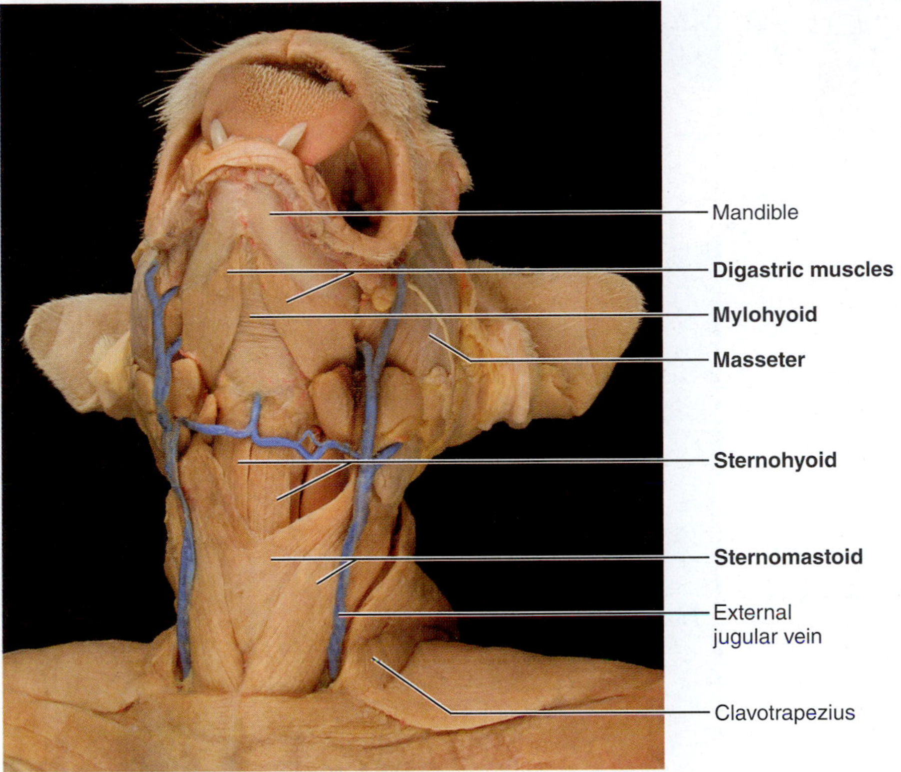

Figure 12.19 Superficial muscles of the anterior neck of the cat.

Preparing the Dissection Animal for Storage

Before leaving the lab, prepare your animal for storage as follows:

1. To prevent the internal organs from drying out, dampen a layer of folded paper towels with embalming fluid, and wrap them snugly around the animal's torso. (Do not use *water-soaked* paper towels because this will encourage the growth of mold.) Make sure the dissected areas are completely enveloped.

2. Return the animal's skin flaps to their normal position over the ventral cavity body organs.

3. Place the animal in a plastic storage bag. Add more embalming fluid if necessary, press out excess air, and securely close the bag with a rubber band or twine.

4. Make sure your name tag is securely attached, and place the animal in the designated storage container.

5. Clean all dissecting equipment with soapy water, and rinse and dry it for return to the storage area. Wash down the lab bench, and properly dispose of organic debris and your gloves before leaving the laboratory.

ACTIVITY 8

Dissecting Trunk and Neck Muscles

To dissect muscles properly, you must be sure to carefully separate one muscle from another and transect the superficial muscles in order to study those lying deeper. In general, when you are directed to transect a muscle, make sure the muscle is completely freed from all adhering connective tissue, and then cut through it about halfway between its origin and insertion points. *Use caution when working around points of muscle origin or insertion, and do not remove the fascia associated with such attachments.*

As a rule, all the fibers of one muscle are held together by a connective tissue sheath (epimysium) and run in the same general direction. Before you begin dissection, observe your skinned cat. If you look carefully, you can see changes in the direction of the muscle fibers, which will help you to locate the muscle borders. Pulling in slightly different directions on two adjacent muscles will usually allow you to expose the subtle white cleavage line between them. Once you have identified cleavage lines, *use a blunt probe* to break the connective tissue between muscles and to separate them. If the muscles separate as clean, distinct bundles, your procedure is probably correct. If they appear ragged or chewed up, you are probably tearing a muscle apart rather than separating it from adjacent muscles. Because of time considerations, in this exercise you will identify only the muscles that are easiest to identify and separate out.

Anterior Neck Muscles

1. Examine the anterior neck surface of the cat, and identify the following superficial neck muscles. The platysma belongs in this group but was probably removed during the skinning process. (Refer to **Figure 12.19** as you work.) The **sternomastoid** muscle and the more lateral and deeper **cleidomastoid** muscle (not visible in Figure 12.19) are joined in humans to form the sternocleidomastoid. The large external jugular vein, which drains the head, should be obvious crossing the anterior aspect of these muscles. The **mylohyoid** muscle parallels the bottom aspect of the chin, and the **digastric** muscles form a V over the mylohyoid muscle. Although it is not one of the neck muscles, you can now identify the fleshy **masseter** muscle, which flanks the digastric muscle laterally. Finally,

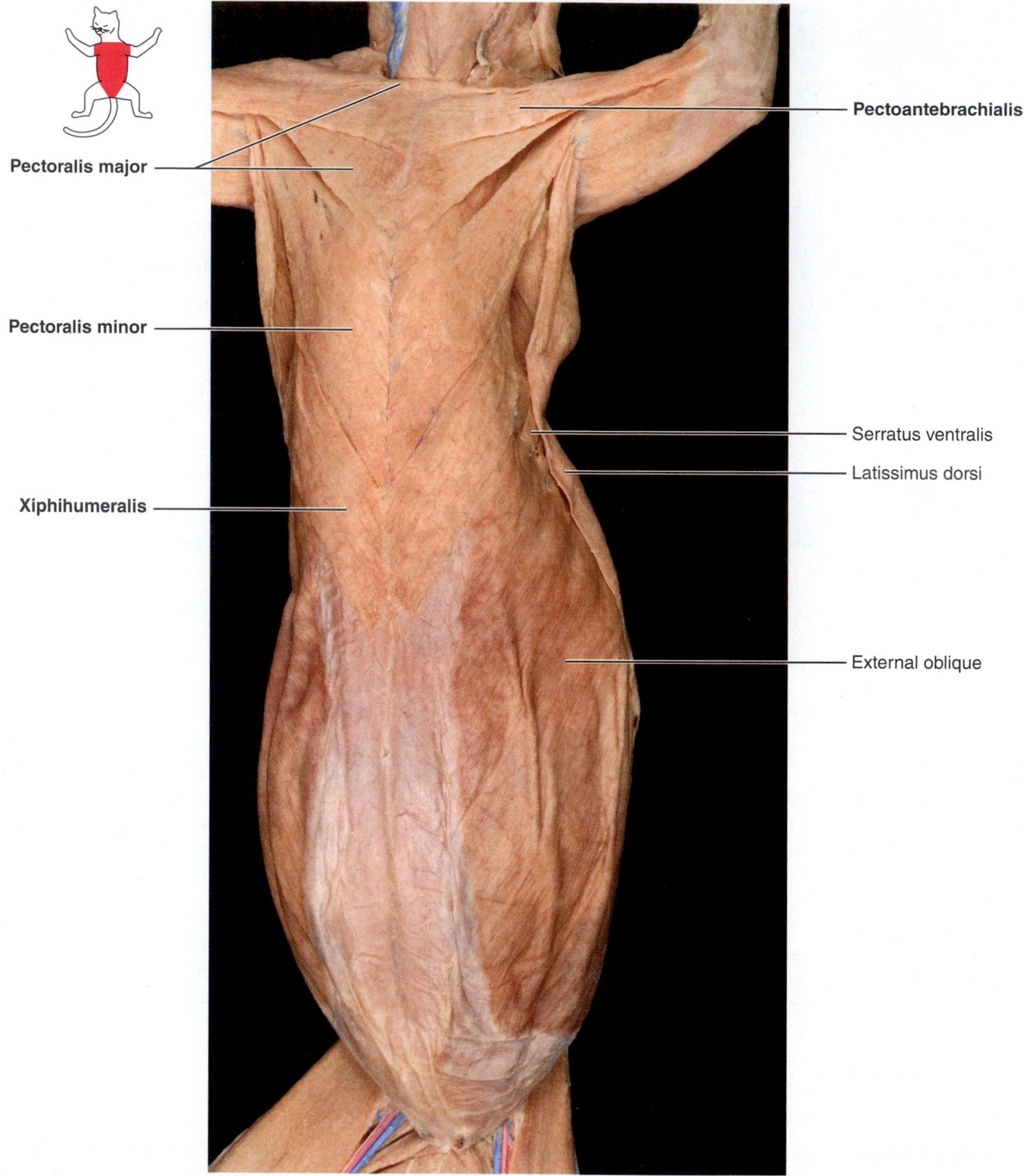

Figure 12.20 Superficial thoracic and abdominal muscles, ventral view. Latissimus dorsi is reflected away from the thorax.

the **sternohyoid** is a narrow muscle between the mylohyoid (superiorly) and the inferior sternomastoid.

2. The deeper muscles of the anterior neck of the cat are small and straplike and hardly worth the effort of dissection. However, one of these deeper muscles can be seen with a minimum of extra effort. Transect the sternomastoid and sternohyoid muscles approximately at midbelly. Reflect the cut ends to reveal the **sternothyroid** muscle (not visible in Figure 12.19), which runs along the anterior surface of the throat just deep and lateral to the sternohyoid muscle. The cleidomastoid muscle, which lies deep to the sternomastoid, is also more easily identified now.

Superficial Chest Muscles

In the cat, the chest or pectoral muscles adduct the arm, just as they do in humans. However, humans have only two pectoral muscles, and cats have four—the pectoralis major, pectoralis minor, xiphihumeralis, and pectoantebrachialis **(Figure 12.20)**. However, because of their relatively great degree of fusion, the cat's pectoral muscles appear to be a single muscle. The pectoral muscles are rather difficult to dissect and identify because they do not separate from one another easily.

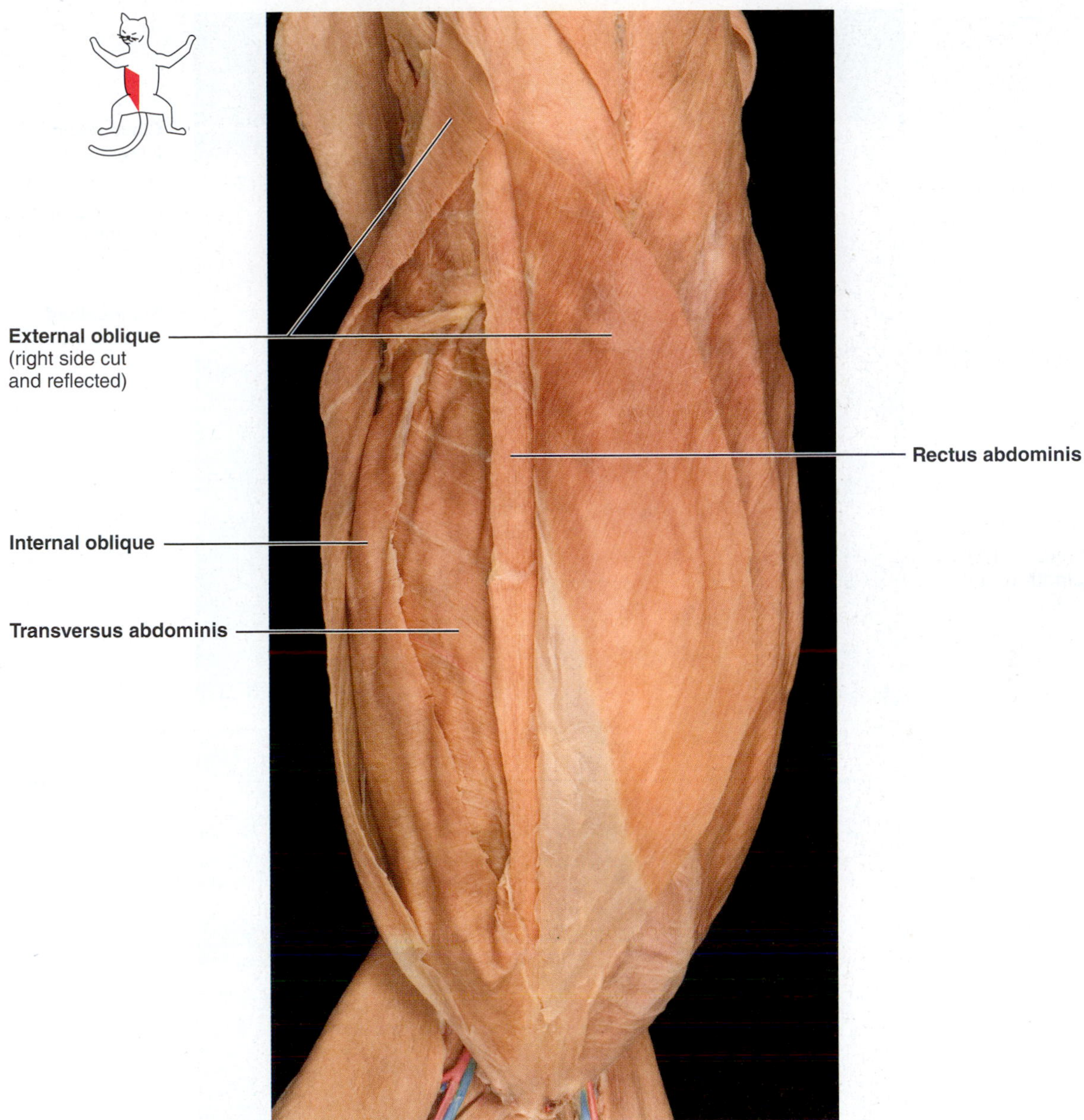

Figure 12.21 **Muscles of the abdominal wall of the cat.**

The **pectoralis major** is 5 to 8 cm (2 to 3 inches) wide and can be seen arising on the manubrium, just inferior to the sternomastoid muscle of the neck, and running to the humerus. Its fibers run at right angles to the longitudinal axis of the cat's body.

The **pectoralis minor** lies beneath the pectoralis major and extends posterior to it on the abdominal surface. It originates on the sternum and inserts on the humerus. Its fibers run obliquely to the long axis of the body, which helps to distinguish it from the pectoralis major. Contrary to what its name implies, the pectoralis minor is a larger and thicker muscle than the pectoralis major.

The **xiphihumeralis** can be distinguished from the posterior edge of the pectoralis minor only by virtue of the fact that its origin is lower—on the xiphoid process of the sternum. Its fibers run parallel to and are fused with those of the pectoralis minor.

The **pectoantebrachialis** is a thin, straplike muscle, about 1.3 cm (½ inch) wide, lying over the pectoralis major. It originates from the manubrium, passes laterally over the pectoralis major, and merges with the muscles of the forelimb approximately halfway down the humerus. It has no homologue in humans.

Identify, free, and trace out the origin and insertion of the cat's chest muscles. (Refer to Figure 12.20 as you work.)

Muscles of the Abdominal Wall

The superficial trunk muscles include those of the abdominal wall (see Figure 12.20 and **Figure 12.21**). Cat musculature in this area is quite similar in function to that of humans.

1. Complete the dissection of the more superficial anterior trunk muscles of the cat by identifying the origins and insertions of the muscles of the abdominal wall. Work carefully here. These muscles are very thin, and it is easy to miss their

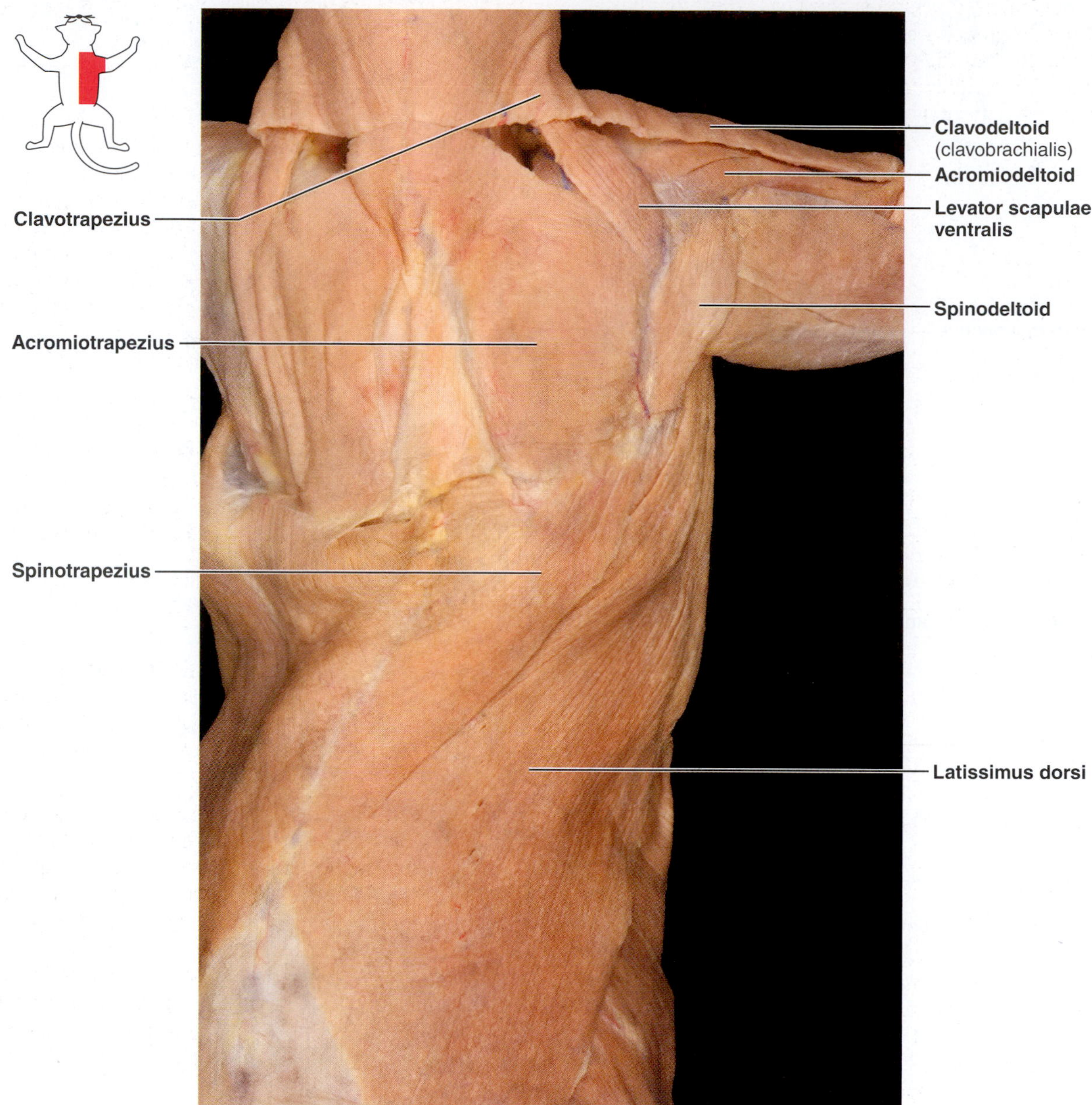

Figure 12.22 Superficial muscles of the anterodorsal aspect of the shoulder, trunk, and neck of the cat.

boundaries. Begin with the **rectus abdominis,** a long band of muscle approximately 2½ cm (1 inch) wide running immediately lateral to the midline of the body on the abdominal surface. Humans have four transverse *tendinous intersections* in the rectus abdominis (Figure 12.8a), but they are absent or difficult to identify in the cat. Identify the **linea alba,** which separates the rectus abdominis muscles. Note the relationship of the rectus abdominis to the other abdominal muscles and their fascia.

2. The **external oblique** is a sheet of muscle immediately beside the rectus abdominis (Figure 12.21). Carefully free and then transect the external oblique to reveal the anterior attachment of the rectus abdominis. Reflect the external oblique; observe the **internal oblique,** the deeper muscle. Notice which way the fibers run.

How does the fiber direction of the internal oblique compare to that of the external oblique?

____

____

3. Free and then transect the internal oblique muscle to reveal the fibers of the **transversus abdominis,** whose fibers run transversely across the abdomen.

Superficial Muscles of the Shoulder and Dorsal Trunk and Neck

Dissect the superficial muscles of the dorsal surface of the trunk. (Refer to **Figure 12.22**.)

1. Turn your cat on its ventral surface, and start your observations with the **trapezius group.** Humans have a single large trapezius muscle, but the cat has three separate muscles—the clavotrapezius, acromiotrapezius, and spinotrapezius—that together perform a similar function. The prefix (*clavo-*, *acromio-*, and *spino-*) in each case reveals the muscle's site of insertion. The **clavotrapezius,** the most superior muscle of the group, is homologous to the part of the human trapezius that inserts into the clavicle. Slip a probe under this muscle and follow it to its apparent origin.

Where does the clavotrapezius appear to originate?

____

Is this similar to its origin in humans? ______________

The fibers of the clavotrapezius are continuous inferiorly with those of the clavicular part of the cat's deltoid muscle (clavodeltoid), and the two muscles work together to flex the humerus. Release the clavotrapezius muscle from adjoining muscles. The **acromiotrapezius** is a large, nearly square muscle easily identified by its aponeurosis, which passes over the vertebral border of the scapula. It originates from the cervical and T_1 vertebrae and inserts into the scapular spine. The triangular **spinotrapezius** runs from the thoracic vertebrae to the scapular spine. This is the most posterior of the trapezius muscles in the cat. Now that you know where they are located, pull on the three trapezius muscles to mimic their action.

Do the trapezius muscles appear to have the same functions

in cats as in humans? ______________

2. The **levator scapulae ventralis,** a flat, straplike muscle, can be located in the triangle created by the division of the fibers of the clavotrapezius and acromiotrapezius. Its anterior fibers run underneath the clavotrapezius from its origin at the base of the skull (occipital bone), and it inserts on the vertebral border of the scapula. In the cat, the levator scapulae ventralis helps to hold the upper edges of the scapulae together (drawing them toward the head).

What is the function of the levator scapulae in humans?

____

3. The **deltoid group:** like the trapezius, the human deltoid muscle is represented by three separate muscles in the cat—the clavodeltoid, acromiodeltoid, and spinodeltoid. The **clavodeltoid** (also called the *clavobrachialis*), the most superficial muscle of the shoulder, is a continuation of the clavotrapezius below the clavicle, which is this muscle's point of origin (see Figure 12.22). Follow its course down the forelimb to the point where it merges along a white line with the pectoantebrachialis. Separate it from the pectoantebrachialis, and then transect it and pull it back.

Where does the clavodeltoid insert? ______________

What do you think the function of this muscle is? ______________

____

The **acromiodeltoid** lies posterior to the clavodeltoid and runs over the top of the shoulder. This small triangular muscle originates on the acromion of the scapula. It inserts into the spinodeltoid (a muscle of similar size) posterior to it. The **spinodeltoid** is covered with fascia near the anterior end of the scapula. Its tendon extends under the acromiodeltoid muscle and inserts on the humerus. Notice that its fibers run obliquely to those of the acromiodeltoid. Like the human deltoid muscle, the acromiodeltoid and clavodeltoid muscles in the cat abduct and rotate the humerus.

4. The **latissimus dorsi** is a large flat muscle covering most of the lateral surface of the posterior trunk. Its anterior edge is covered by the spinotrapezius. As in humans, it inserts into the humerus. But before inserting, its fibers merge with the fibers of many other muscles, among them the xiphihumeralis of the pectoralis group. Latissimus dorsi extends and adducts the arm.

Deep Muscles of the Laterodorsal Trunk and Neck

1. In preparation, transect the latissimus dorsi, the muscles of the pectoralis group, and the spinotrapezius, and reflect them back. Be careful not to damage the large brachial nerve plexus, which lies in the axillary space beneath the pectoralis group.

2. The **serratus ventralis** represents two separate muscles in humans. The posterior portion, homologous to the *serratus anterior* of humans, arises deep to the pectoral muscles and covers the lateral surface of the rib cage. It is easily identified by its fingerlike muscular origins, which arise on the first 9 or 10 ribs. It inserts into the scapula. The anterior portion of the serratus ventralis, which arises from the cervical vertebrae, is homologous to the *levator scapulae* in humans. Both portions pull the scapula toward the sternum. Trace this muscle to its insertion. In general, in the cat, this muscle pulls the scapula posteriorly and downward.

3. Reflect the upper limb to reveal the **subscapularis,** which occupies most of the ventral surface of the scapula **(Figure 12.23)**. Humans have a homologous muscle.

4. Locate the anterior, posterior, and middle **scalene** muscles on the lateral surface of the cat's neck and trunk. The most prominent and longest of these muscles is the middle scalene, which lies between the anterior and posterior members. The scalenes originate on the ribs and run cephalad over the serratus ventralis to insert in common on the cervical vertebrae. These muscles draw the ribs anteriorly and bend the neck downward; thus they are homologous to the human scalene muscles, which elevate the ribs and flex the neck. (Notice that the difference is only one of position. Humans walk erect, but cats are quadrupeds.)

5. Reflect the flaps of the transected latissimus dorsi, spinodeltoid, acromiodeltoid, and levator scapulae ventralis. The **splenius** is a large flat muscle occupying most of the side of the

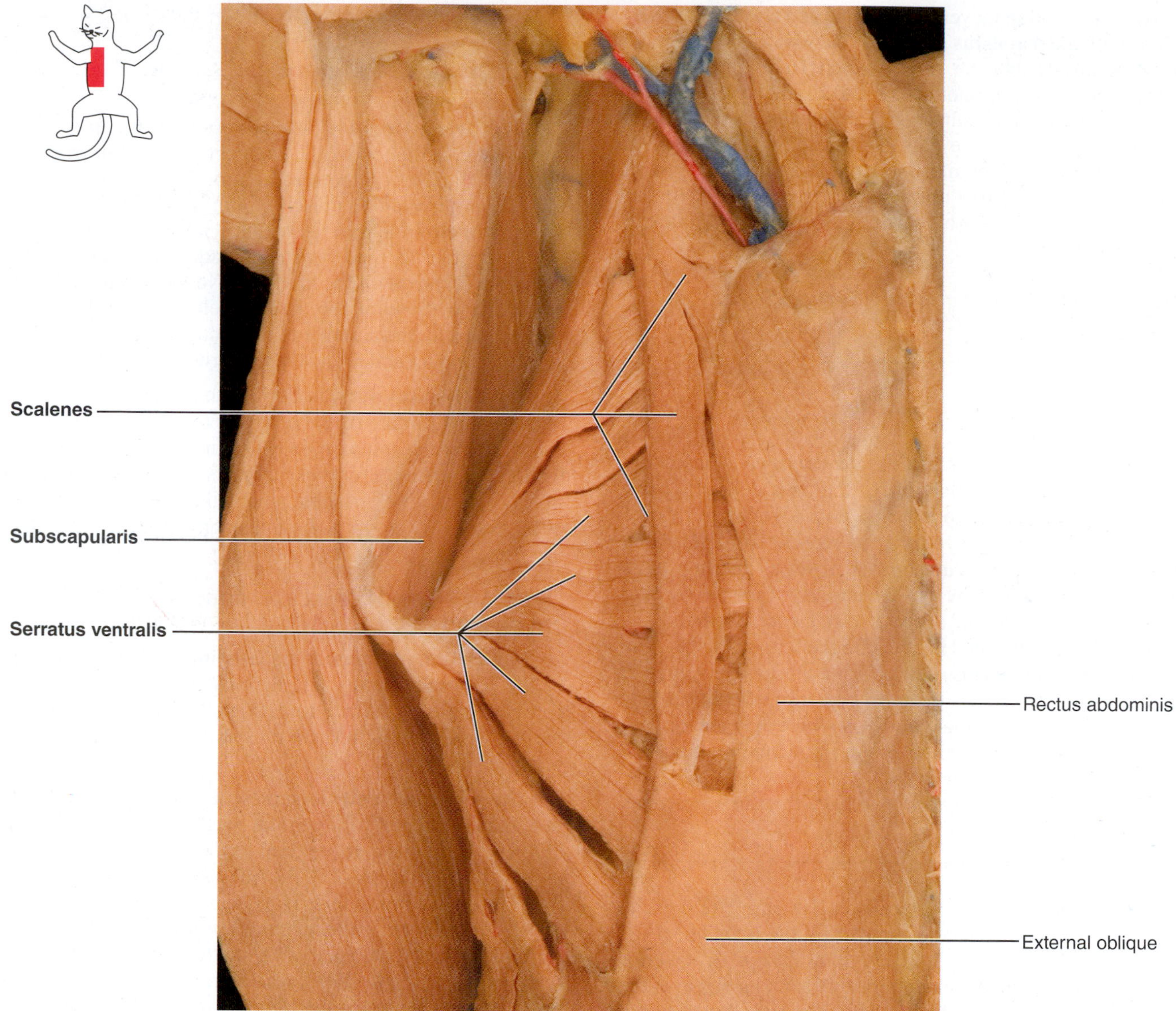

Figure 12.23 **Deep muscles of the inferolateral thorax of the cat.**

neck close to the vertebrae. As in humans, it originates on the ligamentum nuchae at the back of the neck and inserts into the occipital bone. It raises the head. (Refer to **Figure 12.24**.)

6. To view the rhomboid muscles, lay the cat on its side, and hold its forelegs together to spread the scapulae apart. The rhomboid muscles lie between the scapulae and beneath the acromiotrapezius. All the rhomboid muscles originate on the vertebrae and insert on the scapula. They hold the dorsal part of the scapula to the cat's back.

There are three rhomboids in the cat. The ribbonlike **rhomboid capitis,** the most anterolateral muscle of the group, has no counterpart in the human body. The **rhomboid minor,** located posterior to the rhomboid capitis, is much larger. Its fibers run transversely to those of the rhomboid capitis. The most posterior muscle of the group, the **rhomboid major,** is so closely fused to the rhomboid minor that many consider them to be one muscle—the **rhomboideus,** which is homologous to human *rhomboid muscles.*

7. The **supraspinatus** and **infraspinatus** muscles are similar to the same muscles in humans. The supraspinatus can be found under the acromiotrapezius, and the infraspinatus is deep to the spinotrapezius. Both originate on the lateral scapular surface and insert on the humerus. ■

ACTIVITY 9

Dissecting Forelimb Muscles

Cat forelimb muscles fall into the same three categories as human upper limb muscles, but in this section the muscles of the entire forelimb are considered together. (Refer to **Figure 12.25** as you study these muscles.)

Muscles of the Lateral Surface

1. The triceps muscle **(triceps brachii)** of the cat is easily identified if the cat is placed on its side. It is a large, fleshy muscle covering the posterior aspect and much of the side

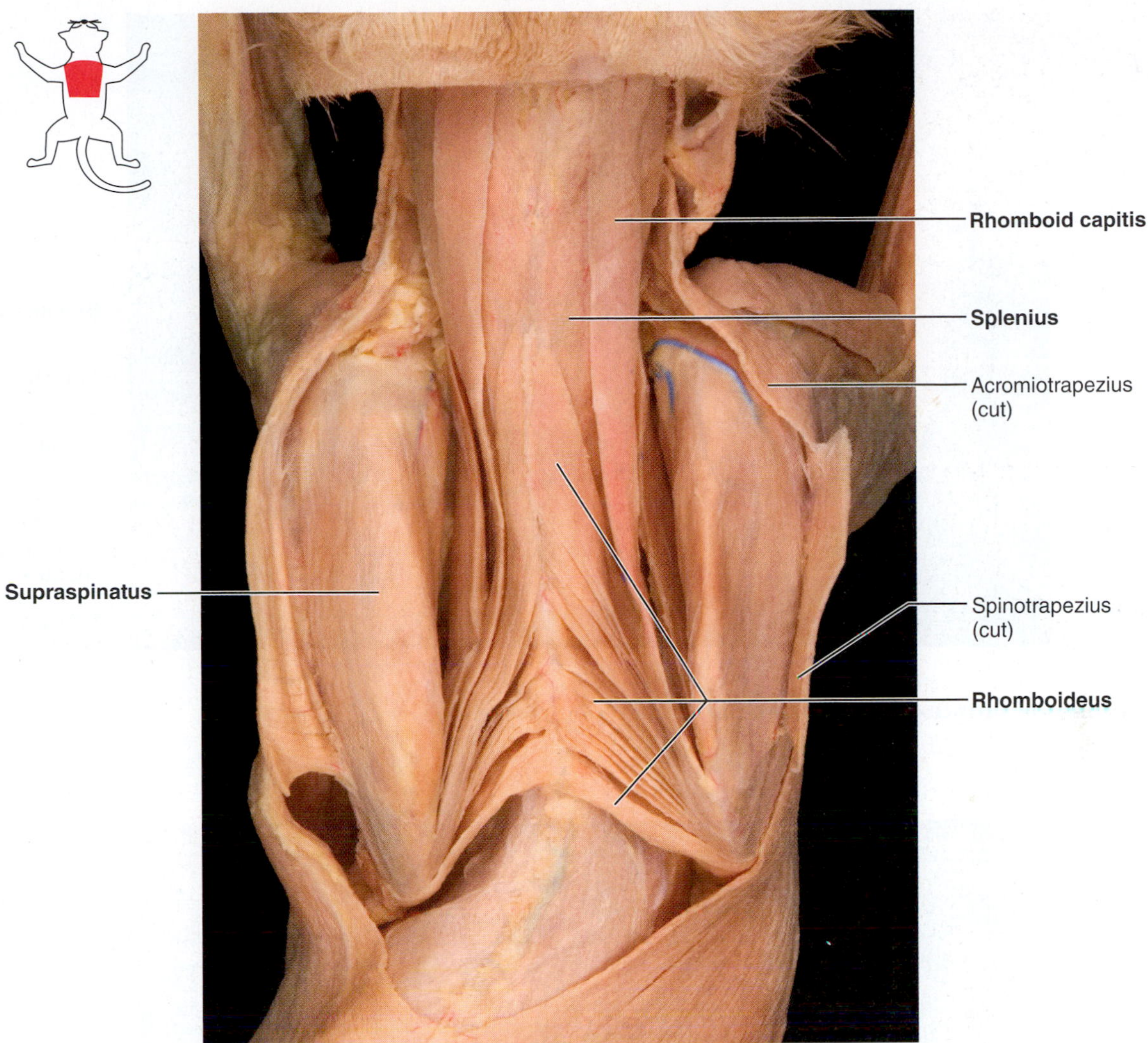

Figure 12.24 Deep muscles of the superior aspect of the dorsal thorax of the cat.

of the humerus. As in humans, this muscle arises from three heads, which originate from the humerus and scapula and insert jointly into the olecranon of the ulna. Remove the fascia from the superior region of the lateral arm surface to identify the lateral and long heads of the triceps. The long head is approximately twice as long as the lateral head and lies medial to it on the posterior arm surface. The medial head can be exposed by transecting the lateral head and pulling it aside. Now pull on the triceps muscle.

How does the function of the triceps muscle compare in cats and in humans?

Anterior and distal to the medial head of the triceps is the tiny **anconeus** muscle (not visible in Figure 12.25), sometimes called the fourth head of the triceps muscle. Notice its darker color and the way it wraps the tip of the elbow.

2. The **brachialis** can be located anterior to the lateral head of the triceps muscle. Identify its origin on the humerus, and trace its course as it crosses the elbow and inserts on the ulna. It flexes the cat's foreleg.

Identifying the forearm muscles is difficult because of the tough fascia sheath that encases them, but give it a try.

3. Remove as much of the connective tissue as possible, and cut through the ligaments that secure the tendons at the wrist (transverse carpal ligaments) so that you will be able to follow the muscles to their insertions. Begin your identification with the muscles at the lateral surface of the forearm. The muscles of this region are very much alike in appearance and are difficult to identify accurately unless you follow a definite order. Thus you will begin with the most anterior muscles and proceed to the posterior aspect. Remember to check carefully the tendons of insertion to verify your muscle identifications.

4. The ribbonlike muscle on the lateral surface of the humerus is the **brachioradialis.** Observe how it passes down the forearm to insert on the radial styloid process. (If you did not remove the fascia carefully, you may have also removed this muscle.)

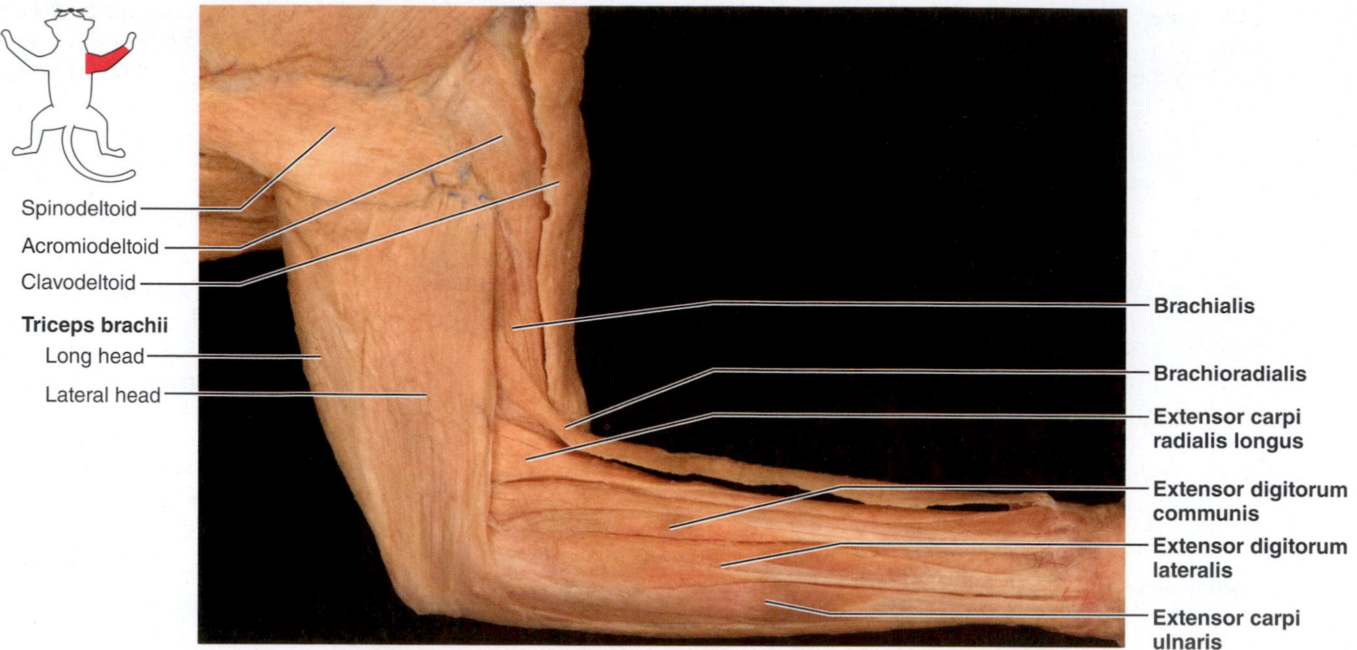

Figure 12.25 Lateral surface of the forelimb of the cat.

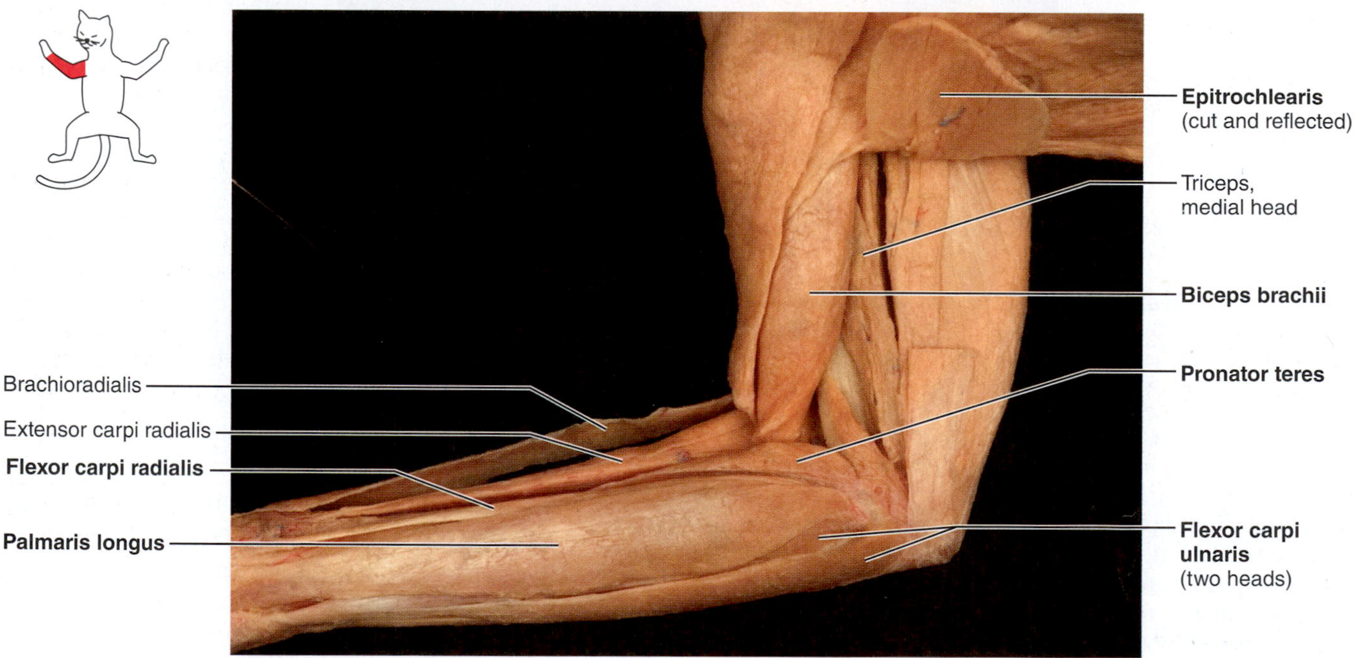

Figure 12.26 Medial surface of the forelimb of the cat.

5. The **extensor carpi radialis longus** has a broad origin and is larger than the brachioradialis. It extends down the anterior surface of the radius (see Figure 12.25). Transect this muscle to view the **extensor carpi radialis brevis** (not shown in Figure 12.25), which is partially covered by and sometimes fused with the extensor carpi radialis longus. Both muscles have origins, insertions, and actions similar to their human counterparts.

6. You can see the entire **extensor digitorum communis** along the lateral surface of the forearm. Trace it to its four tendons, which insert on the second to fifth digits. This muscle extends these digits. The **extensor digitorum lateralis** (absent in humans) also extends the digits. This muscle lies immediately posterior to the extensor digitorum communis.

7. Follow the **extensor carpi ulnaris** from the lateral epicondyle of the humerus to the ulnar side of the fifth metacarpal. Often this muscle has a shiny tendon, which helps you identify it.

Muscles of the Medial Surface

1. The **biceps brachii** (refer to **Figure 12.26**) is a large spindle-shaped muscle medial to the brachialis on the anterior surface of the humerus. Pull back the cut ends of the pectoral muscles to get a good view of the biceps. This muscle is much more prominent in humans, but its origin, insertion, and action are very similar in cats and in humans. Follow the muscle to its origin.

Does the biceps have two heads in the cat? ______________

2. The broad, flat, exceedingly thin muscle on the posteromedial surface of the arm is the **epitrochlearis.** Its tendon originates from the fascia of the latissimus dorsi, and the muscle inserts into the olecranon of the ulna. This muscle extends the forelimb of the cat; it is not found in humans.

3. The **coracobrachialis** of the cat is insignificant (approximately 1.3 cm, or ½ inch long) and can be seen as a very small muscle crossing the ventral aspect of the shoulder joint. It runs beneath the biceps brachii to insert on the humerus and has the same function as the human coracobrachialis.

4. Turn the cat so that the ventral forearm muscles (mostly flexors and pronators) can be observed (refer to Figure 12.26). As in humans, most of these muscles arise from the medial epicondyle of the humerus. The **pronator teres** runs from the medial epicondyle of the humerus and declines in size as it approaches its insertion on the radius. Do not bother to trace it to its insertion.

5. Like its human counterpart, the **flexor carpi radialis** runs from the medial epicondyle of the humerus to insert into the second and third metacarpals.

6. The large flat muscle in the center of the medial surface is the **palmaris longus.** Its origin on the medial epicondyle of the humerus abuts that of the pronator teres and is shared with the flexor carpi radialis. The palmaris longus extends down the forelimb to terminate in four tendons on the digits. Comparatively speaking, this muscle is much larger in cats than in humans.

The **flexor carpi ulnaris** arises from a two-headed origin (medial epicondyle of the humerus and olecranon of the ulna). Its two bellies (fleshy parts) pass downward to the wrist, where they are united by a single tendon that inserts into the carpals of the wrist. As in humans, this muscle flexes the wrist. ■

ACTIVITY 10

Dissecting Hindlimb Muscles

Remove the fat and fascia from all thigh surfaces, but do not cut through or remove the **fascia lata** (or iliotibial band), which is a tough white aponeurosis covering the anterolateral surface of the thigh from the hip to the leg. If the cat is a male, the cordlike sperm duct will be embedded in the fat near the pubic symphysis. Carefully clear around, but not in, this region.

Posterolateral Hindlimb Muscles

1. Turn the cat on its ventral surface, and identify the following superficial muscles of the hip and thigh (refer to **Figure 12.27**). Viewing the lateral aspect of the hindlimb, you will identify these muscles in sequence from the anterior to the posterior aspects of the hip and thigh. Most anterior is the **sartorius,** seen in this view as a thin band (Figure 12.27). Approximately 4 cm (1½ inches) wide, it extends around the lateral aspect of the thigh to the anterior surface, where the major portion of it lies (see Figure 12.29a). Free it from the adjacent muscles, and pass a blunt probe under it to trace its origin and insertion. Homologous to the sartorius muscle in humans, it adducts and rotates the thigh, but in addition, the cat sartorius acts as a knee extensor. Transect this muscle.

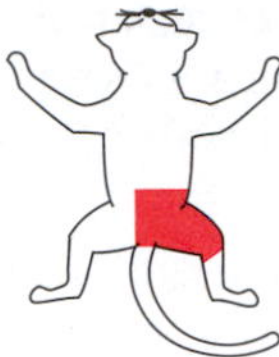

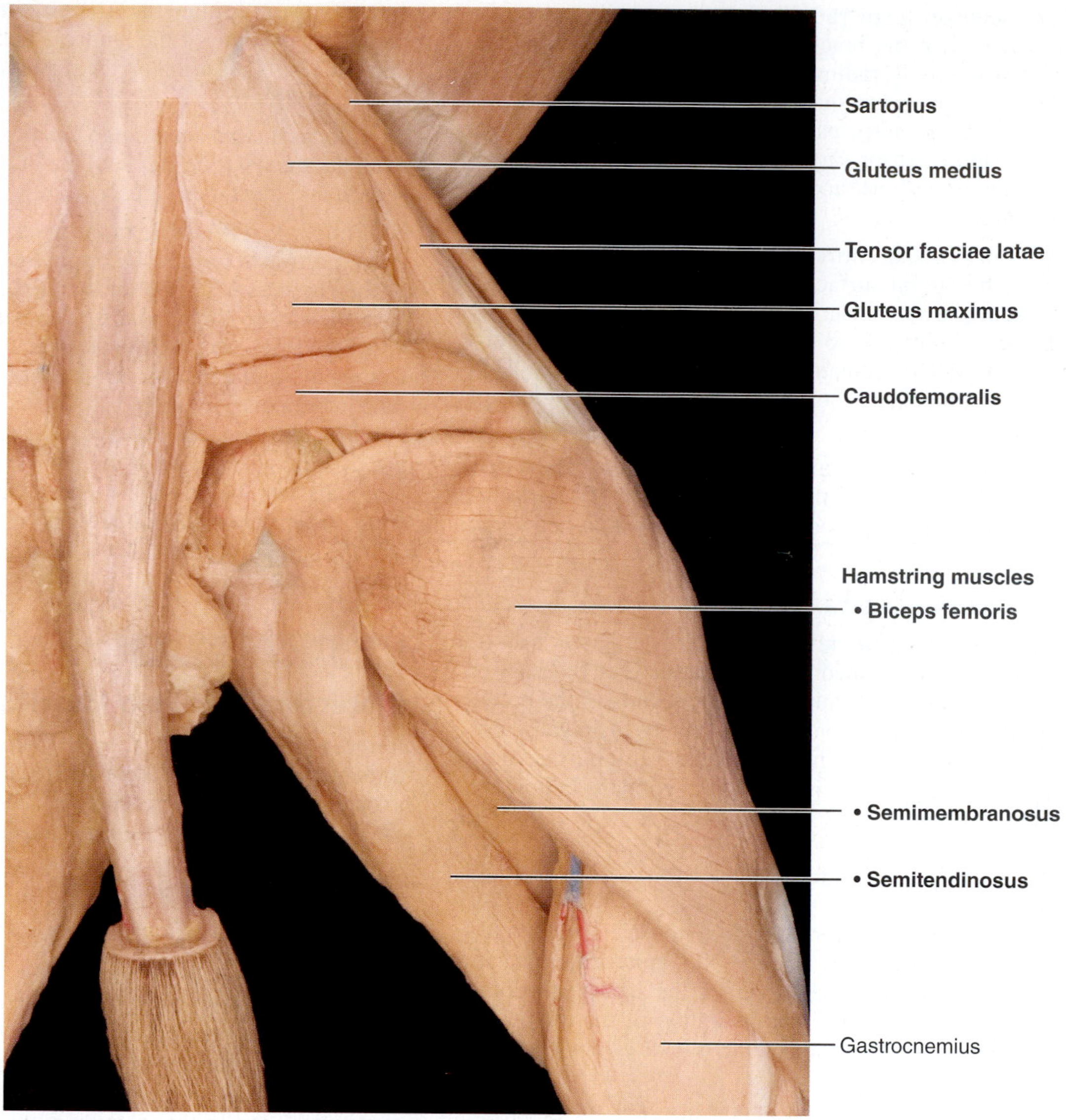

Sartorius
Gluteus medius
Tensor fasciae latae
Gluteus maximus
Caudofemoralis
Hamstring muscles
• Biceps femoris
• Semimembranosus
• Semitendinosus
Gastrocnemius

Figure 12.27 Muscles of the posterolateral thigh in the cat. Superficial view.

2. The **tensor fasciae latae** (Figure 12.27) is posterior to the sartorius. It is wide at its superior end, where it originates on the iliac crest, and narrows as it approaches its insertion into the fascia lata, which runs to the proximal tibial region. Transect its superior end and pull it back to expose the **gluteus medius** lying beneath it. This is the largest of the gluteus muscles in the cat. It originates on the ilium and inserts on the greater trochanter of the femur. The gluteus medius overlays and obscures the gluteus minimus, pyriformis, and gemellus muscles, which will not be identified here.

3. The **gluteus maximus** is a small triangular hip muscle posterior to the superior end of the tensor fasciae latae and paralleling it. In humans the gluteus maximus is a large fleshy muscle forming most of the buttock mass. In the cat it is only about 1.3 cm (½ inch) wide and 5 cm (2 inches) long, and it is smaller than the gluteus medius. The gluteus maximus covers part of the gluteus medius as it extends from the sacral region and the end of the femur. It abducts the thigh.

4. Posterior to the gluteus maximus, identify the triangular **caudofemoralis,** which originates on the caudal vertebrae and inserts into the patella via an aponeurosis. There is no homologue to this muscle in humans; in cats it abducts the thigh and flexes the vertebral column.

5. The **hamstring muscles** of the hindlimb include the biceps femoris, the semitendinosus, and the semimembranosus muscles. The **biceps femoris** is a large, powerful muscle that covers most of the posterolateral surface of the thigh. It is 4–5 cm (1½ to 2 inches) wide throughout its length. Trace it from its origin on the ischial tuberosity to its insertion on

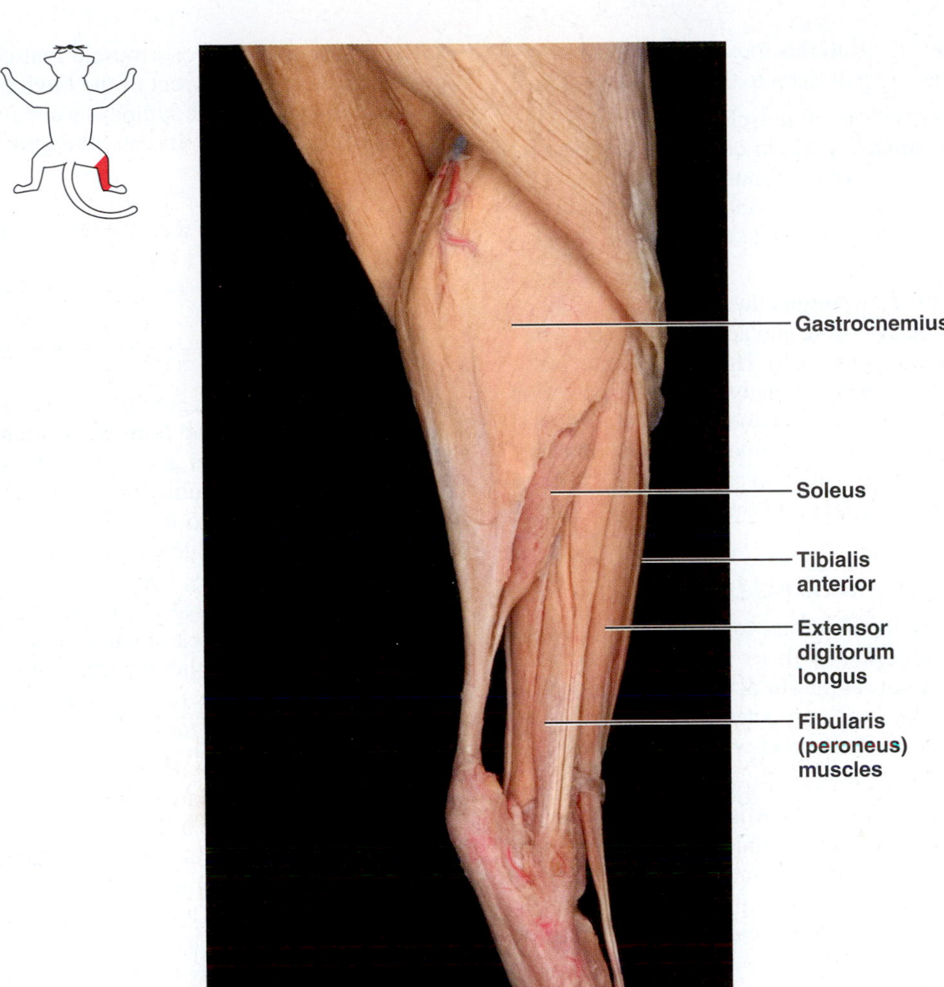

Figure 12.28 Superficial muscles of the posterolateral aspect of the shank (leg).

the tibia. Part of the **semitendinosus** can be seen beneath the posterior border of the biceps femoris. Transect and reflect the biceps muscle to reveal the whole length of the semitendinosus and the large sciatic nerve positioned under the biceps. Contrary to what its name implies ("half-tendon"), this muscle is muscular and fleshy except at its insertion. It is uniformly about 2 cm (¾ inch) wide as it runs down the thigh from the ischial tuberosity to the medial side of the tibia. It flexes the knee. The **semimembranosus,** a large muscle lying medial to the semitendinosus and largely obscured by it, is best seen in an anterior view of the thigh (Figure 12.29b). If desired, however, the semitendinosus can be transected to view it from the posterior aspect. The semimembranosus is larger and broader than the semitendinosus. Like the other hamstrings, it originates on the ischial tuberosity and inserts on the medial epicondyle of the femur and the medial tibial surface.

How does the semimembranosus compare with its human homologue?

___

___

6. Remove the heavy fascia covering the lateral surface of the shank (leg). Moving from the posterior to the anterior aspect, identify the following muscles on the posterolateral shank. (Refer to **Figure 12.28**.) First reflect the lower portion of the biceps femoris to see the origin of the **triceps surae,** the large composite muscle of the calf. Humans also have a triceps surae. The **gastrocnemius,** part of the triceps surae, is the largest muscle on the shank. As in humans, it has two heads and inserts via the calcaneal tendon into the

calcaneus. Run a probe beneath this muscle and then transect it to reveal the **soleus,** which is deep to the gastrocnemius.

7. Another important group of muscles in the leg is the **fibularis (peroneus) muscles,** which collectively appear as a slender, evenly shaped superficial muscle lying anterior to the triceps surae (Figure 12.28). Originating on the fibula and inserting on the digits and metatarsals, the fibularis muscles flex the foot.

8. The **extensor digitorum longus** lies anterior to the fibularis muscles. Its origin, insertion, and action in cats are similar to the homologous human muscle. The **tibialis anterior** is anterior to the extensor digitorum longus. The tibialis anterior is roughly triangular in cross section and heavier at its proximal end. Locate its origin on the proximal fibula and tibia and its insertion on the first metatarsal. You can see the sharp edge of the tibia at the anterior border of this muscle. As in humans, it is a foot flexor.

Anteromedial Hindlimb Muscles

1. Turn the cat onto its dorsal surface to identify the muscles of the anteromedial hindlimb **(Figure 12.29)**. Note once again the straplike sartorius at the surface of the thigh, which you have already identified and transected. It originates on the ilium and inserts on the medial region of the tibia.

2. Reflect the cut ends of the sartorius to identify the **quadriceps** muscles. The **vastus medialis** lies just beneath the sartorius. Resting close to the femur, it arises from the ilium and inserts into the patellar ligament. The small cylindrical muscle anterior and lateral to the vastus medialis is the **rectus femoris.** In cats this muscle originates entirely from the femur.

What is the origin of the rectus femoris in humans?

____

Free the rectus femoris from the most lateral muscle of this group, the large, fleshy **vastus lateralis,** which lies deep to the tensor fasciae latae. The vastus lateralis arises from the lateral femoral surface and inserts, along with the other vasti muscles, into the patellar ligament. Transect this muscle to identify the deep **vastus intermedius,** the smallest of the vasti muscles. It lies medial to the vastus lateralis and merges superiorly with the vastus medialis. The vastus intermedius is not shown in the figure.

3. The **gracilis** is a broad muscle that covers the posterior portion of the medial aspect of the thigh (see Figure 12.29a). It originates on the pubic symphysis and inserts on the medial proximal tibial surface. In cats, the gracilis adducts the leg and draws it posteriorly.

How does this compare with the human gracilis?

____

____

4. Free and transect the gracilis to view the adductor muscles deep to it. The **adductor femoris** is a large muscle that lies beneath the gracilis and abuts the semimembranosus medially. Its origin is the pubic ramus and the ischium, and its fibers pass downward to insert on most of the length of the femoral shaft. The adductor femoris is homologous to the human *adductor magnus, brevis,* and *longus.* Its function is to extend the thigh after it has been drawn forward and to adduct the thigh. A small muscle about 2.5 cm (1 inch) long—the **adductor longus**—touches the superior margin of the adductor femoris. It originates on the pubic bone and inserts on the proximal surface of the femur.

5. Before continuing your dissection, locate the **femoral triangle,** an important area bordered by the proximal edge of the sartorius and the adductor muscles. It is usually possible to identify the femoral artery (injected with red latex) and the femoral vein (injected with blue latex), which span the triangle (Figure 12.29a). (You will identify these vessels again in your study of the circulatory system.) If your instructor wishes you to identify the pectineus and iliopsoas, remove these vessels and go on to steps 6 and 7.

6. Examine the superolateral margin of the adductor longus to locate the small **pectineus.** It is sometimes covered by the gracilis (which you have cut and reflected). The pectineus, which originates on the pubis and inserts on the proximal end of the femur, is similar in all ways to its human homologue.

7. Just lateral to the pectineus you can see a small portion of the **iliopsoas,** a long and cylindrical muscle. Its origin is on the transverse processes of T_1 through T_{12} and the lumbar vertebrae, and it passes posteriorly toward the body wall to insert on the medial aspect of the proximal femur. The iliopsoas flexes and laterally rotates the thigh. It corresponds to the human iliopsoas.

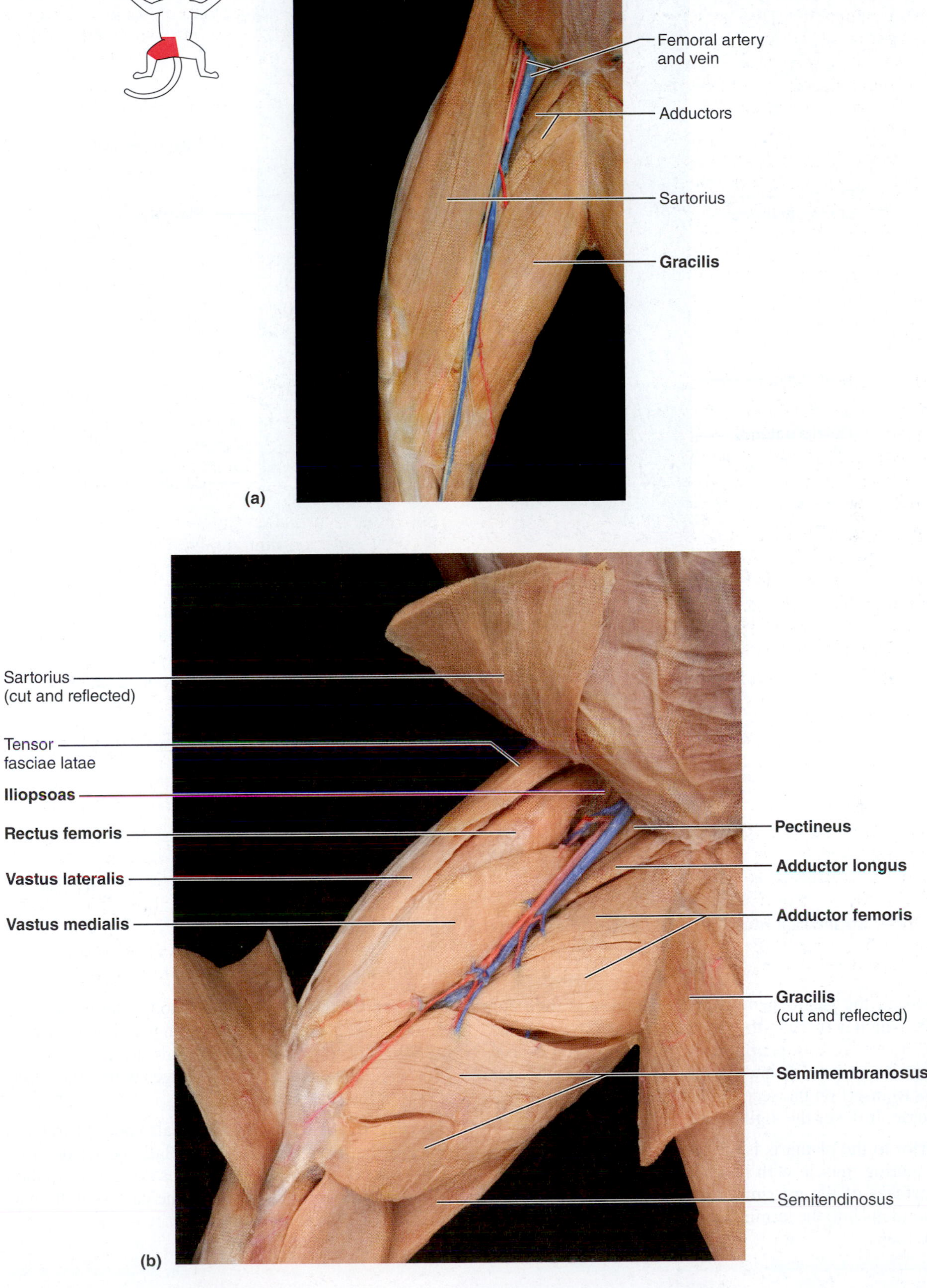

Figure 12.29 Muscles of the anteromedial thigh. (a) The gracilis and sartorius muscles are intact in this superficial view. **(b)** The gracilis and sartorius are transected and reflected to show deeper muscles.

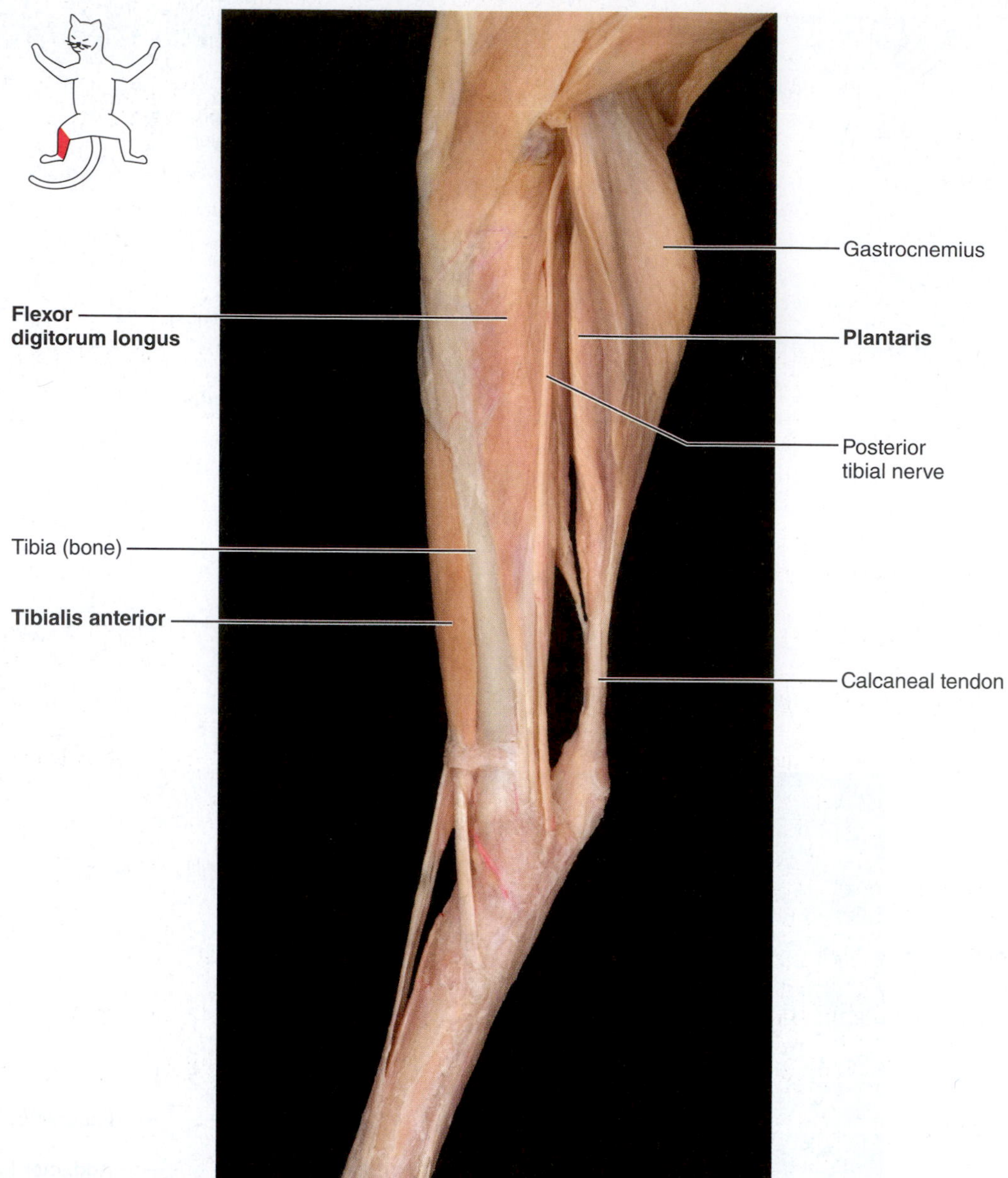

Figure 12.30 Superficial muscles of the anteromedial shank (leg) of the cat.

8. Reidentify the gastrocnemius of the shank and then the **plantaris,** which is fused with the lateral head of the gastrocnemius **(Figure 12.30)**. It originates from the lateral aspect of the femur and patella, and its tendon passes around the calcaneus to insert on the second phalanx. Working with the triceps surae, it flexes the digits and extends the foot.

9. Anterior to the plantaris is the **flexor digitorum longus,** a long, tapering muscle with two heads. It originates on the lateral surfaces of the proximal fibula and tibia and inserts via four tendons into the terminal phalanges. As in humans, it flexes the toes.

10. The **tibialis posterior** is a long, flat muscle lateral and deep to the flexor digitorum longus (not shown in Figure 12.30). It originates on the medial surface of the head of the fibula and the ventral tibia. It merges with a flat, shiny tendon to insert into the tarsals.

11. The **flexor hallucis longus** (also not illustrated) is a long muscle that lies lateral to the tibialis posterior. It originates from the posterior tibia and passes downward to the ankle. It is a uniformly broad muscle in the cat. As in humans, it is a flexor of the great toe.

NAME ______________________

LAB TIME/DATE ______________________

EXERCISE 12

REVIEW SHEET

Gross Anatomy of the Muscular System

Classification of Skeletal Muscles

1. Several criteria were given relative to the naming of muscles. For each muscle name, choose from the key all criteria on which the name is based. Some responses may be used more than once.

		Key:
_______	1. gluteus maximus	a. action of the muscle
_______	2. adductor magnus	b. shape of the muscle
_______	3. biceps femoris	c. location of the origin and/or insertion of the muscle
_______	4. transversus abdominis	d. number of origins
_______	5. extensor carpi ulnaris	e. location of the muscle relative to a bone or body region
_______	6. trapezius	f. direction in which the muscle fibers run relative to some imaginary line
_______	7. rectus femoris	g. relative size of the muscle
_______	8. external oblique	

2. Match the key terms to the muscles and movements described below. Some responses may be used more than once.

Key: a. prime mover (agonist) b. antagonist c. synergist d. fixator

_______________ 1. term for the biceps brachii during elbow flexion

_______________ 2. term that describes the relation of the brachialis to the biceps brachii during elbow flexion

_______________ 3. term for the triceps brachii during elbow flexion

_______________ 4. term for the iliopsoas during hip extension

_______________ 5. term for the gluteus maximus during hip extension when walking up stairs

_______________ 6. terms for the rotator cuff muscles and deltoid when the elbow is flexed and the hand grabs a tabletop to lift the table

Muscles of the Head and Neck

3. Using choices from the key on the right, correctly identify muscles provided with leader lines on the diagram.

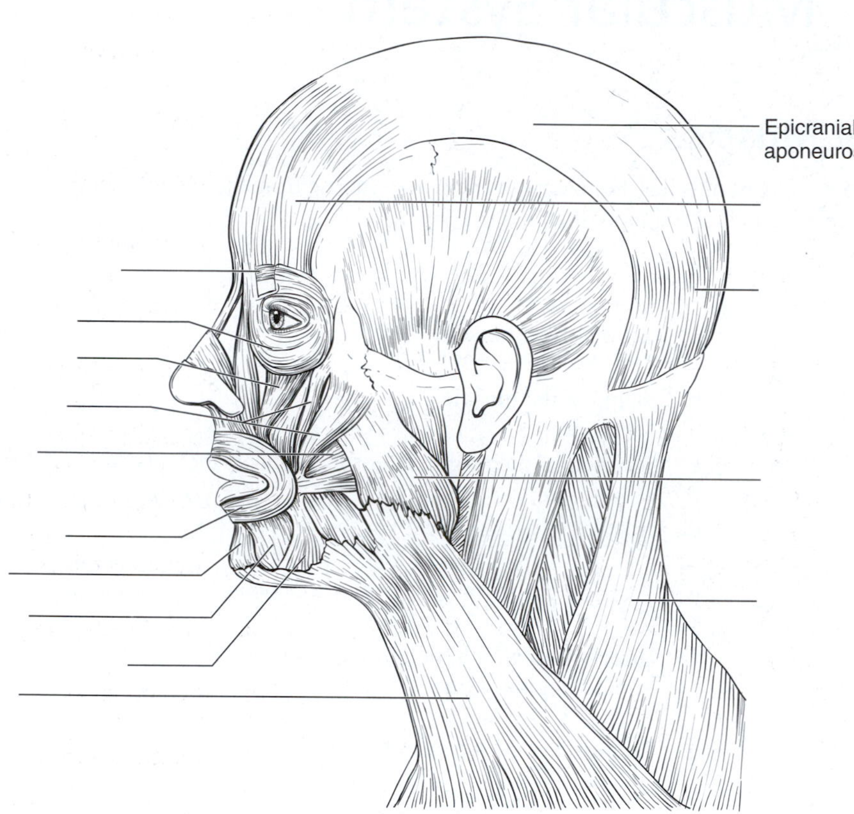

Key:

a. buccinator

b. corrugator supercilii

c. depressor anguli oris

d. depressor labii inferioris

e. frontal belly of epicranius

f. levator labii superioris

g. masseter

h. mentalis

i. occipital belly of epicranius

j. orbicularis oculi

k. orbicularis oris

l. platysma

m. trapezius

n. zygomaticus major and minor

4. Using the key provided in question 3, identify the muscles described below. (Not all the terms from question 3 will be used.)

_______ 1. used in smiling

_______ 2. used to suck in your cheeks

_______ 3. used in blinking and squinting

_______ 4. used to pout (pulls the corners of the mouth downward)

_______ 5. raises your eyebrows for a questioning expression

_______ 6. used to form the vertical frown crease on the forehead

_______ 7. your “kisser”

_______ 8. prime mover to raise the mandible

_______ 9. tenses skin of the neck during shaving

Muscles of the Trunk

5. Correctly identify both intact and transected (cut) muscles depicted in the diagram, using the key on the right. (Not all listed terms will be used in this exercise.)

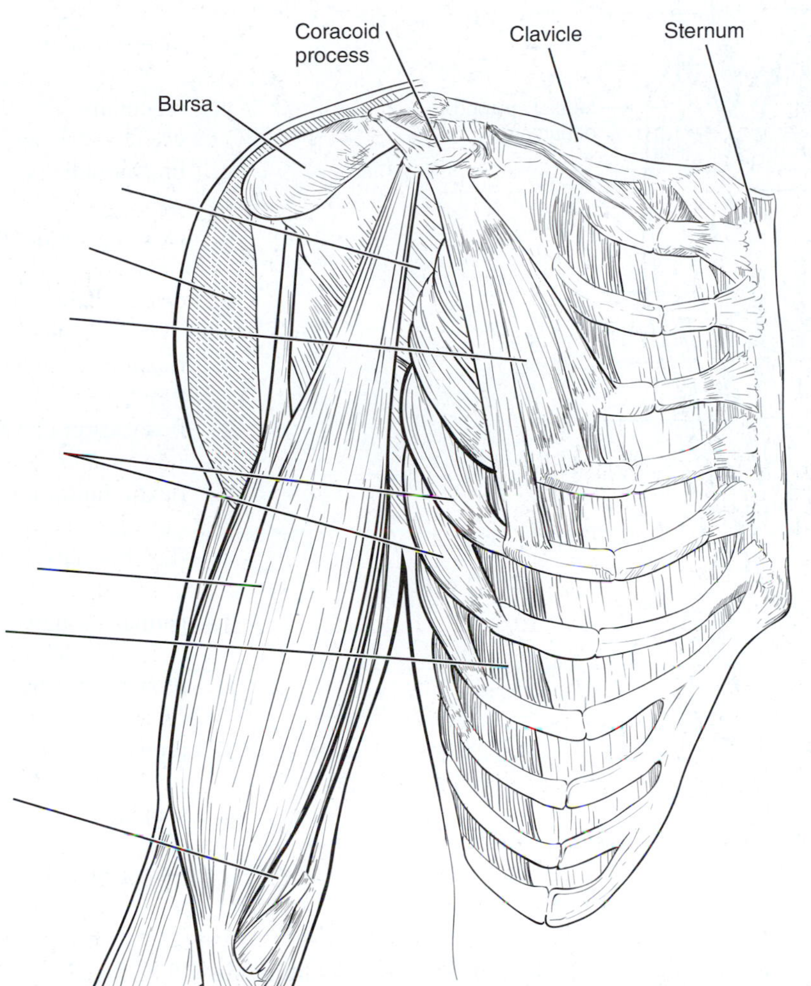

Key:

a. biceps brachii
b. brachialis
c. deltoid (cut)
d. external intercostals
e. external oblique
f. internal oblique
g. latissimus dorsi
h. pectoralis major (cut)
i. pectoralis minor
j. rectus abdominis
k. rhomboids
l. serratus anterior
m. subscapularis
n. transversus abdominis
o. trapezius

6. Using the key provided in question 5, identify the major muscles described below. (Not all terms will be used.)

_______ 1. a major spine flexor

_______ 2. prime mover for pulling the arm posteriorly

_______ 3. prime mover for shoulder flexion

_______ 4. assume major responsibility for forming the abdominal girdle (three pairs of muscles)

_______ 5. pulls the shoulder backward and downward

_______ 6. prime mover of shoulder abduction

_______ 7. important in shoulder adduction; antagonists of the shoulder abductor (two muscles)

_______ 8. moves the scapula forward and downward

_______ 9. small, inspiratory muscles between the ribs; elevate the ribs

_______ 10. extends the head

_______ 11. pull the scapulae medially

Muscles of the Upper Limb

7. Using the terms from the key on the right, correctly identify all muscles provided with leader lines in the diagram below. (Not all listed terms will be used in this exercise.)

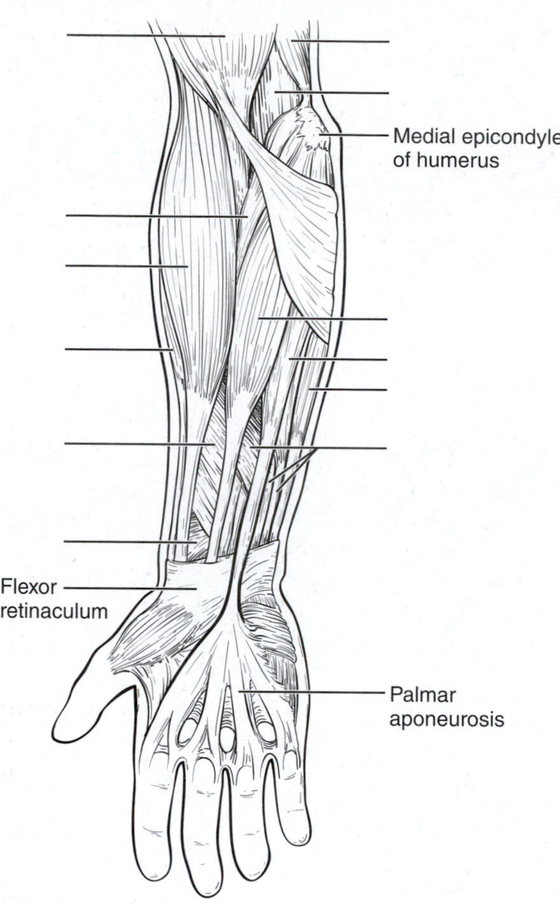

Key:

a. biceps brachii
b. brachialis
c. brachioradialis
d. extensor carpi radialis longus
e. extensor digitorum
f. flexor carpi radialis
g. flexor carpi ulnaris
h. flexor digitorum superficialis
i. flexor pollicis longus
j. palmaris longus
k. pronator quadratus
l. pronator teres
m. supinator
n. triceps brachii

8. Use the key provided in question 7 to identify the muscles described below. (Not all terms will be used.)

______ 1. flexes the forearm and supinates the hand

______ 2. synergist for supinating the hand

______ 3. forearm flexors; no role in supination (two muscles)

______ 4. elbow extensor

______ 5. power wrist flexor and abductor

______ 6. flexes wrist and middle phalanges

______ 7. pronate the hand (two muscles)

______ 8. flexes the thumb

______ 9. extend and abduct the wrist (two muscles)

______ 10. extends the wrist and digits

______ 11. flat muscle that is a weak wrist flexor; tenses skin of palm

Muscles of the Lower Limb

9. Using the terms from the key on the right, correctly identify all muscles provided with leader lines in the diagram below. (Not all listed terms will be used in this exercise.)

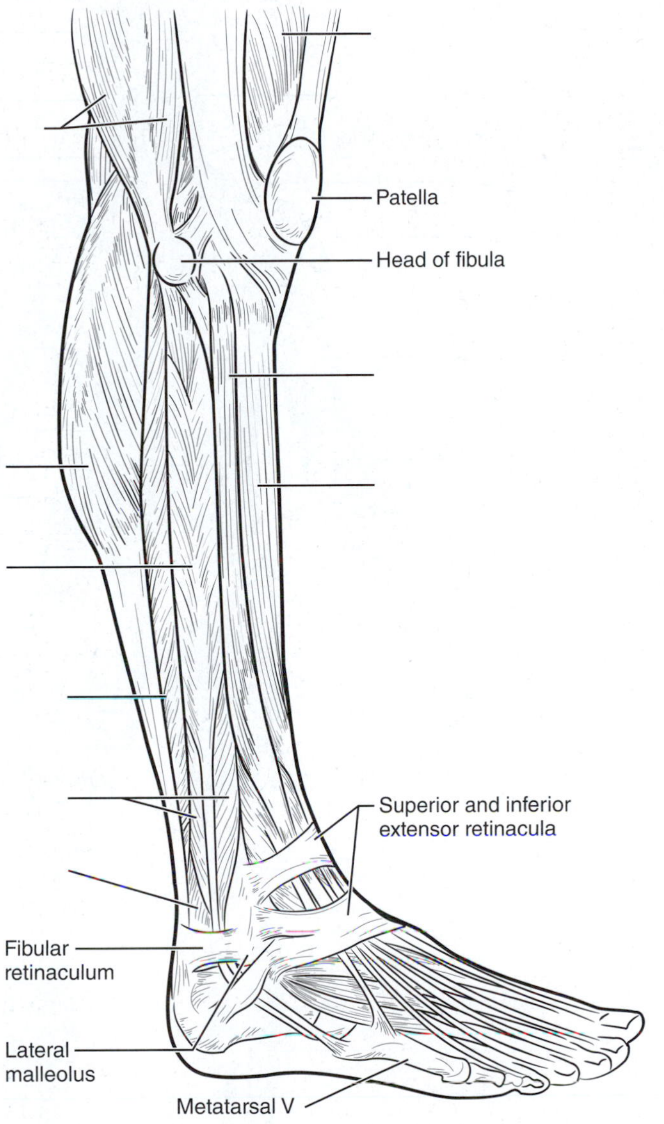

Key:

a. adductor group
b. biceps femoris
c. extensor digitorum longus
d. fibularis brevis
e. fibularis longus
f. flexor hallucis longus
g. gastrocnemius
h. gluteus maximus
i. gluteus medius
j. rectus femoris
k. semimembranosus
l. semitendinosus
m. soleus
n. tensor fasciae latae
o. tibialis anterior
p. tibialis posterior
q. vastus lateralis

10. Use the key provided in question 9 to identify the muscles described below. (Not all terms will be used.)

_______ 1. flexes the great toe and inverts the ankle

_______ 2. lateral compartment muscles that plantar flex and evert the ankle (two muscles)

_______ 3. abduct the thigh to take the "at ease" stance (two muscles)

_______ 4. used to extend the hip when climbing stairs

_______ 5. prime movers of ankle plantar flexion (two muscles)

_______ 6. major foot inverter

_______ 7. prime mover of dorsiflexion of the foot

_______ 8. adduct the thigh, as when standing at attention

_______ 9. extends the toes

_______ 10. extend thigh and flex knee (three muscles)

_______ 11. extends knee and flexes thigh

General Review: Muscle Recognition

11. Identify each lettered muscle in this diagram of the human anterior superficial musculature by matching its letter with one of the following muscle names:

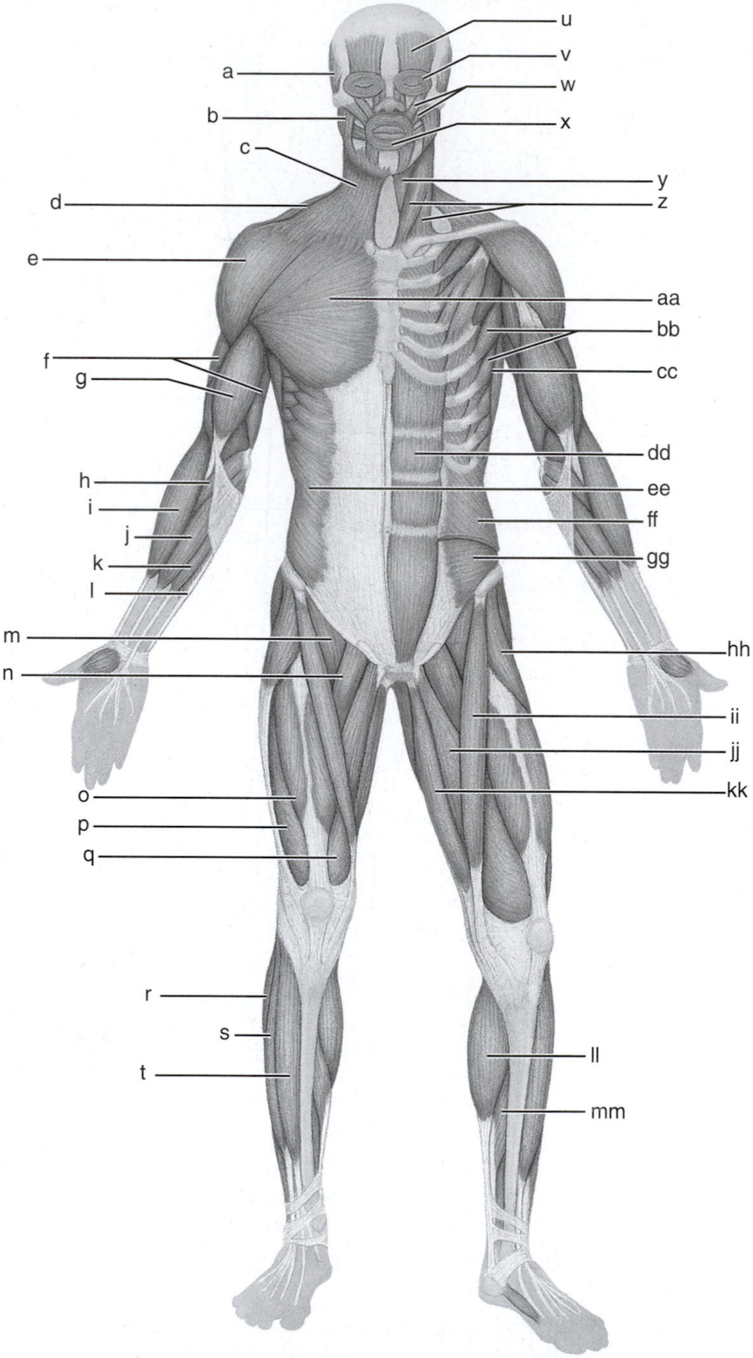

_______ 1. adductor longus

_______ 2. biceps brachii

_______ 3. brachioradialis

_______ 4. deltoid

_______ 5. extensor digitorum longus

_______ 6. external oblique

_______ 7. fibularis longus

_______ 8. flexor carpi radialis

_______ 9. flexor carpi ulnaris

_______ 10. frontal belly of epicranius

_______ 11. gastrocnemius

_______ 12. gracilis

_______ 13. iliopsoas

_______ 14. internal oblique

_______ 15. latissimus dorsi

_______ 16. masseter

_______ 17. orbicularis oculi

_______ 18. orbicularis oris

_______ 19. palmaris longus

_______ 20. pectineus

_______ 21. pectoralis major

_______ 22. platysma

_______ 23. pronator teres

_______ 24. rectus abdominis

_______ 25. rectus femoris

_______ 26. sartorius

_______ 27. serratus anterior

_______ 28. soleus

_______ 29. sternocleidomastoid

_______ 30. sternohyoid

_______ 31. temporalis

_______ 32. tensor fasciae latae

_______ 33. tibialis anterior

_______ 34. transversus abdominis

_______ 35. trapezius

_______ 36. triceps brachii

_______ 37. vastus lateralis

_______ 38. vastus medialis

_______ 39. zygomaticus

12. Identify each lettered muscle in this diagram of the human posterior superficial musculature by matching its letter with one of the following muscle names:

_______ 1. adductor magnus

_______ 2. biceps femoris

_______ 3. brachialis

_______ 4. brachioradialis

_______ 5. deltoid

_______ 6. extensor carpi radialis longus

_______ 7. extensor carpi ulnaris

_______ 8. extensor digitorum

_______ 9. external oblique

_______ 10. flexor carpi ulnaris

_______ 11. gastrocnemius

_______ 12. gluteus maximus

_______ 13. gluteus medius

_______ 14. gracilis

_______ 15. iliotibial tract (tendon)

_______ 16. infraspinatus

_______ 17. latissimus dorsi

_______ 18. occipital belly of epicranius

_______ 19. semimembranosus

_______ 20. semitendinosus

_______ 21. sternocleidomastoid

_______ 22. teres major

_______ 23. trapezius

_______ 24. triceps brachii

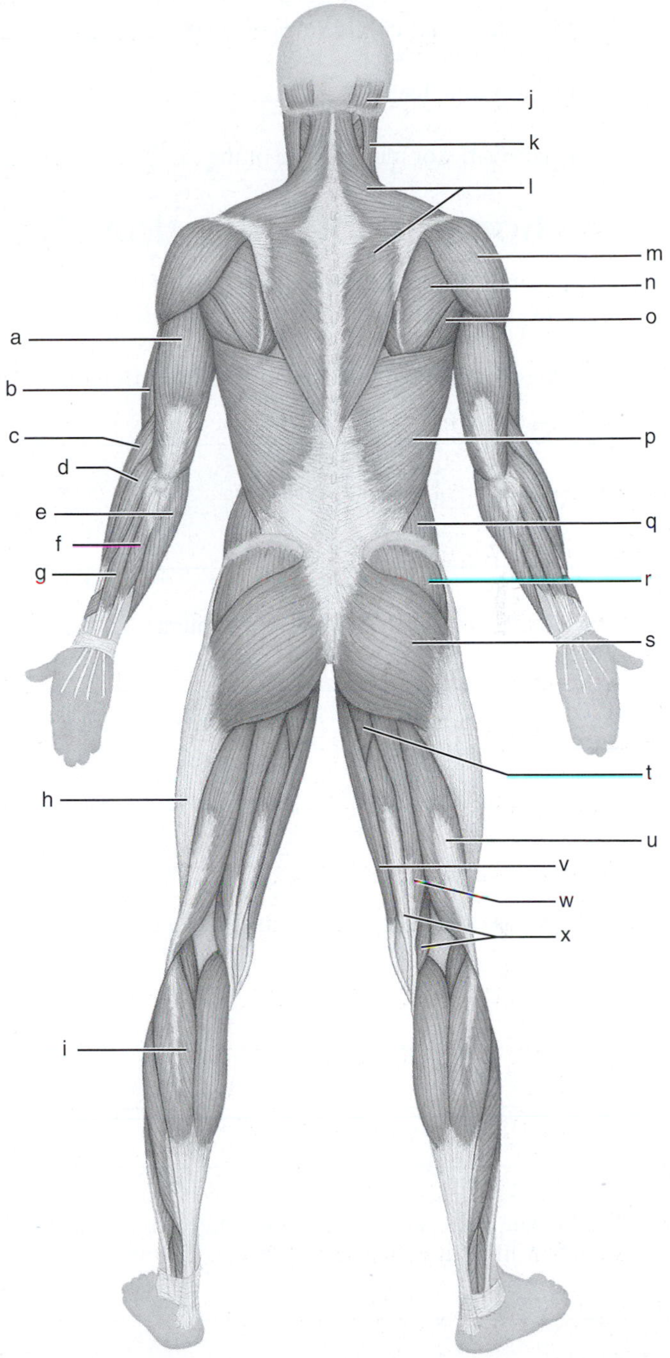

a
b
c
d
e
f
g
h
i
j
k
l
m
n
o
p
q
r
s
t
u
v
w
x

General Review: Muscle Descriptions

13. Identify the muscles described by completing the following statements. Use an appropriate reference as needed.

1. The _______________, _______________, _______________, and _______________ are commonly used for intramuscular injections (four muscles).

2. The insertion tendon of the _______________ group contains a large sesamoid bone, the patella.

3. The triceps surae insert in common into the _______________ tendon.

4. The bulk of the tissue of a muscle tends to lie ____________ to the part of the body it causes to move.

5. The extrinsic muscles of the hand originate on the ____________.

6. Most flexor muscles are located on the ____________ aspect of the body; most extensors are located ____________. An exception to this generalization is the extensor-flexor musculature of the ____________.

Dissection and Identification: Cat Muscles

Many human muscles are modified from those of the cat (or any quadruped) as a result of the requirements of an upright posture. The following questions refer to these differences.

14. How does the human trapezius muscle differ from the cat's?

____

15. How does the deltoid differ?

____

16. How do the size and orientation of the human gluteus maximus muscle differ from that in the cat?

____

____

____

____

17. Explain these differences in cat and human muscles in terms of differences in function.

____

____

____

____

18. The human rectus abdominis is definitely divided by four transverse tendons (tendinous intersections). These tendons are absent or difficult to identify in the cat. How do these tendons affect the human upright posture?

____

____

19. Match the terms in column B to the descriptions in column A.

	Column A	Column B
____________	1. to separate muscles	a. dissect
____________	2. to fold back a muscle	b. embalm
____________	3. to cut through a muscle	c. reflect
____________	4. to preserve tissue	d. transect

13

Histology of Nervous Tissue

EXERCISE

MATERIALS

- ☐ Model of a "typical" neuron (if available)
- ☐ Compound microscope
- ☐ Immersion oil
- ☐ Prepared slides of an ox spinal cord smear and teased myelinated nerve fibers
- ☐ Prepared slides of Purkinje cells (cerebellum), pyramidal cells (cerebrum), and a dorsal root ganglion
- ☐ Prepared slide of a nerve (x.s.)
- ☐ Prepared slides (l.s.) of lamellar corpuscles, tactile corpuscles, tendon organs, and muscle spindles

OBJECTIVES

1. Discuss the functional differences between neurons and neuroglia.
2. List six types of neuroglia, and indicate where each is found.
3. Identify the important anatomical features of a neuron on an appropriate image.
4. List the functions of dendrites, axons, and terminal boutons.
5. Explain how a nerve impulse is transmitted from one neuron to another.
6. State the function of myelin sheaths, and explain how Schwann cells myelinate axons in the peripheral nervous system.
7. Classify neurons structurally and functionally.
8. Differentiate a nerve from a tract, and a ganglion from a CNS nucleus.
9. Identify endoneurium, perineurium, and epineurium microscopically or in an appropriate image, and cite their functions.
10. Recognize and describe the various types of general sensory receptors as studied in the laboratory, and list the functions and locations of each.

PRE-LAB QUIZ

1. ______________________ are the functional units of nervous tissue.
2. Neuroglia of the peripheral nervous system include:
   a. ependymal cells and satellite cells
   b. oligodendrocytes and astrocytes
   c. satellite cells and Schwann cells
3. These branching neuron processes serve as receptive regions and transmit electrical signals toward the cell body. They are:
   a. axons b. collaterals c. dendrites d. neuroglia
4. Most axons are covered with a fatty material called ____________________, which insulates the fibers and increases the speed of neurotransmission.
5. Circle the correct underlined term. Neuron fibers (axons) running through the central nervous system form <u>tracts</u> / <u>nerves</u> of white matter.
6. Neurons can be classified according to structure. ____________________ neurons have many processes that issue from the cell body.
   a. Bipolar b. Multipolar c. Unipolar
7. Within a nerve, each axon is surrounded by a covering called the:
   a. endoneurium b. epineurium c. perineurium
8. Sensory receptors can be classified according to their source of stimulus. ______________________ are found close to the body surface and react to stimuli in the external environment.
   a. Exteroceptors b. Interoceptors c. Proprioceptors d. Visceroceptors
9. Tactile corpuscles respond to light touch. Where would you expect to find tactile corpuscles?
   a. deep within the dermal layer of hairy skin
   b. in the dermal papillae of hairless skin
   c. in the hypodermis of hairless skin
   d. in the uppermost portion of the epidermis
10. ______________________ are sensory receptors that detect stretch in skeletal muscles, tendons, and joints.

The nervous system is the master integrating and coordinating system, continuously monitoring and processing sensory information both from the external environment and from within the body. Every thought, action, and sensation is a reflection of its activity. Like a computer, it processes and integrates new "inputs" with information previously fed into it ("programmed") to produce an appropriate response ("readout"). However, no computer can possibly compare in complexity and scope to the human nervous system.

Two primary divisions make up the nervous system: the central nervous system, or CNS, consisting of the brain and spinal cord; and the peripheral nervous system, or PNS, which includes all the nervous elements located outside the central nervous system. PNS structures include nerves, sensory receptors, and some clusters of nerve cells.

Despite its complexity, nervous tissue is made up of just two principal cell types: neurons and neuroglia.

Neuroglia

The **neuroglia** ("nerve glue"), or **glial cells,** of the CNS include *astrocytes, oligodendrocytes, microglial cells,* and *ependymal cells* **(Figure 13.1)**. The neuroglia found in the PNS include *Schwann cells,* also called neurolemmocytes, and *satellite cells.*

Neuroglia serve the needs of the delicate neurons by bracing and protecting them. In addition, they act as phagocytes (microglial cells), myelinate the cytoplasmic extensions of the neurons (oligodendrocytes and Schwann cells), play a role in capillary-neuron exchanges, and control the chemical environment around neurons (astrocytes). Although neuroglia resemble neurons in some ways (they have fibrous cellular extensions), they are not capable of generating and transmitting nerve impulses, a capability that is highly developed in neurons. In this exercise, we focus on the highly excitable neurons.

Neurons

Neurons, or nerve cells, are the basic functional units of nervous tissue. They are highly specialized to transmit messages from one part of the body to another in the form of nerve impulses. Although neurons differ structurally, they have many identifiable features in common (**Figure 13.2a** and **b**). All have a **cell body** from which slender processes extend. The cell body is both the *biosynthetic center* of the neuron and part of its *receptive region*. Neuron cell bodies make up the gray matter of the CNS and form clusters there that are called **nuclei.** In the PNS, clusters of neuron cell bodies are called **ganglia.**

The neuron cell body contains a large round nucleus surrounded by cytoplasm. Two prominent structures are found in the cytoplasm: (1) cytoskeletal elements called **neurofibrils,** which provide support for the cell and a means to transport substances throughout the neuron; and (2) darkly staining

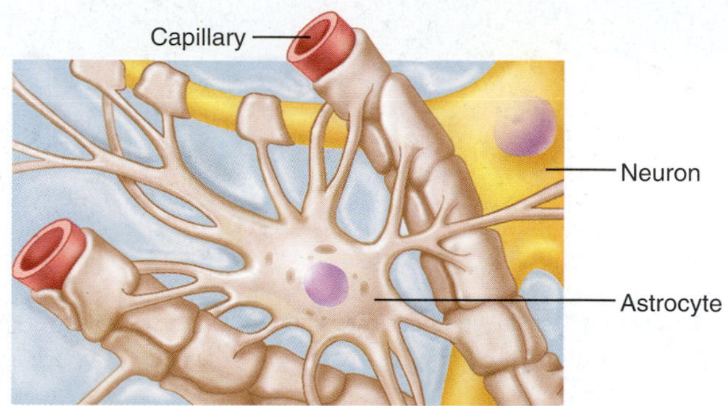

(a) Astrocytes are the most abundant CNS neuroglia.

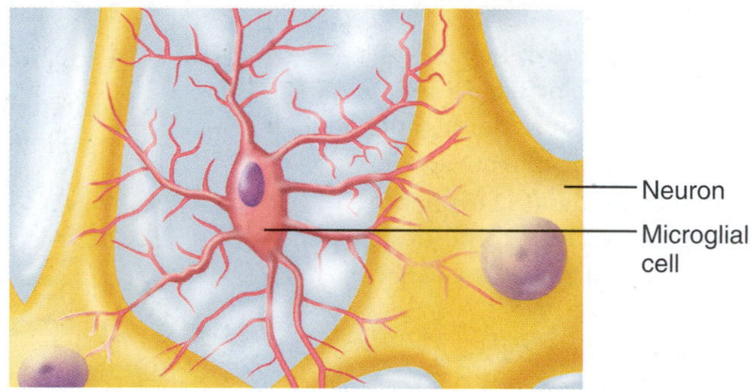

(b) Microglial cells are defensive cells in the CNS.

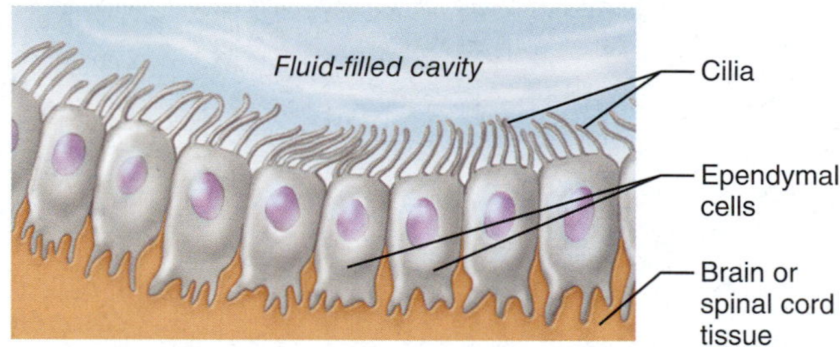

(c) Ependymal cells line cerebrospinal fluid–filled cavities.

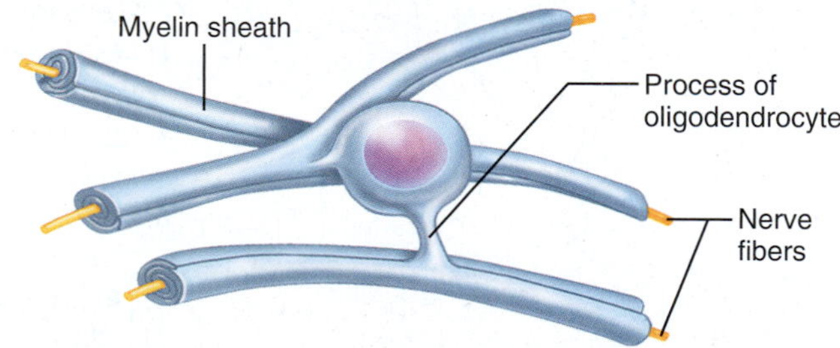

(d) Oligodendrocytes have processes that form myelin sheaths around CNS nerve fibers.

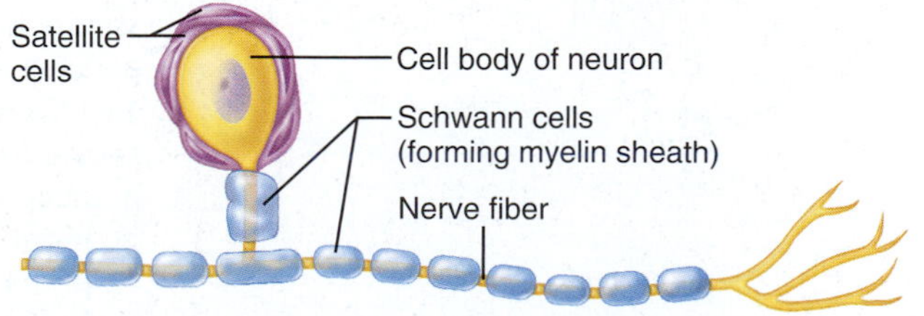

(e) Satellite cells and Schwann cells (which form myelin) surround neurons in the PNS.

Figure 13.1 **Neuroglia. (a–d)** The four types of neuroglia of the central nervous system. **(e)** Neuroglia of the peripheral nervous system.

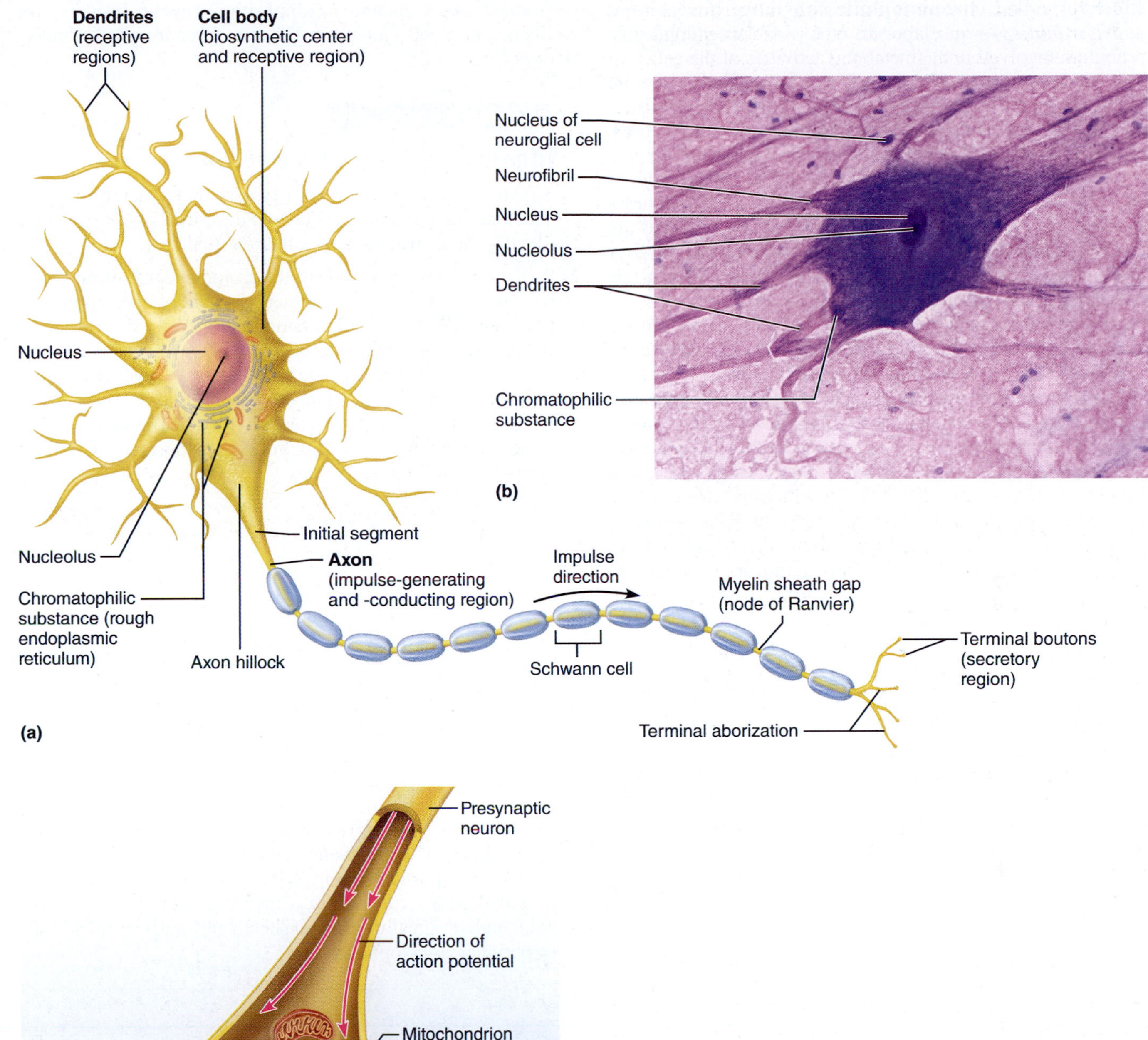

Figure 13.2 Structure of a typical motor neuron. **(a)** Diagram view. **(b)** Photomicrograph (410×). **(c)** Diagram of an enlarged synapse.

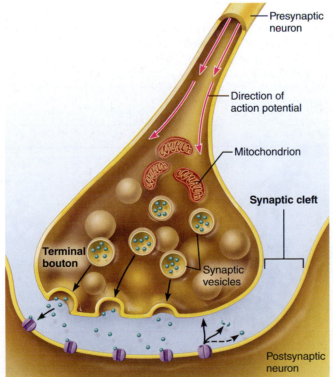

structures called **chromatophilic substance** (also known as *Nissl bodies*)—an elaborate type of rough endoplasmic reticulum involved in the metabolic activities of the cell.

Neurons have two types of processes. **Dendrites** are *receptive regions* that bear receptors for neurotransmitters released by the terminal boutons of other neurons. **Axons,** also called *nerve fibers*, form the *impulse-generating and impulse-conducting region* of the neuron. The white matter of the nervous system is made up of axons. In the CNS, bundles of axons are called **tracts;** in the PNS, bundles of axons are called **nerves.** Neurons may have many dendrites, but they have only a single axon. The axon may branch, forming one or more processes called **axon collaterals.**

In general, a neuron is excited by other neurons when their axons release neurotransmitters close to its dendrites or cell body. The electrical signal produced travels across the cell body, and if it is great enough, it elicits a regenerative electrical signal, an *impulse* or *action potential,* that travels down the axon. The axon in motor neurons begins just distal to a slightly enlarged cell body structure called the **axon hillock** (Figure 13.2a). The point at which the axon hillock narrows to axon diameter is referred to as the *initial segment.* The axon ends in many small structures called **terminal boutons,** or **axon terminals,** which form **synapses,** or junctions, with neurons or effector cells. These terminals store the neurotransmitter chemical in tiny vesicles. Each terminal bouton of the presynaptic neuron is separated from the cell body or dendrites of the next (postsynaptic) neuron by a tiny gap called the **synaptic cleft** (Figure 13.2c). Thus, although they are close, there is no actual physical contact between neurons. When an action potential reaches the terminal bouton, some of the *synaptic vesicles* rupture and release neurotransmitter into the synaptic cleft. The neurotransmitter then diffuses across the synaptic cleft to bind to membrane receptors on the next neuron, initiating an electrical signal or synaptic potential. Specialized synapses between neurons and skeletal muscle cells are called neuromuscular junctions. (They are discussed in Exercise 11.)

Most long nerve fibers are covered with a fatty material called *myelin,* and such fibers are referred to as **myelinated fibers.** Because of its chemical composition, myelin insulates the fibers and greatly increases the speed of neurotransmission by neuron fibers. Axons in the peripheral nervous system are typically heavily myelinated by special supporting cells called **Schwann cells,** which wrap themselves tightly around the axon in jelly-roll fashion **(Figure 13.3)**. During the wrapping process, the cytoplasm is squeezed from between adjacent layers of the Schwann cell membranes, so that when the process is completed a tight core of plasma membrane material (protein-lipoid material) encompasses the axon. This wrapping is the **myelin sheath.** The Schwann cell nucleus and the bulk of its cytoplasm ends up just beneath the outermost portion of its plasma membrane. This peripheral part of the Schwann cell and its exposed plasma membrane is referred to as the **outer collar of perinuclear cytoplasm.** Since the myelin sheath is formed by many individual Schwann cells, it is a discontinuous sheath. The gaps or indentations in the sheath are called **myelin sheath gaps** or *nodes of Ranvier* (see Figure 13.2).

Within the CNS, myelination is accomplished by neuroglia called **oligodendrocytes** (see Figure 13.1d). These CNS sheaths do not exhibit the outer collar of perinuclear cytoplasm seen in fibers myelinated by Schwann cells. Because of its chemical composition, myelin insulates the fibers and greatly increases the transmission speed of nerve impulses.

ACTIVITY 1

Identifying Parts of a Neuron

1. Study the illustration of a typical motor neuron (Figure 13.2), noting the structural details described above, and then identify these structures on a neuron model.

2. Obtain a prepared slide of the ox spinal cord smear, which has large, easily identifiable neurons. Study one representative neuron under oil immersion, and identify the cell body; the nucleus; the large, prominent "owl's eye" nucleolus; and the granular chromatophilic substance. If possible, distinguish the axon from the many dendrites.

Sketch the cell in the space provided below, and label the important anatomical details you have observed. (Compare your sketch to Figure 13.2c.)

3. Obtain a prepared slide of teased myelinated nerve fibers. Identify the following (use **Figure 13.4** as a guide): myelin sheath gaps, axon, Schwann cell nuclei, and myelin sheath.

Sketch a portion of a myelinated nerve fiber in the space provided below, illustrating a myelin sheath gap. Label the axon, myelin sheath, a myelin sheath gap, and the outer collar of perinuclear cytoplasm.

Do the gaps seem to occur at consistent intervals, or are they irregularly distributed? ______________________________

Explain the functional significance of this finding: __________

____

____

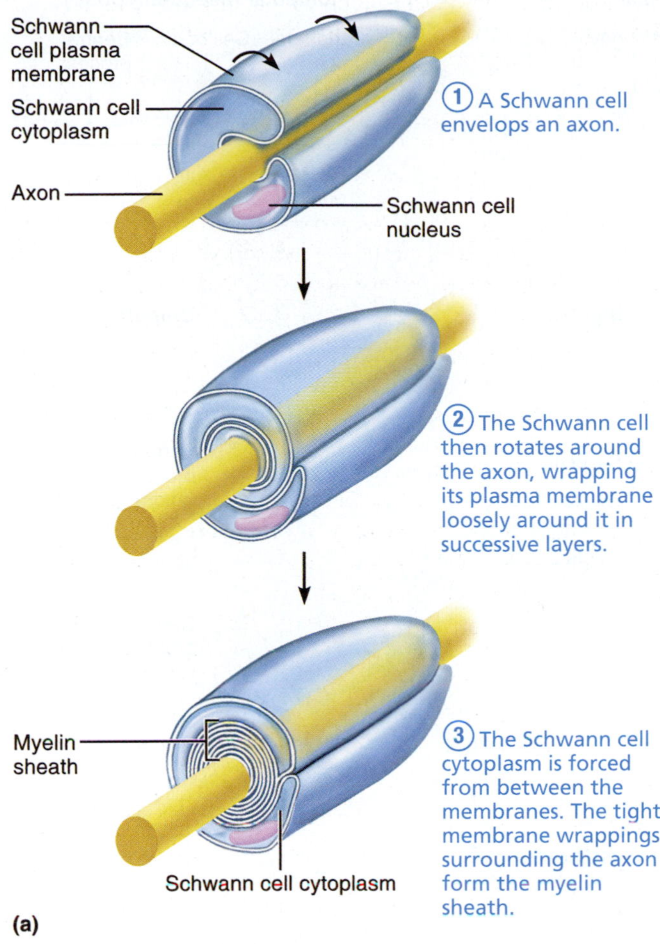

Schwann cell plasma membrane
Schwann cell cytoplasm
Axon
Schwann cell nucleus
① A Schwann cell envelops an axon.
② The Schwann cell then rotates around the axon, wrapping its plasma membrane loosely around it in successive layers.
Myelin sheath
Schwann cell cytoplasm
③ The Schwann cell cytoplasm is forced from between the membranes. The tight membrane wrappings surrounding the axon form the myelin sheath.
(a)

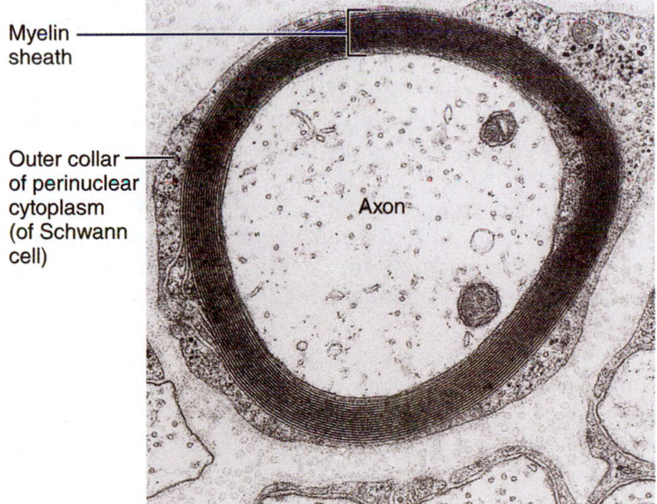

Myelin sheath
Outer collar of perinuclear cytoplasm (of Schwann cell)
Axon
(b)

Figure 13.3 Myelination of a nerve fiber (axon) by individual Schwann cells. (a) Nerve fiber myelination. **(b)** Electron micrograph of cross section through a myelinated axon (11,000×).

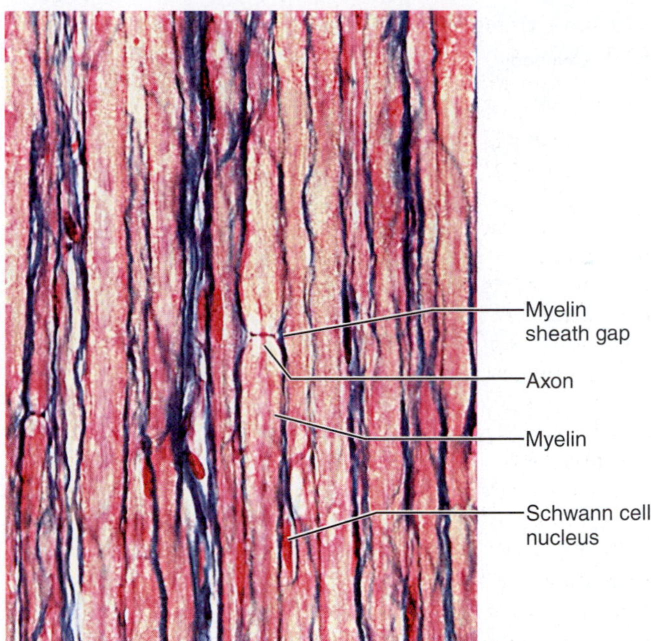

Myelin sheath gap
Axon
Myelin
Schwann cell nucleus

Figure 13.4 Photomicrograph of a small portion of a peripheral nerve in longitudinal section (400×).

Neuron Classification

Neurons may be classified on the basis of structure or of function.

Classification by Structure

Structurally, neurons may be differentiated according to the number of processes attached to the cell body **(Figure 13.5a)**. In **unipolar neurons,** one very short process, which divides into *peripheral* and *central processes,* extends from the cell body. Functionally, only the most distal portions of the peripheral process act as receptive endings; the rest acts as an axon along with the central process. Unipolar neurons are move accurately called **pseudounipolar neurons** because they are derived from bipolar neurons. Nearly all neurons that conduct impulses toward the CNS are unipolar.

Bipolar neurons have two processes attached to the cell body. This neuron type is quite rare, typically found only as part of the receptor apparatus of the eye, ear, and olfactory mucosa.

Many processes issue from the cell body of **multipolar neurons,** all classified as dendrites except for a single axon. Most neurons in the brain and spinal cord (CNS neurons) and those whose axons carry impulses away from the CNS fall into this last category.

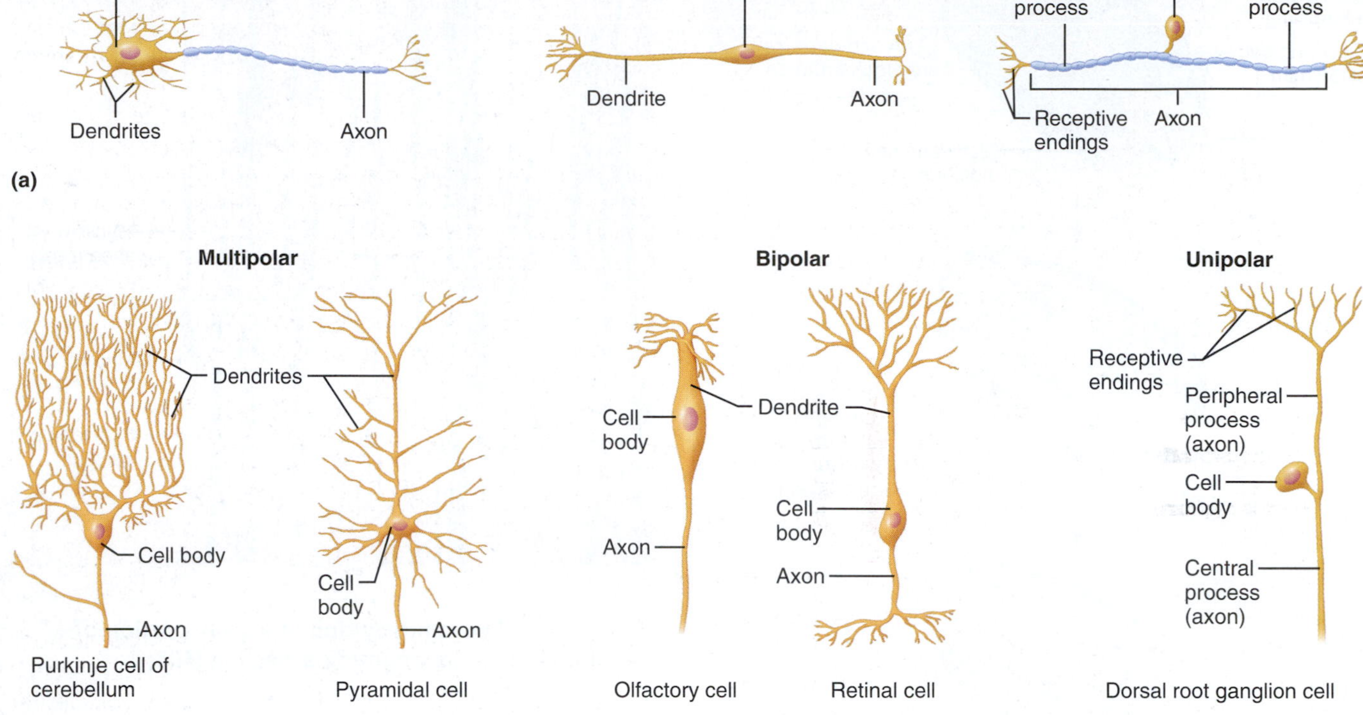

Figure 13.5 Classification of neurons according to structure. (a) Classification of neurons based on structure (number of processes extending from the cell body). **(b)** Structural variations within the classes.

ACTIVITY 2

Studying the Microscopic Structure of Selected Neurons

Obtain prepared slides of pyramidal cells of the cerebral cortex, Purkinje cells of the cerebellar cortex, and a dorsal root ganglion. As you observe them under the microscope, try to pick out the anatomical details (compare the cells to Figure 13.5b and **Figure 13.6**). Notice that the neurons of the cerebral and cerebellar tissues (both brain tissues) are extensively branched; in contrast, the neurons of the dorsal root ganglion are more rounded. The many small nuclei visible surrounding the neurons are those of bordering neuroglia.

Which of these neuron types would be classified as multipolar neurons?

____

Which as unipolar? ______________________________

Classification by Function

In general, neurons carrying impulses from sensory receptors in the internal organs (viscera), the skin, skeletal muscles, joints, or special sensory organs are termed **sensory,** or **afferent, neurons** **(Figure 13.7)**. The receptive endings of sensory neurons are often equipped with specialized receptors that are stimulated by specific changes in their immediate environment. The cell bodies of sensory neurons are always found in a ganglia outside the CNS, and these neurons are typically unipolar.

Neurons carrying impulses from the CNS to the viscera and/or body muscles and glands are termed **motor,** or **efferent, neurons.** Motor neurons are most often multipolar, and their cell bodies are almost always located in the CNS.

The third functional category of neurons is the **interneurons,** which are situated between and contribute to pathways that connect sensory and motor neurons. Their cell bodies are always located within the CNS, and they are multipolar neurons structurally.

Structure of a Nerve

A nerve is a bundle of axons found in the PNS. Wrapped in connective tissue coverings, nerves extend to and/or from the CNS and visceral organs or structures of the body periphery, such as skeletal muscles, glands, and skin.

Like neurons, nerves are classified according to the direction in which they transmit impulses. **Sensory (afferent) nerves** conduct impulses only toward the CNS. A few of

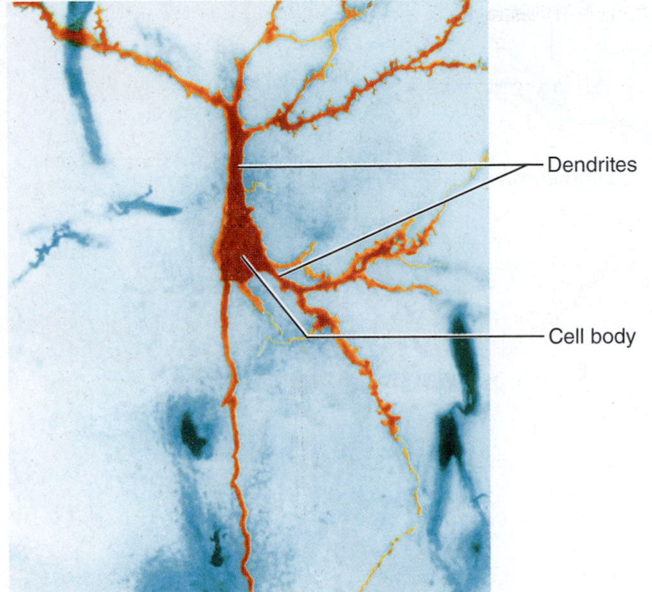

(a)

Dendrites
Cell body

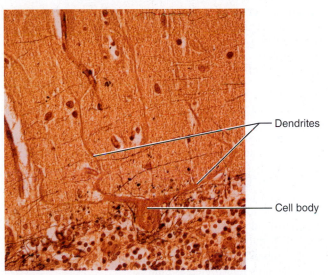

(b)

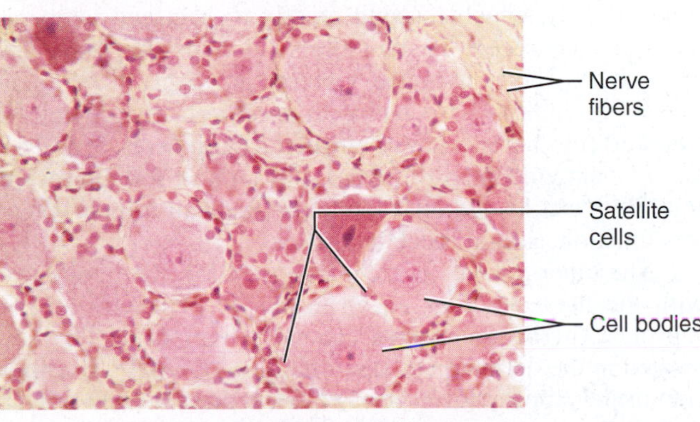

(c)

Figure 13.6 Photomicrographs of neurons. (a) Pyramidal neuron from cerebral cortex (195×). **(b)** Purkinje cell from the cerebellar cortex (200×). **(c)** Dorsal root ganglion cells (235×).

the cranial nerves are pure sensory nerves. **Motor (efferent) nerves** carry impulses only away from the CNS. The ventral roots of the spinal cord are motor nerves. Nerves carrying both sensory (afferent) and motor (efferent) fibers are called **mixed nerves;** most nerves of the body, including all spinal nerves, are mixed nerves.

Within a nerve, each axon is surrounded by a delicate connective tissue sheath called an **endoneurium,** which insulates it from the other neuron processes adjacent to it. The endoneurium is often mistaken for the myelin sheath; it is instead an additional sheath that surrounds the myelin sheath. Groups of axons are bound by a coarser connective tissue, called the **perineurium,** to form bundles of fibers called **fascicles.** Finally, all the fascicles are bound together by a white, fibrous connective tissue sheath called the **epineurium,** forming the cordlike nerve **(Figure 13.8)**. In addition to the connective tissue wrappings, blood vessels and lymphatic vessels serving the fibers also travel within a nerve.

ACTIVITY 3

Examining the Microscopic Structure of a Nerve

Use the compound microscope to examine a prepared cross section of a peripheral nerve. Identify axons, myelin sheaths, fascicles, and endoneurium, perineurium, and epineurium sheaths. If desired, sketch the nerve in the space below. ■

Structure of General Sensory Receptors

You cannot become aware of changes in the environment unless your sensory neurons and their receptors are operating properly. Sensory receptors are either modified dendritic endings or specialized cells associated with the dendrites that are sensitive to certain environmental stimuli. They react to such stimuli by initiating a nerve impulse. Sensory receptors may be classified by the type of stimulus they detect (for example, touch, pain, or temperature), their structure (free nerve endings or complex encapsulated structures), or their body location. **Exteroceptors** respond to stimuli in the external environment, and are typically found close to the body surface. **Interoceptors** respond to stimuli arising within the body (including the visceral organs). **Proprioceptors,** a subclass of interoceptors, are found in skeletal muscles, joints, tendons, and ligaments and report on the degree of stretch in these structures.

The receptors of the special sense organs are complex and deserve considerable study. (The special senses are covered separately in Exercises 17 through 20.) Only the anatomically

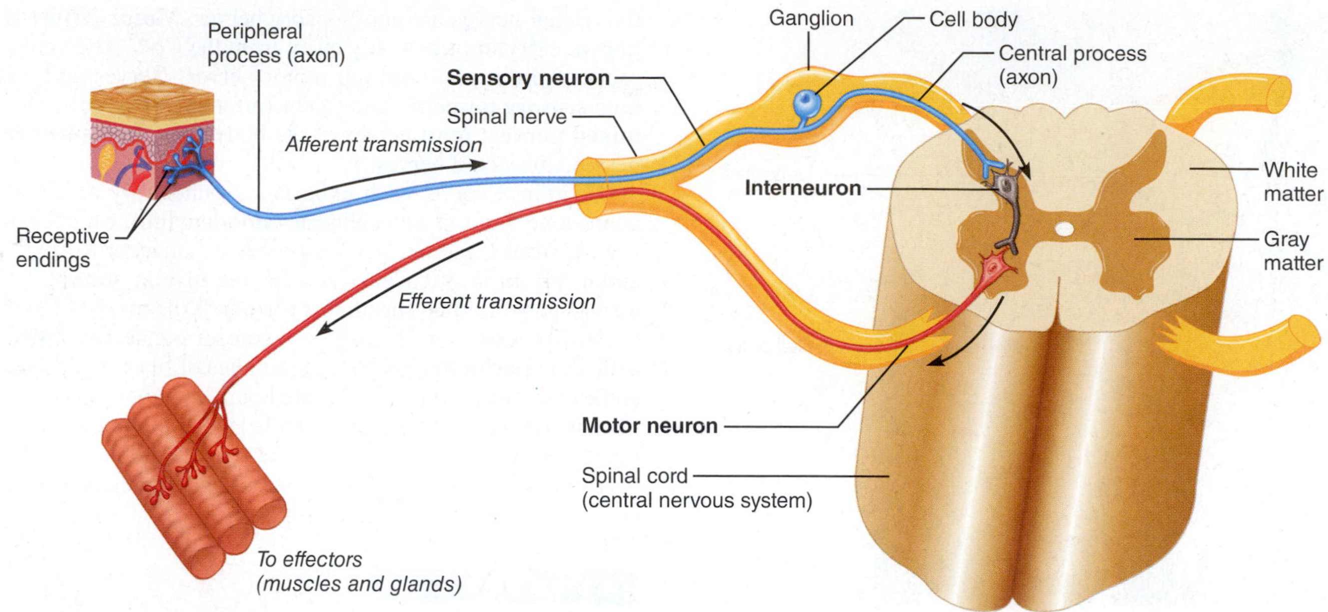

Figure 13.7 Classification of neurons on the basis of function. Sensory (afferent) neurons conduct impulses from the body's sensory receptors to the central nervous system; most are unipolar neurons with their nerve cell bodies in ganglia in the peripheral nervous system (PNS). Motor (efferent) neurons transmit impulses from the CNS to effectors such as muscles and glands. Interneurons complete the communication line between sensory and motor neurons. They are typically multipolar, and their cell bodies reside in the CNS.

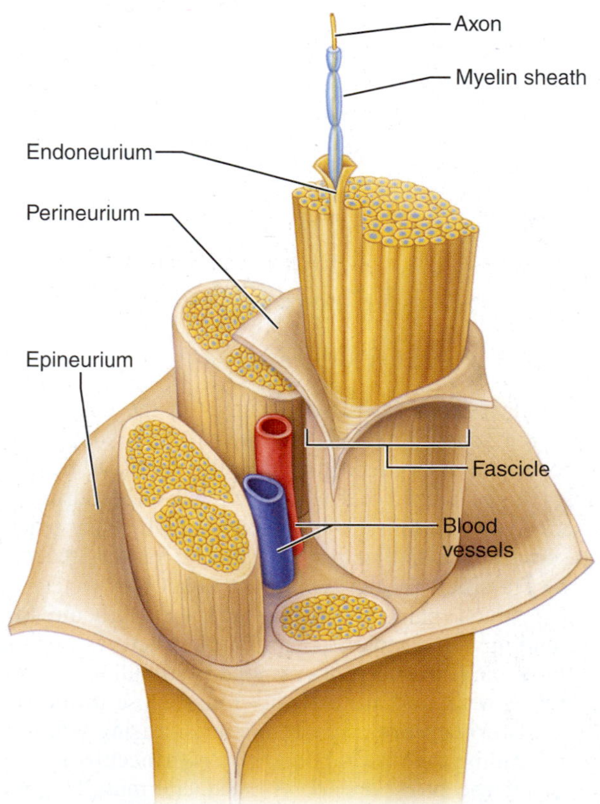

Figure 13.8 Three-dimensional view of a portion of a nerve showing connective tissue wrappings.

simpler **general sensory receptors** (**Figure 13.9** and **Figure 13.10**)—cutaneous receptors and proprioceptors—will be studied in this section. Cutaneous receptors reside in the skin (Figure 13.9). Proprioceptors are located in skeletal muscles, tendons, and joint capsules. Although many references link each type of receptor to specific stimuli, there is still considerable controversy about the precise qualitative function of each receptor. It may be that the responses of all these receptors overlap considerably. Certainly, intense stimulation of any of them is always interpreted as pain.

The least specialized of the cutaneous receptors are the **nonencapsulated (free) nerve endings** of sensory neurons (Figure 13.9), which respond chiefly to pain and temperature. The pain receptors are widespread in the skin and make up a sizable portion of the visceral interoceptors. Certain free nerve endings associate with specific tactile epidermal (Merkel) cells to form **tactile epithelial complexes,** or *Merkel discs,* or entwine in hair follicles to form **hair follicle receptors.** Both tactile epithelial complexes and hair follicle receptors function as light touch receptors.

The other cutaneous receptors are a bit more complex, with the nerve endings encapsulated by connective tissue capsules. **Tactile corpuscles** respond to light touch. They are located in the dermal papillae of hairless skin only. **Bulbous corpuscles** appear to respond to deep pressure and stretch stimuli. As you inspect the illustration of cutaneous receptors (Figure 13.9), notice that all of the encapsulated receptors are quite similar. However, **lamellar corpuscles** are anatomically more distinctive and lie deepest in the dermis. These receptors respond only when deep pressure is first applied. They are best suited to monitor high-frequency vibrations.

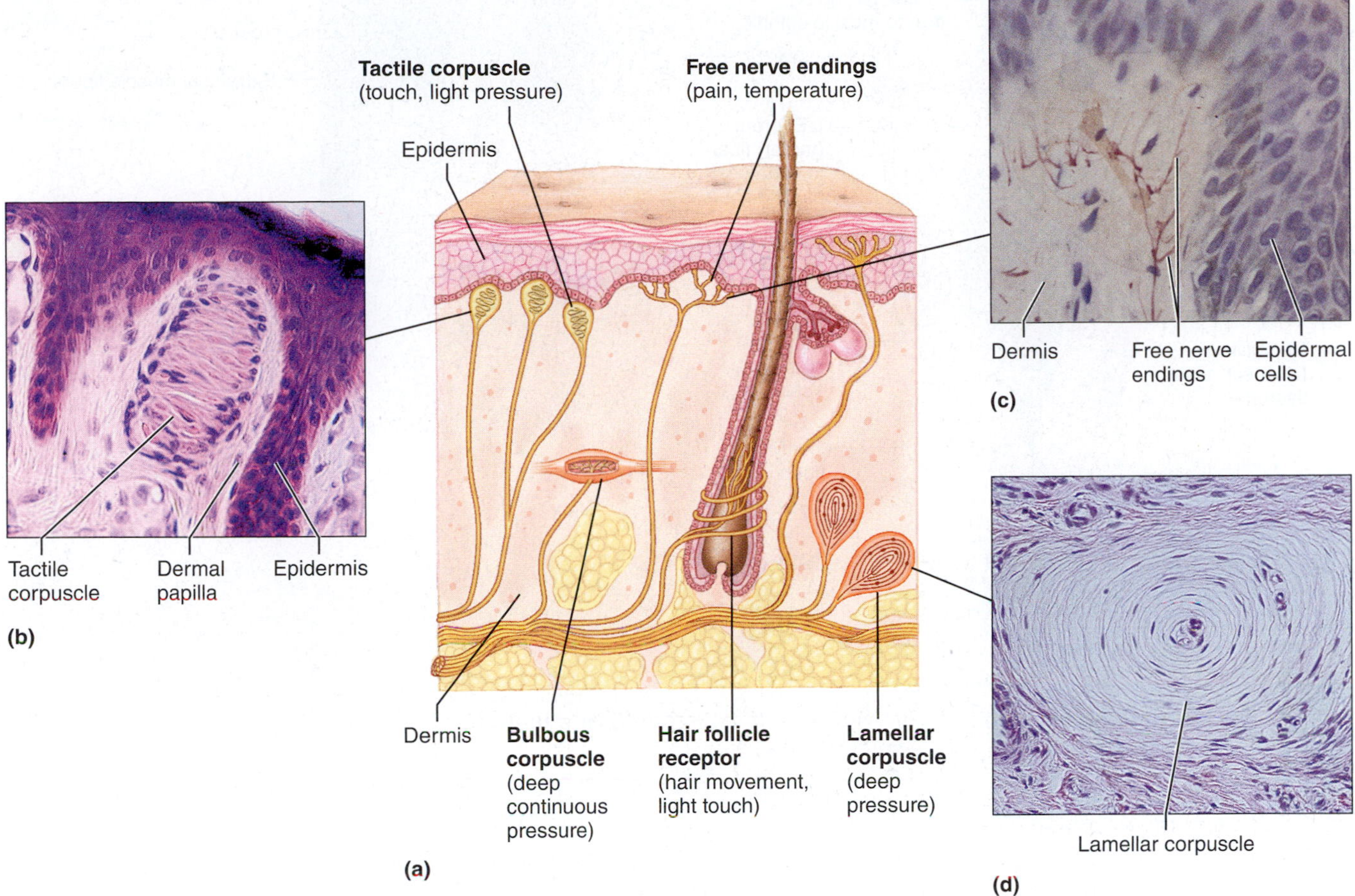

Figure 13.9 Examples of cutaneous receptors. Drawing (a) and photomicrographs (b–d). **(a)** Free nerve endings, hair follicle receptor, tactile corpuscle, lamellar corpuscle, and bulbous corpuscle. Tactile epithelial complexes are not illustrated. **(b)** Tactile corpuscle in a dermal papilla (300×). **(c)** Free nerve endings at dermal-epidermal junction (330×). **(d)** Cross section of a lamellar corpuscle in the dermis (220×).

ACTIVITY 4

Studying the Structure of Selected Sensory Receptors

1. Obtain histologic slides of lamellar and tactile corpuscles. Locate, under low power, a tactile corpuscle in the dermal layer of the skin. As mentioned above, these are usually found in the dermal papillae. Then switch to the oil immersion lens for a detailed study. Notice that the free nerve fibers within the capsule are aligned parallel to the skin surface. Compare your observations to the photomicrographs (Figure 13.9).

2. Next, observe a lamellar corpuscle located much deeper in the dermis. Try to identify the slender free dendrite ending in the center of the receptor and the heavy capsule of connective tissue surrounding it (which looks rather like an onion cut lengthwise). Also, notice how much larger the lamellar corpuscles are than the tactile corpuscles. Compare your observations to the photomicrographs (Figure 13.9).

3. Obtain slides of muscle spindles and tendon organs, the two major types of proprioceptors (see Figure 13.10). In the slide of **muscle spindles,** note that minute extensions of the dendrites of the sensory neurons coil around specialized slender skeletal muscle cells called **intrafusal muscle fibers.** The **tendon organs** are composed of dendrites that ramify through the tendon tissue close to the muscle tendon attachment. Stretching the muscles or tendons excites these receptors, which then transmit impulses that ultimately reach the cerebellum for interpretation. Compare your observations to the photomicrograph (Figure 13.10). ■

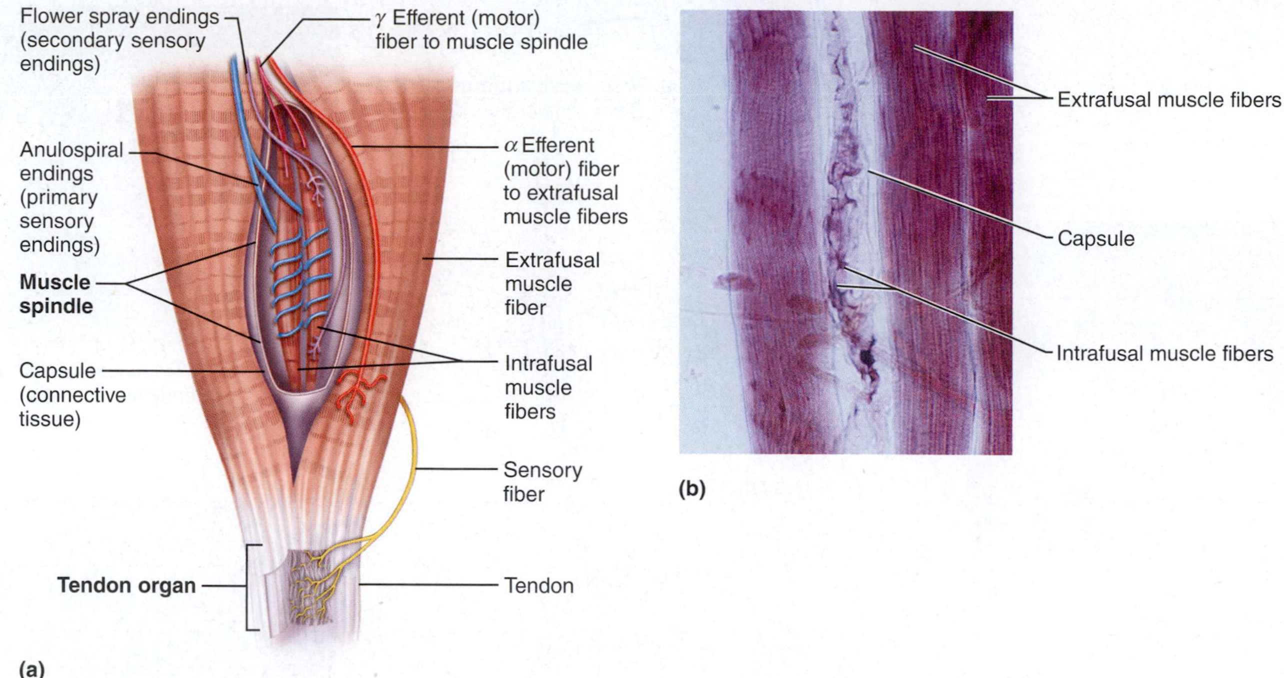

Figure 13.10 Proprioceptors. (a) Diagram view of a muscle spindle and tendon organ. **(b)** Photomicrograph of a muscle spindle (80×).

NAME ______________________

LAB TIME/DATE ______________________

EXERCISE 13

REVIEW SHEET

Histology of Nervous Tissue

1. The basic functional unit of the nervous system is the neuron. What is the major function of this cell type?

____

2. Name four types of neuroglia in the CNS, and list at least one function for each type. (You will need to consult your textbook for this.)

Types	**Functions**
a. Most abundant: ______________________	a. ______________________
b. ______________________	b. ______________________
c. ______________________	c. ______________________
d. ______________________	d. ______________________

Name the PNS neuroglial cell that forms myelin. ______________________

Name the PNS neuroglial cell that surrounds dorsal root ganglion neurons. ______________________

3. For each description, choose the appropriate term from the key. (Not all terms will be used.)

Key:

a. afferent neuron	e. interneuron	i. nuclei
b. central nervous system	f. neuroglia	j. peripheral nervous system
c. efferent neuron	g. neurotransmitters	k. synapse
d. ganglion	h. nerve	l. tract

______________ 1. the brain and spinal cord collectively

______________ 2. specialized supporting cells in the CNS

______________ 3. junction or point of close contact between neurons

______________ 4. a bundle of axons inside the CNS

______________ 5. neuron serving as part of the conduction pathway between sensory and motor neurons

______________ 6. ganglia and spinal and cranial nerves

______________ 7. collection of nerve cell bodies found outside the CNS

______________ 8. neuron that conducts impulses away from the CNS to muscles and glands

______________ 9. neuron that conducts impulses toward the CNS from the body periphery

______________ 10. chemicals released by neurons that stimulate or inhibit other neurons or effectors

7. Circle the correct underlined term. The outer cortex of the brain contains the cell bodies of cerebral neurons and is known as white matter / gray matter.
8. The brain and spinal cord are covered and protected by three connective tissue layers called:
   a. lobes
   b. meninges
   c. sulci
   d. ventricles
9. Circle True or False. Cerebrospinal fluid is produced by the frontal lobe of the cerebrum and is unlike any other body fluid.
10. How many pairs of cranial nerves are there? ___________

When viewed alongside all nature's animals, humans are indeed unique, and the key to our uniqueness is found in the brain. Each of us is a composite reflection of our brain's experience. If all past sensory input could mysteriously and suddenly be "erased," we would be unable to walk, talk, or communicate in any manner. Spontaneous movement would occur, as in a fetus, but no voluntary integrated function of any type would be possible. Clearly we would cease to be the same individuals.

Because of the complexity of the nervous system, its anatomical structures are usually considered in terms of two principal divisions: the central nervous system and the peripheral nervous system. The **central nervous system (CNS)** consists of the brain and spinal cord, which primarily interpret incoming sensory information and issue instructions based on that information and on past experience. The **peripheral nervous system (PNS)** consists of the cranial and spinal nerves, ganglia, and sensory receptors. These structures serve as communication lines as they carry impulses—from the sensory receptors to the CNS and from the CNS to the appropriate glands, muscles, or other effector organs.

The PNS has two major subdivisions: the **sensory portion,** which consists of nerve fibers that conduct impulses toward the CNS, and the **motor portion,** which contains nerve fibers that conduct impulses away from the CNS. The motor arm, in turn, consists of the **somatic division** (sometimes called the *voluntary system*), which controls the skeletal muscles, and the other subdivision, the **autonomic nervous system (ANS),** which controls smooth and cardiac muscles and glands. The ANS is often referred to as the *involuntary nervous system*. Its sympathetic and parasympathetic branches play a major role in maintaining homeostasis.

This exercise focuses on the brain (CNS) and cranial nerves (PNS) because of their close anatomical relationship.

The Human Brain

During embryonic development of all vertebrates, the CNS first makes its appearance as a simple tubelike structure, the **neural tube,** that extends down the dorsal median plane. By the fourth week, the human brain begins to form as an expansion of the anterior or rostral end of the neural tube (the end toward the head). Shortly thereafter, constrictions appear, dividing the developing brain into three major regions—**forebrain, midbrain,** and **hindbrain** (**Figure 14.1**). The remainder of the neural tube becomes the spinal cord.

During fetal development, two anterior outpocketings extend from the forebrain and grow rapidly to form the cerebral hemispheres. The skull imposes space restrictions that force the cerebral hemispheres to grow posteriorly and inferiorly, and they finally end up enveloping and obscuring the rest of the forebrain and most midbrain structures. Somewhat later in development, the dorsal hindbrain also enlarges to produce the cerebellum. The central canal of the neural tube, which remains continuous throughout the brain and cord, enlarges in four regions of the brain, forming chambers called **ventricles** (see Figure 14.8a and b, page 263).

ACTIVITY 1

Identifying External Brain Structures

Identify external brain structures using the figures cited. Also use a model of the human brain and other learning aids as they are mentioned.

Generally, the brain is studied in terms of four major regions: the cerebral hemispheres, diencephalon, brain stem, and cerebellum. It's useful to be aware of the relationship between these four anatomical regions and the structures of the forebrain, midbrain, and hindbrain (Figure 14.1).

Cerebral Hemispheres

The **cerebral hemispheres** are the most superior portion of the brain (**Figure 14.2**). Their entire surface is thrown into elevated ridges of tissue called **gyri** that are separated by shallow grooves called **sulci** or deeper grooves called **fissures.** Many of the fissures and gyri are important anatomical landmarks.

The cerebral hemispheres are divided by a single deep fissure, the **longitudinal fissure.** The **central sulcus** divides the **frontal lobe** from the **parietal lobe,** and the **lateral sulcus** separates the **temporal lobe** from the parietal lobe. The **parieto-occipital sulcus** on the medial surface of each hemisphere divides the **occipital lobe** from the parietal lobe. It is not visible externally. Notice that these cerebral hemisphere lobes are named for the cranial bones that lie over them. A fifth lobe of each cerebral hemisphere, the **insula,** is buried deep within the lateral sulcus and is covered by portions of the temporal, parietal, and frontal lobes.

Some important functional areas of the cerebral hemispheres have also been located (Figure 14.2d). The **primary somatosensory cortex** is located in the **postcentral gyrus** of the parietal lobe. Impulses traveling from the body's sensory receptors (such as those for pressure, pain, and temperature)

(a) Neural tube (contains neural canal)
Anterior (rostral)
Posterior (caudal)
(b) Primary brain vesicles
Prosencephalon (forebrain)
Mesencephalon (midbrain)
Rhombencephalon (hindbrain)
(c) Secondary brain vesicles
Telencephalon
Diencephalon
Mesencephalon
Metencephalon
Myelencephalon
(d) Adult brain structures
Cerebrum: cerebral hemispheres (cortex, white matter, basal nuclei)
Diencephalon (thalamus, hypothalamus, epithalamus), retina
Brain stem: midbrain
Brain stem: pons
Cerebellum
Brain stem: medulla oblongata
Spinal cord
(e) Adult neural canal regions
Lateral ventricles
Third ventricle
Cerebral aqueduct
Fourth ventricle
Central canal

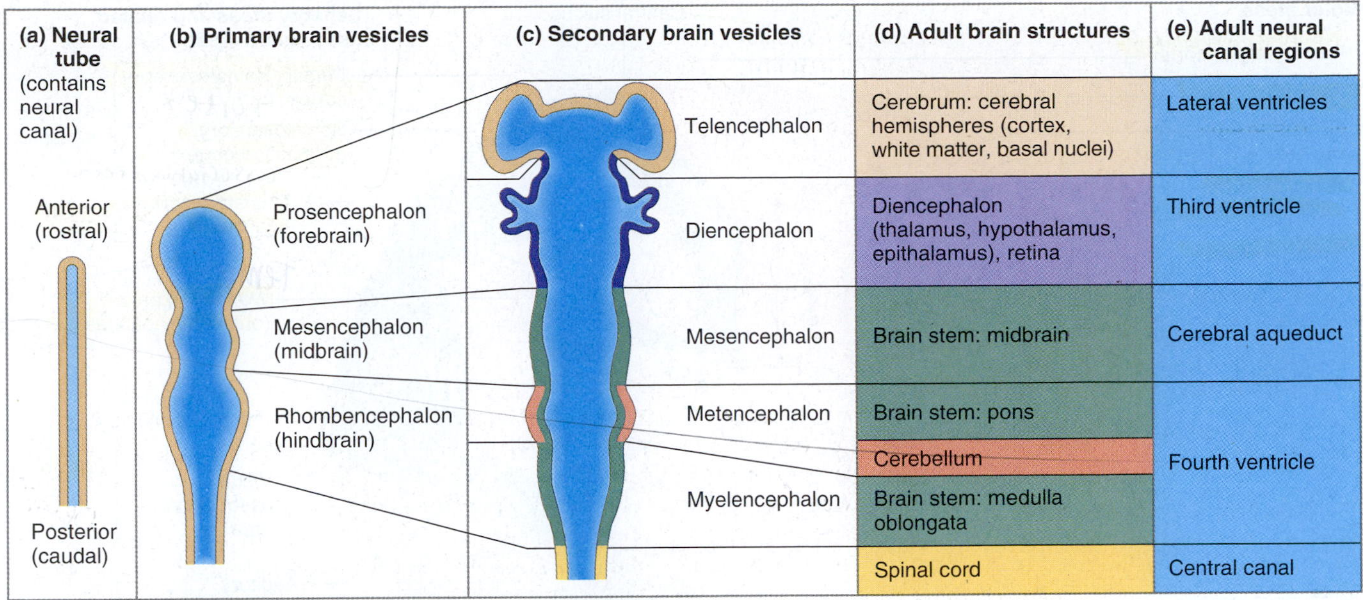

Figure 14.1 Embryonic development of the human brain. (a) The neural tube becomes subdivided into **(b)** the primary brain vesicles, which subsequently form **(c)** the secondary brain vesicles, which differentiate into **(d)** the adult brain structures. **(e)** The adult structures derived from the neural canal.

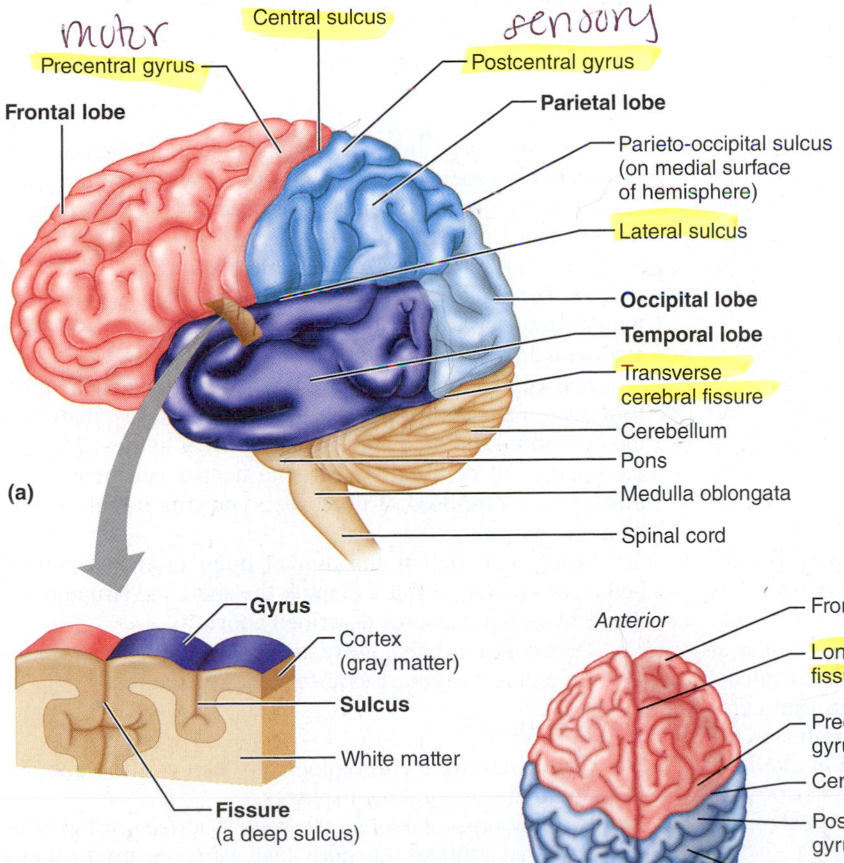

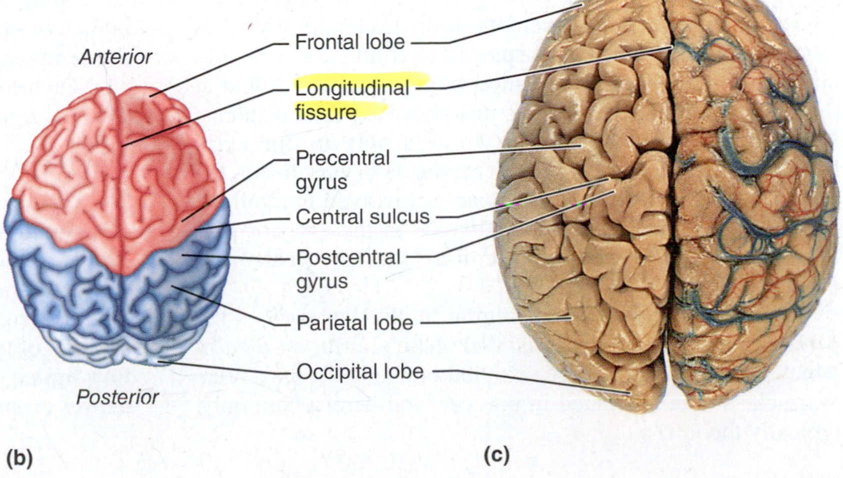

Central sulcus
Precentral gyrus
Postcentral gyrus
Frontal lobe
Parietal lobe
Parieto-occipital sulcus (on medial surface of hemisphere)
Lateral sulcus
Occipital lobe
Temporal lobe
Transverse cerebral fissure
Cerebellum
Pons
Medulla oblongata
Spinal cord
(a)
Gyrus
Cortex (gray matter)
Sulcus
White matter
Fissure (a deep sulcus)

Anterior
Posterior
Frontal lobe
Longitudinal fissure
Precentral gyrus
Central sulcus
Postcentral gyrus
Parietal lobe
Occipital lobe
(b)
(c)

Figure 14.2 External structure (lobes and fissures) of the cerebral hemispheres. (a) Left lateral view of the brain. **(b)** Superior view. **(c)** Photograph of the superior aspect of the human brain.

Figure 14.2 *(continued)* **External structure (lobes and fissures) of the cerebral hemispheres. (d)** Functional areas of the left cerebral cortex. The olfactory area, which is deep to the temporal lobe on the medial hemispheric surface, is not identified here.

are localized in this area of the brain. ("This information is from my big toe.") Immediately posterior to the primary somatosensory cortex is the **somatosensory association cortex,** in which the meaning of incoming stimuli is analyzed. ("Ouch! I have a *pain* there.") Thus, the somatosensory association cortex allows you to become aware of pain, coldness, a light touch, and the like.

Impulses from the special sense organs are interpreted in other specific areas (Figure 14.2d). For example, the visual areas are in the posterior portion of the occipital lobe, and the auditory area is located in the temporal lobe in the gyrus bordering the lateral sulcus. The olfactory area is deep within the temporal lobe along its medial surface, in a region called the **uncus** (Figure 14.4a).

The **primary motor cortex,** which is responsible for conscious or voluntary movement of the skeletal muscles, is located in the **precentral gyrus** of the frontal lobe. A specialized motor speech area called **Broca's area** is found at the base of the precentral gyrus just above the lateral sulcus. Damage to this area (which is located only in one cerebral hemisphere, usually the left) reduces or eliminates the ability to articulate words. Many areas involved in intellect, complex reasoning, and personality lie in the anterior portions of the frontal lobes, in a region called the **prefrontal cortex.**

A rather poorly defined region at the junction of the parietal and temporal lobes is **Wernicke's area,** an area in which unfamiliar words are sounded out. Like Broca's area, Wernicke's area is located in one cerebral hemisphere only, typically the left.

Although there are many similar functional areas in both cerebral hemispheres, such as motor and sensory areas, each hemisphere is also a "specialist" in certain ways. For example, the left hemisphere is the "language brain" in most of us, because it houses centers associated with language skills and speech. The right hemisphere is more specifically concerned with abstract, conceptual, or spatial processes—skills associated with artistic or creative pursuits.

The cell bodies of cerebral neurons involved in these functions are found only in the outermost gray matter of the cerebrum, the area called the **cerebral cortex.** Most of the balance of cerebral tissue—the deeper **cerebral white matter**—is composed of fiber tracts carrying impulses to or from the cortex.

Using a model of the human brain (and a preserved human brain, if available), identify the areas and structures of the cerebral hemispheres described above.

Then continue using the model and preserved brain along with the figures as you read about other structures.

Diencephalon

The **diencephalon** is embryologically part of the forebrain, along with the cerebral hemispheres.

Turn the brain model so the ventral surface of the brain can be viewed. Starting superiorly (and using **Figure 14.3** as a guide), identify the externally visible structures that mark the position of the floor of the diencephalon. These are the **olfactory bulbs** (synapse point of cranial nerve I) and **tracts, optic nerves** (cranial nerve II), **optic chiasma** (where the fibers of

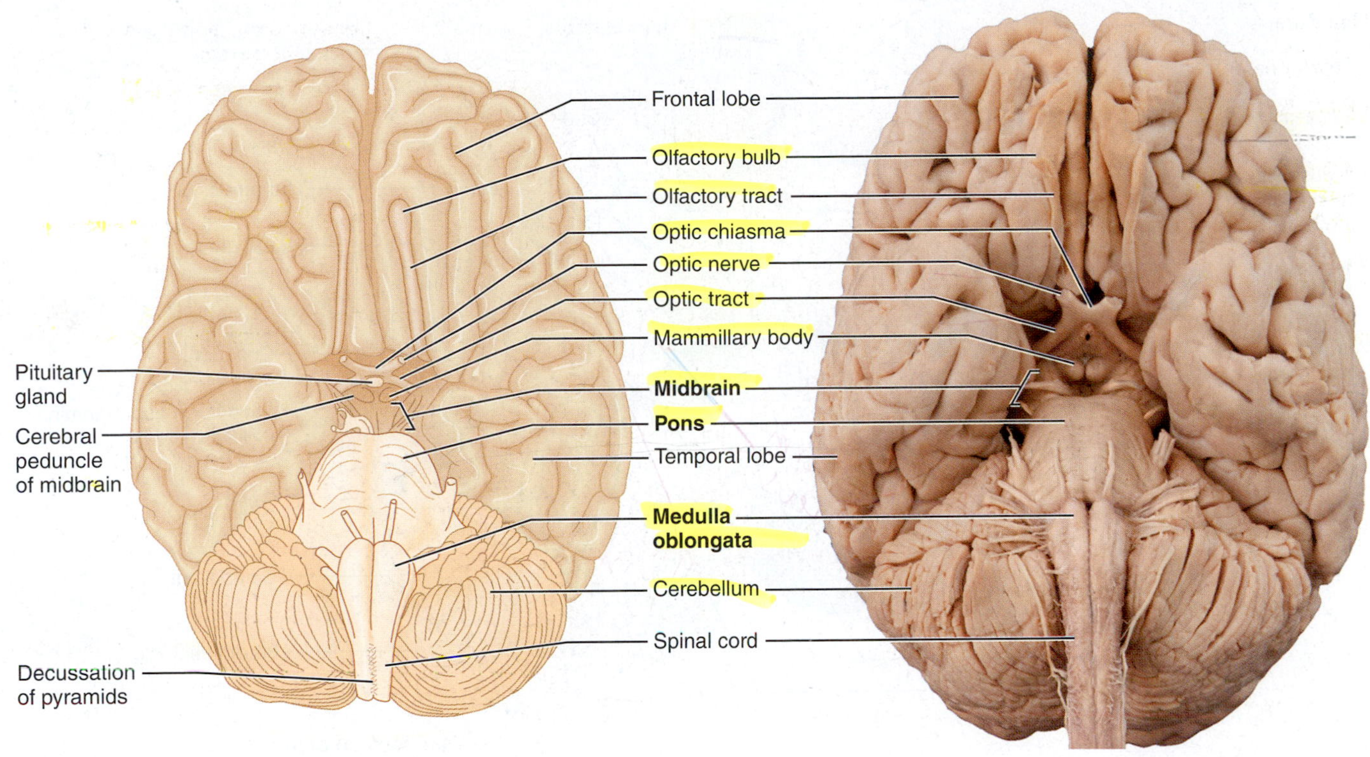

Figure 14.3 Ventral (inferior) aspect of the human brain, showing the three regions of the brain stem. Only a small portion of the midbrain can be seen; the rest is surrounded by other brain regions.

the optic nerves partially cross over), **optic tracts, pituitary gland,** and **mammillary bodies.**

Brain Stem

Continue inferiorly to identify the **brain stem** structures—the **cerebral peduncles** (fiber tracts in the **midbrain** connecting the pons below with the cerebrum above), the pons, and the medulla oblongata. *Pons* means "bridge," and the **pons** consists primarily of motor and sensory fiber tracts connecting the brain with lower CNS centers. The lowest brain stem region, the **medulla oblongata,** is also composed primarily of fiber tracts. You can see the **decussation of pyramids,** a crossover point for the major motor tracts (pyramidal tracts) descending from the motor areas of the cerebrum to the cord, on the medulla's surface. The medulla also houses many vital autonomic centers involved in the control of heart rate, respiratory rhythm, and blood pressure as well as involuntary centers involved in vomiting, swallowing, and so on.

Cerebellum

1. Turn the brain model so you can see the dorsal aspect. Identify the large cauliflower-like **cerebellum,** which projects dorsally from under the occipital lobe of the cerebrum. Notice that, like the cerebrum, the cerebellum has two major hemispheres and a convoluted surface (see Figure 14.6). It also has an outer cortex made up of gray matter with an inner region of white matter.

2. Remove the cerebellum to view the **corpora quadrigemina (Figure 14.4),** located on the posterior aspect of the midbrain, a brain stem structure. The two superior prominences are the **superior colliculi** (visual reflex centers); the two smaller inferior prominences are the **inferior colliculi** (auditory reflex centers).

ACTIVITY 2

Identifying Internal Brain Structures

The deeper structures of the brain have also been well mapped. Like the external structures, these can be studied in terms of the four major regions. As the internal brain areas are described, identify them on the figures cited. Also, use the brain model as indicated to help you in this study.

Cerebral Hemispheres

1. Take the brain model apart so you can see a median sagittal view of the internal brain structures (see Figure 14.4). Observe the model closely to see the extent of the outer cortex (gray matter), which contains the cell bodies of cerebral neurons. The pyramidal cells of the cerebral motor cortex (Exercise 13 and Figure 13.5) are representative of the neurons seen in the precentral gyrus.

2. Observe the deeper area of white matter, which is made up of fiber tracts. The fiber tracts found in the cerebral hemisphere white matter are called *association tracts* if they connect two portions of the same hemisphere, *projection tracts*

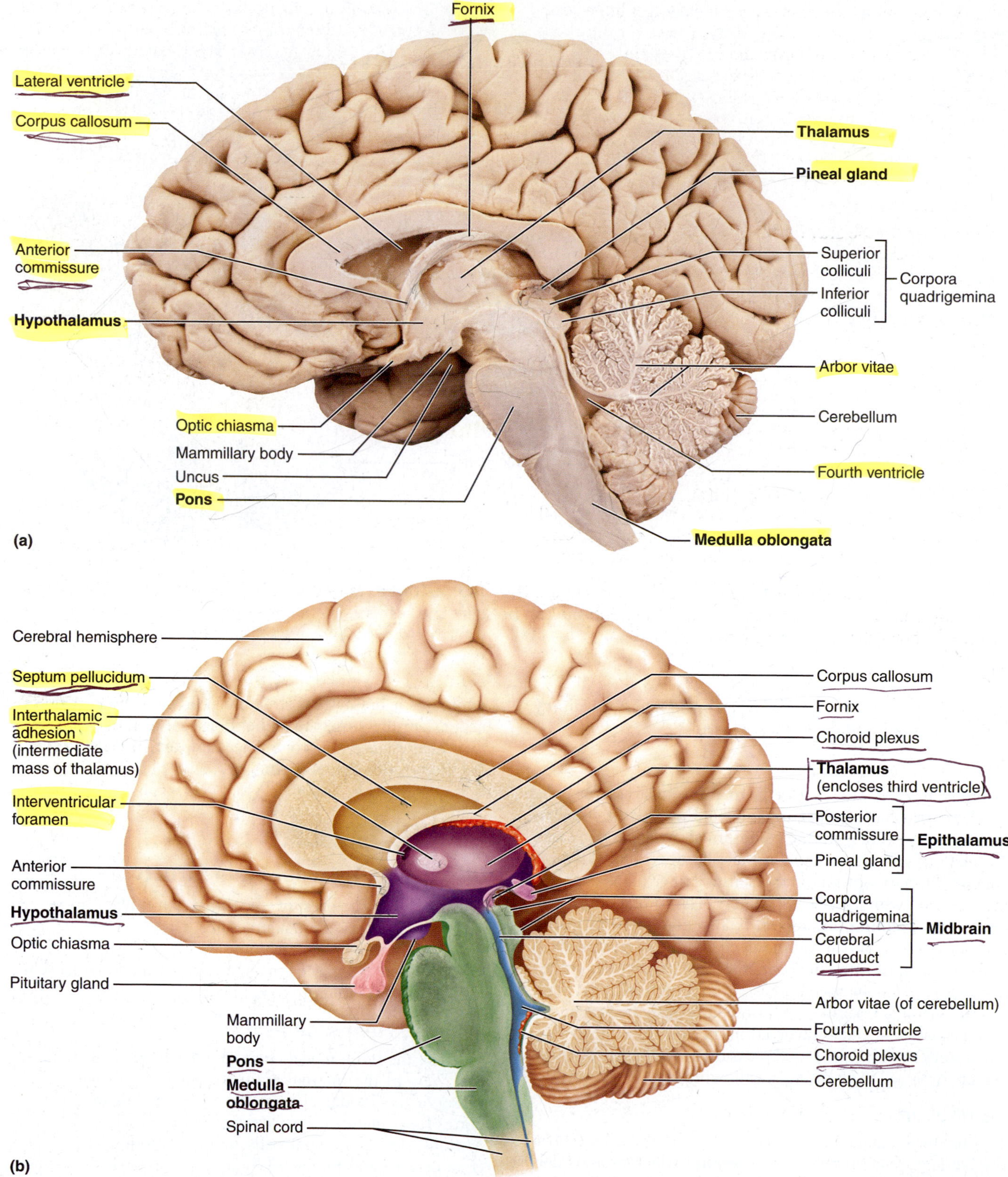

Figure 14.4 Diencephalon and brain stem structures as seen in a midsagittal section of the brain. (a) Photograph. **(b)** Diagram view.

if they run between the cerebral cortex and lower brain structures or spinal cord, and *commissures* if they run from one hemisphere to another. Observe the large **corpus callosum,** the major commissure connecting the cerebral hemispheres. The corpus callosum arches above the structures of the diencephalon and roofs over the lateral ventricles. Notice also the **fornix,** a bandlike fiber tract concerned with olfaction as well as limbic system functions, and the membranous **septum pellucidum,** which separates the lateral ventricles of the cerebral hemispheres.

3. In addition to the gray matter of the cerebral cortex, there are several clusters of neuron cell bodies called **nuclei** buried deep within the white matter of the cerebral hemispheres. One important group of cerebral nuclei, called the **basal nuclei,** or **basal ganglia,*** flank the lateral and third ventricles. You can see these nuclei if you have an appropriate dissectible model or a coronally or cross-sectioned human brain slice. Otherwise, the figure **(Figure 14.5)** will suffice.

The basal nuclei, part of the *indirect pathway,* are involved in regulating voluntary motor activities. The most important of them are the arching, comma-shaped **caudate nucleus,** the **putamen,** and **globus pallidus.**

The **corona radiata,** a spray of projection fibers coursing down from the precentral (motor) gyrus, combines with sensory fibers traveling to the primary somatosensory cortex to form a broad band of fibrous material called the **internal capsule.** The internal capsule passes between the diencephalon and the basal nuclei and through the caudate and putamen, giving them a striped appearance. Hence they are called the **striatum,** or "striped" body (Figure 14.5a).

4. Examine the relationship of the lateral ventricles and corpus callosum to the diencephalon structures; that is, thalamus and third ventricle—from the cross-sectional viewpoint (see Figure 14.5b).

Diencephalon

1. The major internal structures of the diencephalon are the thalamus, hypothalamus, and epithalamus (see Figure 14.4). The **thalamus** consists of two large lobes of gray matter that laterally enclose the shallow third ventricle of the brain. A slender stalk of thalamic tissue, the **interthalamic adhesion,** or **intermediate mass,** connects the two thalamic lobes and bridges the ventricle. The thalamus is a major integrating and relay station for sensory impulses passing upward to the cortical sensory areas for localization and interpretation. Locate also the **interventricular foramen,** a tiny opening connecting the third ventricle with the lateral ventricle on the same side.

2. The **hypothalamus** makes up the floor and the inferolateral walls of the third ventricle. It is an important autonomic center involved in regulating body temperature, water balance, and fat and carbohydrate metabolism as well as in many other activities and drives (sex, hunger, thirst). Locate again the pituitary gland, which hangs from the anterior floor of the hypothalamus by a slender stalk, the **infundibulum.** (The pituitary gland is usually not present in preserved brain specimens.) The pituitary rests in the hypophyseal fossa of the sella turcica of the sphenoid bone. (Its function is discussed in Exercise 21.)

Anterior to the pituitary, identify the optic chiasma portion of the optic pathway to the brain. The **mammillary bodies,** relay stations for olfaction, bulge exteriorly from the floor of the hypothalamus just posterior to the pituitary gland.

3. The **epithalamus** forms the roof of the third ventricle and is the most dorsal portion of the diencephalon. Important structures in the epithalamus are the **pineal gland** (a neuroendocrine structure), and the **choroid plexus** of the third ventricle. The choroid plexuses, knotlike collections of capillaries within each ventricle, produce the cerebrospinal fluid.

Brain Stem

1. Now trace the short midbrain from the mammillary bodies to the rounded pons below. (Continue to refer to Figure 14.4.) The **cerebral aqueduct** is a slender canal traveling through the midbrain; it connects the third ventricle to the fourth ventricle in the hindbrain below. The cerebral peduncles and the rounded corpora quadrigemina make up the midbrain tissue anterior and posterior (respectively) to the cerebral aqueduct.

2. Locate the hindbrain structures. Trace the rounded pons to the medulla oblongata below, and identify the fourth ventricle posterior to these structures. Attempt to identify the single median aperture and the two lateral apertures, three openings found in the walls of the fourth ventricle. These apertures serve as passageways for cerebrospinal fluid to circulate into the subarachnoid space from the fourth ventricle.

Cerebellum

Examine the cerebellum. Notice that it is composed of two lateral hemispheres, each with three lobes (*anterior, posterior,* and a deep *flocculonodular*) connected by a midline lobe called the **vermis (Figure 14.6).** As in the cerebral hemispheres, the cerebellum has an outer cortical area of gray matter and an inner area of white matter. The treelike branching of the cerebellar white matter is referred to as the **arbor vitae,** or "tree of life." The cerebellum is concerned with unconscious coordination of skeletal muscle activity and control of balance and equilibrium. Fibers converge on the cerebellum from the equilibrium apparatus of the internal ear, visual pathways, proprioceptors of tendons and skeletal muscles, and from many other areas. Thus the cerebellum remains constantly aware of the position and state of tension of the various body parts. ■

*The historical term for these nuclei, *basal ganglia,* is misleading because ganglia are PNS structures. Although technically not the correct anatomical term, "basal ganglia" is included here because it is widely used in clinical settings.

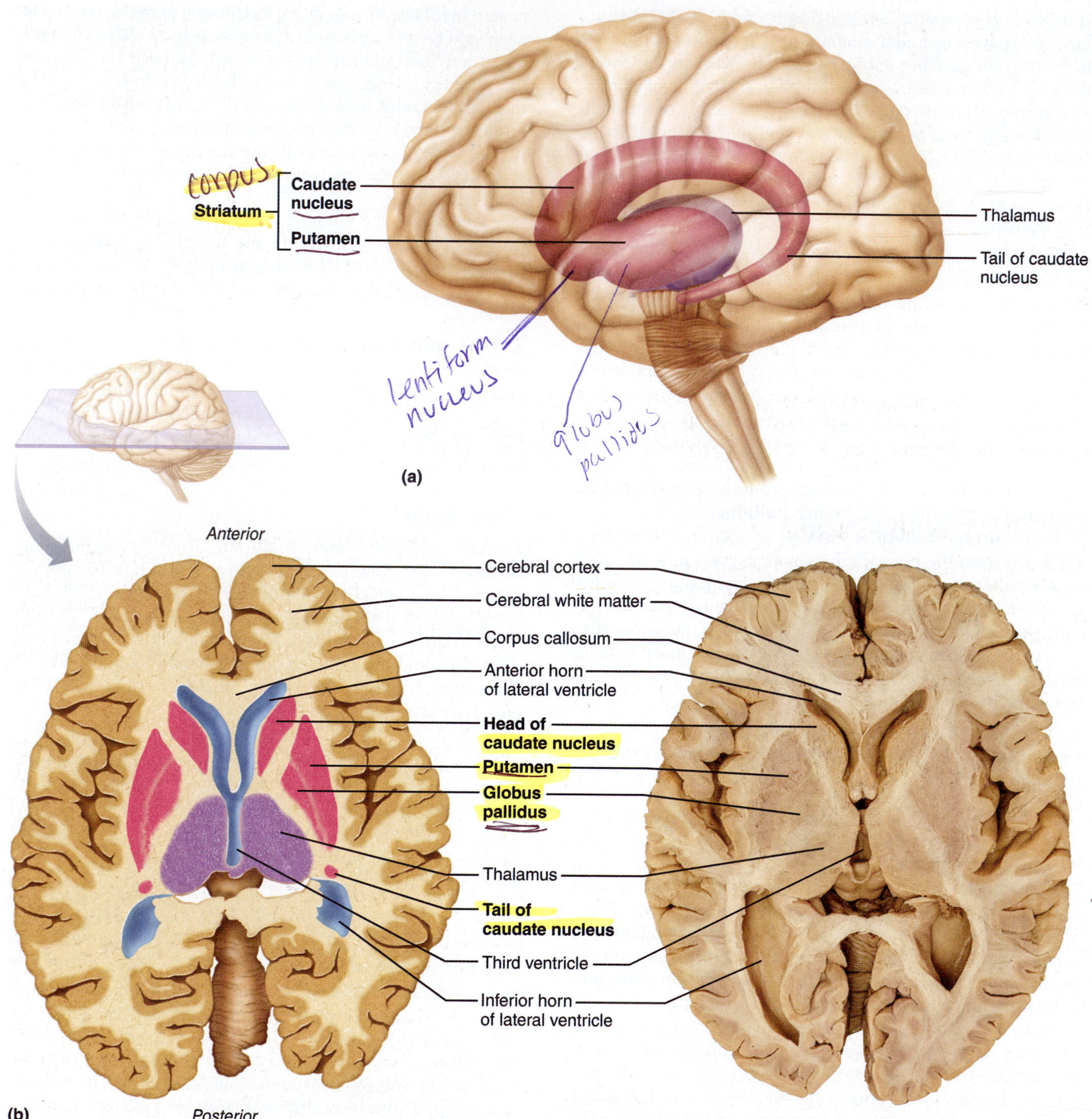

Figure 14.5 Basal nuclei. (a) Three-dimensional view of the basal nuclei showing their positions within the cerebrum. Globus pallidus lies medial and deep to the putamen. **(b)** A transverse section of the cerebrum and diencephalon showing the relationship of the basal nuclei to the thalamus and the lateral and third ventricles.

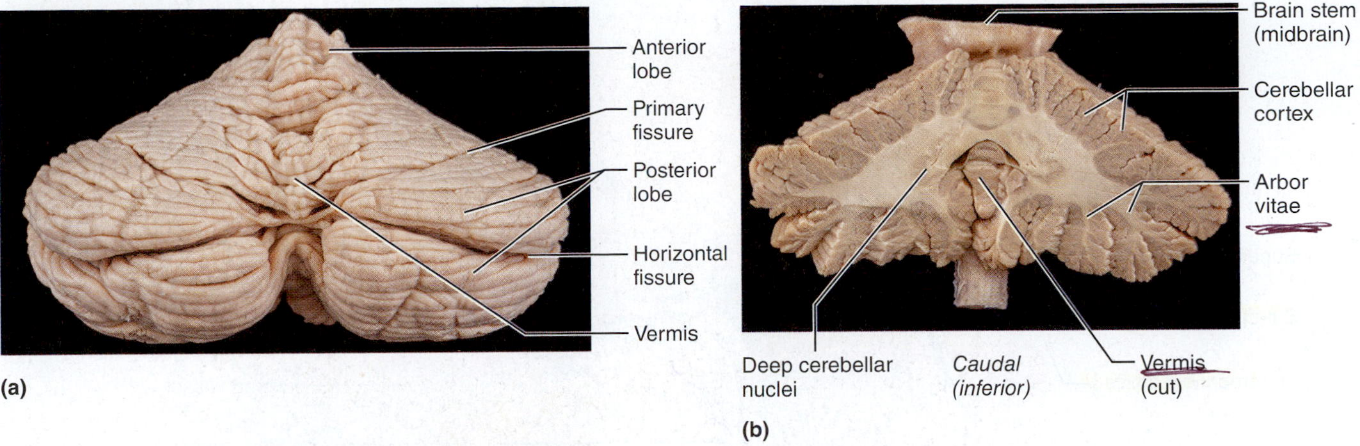

Figure 14.6 Cerebellum. (a) Posterior (dorsal) view. **(b)** The cerebellum, sectioned to reveal its cortex and medullary regions. (Note that the cerebellum is sectioned frontally and the brain stem is sectioned horizontally in this posterior view.)

Meninges of the Brain

The brain and spinal cord are covered and protected by three connective tissue membranes called **meninges** (Figure 14.7). The outermost meninx is the leathery **dura mater,** a double-layered membrane. One of its layers (the *periosteal layer*) is attached to the inner surface of the skull, forming the periosteum. The other layer (the *meningeal layer*) forms the outermost brain covering and is continuous with the dura mater of the spinal cord.

The dural layers are fused together except in three places where the inner membrane extends inward to form a septum that secures the brain to structures inside the cranial cavity. One such extension, the **falx cerebri,** dips into the longitudinal fissure between the cerebral hemispheres to attach to the crista galli of the ethmoid bone of the skull (Figure 14.7a). The cavity created at this point is the large **superior sagittal sinus,** which collects blood draining from the brain tissue. The **falx cerebelli,** separating the two cerebellar hemispheres, and the **tentorium cerebelli,** separating the cerebrum from the cerebellum below, are two other important inward folds of the inner dural membrane.

The middle meninx, the weblike **arachnoid mater,** underlies the dura mater and is partially separated from it by the **subdural space.** Threadlike projections bridge the **subarachnoid space** to attach the arachnoid to the innermost meninx, the **pia mater.** The delicate pia mater is highly vascular and clings tenaciously to the surface of the brain, following its convolutions.

The subarachnoid space is filled with cerebrospinal fluid. Specialized projections of the arachnoid tissue called **arachnoid granulations,** or *arachnoid villi,* protrude through the dura mater. These granulations allow the cerebrospinal fluid to drain back into the venous circulation via the superior sagittal sinus and other dural sinuses.

Meningitis, inflammation of the meninges, is a serious threat to the brain because of the intimate association between the brain and meninges. Should infection spread to the neural tissue of the brain itself, life-threatening **encephalitis** may occur. Meningitis is often diagnosed by taking a sample of cerebrospinal fluid from the subarachnoid space. +

Cerebrospinal Fluid

The cerebrospinal fluid (CSF), much like plasma in composition, is continually formed by the **choroid plexuses,** small capillary knots hanging from the roof of the ventricles of the brain. The cerebrospinal fluid in and around the brain forms a watery cushion that protects the delicate brain tissue against blows to the head.

Within the brain, the cerebrospinal fluid circulates from the two lateral ventricles (in the cerebral hemispheres) into the third ventricle via the **interventricular foramina,** and then through the cerebral aqueduct of the midbrain into the fourth ventricle in the hindbrain (Figure 14.8). CSF enters the subarachnoid space through the three foramina in the walls of the fourth ventricle. There it bathes the outer surfaces of the brain and spinal cord. The fluid returns to the blood in the dural sinuses via the arachnoid granulations.

Ordinarily, cerebrospinal fluid forms and drains at a constant rate. However, under certain conditions—for example, obstructed drainage or circulation resulting from tumors or anatomical deviations—cerebrospinal fluid accumulates and exerts increasing pressure on the brain which, uncorrected, causes neurological damage in adults. In infants, **hydrocephalus** (literally, "water on the brain") is indicated by a gradually enlarging head. The infant's skull is still flexible and contains fontanelles, so it can expand to accommodate the increasing size of the brain. +

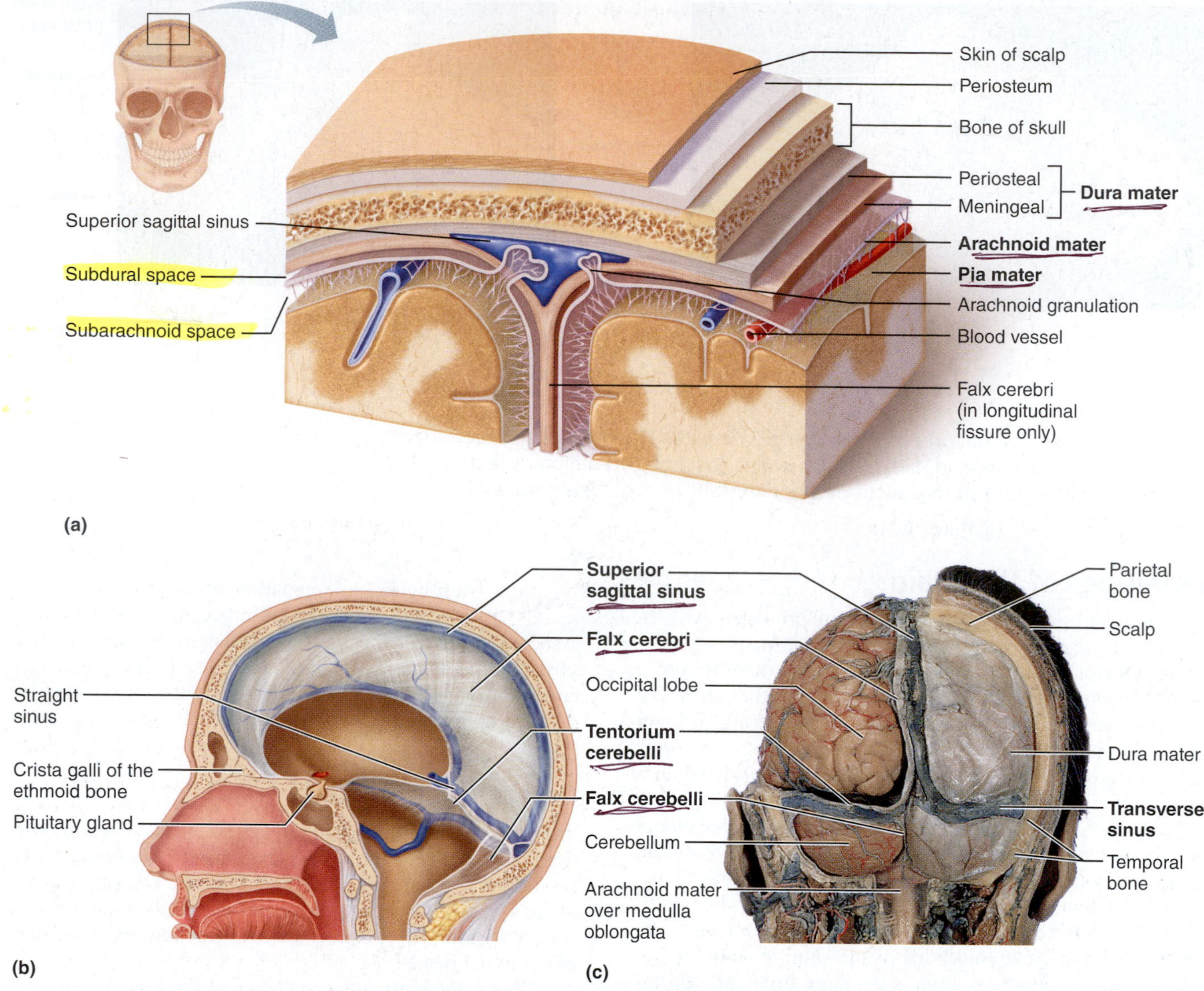

Figure 14.7 Meninges of the brain. (a) Three-dimensional frontal section showing the relationship of the dura mater, arachnoid mater, and pia mater. The meningeal dura forms the falx cerebri fold, which extends into the longitudinal fissure and attaches the brain to the ethmoid bone of the skull. The superior sagittal sinus is enclosed by the dural membranes superiorly. Arachnoid granulations, which return cerebrospinal fluid to the dural sinus, are also shown. **(b)** Midsagittal view showing the position of the dural folds, the falx cerebri, tentorium cerebelli, and falx cerebelli. **(c)** Posterior view of the brain in place, surrounded by the dura mater. Sinuses between periosteal and meningeal dura contain venous blood.

Cranial Nerves

The **cranial nerves (Figure 14.9)** are part of the peripheral nervous system and not part of the brain proper, but they are most appropriately identified while studying brain anatomy. The 12 pairs of cranial nerves primarily serve the head and neck. Only one pair, the vagus nerves, extends into the thoracic and abdominal cavities. All but the first two pairs (olfactory and optic nerves) arise from the brain stem and pass through foramina in the base of the skull to reach their destination.

The cranial nerves are numbered consecutively, and in most cases their names reflect the major structures they control. The cranial nerves are described by name, number (Roman numeral), origin, course, and function in the table **(Table 14.1)**. You should memorize this information. A mnemonic device that might be helpful for remembering the cranial nerves in order is "**O**n **O**ccasion, **O**ur **T**rusty **T**ruck **A**cts **F**unny—**V**ery **G**ood **V**ehicle **A**ny**H**ow." The first letter of each word and the "a" and "h" of the final word "anyhow" will remind you of the first letter of each cranial nerve name.

(Text continues on page 267.)

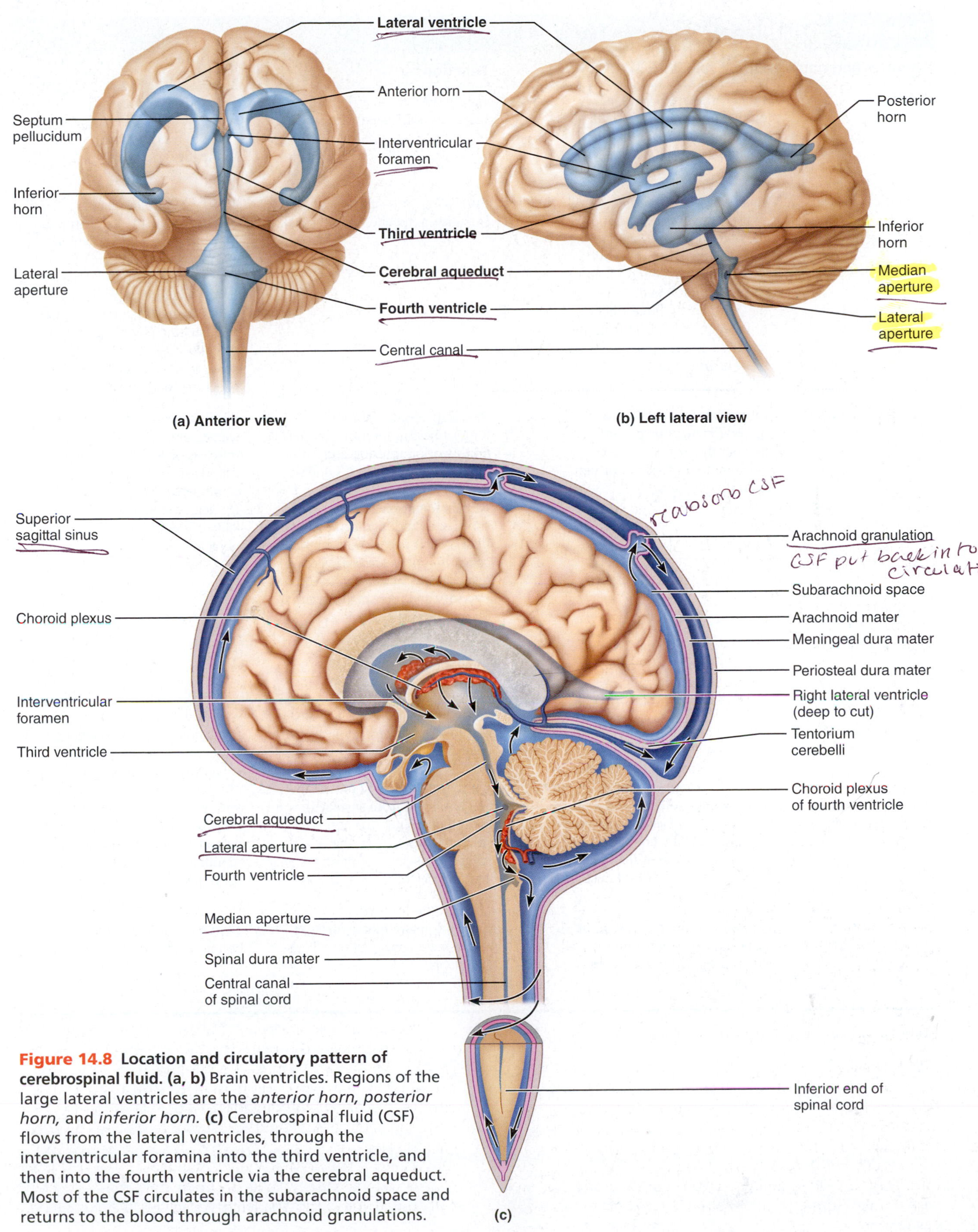

Figure 14.8 Location and circulatory pattern of cerebrospinal fluid. (a, b) Brain ventricles. Regions of the large lateral ventricles are the *anterior horn, posterior horn,* and *inferior horn.* **(c)** Cerebrospinal fluid (CSF) flows from the lateral ventricles, through the interventricular foramina into the third ventricle, and then into the fourth ventricle via the cerebral aqueduct. Most of the CSF circulates in the subarachnoid space and returns to the blood through arachnoid granulations.

Table 14.1 The Cranial Nerves (see Figure 14.9)

Number and name	Origin and course	Function*	Testing
I. Olfactory	Fibers arise from olfactory epithelium and run through cribriform plate of ethmoid bone to synapse in olfactory bulbs.	Purely sensory—carries afferent impulses associated with sense of smell.	Person is asked to sniff aromatic substances, such as oil of cloves and vanilla, and to identify each.
II. Optic	Fibers arise from retina of eye to form the optic nerve and pass through optic canal of the orbit. Fibers partially cross over at the optic chiasma and continue on to the thalamus as the optic tracts. Final fibers of this pathway travel from the thalamus to the optic cortex as the optic radiation.	Purely sensory—carries afferent impulses associated with vision.	Vision and visual field are determined with eye chart and by testing the point at which the person first sees an object (finger) moving into the visual field. Fundus of eye viewed with ophthalmoscope to detect papilledema (swelling of optic disc, the point at which optic nerve leaves the eye) and to observe blood vessels.
III. Oculomotor	Fibers emerge from dorsal midbrain and course ventrally to enter the orbit. They exit from skull via superior orbital fissure.	Primarily motor—somatic motor fibers to inferior oblique and superior, inferior, and medial rectus muscles, which direct eyeball, and to levator palpebrae muscles of the superior eyelid; parasympathetic fibers to iris and smooth muscle controlling lens shape (reflex responses to varying light intensity and focusing of eye for near vision).	Pupils are examined for size, shape, and equality. Pupillary reflex is tested with penlight (pupils should constrict when illuminated). Convergence for near vision is tested, as is subject's ability to follow objects with the eyes.
IV. Trochlear	Fibers emerge from midbrain and exit from skull via superior orbital fissure to run to eye.	Primarily motor—provides somatic motor fibers to superior oblique muscle that moves the eyeball.	Tested in common with cranial nerve III.
V. Trigeminal	Fibers run from face to pons and form three divisions, which exit separately from skull: mandibular division fibers pass through foramen ovale in sphenoid bone, maxillary division fibers pass via foramen rotundum in sphenoid bone, and ophthalmic division fibers pass through superior orbital fissure of eye socket.	Mixed—major sensory nerve of face; conducts sensory impulses from skin of face and anterior scalp, from mucosae of mouth and nose, and from surface of eyes; mandibular division also contains motor fibers that innervate muscles of mastication and muscles of floor of mouth.	Sensations of pain, touch, and temperature are tested with safety pin and hot and cold objects. Corneal reflex tested with wisp of cotton. Motor branch assessed by asking person to clench the teeth, open mouth against resistance, and move jaw side to side.
VI. Abducens	Fibers leave inferior region of pons and exit from skull via superior orbital fissure to run to eye.	Carries somatic motor fibers to lateral rectus muscle that moves the eyeball.	Tested in common with cranial nerve III.

*Does not include sensory impulses from proprioceptors.

Table 14.1 *(continued)*

Number and name	Origin and course	Function*	Testing
VII. Facial	Fibers leave pons and travel through temporal bone via internal acoustic meatus, exiting via stylomastoid foramen to reach the face.	Mixed—supplies somatic motor fibers to muscles of facial expression and parasympathetic motor fibers to lacrimal and salivary glands; carries sensory fibers from taste receptors of anterior portion of tongue.	Anterior two-thirds of tongue is tested for ability to taste sweet (sugar), salty, sour (vinegar), and bitter (quinine) substances. Symmetry of face is checked. Subject is asked to close eyes, smile, whistle, and so on. Tearing is assessed with ammonia fumes.
VIII. Vestibulocochlear	Fibers run from internal-ear equilibrium and hearing apparatus, housed in temporal bone, through internal acoustic meatus to enter pons.	Primarily sensory—vestibular branch transmits impulses associated with sense of equilibrium from vestibular apparatus and semicircular canals; cochlear branch transmits impulses associated with hearing from cochlea.	Hearing is checked by air and bone conduction using tuning fork.
IX. Glossopharyngeal	Fibers emerge from medulla and leave skull via jugular foramen to run to throat.	Mixed—somatic motor fibers serve pharyngeal muscles, and parasympathetic motor fibers serve salivary glands; sensory fibers carry impulses from pharynx, tonsils, posterior tongue (taste buds), and pressure receptors of carotid artery.	A tongue depressor is used to check the position of the uvula. Gag and swallowing reflexes are checked. Subject is asked to speak and cough. Posterior third of tongue may be tested for taste.
X. Vagus	Fibers emerge from medulla and pass through jugular foramen and descend through neck region into thorax and abdomen.	Mixed—fibers carry somatic motor impulses to pharynx and larynx and sensory fibers from same structures; very large portion is composed of parasympathetic motor fibers, which supply heart and smooth muscles of abdominal visceral organs; transmits sensory impulses from viscera.	As for cranial nerve IX (IX and X are tested in common, because they both innervate muscles of throat and mouth).
XI. Accessory	Fibers arise from medulla and superior aspect of spinal cord and travel through jugular foramen to reach muscles of neck and back.	Mixed (but primarily motor in function)—provides somatic motor fibers to sternocleidomastoid and trapezius muscles and to muscles of soft palate, pharynx, and larynx (spinal and medullary fibers respectively).	Sternocleidomastoid and trapezius muscles are checked for strength by asking person to rotate head and shoulders against resistance.
XII. Hypoglossal	Fibers arise from medulla and exit from skull via hypoglossal canal to travel to tongue.	Mixed (but primarily motor in function)—carries somatic motor fibers to muscles of tongue.	Person is asked to protrude and retract tongue. Any deviations in position are noted.

*Does not include sensory impulses from proprioceptors.

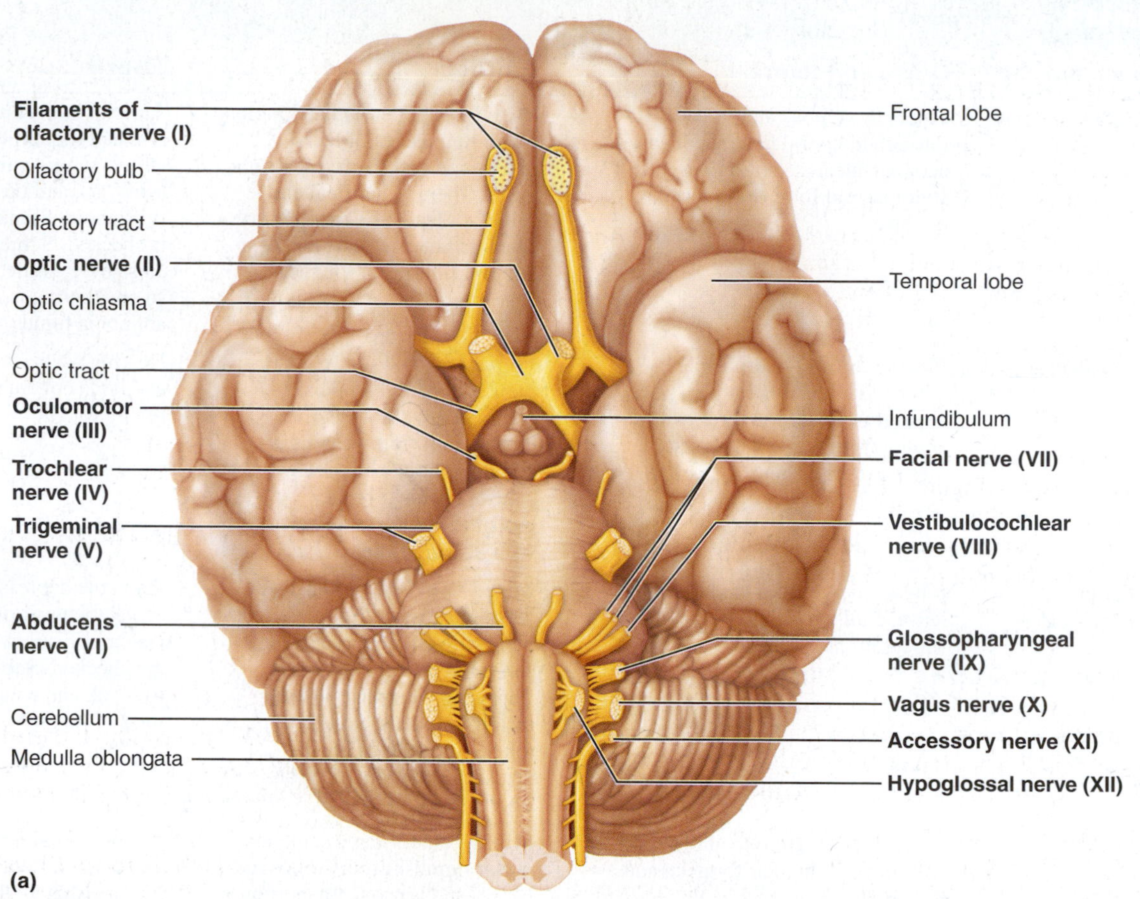

Cranial nerves	Sensory function		Motor function	
	Somatic sensory (SS)	Visceral sensory (VS)	Somatic motor (SM)	Visceral motor: parasympathetic (VM)
I Olfactory		Smell		
II Optic	Vision			
III Oculomotor			SM	VM
IV Trochlear			SM	
V Trigeminal	General		SM	
VI Abducens			SM	

Cranial nerves	Sensory function		Motor function	
	Somatic sensory (SS)	Visceral sensory (VS)	Somatic motor (SM)	Visceral motor: parasympathetic (VM)
VII Facial	General	General; taste	SM	VM
VIII Vestibulocochlear	Hearing; equilibrium		Some	
IX Glossopharyngeal	General	General; taste	SM	VM
X Vagus	General	General; taste	SM	VM
XI Accessory			SM	
XII Hypoglossal			SM	

(b)

Figure 14.9 Ventral aspect of the human brain, showing the cranial nerves. (See also Figure 14.3.)

Most cranial nerves are mixed nerves (containing both motor and sensory fibers). But close scrutiny of the table (Table 14.1) will reveal that three pairs of cranial nerves (optic, olfactory, and vestibulocochlear) are primarily or exclusively sensory in function.

Recall that the cell bodies of neurons are always located within the central nervous system (cortex or nuclei) or in specialized collections of cell bodies (ganglia) outside the CNS. Neuron cell bodies of the sensory cranial nerves are located in ganglia; those of the mixed cranial nerves are found both within the brain and in peripheral ganglia.

ACTIVITY 3

Identifying and Testing the Cranial Nerves

1. Observe the anterior surface of the brain model to identify the cranial nerves (Figure 14.9). Notice that the first (olfactory) cranial nerves are not visible on the model because they consist only of short axons that run from the nasal mucosa through the cribriform plate of the ethmoid bone. (However, the synapse points of the first cranial nerves, the *olfactory bulbs,* are visible on the model.)

2. Testing cranial nerves is an important part of any neurological examination. (See the last column of Table 14.1 for techniques you can use for such tests.) The results may help you understand cranial nerve function, especially as it pertains to some aspects of brain function.

3. Several cranial nerve ganglia are named in the accompanying chart. *Using your textbook or another appropriate reference,* fill in the **Activity 3 chart** by naming the cranial nerve the ganglion is associated with and stating its location. ■

Activity 3: Cranial Nerve Ganglia

Cranial nerve ganglion	Cranial nerve	Site of ganglion
Trigeminal		
Geniculate		
Inferior		
Superior		
Spiral		
Vestibular		

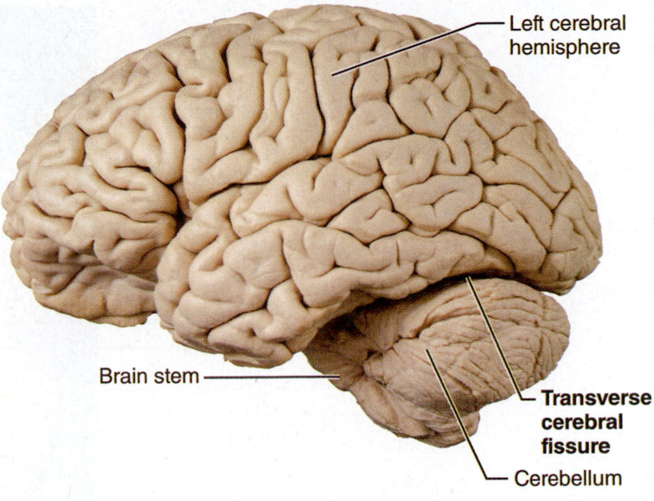

Figure 14.10 Photo of lateral aspect of the human brain.

DISSECTION

The Sheep Brain

The sheep brain is enough like the human brain to warrant comparison. Obtain a sheep brain, disposable gloves, dissecting tray, and instruments, and bring them to your laboratory bench.

1. If the dura mater is present, remove it as described here. Don disposable gloves. Place the intact sheep brain ventral surface down on the dissecting pan, and observe the dura mater. Feel its consistency, and note its toughness. Cut through the dura mater along the line of the longitudinal fissure (which separates the cerebral hemispheres) to enter the superior sagittal sinus. Gently force the cerebral hemispheres apart laterally to expose the corpus callosum deep to the longitudinal fissure.

2. Carefully remove the dura mater, and examine the superior surface of the brain. Notice that like the human brain, its surface is thrown into convolutions (fissures and gyri). Locate the arachnoid mater, which appears on the brain surface as a delicate "cottony" material spanning the fissures. In contrast, the innermost meninx, the pia mater, closely follows the cerebral contours.

3. Before beginning the dissection, turn your sheep brain so that you are viewing its left lateral aspect. Compare the various areas of the sheep brain (cerebrum, brain stem, cerebellum) to the photo of the human brain **(Figure 14.10)**. Relatively speaking, which of these structures is obviously much larger in the human brain?

____

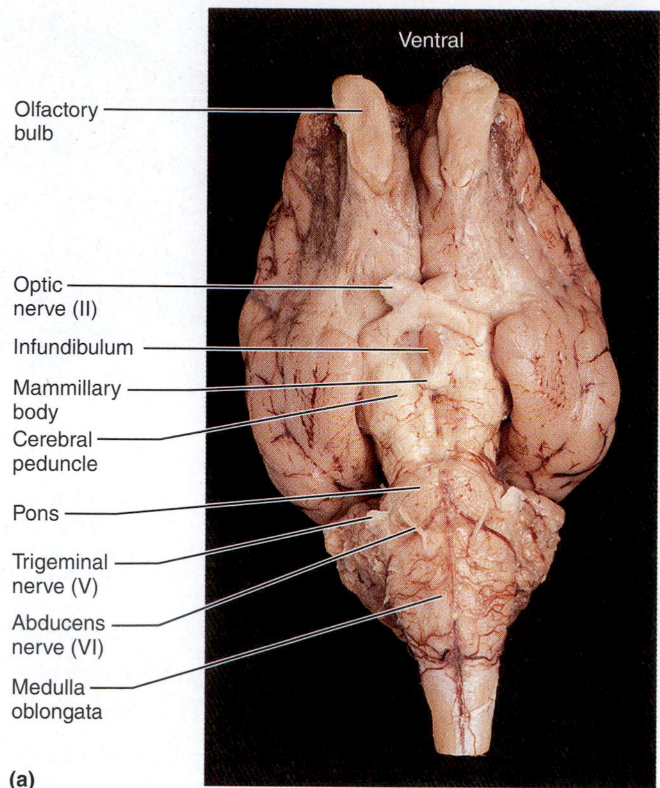

Figure 14.11 Intact sheep brain. (a) Photograph of ventral view.

Ventral Structures

Turn the brain so that its ventral surface is uppermost. (**Figure 14.11a** and **b** show the important features of the ventral surface of the brain.)

1. Look for the clublike olfactory bulbs anteriorly, on the inferior surface of the frontal lobes of the cerebral hemispheres. Axons of olfactory neurons run from the nasal mucosa through the perforated cribriform plate of the ethmoid bone to synapse with the olfactory bulbs.

How does the size of these olfactory bulbs compare with those of humans?

___

Is the sense of smell more important as a protective and a food-getting sense in sheep or in humans?

___

2. The optic nerve (II) carries sensory impulses from the retina of the eye. Thus this cranial nerve is involved in the sense of vision. Identify the optic nerves, optic chiasma, and optic tracts.

3. Posterior to the optic chiasma, two structures protrude from the ventral aspect of the hypothalamus—the infundibulum (stalk of the pituitary gland) immediately posterior to the optic chiasma and the mammillary body. Notice that the sheep's mammillary body is a single rounded eminence. In humans it is a double structure.

4. Identify the cerebral peduncles on the ventral aspect of the midbrain, just posterior to the mammillary body of the hypothalamus. The cerebral peduncles are fiber tracts connecting the cerebrum and medulla oblongata. Identify the large oculomotor nerves (III), which arise from the ventral midbrain surface, and the tiny trochlear nerves (IV), which can be seen at the junction of the midbrain and pons. Both of these cranial nerves provide motor fibers to extrinsic muscles of the eyeball.

5. Move posteriorly from the midbrain to identify first the pons and then the medulla oblongata, both hindbrain structures composed primarily of ascending and descending fiber tracts.

6. Return to the junction of the pons and midbrain, and proceed posteriorly to identify the following cranial nerves, all arising from the pons. Check them off as you locate them. (Figure 14.11b):

☐ Trigeminal nerves (V), which are involved in chewing and sensations of the head and face.

☐ Abducens nerves (VI), which abduct the eye (and thus work in conjunction with cranial nerves III and IV).

☐ Facial nerves (VII), large nerves involved in taste sensation, gland function (salivary and lacrimal glands), and facial expression.

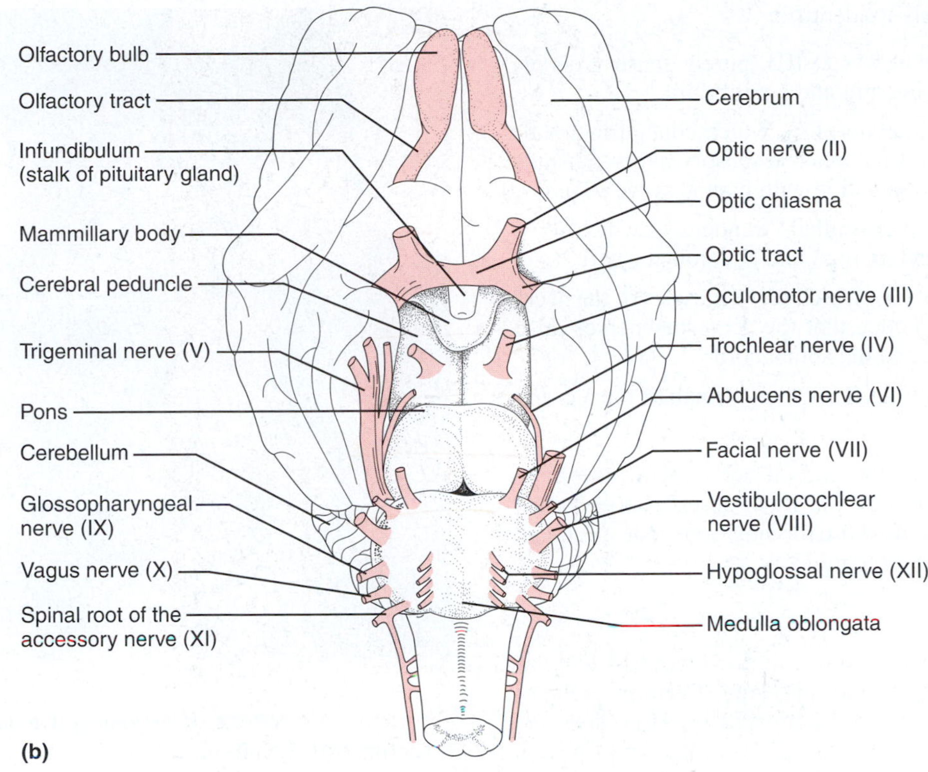

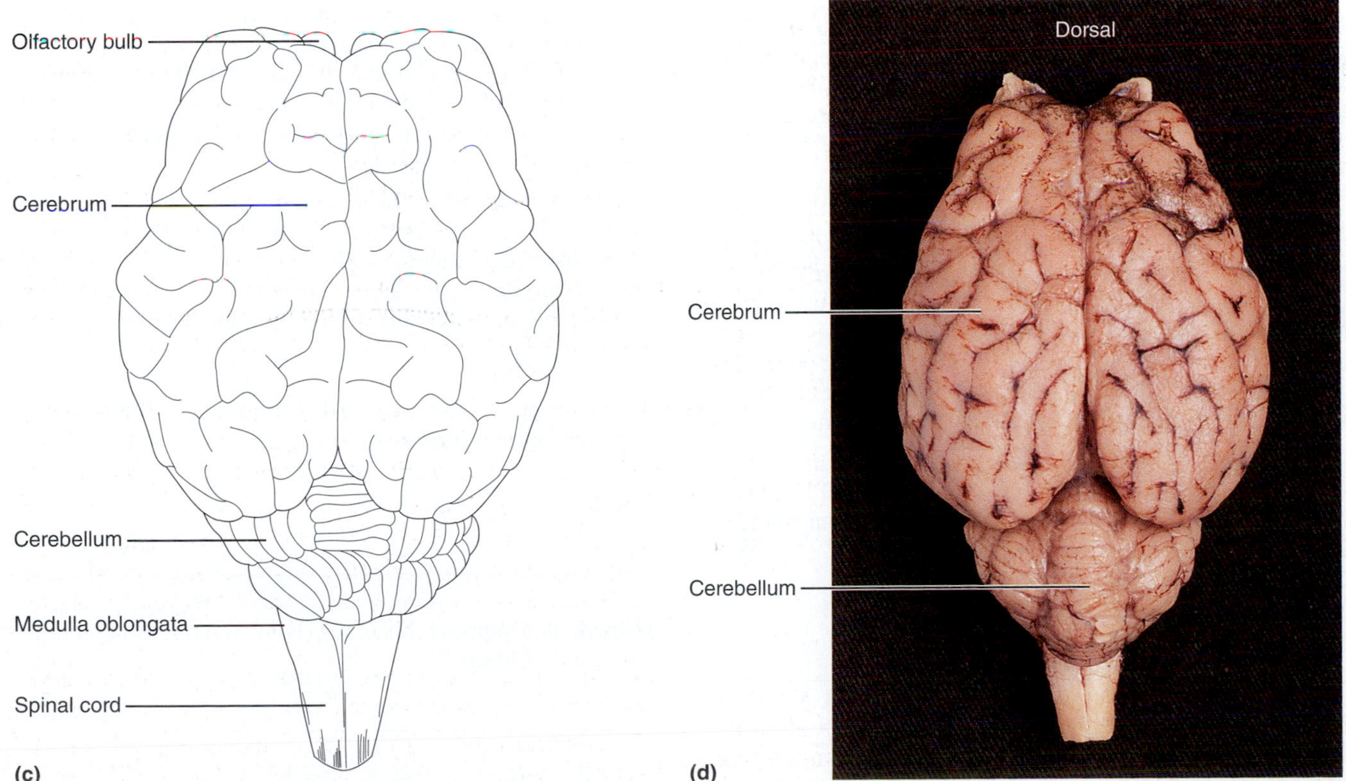

Figure 14.11 *(continued)* **Intact sheep brain. (b)** Diagrammatic ventral view. **(c)** Diagrammatic dorsal view. **(d)** Photograph of dorsal view.

7. Continue posteriorly to identify:

☐ Vestibulocochlear nerves (VIII), purely sensory nerves that are involved with hearing and equilibrium.

☐ Glossopharyngeal nerves (IX), which contain motor fibers innervating throat structures and sensory fibers transmitting taste stimuli (in conjunction with cranial nerve VII).

☐ Vagus nerves (X), often called "wanderers," which serve many organs of the head, thorax, and abdominal cavity.

☐ Accessory nerves (XI), which serve muscles of the neck, larynx, and shoulder; notice that the accessory nerves arise from both the medulla and the spinal cord.

☐ Hypoglossal nerves (XII), which stimulate tongue and neck muscles.

It is likely that some of the cranial nerves will have been broken off during brain removal. If so, observe sheep brains of other students to identify those missing from your specimen, using your check marks as a guide.

Dorsal Structures

1. Refer to the dorsal view illustrations (Figure 14.11c and d) as a guide in identifying the following structures. Reidentify the now exposed cerebral hemispheres. How does the depth of the fissures in the sheep's cerebral hemispheres compare to that in the human brain?

___

2. Examine the cerebellum. Notice that in contrast to the human cerebellum, it is not divided longitudinally and that its fissures are oriented differently. What dural falx (falx cerebri or falx cerebelli) is missing that is present in humans?

___

3. Locate the three pairs of cerebellar peduncles, fiber tracts that connect the cerebellum to other brain structures, by lifting the cerebellum dorsally away from the brain stem. The most posterior pair, the inferior cerebellar peduncles, connect the cerebellum to the medulla. The middle cerebellar peduncles attach the cerebellum to the pons, and the superior cerebellar peduncles run from the cerebellum to the midbrain.

4. To expose the dorsal surface of the midbrain, gently separate the cerebrum and cerebellum **(Figure 14.12)**. Identify the corpora quadrigemina, which appear as four rounded prominences on the dorsal midbrain surface.

What is the function of the corpora quadrigemina?

___

___

Also locate the pineal gland, which appears as a small oval protrusion in the midline just anterior to the corpora quadrigemina.

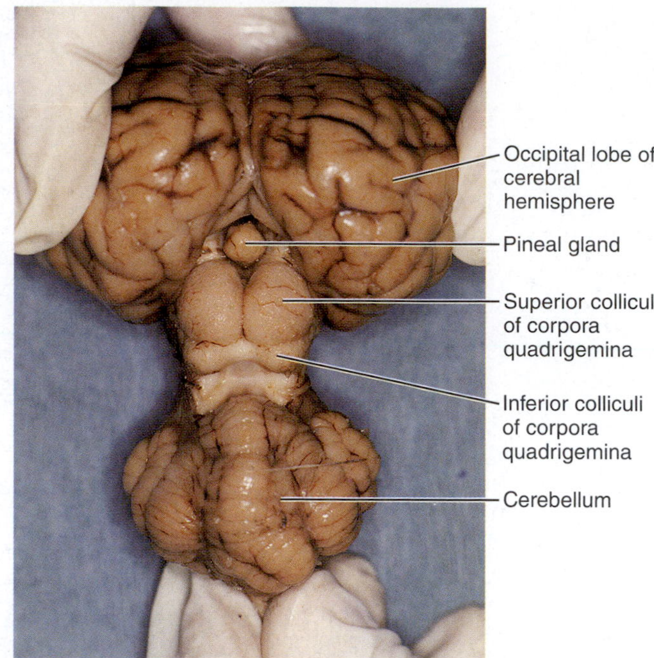

Figure 14.12 Means of exposing the dorsal midbrain structures of the sheep brain.

Internal Structures

1. The internal structure of the brain can be examined only after further dissection. Place the brain ventral side down on the dissecting tray, and make a cut completely through it in a superior-to-inferior direction. Cut through the longitudinal fissure, corpus callosum, and midline of the cerebellum. (Refer to **Figure 14.13** as you work.)

2. The thin nervous tissue membrane immediately ventral to the corpus callosum that separates the lateral ventricles is the septum pellucidum. Pierce this membrane, and probe the lateral ventricle cavity. The fiber tract ventral to the septum pellucidum and anterior to the third ventricle is the fornix.

How does the relative size of the fornix in this brain compare with the human fornix?

___

___

Why do you suppose this is so? (Hint: What is the function of this band of fibers?)

___

___

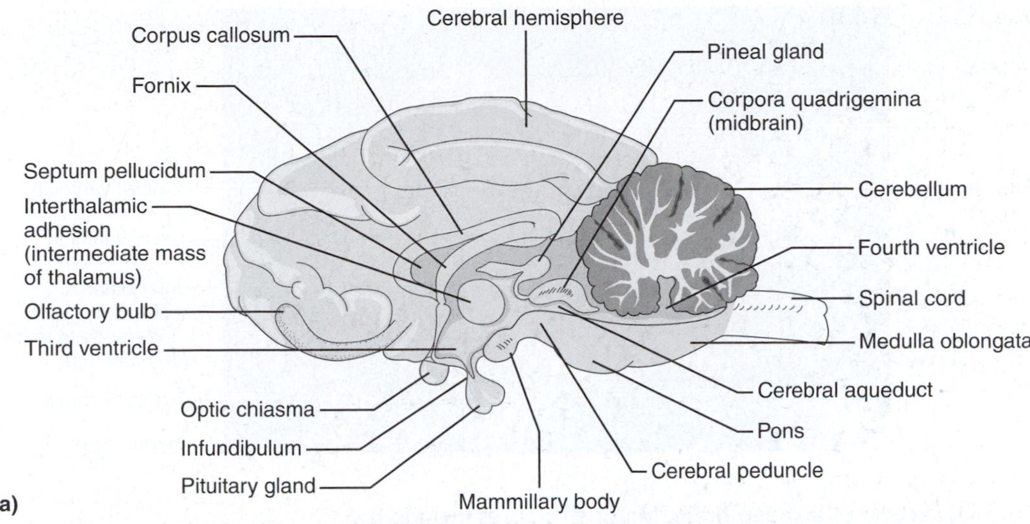

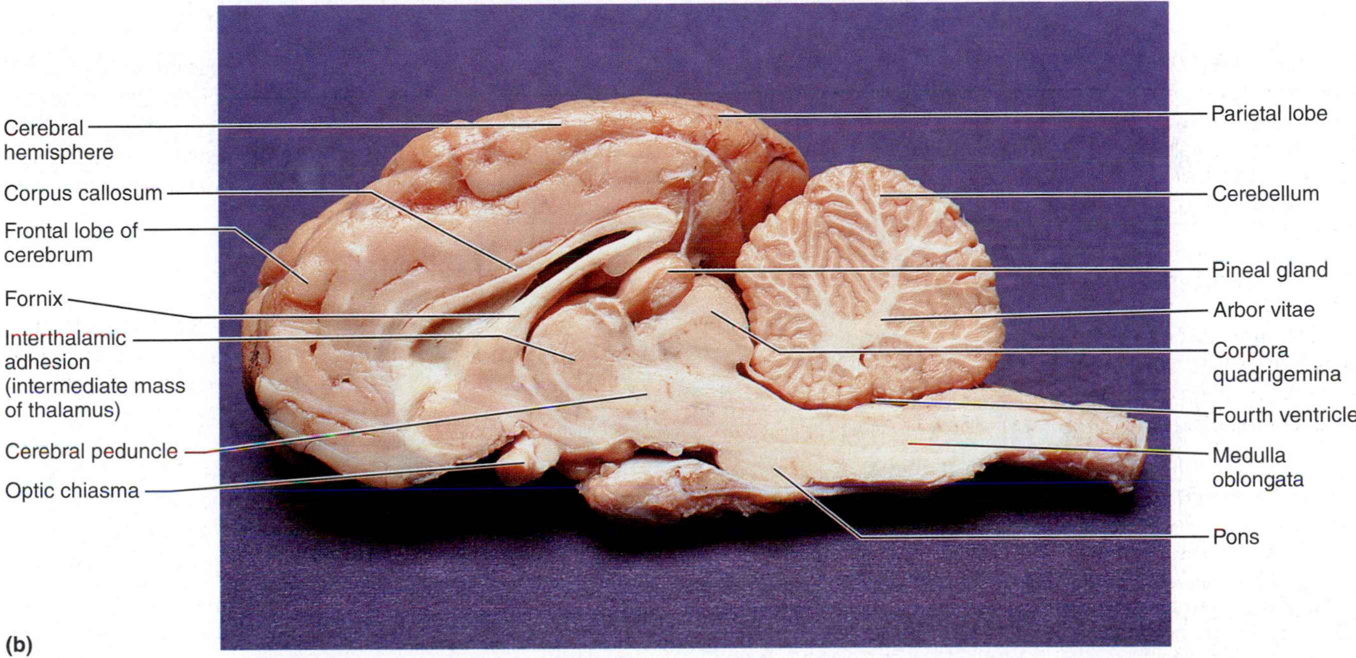

Figure 14.13 Sagittal section of the sheep brain showing internal structures. (a) Diagram view. **(b)** Photograph.

3. Identify the thalamus, which forms the walls of the third ventricle and is located posterior and ventral to the fornix. The interthalamic adhesion spanning the ventricular cavity appears as an oval protrusion of the thalamic wall. Anterior to the interthalamic adhesion, locate the interventricular foramen, a canal connecting the lateral ventricle on the same side with the third ventricle.

4. The hypothalamus forms the floor of the third ventricle. Identify the optic chiasma, infundibulum, and mammillary body on its exterior surface. You can see the pineal gland at the superoposterior end of the third ventricle, just beneath the junction of the corpus callosum and fornix.

5. Locate the midbrain by identifying the corpora quadrigemina that form its dorsal roof. Follow the cerebral aqueduct (the narrow canal connecting the third and fourth ventricles) through the midbrain tissue to the fourth ventricle. Identify the cerebral peduncles, which form its anterior walls.

6. Identify the pons and medulla oblongata, which lie anterior to the fourth ventricle. The medulla continues into the spinal cord without any obvious anatomical change, but the point at which the fourth ventricle narrows to a small canal is generally accepted as the beginning of the spinal cord.

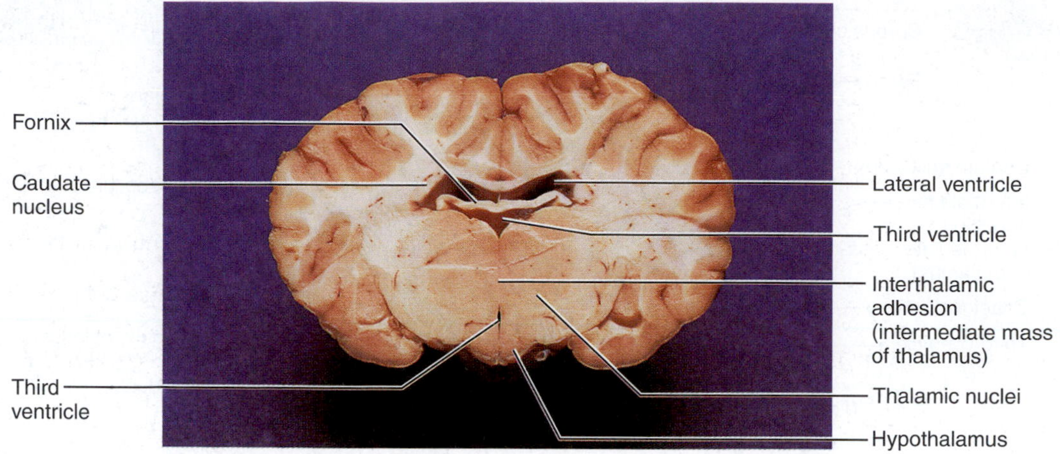

Figure 14.14 Coronal section of a sheep brain. Major structures include the thalamus, hypothalamus, and lateral and third ventricles.

7. Identify the cerebellum posterior to the fourth ventricle. Notice its internal treelike arrangement of white matter, the arbor vitae.

8. If time allows, obtain another sheep brain, and section it along the coronal plane so that the cut passes through the infundibulum. Compare your specimen to the photograph **(Figure 14.14)**, and attempt to identify all the structures shown in the figure.

9. Ask your instructor whether you should save a small portion of spinal cord from your brain specimen for the spinal cord studies (Exercise 15). Otherwise, dispose of all the organic debris in the appropriate laboratory containers, and clean the laboratory bench, the dissection instruments, and the tray before leaving the laboratory. ■

GROUP CHALLENGE

Odd (Cranial) Nerve Out

The following boxes each contain four cranial nerves. One of the listed nerves does not share a characteristic with the other three. Circle the cranial nerve that doesn't belong with the others, and explain why it is singled out. What characteristic is missing? Sometimes there may be multiple reasons why the cranial nerve doesn't belong with the others.

1. Which is the "odd" nerve?	Why is it the odd one out?
Optic nerve (II) Oculomotor nerve (III) Olfactory nerve (I) Vestibulocochlear nerve (VIII)	

2. Which is the "odd" nerve?	Why is it the odd one out?
Oculomotor nerve (III) Trochlear nerve (IV) Abducens nerve (VI) Hypoglossal nerve (XII)	

3. Which is the "odd" nerve?	Why is it the odd one out?
Facial nerve (VII) Hypoglossal nerve (XII) Trigeminal nerve (V) Glossopharyngeal nerve (IX)	

NAME ______________________________

LAB TIME/DATE ______________________

EXERCISE 14

REVIEW SHEET

Gross Anatomy of the Brain and Cranial Nerves

The Human Brain

1. Match the letters on the diagram of the human brain (right lateral view) to the appropriate terms listed on the left.

____________ 1. frontal lobe

____________ 2. parietal lobe

____________ 3. temporal lobe

____________ 4. precentral gyrus

____________ 5. parieto-occipital sulcus

____________ 6. postcentral gyrus

____________ 7. lateral sulcus

____________ 8. central sulcus

____________ 9. cerebellum

____________ 10. medulla

____________ 11. occipital lobe

____________ 12. pons

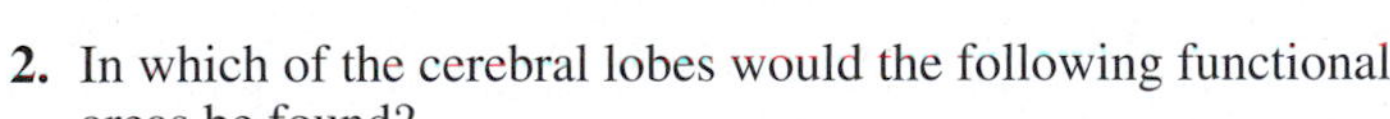

2. In which of the cerebral lobes would the following functional areas be found?

auditory cortex: ____

primary motor cortex: ____

primary somatosensory cortex: ____

olfactory cortex: ____

primary visual cortex: ____

Broca's area: ____

3. Which of the following structures are *not* part of the brain stem? (Circle the appropriate response or responses.)

cerebral hemispheres   pons   midbrain   cerebellum   medulla   diencephalon

4. Complete the following statements by writing the proper word or phrase on the corresponding blanks on the right.

A(n) __1__ is an elevated ridge of cerebral tissue. The convolutions seen in the cerebrum are important because they increase the __2__. Gray matter is composed of __3__. White matter is composed of __4__. A fiber tract that provides for communication between different parts of the same cerebral hemisphere is called a(n) __5__ tract, whereas one that carries impulses from the cerebrum to lower CNS areas is called a(n) __6__ tract. The caudate, putamen, and globus pallidus are collectively called the __7__.

1. ______________________
2. ______________________
3. ______________________
4. ______________________
5. ______________________
6. ______________________
7. ______________________

5. Identify the structures on the following sagittal view of the human brain by matching the numbered areas to the proper terms in the list.

__________	a.	cerebellum	__________	k.	interthalamic adhesion
__________	b.	cerebral aqueduct	__________	l.	mammillary bodies
__________	c.	(small part of) cerebral hemisphere	__________	m.	medulla oblongata
__________	d.	cerebral peduncle	__________	n.	optic chiasma
__________	e.	choroid plexus	__________	o.	pineal gland
__________	f.	corpora quadrigemina	__________	p.	pituitary gland
__________	g.	corpus callosum	__________	q.	pons
__________	h.	fornix	__________	r.	septum pellucidum
__________	i.	fourth ventricle	__________	s.	thalamus
__________	j.	hypothalamus			

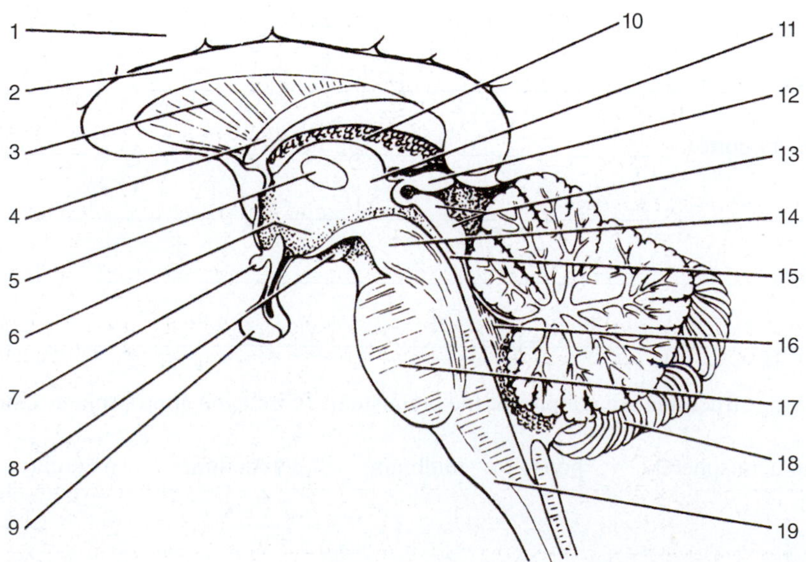

6. Using the terms from question 5, match the appropriate structures with the descriptions given below. (Not all terms will be used.)

__________ 1. site of regulation of body temperature and water balance; most important autonomic center

__________ 2. consciousness depends on the function of this part of the brain

__________ 3. located in the midbrain; contains reflex centers for vision and hearing

__________ 4. responsible for regulation of posture and coordination of complex muscular movements

__________ 5. important synapse site for afferent fibers traveling to the primary somatosensory cortex

__________ 6. contains autonomic centers regulating blood pressure, heart rate, and respiratory rhythm, as well as coughing, sneezing, and swallowing centers

__________ 7. large commissure connecting the cerebral hemispheres

__________ 8. fiber tract involved with olfaction

__________ 9. connects the third and fourth ventricles

__________ 10. encloses the third ventricle

7. Embryologically, the brain arises from the rostral end of a tubelike structure that quickly becomes divided into three major regions. Groups of structures that develop from the embryonic brain are listed below. Designate the embryonic origin of each group as the hindbrain, midbrain, or forebrain.

______________ 1. the diencephalon, including the thalamus, optic chiasma, and hypothalamus

______________ 2. the medulla, pons, and cerebellum

______________ 3. the cerebral hemispheres

8. What is the function of the basal nuclei? ______________________________

____

9. What is the striatum, and how is it related to the fibers of the internal capsule? ______________________________

____

____

10. A brain hemorrhage within the region of the right internal capsule results in paralysis of the left side of the body.

Explain why the left side (rather than the right side) is affected. ______________________________

____

11. Explain why trauma to the base of the brain is often much more dangerous than trauma to the frontal lobes. (Hint: Think about the relative functioning of the cerebral hemispheres and the brain stem structures. Which contain centers more vital to life?)

____

____

____

12. In "split brain" experiments, the main commissure connecting the cerebral hemispheres is cut. Name this commissure.

____

Meninges of the Brain

13. Identify the meningeal (or associated) structures described below:

______________ 1. outermost meninx covering the brain; composed of tough fibrous connective tissue

______________ 2. innermost meninx covering the brain; delicate and highly vascular

______________ 3. structures instrumental in returning cerebrospinal fluid to the venous blood in the dural sinuses

______________ 4. structure that forms the cerebrospinal fluid

______________ 5. middle meninx; like a cobweb in structure

______________ 6. its outer layer forms the periosteum of the skull

______________ 7. a dural fold that attaches the cerebrum to the crista galli of the skull

______________ 8. a dural fold separating the cerebrum from the cerebellum

Cerebrospinal Fluid

14. Label on the accompanying diagram the structures involved with circulation of cerebrospinal fluid.

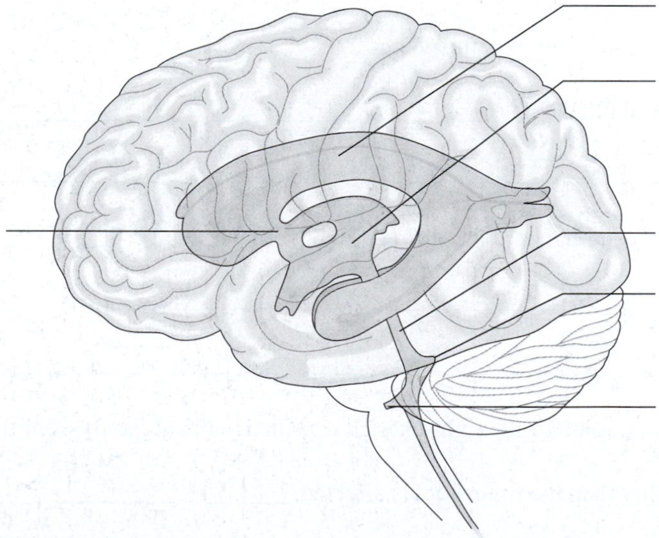

Add arrows to the figure above to indicate the flow of cerebrospinal fluid from its formation in the lateral ventricles to the site of its exit from the fourth ventricle. Then fill in the blanks in the following paragraph.

Cerebrospinal fluid flows from the fourth ventricle into the central canal of the spinal cord and the __1__ space surrounding the brain and spinal cord. From this space it drains through the __2__ into the __3__.

1. ______________

2. ______________

3. ______________

Cranial Nerves

15. Using the terms below, correctly identify all structures indicated by leader lines on the diagram.

a. abducens nerve (VI)
b. accessory nerve (XI)
c. cerebellum
d. cerebral peduncle
e. decussation of the pyramids
f. facial nerve (VII)
g. frontal lobe of cerebral hemisphere
h. glossopharyngeal nerve (IX)
i. hypoglossal nerve (XII)
j. longitudinal fissure
k. mammillary body
l. medulla oblongata
m. oculomotor nerve (III)
n. olfactory bulb
o. olfactory tract
p. optic chiasma
q. optic nerve
r. optic tract
s. pituitary gland
t. pons
u. spinal cord
v. temporal lobe of cerebral hemisphere
w. trigeminal nerve (V)
x. trochlear nerve (IV)
y. vagus nerve (X)
z. vestibulocochlear nerve (VIII)

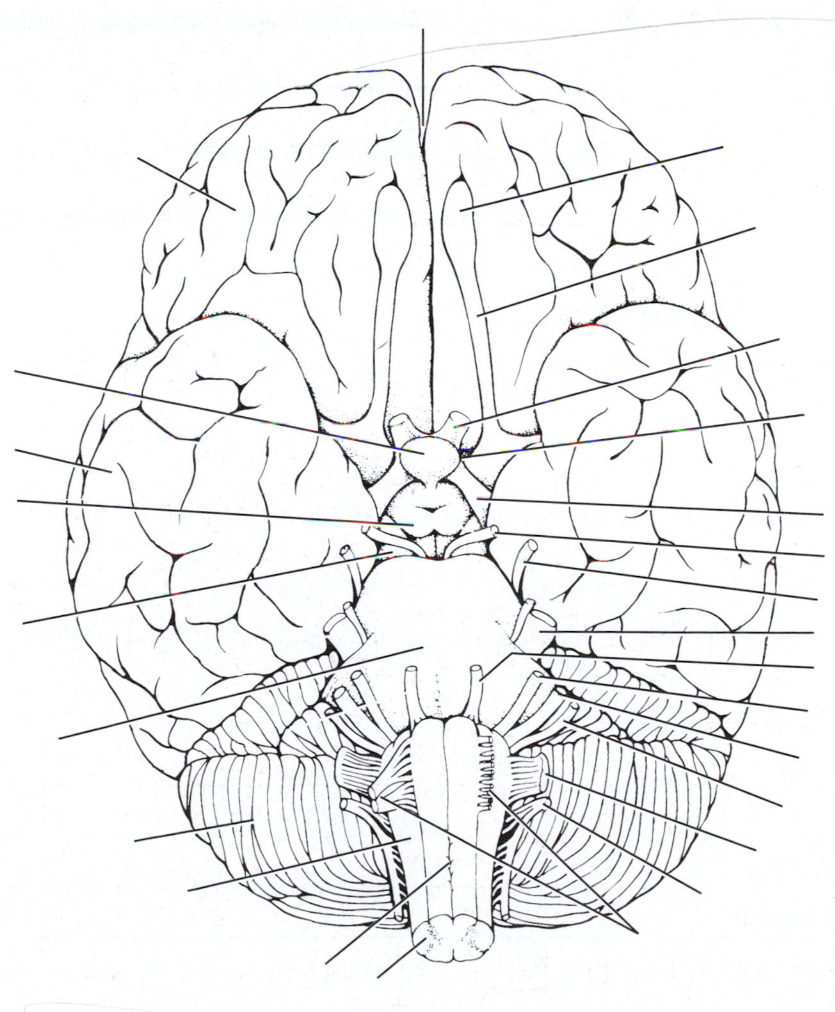

16. Provide the name and number of the cranial nerves involved in each of the following activities, sensations, or disorders.

______________________ 1. rotating the head

______________________ 2. smelling a flower

______________________ 3. raising the eyelids; constricting the pupils of the eye

______________________ 4. slowing the heart; increasing the motility of the digestive tract

______________________ 5. involved in Bell's palsy (facial paralysis)

______________________ 6. chewing food

______________________ 7. listening to music; seasickness

______________________ 8. secreting saliva; tasting well-seasoned food

______________________ 9. involved in "rolling" the eyes (three nerves—provide numbers only)

______________________ 10. feeling a toothache

______________________ 11. reading the newspaper

______________________ 12. primarily sensory in function (three nerves—provide numbers only)

Dissection of the Sheep Brain

17. Describe the firmness and texture of the sheep brain tissue as observed when you cut into it. ______________________

____

Given that formalin hardens all tissue, what conclusions might you draw about the firmness and texture of living brain tissue?

____

18. When comparing human and sheep brains, you observe some profound differences between them. Record your observations in the chart below.

Structure	Human	Sheep
Olfactory bulb		
Pons-medulla relationship		
Location of cranial nerve III		
Mammillary body		
Corpus callosum		
Interthalamic adhesion		
Relative size of superior and inferior colliculi		
Pineal gland		

EXERCISE 15

The Spinal Cord and Spinal Nerves

MATERIALS

- ☐ Spinal cord model (cross section)
- ☐ Three-dimensional models or laboratory charts of the spinal cord and spinal nerves
- ☐ Red and blue pencils
- ☐ Preserved cow spinal cord sections with meninges and nerve roots intact (or spinal cord segment saved from the brain dissection in Exercise 14)
- ☐ Dissecting instruments and tray
- ☐ Disposable gloves
- ☐ Stereomicroscope
- ☐ Prepared slide of spinal cord (x.s.)
- ☐ Compound microscope
- ☐ Animal specimen from previous dissections
- ☐ Embalming fluid
- ☐ Organic debris container

OBJECTIVES

1. List two major functions of the spinal cord.
2. Define *conus medullaris, cauda equina,* and *filum terminale*.
3. Name the meningeal coverings of the spinal cord, and state their functions.
4. Indicate two major areas where the spinal cord is enlarged, and explain the reasons for the enlargement.
5. Identify important anatomical areas on a model or image of a cross section of the spinal cord. Where applicable, name the neuron type found in these areas.
6. Locate on a diagram the fiber tracts in the spinal cord, and state their functions.
7. Note the number of pairs of spinal nerves that arise from the spinal cord, describe their division into groups, and identify the number of pairs in each group.
8. Describe the origin and fiber composition of the spinal nerves, differentiating among roots, the spinal nerve proper, and rami. Discuss the result of transecting these structures.
9. Discuss the distribution of the dorsal and ventral rami of the spinal nerves.
10. Identify the four major nerve plexuses on a model or image, name the major nerves of each plexus, and describe the destination and function of each.
11. Identify on a dissected animal the musculocutaneous, radial, median, and ulnar nerves of the upper limb and the femoral, saphenous, sciatic, common fibular, and tibial nerves of the lower limb.

PRE-LAB QUIZ

1. The spinal cord extends from the foramen magnum of the skull to the first or second lumbar vertebra, where it terminates in the:
   a. conus medullaris
   b. denticulate ligament
   c. filum terminale
   d. gray matter
2. How many pairs of spinal nerves do humans have?
   a. 10 c. 31
   b. 12 d. 47
3. Circle the correct underlined term. In cross section, the gray / white matter of the spinal cord looks like a butterfly or the letter H.
4. Circle True or False. The cell bodies of sensory neurons are found in an enlarged area of the dorsal root called the gray commissure.
5. Circle the correct underlined term. Fiber tracts conducting impulses to the brain are called ascending or sensory / motor tracts.

(Text continues on next page.)

6. Circle True or False. Because the spinal nerves arise from fusion of the ventral and dorsal roots of the spinal cord and contain motor and sensory fibers, all spinal nerves are considered mixed nerves.
7. The ventral rami of all spinal nerves except for T_2 through T_{12} form complex networks of nerves known as:
   a. fissures c. plexuses
   b. ganglia d. sulci
8. Severe injuries to the ___________ plexus cause weakness or paralysis of the entire upper limb.
   a. brachial c. lumbar
   b. cervical d. sacral
9. Circle True or False. The femoral nerve is the largest nerve from the sacral plexus.
10. Circle the correct underlined term. The sciatic nerve divides into the tibial and posterior femoral cutaneous / common fibular nerves.

The cylindrical **spinal cord,** a continuation of the brain stem, is an association and communication center. It plays a major role in spinal reflex activity and provides neural pathways to and from higher nervous centers.

Anatomy of the Spinal Cord

Enclosed within the vertebral canal of the spinal column, the spinal cord extends from the foramen magnum of the skull to the first or second lumbar vertebra, where it terminates in the cone-shaped **conus medullaris** (**Figure 15.1**). Like the brain, the spinal cord is cushioned and protected by meninges. The dura mater and arachnoid meningeal coverings extend beyond the conus medullaris, approximately to the level of S_2, and the **filum terminale,** a fibrous extension of the pia mater, extends even farther into the coccygeal canal to attach to the posterior coccyx. **Denticulate ligaments,** saw-toothed shelves of pia mater, secure the spinal cord to the bony wall of the vertebral column all along its length (see Figure 15.1c).

The cerebrospinal fluid–filled meninges extend well beyond the end of the spinal cord, providing an excellent site for removing cerebrospinal fluid without endangering the delicate spinal cord. Analysis of the fluid can provide important information about suspected bacterial or viral infections of the spinal cold or meninges. This procedure, called a *lumbar tap,* is usually performed below L_3. Additionally, "saddle block," or caudal anesthesia for childbirth, is normally administered (injected) between L_3 and L_5.

In humans, 31 pairs of spinal nerves arise from the spinal cord and pass through intervertebral foramina to serve the body area at their approximate level of emergence. The cord is about the size of a finger in circumference for most of its length, but there are obvious enlargements in the cervical and lumbar areas where the nerves serving the upper and lower limbs issue from the cord.

Because the spinal cord does not extend to the end of the vertebral column, the spinal nerves emerging from the inferior end of the cord must travel through the vertebral canal for some distance before exiting at the appropriate intervertebral foramina. This collection of spinal nerves passing through the inferior end of the vertebral canal is called the **cauda equina** (Figure 15.1a and d) because of its similarity to a horse's tail (the literal translation of *cauda equina*).

ACTIVITY 1

Identifying Structures of the Spinal Cord

Obtain a three-dimensional model or laboratory chart of a cross section of a spinal cord, and identify its structures as they are described next. ■

Gray Matter

In cross section, the **gray matter** of the spinal cord looks like a butterfly or the letter H (**Figure 15.2**). The two posterior projections are called the **dorsal (posterior) horns.** The two anterior projections are the **ventral (anterior) horns.** The tips of the ventral horns are broader and less tapered than those of the dorsal horns. In the thoracic and lumbar regions of the cord, there is also a lateral outpocketing of gray matter on each side referred to as the **lateral horn.** The central area of gray matter connecting the two vertical regions is the **gray commissure,** which surrounds the **central canal** of the cord.

Neurons with specific functions can be localized in the gray matter. The dorsal horns contain interneurons and sensory fibers that enter the cord from the body periphery via the **dorsal root.** The cell bodies of these sensory neurons are found in an enlarged area of the dorsal root called the **dorsal root ganglion.** The ventral horns mainly contain cell bodies of motor neurons of the somatic nervous system, which send their axons out via the **ventral root** of the cord to enter the adjacent spinal nerve. Since they are formed by the fusion of the dorsal and ventral roots, all **spinal nerves** are **mixed nerves,** containing both sensory and motor fibers. The lateral horns, where present, contain nerve cell bodies of motor neurons of the autonomic nervous system (sympathetic division). Their axons also leave the cord via the ventral roots, along with those of the motor neurons of the ventral horns.

White Matter

The **white matter** of the spinal cord is nearly bisected by fissures (see Figure 15.2). The more open anterior fissure is the **ventral median fissure,** and the shallow posterior one is the **dorsal median sulcus.** The white matter is composed of myelinated and nonmyelinated fibers—some running to higher centers, some traveling from the brain to the cord, and

(Text continues on page 283.)

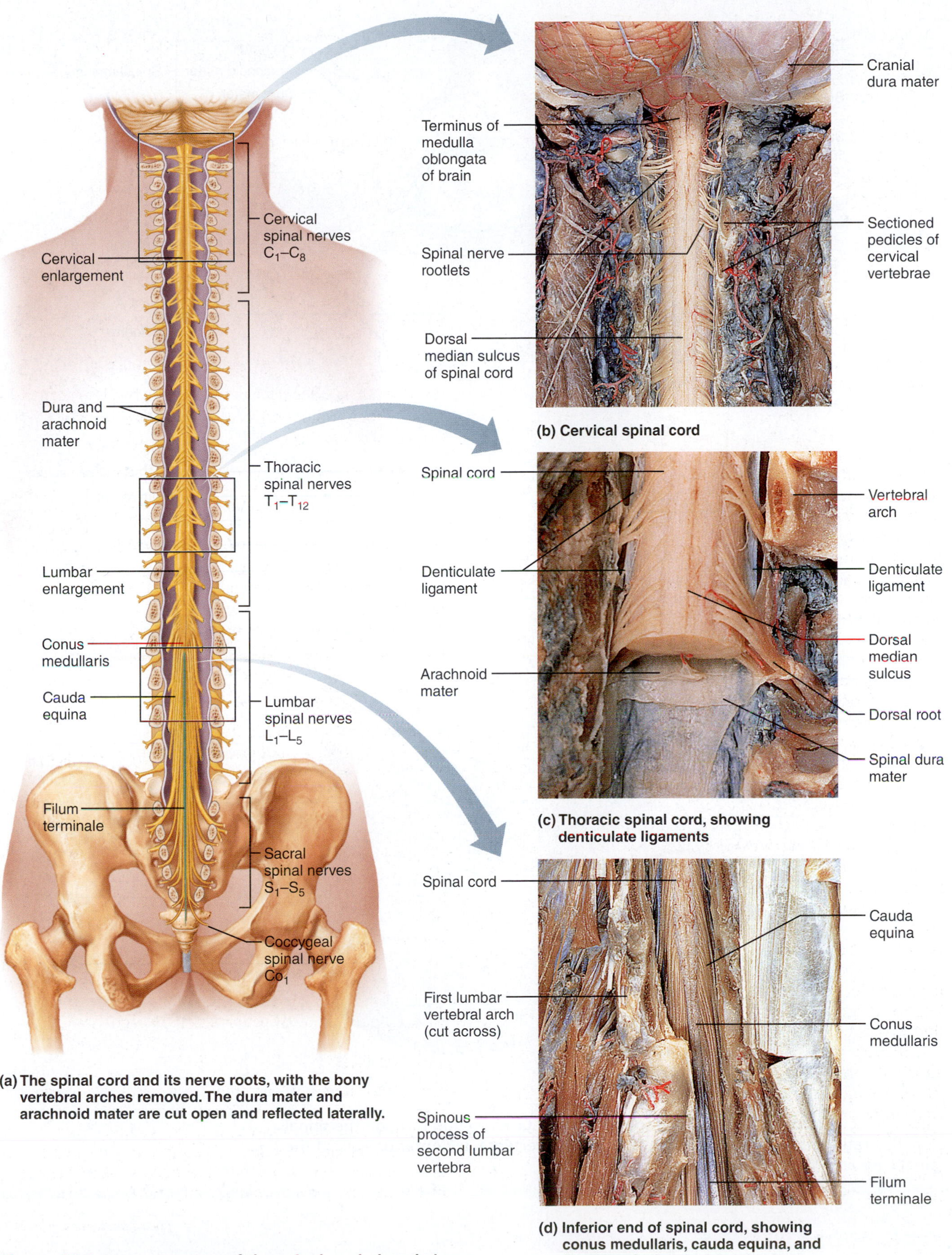

(a) The spinal cord and its nerve roots, with the bony vertebral arches removed. The dura mater and arachnoid mater are cut open and reflected laterally.

(b) Cervical spinal cord

(c) Thoracic spinal cord, showing denticulate ligaments

(d) Inferior end of spinal cord, showing conus medullaris, cauda equina, and filum terminale

Figure 15.1 Gross structure of the spinal cord, dorsal view.

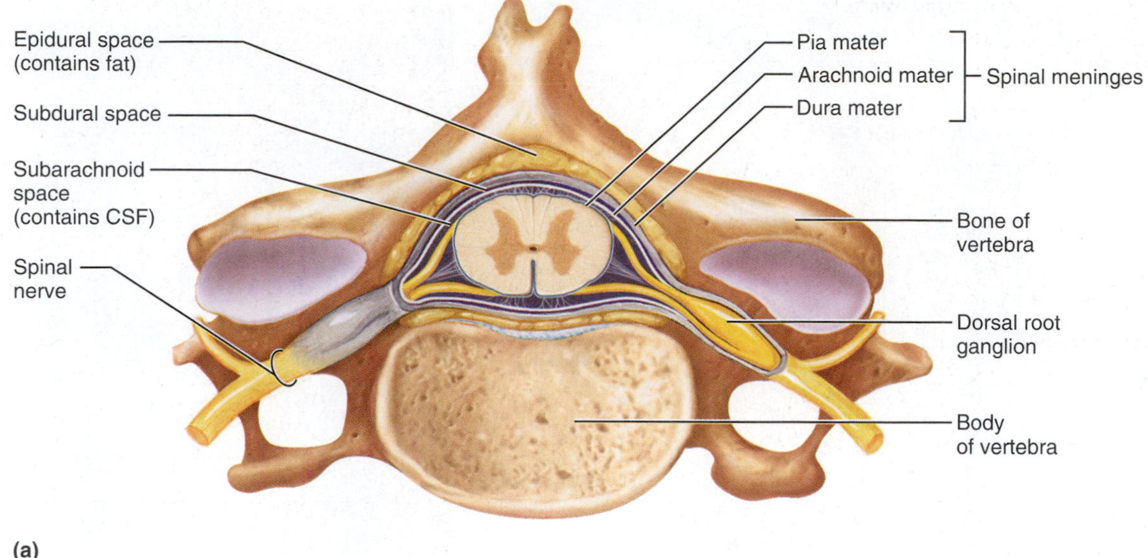

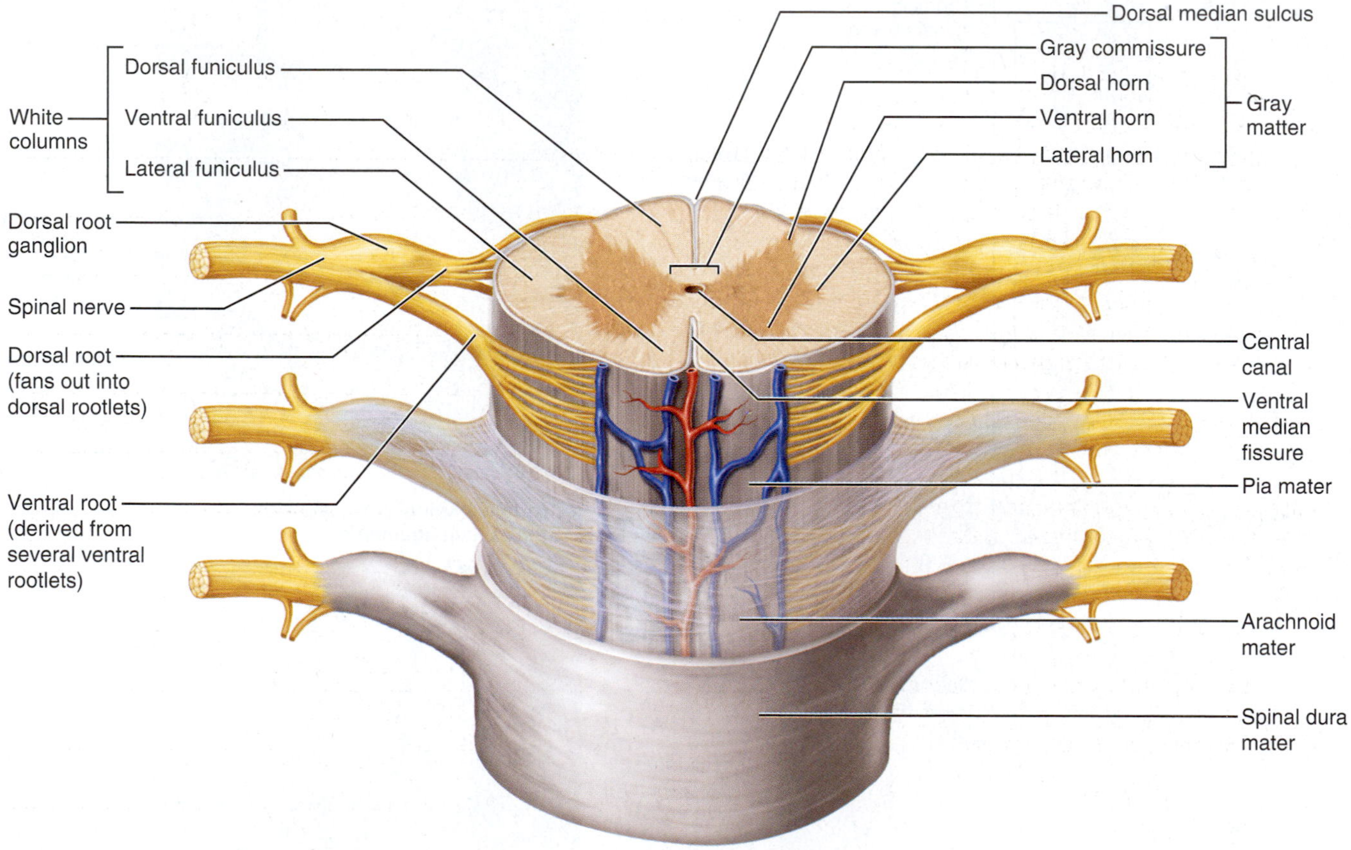

Figure 15.2 Anatomy of the human spinal cord. (a) Cross section through the spinal cord illustrating its relationship to the surrounding vertebra. **(b)** Anterior view of the spinal cord and its meningeal coverings.

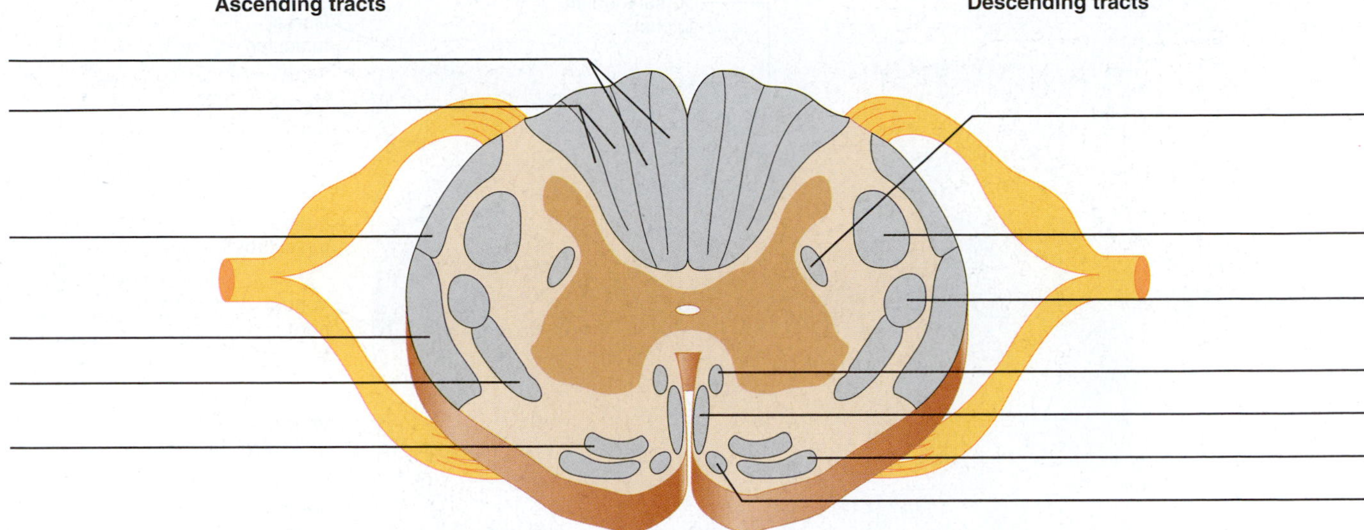

Figure 15.3 Cross section of the spinal cord showing the relative positioning of its major tracts.

some conducting impulses from one side of the cord to the other (see Figure 15.2).

Because of the irregular shape of the gray matter, the white matter on each side of the cord can be divided into three primary regions, or **white columns:** the **dorsal, lateral,** and **ventral funiculi.** Each funiculus contains a number of fiber **tracts** composed of axons with the same origin, terminus, and function. Tracts conducting sensory impulses to the brain are called *ascending,* or *sensory, tracts;* those carrying impulses from the brain to the skeletal muscles are *descending,* or *motor, tracts.*

Because it serves as the transmission pathway between the brain and the body periphery, the spinal cord is extremely important functionally. Even though it is protected by meninges and cerebrospinal fluid in the vertebral canal, it is highly vulnerable to traumatic injuries, such as might occur in an automobile accident.

When the cord is transected (or severely traumatized), both motor and sensory functions are lost in body areas normally served by that region and lower regions of the spinal cord. Injury to certain spinal cord areas may even result in a permanent flaccid paralysis of both legs, called **paraplegia,** or of all four limbs, called **quadriplegia.** +

ACTIVITY 2

Identifying Spinal Cord Tracts

With the help of your textbook, label the spinal cord diagram **(Figure 15.3)** with the tract names that follow. Each tract is represented on both sides of the cord, but for clarity, label the motor tracts on the right side of the diagram and the sensory tracts on the left side of the diagram. *Color ascending tracts blue and descending tracts red.* Then fill in the functional importance of each tract beside its name in the list that follows. As you work, try to be aware of how the naming of the tracts is related to their anatomical distribution.

Dorsal columns

Fasciculus gracilis ______________________

Fasciculus cuneatus ______________________

Dorsal spinocerebellar ______________________

Ventral spinocerebellar ______________________

Lateral spinothalamic ______________________

Ventral spinothalamic ______________________

Lateral corticospinal ______________________

Ventral corticospinal ______________________

Rubrospinal ______________________

Tectospinal ______________________

Vestibulospinal ______________________

Medial reticulospinal ______________________

Lateral reticulospinal ______________________

DISSECTION Spinal Cord

1. Obtain a dissecting tray and instruments, disposable gloves, and a segment of preserved spinal cord (from a cow or saved from the brain specimen used in Exercise 14). Identify the tough outer meninx (dura mater) and the weblike arachnoid mater.

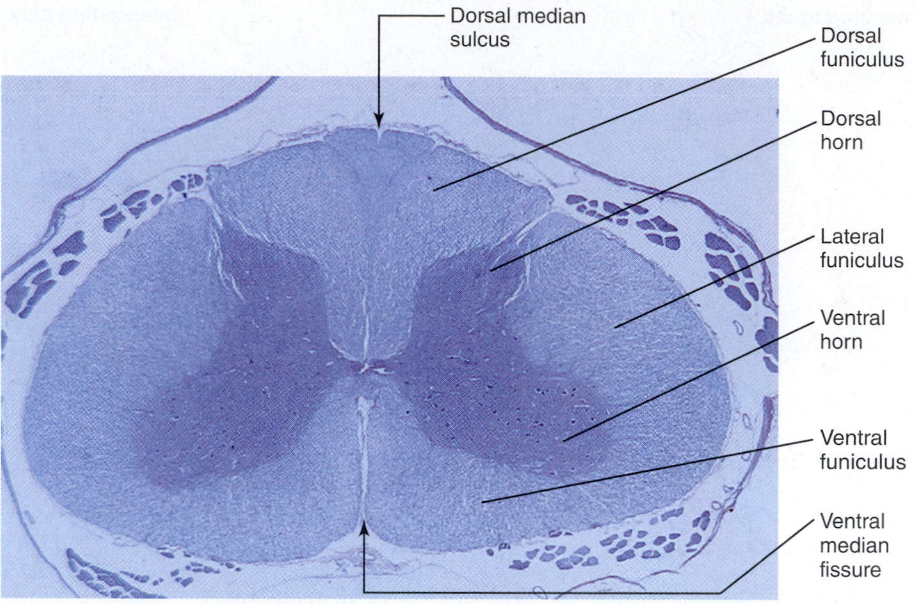

Figure 15.4 Cross section of the spinal cord (10×).

What name is given to the third meninx, and where is it found?

___

___

Peel back the dura mater, and observe the fibers making up the dorsal and ventral roots. If possible, identify a dorsal root ganglion.

2. Cut a thin cross section of the cord, and identify the ventral and dorsal horns of the gray matter with the naked eye or with the aid of a dissecting microscope.

How can you be certain that you are correctly identifying the ventral and dorsal horns?

___

___

Also identify the central canal, white matter, ventral median fissure, dorsal median sulcus, and dorsal, ventral, and lateral funiculi.

3. Obtain a prepared slide of the spinal cord (cross section) and a compound microscope. Examine the slide carefully under low power (refer to **Figure 15.4** to identify spinal cord features). Observe the shape of the central canal.

Is it basically circular or oval? ___

Name the neuroglial cell type that lines this canal. ___

Can you see any neuron cell bodies? ___

If so, where and what type of neuron would these most likely be—motor neurons, sensory neurons, or interneurons?

___

___ ■

Spinal Nerves and Nerve Plexuses

The 31 pairs of human spinal nerves arise from the fusions of the ventral and dorsal roots of the spinal cord (see Figure 15.2a). There are 8 pairs of cervical nerves (C_1–C_8), 12 pairs of thoracic nerves (T_1–T_{12}), 5 pairs of lumbar nerves (L_1–L_5), 5 pairs of sacral nerves (S_1–S_5), and 1 pair of coccygeal nerves (Co_1) **(Figure 15.5a)**. The first pair of spinal nerves leaves the vertebral canal between the base of the occiput and the atlas, but all the rest exit via the intervertebral foramina. The first through seventh pairs of cervical nerves emerge *above* the vertebra for which they are named; C_8 emerges between C_7 and T_1. (Notice that there are 7 cervical vertebrae, but 8 pairs of cervical nerves.) The remaining spinal nerve pairs emerge from the spinal cord area *below* the same-numbered vertebra.

Almost immediately after emerging, each nerve divides into **dorsal** and **ventral rami.** Thus, each spinal nerve is only about 1 or 2 cm long. The rami, like the spinal nerves, contain both motor and sensory fibers. The smaller dorsal rami serve the skin and musculature of the posterior body trunk at their approximate level of emergence. The ventral rami of spinal nerves T_2 through T_{12} pass anteriorly as the **intercostal nerves** to supply the muscles of intercostal spaces, and the skin and muscles of the anterior and lateral trunk (see Figure 15.5b). The ventral rami of all other spinal nerves form complex networks of nerves called **nerve plexuses.** These plexuses primarily serve the muscles and skin of the limbs. The fibers

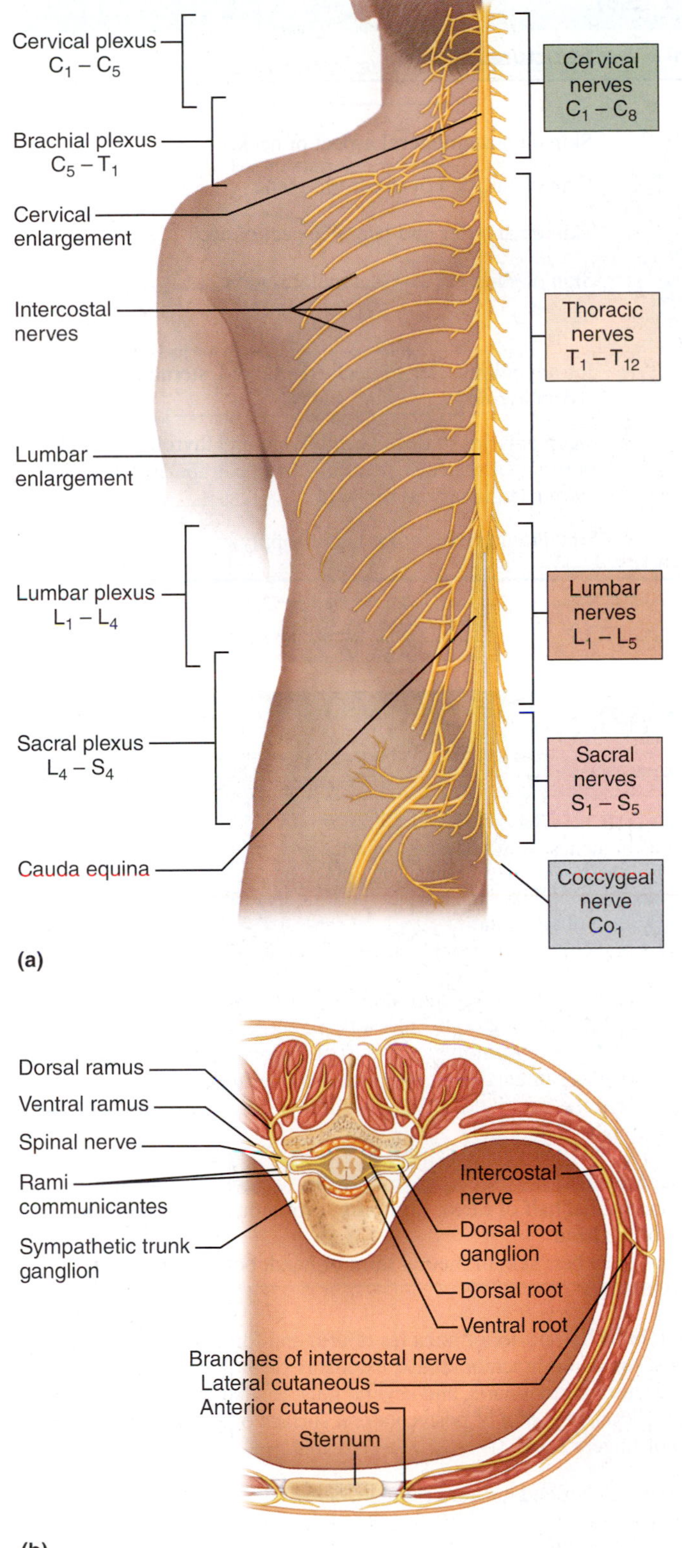

Figure 15.5 Human spinal nerves. (a) Spinal nerves are shown at right; ventral rami and the major nerve plexuses are shown at left. **(b)** Relative distribution of the ventral and dorsal rami of a spinal nerve (cross section of left thorax).

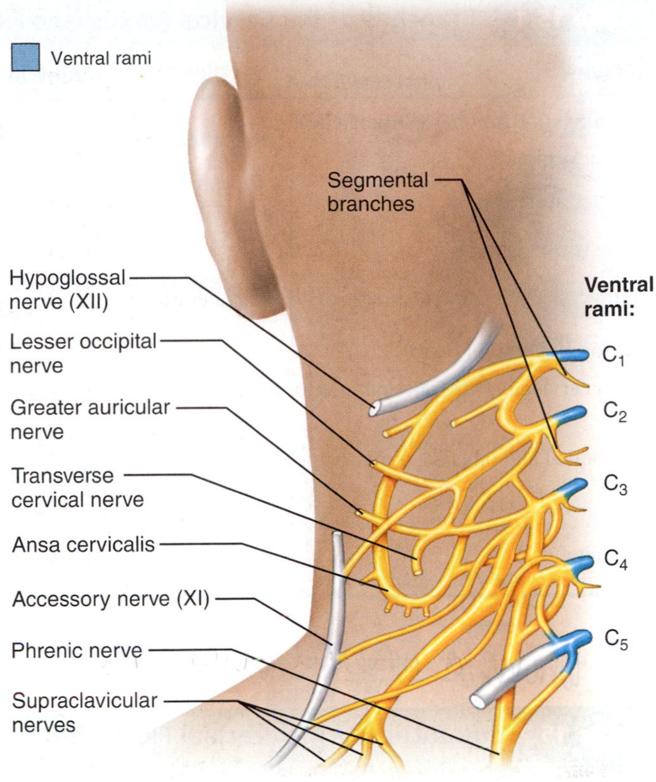

Figure 15.6 The cervical plexus. The nerves colored gray connect to the plexus but do not belong to it. (See Table 15.1.)

of the ventral rami unite in the plexuses (with a few rami supplying fibers to more than one plexus). From the plexuses the fibers diverge again to form peripheral nerves, each of which contains fibers from more than one spinal nerve. The four major nerve plexuses and their chief peripheral nerves are described in the tables (**Tables 15.1**, **15.2**, **15.3**, and **15.4**) and illustrated in the figures (**Figures 15.6**, **15.7**, **15.8**, and **15.9**). Their names and sites of origin should be committed to memory. The tiny S_5 and Co_1 spinal nerves contribute to a small plexus that serves part of the pelvic floor.

Cervical Plexus and the Neck

The **cervical plexus** (Figure 15.6 and Table 15.1) arises from the ventral rami of C_1 through C_5 to supply muscles of the shoulder and neck. The major motor branch of this plexus is the **phrenic nerve,** which arises from C_3 through C_4 (plus some fibers from C_5) and passes into the thoracic cavity in front of the first rib to innervate the diaphragm. The primary danger of a broken neck is that the phrenic nerve may be severed, leading to paralysis of the diaphragm and cessation of breathing. A jingle to help you remember the rami (roots) forming the phrenic nerves is "C_3, C_4, C_5 keep the diaphragm alive."

Brachial Plexus and the Upper Limb

The **brachial plexus** is large and complex, arising from the ventral rami of C_5 through C_8 and T_1 (Table 15.2). The plexus,

(Text continues on page 288.)

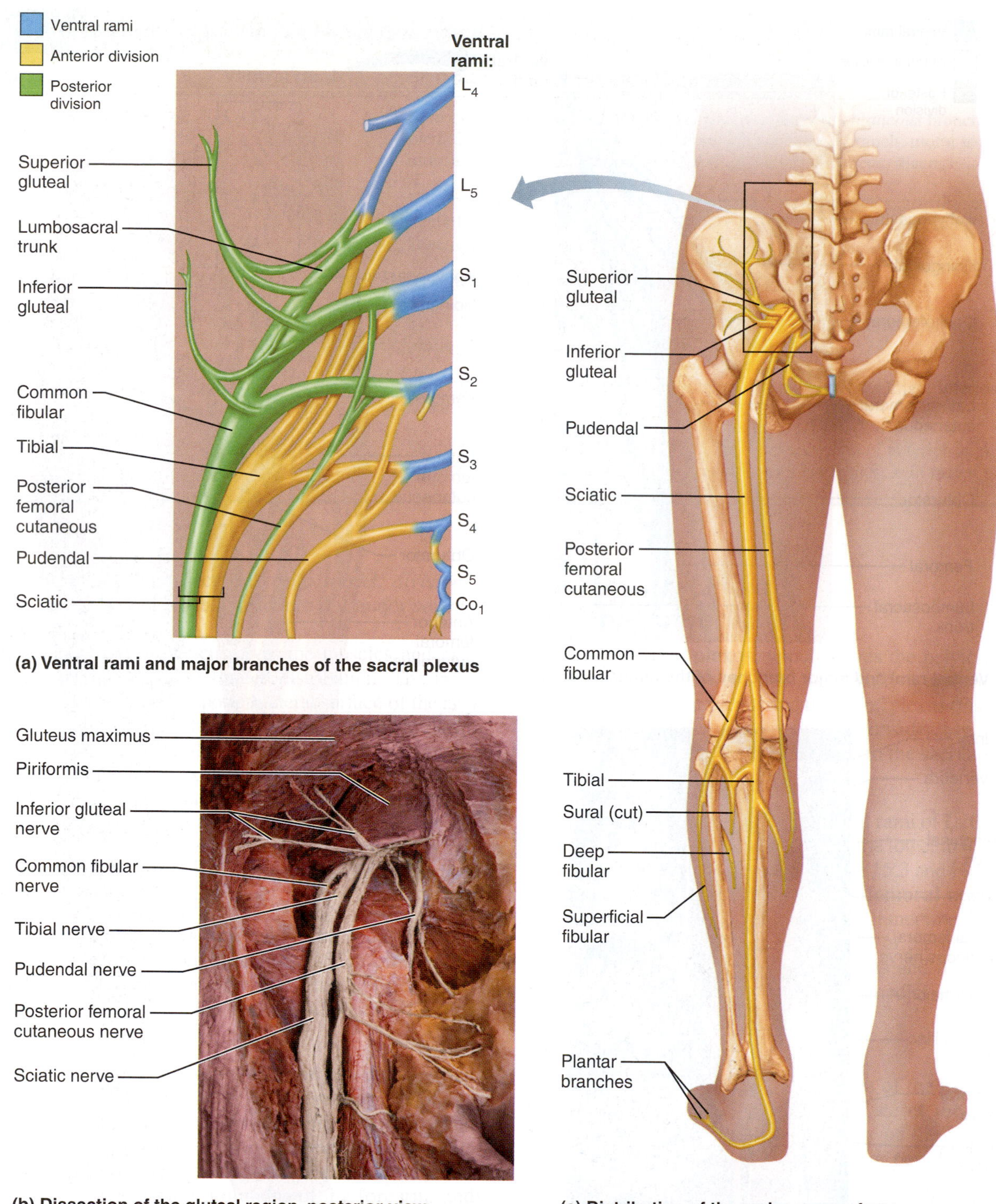

(a) Ventral rami and major branches of the sacral plexus

(b) Dissection of the gluteal region, posterior view

(c) Distribution of the major nerves from the sacral plexus to the lower limb

Figure 15.9 The sacral plexus (posterior view). Illustrations **(a)** and **(c)**. **(b)** Photograph of the sacral plexus from a cadaver. (See Table 15.4)

Table 15.4 Branches of the Sacral Plexus (see Figure 15.9)

Nerves	Ventral rami	Structures served
Sciatic nerve	L_4, L_5, S_1–S_3	Composed of two nerves (tibial and common fibular) in a common sheath; they diverge just proximal to the knee
• Tibial (including sural branch and medial and lateral plantar branches)	L_4–S_3	Cutaneous branches: to skin of posterior surface of leg and sole of foot Motor branches: to muscles of back of thigh, leg, and foot (hamstrings [except short head of biceps femoris], posterior part of adductor magnus, triceps surae, tibialis posterior, popliteus, flexor digitorum longus, flexor hallucis longus, and intrinsic muscles of foot)
• Common fibular (superficial and deep branches)	L_4–S_2	Cutaneous branches: to skin of anterior and lateral surface of leg and dorsum of foot Motor branches: to short head of biceps femoris of thigh, fibularis muscles of lateral leg, tibialis anterior, and extensor muscles of toes (extensor hallucis longus, extensors digitorum longus and brevis)
Superior gluteal	L_4, L_5, S_1	Motor branches: to gluteus medius and minimus and tensor fasciae latae
Inferior gluteal	L_5–S_2	Motor branches: to gluteus maximus
Posterior femoral cutaneous	S_1–S_3	Skin of buttock, posterior thigh, and popliteal region; length variable; may also innervate part of skin of calf and heel
Pudendal	S_2–S_4	Supplies most of skin and muscles of perineum (region encompassing external genitalia and anus and including clitoris, labia, and vaginal mucosa in females, and scrotum and penis in males); external anal sphincter

ACTIVITY 3

Identifying the Major Nerve Plexuses and Peripheral Nerves

Identify each of the four major nerve plexuses and its major nerves (Figures 15.6 to 15.9) on a large laboratory chart or model. Trace the courses of the nerves, and relate those observations to the information provided (Tables 15.1 to 15.4). ■

DISSECTION AND IDENTIFICATION

Cat Spinal Nerves

The cat has 38 or 39 pairs of spinal nerves (as compared to 31 in humans). Of these, 8 are cervical, 13 thoracic, 7 lumbar, 3 sacral, and 7 or 8 caudal.

A complete dissection of the cat's spinal nerves would be extraordinarily time-consuming and exacting and is not warranted in a basic anatomy course. However, it is desirable for you to have some dissection work to complement your study of the anatomical charts. Thus, at this point you will carry out a partial dissection of the brachial plexus and lumbosacral plexus and identify some of the major nerves. ■

ACTIVITY 4

Dissecting Nerves of the Brachial Plexus

1. Don disposable gloves. Place your cat specimen on the dissecting tray, dorsal side down. Reflect the cut ends of the left pectoralis muscles to expose the large brachial plexus in the axillary region **(Figure 15.10)**. Carefully clean the exposed nerves as far back toward their points of origin as possible.

2. The **musculocutaneous nerve** is the most superior nerve of this group. It splits into two subdivisions that run under the margins of the coracobrachialis and biceps brachii muscles. Trace its fibers into the ventral muscles of the arm it serves.

3. Locate the large **radial nerve** inferior to the musculocutaneous nerve. The radial nerve serves the dorsal muscles of the arm and forearm. Follow it into the three heads of the triceps brachii muscle.

4. In the cat, the **median nerve** is closely associated with the brachial artery and vein (Figure 15.10). It courses through the arm to supply the ventral muscles of the forearm (with the exception of the flexor carpi ulnaris and the ulnar head of the flexor digitorum profundus). It also innervates some of the intrinsic hand muscles, as in humans.

5. The **ulnar nerve** is the most posterior of the large brachial plexus nerves. Follow it as it travels down the forelimb, passing over the medial epicondyle of the humerus, to supply the flexor carpi ulnaris, the ulnar head of the flexor digitorum profundus, and the hand muscles. ■

ACTIVITY 5

Dissecting Nerves of the Lumbosacral Plexus

1. To locate the **femoral nerve** arising from the lumbar plexus, first identify the right *femoral triangle,* which is bordered

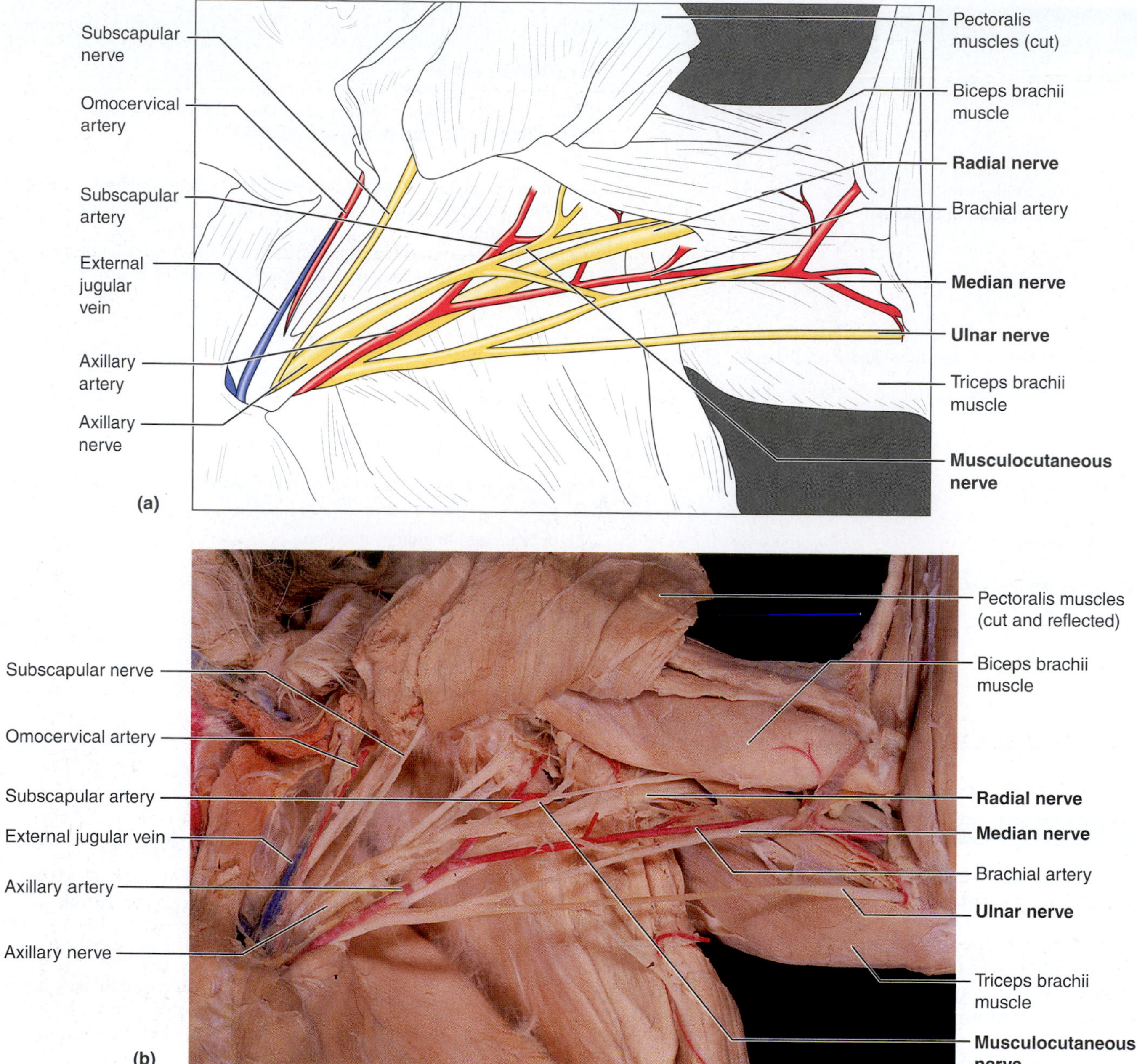

Figure 15.10 Brachial plexus and major blood vessels of the left forelimb of the cat, ventral aspect. (a) Diagram view. **(b)** Photograph.

by the sartorius and adductor muscles of the anterior thigh **(Figure 15.11)**. The large femoral nerve travels through this region after emerging from the psoas major muscle in close association with the femoral artery and vein. Follow the nerve into the muscles and skin of the anterior thigh, which it supplies. Notice also its cutaneous branch in the cat, the **saphenous nerve,** which continues down the anterior medial surface of the thigh with the great saphenous artery and vein to supply the skin of the anterior shank and foot.

2. Turn the cat ventral side down so you can view the posterior aspect of the lower limb **(Figure 15.12)**. Reflect the ends of the transected biceps femoris muscle to view the large cordlike sciatic nerve. The **sciatic nerve** arises from the sacral plexus and serves the dorsal thigh muscles and all the muscles of the leg and foot. Follow the nerve as it travels down the posterior thigh lateral to the semimembranosus muscle. Note that just superior to the gastrocnemius muscle of the calf, it divides into its two major branches, which serve the leg.

3. Identify the **tibial nerve** medially and the **common fibular (peroneal) nerve,** which curves over the lateral surface of the gastrocnemius.

4. Before you leave the laboratory, follow the boxed instructions to prepare your cat for storage and to clean the area (page 217). ■

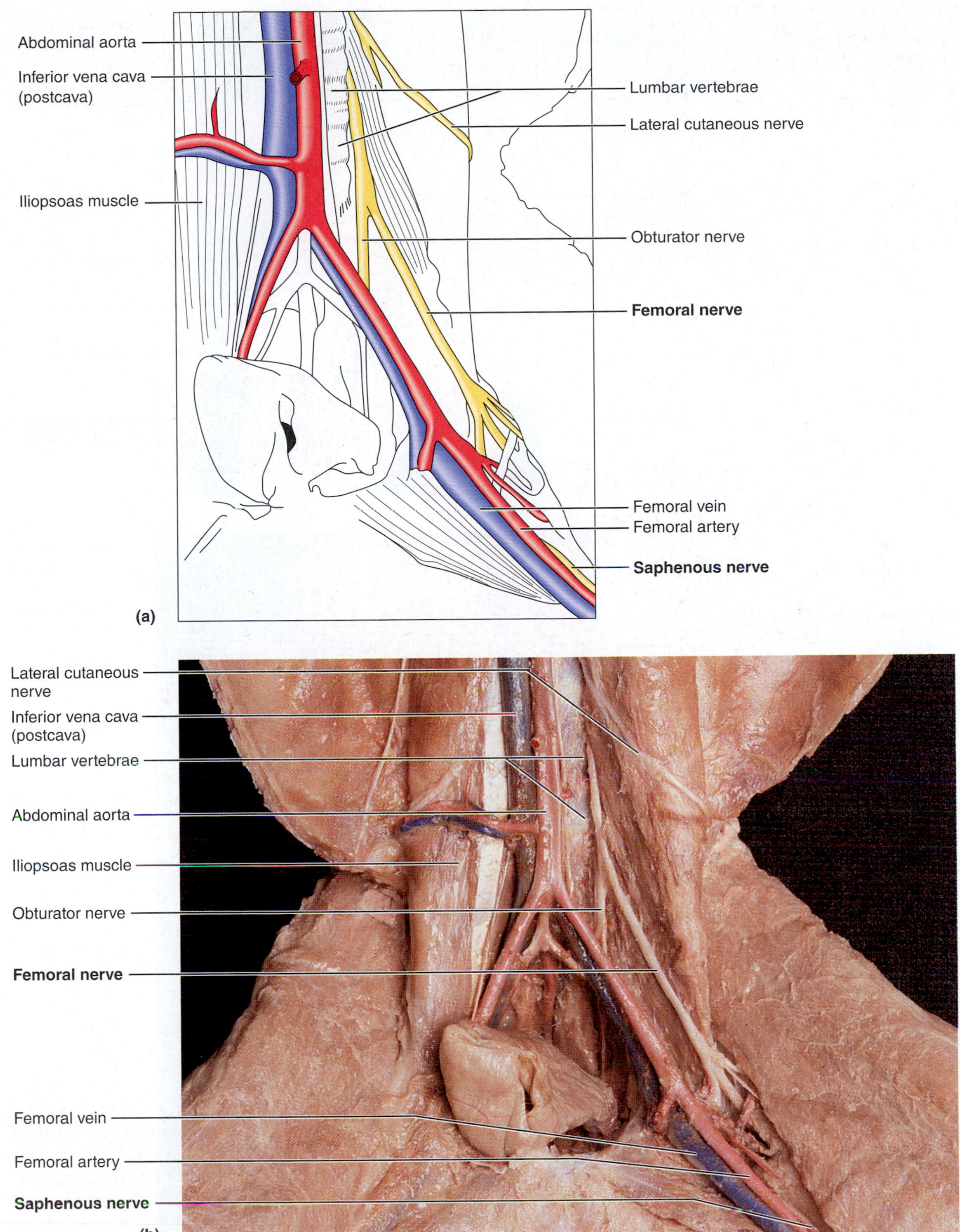

Figure 15.11 Lumbar plexus of the cat, ventral aspect. (a) Diagram view. **(b)** Photograph.

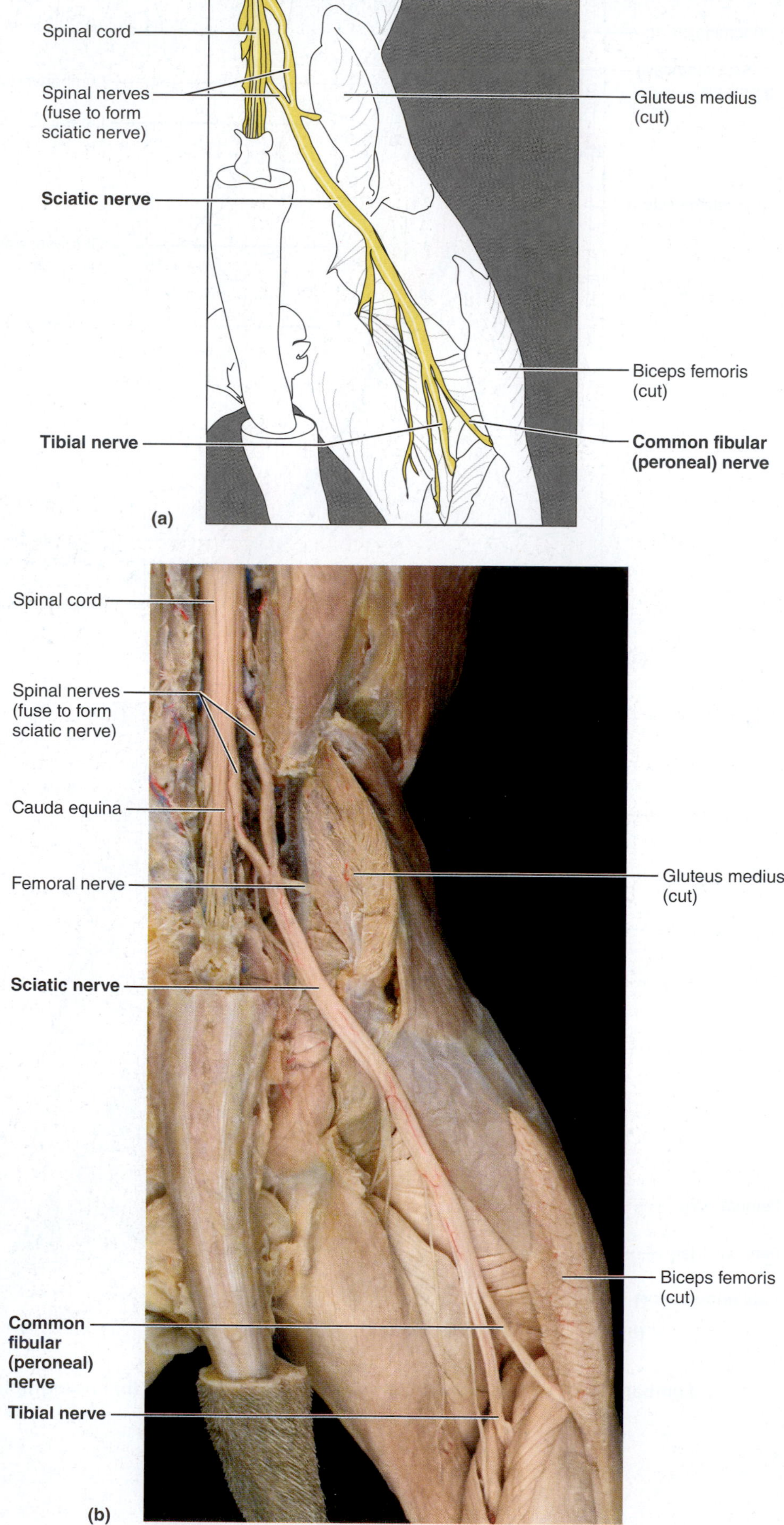

Figure 15.12 Sacral plexus of the cat, dorsal aspect. **(a)** Diagram view. **(b)** Photograph.

NAME ___________________________

LAB TIME/DATE ___________________

EXERCISE 15

REVIEW SHEET

The Spinal Cord and Spinal Nerves

Anatomy of the Spinal Cord

1. Match each description with the proper anatomical term from the key.

Key: a. cauda equina b. conus medullaris c. filum terminale d. foramen magnum

__________ 1. most superior boundary of the spinal cord

__________ 2. meningeal extension beyond the spinal cord terminus

__________ 3. spinal cord terminus

__________ 4. collection of spinal nerves traveling in the vertebral canal below the terminus of the spinal cord

2. Match the letters on the diagram with the following terms.

________ 1. arachnoid mater

________ 2. central canal

________ 3. dorsal horn

________ 4. dorsal ramus of spinal nerve

________ 5. dorsal root ganglion

________ 6. dorsal root of spinal nerve

________ 7. dura mater

________ 8. gray commissure

________ 9. lateral horn

________ 10. pia mater

________ 11. spinal nerve

________ 12. ventral horn

________ 13. ventral ramus of spinal nerve

________ 14. ventral root of spinal nerve

________ 15. white matter

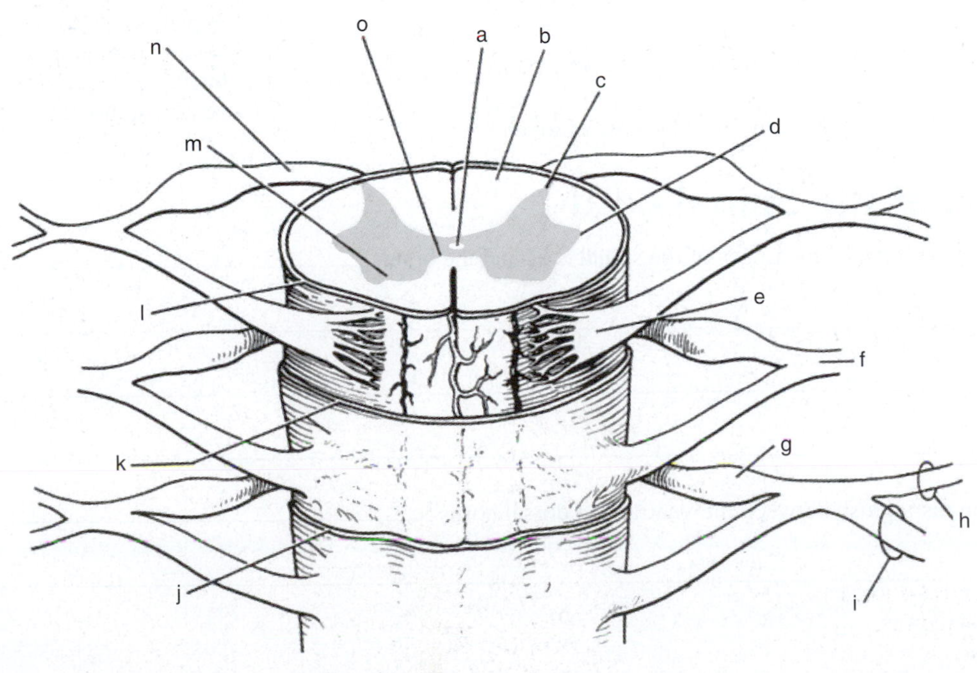

3. Choose the proper answer from the key to respond to the descriptions relating to spinal cord anatomy. (Some terms are used more than once.)

Key: a. sensory b. motor c. both sensory and motor d. interneurons

_____________ 1. neuron type found in dorsal horn

_____________ 2. neuron type found in ventral horn

_____________ 3. neuron type in dorsal root ganglion

_____________ 4. fiber type in ventral root

_____________ 5. fiber type in dorsal root

_____________ 6. fiber type in spinal nerve

4. Where in the vertebral column is a lumbar puncture generally done? _____________

Why is this the site of choice? _____________

____

5. The spinal cord is enlarged in two regions, the _____________ and the _____________ regions.

What is the significance of these enlargements? _____________

____

6. How does the position of the gray and white matter differ in the spinal cord and the cerebral hemispheres?

____

____

7. From the key on the right, choose the name of the tract that might be damaged when the following conditions are observed. (More than one choice may apply. Some terms are used more than once.)

_____________ 1. uncoordinated movement

_____________ 2. lack of voluntary movement

_____________ 3. tremors, jerky movements

_____________ 4. diminished pain perception

_____________ 5. diminished sense of touch

Key:
a. dorsal columns (fasciculus cuneatus and fasciculus gracilis)
b. lateral corticospinal tract
c. ventral corticospinal tract
d. tectospinal tract
e. rubrospinal tract
f. vestibulospinal tract
g. lateral spinothalamic tract
h. ventral spinothalamic tract

Dissection of the Spinal Cord

8. Compare and contrast the meninges of the spinal cord and the brain. _____________

____

____

____

9. How can you distinguish between the ventral and dorsal horns? _____________

____

Spinal Nerves and Nerve Plexuses

10. In the human, there are 31 pairs of spinal nerves, named according to the region of the vertebral column from which they issue. The spinal nerves are named below. Indicate how they are numbered.

cervical nerves ______________________ sacral nerves ______________________

lumbar nerves ______________________ thoracic nerves ______________________

11. The ventral rami of spinal nerves C_1 through T_1 and T_{12} through S_4 take part in forming ______________________, which serve the ______________________ of the body. The ventral rami of T_2 through T_{12} run between the ribs to serve the ______________________. The dorsal rami of the spinal nerves serve ______________________.

12. What would happen if the following structures were damaged or transected? (Use the key choices for responses.)

Key: a. loss of motor function b. loss of sensory function c. loss of both motor and sensory function

__________ 1. dorsal root of a spinal nerve

__________ 2. ventral root of a spinal nerve

__________ 3. ventral ramus of a spinal nerve

13. Define *plexus.* ______________________

____

14. Name the major nerves that serve the following body areas.

______________________ 1. head, neck, shoulders (name plexus only)

______________________ 2. diaphragm

______________________ 3. posterior thigh

______________________ 4. leg and foot (name two)

______________________ 5. anterior forearm muscles (name two)

______________________ 6. arm muscles (name two)

______________________ 7. abdominal wall (name plexus only)

______________________ 8. anterior thigh

______________________ 9. medial side of the hand

Dissection and Identification: Cat Spinal Nerves

15. From anterior to posterior, put the nerves issuing from the brachial plexus in their proper order (i.e., the median, musculocutaneous, radial, and ulnar nerves).

____

16. Which of the nerves named above serves the cat's forearm extensor muscles? ____________________

Which serves the forearm flexors? ____________________

17. Just superior to the gastrocnemius muscle, the sciatic nerve divides into its two main branches, the ____________________

and ____________________ nerves.

18. What name is given to the cutaneous nerve of the cat's thigh? ____________________

16

EXERCISE

The Autonomic Nervous System

MATERIALS

- ☐ Laboratory chart or three-dimensional model of the sympathetic trunk (chain)

OBJECTIVES

1. Identify the site of origin and the function of the sympathetic and parasympathetic divisions of the autonomic nervous system.
2. State how the autonomic nervous system differs from the somatic nervous system.
3. Identify the neurotransmitters associated with the sympathetic and parasympathetic fibers.

PRE-LAB QUIZ

1. The _____________ nervous system is the subdivision of the peripheral nervous system that regulates body activities that are generally *not* under conscious control.
   a. autonomic c. somatic
   b. cephalic d. vascular
2. Circle the correct underlined term. The parasympathetic division of the autonomic nervous system is also known as the craniosacral / thoracolumbar division.
3. Circle True or False. Cholinergic fibers release epinephrine.
4. The _____________ division of the autonomic nervous system is responsible for the "fight-or-flight" response because it adapts the body for extreme conditions such as exercise.
5. Circle True or False. Preganglionic fibers of the autonomic nervous system release acetylcholine.

The **autonomic nervous system (ANS)** is the subdivision of the peripheral nervous system (PNS) that regulates body activities that are generally not under conscious control. It is composed of a special group of motor neurons serving smooth muscle, cardiac muscle, and glands. The ANS is also called the *involuntary nervous system*, which reflects its subconscious control.

There is a basic anatomical difference between the motor pathways of the **somatic** (voluntary) **nervous system,** which innervates the skeletal muscles, and those of the autonomic nervous system. In the somatic division, the cell bodies of the motor neurons reside in the brain stem or ventral horns of the spinal cord, and their axons, sheathed in cranial or spinal nerves, extend directly to the skeletal muscles they serve. However, the autonomic nervous system consists of chains of two motor neurons. The first motor neuron of each pair, called the *preganglionic neuron,* resides in the brain stem or lateral horn of the spinal cord. Its axon leaves the central nervous system to synapse with the second motor neuron *(postganglionic neuron),* whose cell body is located in a ganglion outside the CNS. The axon of the postganglionic neuron then extends to the organ it serves.

The ANS has two major functional subdivisions **(Figure 16.1)**, the sympathetic and parasympathetic divisions. Both serve most of the same organs, but generally cause opposing or antagonistic effects.

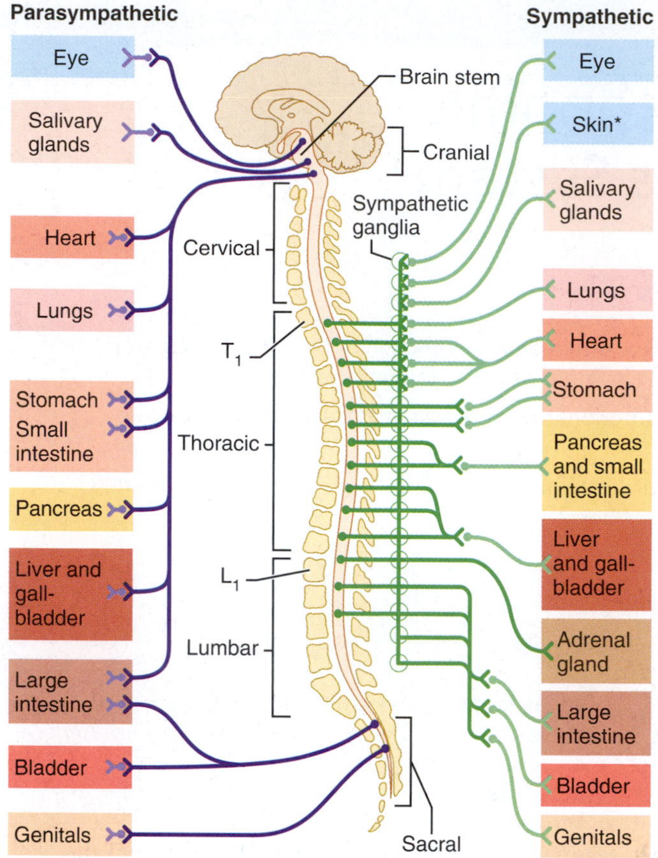

Figure 16.1 Overview of the subdivisions of the autonomic nervous system. The parasympathetic and sympathetic divisions differ anatomically in (1) the sites of origin of their nerves, (2) the relative lengths of preganglionic and postganglionic fibers, and (3) the locations of the ganglia. Although sympathetic innervation to the skin (*) is shown only for the cervical area, all nerves to the periphery carry postganglionic sympathetic fibers.

Autonomic Functioning

Most body organs served by the autonomic nervous system receive fibers from both the sympathetic and parasympathetic divisions. The only exceptions are the structures of the skin (sweat glands and arrector pili muscles attached to the hair follicles), the adrenal medulla, and essentially all blood vessels except those of the external genitalia, all of which receive sympathetic innervation only. When both divisions serve an organ, they usually have opposite effects. This is because their postganglionic axons release different neurotransmitters. The parasympathetic fibers, called **cholinergic fibers,** release acetylcholine; the sympathetic postganglionic fibers, called **adrenergic fibers,** release norepinephrine. However, there are isolated examples of postganglionic sympathetic fibers, such as those serving sweat glands and some blood vessels, that release acetylcholine. The preganglionic fibers of both divisions release acetylcholine.

The parasympathetic division is often referred to as the housekeeping, or "rest-and-digest," system because it maintains the visceral organs in a state most suitable for normal functions and internal homeostasis; that is, it promotes normal digestion and elimination. In contrast, activation of the sympathetic division is referred to as the "fight-or-flight" response because it readies the body to cope with situations that threaten homeostasis. Under such emergency conditions, the sympathetic nervous system induces an increase in heart rate and blood pressure, dilates the bronchioles of the lungs, increases blood sugar levels, and promotes many other effects that help the individual cope with a stressor.

As we grow older, our sympathetic nervous system gradually becomes less and less efficient, particularly in causing vasoconstriction of blood vessels. When elderly people stand up quickly after sitting or lying down, they often become light-headed or faint. This is because the sympathetic nervous system is not able to react quickly enough to counteract the pull of gravity by activating the vasoconstrictor fibers. So, blood pools in the feet. This condition, **orthostatic hypotension,** is a type of low blood pressure resulting from changes in body position as described. Orthostatic hypotension can be prevented to some degree if the person changes position *slowly.* This gives the sympathetic nervous system a little more time to react and adjust. +

Parasympathetic (Craniosacral) Division

The preganglionic neurons of the **parasympathetic,** or **craniosacral,** division are located in brain stem nuclei of cranial nerves III, VII, IX, X and in the S_2 through S_4 level of the spinal cord. The axons of preganglionic neurons of the cranial region travel in their respective cranial nerves to the *immediate area* of the head and neck organs to be stimulated. There they synapse with postganglionic neurons. The postganglionic neuron then sends out a very short postganglionic axon to the organ it serves. In the sacral region, the preganglionic axons leave the ventral roots of the spinal cord and collectively form the **pelvic splanchnic nerves,** which travel to the pelvic cavity. In the pelvic cavity, the preganglionic axons synapse with the postganglionic neurons in ganglia located on or close to the organs served.

Sympathetic (Thoracolumbar) Division

The preganglionic neurons of the **sympathetic,** or **thoracolumbar,** division are located in the lateral horns of the gray matter of the spinal cord from T_1 through L_2. The preganglionic axons leave the cord via the ventral root with the axons of the somatic motor neurons. They enter the spinal nerve, and then travel briefly in the ventral ramus **(Figure 16.2)**. From the ventral ramus, they pass through a small branch called the **white ramus communicans** to enter a **sympathetic trunk ganglion.** These two trunks, or *chains,* lie alongside the vertebral column and are also called *paravertebral ganglia.*

Having reached the ganglion, a preganglionic axon may take one of three main courses (see Figure 16.2b). First, it may synapse with a postganglionic neuron in the sympathetic trunk at that level. Second, the axon may travel upward or downward through the sympathetic trunk to synapse with a

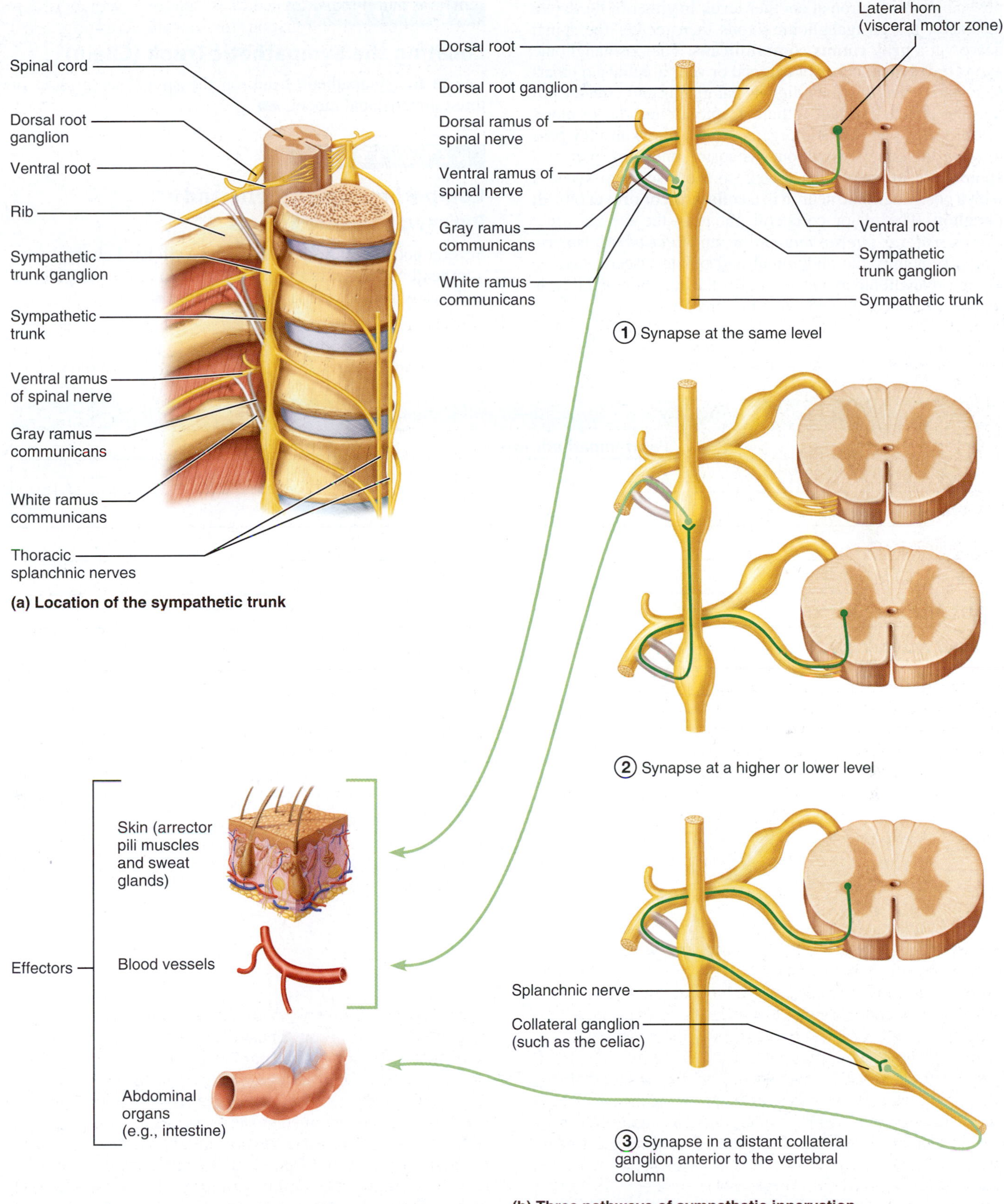

Figure 16.2 Sympathetic trunks and pathways. (a) Diagram of the right sympathetic trunks in the posterior thorax. **(b)** Synapses between preganglionic and postganglionic neurons can occur at three different places.

postganglionic neuron at another level. In either of these two instances, the postganglionic axons then reenter the spinal nerve via a **gray ramus communicans.** The postganglionic axons travel in branches of a dorsal or ventral ramus to innervate skin structures (including sweat glands and arrector pili muscles attached to hair follicles), smooth muscles of blood vessel walls, and thoracic organs. Third, the axon may pass through the ganglion without synapsing and form part of a **splanchnic nerve,** which travels to the viscera to synapse with a postganglionic neuron in a **collateral,** or **prevertebral, ganglion.** The major collateral ganglia—the *celiac, superior mesenteric, inferior mesenteric,* and *inferior hypogastric ganglia*—supply the abdominal and pelvic visceral organs. The postganglionic axon then leaves the ganglion and travels to a nearby visceral organ which it innervates.

ACTIVITY 1

Locating the Sympathetic Trunk (Chain)

Locate the sympathetic trunk on the spinal nerve chart or three-dimensional model.

ACTIVITY 2

Comparing Sympathetic and Parasympathetic Effects

Several body organs are listed in the **Activity 2 chart.** Using your textbook as a reference, list the effect of the sympathetic and parasympathetic divisions on each.

Activity 2: Parasympathetic and Sympathetic Effects

Organ	Parasympathetic effect	Sympathetic effect
Heart		
Bronchioles of lungs		
Digestive tract		
Urinary bladder		
Iris of the eye		
Blood vessels (most)		
Penis/clitoris		
Sweat glands		
Adrenal medulla		
Pancreas		

NAME ______________________

LAB TIME/DATE ______________________

EXERCISE 16

REVIEW SHEET

The Autonomic Nervous System

The Autonomic Nervous System

1. For the most part, sympathetic and parasympathetic fibers serve the same organs and structures. How can they exert antagonistic effects? (After all, nerve impulses are nerve impulses—aren't they?)

____

2. Name three structures that receive sympathetic but *not* parasympathetic innervation.

____

3. A pelvic splanchnic nerve contains (circle one):

   a. preganglionic sympathetic fibers

   b. postganglionic sympathetic fibers

   c. preganglionic parasympathetic fibers

   d. postganglionic parasympathetic fibers

4. The following chart states a number of conditions. Use a check mark to show which division of the autonomic nervous system is involved in each.

Sympathetic division	Condition	Parasympathetic division
	Secretes norepinephrine; adrenergic fibers	
	Secretes acetylcholine; cholinergic fibers	
	Long preganglionic axon; short postganglionic axon	
	Short preganglionic axon; long postganglionic axon	
	Arises from cranial and sacral nerves	
	Arises from spinal nerves T_1 through L_3	
	Normally in control	
	"Fight-or-flight" system	
	Has more specific control (Look it up!)	

17

EXERCISE

Special Senses: Anatomy of the Visual System

MATERIALS

- ☐ Chart of eye anatomy
- ☐ Dissectible eye model
- ☐ Prepared slide of longitudinal section of an eye showing retinal layers
- ☐ Compound microscope
- ☐ Preserved cow or sheep eye
- ☐ Dissecting instruments and tray
- ☐ Disposable gloves

OBJECTIVES

1. Identify the external, internal, and accessory anatomical structures of the eye on a model or appropriate image, and list the function(s) of each; identify the structural components that are present in a preserved sheep or cow eye (if available).
2. Define *conjunctivitis, cataract,* and *glaucoma*.
3. Describe the cellular makeup of the retina.
4. Explain the difference between rods and cones with respect to visual perception and retinal localization.
5. Trace the visual pathway to the primary visual cortex, and indicate the effects of damage to various parts of this pathway.

PRE-LAB QUIZ

1. Name the mucous membrane that lines the internal surface of the eyelids and continues over the anterior surface of the eyeball. ________________
2. How many extrinsic eye muscles are attached to the exterior surface of each eyeball?
   a. three   c. five
   b. four   d. six
3. The wall of the eye has three layers. The outermost fibrous layer is made up of the opaque white sclera and the transparent:
   a. choroids   c. cornea
   b. ciliary gland   d. lacrima
4. Circle the correct underlined term. The aqueous humor / vitreous humor is a clear, watery fluid that helps to maintain the intraocular pressure of the eye and provides nutrients for the avascular lens and cornea.
5. Circle True or False. At the optic chiasma, the fibers from the medial side of each eye cross over to the opposite side.

Anatomy of the Eye

External Anatomy and Accessory Structures

The adult human eye is a sphere measuring about 2.5 cm (1 inch) in diameter. Only about one-sixth of the eye's anterior surface is observable **(Figure 17.1)**; the remainder is enclosed and protected by a cushion of fat and the walls of the bony orbit.

The **lacrimal apparatus** consists of the lacrimal gland, lacrimal canaliculi, lacrimal sac, and the nasolacrimal duct. The **lacrimal glands** are situated superior to the lateral aspect of each eye. They continually liberate a dilute salt solution (tears) that flows onto the anterior surface of the eyeball through several small ducts. The tears flush across the eyeball and through the **lacrimal puncta,** the tiny openings of the **lacrimal canaliculi** medially, then into the **lacrimal sac,** and finally into the **nasolacrimal duct,** which empties into the nasal cavity. The

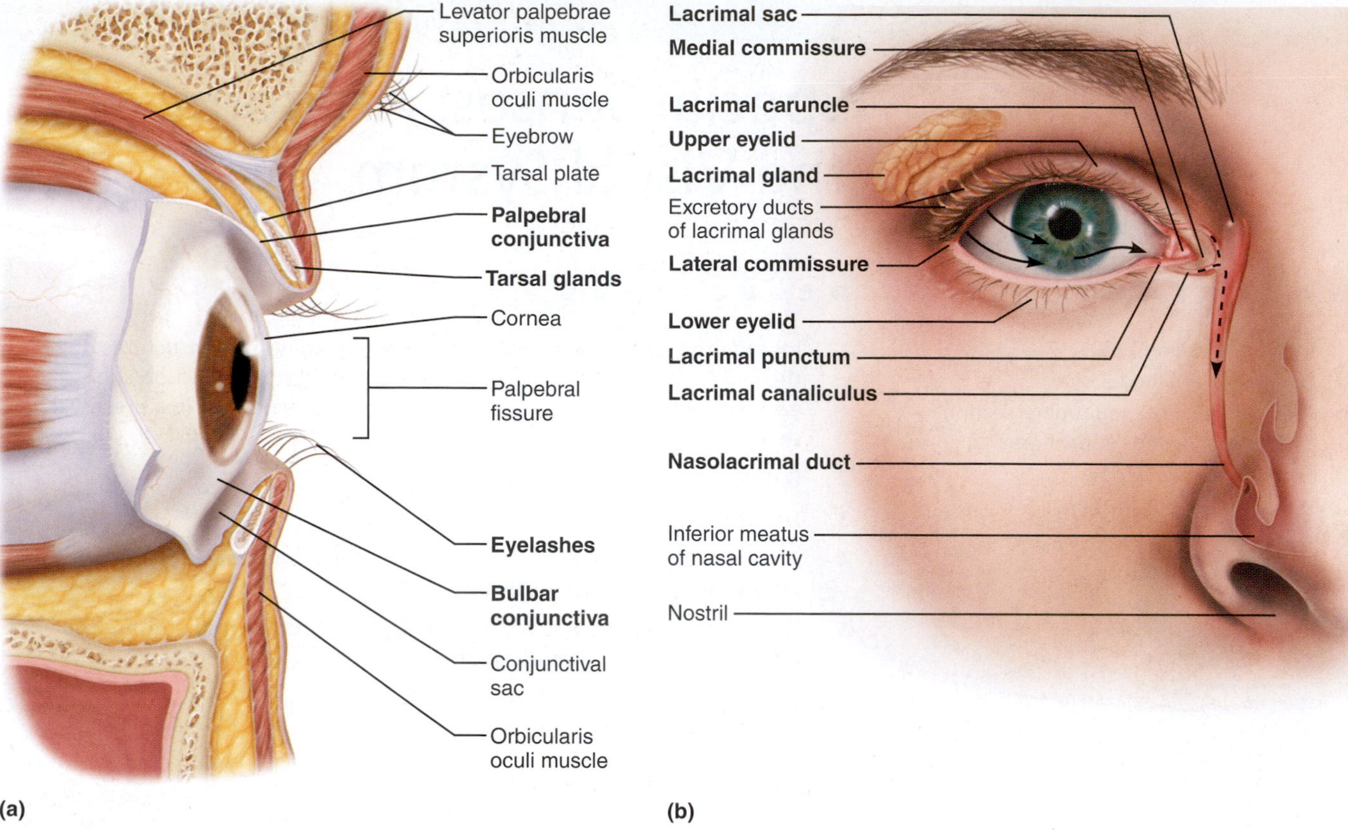

Figure 17.1 External anatomy of the eye and accessory structures. (a) Lateral view; some structures shown in sagittal section. **(b)** Anterior view with lacrimal apparatus.

lacrimal secretion also contains **lysozyme,** an antibacterial enzyme. Because it constantly flushes the eyeball, the lacrimal fluid cleanses and protects the eye surface as it moistens and lubricates it. As we age, tear production decreases, causing our eyes to become dry and making them more vulnerable to bacterial invasion and irritation.

The anterior surface of each eye is protected by the **eyelids, or palpebrae.** (See Figure 17.1.) The medial and lateral junctions of the upper and lower eyelids are referred to as the **medial** and **lateral commissures,** respectively. The **caruncle,** a fleshy elevation at the medial commissure, produces a whitish oily secretion. A mucous membrane, the **conjunctiva,** lines the internal surface of the eyelids (as the *palpebral conjunctiva*) and continues over the anterior surface of the eyeball to its junction with the corneal epithelium (as the *bulbar* or *ocular conjunctiva*). The conjunctiva secretes mucus, which aids in lubricating the eyeball. Inflammation of the conjunctiva, often accompanied by redness of the eye, is called **conjunctivitis.**

Projecting from the border of each eyelid is a row of short hairs, the **eyelashes.** The **ciliary glands,** modified sweat glands, lie between the eyelash hair follicles and help lubricate the eyeball. Small sebaceous glands associated with the hair follicles and the larger **tarsal glands,** located posterior to the eyelashes, secrete an oily substance. An inflammation of one of the ciliary glands or a small oil gland is called a **sty.**

Six **extrinsic eye muscles** attached to the exterior surface of each eyeball control eye movement and make it possible for the eye to follow a moving object. The names and positioning of these extrinsic muscles are noted in the figure **(Figure 17.2)**. Their actions are given in the chart accompanying that figure (Figure 17.2c).

ACTIVITY 1

Identifying Accessory Eye Structures

Using a chart of eye anatomy or the figure of the eye and accessory structures (Figure 17.1), observe the eyes of another student, and identify as many of the accessory structures as possible. Ask the student to look to the left. What extrinsic eye muscles are responsible for this action?

Right eye: ______________________________

Left eye: ______________________________

Internal Anatomy of the Eye

Anatomically, the wall of the eye is constructed of three layers **(Figure 17.3)**. The outermost **fibrous layer** is a protective layer composed of dense avascular connective tissue. It has two obviously different regions: the sclera and the cornea. The opaque white **sclera** forms the bulk of the fibrous layer and is observable anteriorly as the "white of the eye." Its anteriormost portion is modified structurally to form the transparent **cornea,** through which light enters the eye.

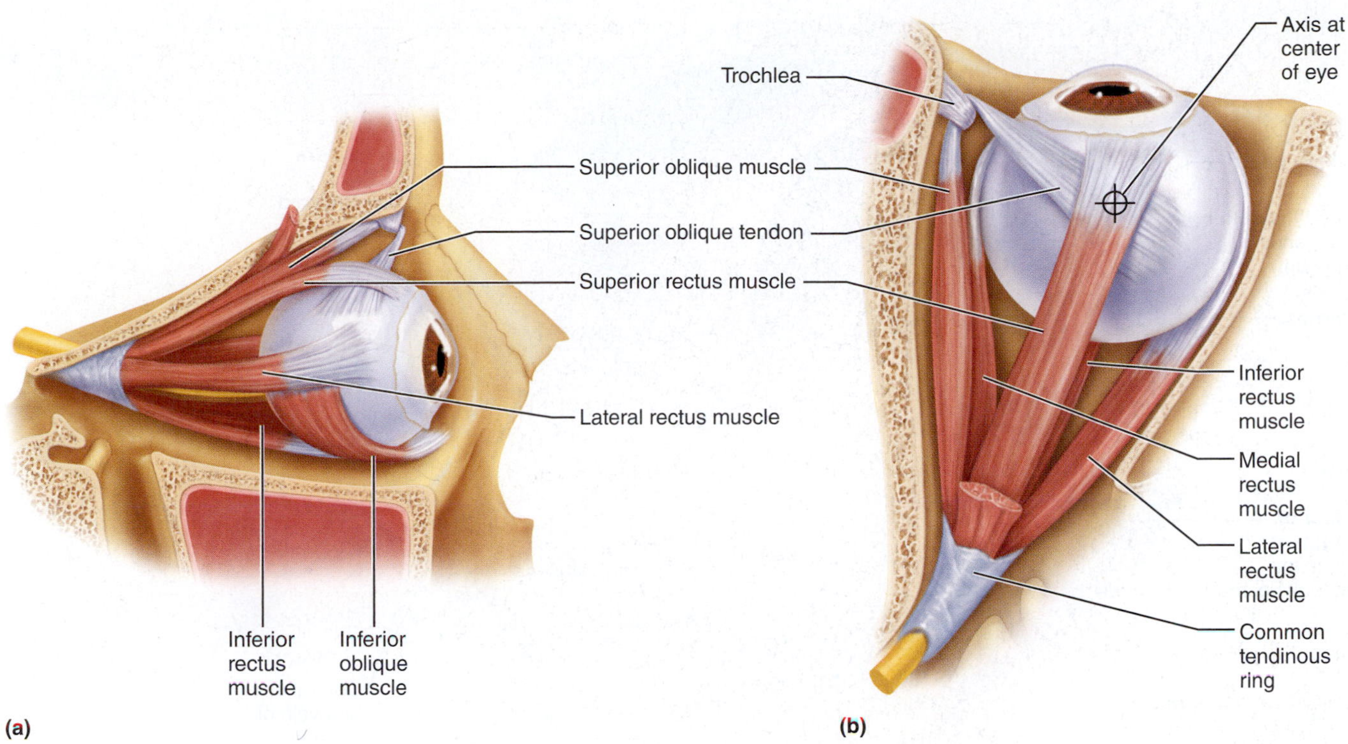

Muscle	Action	Controlling cranial nerve
Lateral rectus	Moves eye laterally	VI (abducens)
Medial rectus	Moves eye medially	III (oculomotor)
Superior rectus	Elevates eye and turns it medially	III (oculomotor)
Inferior rectus	Depresses eye and turns it medially	III (oculomotor)
Inferior oblique	Elevates eye and turns it laterally	III (oculomotor)
Superior oblique	Depresses eye and turns it laterally	IV (trochlear)

(c)

Figure 17.2 Extrinsic muscles of the eye. (a) Lateral view of the right eye. **(b)** Superior view of the right eye. **(c)** Summary of actions of the extrinsic eye muscles and cranial nerves that control them.

The middle layer is the **vascular layer,** also called the *uvea*. Its posteriormost part, the **choroid,** is a blood-rich nutritive layer containing a dark pigment that prevents light scattering within the eye. Anteriorly, the choroid is modified to form the **ciliary body,** which is chiefly composed of *ciliary muscles,* smooth muscles important in controlling lens shape, and **ciliary processes.** The ciliary processes secrete aqueous humor. The most anterior part of the vascular layer is the pigmented **iris.** The iris is incomplete, resulting in a rounded opening, the **pupil,** through which light passes.

The iris is composed of circularly and radially arranged smooth muscle fibers and acts as a reflexively activated diaphragm to regulate the amount of light entering the eye. In close vision and bright light, the **sphincter pupillae** *(circular)* muscles of the iris contract, and the pupil constricts. In distant vision and in dim light, the **dilator pupillae** *(radial)* muscles contract, enlarging (dilating) the pupil and allowing more light to enter the eye.

The innermost **sensory layer** of the eye is the delicate, two-layered **retina** (Figure 17.3 and **Figure 17.4**). The outer **pigmented layer** abuts the choroid and extends anteriorly to cover the ciliary body and the posterior side of the iris. The pigment cells, like those of the choroid, absorb light and prevent it from scattering in the eye. They also participate in renewing photoreceptor cells by acting as phagocytes, and they store vitamin A needed by the photoreceptor cells. The transparent inner **neural layer** extends anteriorly only to the ciliary body. It contains the photoreceptors, *rods* and *cones,* which begin the chain of electrical events that ultimately result in the transduction of light energy into nerve impulses that are transmitted to the primary visual cortex of the brain. Vision is the result. The photoreceptor cells are distributed over the entire neural retina, except where the optic nerve leaves the eyeball. This site, called the **optic disc,** or *blind spot,* is located in a weak spot in the **fundus** (posterior wall). Lateral to each blind spot, and directly posterior to the lens, is an area called the **macula lutea,** "yellow spot," an area of

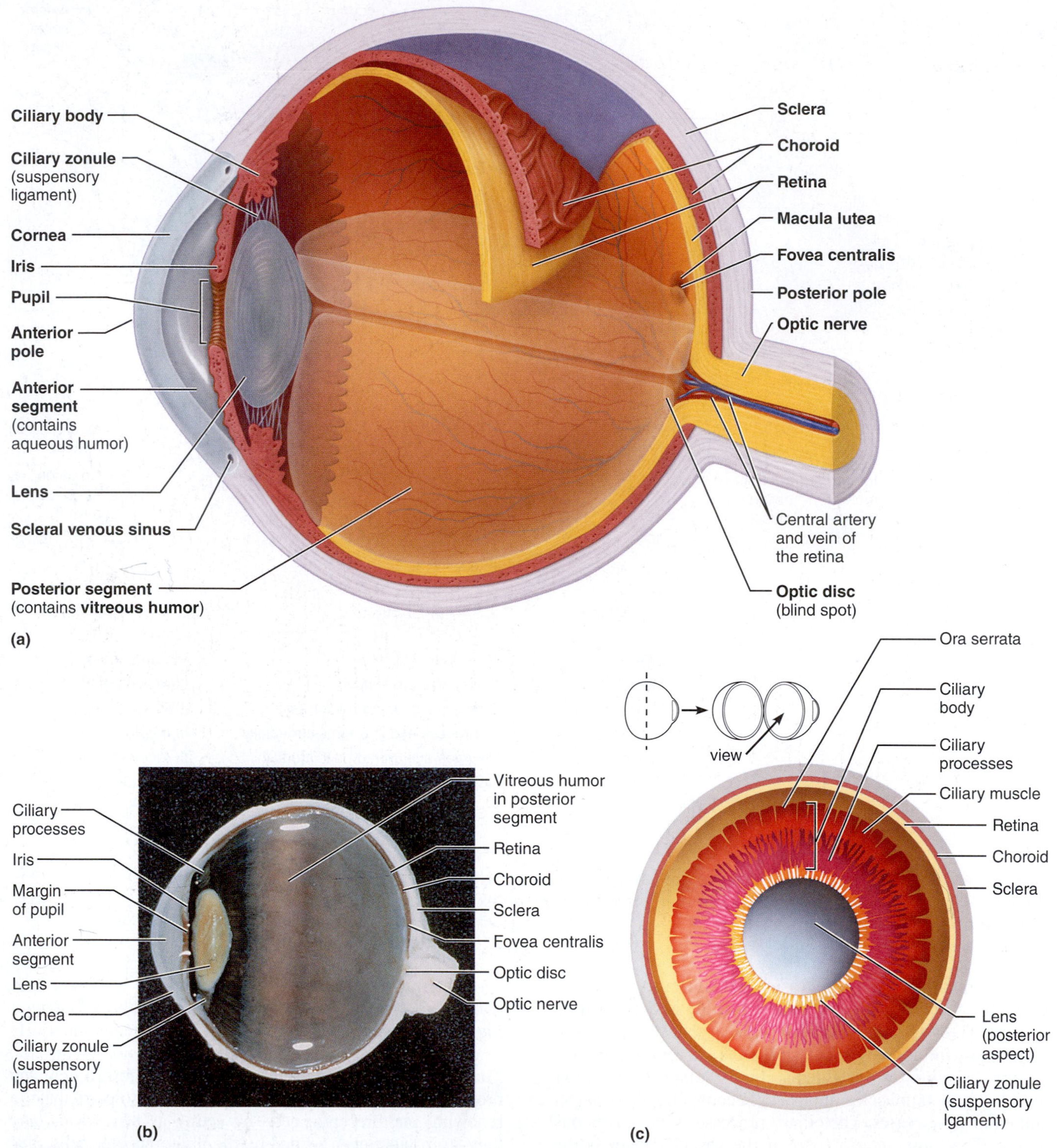

Figure 17.3 Internal anatomy of the eye. (a) Sagittal section, diagram view. The vitreous humor is illustrated only in the bottom half of the eyeball. **(b)** Photograph of a sagittal section of the human eye. **(c)** Posterior view of anterior half of the eye.

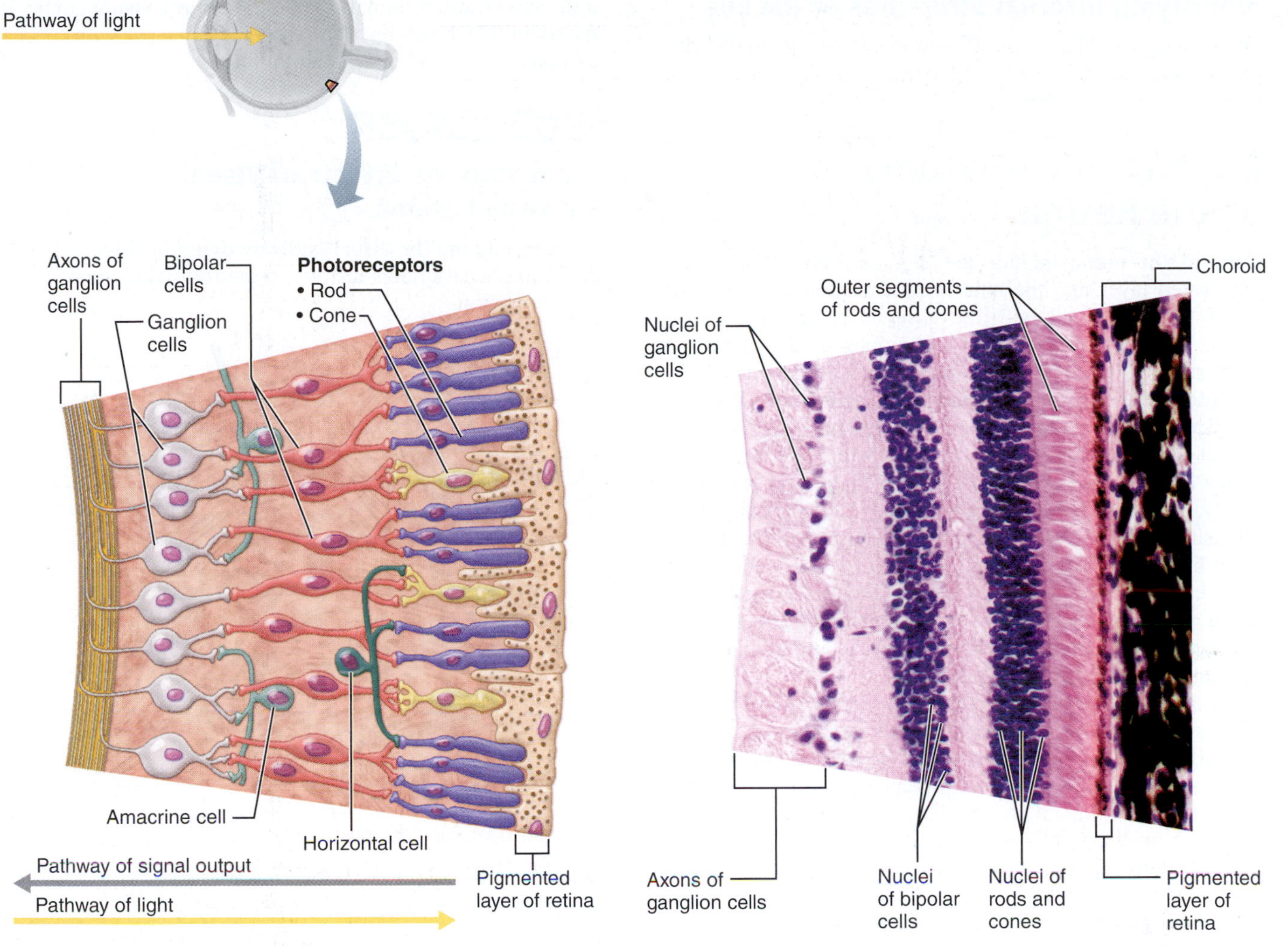

Figure 17.4 Microscopic anatomy of the retina. (a) Diagram view of cells of the neural retina. Note the pathway of light through the retina. Neural signals (output of the retina) flow in the opposite direction. **(b)** Photomicrograph of the retina (185×).

high cone density. In its center is the **fovea centralis,** a small pit about 0.4 mm in diameter, which contains only cones and is the area of greatest visual acuity. Focusing for discriminative vision occurs in the fovea centralis.

Light entering the eye is focused on the retina by the **lens,** a flexible crystalline structure held vertically in the eye's interior by the **ciliary zonule** *(suspensory ligament),* attached to the ciliary body. Activity of the ciliary muscle, which accounts for the bulk of ciliary body tissue, changes lens thickness to allow light to be properly focused on the retina.

In the elderly, the lens becomes increasingly hard and opaque. **Cataracts,** which often result from this process, cause vision to become hazy or entirely obstructed. +

The lens divides the eye into two segments: the **anterior segment** anterior to the lens, which contains a clear watery fluid called the **aqueous humor,** and the **posterior segment** behind the lens, filled with a gel-like substance, the **vitreous humor,** or **vitreous body.** The anterior segment is further divided into **anterior** and **posterior chambers,** located before and after the iris, respectively. The aqueous humor is continually formed by the capillaries of the **ciliary processes** of the ciliary body. It helps to maintain the intraocular pressure of the eye and provides nutrients for the avascular lens and cornea. The aqueous humor is reabsorbed into the **scleral venous sinus.** The vitreous humor provides the major internal reinforcement of the posterior part of the eyeball, and helps to keep the retina pressed firmly against the wall of the eyeball. It is formed *only* before birth.

Anything that interferes with drainage of the aqueous fluid increases intraocular pressure. When intraocular pressure reaches dangerously high levels, the retina and optic nerve are compressed, resulting in pain and possible blindness, a condition called **glaucoma.** +

ACTIVITY 2

Identifying Internal Structures of the Eye

Obtain a dissectible eye model and identify its internal structures described previously. As you work, also refer to the figure (Figure 17.3).

Microscopic Anatomy of the Retina

Cells of the retina include the pigment cells of the outer pigmented layer and the inner photoreceptors and *neurons,* which are in contact with the vitreous humor (see Figure 17.4). The inner neural layer is composed of three major populations. These are, from outer to inner aspect, the **photoreceptors,** the **bipolar cells,** and the **ganglion cells.**

The **rods** are the specialized receptors for dim light. Visual interpretation of their activity is in gray tones. The **cones** are color receptors that permit high levels of visual acuity, but they function only under conditions of high light intensity; thus, for example, no color vision is possible in moonlight. The fovea contains only cones; the macula contains mostly cones; and from the edge of the macula to the retina periphery, cone density declines gradually. By contrast, rods are most numerous in the periphery, and their density decreases as the macula is approached.

Light must pass through the ganglion cell layer and the bipolar cells to reach and excite the rods and cones. As a result of a light stimulus, the photoreceptors undergo changes in their membrane potential that influence the bipolar neurons. These in turn stimulate the ganglion cells, whose axons leave the retina in the tight bundle of fibers known as the **optic nerve** (Figure 17.3). The retinal layer is thickest where the optic nerve attaches to the eyeball because increasing numbers of ganglion cell axons converge at this point. It thins as it approaches the ciliary body. In addition to these three major cell types, the retina also contains other types of neurons, the *horizontal cells* and *amacrine cells,* which play a role in visual processing.

ACTIVITY 3

Studying the Microscopic Anatomy of the Retina

Use a compound microscope to examine a histologic slide of a longitudinal section of the eye. Identify the retinal layers by comparing your view to the photomicrograph (Figure 17.4b).

Visual Pathways to the Brain

The axons of the ganglion cells of the retina converge at the posterior aspect of the eyeball and exit from the eye as the optic nerve. At the **optic chiasma,** the fibers from the medial side of each eye cross over to the opposite side **(Figure 17.5)**. The fiber tracts thus formed are called the **optic tracts.** Each optic tract contains fibers from the lateral side of the eye on the same side and from the medial side of the opposite eye.

The optic tract fibers synapse with neurons in the **lateral geniculate nucleus** of the thalamus, whose axons form the **optic radiation,** terminating in the **primary visual cortex** in the occipital lobe of the brain. Here they synapse with cortical neurons, and visual interpretation begins.

ACTIVITY 4

Predicting the Effects of Visual Pathway Lesions

After examining the visual pathway diagram (Figure 17.5a), determine what effects lesions in the following areas would have on vision:

In the right optic nerve: ______________________________

____

Through the optic chiasma: ____________________________

____

In the left optic tract: _______________________________

____

In the right cerebral cortex (visual area): _______________

____

DISSECTION

The Cow (Sheep) Eye

1. Obtain a preserved cow or sheep eye, dissecting instruments, and a dissecting tray. Don disposable gloves.

2. Examine the external surface of the eye, noting the thick cushion of adipose tissue. Identify the optic nerve (cranial nerve II) as it leaves the eyeball, the remnants of the extrinsic eye muscles, the conjunctiva, the sclera, and the cornea. The normally transparent cornea is opalescent or opaque if the eye has been preserved. (Refer to **Figure 17.6** as you work).

3. Trim away most of the fat and connective tissue, but leave the optic nerve intact. Holding the eye with the cornea facing downward, carefully make an incision with a sharp scalpel into the sclera about 6 mm (¼ inch) above the cornea. (The sclera of the preserved eyeball is very tough, so you will have to apply substantial pressure to penetrate it.) Using scissors, complete the incision around the circumference of the eyeball paralleling the corneal edge.

4. Carefully lift the anterior part of the eyeball away from the posterior portion. If you have worked carefully, the vitreous body should remain with the posterior part of the eyeball.

5. Examine the anterior part of the eye, and identify the following structures:

Ciliary body: Black pigmented body that appears to be a halo encircling the lens.

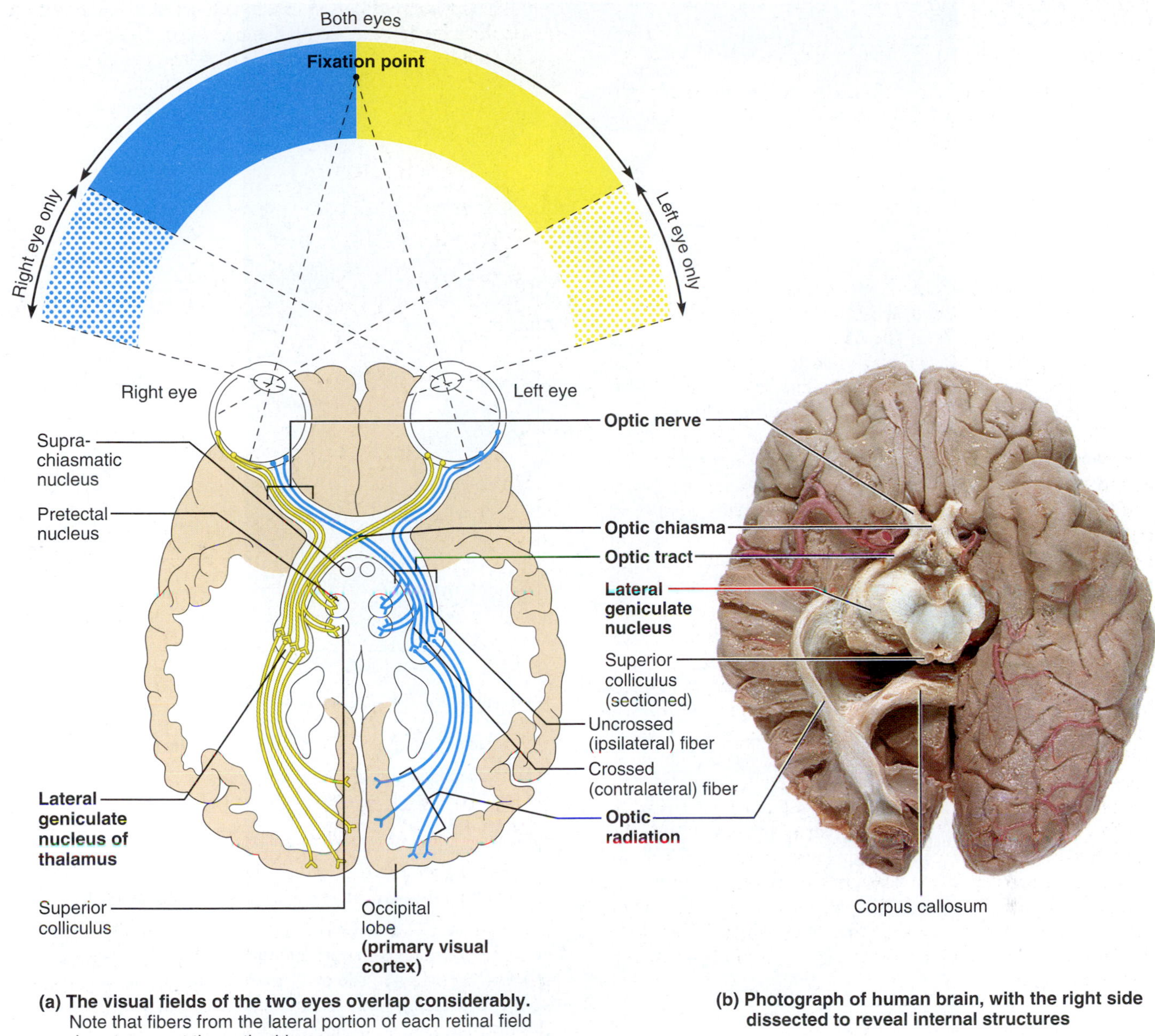

(a) The visual fields of the two eyes overlap considerably. Note that fibers from the lateral portion of each retinal field do not cross at the optic chiasma.

(b) Photograph of human brain, with the right side dissected to reveal internal structures

Figure 17.5 Visual pathway to the brain and visual fields, inferior view.

Lens: Biconvex structure that is opaque in preserved specimens.

Carefully remove the lens and identify the adjacent structures:

Iris: Anterior continuation of the ciliary body penetrated by the pupil.

Cornea: More convex anteriormost portion of the sclera; normally transparent, but cloudy in preserved specimens.

6. Examine the posterior portion of the eyeball. Carefully remove the vitreous humor, and identify the following structures:

Retina: The neural layer of the retina appears as a delicate tan, probably crumpled membrane that separates easily from the pigmented choroid.

Note its point of attachment. What is this point called?

____

Pigmented choroid coat: Appears iridescent in the cow or sheep eye owing to a special reflecting surface called the **tapetum lucidum.** This specialized surface reflects the light within the eye and is found in the eyes of animals that live under conditions of low-intensity light. It is not found in humans. ■

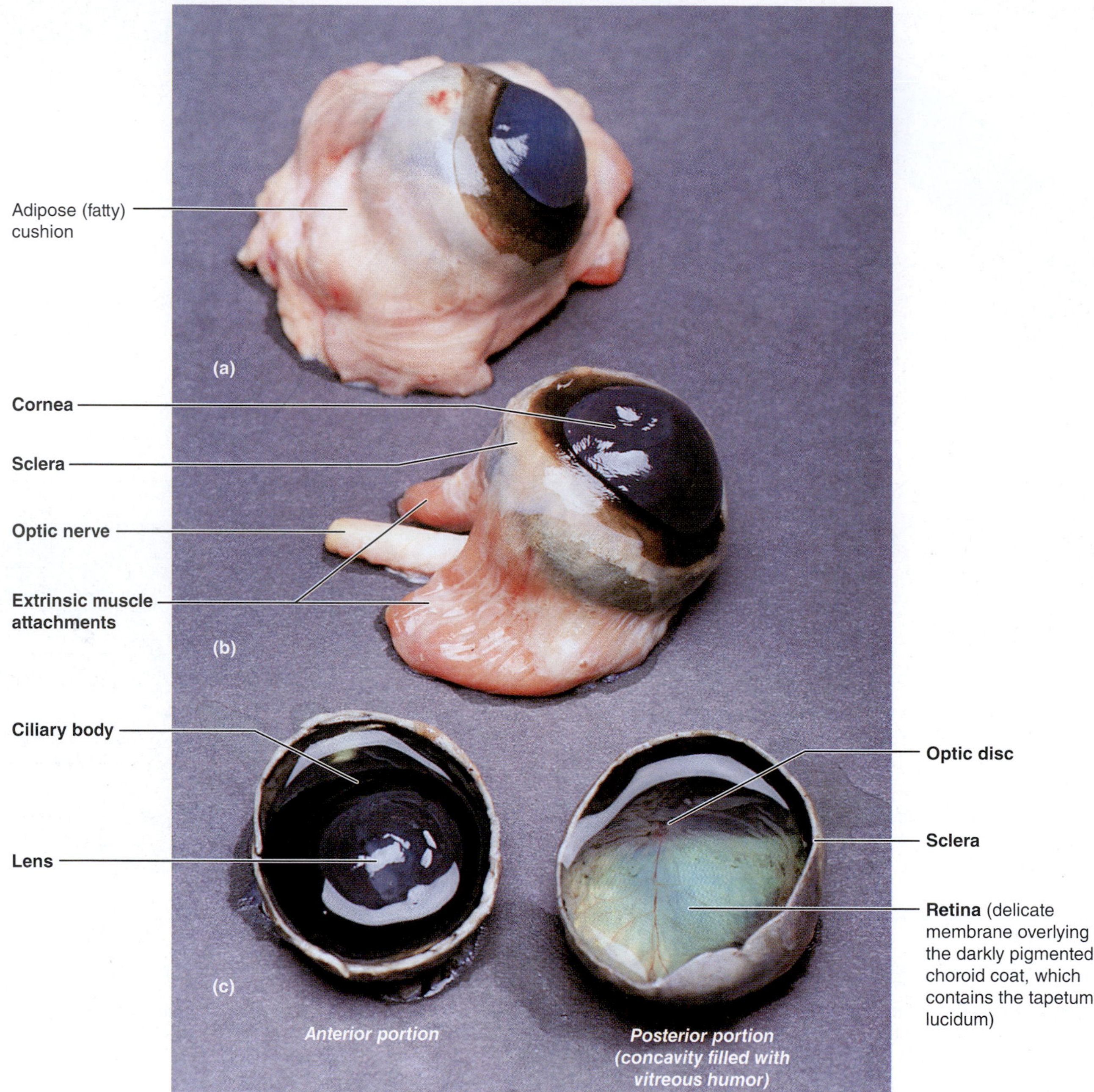

Figure 17.6 Anatomy of the cow eye. (a) Cow eye (entire) removed from orbit (notice the large amount of fat cushioning the eyeball). **(b)** Cow eye (entire) with fat removed to show the extrinsic muscle attachments and optic nerve. **(c)** Cow eye cut along the frontal plane to reveal internal structures.

Name ______________________

Lab Time/Date ______________________

EXERCISE 17

REVIEW SHEET

Special Senses: Anatomy of the Visual System

Anatomy of the Eye

1. Name five accessory eye structures that contribute to the formation of tears and/or aid in lubrication of the eyeball, and then name the major secretory product of each. Indicate which has antibacterial properties by circling the correct secretory product.

Accessory structures	Product

2. The eyeball is wrapped in adipose tissue within the orbit. What is the function of the adipose tissue?

____

3. Why does one often have to blow one's nose after crying? ______________________

____

4. Identify the extrinsic eye muscle predominantly responsible for the actions described below.

______________ 1. turns the eye laterally

______________ 2. turns the eye medially

______________ 3. turns the eye up and laterally

______________ 4. turns the eye down and medially

______________ 5. turns the eye up and medially

______________ 6. turns the eye down and laterally

5. What is a sty? ______________________

____

What is conjunctivitis? ______________________

____

6. For each numbered term in the key, correctly identify the corresponding structure provided with a letter in the diagram.

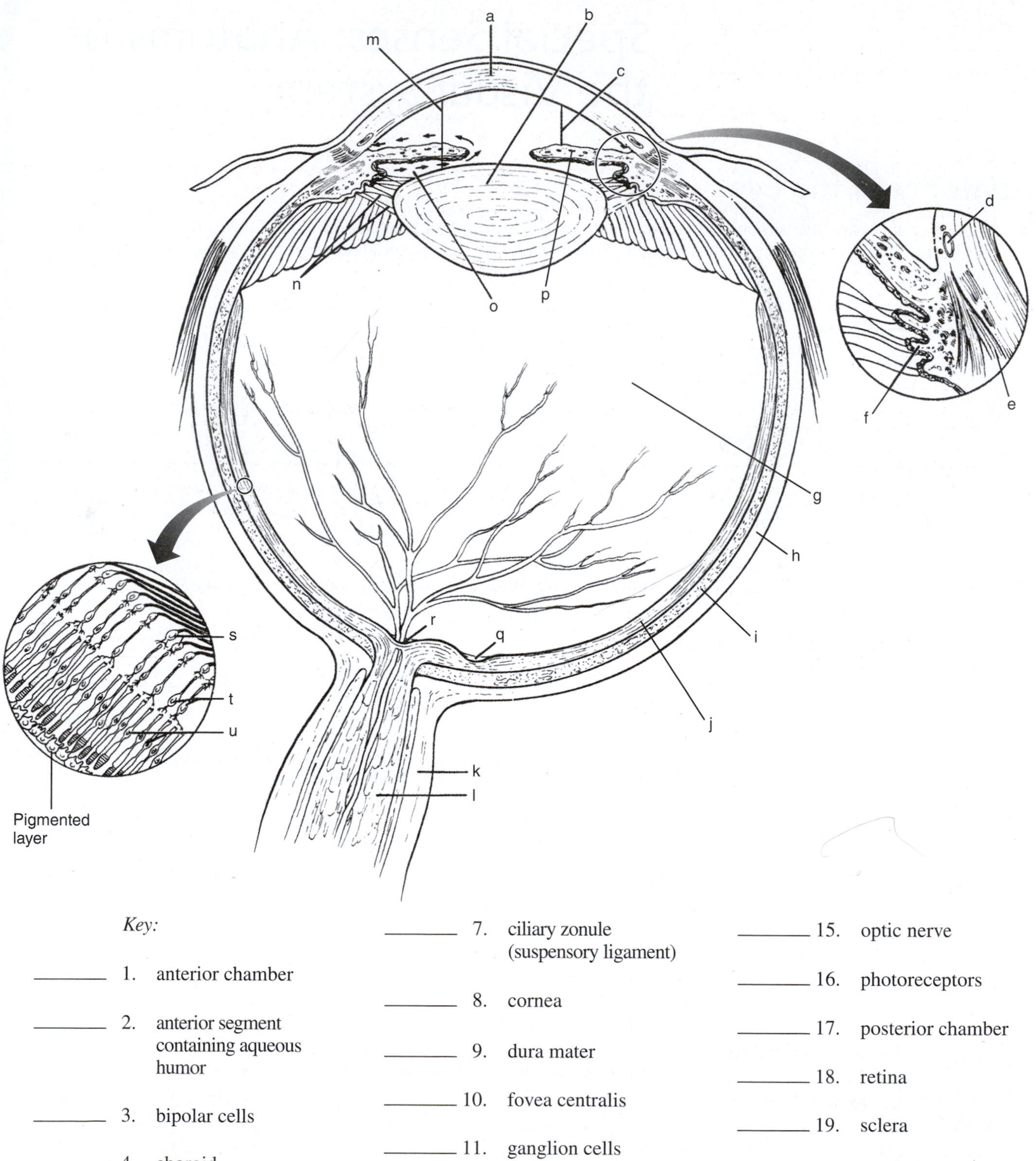

Key:

_______ 1. anterior chamber

_______ 2. anterior segment containing aqueous humor

_______ 3. bipolar cells

_______ 4. choroid

_______ 5. ciliary body and processes

_______ 6. ciliary muscle

_______ 7. ciliary zonule (suspensory ligament)

_______ 8. cornea

_______ 9. dura mater

_______ 10. fovea centralis

_______ 11. ganglion cells

_______ 12. iris

_______ 13. lens

_______ 14. optic disc

_______ 15. optic nerve

_______ 16. photoreceptors

_______ 17. posterior chamber

_______ 18. retina

_______ 19. sclera

_______ 20. scleral venous sinus

_______ 21. vitreous body in posterior segment

Notice the arrows drawn close to the left side of the iris (on page 314). What do they indicate?

____

____

7. The iris is composed primarily of two smooth muscle layers, one arranged radially and the other circularly.

Which of these dilates the pupil? ______________________________

8. In which of the following circumstances would you expect the pupil to be dilated?

a. in bright light
b. in dim light
c. focusing for near vision
d. observing distant objects

9. Which of the following controls the intrinsic eye muscles?

autonomic nervous system somatic nervous system

10. Match the key terms with the descriptions that follow. (Some choices will be used more than once.)

Key:
a. aqueous humor
b. choroid
c. ciliary body
d. ciliary processes of the ciliary body
e. cornea
f. fovea centralis
g. iris
h. lens
i. optic disc
j. retina
k. sclera
l. scleral venous sinus
m. vitreous humor

_______ 1. fluid filling the anterior segment of the eye

_______ 2. the "white" of the eye

_______ 3. part of the retina that lacks photoreceptors

_______ 4. modification of the choroid that controls the shape of the crystalline lens and contains the ciliary muscle

_______ 5. drains the aqueous humor from the eye

_______ 6. layer containing the rods and cones

_______ 7. substance occupying the posterior segment of the eyeball

_______ 8. forms the bulk of the heavily pigmented vascular layer

_______, _______ 9. smooth muscle structures (2)

_______ 10. area of critical focusing and discriminatory vision

_______ 11. form (by filtration) the aqueous humor

_______, _______, _______, _______ 12. light-bending media of the eye (4)

_______ 13. anterior continuation of the sclera—your "window on the world"

_______ 14. composed of tough, white, opaque, fibrous connective tissue

Microscopic Anatomy of the Retina

11. The two major layers of the retina are the pigmented and neural layers. In the neural layer, the neuron populations are arranged as follows from the pigmented layer to the vitreous humor. (Circle the proper response.)

bipolar cells, ganglion cells, photoreceptors

photoreceptors, ganglion cells, bipolar cells

ganglion cells, bipolar cells, photoreceptors

photoreceptors, bipolar cells, ganglion cells

12. The axons of the ________________ cells form the optic nerve, which exits from the eyeball.

13. Complete the following statements by writing either *rods* or *cones* on each blank.

The dim light receptors are the ____________. Only ____________ are found in the fovea centralis, whereas mostly ____________ are found in the periphery of the retina. ____________ are the photoreceptors that operate best in bright light and allow for color vision.

Visual Pathways to the Brain

14. The visual pathway to the occipital lobe of the brain consists most simply of a chain of five neurons. Beginning with the photoreceptor cell of the retina, name them, and note their location in the pathway.

1. ________________________
2. ________________________
3. ________________________
4. ________________________
5. ________________________

15. Visual field tests are done to reveal destruction along the visual pathway from the retina to the optic region of the brain. Note where the lesion is likely to be in the following cases.

Normal vision in left eye visual field; absence of vision in right eye visual field: ________________________

Normal vision in both eyes for right half of the visual field; absence of vision in both eyes for left half of the visual field:

____

16. How is the right optic *tract* anatomically different from the right optic *nerve*? ________________________

____

____

Dissection of the Cow (Sheep) Eye

17. What modification of the choroid that is *not* present in humans is found in the cow eye? ________________________

________________________ What is its function? ________________________

18. What does the retina look like? ________________________

____

At what point is the retina attached to the posterior aspect of the eyeball? ________________________

18

EXERCISE

Special Senses: Visual Tests and Experiments

MATERIALS

- ☐ Metric ruler; meter stick
- ☐ Common straight pins
- ☐ Snellen eye chart, floor marked with chalk or masking tape to indicate 20-ft. distance from posted Snellen chart
- ☐ Ishihara's color plates
- ☐ Two pencils
- ☐ Test tubes, each large enough to accommodate a pencil
- ☐ Laboratory lamp or penlight
- ☐ Ophthalmoscope (if available)

OBJECTIVES

1. Discuss the mechanism by which images form on the retina.
2. Define the following terms: *accommodation, astigmatism, emmetropic, hyperopia, myopia, refraction,* and *presbyopia;* describe several simple visual tests to which the terms apply.
3. Discuss the benefits of binocular vision.
4. Define *convergence,* and discuss the importance of the pupillary and convergence reflexes.
5. State the importance of an ophthalmoscopic examination.

PRE-LAB QUIZ

1. Circle the correct underlined term. Photoreceptors are distributed over the entire neural retina, except where the optic nerve leaves the eyeball. This site is called the macula lutea / optic disc.
2. Circle True or False. Persons with difficulty seeing objects at a distance are said to have myopia.
3. A condition that results in the loss of elasticity of the lens and difficulty focusing on a close object is called:
   a. myopia c. hyperopia
   b. presbyopia d. astigmatism
4. Photoreceptors of the eye include rods and cones. Which is responsible for interpreting color; which can function only under conditions of high light intensity? ______________________________
5. Circle the correct underlined term. Extrinsic / Intrinsic eye muscles are controlled by the autonomic nervous system.

In this exercise, you will perform several visual tests and experiments focusing on the physiology of vision. The first test involves demonstrating the blind spot (optic disc), the site where the optic nerve exits the eyeball.

The Optic Disc

ACTIVITY 1

Demonstrating the Blind Spot

1. Hold the figure for the blind spot test **(Figure 18.1)** about 46 cm (18 inches) from your eyes. Close your left eye, and focus your right eye on the X, which should be positioned so that it is directly in line with your right eye. Move the figure slowly toward your face, keeping your right eye focused on the X. When the dot focuses on the blind spot, which lacks photoreceptors, it will disappear.

Figure 18.1 Bind spot test figure.

2. Have your laboratory partner obtain a metric ruler and record in metric units the distance at which this occurs. The dot will reappear as the figure is moved closer. Distance at which the dot disappears:

Right eye ______________________________

Repeat the test for the left eye, this time closing the right eye and focusing the left eye on the dot. Record the distance at which the X disappears:

Left eye ______________________________

Refraction, Visual Acuity, and Astigmatism

When light rays pass from one medium to another, their velocity, or speed of transmission, changes, and the rays are bent, or **refracted.** Thus, the light rays in the visual field are refracted as they encounter the cornea, lens, and aqueous and vitreous humors of the eye.

The refractive index (bending power) of the cornea and humors are constant. But the lens's refractive index can be varied by changing the lens's shape—that is, by making it more or less convex so that the light is properly converged and focused on the retina. The greater the lens convexity, or bulge, the more the light will be bent, and the stronger the lens. Conversely, the less the lens convexity (the flatter it is), the less it bends the light.

In general, light from a distant source (over 6 m, or 20 feet) approaches the eye as parallel rays, and no change in lens convexity is necessary for it to focus properly on the retina. However, light from a close source tends to diverge, and the convexity of the lens must increase to make close vision possible. To achieve this, the ciliary muscle contracts, decreasing the tension on the ciliary zonule attached to the lens and allowing the elastic lens to "round up." Thus, a lens capable of bringing a *close* object into sharp focus is stronger (more convex) than a lens focusing on a more distant object. The ability of the eye to focus differentially for objects of near vision (less than 6 m, or 20 feet) is called **accommodation.** It should be noted that the image formed on the retina as a result of the refractory activity of the lens **(Figure 18.2)** is a **real**

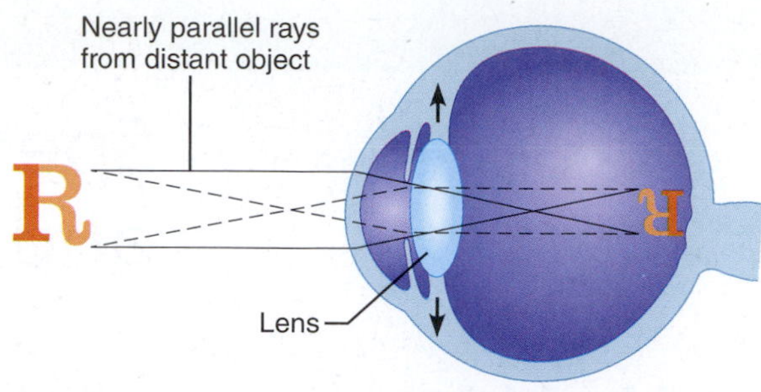

Figure 18.2 Refraction and real images. The refraction of light in the eye produces a real image (reversed, inverted, and reduced) on the retina.

image (reversed from left to right, inverted, and smaller than the object).

The normal eye, or **emmetropic eye,** is able to accommodate properly **(Figure 18.3a)**. However, visual problems may result from (1) lenses that are too strong or too "lazy" (overconverging and underconverging, respectively), (2) from structural problems such as an eyeball that is too long or too short to provide for proper focusing by the lens, or (3) a cornea or lens with improper curvatures.

Individuals in whom the image normally focuses in front of the retina are said to have **myopia,** or nearsightedness (Figure 18.3b); they can see close objects without difficulty, but distant objects are blurred or seen indistinctly. Correction requires a concave lens, which causes the light reaching the eye to diverge.

If the image focuses behind the retina, the individual is said to have **hyperopia,** or farsightedness. Such persons have no problems with distant vision but need glasses with convex lenses to augment the converging power of the lens for close vision (Figure 18.3c).

Irregularities in the curvatures of the lens and/or the cornea lead to a blurred vision problem called **astigmatism.** Cylindrically ground lenses, which compensate for inequalities in the curvatures of the refracting surfaces, are prescribed to correct the condition. +

Near Point of Accommodation

The elasticity of the lens decreases dramatically with age, resulting in difficulty in focusing for near or close vision. This condition is called **presbyopia**—literally, "old vision." Lens elasticity can be tested by measuring the **near point of accommodation.** The near point of vision is about 10 cm from the eye in young adults. It is closer in children and farther in elderly people.

ACTIVITY 2

Determining Near Point of Accommodation

To determine your near point of accommodation, hold a common straight pin at arm's length in front of one eye. (If desired, the text in the lab manual can be used rather than a pin.) Slowly move the pin toward that eye until the pin image becomes distorted. Have your lab partner use a metric ruler to measure the distance in centimeters from your eye to the pin at this point, and record the distance on the next page. Repeat the procedure for the other eye.

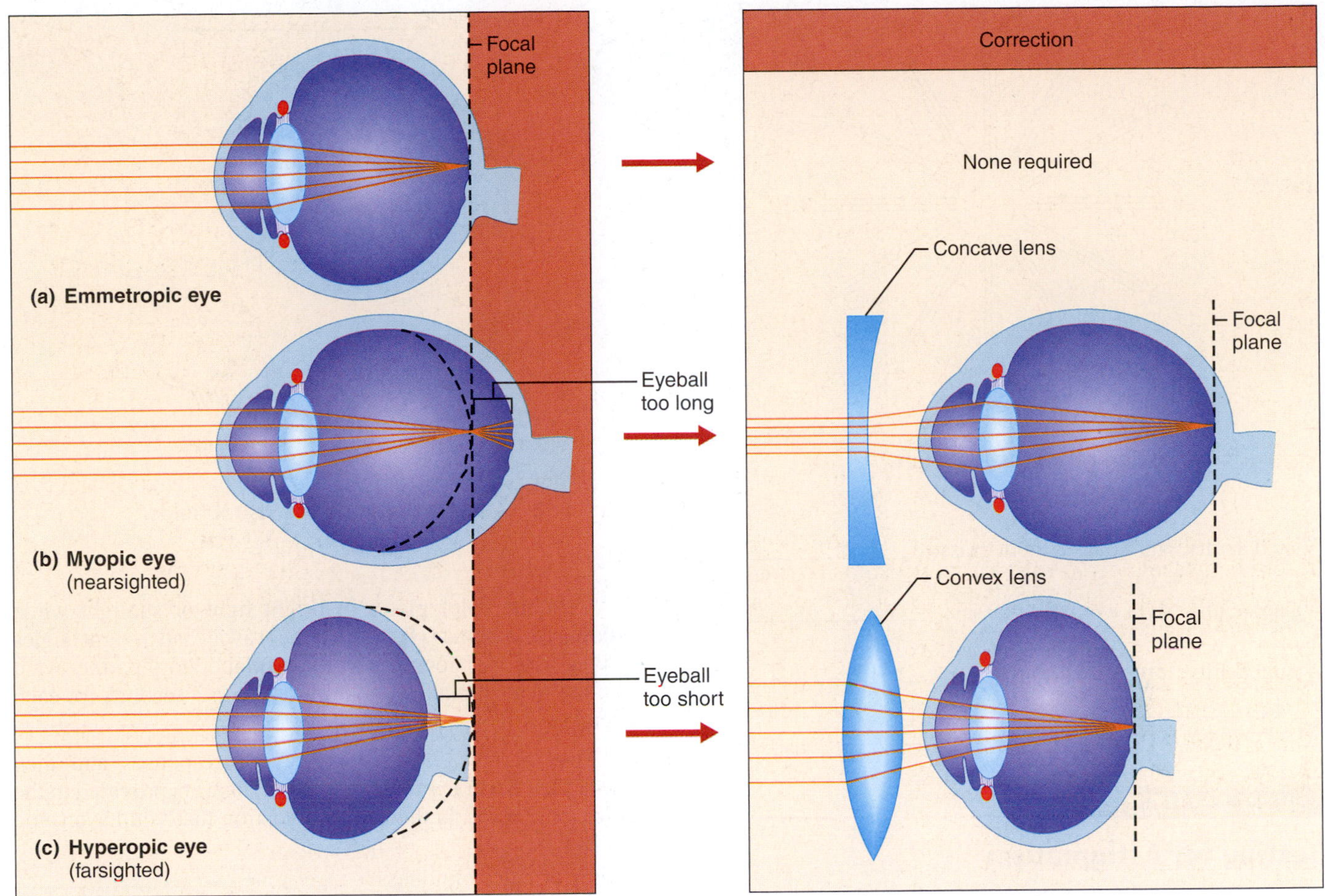

Figure 18.3 Problems of refraction. (a) In the emmetropic (normal) eye, light from both near and far objects is focused properly on the retina. **(b)** In a myopic eye, light from distant objects is brought to a focal point before reaching the retina. It then diverges. Applying a concave lens focuses objects properly on the retina. **(c)** In the hyperopic eye, light from a near object is brought to a focal point behind the retina. Applying a convex lens focuses objects properly on the retina. (Refractory effect of cornea is ignored here.)

Near point for right eye: ______________________

Near point for left eye: ______________________

Visual Acuity

Visual acuity, or sharpness of vision, is generally tested with a Snellen eye chart, which consists of letters of various sizes printed on a white card. This test is based on the fact that letters of a certain size can be seen clearly by eyes with normal vision at a specific distance. The distance at which the normal, or emmetropic, eye can read a line of letters is printed at the end of that line.

ACTIVITY 3

Testing Visual Acuity

1. Have your partner stand 6 m (20 feet) from the posted Snellen eye chart and cover one eye with a card or hand. As your partner reads each consecutive line aloud, check for accuracy. If this individual wears glasses, give the test twice—first with glasses off and then with glasses on. *Do not remove contact lenses, but note that they were in place during the test.*

2. Record the number of the line with the smallest-sized letters read. If it is 20/20, the person's vision for that eye is normal. If it is 20/40, or any ratio with a value less than one, he or she has less than the normal visual acuity. (Such an individual is myopic.) If the visual acuity is 20/15, vision is better than normal, because this person can stand at 6 m (20 feet) from the chart and read letters that are only discernible by the normal eye at 4.5 m (15 feet). Give your partner the number of the line corresponding to the smallest letters read, to record in step 4.

3. Repeat the process for the other eye.

4. Have your partner test and record your visual acuity. If you wear glasses, the test results *without* glasses should be recorded first.

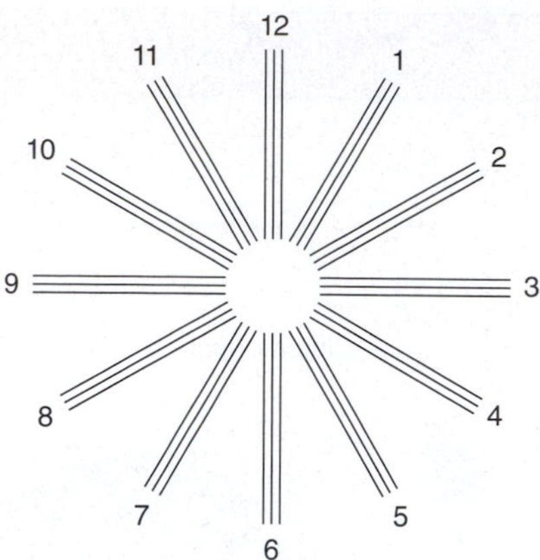

Figure 18.4 Astigmatism testing chart.

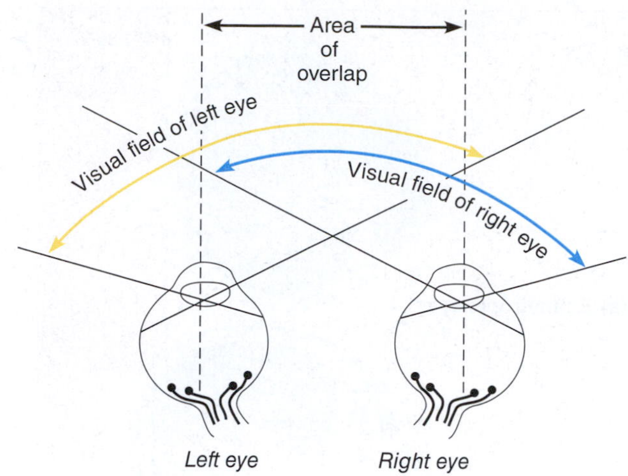

Figure 18.5 Overlapping of the visual fields.

Visual acuity, right eye without glasses: ________________

Visual acuity, right eye with glasses: ________________

Visual acuity, left eye without glasses: ________________

Visual acuity, left eye with glasses: ________________

ACTIVITY 4

Testing for Astigmatism

The astigmatism chart **(Figure 18.4)** is designed to test for defects in the refracting surface of the lens and/or cornea.

View the chart first with one eye and then with the other, focusing on the center of the chart. If all the radiating lines appear equally dark and distinct, there is no distortion of your refracting surfaces. If some of the lines are blurred or appear less dark than others, at least some degree of astigmatism is present.

Is astigmatism present in your left eye? ________________

Is it present in your right eye? ________________

Color Blindness

Ishihara's color plates are designed to test for deficiencies in the cones or color photoreceptor cells. There are three cone types, each containing a different light-absorbing pigment. One type primarily absorbs the red wavelengths of the visible light spectrum, another the blue wavelengths, and a third the green wavelengths. Nerve impulses reaching the brain from these different photoreceptor types are then interpreted (seen) as red, blue, and green, respectively. Interpretation of the intermediate colors of the visible light spectrum is a result of overlapping input from more than one cone type.

ACTIVITY 5

Testing for Color Blindness

1. Find the interpretation table that accompanies the Ishihara color plates, and prepare a sheet to record data for the test. Note which plates are patterns rather than numbers.

2. View the color plates in bright light or sunlight while holding them about 0.8 m (30 inches) away and at right angles to your line of vision. Report to your laboratory partner what you see in each plate. Take no more than 3 seconds for each decision.

3. Your partner should record your responses and then check their accuracy with the correct answers provided in the color plate book. Is there any indication that you have some degree of color blindness? ________________

If so, what type? ________________

____

Repeat the procedure to test your partner's color vision.

Binocular Vision

Humans, cats, predatory birds, and most primates are endowed with *binocular vision.* Their visual fields, each about 170 degrees, overlap to a considerable extent, and each eye sees a slightly different view **(Figure 18.5)**. The primary visual cortex fuses the slightly different images, providing **depth perception** (or **three-dimensional vision**). This provides an accurate means of locating objects in space.

In contrast, the eyes of rabbits, pigeons, and others are on the sides of their head. Such animals see in two different directions and thus have a panoramic field of view and panoramic vision. A mnemonic device to keep these straight is "Eyes in the front—likes to hunt. Eyes on the side—likes to hide."

ACTIVITY 6

Testing for Depth Perception

1. To demonstrate that each eye sees a slightly different view, perform the following simple experiment.

Close your left eye. Hold a pencil at arm's length directly in front of your right eye. Position another pencil directly beneath

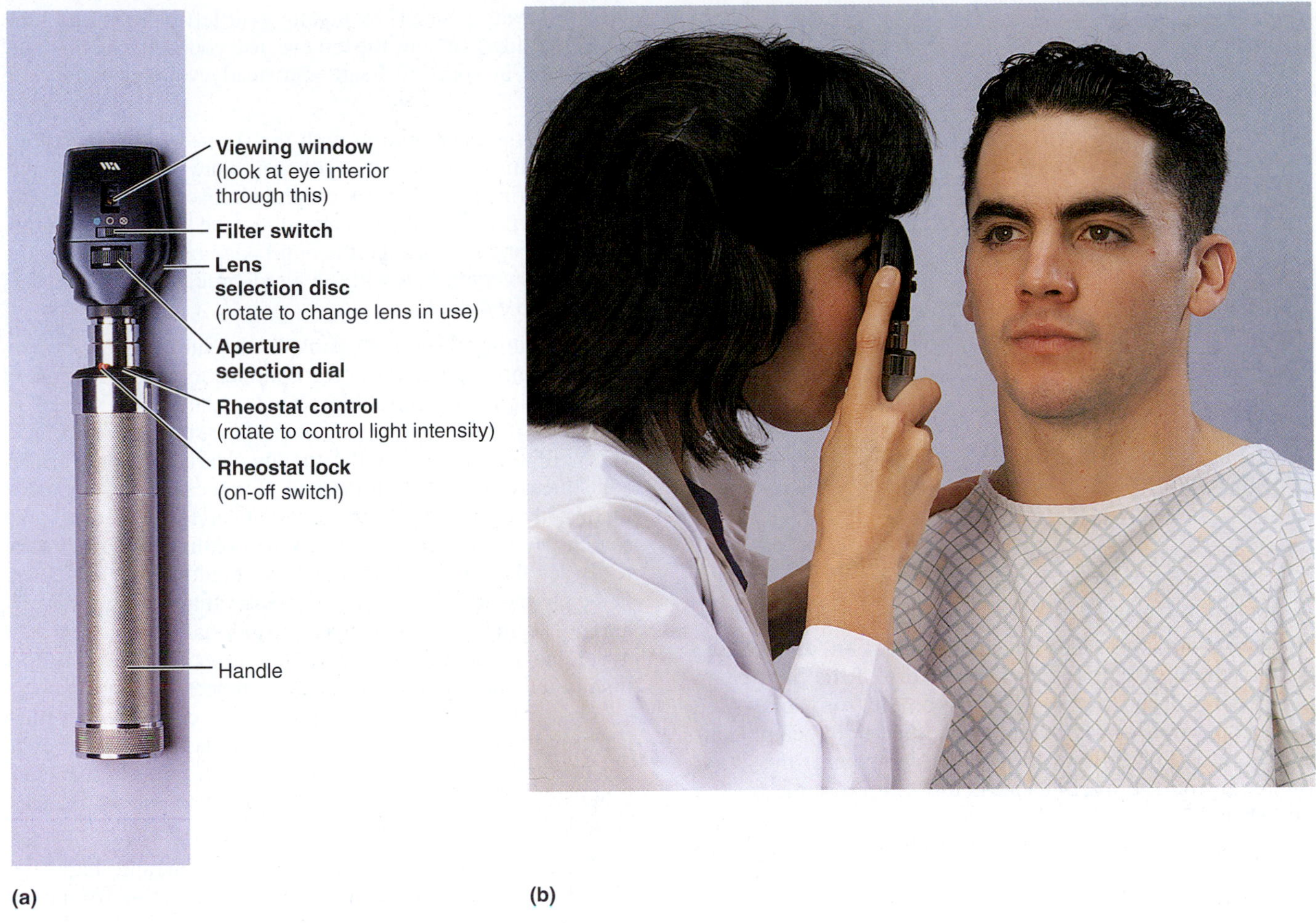

Figure 18.6 Structure and use of an ophthalmoscope. (a) Photograph of an ophthalmoscope. **(b)** Proper position for beginning to examine the right eye with an ophthalmoscope.

it, and then move the lower pencil about half the distance toward you. As you move the lower pencil, make sure it remains in the *same plane* as the stationary pencil, so that the two pencils continually form a straight line. Then, without moving the pencils, close your right eye, and open your left eye. Notice that with only the right eye open, the moving pencil stays in the same plane as the fixed pencil, but that when viewed with the left eye, the moving pencil is displaced laterally away from the plane of the fixed pencil.

2. To demonstrate the importance of two-eyed binocular vision for depth perception, perform this second simple experiment.

Have your laboratory partner hold a test tube erect about arm's length in front of you. With both eyes open, quickly insert a pencil into the test tube. Remove the pencil, bring it back close to your body, close one eye, and quickly and without hesitation insert the pencil into the test tube. *(Do not feel for the test tube with the pencil!)* Repeat with the other eye closed.

Was it as easy to dunk the pencil with one eye closed as with both eyes open?

____

Ophthalmoscopic Examination of the Eye (Optional)

The ophthalmoscope is an instrument used to examine the *fundus,* or eyeball interior, to determine visually the condition of the retina, optic disc, and internal blood vessels. Such an examination can detect certain pathological conditions such as diabetes mellitus, arteriosclerosis, and degenerative changes of the optic nerve and retina. The ophthalmoscope consists of a set of lenses mounted on a rotating disc (the **lens selection disc**), a light source regulated by a **rheostat control,** and a mirror that reflects the light so that the eye interior can be illuminated.

The lens selection disc is positioned in a small slit in the mirror, and the examiner views the eye interior through this slit, appropriately called the **viewing window (Figure 18.6a)**. The focal length of each lens is indicated in diopters preceded by a plus (+) sign if the lens is convex and by a negative (–) sign if the lens is concave. When the zero (0) is seen in the **diopter window,** on the examiner's side of the instrument, there is no lens positioned in the slit. Changing the lens will change the depth of focus for viewing the eye interior.

The light is turned on by depressing the red **rheostat lock button** and then rotating the rheostat control in the

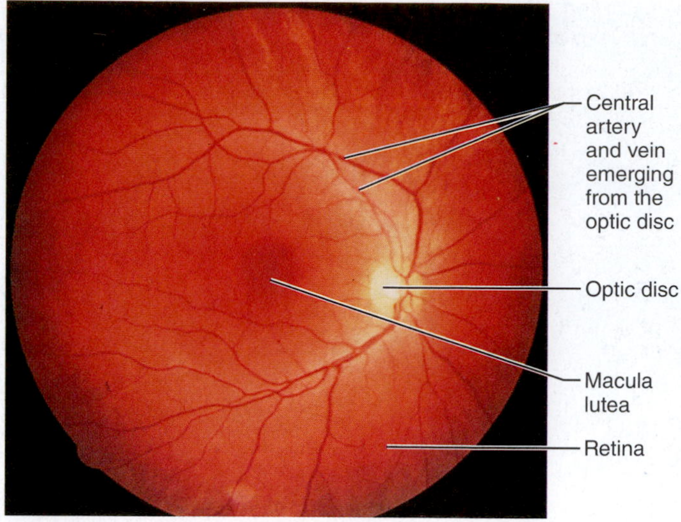

Figure 18.7 Posterior wall (fundus) of right retina.

clockwise direction. The **aperture selection dial** on the front of the instrument allows the nature of the light beam to be altered. The **filter switch,** also on the front, allows the choice of a green, unfiltered, or polarized light beam. Generally, green light allows for clearest viewing of the blood vessels in the eye interior and is most comfortable for the subject.

Once you have examined the ophthalmoscope and have become familiar with it, you are ready to conduct an eye examination.

ACTIVITY 7

Conducting an Ophthalmoscopic Examination

1. Conduct the examination in a dimly lighted or darkened room with the subject comfortably seated and gazing straight ahead. To examine the right eye, sit face-to-face with the subject, hold the instrument in your right hand, and use your right eye to view the eye interior (Figure 18.6b). You may want to steady yourself by resting your left hand on the subject's shoulder. To view the left eye, use your left eye, hold the instrument in your left hand, and steady yourself with your right hand.

2. Begin the examination with the 0 (no lens) in position. Grasp the instrument so that you can rotate the lens disc with the index finger. Holding the ophthalmoscope about 15 cm (6 inches) from the subject's eye, direct the light into the pupil at a slight angle—through the pupil edge rather than directly through its center. You will see a red circular area that is the illuminated eye interior.

3. Move in as close as possible to the subject's cornea (to within 5 cm, or 2 inches) as you continue to observe the area. Steady your instrument-holding hand on the subject's cheek if necessary. If both your eye and that of the subject are normal, you will be able to view the fundus clearly without further adjustment of the ophthalmoscope. If you cannot focus on the fundus, slowly rotate the lens disc counterclockwise until you can clearly see the fundus. When the ophthalmoscope is correctly set, the fundus of the right eye should appear as shown in the photograph **(Figure 18.7)**. (**Note:** If a positive [convex] lens is required and your eyes are normal, the subject has hyperopia. If a negative [concave] lens is necessary to view the fundus and your eyes are normal, the subject is myopic.)

When the examination is proceeding correctly, the subject can often see images of retinal vessels in his or her own eye that appear rather like cracked glass. If you are unable to achieve a sharp focus or to see the optic disc, move medially or laterally and begin again.

4. Examine the optic disc for color, elevation, and sharpness of outline, and observe the blood vessels radiating from near its center. Locate the macula, lateral to the optic disc. It is a darker area in which blood vessels are absent, and the fovea appears to be a slightly lighter area in its center. The macula is most easily seen when the subject looks directly into the light of the ophthalmoscope.

Do not examine the macula for longer than 1 second at a time.

5. When you have finished examining your partner's retina, shut off the ophthalmoscope. Change places with your partner (become the subject), and repeat steps 1–4.

Name ______________________________

Lab Time/Date______________________

EXERCISE 18

REVIEW SHEET

Special Senses: Visual Tests and Experiments

Blind Spot

1. Explain why vision is lost when light hits the blind spot. ______________________________

____

Refraction, Visual Acuity, and Astigmatism

2. Match the terms in column B with the descriptions in column A.

	Column A	Column B
______________	1. light bending	a. accommodation
______________	2. ability to focus for close (less than 20 feet) vision	b. astigmatism
______________	3. normal vision	c. convergence
______________	4. inability to focus well on close objects (farsightedness)	d. emmetropia
______________	5. nearsightedness	e. hyperopia
______________	6. blurred vision due to unequal curvatures of the lens or cornea	f. myopia
______________	7. medial movement of the eyes during focusing on close objects	g. refraction

3. Complete the following statements:

In farsightedness, the light is focused _1_ the retina. The lens required to treat myopia is a _2_ lens. The "near point" increases with age because the _3_ of the lens decreases as we get older. A convex lens, like that of the eye, produces an image that is upside down and reversed from left to right. Such an image is called a _4_ image.

1. ______________________________
2. ______________________________
3. ______________________________
4. ______________________________

4. Use terms from the key to complete the statements concerning near and distance vision. (Some terms will be used more than once.)

Key: a. contracted b. decreased c. increased d. relaxed e. taut

During distance vision, the ciliary muscle is __________, the ciliary zonule is __________, the convexity of the lens is __________, and light refraction is __________. During close vision, the ciliary muscle is __________, the ciliary zonule is __________, lens convexity is __________, and light refraction is __________.

5. Using your Snellen eye test results, answer the following questions.

Is your visual acuity normal, less than normal, or better than normal? ______

Explain your answer. ______

______

Explain why each eye is tested separately when the Snellen eye chart is used. ______

______

Explain what "20/40 vision" means. ______

______

Explain what "20/10 vision" means. ______

______

6. Define *astigmatism*. ______

______

How can it be corrected? ______

7. Define *presbyopia*. ______

______

What causes it? ______

Color Blindness

8. To which wavelengths of light do the three cone types of the retina respond maximally?

______, ______, and ______

9. How can you explain the fact that we see a great range of colors even though only three cone types exist?

______

Binocular Vision

10. Explain the difference between binocular and panoramic vision. ______

______

______

What is the advantage of binocular vision? ______

What factor(s) are responsible for binocular vision? ______

______

Ophthalmoscopic Examination

11. Why is the ophthalmoscopic examination an important diagnostic tool? ______________________________

____

12. Many college students struggling through mountainous reading assignments are told that they need glasses for "eyestrain." Why is it more of a strain on the extrinsic and intrinsic eye muscles to look at close objects than at far objects?

____

____

EXERCISE

19

Special Senses: Hearing and Equilibrium

MATERIALS

- ☐ Three-dimensional dissectible ear model and/or chart of ear anatomy
- ☐ Otoscope (if available)
- ☐ Disposable otoscope tips (if available) and autoclave bag
- ☐ Alcohol swabs
- ☐ Compound microscope
- ☐ Prepared slides of the cochlea of the ear
- ☐ Absorbent cotton
- ☐ Pocket watch or clock that ticks
- ☐ Metric ruler
- ☐ Tuning forks (range of frequencies)
- ☐ Rubber mallet
- ☐ Demonstration: Microscope focused on a slide of a crista ampullaris receptor of a semicircular canal
- ☐ Blackboard and chalk, or whiteboard and markers

OBJECTIVES

1. Identify the anatomical structures of the external, middle, and internal ear on a model or appropriate diagram, and explain their functions.
2. Describe the anatomy of the organ of hearing (spiral organ in the cochlea), and explain its function in sound reception.
3. Discuss how one is able to localize the source of sounds.
4. Define *sensorineural deafness* and *conduction deafness*, and relate these conditions to the Weber and Rinne tests.
5. Describe the anatomy of the organs of equilibrium in the internal ear (cristae ampullares and maculae), and explain their relative function in maintaining equilibrium.
6. State the locations and functions of endolymph and perilymph.
7. Discuss the effects of acceleration on the semicircular canals.
8. Define *nystagmus*, and relate this event to the balance test.
9. State the purpose of the Romberg test.
10. Explain the role of vision in maintaining equilibrium.

PRE-LAB QUIZ

1. Circle the correct underlined term. The ear is divided into <u>three</u> / <u>four</u> major areas.
2. The external ear is composed primarily of the _______ and the external acoustic meatus.
   a. auricle c. eardrum
   b. cochlea d. stapes
3. Circle the correct underlined term. Sound waves that enter the external acoustic meatus eventually encounter the <u>tympanic membrane</u> / <u>oval window</u>, which then vibrates at the same frequency as the sound waves hitting it.
4. Three small bones found within the middle ear are the malleus, incus, and:
   a. auricle c. eardrum
   b. cochlea d. stapes
5. The snail-like _______, found in the internal ear, contains sensory receptors for hearing.
   a. cochlea c. semicircular canals
   b. lobule d. vestibule
6. Circle the correct underlined term. Today you will use an <u>ophthalmoscope</u> / <u>otoscope</u> to examine the ear.
7. The _______ test is used for comparing bone and air-conduction hearing.
   a. balance b. Rinne c. Weber
8. The equilibrium apparatus of the ear, the vestibular apparatus, is found in the:
   a. external ear b. internal ear c. middle ear

(Text continues on next page.)

9. Circle the correct underlined terms. The crista ampullaris / macula located in the semicircular duct / vestibule is essential for detecting static equilibrium.
10. Nystagmus is:
   a. the ability to hear only high-frequency tones
   b. the ability to hear only low-frequency tones
   c. involuntary trailing of eyes in one direction, then rapid movement in the other
   d. the sensation of dizziness

Anatomy of the Ear

Gross Anatomy

The ear is a complex structure containing sensory receptors for hearing and equilibrium. The ear is divided into three major areas: the *external ear,* the *middle ear,* and the *internal ear* **(Figure 19.1)**. The external and middle ear structures serve the needs of the sense of hearing *only,* whereas internal ear structures function both in equilibrium and hearing reception.

ACTIVITY 1

Identifying Structures of the Ear

Obtain a dissectible ear model or chart of ear anatomy and identify the structures described below. Refer to the figure (Figure 19.1) as you work.

The **external (outer) ear** is composed primarily of the auricle and the external acoustic meatus. The **auricle (pinna)** is the skin-covered cartilaginous structure encircling the auditory canal opening. In many animals, it collects and directs sound waves into the external acoustic meatus. In humans, this function of the auricle is largely lost. The portion of the auricle lying inferior to the external acoustic meatus is the **lobule.**

The **external acoustic meatus (external auditory canal)** is a short, narrow (about 2.5 cm long by 0.6 cm wide) chamber carved into the temporal bone. In its skin-lined walls are wax-secreting glands called **ceruminous glands.** Sound waves that enter the external acoustic meatus eventually encounter the **tympanic membrane,** or **eardrum,** which vibrates at exactly the same frequency as the sound wave(s) hitting it. The membranous eardrum separates the external ear from the middle ear.

The **middle ear** is essentially a small chamber—the **tympanic cavity**—found within the temporal bone. The cavity is spanned by three small bones, collectively called the **auditory ossicles** (**malleus, incus,** and **stapes**), which articulate to form a lever system that amplifies and transmits the vibratory motion of the eardrum to the fluids of the internal ear via the **oval window.** The ossicles are often referred to by their common names: hammer, anvil, and stirrup, respectively.

Connecting the middle ear chamber with the nasopharynx is the **pharyngotympanic (auditory) tube** (formerly known as the eustachian tube). Normally this tube is flattened and closed, but swallowing or yawning can cause it to open temporarily to equalize the pressure of the middle ear cavity with external air pressure. This is an important function. The eardrum does not vibrate properly unless the pressure on both of its surfaces is the same.

Because the mucosal membranes of the middle ear cavity and nasopharynx are continuous through the

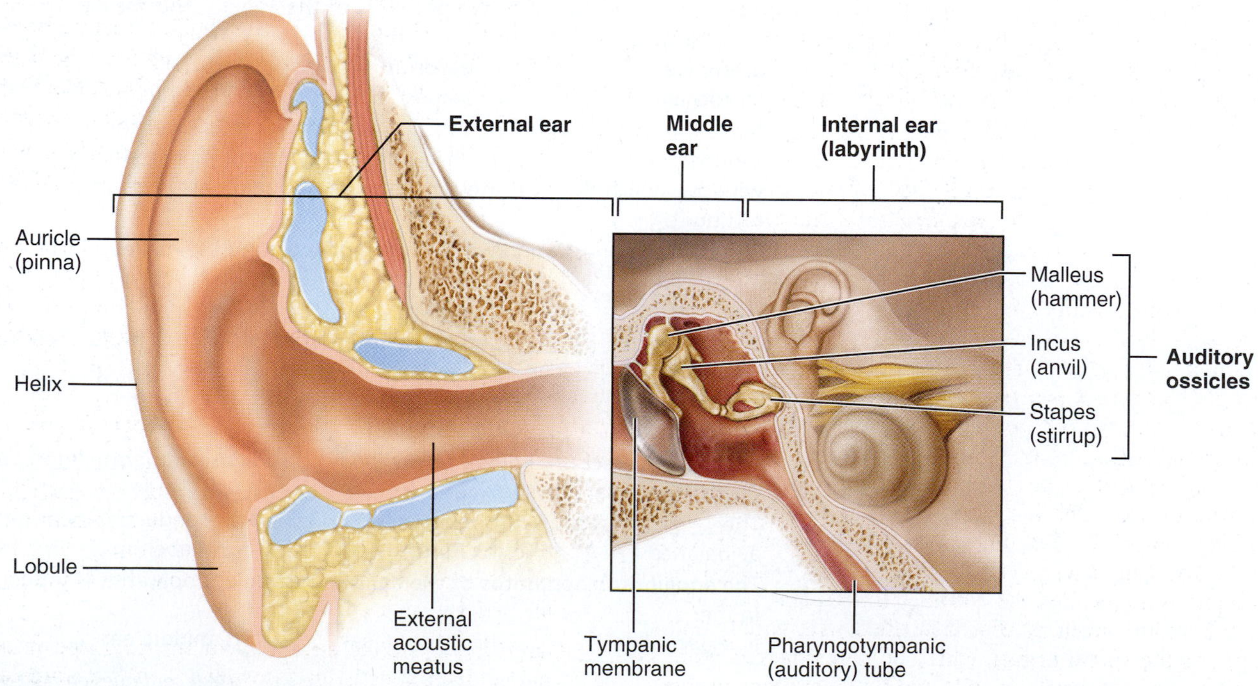

Figure 19.1 Anatomy of the ear.

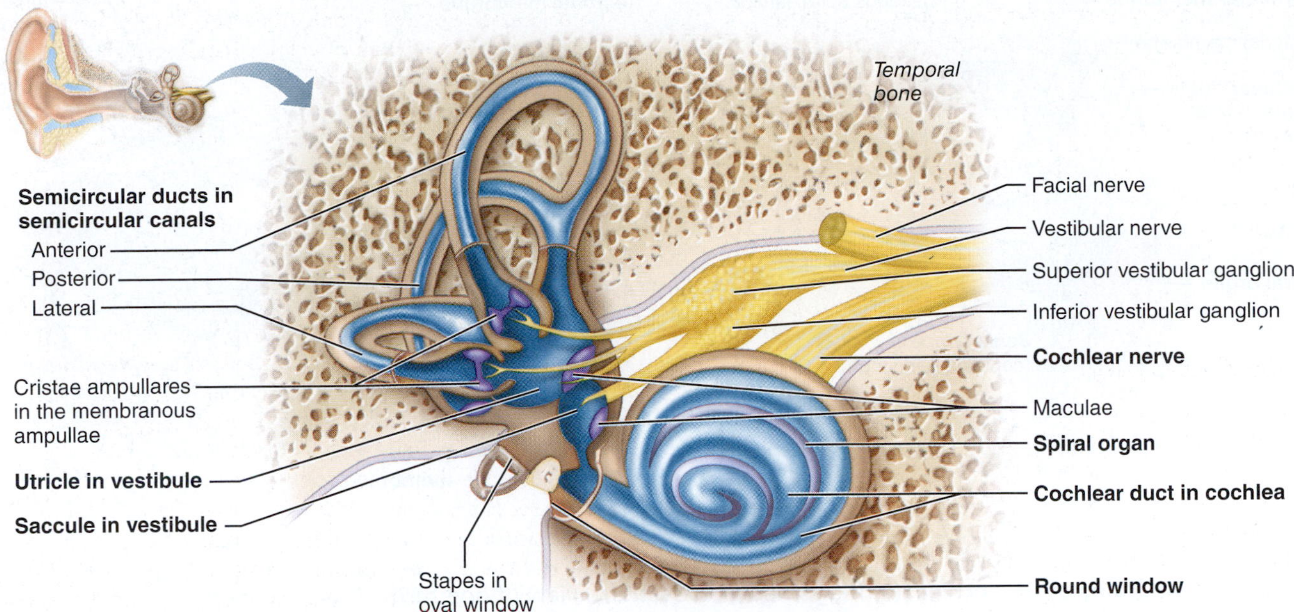

Figure 19.2 **Internal ear.** Right membranous labyrinth (blue) shown within the bony labyrinth (tan). The locations for sensory organs for hearing and equilibrium are shown in purple.

pharyngotympanic tube, **otitis media,** or inflammation of the middle ear, is a fairly common condition, especially among children prone to sore throats. In cases where large amounts of fluid or pus accumulate in the middle ear cavity, an emergency myringotomy (lancing of the eardrum) may be necessary to relieve the pressure. Frequently, tiny ventilating tubes are put in during the procedure. +

The **internal ear** consists of a system of bony and rather tortuous chambers called the **bony labyrinth,** which is filled with an aqueous fluid called **perilymph** (**Figure 19.2**). Suspended in the perilymph is the **membranous labyrinth,** a system that mostly follows the contours of the bony labyrinth. The membranous labyrinth is filled with a more viscous fluid called **endolymph.** The three subdivisions of the bony labyrinth are the cochlea, the vestibule, and the semicircular canals, with the vestibule situated between the cochlea and semicircular canals. The **vestibule** and the **semicircular canals** are involved with equilibrium.

The snail-like **cochlea** (see Figure 19.2 and **Figure 19.3**) contains the sensory receptors for hearing. The membranous **cochlear duct** is a soft wormlike tube about 3.8 cm long. It winds through the full two and three-quarter turns of the cochlea and separates the perilymph-containing cochlear cavity into upper and lower chambers, the **scala vestibuli** and **scala tympani.** The scala vestibuli terminates at the oval window, which "seats" the foot plate of the stirrup located laterally in the tympanic cavity. The scala tympani is bounded by a membranous area called the **round window.** The cochlear duct is the middle **scala media.** It is filled with endolymph and supports the **spiral organ,** which contains the receptors for hearing—the sensory hair cells and nerve endings of the **cochlear nerve,** a division of the vestibulocochlear nerve (VIII).

ACTIVITY 2

Examining the Ear with an Otoscope (Optional)

1. Obtain an otoscope and two alcohol swabs. Inspect your partner's ear canal and then select the largest-*diameter* (not length!) speculum that will fit comfortably into his or her ear to permit full visibility. Clean the speculum thoroughly with an alcohol swab, and then attach the speculum to the battery-containing otoscope handle. Before beginning, check that the otoscope light beam is strong. If not, obtain another otoscope or new batteries. Some otoscopes come with disposable tips. Be sure to use a new tip for each ear examined. Dispose of these tips in an autoclave bag after use.

! 2. When you are ready to begin the examination, hold the lighted otoscope securely between your thumb and forefinger (like a pencil), and rest the little finger of the otoscope-holding hand against your partner's head. This maneuver forms a brace that allows the speculum to move as your partner moves and prevents the speculum from penetrating too deeply into the ear canal during unexpected movements.

3. Grasp the ear auricle firmly and pull it up, back, and slightly laterally. If your partner experiences pain or discomfort when the auricle is manipulated, an inflammation or infection of the external ear may be present. If this occurs, do not attempt to examine the ear canal.

4. Carefully insert the speculum of the otoscope into the external acoustic meatus in a downward and forward direction only far enough to permit examination of the tympanic membrane or eardrum. Note its shape, color, and vascular

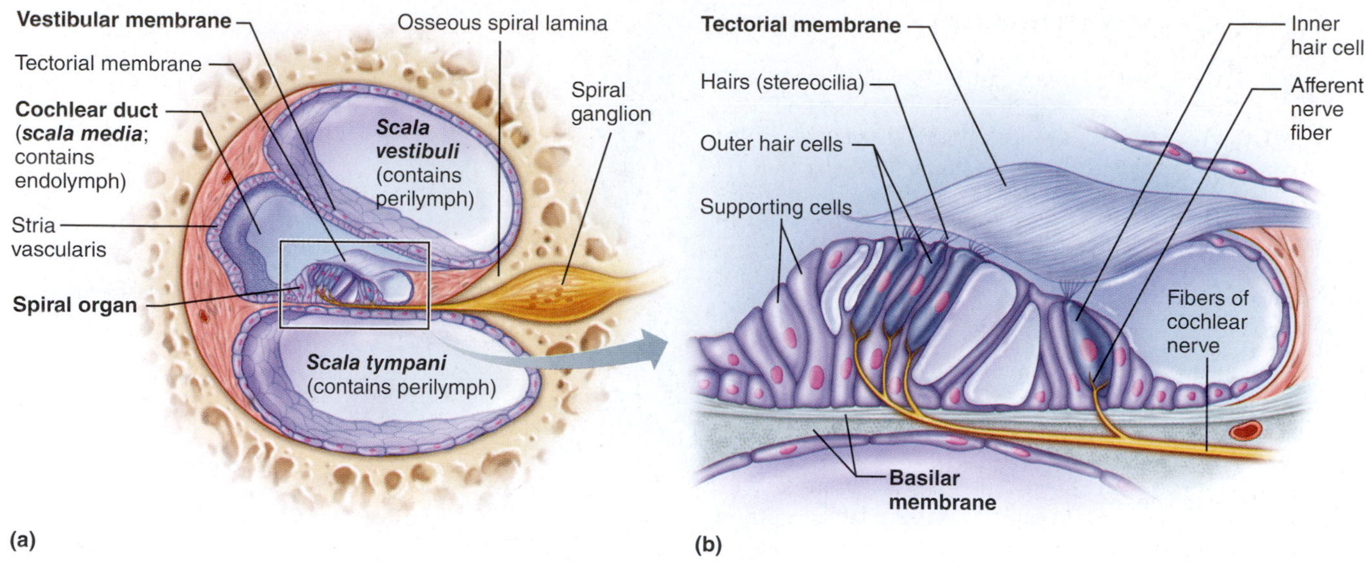

Figure 19.3 Anatomy of the cochlea. (a) Magnified cross-sectional view of one turn of the cochlea, showing the relationship of the three scalae. The scalae vestibuli and tympani contain perilymph; the cochlear duct (scala media) contains endolymph. **(b)** Detailed structure of the spiral organ.

network. The healthy tympanic membrane is pearly white. During the examination, notice whether there is any discharge or redness in the canal, and identify earwax.

5. After the examination, thoroughly clean the speculum with the second alcohol swab before returning the otoscope to the supply area.

Microscopic Anatomy of the Spiral Organ and the Mechanism of Hearing

In the spiral organ, the auditory receptors are hair cells that rest on the **basilar membrane,** which forms the floor of the cochlear duct (Figure 19.3). Their "hairs" are stereocilia that project into a gelatinous membrane, the **tectorial membrane,** that overlies them. The roof of the cochlear duct is called the **vestibular membrane.** The endolymph-filled chamber of the cochlear duct is the **scala media.**

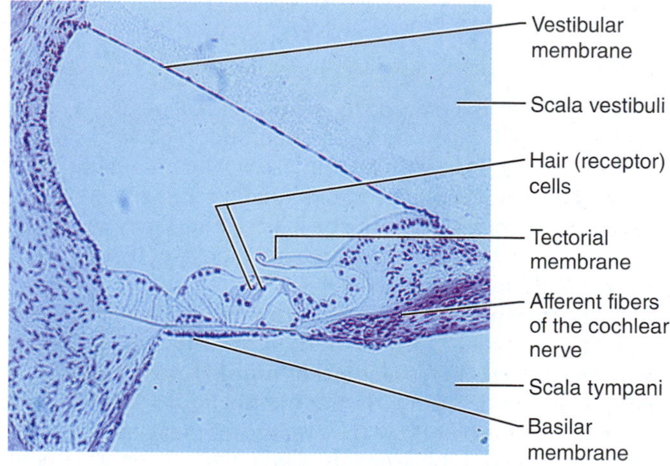

Figure 19.4 Histologic image of the spiral organ (100×).

ACTIVITY 3

Examining the Microscopic Structure of the Cochlea

Obtain a compound microscope and a prepared microscope slide of the cochlea, and identify the areas shown in the photomicrograph **(Figure 19.4)**.

The mechanism of hearing begins as sound waves pass through the external acoustic meatus and through the middle ear into the internal ear, where the vibration eventually reaches the spiral organ, which contains the receptors for hearing.

Vibration of the stirrup at the oval window initiates traveling pressure waves in the perilymph that cause maximal displacements of the basilar membrane, where they peak and stimulate the hair cells of the spiral organ in that region. Because the area at which the traveling waves peak is a high-pressure area, the vestibular membrane is compressed at this point and, in turn, compresses the endolymph and the basilar membrane of the cochlear duct. The resulting pressure on the perilymph in the scala tympani causes the membrane of the round window to bulge outward into the middle ear chamber, thus acting as a relief valve for the compressional wave. High-frequency waves (high-pitched sounds) peak close to the oval window, and low-frequency waves (low-pitched sounds) peak farther up the basilar membrane near the apex of the cochlea. The mechanism of sound reception by the spiral organ is complex. Hair cells at a given spot on the basilar membrane are stimulated by sounds of a specific frequency and amplitude. Once stimulated, they depolarize and begin the chain of nervous impulses that travel along the cochlear nerve to the auditory centers of the temporal lobe cortex. This series of events results in the phenomenon we call hearing **(Figure 19.5)**.

Sensorineural deafness results from damage to neural structures anywhere from the cochlear hair cells through neurons of the auditory cortex. **Presbycusis** is a type of sensorineural deafness that occurs commonly in people by the time they are in their 60s. It results from a gradual deterioration and atrophy of the spiral organ, leading to a loss in the ability to hear high tones and speech sounds. Because many elderly people refuse to accept their hearing loss and resist using hearing aids, they begin to rely more and more on their vision for clues as to what is going on around them and may be accused of ignoring people.

Although presbycusis is considered to be a disability of old age, it is becoming much more common in younger people as our world grows noisier. Prolonged or excessive noise tears the cilia from hair cells, and the damage is progressive and cumulative. Each assault causes a bit more damage. Music played and listened to at deafening levels definitely contributes to the deterioration of hearing receptors. +

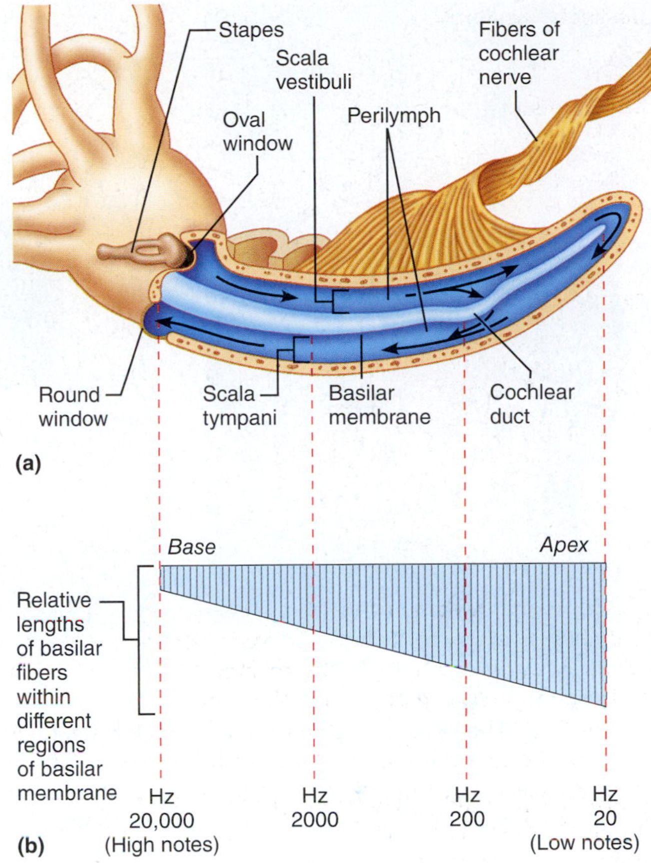

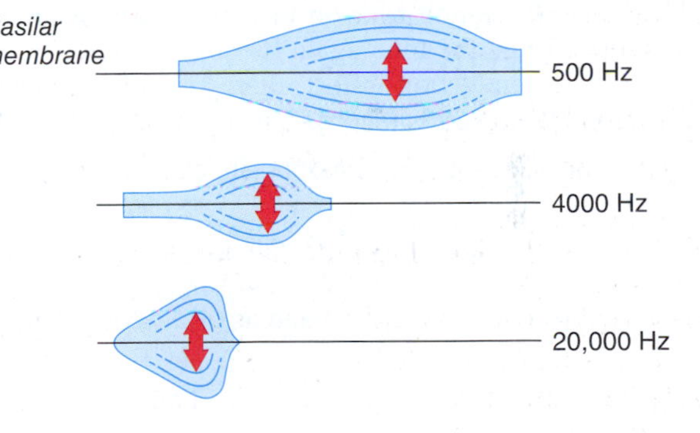

Figure 19.5 Resonance of the basilar membrane. The cochlea is depicted as if it has been uncoiled. **(a)** Fluid movement in the cochlea following the stirrup thrust at the oval window. The compressional wave thus created causes the round window to bulge into the middle ear. Pressure waves set up vibrations in the basilar membrane. **(b)** Fibers span the basilar membrane. The length of the fibers "tunes" specific regions to vibrate at specific frequencies. **(c)** Different frequencies of pressure waves in the cochlea stimulate particular hair cells and neurons.

ACTIVITY 4

Conducting Laboratory Tests of Hearing

Perform the following hearing tests in a quiet area.

Weber Test to Determine Conduction and Sensorineural Deafness

Strike a tuning fork on the heel of your hand or with a rubber mallet, and place the handle of the tuning fork medially on your partner's head **(Figure 19.6a)**. Is the tone equally loud in both ears, or is it louder in one ear?

____

If it is equally loud in both ears, the subject has equal hearing or equal loss of hearing in both ears. If sensorineural deafness is present in one ear, the subject will hear the tone in the unaffected ear, but not in the ear with sensorineural deafness. *Conduction deafness* occurs when something prevents sound waves from reaching the fluids of the internal ear. Compacted earwax, a perforated eardrum, inflammation of the middle ear (otitis media), and damage to the ossicles are all causes of conduction deafness. If conduction deafness is present, the subject will hear the sound more strongly in the ear in which there is a hearing loss, because the bone of the skull is conducting the sound. Conduction deafness can be simulated by plugging one ear with cotton.

Rinne Test for Comparing Bone- and Air-Conduction Hearing

1. Strike the tuning fork, and place its handle on your partner's mastoid process (Figure 19.6b).

2. When your partner indicates that the sound is no longer audible, hold the still-vibrating prongs close to his or her acoustic meatus (Figure 19.6c). If your partner hears the fork again (by air conduction) when it is moved to that position, hearing is not impaired, and the test result is to be recorded as positive (+). Record in step 5 below.

3. Repeat the test, but this time test air-conduction hearing first.

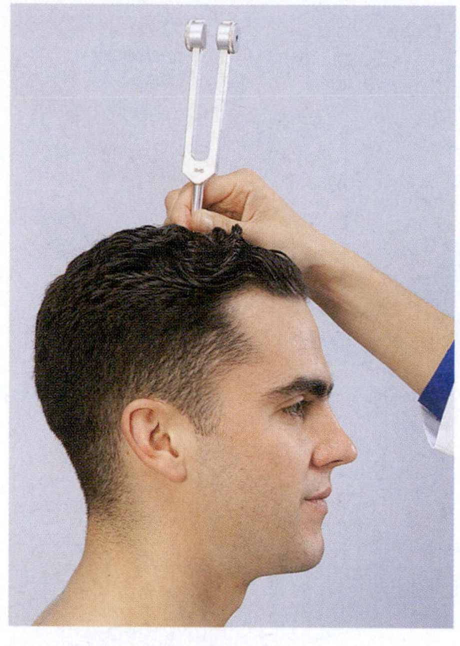

(a)

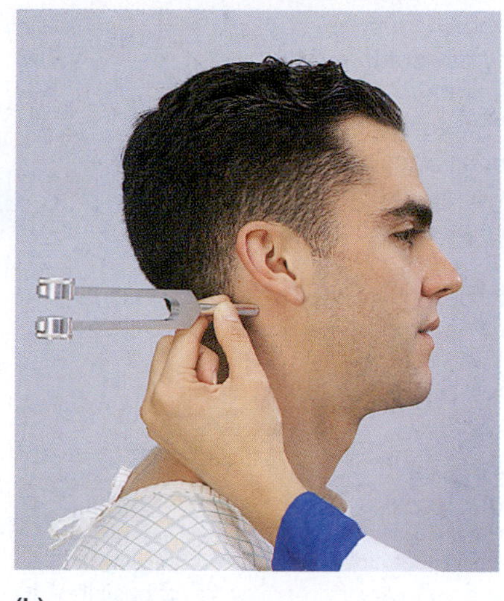

(b)

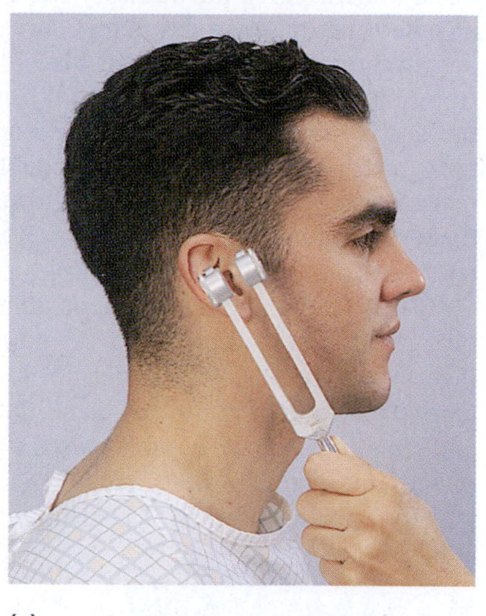

(c)

Figure 19.6 The Weber and Rinne tuning fork tests. (a) The Weber test to evaluate whether the sound remains centralized (normal) or lateralizes to one side or the other (indicative of some degree of conduction or sensorineural deafness). **(b, c)** The Rinne test to compare bone conduction and air conduction.

4. After the tone is no longer heard by air conduction, hold the handle of the tuning fork on the bony mastoid process. If the subject hears the tone again by bone conduction after hearing by air conduction is lost, there is some conduction deafness, and the result is recorded as negative (–).

5. Repeat the sequence for the opposite ear.

Right ear: ______________ Left ear: ______________

Does the subject hear better by bone or by air conduction?

____

Acuity Test

Have your lab partner pack one ear with cotton and sit quietly with eyes closed. Obtain a ticking clock or pocket watch, and hold it very close to his or her *unpacked* ear. Then slowly move it away from the ear until your partner signals that the ticking is no longer audible. Record the distance in centimeters at which ticking is inaudible, and then remove the cotton from the packed ear.

Right ear: ______________ Left ear: ______________

Is the threshold of audibility sharp or indefinite?

____

Sound Localization

Ask your partner to close both eyes. Hold the pocket watch at an audible distance (about 15 cm) from the ear, and move it to various locations (front, back, sides, and above the head). Have your partner locate the position by pointing in each instance. Can the sound be localized equally well at all positions? ______________________________

If not, at what position(s) was the sound less easily located?

____

____

The ability to localize the source of a sound depends on two factors—the difference in the loudness of the sound reaching each ear and the time of arrival of the sound at each ear. How does this information help to explain your findings?

____

____

Frequency Range of Hearing

Obtain three tuning forks: one with a low frequency (75 to 100 hertz [Hz; 1 Hz = 1 cycle per second, or cps]), one with a frequency of approximately 1000 Hz, and one with a frequency of 4000 to 5000 Hz. Strike the lowest-frequency fork, and hold it close to your partner's ear. Repeat with the other two forks.

Which fork was heard most clearly and comfortably?

______________ Hz

Which was heard least well? ______________ Hz

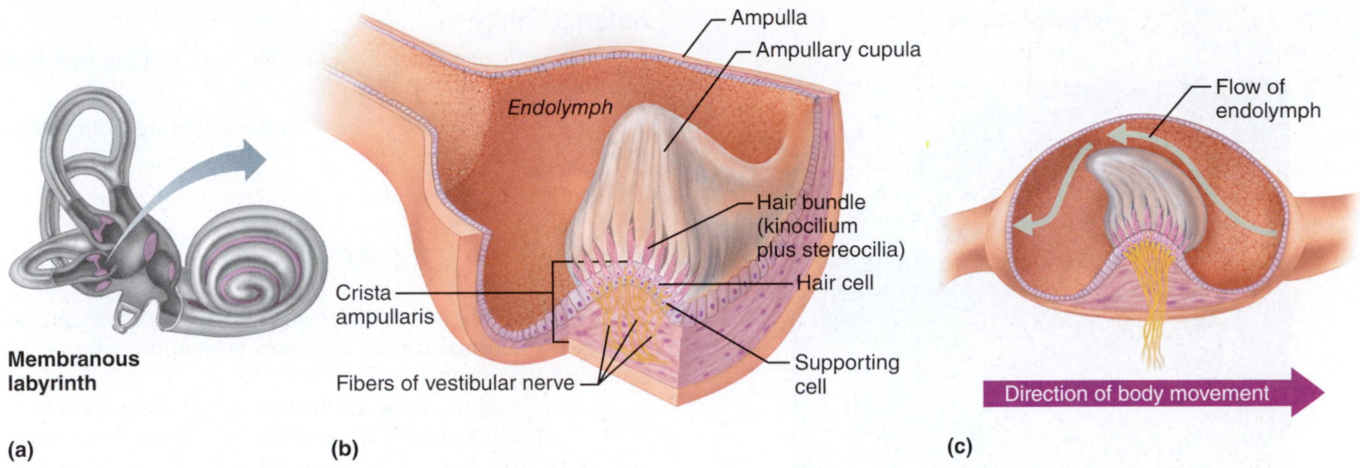

Figure 19.7 Structure and function of the crista ampullaris. (a) Arranged in the three spatial planes, the semicircular ducts in the semicircular canals each have a swelling called an ampulla at their base. **(b)** Each ampulla contains a crista ampullaris, a receptor that is essentially a cluster of hair cells with hairs projecting into a gelatinous cap called the ampullary cupula. **(c)** Movement of the ampullary cupula during angular acceleration of the head.

Microscopic Anatomy of the Equilibrium Apparatus and Mechanisms of Equilibrium

The equilibrium receptors of the internal ear are collectively called the **vestibular apparatus,** and are found in the vestibule and semicircular canals of the bony labyrinth. Their chambers are filled with perilymph, in which membranous labyrinth structures are suspended. The vestibule contains the saclike **utricle** and **saccule,** and the semicircular chambers contain **membranous semicircular ducts** (see Figure 19.2). Like the cochlear duct, these membranes are filled with endolymph and contain receptor cells that are activated by the bending of their cilia.

The semicircular canals monitor angular movements of the head. This process is called **dynamic equilibrium.** The canals are 1.2 cm in circumference and are oriented in three planes—horizontal, frontal, and sagittal. At the base of each semicircular duct is an enlarged region, the **ampulla,** which communicates with the utricle of the vestibule. Within each ampulla is a receptor region called a **crista ampullaris,** which consists of a tuft of hair cells covered with a gelatinous cap, or **ampullary cupula** (Figure 19.7).

The cristae respond to changes in the velocity of rotational head movements. For example, consider what happens when you twirl around. During acceleration, when you begin to twirl around, inertia causes the endolymph in the canal to lag behind the head movement, pushing the ampullary cupula—like a swinging door—in the opposite direction. The head movement depolarizes the hair cells and results in enhanced impulse transmission in the vestibular division of the eighth cranial nerve to the brain (Figure 19.7c). If the body continues to rotate at a constant rate, the endolymph eventually comes to rest and moves at the same speed as the body. The ampullary cupula returns to its upright position, hair cells are no longer stimulated, and you lose the sensation of spinning. When rotational movement stops suddenly, the endolymph keeps on going in the direction of head movement. This pushes the ampullary cupula in the *same* direction as the previous head movement and hyperpolarizes the hair cells; as a result, fewer impulses are transmitted to the brain. This tells the brain that you have stopped moving and accounts for the reversed motion sensation you feel when you stop twirling suddenly.

ACTIVITY 5

Examining the Microscopic Structure of the Crista Ampullaris

Go to the demonstration area and examine the slide of a crista ampullaris. Identify the areas depicted in the photomicrograph (Figure 19.8) and the labeled diagram (Figure 19.7b and c). ■

Maculae in the vestibule contain another set of **hair cells,** receptors that in this case monitor head position and acceleration in a straight line. This monitoring process is called **static equilibrium.** The maculae respond to gravitational pull, thus providing information on which way is up or down, and to linear or straightforward changes in speed. They are located on the walls of the saccule and utricle. The hair cells in each macula are embedded in the **otolith membrane,** a gelatinous material containing small grains of calcium carbonate called **otoliths.** When the head moves, the otoliths move in response to variations in gravitational pull. As they deflect different hair cells, they trigger hyperpolarization or depolarization of the hair cells and modify the rate of impulse transmission along the vestibular nerve (Figure 19.9).

Although the receptors of the semicircular canals and the vestibule are responsible for dynamic and static equilibrium respectively, they rarely act independently. Complex interaction of many of the receptors is the rule. Processing is also

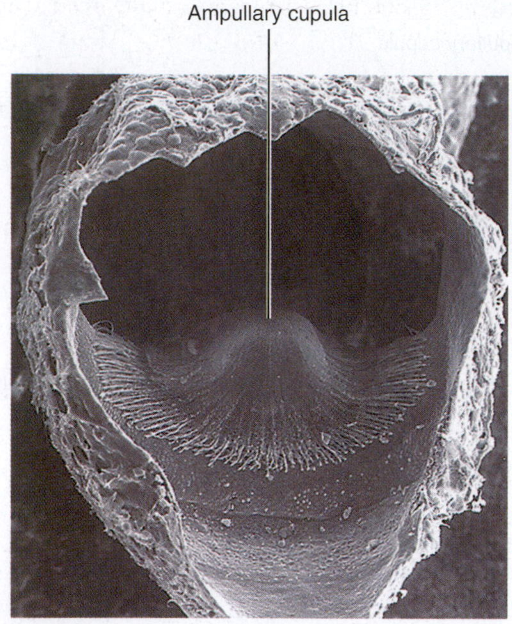

Figure 19.8 Scanning electron micrograph of a crista ampullaris (20×).

complex and involves the brain stem and cerebellum as well as input from proprioceptors and the eyes.

ACTIVITY 6

Conducting Laboratory Tests on Equilibrium

The function of the semicircular canals and vestibule is not routinely tested in the laboratory, but the following simple tests illustrate normal equilibrium apparatus function as well as some of the complex processing interactions.

Balance Tests

1. Have your partner walk a straight line, placing one foot directly in front of the other.

Is he or she able to walk without undue wobbling from side to side? ______________________

Did he or she experience any dizziness? ______________

The ability to walk with balance and without dizziness, unless subject to rotational forces, indicates normal function of the equilibrium apparatus.

Nystagmus is the involuntary rolling of the eyes in any direction or the trailing of the eyes slowly in one direction, followed by their rapid movement in the opposite direction. It is normal after rotation, abnormal otherwise. The direction of nystagmus is that of its quick phase on acceleration. ✚

Was nystagmus present? ______________________

2. Place three coins of different sizes on the floor. Ask your lab partner to pick up the coins, and carefully observe his or her muscle activity and coordination.

Did your lab partner have any difficulty locating and picking up the coins? ______________________

Describe your observations and your lab partner's observations during the test.

____

____

____

____

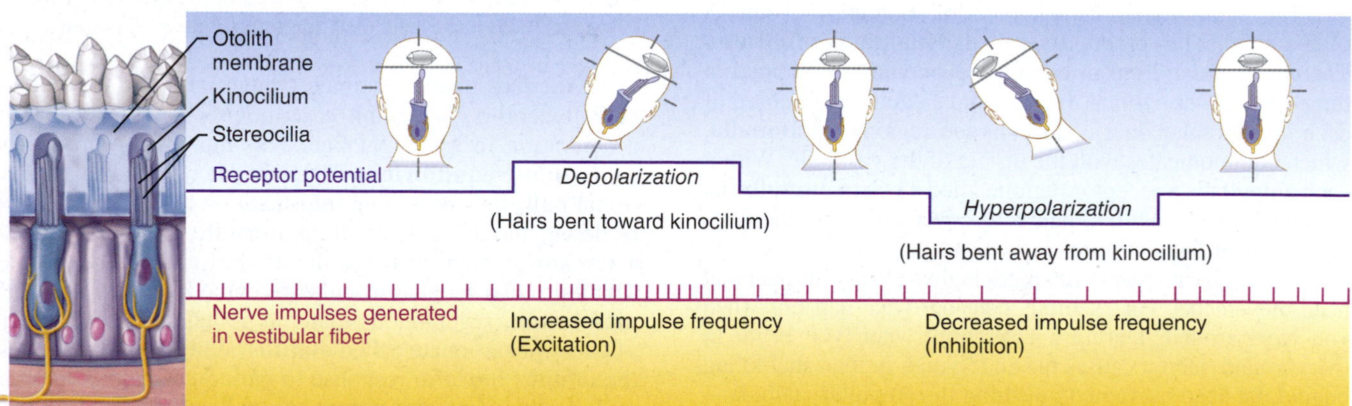

Figure 19.9 The effect of gravitational pull on a macula receptor in the utricle. When movement of the otolith membrane bends the hair cells in the direction of the kinocilium, the hair cells depolarize, exciting the nerve fibers, which generate action potentials more rapidly. When the hairs are bent in the direction away from the kinocilium, the hair cells become hyperpolarized, inhibiting the nerve fibers and decreasing the action potential rate (i.e., below the resting rate of discharge).

What kinds of interactions involving balance and coordination must occur for a person to move fluidly during this test?

___

___

3. If a person has a depressed nervous system, mental concentration may result in a loss of balance. Ask your lab partner to stand up and count backward from 10 as rapidly as possible.

Did your lab partner lose balance? ___

Romberg Test

The Romberg test determines the integrity of the dorsal white column of the spinal cord, which transmits impulses to the brain from the proprioceptors involved with posture.

1. Have your partner stand with his or her back to the blackboard or whiteboard.

2. Draw one line parallel to each side of your partner's body. He or she should stand erect, with eyes open and staring straight ahead for 2 minutes while you observe any movements. Did you see any gross swaying movements?

___

3. Repeat the test. This time the subject's eyes should be closed. Note and record the degree of side-to-side movement.

___

4. Repeat the test with the subject's eyes first open and then closed. This time, however, the subject should be positioned with the left shoulder toward, but not touching, the board so that you may observe and record the degree of front-to-back swaying.

___

___

___

Do you think the equilibrium apparatus of the internal ear was operating equally well in all these tests?

___

The proprioceptors? ___

Why was the observed degree of swaying greater when the eyes were closed?

___

___

What conclusions can you draw regarding the factors necessary for maintaining body equilibrium and balance?

___

___

___

___

Role of Vision in Maintaining Equilibrium

To further demonstrate the role of vision in maintaining equilibrium, perform the following experiment. (Ask your lab partner to record observations and act as a "spotter.") Stand erect, with your eyes open. Raise your left foot approximately 30 cm off the floor, and hold it there for 1 minute.

1. Record the observations: ___

___

___

2. Rest for 1 or 2 minutes, and then repeat the experiment with the same foot raised but with your eyes closed. Record the observations:

___

___

Name ______________________________

Lab Time/Date ______________________

EXERCISE 19

REVIEW SHEET

Special Senses: Hearing and Equilibrium

Anatomy of the Ear

1. Select the terms from column B that apply to the column A descriptions. (Some terms are used more than once.)

Column A	Column B
________, ________, ________ 1. structures composing the external ear (3)	a. auricle
________, ________, ________ 2. structures composing the internal ear (3)	b. cochlea
________, ________, ________ 3. collectively called the auditory ossicles (3)	c. endolymph
________ 4. involved in equalizing the pressure in the middle ear with atmospheric pressure	d. external acoustic meatus
________ 5. vibrates at the same frequency as sound waves hitting it; transmits the vibrations to the auditory ossicles	e. incus (anvil)
________ 6. transmits the vibratory motion of the stirrup to the fluid in the scala vestibuli of the internal ear	f. malleus (hammer)
________ 7. acts as a pressure relief valve for the increased fluid pressure in the scala tympani; bulges into the tympanic cavity	g. oval window
________ 8. passage between the throat and the tympanic cavity	h. perilymph
________, ________ 9. contain receptors for the sense of balance (2)	i. pharyngotympanic (auditory) tube
________ 10. fluid contained within the membranous labyrinth	j. round window
________ 11. fluid contained within the bony labyrinth and bathing the membranous labyrinth	k. semicircular canals
	l. stapes (stirrup)
	m. tympanic membrane
	n. vestibule

2. Identify all indicated structures and ear regions in the following diagram.

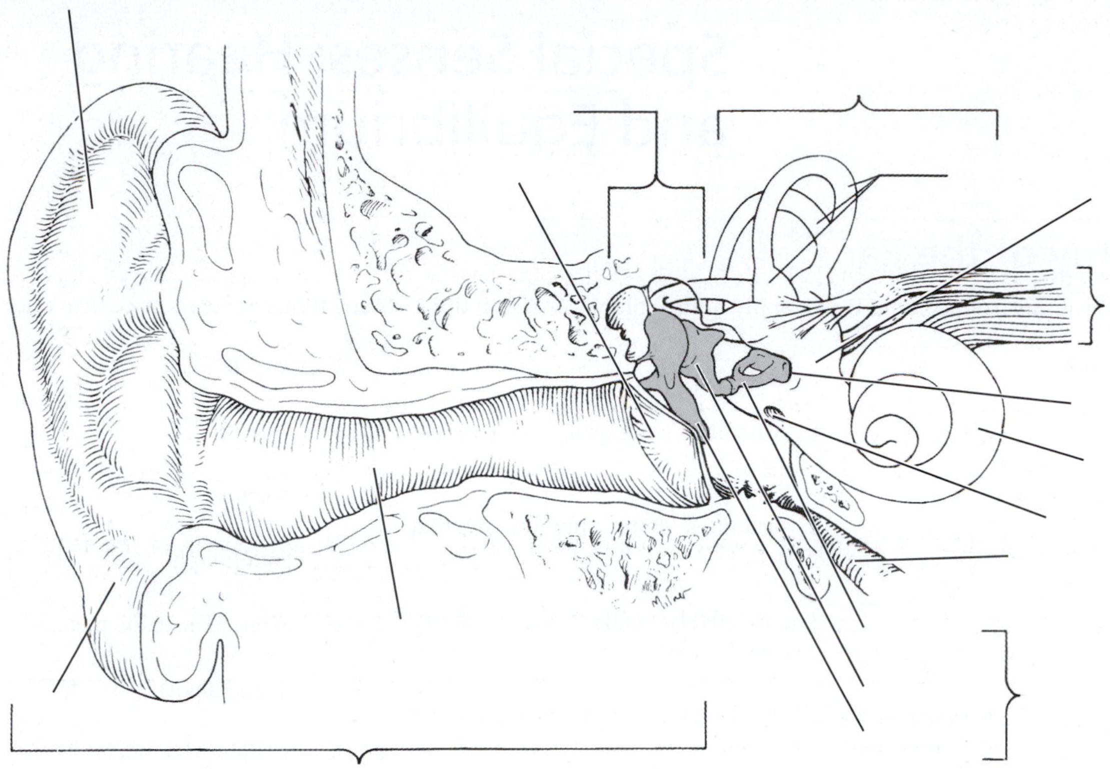

3. Match the membranous labyrinth structures listed in column B with the descriptive statements in column A. (Some terms are used more than once.)

Column A	Column B
________, ________ 1. sacs found within the vestibule (2)	a. ampulla
________ 2. contains the spiral organ	b. ampullary cupula
________, ________ 3. sites of the maculae (2)	c. basilar membrane
________ 4. positioned in all spatial planes	d. cochlear duct
________ 5. hair cells of spiral organ rest on this membrane	e. cochlear nerve
________ 6. gelatinous membrane overlying the hair cells of the spiral organ	f. otoliths
________ 7. contains the crista ampullaris	g. saccule
________, ________, ________, ________ 8. function in static equilibrium (4)	h. semicircular ducts
________, ________, ________, ________ 9. function in dynamic equilibrium (4)	i. tectorial membrane
________ 10. carries auditory information to the brain	j. utricle
________ 11. gelatinous cap overlying hair cells of the crista ampullaris	k. vestibular nerve
________ 12. grains of calcium carbonate in the maculae	

4. Sound waves hitting the tympanic membrane initiate its vibratory motion. Trace the pathway through which vibrations and fluid currents are transmitted to finally stimulate the hair cells in the spiral organ. (Name the appropriate ear structures in their correct sequence.)

Tympanic membrane → ______________________________

____

____

5. Describe how sounds of different frequency (pitch) are differentiated in the cochlea. ______________________________

____

____

____

6. Explain the role of the endolymph of the semicircular canals in activating the receptors during angular motion.

____

____

____

7. Explain the role of the otoliths in perception of static equilibrium (head position). ______________________________

____

____

____

Laboratory Tests

8. Was the auditory acuity measurement made in Activity 4 (on page 332) the same or different for both ears?

______________________________ What factors might account for a difference in the acuity of the two ears?

____

____

9. During the sound localization experiment in Activity 4 (on page 332), in which position(s) was the sound least easily located?

____

How can this phenomenon be explained? ______________________________

____

10. In the frequency experiment in Activity 4 (on page 332), which tuning fork was the most difficult to hear? __________ Hz

What conclusion can you draw? ______________________________

____

11. When the tuning fork handle was pressed to your forehead during the Weber test, where did the sound seem to originate?

____

Where did it seem to originate when one ear was plugged with cotton? ______________________________

How do sound waves reach the cochlea when conduction deafness is present? ______________________________

____

12. Indicate whether the following conditions relate to conduction deafness (C), or sensorineural deafness (S), or both (C and S).

_______ 1. can result from the fusion of the ossicles

_______ 2. can result from a lesion on the cochlear nerve

_______ 3. sound heard in one ear but not in the other during bone and air conduction

_______ 4. can result from otitis media

_______ 5. can result from impacted cerumen or a perforated eardrum

_______ 6. can result from a blood clot in the auditory cortex

13. The Rinne test evaluates an individual's ability to hear sounds conducted by air or bone. Which is more indicative of normal hearing? ______________________________

14. Define *nystagmus*. ______________________________

____

15. What is the usual reason for conducting the Romberg test? ______________________________

____

Was the degree of sway greater with the eyes open or closed? Why? ______________________________

____

16. Normal balance, or equilibrium, depends on input from a number of sensory receptors. Name them.

____

____

20

EXERCISE

Special Senses: Olfaction and Taste

MATERIALS

- ☐ Prepared slides: the tongue showing taste buds; nasal olfactory epithelium (l.s.)
- ☐ Compound microscope
- ☐ Small mirror
- ☐ Paper towels
- ☐ Packets of granulated sugar
- ☐ Disposable autoclave bag
- ☐ Cotton-tipped swabs
- ☐ Prepared vials of oil of cloves, oil of peppermint, and oil of wintergreen, or corresponding flavors found in the condiment section of a supermarket
- ☐ Nose clips
- ☐ Paper cups
- ☐ Flask of distilled or tap water
- ☐ Absorbent cotton
- ☐ Paper plates
- ☐ Foil-lined egg carton containing equal-sized food cubes of cheese, apple, raw potato, dried prunes, banana, raw carrot, and hard-cooked egg white
- ☐ Toothpicks
- ☐ Disposable gloves
- ☐ Chipped ice

OBJECTIVES

1. State the location and cellular composition of the olfactory epithelium.
2. Describe the structure of olfactory sensory neurons, and state their function.
3. Discuss the locations and cellular composition of taste buds.
4. Describe the structure of gustatory epithelial cells, and state their function.
5. Identify the cranial nerves that carry the sensations of olfaction and taste.
6. Name five basic qualities of taste sensation.
7. Explain the interdependence between the senses of smell and taste.
8. Name two factors other than olfaction that influence taste appreciation of foods.
9. Define *olfactory adaptation*.

PRE-LAB QUIZ

1. Circle True or False. Receptors for olfaction and taste are classified as chemoreceptors because they respond to dissolved chemicals.
2. The organ of smell is the ___________, located in the roof of the nasal cavity.
   a. nares c. olfactory epithelium
   b. nostrils d. olfactory nerve
3. Circle the correct underlined term. Olfactory receptors are bipolar / unipolar sensory neurons whose olfactory cilia extend outward from the epithelium.
4. Most taste buds are located in ___________, peglike projections of the tongue mucosa.
   a. cilia c. papillae
   b. conchae d. supporting cells
5. Circle the correct underlined term. Vallate papillae are arranged in a V formation on the anterior / posterior surface of the tongue.
6. Circle the correct underlined term. Most taste buds are made of two / three types of modified epithelial cells.
7. There are five basic taste sensations. Name one. ___________
8. Circle True or False. Taste buds typically respond optimally to one of the five basic taste sensations.
9. Circle True or False. Texture, temperature, and smell have little or no effect on the sensation of taste.
10. You will use absorbent cotton and oil of wintergreen, peppermint, or cloves to test for olfactory:
    a. accommodation c. identification
    b. adaptation d. recognition

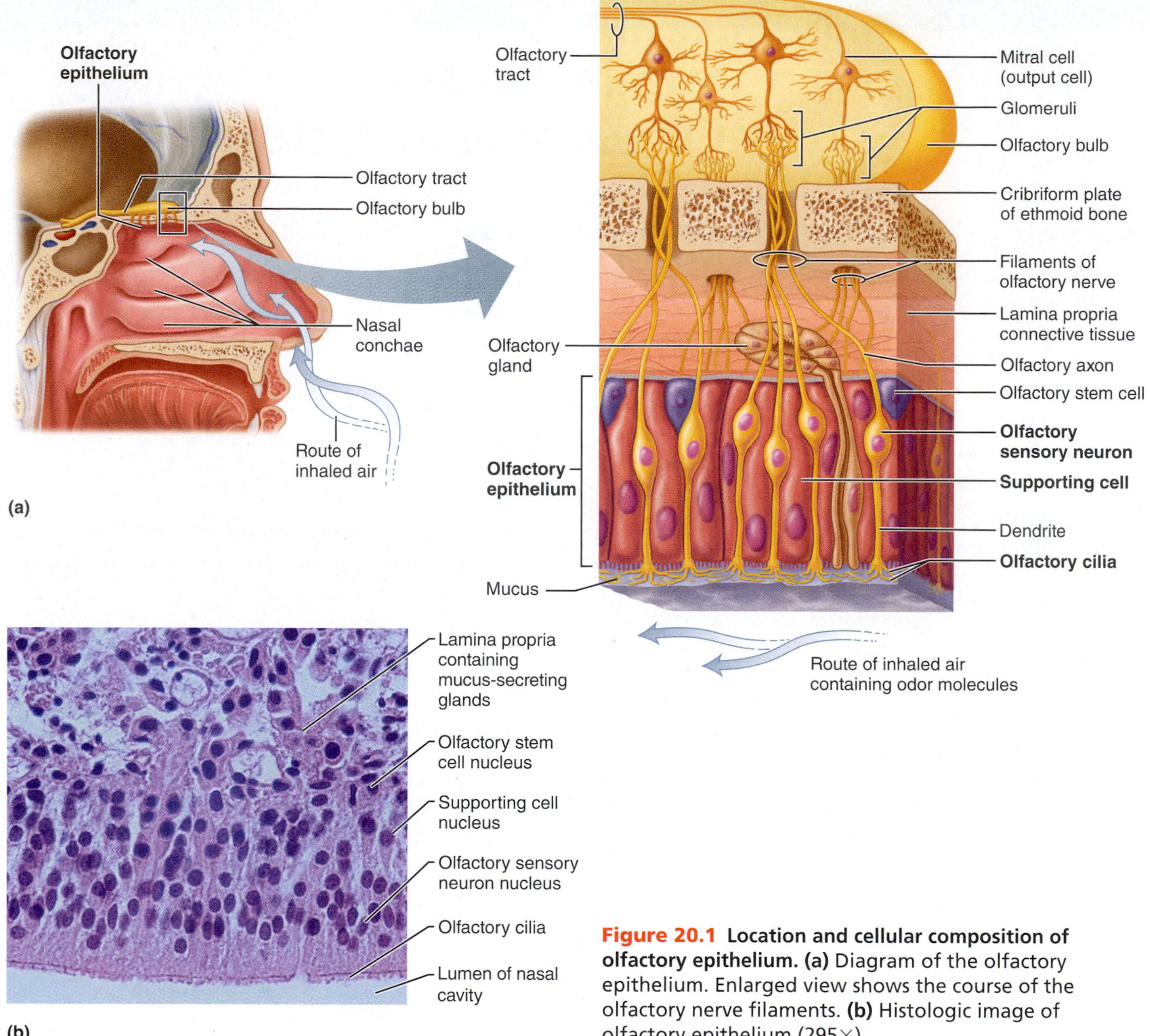

Figure 20.1 Location and cellular composition of olfactory epithelium. (a) Diagram of the olfactory epithelium. Enlarged view shows the course of the olfactory nerve filaments. **(b)** Histologic image of olfactory epithelium (295×).

The receptors for olfaction and taste are classified as **chemoreceptors** because they respond to chemicals in solution. Although five relatively specific types of taste receptors have been identified, the olfactory receptors are considered sensitive to a much wider range of chemical sensations. The sense of smell is the least understood of the special senses.

Localization and Anatomy of the Olfactory Receptors

The **olfactory epithelium** is the organ of smell. It occupies an area of about 5 cm^2 in the roof of the nasal cavity **(Figure 20.1a)**. Because the air entering the human nasal cavity must make a hairpin turn to enter the respiratory passages below, the nasal epithelium is in a rather poor position for performing its function. This is why sniffing, which brings more air into contact with the receptors, increases your ability to detect odors.

The specialized receptor cells in the olfactory epithelium are **olfactory sensory neurons.** They are surrounded by epithelial supporting cells. The bipolar neurons have **olfactory cilia** that extend outward from the epithelium (Figure 20.1a and b). Axons emerging from their basal ends penetrate the cribriform plate of the ethmoid bone and proceed as the *olfactory filaments* (cranial nerve I) to synapse in the olfactory bulbs lying on either side of the crista galli of the ethmoid bone. Impulses from neurons of the olfactory bulbs are then conveyed to the olfactory portion of the cortex without synapsing in the thalamus.

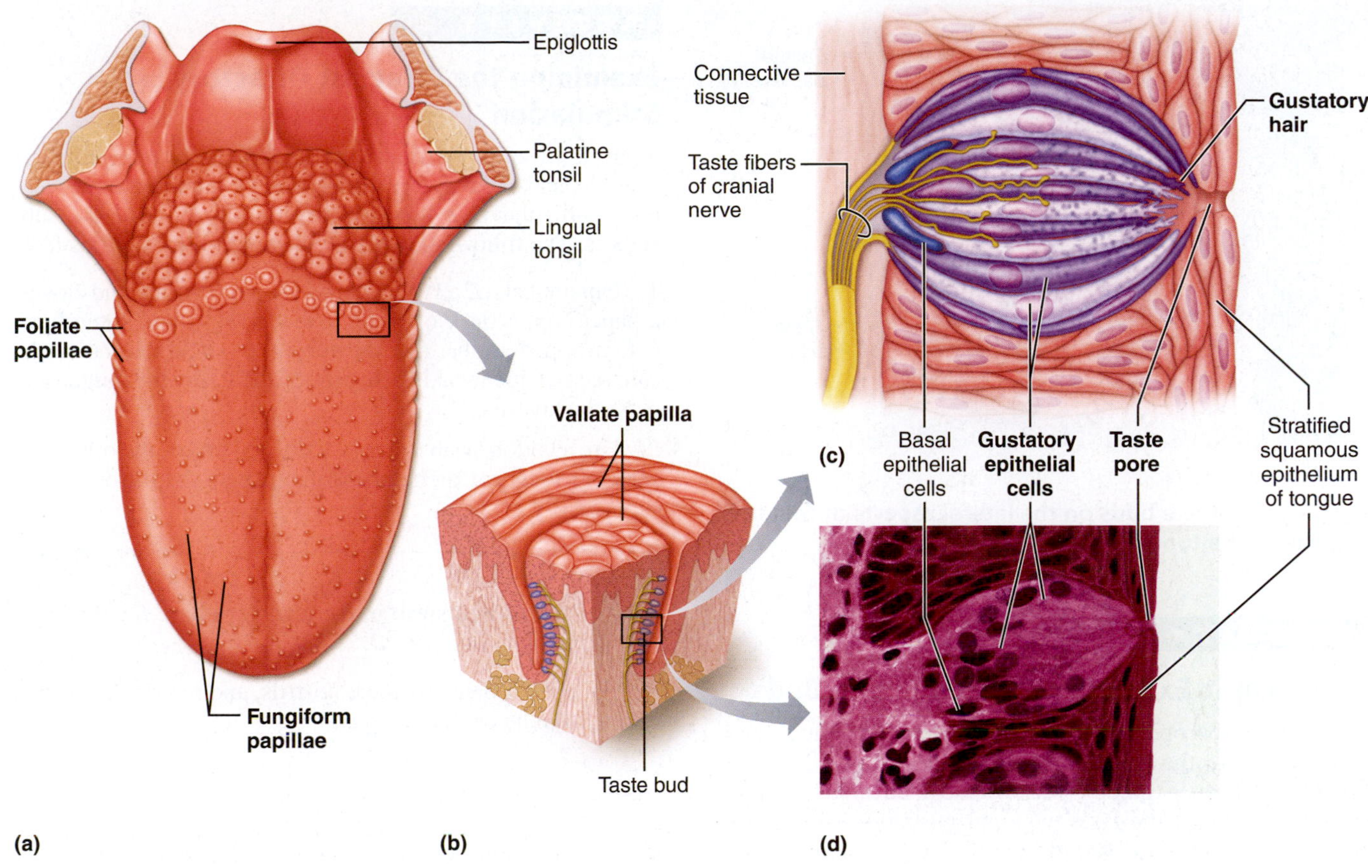

Figure 20.2 Location and structure of taste buds. (a) Taste buds on the tongue are associated with papillae, projections of the tongue mucosa. **(b)** A sectioned vallate papilla shows the position of the taste buds in its lateral walls. **(c)** An enlarged view of a taste bud. **(d)** Photomicrograph of a taste bud (510×). (See also Figure 20.3.)

ACTIVITY 1

Microscopic Examination of the Olfactory Epithelium

Obtain a longitudinal section of olfactory epithelium. Examine it closely using a compound microscope, comparing it to the photomicrograph (Figure 20.1b).

Localization and Anatomy of the Taste Buds

The **taste buds,** containing specific receptors for the sense of taste, are widely but not uniformly distributed in the oral cavity. Most are located in **papillae,** peglike projections of the mucosa, on the dorsal surface of the tongue (as described next). A few are found on the soft palate, epiglottis, pharynx, and inner surface of the cheeks.

The taste buds are located primarily on the sides of the large, **vallate papillae** (arranged in a V formation on the posterior surface of the tongue); in the side walls of the **foliate papillae;** and on the tops of the more numerous **fungiform papillae.** The latter look rather like minute mushrooms and are widely distributed on the tongue **(Figure 20.2)**.

☐ Use a mirror to examine your tongue. Which of the various papillae types can you pick out?

____

____

Each taste bud consists largely of a globular arrangement of two types of modified epithelial cells: the **gustatory epithelial cells,** which are the actual receptor cells, and **basal epithelial cells.** Several nerve fibers enter each taste bud and supply sensory nerve endings to each of the gustatory epithelial cells. The long microvilli of the receptor cells penetrate the epithelial surface through an opening called the **taste pore.** When these microvilli, called **gustatory hairs,** contact specific chemicals in the solution, the receptor cells depolarize. The afferent fibers from the taste buds to the primary somatosensory cortex in the postcentral gyrus of the brain are carried in three cranial nerves: the *facial nerve (VII)* serves the anterior two-thirds of the tongue; the *glossopharyngeal nerve (IX)* serves the posterior third of the tongue; and the *vagus nerve (X)* carries a few fibers from the pharyngeal region.

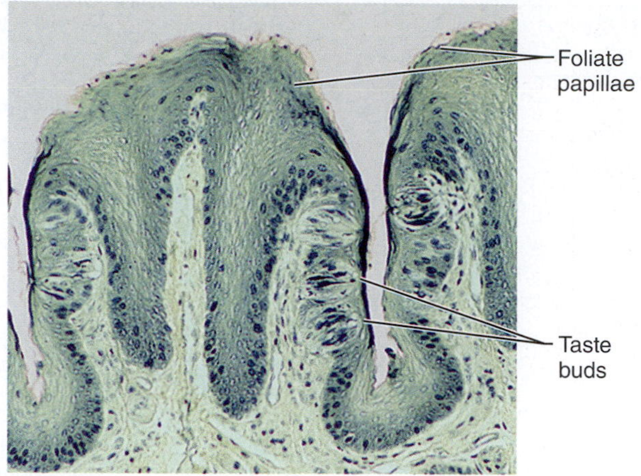

Figure 20.3 Taste buds on the lateral aspects of foliate papillae of the tongue (180×).

ACTIVITY 2

Microscopic Examination of the Taste Buds

Obtain a microscope and a prepared slide of a tongue cross section. Locate the taste buds on the tongue papillae (use Figure 20.2b as a guide). Make a detailed study of one taste bud. Identify the taste pore and gustatory hairs if observed. Compare your observations to the photomicrograph **(Figure 20.3)**. ■

When taste is tested with pure chemical compounds, most taste sensations can be grouped into one of five basic qualities—sweet, sour, bitter, salty, or umami (oo-mom-ē, "delicious"). Although all taste buds are believed to respond in some degree to all five classes of chemical stimuli, each type responds optimally to only one.

Laboratory Experiments

Notify instructor of any food or scent allergies before beginning experiments.

ACTIVITY 3

Stimulating Taste Buds

1. Obtain several paper towels, a sugar packet, and a disposable autoclave bag and bring them to your bench.
2. With a paper towel, dry the dorsal surface of your tongue.

Immediately dispose of the paper towel in the autoclave bag.

3. Tear off a corner of the sugar packet, and shake a few sugar crystals on your dried tongue. Do *not* close your mouth.

Time how long it takes to taste the sugar. __________ sec

Why couldn't you taste the sugar immediately?

____ ■

ACTIVITY 4

Examining the Effect of Olfactory Stimulation

There is no question that what is commonly referred to as taste depends heavily on stimulation of the olfactory receptors, particularly in the case of strongly odoriferous substances. The following experiment should illustrate this fact.

1. Obtain vials of oil of wintergreen, peppermint, and cloves, a paper cup, a flask of water, paper towels, and some fresh cotton-tipped swabs. Ask the subject to sit so that he or she cannot see which vial is being used, and to dry the tongue and close the nostrils.
2. Use a cotton swab to apply a drop of one of the oils to the subject's tongue. Can he or she distinguish the flavor?

____

Put the used swab in the autoclave bag. *Do not redip the swab into the oil.*

3. Have the subject open the nostrils, and record the change in sensation he or she reports.

____

4. Have the subject rinse the mouth well and dry the tongue.
5. Prepare two swabs, each with one of the two remaining oils.
6. Hold one swab under the subject's open nostrils while touching the second swab to the tongue.

Record the reported sensations. ______________________

____

____

____

7. Dispose of the used swabs and paper towels in the autoclave bag before continuing.

Which sense, taste or smell, appears to be more important in proper identification of a strongly flavored volatile substance?

____ ■

ACTIVITY 5

Demonstrating Olfactory Adaptation

When olfactory receptors are subjected to the same odor for a prolonged period, they eventually stop responding to that stimulus, a phenomenon called **olfactory adaptation.** This activity allows you to demonstrate olfactory adaptation.

Obtain some absorbent cotton and two of the following oils (oil of wintergreen, peppermint, or cloves). Place several

Activity 6: Identification by Texture and Smell				
Food tested	Texture only	Chewing with nostrils pinched	Chewing with nostrils open	Identification not made

drops of one oil on the absorbent cotton. Press one nostril shut. Hold the cotton under the open nostril, and exhale through the mouth. Record the time required for the odor to disappear (for olfactory adaptation to occur).

________________ sec

Repeat the procedure with the other nostril.

________________ sec

Immediately test another oil with the nostril that has just experienced olfactory adaptation. What are the results?

____

What conclusions can you draw? ________________

____

ACTIVITY 6

Examining the Combined Effects of Smell, Texture, and Temperature on Taste

Effects of Smell and Texture

1. Ask the subject to sit with eyes closed and to pinch his or her nostrils shut.

2. Using a paper plate, obtain samples of the food items provided by your laboratory instructor. Do not let the subject see the foods being tested. Wear plastic gloves, and use toothpicks to handle food.

3. For each test, place a cube of food in the subject's mouth, and ask him or her to identify the food by using the following sequence of activities:

- First, manipulate the food with the tongue.
- Second, chew the food.
- Third, if the subject cannot make a positive identification with the first two techniques and the taste sense, ask him or her to release the pinched nostrils and to continue chewing with the nostrils open to make a positive identification.

In the **Activity 6 chart,** record the type of food, and then put a check mark in the appropriate column for the result.

Was the sense of smell equally important in all cases?

____

When did it seem to be important, and why?

____

____

Discard gloves in the autoclave bag.

Effect of Temperature

In addition to the roles that olfaction and food texture play in determining our taste sensations, the temperature of foods also helps determine whether the food is appreciated or even tasted. To illustrate this, have your partner hold some chipped ice on the tongue for approximately a minute and then close his or her eyes. Immediately place in the person's mouth any of the foods previously identified, and ask for an identification.

What are the results?

____

____

____

____

NAME ___________________________

LAB TIME/DATE ___________________

EXERCISE 20

REVIEW SHEET

Special Senses: Olfaction and Taste

Localization and Anatomy of the Olfactory Receptors

1. Describe the location and cellular composition of the olfactory epithelium. ___________________________

____

____

2. How and why does sniffing increase your ability to detect an odor? ___________________________

____

Localization and Anatomy of Taste Buds

3. Name five sites where receptors for taste are found, and circle the predominant site.

_______________, _______________, _______________,

_______________, and _______________

4. Describe the cellular makeup and arrangement of a taste bud. ___________________________

____

Laboratory Experiments

5. Taste and smell receptors are both classified as _______________, because they both respond to _______________.

6. Why is it impossible to taste substances with a dry tongue? ___________________________

____

7. Name the five basic taste sensations.

1. _______________
2. _______________
3. _______________
4. _______________
5. _______________

8. Name three factors that influence our appreciation of foods. Substantiate each choice with an example from the lab experience.

1. ______________ Substantiation ______________

____

2. ______________ Substantiation ______________

____

3. ______________ Substantiation ______________

____

Expand on your choices by explaining why a cold, greasy hamburger is unappetizing to most people. ______________

____

9. How palatable is food when you have a cold? ______________

Explain your answer. ______________

____

10. In your opinion, is olfactory adaptation desirable? ______________

Explain your answer. ______________

____

EXERCISE 21

Functional Anatomy of the Endocrine Glands

MATERIALS

- ☐ Human torso model
- ☐ Anatomical chart of the human endocrine system
- ☐ Compound microscope
- ☐ Prepared slides of the anterior pituitary and pancreas (with differential staining), posterior pituitary, thyroid gland, parathyroid glands, and adrenal gland
- ☐ Dissection animal, tray, and instruments
- ☐ Bone cutters
- ☐ Embalming fluid
- ☐ Disposable gloves
- ☐ Organic debris container

OBJECTIVES

1. Identify the major endocrine glands and tissues of the body when provided with an appropriate image.
2. List the hormones produced by the endocrine glands, and discuss the general function of each.
3. Explain how hormones contribute to body homeostasis by giving appropriate examples of hormonal actions.
4. Discuss some mechanisms that stimulate release of hormones from endocrine glands.
5. Describe the structural and functional relationship between the hypothalamus and the pituitary gland.
6. For several of the hormones studied, name a major pathological consequence of hypersecretion and of hyposecretion.
7. Correctly identify the histologic structure of the thyroid, parathyroid, pancreas, anterior and posterior pituitary, adrenal cortex, and adrenal medulla by microscopic inspection or in an image.
8. Name and point out the specialized hormone-secreting cells in the tissues listed above.
9. Identify and name the major endocrine organs on a dissected cat.

PRE-LAB QUIZ

1. Define *hormone*. ____
____
2. Circle the correct underlined term. An <u>endocrine</u> / <u>exocrine</u> gland is a ductless gland that empties its hormone into the extracellular fluid, from which it enters the blood.
3. The pituitary gland, also known as the ____________, is located in the sella turcica of the sphenoid bone.
   a. hypophysis
   b. hypothalamus
   c. thalamus
4. Circle True or False. The anterior pituitary gland is also referred to as the master endocrine gland because it controls the activity of many other endocrine glands.
5. The ____________ gland, composed of two lobes, is located in the throat, just inferior to the larynx.
   a. pancreas — c. thymus
   b. posterior pituitary — d. thyroid
6. The pancreas produces two hormones that are responsible for regulating blood sugar levels. Name the hormone that increases blood glucose levels. ______________________
7. Circle True or False. The gonads are considered to be both endocrine and exocrine glands.

(Text continues on next page.)

8. This gland is rather large in an infant, but begins to atrophy at puberty and is relatively inconspicuous by old age. It produces hormones that direct the maturation of T cells. It is the ________ gland.
   a. pineal
   b. testes
   c. thymus
   d. thyroid

9. Circle the correct underlined term. Pancreatic islets / Acinar cells form the endocrine portion of the pancreas.

10. The outer cortex of the adrenal gland is divided into three areas or regions. Which one produces aldosterone?
   a. zona fasciculata
   b. zona glomerulosa
   c. zona reticularis

The **endocrine system** is the second major control system of the body. Acting with the nervous system, it helps coordinate and integrate the activity of the body. The nervous system uses electrochemical impulses to bring about rapid control, whereas the more slowly acting endocrine system uses chemical messengers, or **hormones,** which ultimately enter the blood to be transported throughout the body.

The term *hormone* comes from a Greek word meaning "to arouse." The body's hormones, which are steroids or amino acid–based molecules, arouse the body's tissues and cells by stimulating changes in their metabolic activity. These changes lead to growth and development and to the physiological homeostasis of many body systems. Although all hormones are bloodborne, a given hormone affects the biochemical activity of only a specific organ or organs. Organs that respond to a particular hormone are referred to as the **target organs** of that hormone. The ability of the target tissue to respond depends on the ability of the hormone to bind with specific receptors occurring on the cells' plasma membranes or within the cells.

Although the function of some hormone-producing glands (the anterior lobe of the pituitary, thyroid, adrenals, parathyroids) is purely endocrine, the function of others (the pancreas and gonads) is mixed—both endocrine and exocrine. Both types of glands are derived from epithelium, but the endocrine glands release their hormones directly into the extracellular fluid, from which it enters blood or lymph. The exocrine glands release their products at the body's surface or upon an epithelial membrane via ducts. In addition, there are hormone-producing cells in the heart, gastrointestinal tract, kidney, skin, adipose tissue, skeleton, and placenta—organs whose functions are primarily nonendocrine. Here we consider only the major endocrine organs.

Gross Anatomy and Basic Function of the Endocrine Glands

Pituitary Gland (Hypophysis)

The **pituitary gland,** or **hypophysis,** is located in the hypophyseal fossa of the sella turcica of the sphenoid bone. It consists largely of two functional lobes, the **anterior lobe,** or **adenohypophysis,** and the **posterior lobe,** or **neurohypophysis** and **infundibulum**—the stalk that attaches the pituitary gland to the hypothalamus **(Figure 21.1)**.

Anterior Lobe Hormones

The anterior lobe of the pituitary produces and secretes a number of hormones, four of which are **tropic hormones.** The target organ of a tropic hormone is another endocrine gland, which secretes its hormone in response to stimulation. Hormones from these target glands exert their effects on other body organs and tissues.

Because the anterior pituitary controls the activity of many other endocrine glands, it has been called the *master endocrine gland*. However, it is now recognized that *releasing* or *inhibiting* hormones from neurons of the ventral hypothalamus control anterior pituitary cells; thus, the hypothalamus has superseded the anterior pituitary as the major controller of endocrine glands.

The anterior pituitary *tropic hormones* include the following:

- **Gonadotropins—follicle-stimulating hormone (FSH)** and **luteinizing hormone (LH)**—regulate gamete production and hormonal activity of the gonads (ovaries and testes).
- **Adrenocorticotropic hormone (ACTH)** regulates the endocrine activity of the adrenal cortex.
- **Thyroid-stimulating hormone (TSH),** or **thyrotropin,** influences the growth and activity of the thyroid gland.

The two other important hormones produced by the anterior lobe, growth hormone and prolactin, are not directly involved in regulating other endocrine glands of the body. They are:

- **Growth hormone (GH)** is a general metabolic hormone that plays an important role in determining body size. It affects many tissues of the body; however, it mainly affects the growth of muscle and the long bones of the body.

Hyposecretion results in pituitary dwarfism in children. Hypersecretion causes gigantism in children and **acromegaly** (overgrowth of bones in hands, feet, and face) in adults. +

- **Prolactin (PRL)** stimulates milk production by the breasts. The role of prolactin in males is not well understood.

The ventral hypothalamic hormones control production and secretion of the tropic hormones GH and PRL. The hypothalamic hormones reach the cells of the anterior pituitary through the **hypophyseal portal system** (Figure 21.1), a complex vascular arrangement of two capillary beds that are connected by the hypophyseal portal veins.

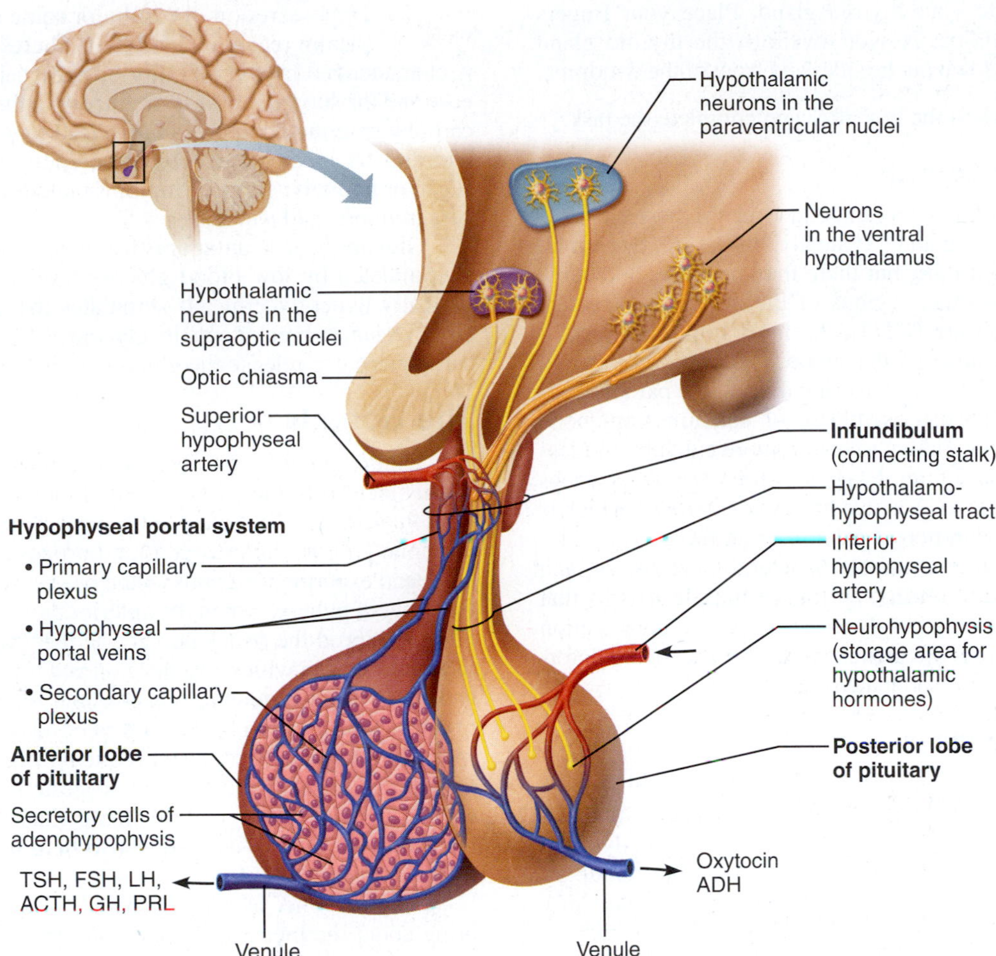

Figure 21.1 Hypothalamus and pituitary gland. Neural and vascular relationships between the hypothalamus and the anterior and posterior lobes of the pituitary are depicted.

Posterior Lobe Hormones

The posterior lobe is not an endocrine gland because it does not synthesize the hormones it releases. Instead, it acts as a storage area for two *neurohormones* transported to it via the axons of neurons in the paraventricular and supraoptic nuclei of the hypothalamus. These axons form the hypothalamo-hypophyseal tract. The hormones are released in response to nerve impulses in these neurons. The first of these hormones is **oxytocin,** which stimulates powerful uterine contractions during childbirth and also causes milk ejection in the lactating mother. The second, **antidiuretic hormone (ADH),** causes the tubules of the kidneys to resorb more water from the urinary filtrate, thereby reducing urine output and conserving body water. It also plays a minor role in increasing blood pressure because of its vasoconstrictor effect on the arterioles.

Hyposecretion of ADH results in dehydration from excessive urine output, a condition called **diabetes insipidus.** Individuals with this condition experience an insatiable thirst. Hypersecretion results in edema, headache, and disorientation. ✚

Thyroid Gland

The *thyroid gland* is composed of two lobes joined by a central mass, or isthmus. It is located in the throat, just inferior to the larynx. It produces two major hormones, thyroid hormone and calcitonin.

Thyroid hormone (TH) is actually two physiologically active hormones known as **T_4 (thyroxine)** and **T_3 (triiodothyronine).** Because its primary function is to control the rate of body metabolism and cellular oxidation, TH affects virtually every cell in the body.

Hyposecretion of thyroxine leads to a condition of mental and physical sluggishness, which is called **myxedema** in adults. Hypersecretion causes elevated metabolic rate, nervousness, weight loss, sweating, and irregular heartbeat. ✚

Calcitonin is released in response to high blood calcium levels. Although it decreases blood calcium levels by stimulating calcium salt deposit in the bones, it is not involved in day-to-day control of calcium homeostasis.

☐ Try to palpate your thyroid gland. Place your fingers against your windpipe. As you swallow, the thyroid gland will move up and down on the sides and front of the windpipe.

Place a check mark in the box after you complete the task.

Parathyroid Glands

The *parathyroid glands* are embedded in the posterior surface of the thyroid gland. Typically, there are two small oval glands on each lobe, but there may be more, and some may be located in other regions of the neck. They secrete **parathyroid hormone (PTH),** the most important regulator of calcium balance of the blood. When blood calcium levels decrease below a certain critical level, the parathyroids release PTH, which causes release of calcium from bone matrix and prods the kidney to resorb more calcium and less phosphate from the filtrate. PTH also stimulates the kidneys to convert vitamin D to its active D_3 form, *calcitriol,* which is required for the absorption of calcium from food.

Hyposecretion increases neural excitability and may lead to **tetany,** prolonged muscle spasms that can result in respiratory paralysis and death. Hypersecretion of PTH results in loss of calcium from bones, causing deformation, softening, and spontaneous fractures. ✚

Adrenal Glands

The two *adrenal,* or *suprarenal, glands* are located atop or close to the kidneys. Anatomically, the **adrenal medulla** develops from neural crest tissue, and it is directly controlled by the sympathetic nervous system. The medullary cells respond to this stimulation by releasing a hormone mix of **epinephrine** (80%) and **norepinephrine** (20%), which act in conjunction with the sympathetic nervous system to elicit the fight-or-flight response to stressors.

The **adrenal cortex** produces three major groups of steroid hormones, collectively called **corticosteroids.** The **mineralocorticoids,** chiefly *aldosterone,* regulate water and electrolyte balance in the extracellular fluids, mainly by regulating sodium ion resorption by kidney tubules. The **glucocorticoids** include *cortisol (hydrocortisone), cortisone,* and *corticosterone,* but only cortisol is secreted in significant amounts in humans. It enables the body to resist long-term stressors, primarily by increasing blood glucose levels. The **gonadocorticoids,** or **sex hormones,** produced by the adrenal cortex are chiefly *androgens* (male sex hormones), but some *estrogens* (female sex hormones) are also formed.

The gonadocorticoids are produced throughout life in relatively insignificant amounts; however, hypersecretion of these hormones produces abnormal hairiness **(hirsutism),** and masculinization occurs. ✚

Pancreas

The *pancreas,* located partially behind the stomach in the abdomen, functions as both an endocrine and exocrine gland. It produces digestive enzymes as well as insulin and glucagon, important hormones concerned with regulating blood sugar levels.

Elevated blood glucose levels stimulate release of **insulin,** which decreases blood sugar levels, primarily by accelerating the transport of glucose into the body cells, where it is oxidized for energy or converted to glycogen or fat for storage.

Hyposecretion of insulin or some deficiency in the insulin receptors leads to **diabetes mellitus,** which is characterized by the inability of body cells to utilize glucose and the subsequent loss of glucose in the urine. Impaired carbohydrate metabolism can also lead to alterations of protein and fat metabolism. Hypersecretion causes low blood sugar, or **hypoglycemia.** Symptoms include anxiety, nervousness, tremors, and weakness. ✚

Glucagon acts antagonistically to insulin. Its release is stimulated by low blood glucose levels, and its action is basically hyperglycemic. It stimulates the liver, its primary target organ, to break down its glycogen stores to glucose and subsequently to release the glucose to the blood.

The Gonads

The *female gonads,* or *ovaries,* are paired, almond-sized organs located in the pelvic cavity. In addition to producing the female sex cells (ova), the ovaries produce two steroid hormone groups, the estrogens and progesterone. The endocrine and exocrine functions of the ovaries do not begin until the onset of puberty, when the anterior pituitary gonadotropic hormones prod the ovary into action that produces rhythmic ovarian cycles in which ova develop and hormonal levels rise and fall. The **estrogens** are responsible for the development of the secondary sex characteristics of the female at puberty (primarily maturation of the reproductive organs and development of the breasts) and act with progesterone to bring about cyclic changes of the uterine lining that occur during the menstrual cycle. The estrogens also help prepare the mammary glands for lactation.

Progesterone, as already noted, acts with estrogen to bring about the menstrual cycle. During pregnancy it maintains the uterine musculature in a quiescent state and helps to prepare the breast tissue for lactation.

The paired oval *testes* of the male are suspended in a pouchlike sac, the scrotum, outside the pelvic cavity. In addition to the male sex cells (sperm), the testes produce the male sex hormone, **testosterone.** Testosterone promotes the maturation of the reproductive system accessory structures, brings about the development of the male secondary sex characteristics, and is responsible for sexual drive, or libido. Both the endocrine and exocrine functions of the testes begin at puberty under the influence of the anterior pituitary gonadotropins. (For a more detailed discussion of the function and histology of the ovaries and testes, see Exercise 29.)

Two glands not mentioned earlier as major endocrine glands should also be briefly considered here, the thymus and the pineal gland.

Thymus

The *thymus* is a bilobed gland situated in the superior thorax, posterior to the sternum and anterior to the heart and lungs. Conspicuous in the infant, it begins to atrophy at puberty, and by old age it is relatively inconspicuous. The thymus produces several different families of hormones, including **thymulin, thymosins,** and **thymopoietins.** These hormones are thought to be involved in the development of T lymphocytes and the immune response. Their role is poorly understood, and they act locally as paracrines.

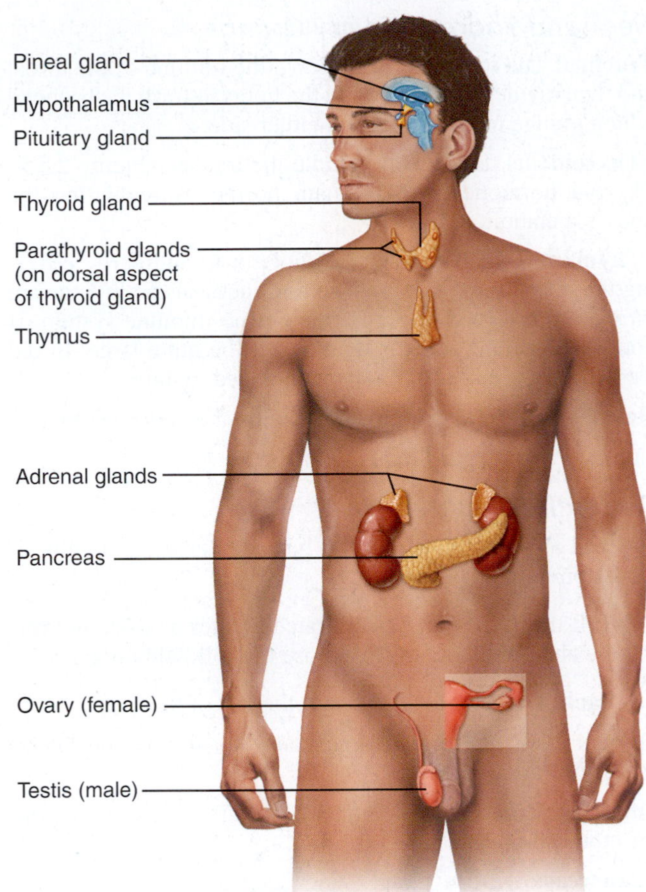

Figure 21.2 Human endocrine organs.

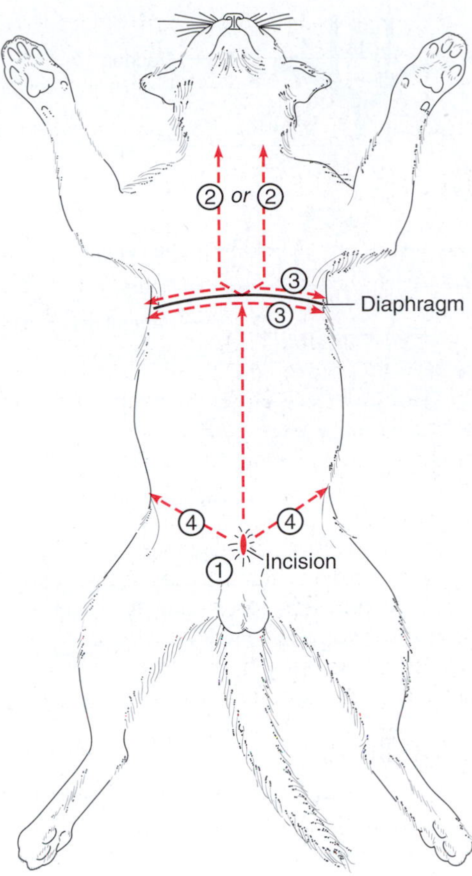

Figure 21.3 Endocrine organs of the cat. Incisions to be made in opening the ventral body cavity of a cat are shown. Numbers indicate sequence.

Pineal Gland

The *pineal gland* is a small, cone-shaped gland located in the roof of the third ventricle of the brain. Its major endocrine product is **melatonin.** Melatonin exhibits a diurnal (daily) cycle. It peaks at night, causing drowsiness, and is lowest around noon.

The endocrine role of the pineal gland in humans is still controversial, but it is known to play a role in mating and migratory behavior of other animals. In humans, melatonin appears to exert some inhibitory effect on the reproductive system that prevents precocious sexual maturation. Changing levels of melatonin may also affect biological rhythms associated with body temperature, sleep, and appetite.

ACTIVITY 1

Identifying the Endocrine Organs

Locate the endocrine organs in the figure of the body **(Figure 21.2)**. Also locate these organs on the anatomical charts or torso model. ■

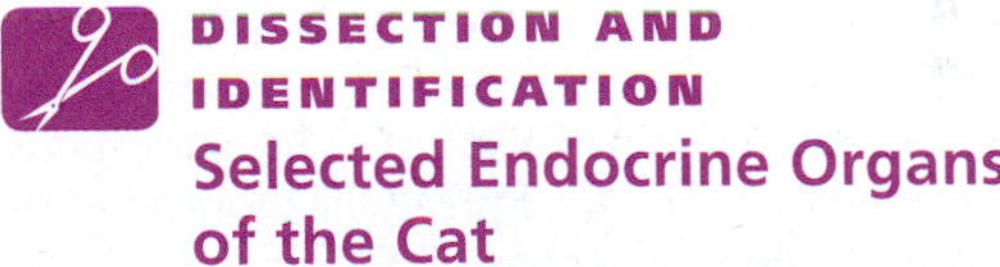

DISSECTION AND IDENTIFICATION

Selected Endocrine Organs of the Cat

If you have not previously opened the ventral body cavity, follow the directions provided in Activity 2. Otherwise, begin with Activity 3, "Identifying Organs." ■

ACTIVITY 2

Opening the Ventral Body Cavity

1. Don gloves, and then obtain your dissection animal. Place the animal on the dissecting tray, ventral side up. Using scissors, make a longitudinal median incision through the ventral body wall. Begin your cut just superior to the midline of the pubis, and continue it anteriorly to the rib cage. Check the incision guide provided **(Figure 21.3)** as you work.

2. Angle the scissors slightly (1.3 cm or ½ in.) to the right or left of the sternum, and continue the cut through the rib cartilages, just lateral to the body midline, to the base of the throat.

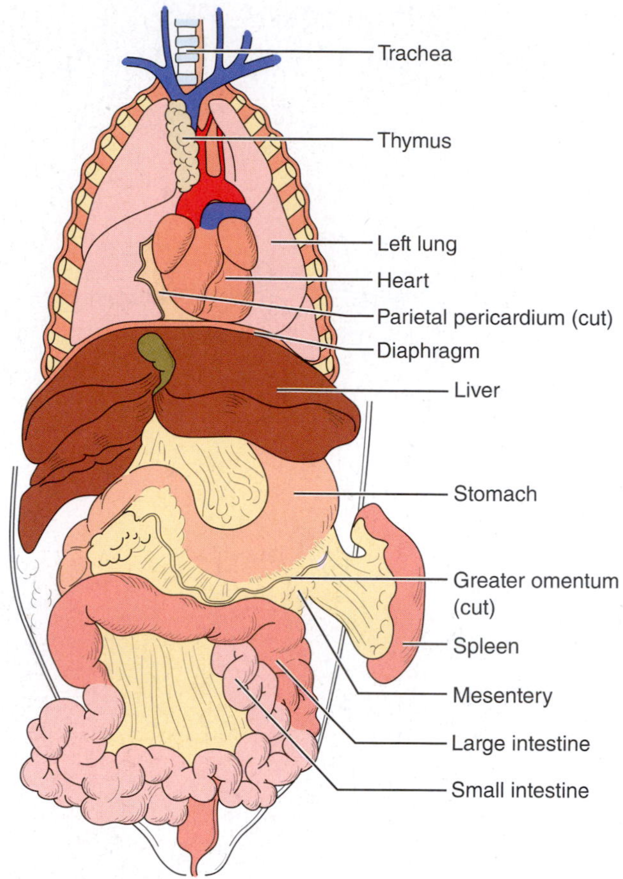

Figure 21.4 Ventral body cavity organs of the cat. Superficial view with greater omentum removed. (Also see Figure 27.20, page 476.)

3. Make two lateral cuts on both sides of the ventral body surface, anterior and posterior to the diaphragm, which separates the thoracic and abdominal parts of the ventral body cavity. *Leave the diaphragm intact.* Spread the thoracic walls laterally to expose the thoracic organs.

4. Make an angled lateral cut on each side of the median incision line just superior to the pubis, and spread the flaps to expose the abdominal cavity organs. ■

ACTIVITY 3

Identifying Organs

A helpful adjunct to identifying selected endocrine organs of the cat is a general overview of ventral body cavity organs **(Figure 21.4)**. You will study the organ systems housed in the ventral body cavity in later exercises, so the objective here is simply to identify the most important organs and those that will help you to locate the desired endocrine organs (marked with a * in the following lists). Examine the schematic and photographic cat images showing the relative positioning of several of the animal's endocrine organs **(Figure 21.5)**.

Neck and Thoracic Cavity Organs

Trachea: The windpipe; runs down the midline of the throat and then divides just anterior to the lungs to form the bronchi, which plunge into the lungs on either side.

***Thyroid:** Its dark lobes straddle the trachea (Figure 21.5). Thyroid hormones are the main hormones regulating the body's metabolic rate.

***Thymus:** Glandular structure superior to and partly covering the heart (Figure 21.4). The hormones of the thymus are intimately involved in programming the immune system. If you have a young cat, the thymus will be quite large. In old cats, most of this organ has been replaced by fat.

Heart: In the mediastinum enclosed by the pericardium.

Lungs: Paired organs flanking the heart.

Abdominal Cavity Organs

Liver: Large multilobed organ lying under the umbrella of the diaphragm.

Lift the drapelike, fat-infiltrated greater omentum covering the abdominal organs to expose the following organs:

Stomach: Dorsally located sac to the left of the liver.

Spleen: Flattened brown organ curving around the lateral aspect of the stomach.

Small intestine: Tubelike organ continuing posteriorly from the stomach.

***Pancreas:** Diffuse gland lying deep to and between the small intestine and stomach (Figure 21.5). Lift the first section of the small intestine with your forceps; you should see the pancreas situated in the delicate mesentery behind the stomach. This gland is extremely important in regulating blood sugar levels.

Large intestine: Takes a U-shaped course around the small intestine to terminate in the rectum.

Push the intestines to one side with a probe to reveal the deeper organs in the abdominal cavity.

Kidneys: Bean-shaped organs located toward the dorsal body wall surface and behind the peritoneum (see Figure 21.5).

***Adrenal glands:** Seen above and medial to each kidney, these small glands produce corticosteroids important in the stress response and in preventing abnormalities of water and electrolyte balance in the body (Figure 21.5).

***Gonads (ovaries or testes):** Sex organs producing sex hormones. The location of the gonads is illustrated (Figure 21.5a), but you will not identify them until the reproductive system organs are considered (Exercise 29).

Before you leave the laboratory, follow the boxed instructions to prepare your cat for storage and clean the area (page 217). ■

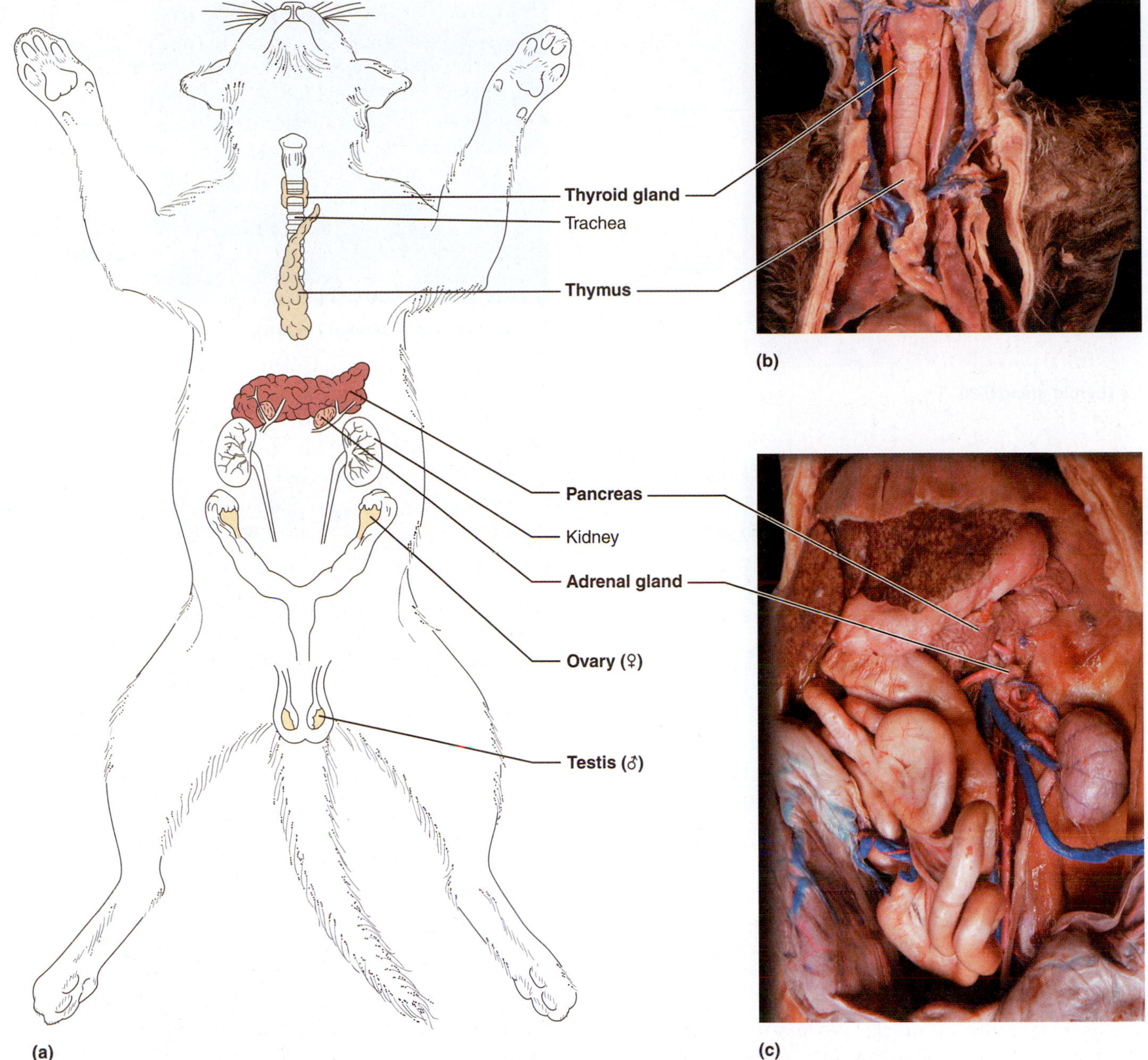

Figure 21.5 Endocrine organs in the cat. (a) Drawing. **(b, c)** Photographs.

Microscopic Anatomy of Selected Endocrine Glands

ACTIVITY 4

Examining the Microscopic Structure of Endocrine Glands

To prepare for the histologic study of the endocrine glands, obtain a microscope and one of each slide on the materials list (page 349). We will study only organs in which it is possible to identify the endocrine-producing cells. Compare your observations with the histology images **(Figure 21.6a–f)**.

Thyroid Gland

1. Scan the thyroid under low power, noting the **follicles,** spherical sacs containing a pink-stained material *(colloid)*. Stored T_3 and T_4 are attached to the protein colloidal material stored in the follicles as **thyroglobulin** and are released gradually to the blood. Compare the tissue viewed to the photomicrograph of thyroid tissue (Figure 21.6a).

2. Observe the tissue under high power. Notice that the walls of the follicles are formed by simple cuboidal or squamous

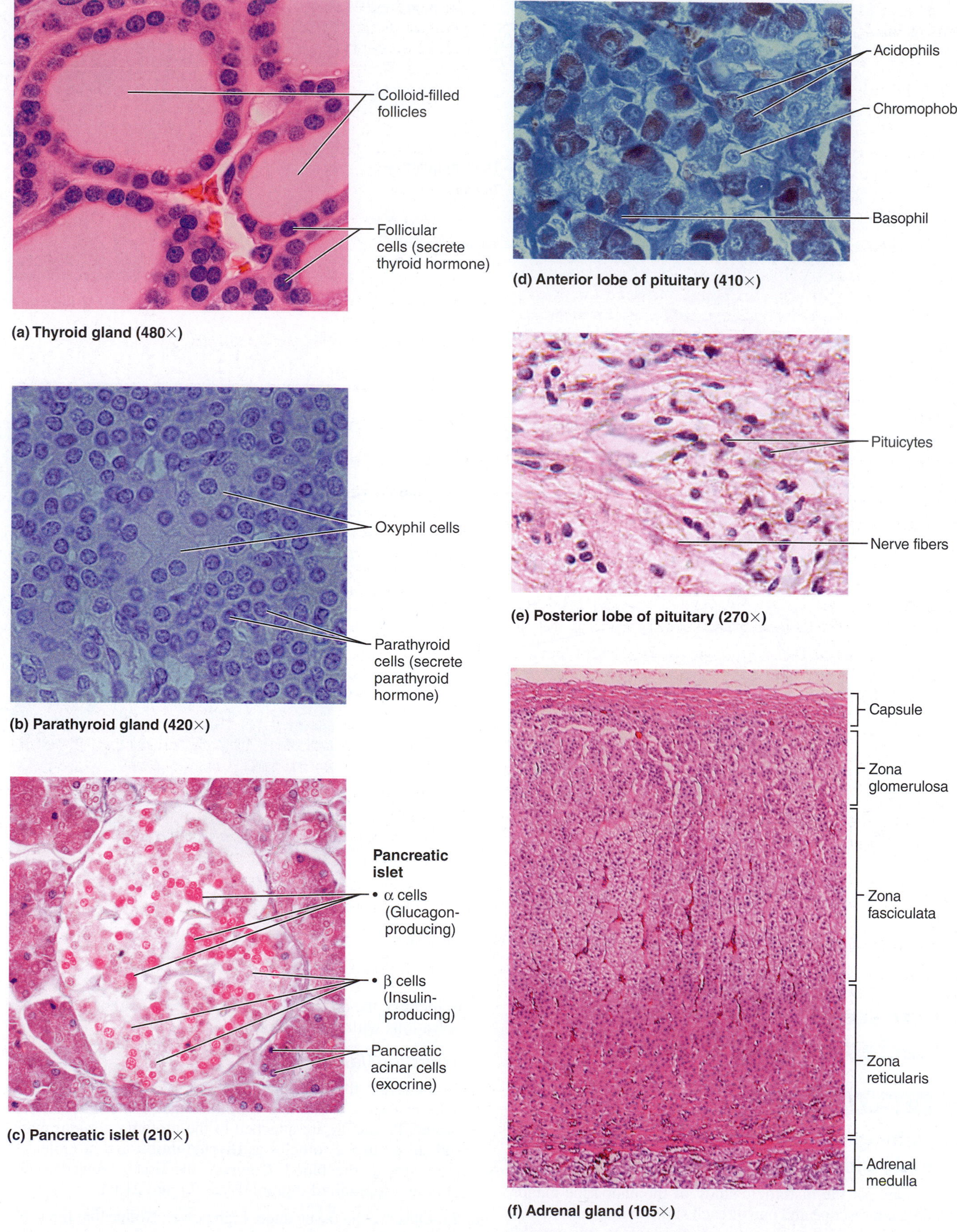

Figure 21.6 Microscopic anatomy of selected endocrine organs.

epithelial cells that synthesize the follicular products. The **parafollicular,** or **C, cells** you see between the follicles are responsible for calcitonin production.

When the thyroid gland is actively secreting, the follicles appear small, and the colloidal material has a ruffled border. When the thyroid is hypoactive or inactive, the follicles are large and plump, and the follicular epithelium appears to be squamouslike. What is the physiological state of the tissue

you have been viewing? ______________________

Parathyroid Glands

Observe the parathyroid tissue under low power to view its two major cell types, the parathyroid cells and the oxyphil cells. Compare your observations to the photomicrograph of parathyroid tissue (Figure 21.6b). The **parathyroid cells,** which synthesize parathyroid hormone (PTH), are small and abundant, and arranged in thick branching cords. The function of the scattered, much larger **oxyphil cells** is unknown.

Pancreas

1. Observe pancreas tissue under low power to identify the roughly circular **pancreatic islets** (also called *islets of Langerhans*), the endocrine portions of the pancreas. The islets are scattered amid the more numerous acinar cells and stain differently (usually lighter), which makes their identification possible. (See Figure 27.14.) The acinar cells produce the enzymatic exocrine product of the pancreas that is released to the duodenum via the pancreatic duct. Alkaline fluid produced by the duct cells accompanies the hydrolytic enzymes.

2. Focus on islet cells under high power. Notice that they are densely packed and have no definite arrangement. In contrast, the cuboidal acinar cells are arranged around secretory ducts. If special stains are used, it will be possible to distinguish the **alpha (α) cells,** which tend to cluster at the periphery of the islets and produce glucagon, from the **beta (β) cells,** which synthesize insulin. With these specific stains, the beta cells are larger and stain gray-blue, and the alpha cells are smaller and appear bright pink, as seen in the photomicrograph (Figure 21.6c).

Pituitary Gland

1. Observe the general structure of the pituitary gland under low power to differentiate the glandular anterior lobe and the neural posterior lobe.

2. Using the high-power lens, focus on the nests of cells of the anterior lobe. It is possible to identify the specialized cell types that secrete the specific hormones when differential stains are used. Using the photomicrograph of the anterior pituitary (Figure 21.6d) as a guide, locate the reddish brown–stained **acidophil cells,** which produce growth hormone and prolactin, and the **basophil cells,** whose deep-blue granules are responsible for the production of the tropic hormones (TSH, ACTH, FSH, and LH). **Chromophobes,** the third cellular population, do not take up the stain and appear rather dull and colorless. The role of the chromophobes is controversial, but they do not appear to be involved in hormone production.

3. Switch your focus to the posterior lobe. Observe the nerve fibers (axons of hypophyseal neurons) that make up most of this portion of the pituitary. Also note the **pituicytes,** neuroglia that are randomly distributed among the nerve fibers. (Refer to Figure 21.6e as you scan the slide.)

Which two hormones are stored here?

______________________ and ______________________

What is their source? ______________________

Adrenal Gland

1. Hold the slide of the adrenal gland up to the light to distinguish the outer cortex and inner medulla areas. Then scan the cortex under low power to distinguish the differences in cell appearance and arrangement in the three cortical areas. Refer to the photomicrograph (Figure 21.6f) and the following descriptions to identify the cortical areas:

- Connective tissue capsule of the adrenal gland.
- The outermost **zona glomerulosa,** where most mineralocorticoid production occurs and where the tightly packed cells are arranged in spherical clusters.
- The deeper intermediate **zona fasciculata,** which produces glucocorticoids. This is the thickest part of the cortex. Its cells are arranged in parallel cords.
- The innermost cortical zone, the **zona reticularis** abutting the medulla, which produces sex hormones and some glucocorticoids. The cells here stain intensely and form a branching network.

2. Switch focus to view the large, lightly stained cells of the adrenal medulla under high power. Notice their clumped arrangement.

Which hormones are produced by the medulla?

______________________ and ______________________ ■

GROUP CHALLENGE

Odd Hormone Out

Each box below contains four hormones. One of the listed hormones does *not* share a characteristic that the other three do. Circle the hormone that doesn't belong with the others, and explain why it is singled out. What characteristic is it missing? Sometimes there may be multiple reasons why the hormone doesn't belong with the others.

1. Which is the "odd hormone"?	Why is it the odd one out?
ACTH Oxytocin LH FSH	
2. Which is the "odd hormone"?	**Why is it the odd one out?**
Aldosterone Cortisol Epinephrine ADH	
3. Which is the "odd hormone"?	**Why is it the odd one out?**
PTH Testosterone LH FSH	
4. Which is the "odd hormone"?	**Why is it the odd one out?**
Insulin Cortisol Calcitonin Glucagon	

Name ______________________________

Lab Time/Date ______________________

EXERCISE 21

REVIEW SHEET

Functional Anatomy of the Endocrine Glands

Gross Anatomy and Basic Function of the Endocrine Glands

1. Both the endocrine and nervous systems are major regulating systems of the body; however, the nervous system has been compared to an airmail delivery system and the endocrine system to the Pony Express. Briefly explain this comparison.

____

____

____

2. Define *hormone.* ____

____

3. Chemically, hormones belong chiefly to two molecular groups: ________________ and ________________.

4. Define *target organ.* ____

____

5. Given that hormones travel in the bloodstream, why don't all tissues respond to all hormones? ________________

____

6. Identify the endocrine organ described by each of the following statements.

________________ 1. located in the throat; bilobed gland connected by an isthmus

________________ 2. found atop the kidney

________________ 3. a mixed gland, located close to the stomach and small intestine

________________ 4. paired glands suspended in the scrotum

________________ 5. ride "horseback" on the thyroid gland

________________ 6. found in the pelvic cavity of the female, concerned with production of ova and female hormones

________________ 7. found in the upper thorax overlying the heart; large during youth

________________ 8. found in the roof of the third ventricle of the brain

7. The table below lists the functions of many of the hormones you have studied. From the keys below, fill in the hormones responsible for each function, and the endocrine glands that produce each hormone. Glands may be used more than once.

Hormones Key:

ACTH
ADH
aldosterone
cortisol
epinephrine
estrogens
FSH
glucagon
insulin
LH
oxytocin
progesterone
prolactin
PTH
T_3 / T_4
testosterone
TSH

Glands Key:

adrenal cortex
adrenal medulla
anterior lobe of pituitary
hypothalamus
ovaries
pancreas
parathyroid gland
posterior lobe of pituitary
testes
thyroid gland

Function	Hormone(s)	Gland(s)
Regulate the function of another endocrine gland	1.	
	2.	
	3.	
	4.	
Maintain salt and water balance in the extracellular fluid	1.	
	2.	
Directly involved in milk production and ejection	1.	
	2.	
Controls the rate of body metabolism and cellular oxidation	1.	
Regulates blood calcium levels	1.	
Regulate blood glucose levels; produced by the same "mixed" gland	1.	
	2.	
Released in response to stressors	1.	
	2.	
Drives development of secondary sex characteristics in males	1.	
Directly responsible for regulating the menstrual cycle	1.	
	2.	

8. Although the pituitary gland is often referred to as the master gland of the body, the hypothalamus exerts control over the pituitary gland. How does the hypothalamus control the functioning of both the anterior and posterior lobes?

___

___

9. Name the hormone(s) produced in *inadequate* amounts that directly result in the following conditions. (Use your textbook as necessary.)

___ 1. tetany

___ 2. excessive urination without high blood glucose levels

___ 3. loss of glucose in urine

___ 4. abnormally small stature, normal proportions

___ 5. lethargy, hair loss, low metabolic rate, mental and physical sluggishness

10. Name the hormone(s) produced in *excessive* amounts that directly result in the following conditions. (Use your textbook as necessary.)

___ 1. large facial bones, hands, and feet in the adult

___ 2. bulging eyeballs, nervousness, increased pulse rate, sweating

___ 3. demineralization of bones, spontaneous fractures

Microscopic Anatomy of Selected Endocrine Glands

11. Choose a response from the key below to name the hormone(s) produced by the cell types listed.

Key:

a. calcitonin
b. GH, prolactin
c. glucagon
d. glucocorticoids
e. insulin
f. mineralocorticoids
g. PTH
h. T_4 / T_3
i. TSH, ACTH, FSH, LH

___ 1. parafollicular cells of the thyroid

___ 2. follicular cells of the thyroid

___ 3. beta cells of the pancreatic islets (islets of Langerhans)

___ 4. alpha cells of the pancreatic islets (islets of Langerhans)

___ 5. basophil cells of the anterior lobe of the pituitary

___ 6. zona fasciculata cells

___ 7. zona glomerulosa cells

___ 8. parathyroid cells

___ 9. acidophil cells of the anterior lobe of the pituitary

12. Six diagrams of the microscopic structures of the endocrine glands are presented here. Identify each, and name all structures indicated by a leader line or bracket.

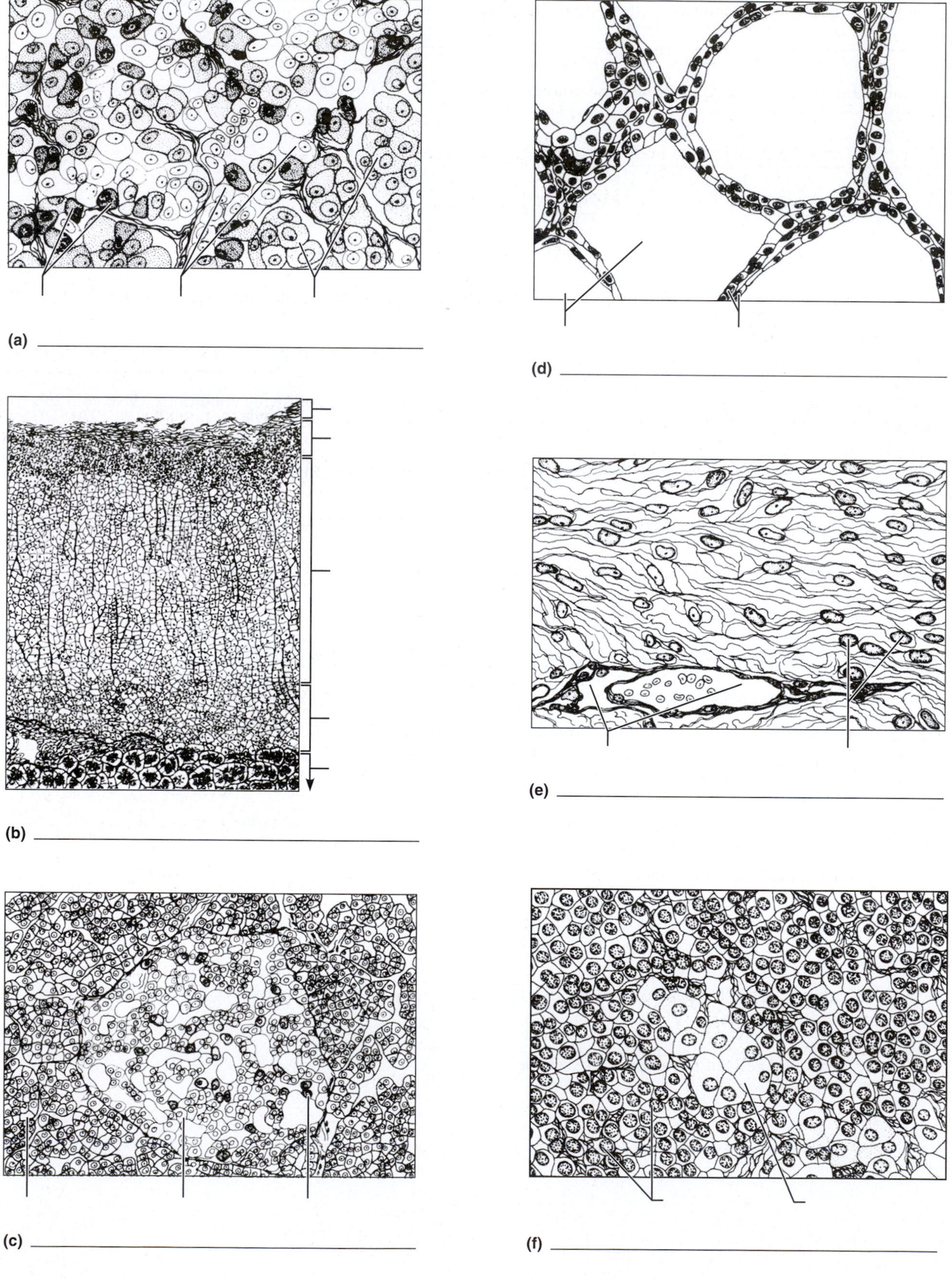

Dissection and Identification: Selected Endocrine Organs of the Cat

13. How do the locations of the endocrine organs in the cat compare with those in the human?

____

____

14. Name two endocrine organs located in the throat region: ______________________ and ______________________

15. Name three endocrine organs located in the abdominal cavity.

____

16. Given the assumption (not necessarily true) that human beings have more stress than cats, which endocrine organs would you expect to be relatively larger in humans?

____

17. Cats are smaller animals than humans. Which would you expect to have a (relatively speaking) more active thyroid gland—cats or humans? ______________ Why? (We know we are asking a lot with this one, but give it a whirl.)

____

____

____

____

22

EXERCISE

Blood

MATERIALS

General Supply Area*

- ☐ Disposable gloves
- ☐ Safety glasses (student-provided)
- ☐ Bucket or large beaker containing 10% household bleach solution for slide and glassware disposal
- ☐ Spray bottles containing 10% bleach solution
- ☐ Autoclave bag
- ☐ Designated lancet (sharps) disposal container
- ☐ Plasma (obtained from an animal hospital or prepared by centrifuging animal blood [for example, cattle or sheep blood] obtained from a biological supply house)
- ☐ Test tubes and test tube racks
- ☐ Wide-range pH paper
- ☐ Stained smears of human blood from a biological supply house or, if desired by the instructor, heparinized animal blood obtained from a biological supply house or an animal hospital (for example, dog blood), or EDTA-treated red cells (reference cells†) with blood type labels obscured (available from Immucor, Inc.)
- ☐ Clean microscope slides
- ☐ Glass stirring rods
- ☐ Wright's stain in a dropper bottle
- ☐ Distilled water in a dropper bottle
- ☐ Sterile lancets
- ☐ Absorbent cotton balls
- ☐ Alcohol swabs (wipes)
- ☐ Paper towels

***Note to the Instructor:** See directions for handling of soiled glassware and disposable items (page 27).

†The blood in these kits (each containing four blood cell types—A1, A2, B, and O—individually supplied in 10-ml vials) is used to calibrate cell counters and other automated clinical laboratory equipment. This blood has been carefully screened and can be safely used by students for blood typing and determining hematocrits. It is not usable for hemoglobin determinations or coagulation studies.

(Text continues on next page.)

OBJECTIVES

1. Name the two major components of blood, and state their average percentages in whole blood.
2. Describe the composition and functional importance of plasma.
3. Define *formed elements*. Name the cell types that comprise the formed elements, state their relative percentages, and describe their major functions.
4. Identify erythrocytes, basophils, eosinophils, monocytes, lymphocytes, and neutrophils on a microscopic preparation or an appropriate image.
5. Provide the normal values for a total white blood cell count and a total red blood cell count, and explain why these tests are important.
6. Conduct the following blood tests in the laboratory, state their norms, and explain the importance of each: differential white blood cell count, hematocrit, hemoglobin determination, clotting time, and plasma cholesterol concentration.
7. Perform an ABO and Rh blood typing test in the laboratory, and discuss why administering mismatched blood causes transfusion reactions.
8. Define *leukocytosis, leukopenia, leukemia, polycythemia,* and *anemia,* and cite a possible cause for each condition.

PRE-LAB QUIZ

1. Circle True or False. There are no special precautions that I need to observe when performing today's lab.
2. Three types of formed elements found in blood include erythrocytes, leukocytes, and ________.
   a. electrolytes b. fibers c. platelets d. sodium salts
3. Circle the correct underlined term. Mature erythrocytes / leukocytes are the most numerous blood cells and do not have a nucleus.
4. The least numerous but largest of all agranulocytes is the:
   a. basophil b. lymphocyte c. monocyte d. neutrophil
5. ________ are the leukocytes responsible for releasing histamine and other mediators of inflammation.
   a. Basophils b. Eosinophils c. Monocytes d. Neutrophils
6. ________ are essential for blood clotting.
7. Circle the correct underlined term. When determining the hematocrit / hemoglobin, you will centrifuge whole blood in order to allow the formed elements to sink to the bottom of the sample.
8. Circle the correct underlined term. The normal hematocrit value for females / males is generally higher than that of the opposite sex.
9. Circle the correct underlined term. Blood typing is based on the presence of proteins known as antigens / antibodies on the outer surface of the red blood cell plasma membrane.
10. Circle True or False. If an individual is transfused with the wrong type blood, the recipient's antibodies react with the donor's antigens, eventually clumping and hemolyzing the donated RBCs.

Table 22.1 Summary of Formed Elements of the Blood

Cell type	Illustration	Description*	Number of cells/mm³ (μl) of blood	Duration of development (D) and life span (LS)	Function
Erythrocytes (red blood cells, RBCs)		Biconcave, anucleate disc; salmon-colored; diameter 7–8 μm	4–6 million	D: about 15 days LS: 100–120 days	Transport oxygen and carbon dioxide
Leukocytes (white blood cells, WBCs)		Spherical, nucleated cells	4800–10,800		
Granulocytes Neutrophil		Nucleus multilobed; inconspicuous cytoplasmic granules; diameter 10–12 μm	3000–7000	D: about 14 days LS: 6 hours to a few days	Phagocytize bacteria
Eosinophil		Nucleus bilobed; red cytoplasmic granules; diameter 10–14 μm	100–400	D: about 14 days LS: 5 days	Kill parasitic worms; destroy antigen-antibody complexes; inactivate some inflammatory chemicals of allergy
Basophil		Nucleus lobed; large blue-purple cytoplasmic granules; diameter 10–14 μm	20–50	D: 3–7 days LS: a few hours to a few days	Release histamine and other mediators of inflammation; contain heparin, an anticoagulant
Agranulocytes Lymphocyte		Nucleus spherical or indented; pale blue cytoplasm; diameter 5–17 μm	1500–3000	D: days to weeks LS: hours to years	Mount immune response by direct cell attack or via antibodies
Monocyte		Nucleus U- or kidney-shaped; gray-blue cytoplasm; diameter 14–24 μm	100–700	D: 2–3 days LS: months	Phagocytosis; develop into macrophages in tissues
Platelets		Discoid cytoplasmic fragments containing granules; stain deep purple; diameter 2–4 μm	150,000–400,000	D: 4–5 days LS: 5–10 days	Seal small tears in blood vessels; instrumental in blood clotting

*Appearance when stained with Wright's stain.

How would you describe the consistency of plasma (slippery, watery, sticky, granular)? Record your observations.

____

ACTIVITY 2

Examining the Formed Elements of Blood Microscopically

In this section, you will observe blood cells on an already prepared (purchased) blood slide or on a slide prepared from your own blood or blood provided by your instructor.

- If you are using a purchased blood slide, obtain a slide and begin your observations at step 6.
- If you are testing blood provided by a biological supply source or an animal hospital, obtain a tube of the supplied blood, disposable gloves, and the supplies listed in step 1, except for the lancets and alcohol swabs. After donning gloves, go to step 3b to begin your observations.
- If you are examining your own blood, you will perform all the steps described below *except* step 3b.

1. Obtain two glass slides, a glass stirring rod, dropper bottles of Wright's stain and distilled water, two or three lancets, cotton balls, and alcohol swabs. Bring this equipment

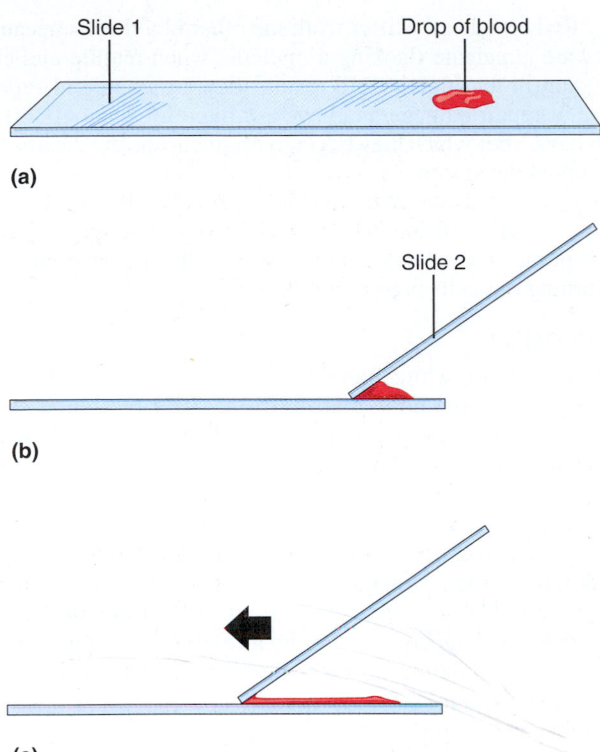

Figure 22.2 Procedure for making a blood smear. **(a)** Place a drop of blood on slide 1 approximately ½ inch from one end. **(b)** Hold slide 2 at a 30° to 40° angle to slide 1 (it should touch the drop of blood) and allow blood to spread along entire bottom edge of angled slide. **(c)** Smoothly advance slide 2 to end of slide 1 (blood should run out before reaching the end of slide 1). Then lift slide 2 away from slide 1, and place it on a paper towel.

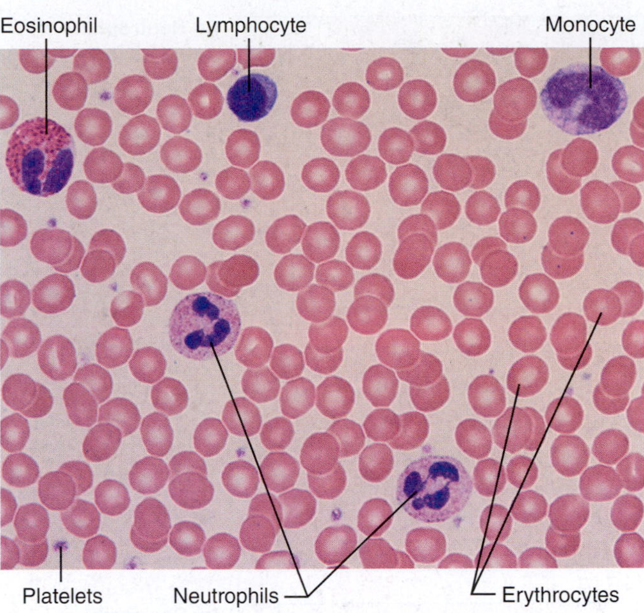

Figure 22.3 Photomicrograph of a human blood smear stained with Wright's stain (715×).

to the laboratory bench. Clean the slides thoroughly, and dry them.

2. Open the alcohol swab packet, and scrub your third or fourth finger with the swab. (Because the pricked finger may be a little sore later, it is better to prepare a finger on the hand used less often.) Circumduct your hand (swing it in a cone-shaped path) for 10 to 15 seconds. This will dry the alcohol and cause your fingers to become engorged with blood. Then, open the lancet packet and grasp the lancet by its blunt end. Quickly jab the pointed end into the prepared finger to produce a free flow of blood. It is *not* a good idea to squeeze or "milk" the finger, because this forces out tissue fluid as well as blood. If the blood is not flowing freely, make another puncture.

! *Under no circumstances are you to use a lancet for more than one puncture.* Dispose of the lancets in the designated disposal container immediately after use.

3a. With a cotton ball, wipe away the first drop of blood; then allow another large drop of blood to form. Touch the blood to one of the cleaned slides approximately 1.3 cm, or ½ inch, from the end. Then quickly (to prevent clotting) use the second slide to form a blood smear **(Figure 22.2)**. When properly prepared, the blood smear is uniformly thin. If the blood smear appears streaked, the blood probably began to clot or coagulate before the smear was made, and another slide should be prepared. Continue at step 4.

3b. Dip a glass rod in the blood provided, and transfer a generous drop of blood to the end of a cleaned microscope slide. For the time being, lay the glass rod on a paper towel on the bench. Then, as described in step 3a (Figure 22.2), use the second slide to make your blood smear.

4. Dry the slide by waving it in the air. When it is completely dry, it will look dull. Place it on a paper towel, and flood it with Wright's stain. Count the number of drops of stain used. Allow the stain to remain on the slide for 3 to 4 minutes, and then flood the slide with an equal number of drops of distilled water. Allow the water and Wright's stain mixture to remain on the slide for 4 or 5 minutes or until a metallic green film or scum is apparent on the fluid surface. Blow on the slide gently every minute or so to keep the water and stain mixed during this interval.

5. Rinse the slide with a stream of distilled water. Then flood it with distilled water, and allow it to lie flat until the slide becomes translucent and takes on a pink cast. Then stand the slide on its long edge on the paper towel, and allow it to dry completely. Once the slide is dry, you can begin your observations.

6. Obtain a microscope and scan the slide under low power to find the area where the blood smear is the thinnest. After scanning the slide under low power to find the areas with the largest numbers of nucleated WBCs, read the following descriptions of cell types, and find each one in the art illustrating blood cell types (Figure 22.1, Table 22.1, **Figure 22.3**, and **Figure 22.4**). Then, switch to the oil immersion lens, and observe the slide carefully to identify each cell type.

7. Set your prepared slide aside for use in Activity 3.

Erythrocytes

Erythrocytes, or red blood cells, which average 7.5 μm in diameter, vary in color from a salmon red color to pale pink,

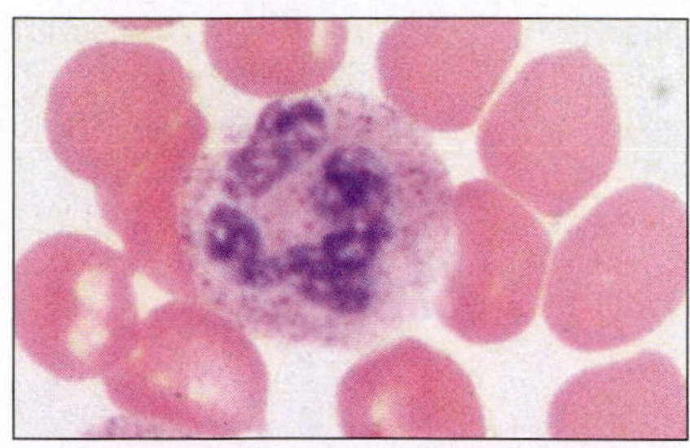

(a) Neutrophil: multilobed nucleus

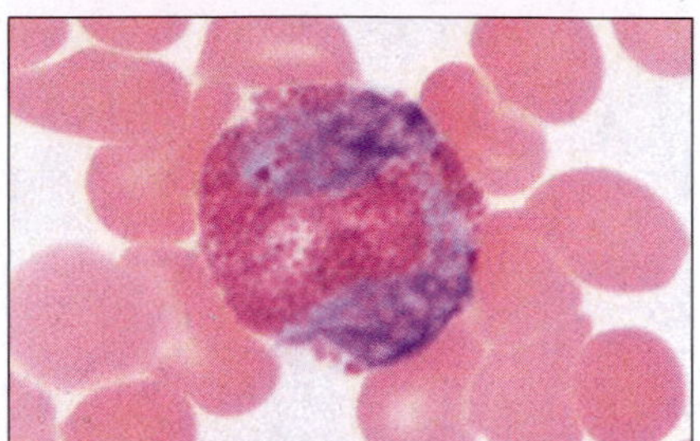

(b) Eosinophil: bilobed nucleus, red cytoplasmic granules

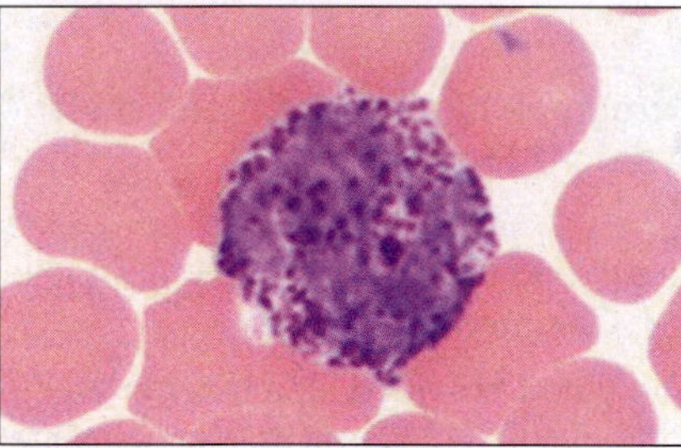

(c) Basophil: bilobed nucleus, purplish black cytoplasmic granules

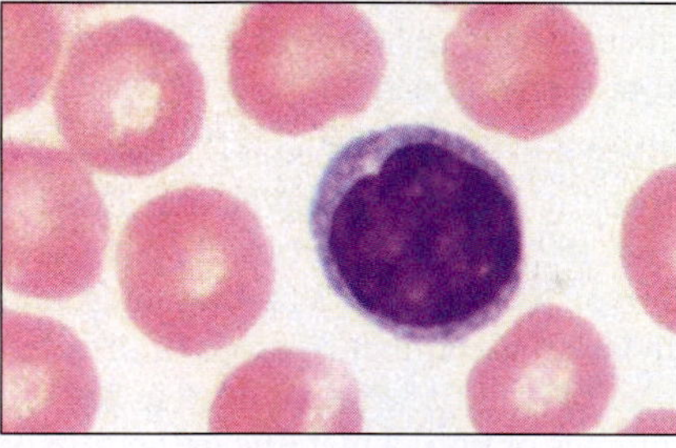

(d) Small lymphocyte: large spherical nucleus

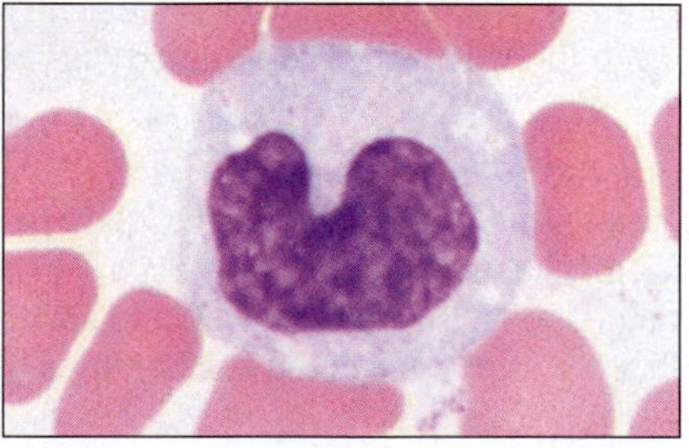

(e) Monocyte: kidney-shaped nucleus

Figure 22.4 Leukocytes. In each case, the leukocytes are surrounded by erythrocytes (all 1750×, Wright's stain).

depending on the effectiveness of the stain. They have a distinctive biconcave disk shape and appear paler in the center than at the edge (Figure 22.3).

As you observe the slide, notice that the red blood cells are by far the most numerous blood cells seen in the field. Their number averages 4.5 million to 5.5 million cells per cubic millimeter of blood (for women and men, respectively).

Red blood cells differ from the other blood cells because they are anucleate (lacking a nucleus) when mature and circulating in the blood. As a result, they are unable to reproduce or repair damage and have a limited life span of 100 to 120 days, after which they begin to fragment and are destroyed, mainly in the spleen.

In various anemias, the red blood cells may appear pale (an indication of decreased hemoglobin content) or may be nucleated (an indication that the bone marrow is turning out cells prematurely). ✚

Leukocytes

Leukocytes, or white blood cells, are nucleated cells that are formed in the bone marrow from the same blood stem cells *(hematopoietic stem cells)* as red blood cells. They are much less numerous than the red blood cells, averaging from 4800 to 10,800 cells per cubic millimeter. Basically, white blood cells are protective, pathogen-destroying cells that are transported to all parts of the body in the blood or lymph. Important to their protective function is their ability to move in and out of blood vessels, a process called **diapedesis,** and to wander through body tissues by **amoeboid motion** to reach sites of inflammation or tissue destruction. They are classified into two major groups, depending on whether or not they contain conspicuous granules in their cytoplasm.

Granulocytes make up the first group. The granules in their cytoplasm stain differentially with Wright's stain, and they have peculiarly lobed nuclei, which often consist of expanded nuclear regions connected by thin strands of nucleoplasm. There are three types of granulocytes:

Neutrophil: The most abundant of the white blood cells (50% to 70% of the leukocyte population); nucleus consists of three to six lobes, and the pale lilac cytoplasm contains fine cytoplasmic granules, which are generally indistinguishable and take up both the acidic (red) and basic (blue) dyes (*neutrophil* = neutral loving); functions as an active phagocyte. The number of neutrophils increases exponentially during acute infections (Figure 22.4a).

Eosinophil: Represents 2% to 4% of the leukocyte population; nucleus is generally figure-8 or bilobed in shape; contains large cytoplasmic granules (elaborate lysosomes) that stain red-orange with the acid dyes in Wright's stain (Figure 22.4b). Eosinophils are about the size of neutrophils and play a role in counterattacking parasitic worms. Eosinophils have complex roles in fighting many other diseases, especially in allergy and asthma.

Basophil: Least abundant leukocyte type, representing less than 1% of the population; large U- or S-shaped nucleus with two or more indentations. Cytoplasm contains coarse, sparse granules that are stained deep purple by the basic dyes in Wright's stain (Figure 22.4c). The granules contain several chemicals, including histamine, a vasodilator that is discharged on exposure to antigens and helps mediate the inflammatory response. Basophils are about the size of neutrophils.

The second group, **agranulocytes,** or **agranular leukocytes,** contains no *visible* cytoplasmic granules. Although found in the bloodstream, they are much more abundant in lymphoid tissues. Their nuclei tend to be closer to the norm, that is, spherical, oval, or kidney-shaped. Specific characteristics of the two types of agranulocytes are described next.

Lymphocyte: The smallest of the leukocytes, approximately the size of a red blood cell (Figure 22.4d). The nucleus stains dark blue to purple, is generally spherical or slightly indented, and accounts for most of the cell mass. Sparse cytoplasm appears as a thin blue rim around the nucleus. Concerned with immunologic responses in the body; one population, the *B lymphocytes,* gives rise to *plasma cells,* which produce antibodies that are released to blood. The second population, *T lymphocytes,* plays a regulatory role and destroys grafts, tumors, and virus-infected cells. Represents 25% or more of the WBC population.

Monocyte: The largest of the leukocytes; approximately twice the size of red blood cells (Figure 22.4e). Represents 3% to 8% of the leukocyte population. Dark blue nucleus is generally kidney-shaped; abundant cytoplasm stains gray-blue. Once in the tissues, monocytes convert to macrophages, active phagocytes (the "long-term cleanup team"), increasing dramatically in number during chronic infections such as tuberculosis.

Students are often asked to list the leukocytes in order from the most abundant to the least abundant. The following silly phrase may help you with this task: *N*ever *l*et *m*onkeys *e*at *b*ananas (neutrophils, lymphocytes, monocytes, eosinophils, basophils).

Platelets

Platelets are cell fragments of large multinucleate cells **(megakaryocytes)** formed in the bone marrow. They appear as darkly staining, irregularly shaped bodies interspersed among the blood cells (see Figure 22.3). The normal platelet count in blood ranges from 150,000 to 400,000 per cubic millimeter. Platelets are instrumental in the clotting process that occurs in plasma when blood vessels are ruptured.

After you have identified these cell types on your slide, observe charts and three-dimensional models of blood cells if these are available. Do not dispose of your slide, because you will use it later for the differential white blood cell count. ■

Hematologic Tests

When someone enters a hospital as a patient, several hematologic tests are routinely done to determine general level of health as well as the presence of pathological conditions. You will be conducting the most common of these tests in this exercise.

! Materials such as cotton balls, lancets, and alcohol swabs are used in nearly all of the following diagnostic tests. These supplies are at the general supply area and should be properly disposed of (glassware to the bleach bucket, lancets in a designated disposal container, and disposable items to the autoclave bag) immediately after use.

Other necessary supplies and equipment are at specific supply areas marked according to the test with which they are used. Because nearly all of the tests require a finger stab, if you will be using your own blood it might be wise to quickly read through the tests to determine in which instances more than one preparation can be done from the same finger stab. For example, the hematocrit capillary tubes and sedimentation rate samples might be prepared at the same time. A little planning will save you the discomfort of a multiple-punctured finger.

An alternative to using blood obtained from the finger stab technique is to use heparinized blood samples supplied by your instructor. The purpose of using heparinized tubes is to prevent the blood from clotting. Thus blood collected and stored in such tubes will be suitable for all tests except coagulation time testing.

Total White and Red Blood Cell Counts

A **total WBC count** or **total RBC count** determines the total number of that cell type per unit volume of blood. Total WBC and RBC counts are a routine part of any physical exam. Most clinical agencies use computers to conduct these counts. Because the hand counting technique typically done in college labs is rather outdated, total RBC and WBC counts will not be done here, but the importance of such counts (both normal and abnormal values) is briefly described below.

Total White Blood Cell Count

Because white blood cells are an important part of the body's defense system, it is essential to note any abnormalities in them.

Leukocytosis, an abnormally high WBC count, may indicate bacterial or viral infection, metabolic disease, hemorrhage, or poisoning by drugs or chemicals. A decrease in the white cell number below 4000/mm^3 **(leukopenia)** may indicate typhoid fever, measles, infectious hepatitis or cirrhosis, tuberculosis, or excessive antibiotic or X-ray therapy. A person with leukopenia lacks the usual protective mechanisms. **Leukemia,** a malignant disorder of the lymphoid tissues characterized by uncontrolled proliferation of abnormal WBCs accompanied by a reduction in the number of RBCs and platelets, is detectable not only by a total WBC count, but also by a differential WBC count. +

Total Red Blood Cell Count

Because RBCs are absolutely necessary for oxygen transport, a doctor typically investigates any excessive change in their number immediately.

An increase in the number of RBCs **(polycythemia)** may result from bone marrow cancer or from living at high altitudes where less oxygen is available. A decrease in the number of RBCs results in anemia. The term **anemia** simply indicates a decreased oxygen-carrying capacity of blood that may result from a decrease in RBC number or size, or a decreased hemoglobin content of the RBCs. A decrease in RBCs may result suddenly from hemorrhage or more gradually from conditions that destroy RBCs or hinder RBC production. +

Differential White Blood Cell Count

To make a **differential white blood cell count,** 100 WBCs are counted and classified according to type. Such a count is routine in a physical examination and in diagnosing illness, because any abnormality or significant elevation in percentages of WBC types may indicate a problem or the source of pathology.

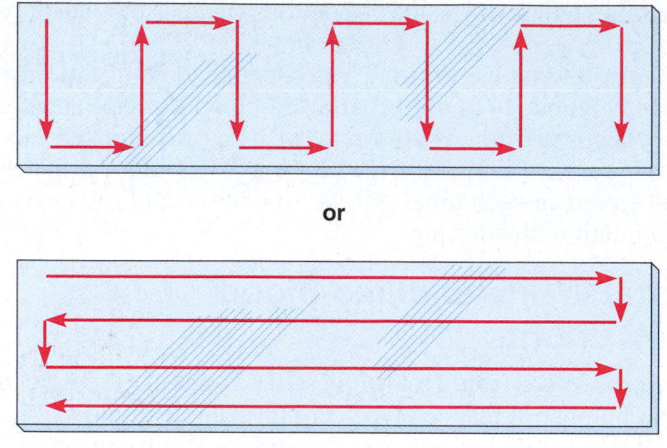

Figure 22.5 Alternative methods of moving the slide for a differential WBC count.

ACTIVITY 3

Conducting a Differential White Blood Cell Count

1. Use the slide prepared for identification of the blood cells in Activity 2. Begin at the edge of the smear, and move the slide in a systematic manner on the microscope stage—either up and down or from side to side as indicated in the figure **(Figure 22.5)**.

2. Record each type of white blood cell you observe by making a count on the **Activity 3** chart (for example, 卌 || = 7 cells) until you have observed and recorded a total of 100 WBCs. Using the following equation, compute the percentage of each WBC type counted, and record the percentages on the Hematologic Test Data Sheet (page 374).

$$\text{Percent (\%)} = \frac{\text{\# observed}}{\text{Total \# counted (100)}} \times 100$$

3. Select a slide marked "Unknown sample," record the slide number, and again use the Activity 3 chart to conduct a differential count. Record the percentages on the Test Data Sheet (page 374).

How does the differential count from the unknown sample slide compare to the normal percentages given for each type in the "Leukocytes" section (pages 370–371)?

___

___

Using the text and other references, try to determine the blood pathology on the unknown slide. Defend your answer.

___

___

4. How does your differential white blood cell count compare to the percentages given in the "Leukocytes" section in Activity 2?

___

Activity 3: Count of 100 WBCs

Cell type	Number observed	
	Student smear	Unknown sample # ____
Neutrophils		
Eosinophils		
Basophils		
Lymphocytes		
Monocytes		

Hematocrit

The **hematocrit,** or **packed cell volume (PCV),** is routinely determined when anemia is suspected. Centrifuging whole blood spins the formed elements to the bottom of the tube, with plasma forming the top layer (see Figure 22.1). Because the blood cell population is primarily RBCs, the PCV is generally considered equivalent to the RBC volume, and this is the only value reported. However, the relative percentage of WBCs can be differentiated, and both WBC and plasma volume will be reported here. Normal hematocrit values for the male and female, respectively, are 47.0 ± 7 and 42.0 ± 5.

ACTIVITY 4

Determining the Hematocrit

The hematocrit is determined by the micromethod, so only a drop of blood is needed. If possible (and the centrifuge allows), all members of the class should prepare their capillary tubes at the same time so the centrifuge can be run only once.

1. Obtain two heparinized capillary tubes, capillary tube sealer or modeling clay, a lancet, alcohol swabs, and some cotton balls.

2. If you are using your own blood, cleanse a finger, and allow the blood to flow freely. Wipe away the first few drops and, holding the red-line-marked end of the capillary tube to the blood drop, allow the tube to fill at least three-fourths full by capillary action **(Figure 22.6a)**. If the blood is not flowing freely, the end of the capillary tube will not be completely submerged in the blood during filling, air will enter, and you will have to prepare another sample.

If you are using instructor-provided blood, simply immerse the red-marked end of the capillary tube in the blood sample, and fill it three-quarters full as just described.

3. Plug the blood-containing end by pressing it into the capillary tube sealer or clay (Figure 22.6b). Prepare a second tube in the same manner.

4. Place the prepared tubes opposite one another in the radial grooves of the microhematocrit centrifuge with the sealed ends abutting the rubber gasket at the centrifuge periphery (Figure 22.6c). This loading procedure balances the centrifuge and prevents blood from spraying everywhere by centrifugal force. *Make a note of the numbers of the grooves your*

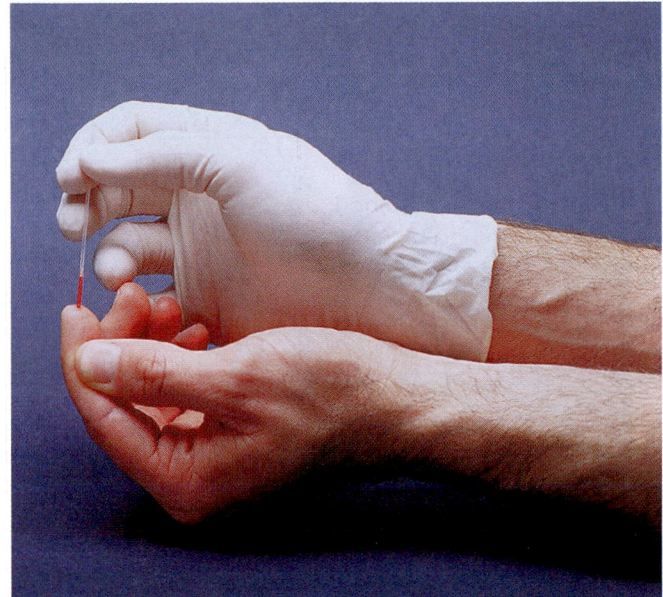

(a)

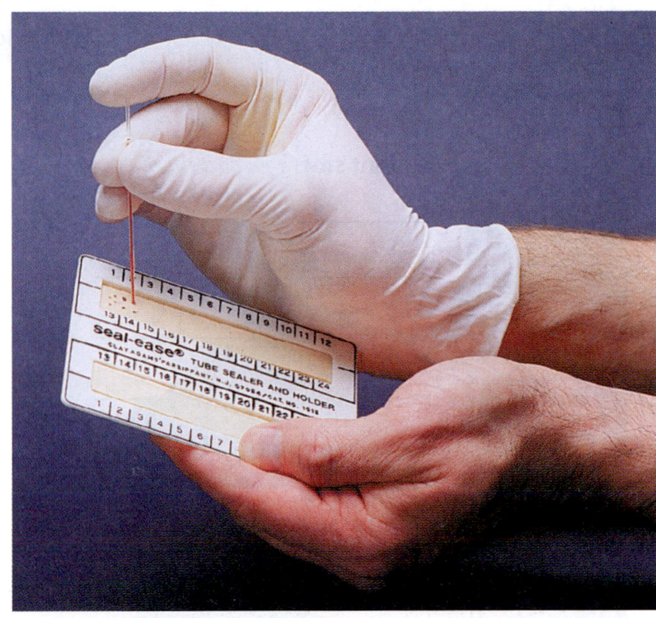

(b)

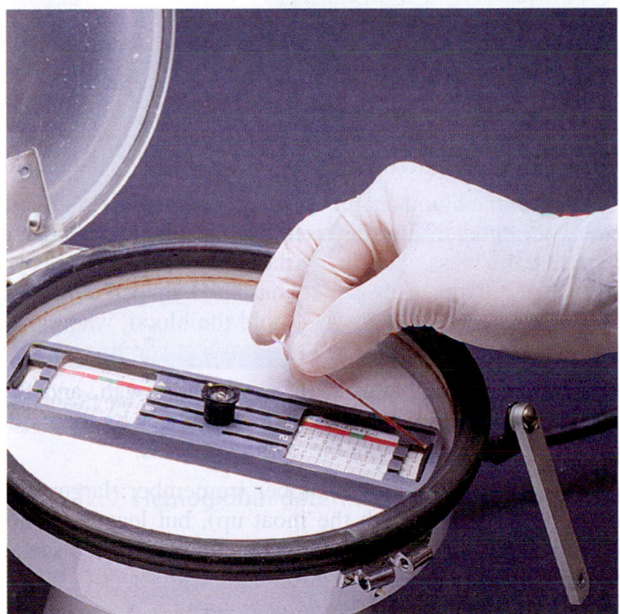

(c)

Figure 22.6 Steps in a hematocrit determination. **(a)** Load a heparinized capillary tube with blood. **(b)** Plug the blood-containing end of the tube with clay. **(c)** Place the tube in a microhematocrit centrifuge. (Centrifuge must be balanced.)

tubes are in. When all the tubes have been loaded, make sure the centrifuge is properly balanced, and secure the centrifuge cover. Turn the centrifuge on, and set the timer for 4 or 5 minutes.

5. Determine the percentage of RBCs, WBCs, and plasma by using the microhematocrit reader. The RBCs are the bottom layer, the plasma is the top layer, and the WBCs are the buff-colored layer between the two. If the reader is not available, use a millimeter ruler to measure the length of the filled capillary tube occupied by each element, and compute its percentage by using the following formula:

$$\frac{\text{Height of the column composed of the element (mm)}}{\text{Height of the original column of whole blood (mm)}} \times 100$$

Record your calculations below and on the Test Data Sheet (page 374).

% RBC ________ % WBC __________ % plasma _________

Usually WBCs constitute 1% of the total blood volume. How do your blood values compare to this figure and to the normal percentages for RBCs and plasma? (See Figure 22.1.)

____

As a rule, a hematocrit is considered a more accurate test than the total RBC count for determining the RBC composition of the blood. A hematocrit within the normal range generally indicates a normal RBC number, whereas an abnormally high or low hematocrit is cause for concern. ■

Hemoglobin Concentration

As noted earlier, a person can be anemic even with a normal RBC count. Because hemoglobin (Hb) is the RBC protein responsible for oxygen transport, perhaps the most accurate way of measuring the oxygen-carrying capacity of the blood is to determine its hemoglobin content. Oxygen, which combines reversibly with the heme (iron-containing portion) of the hemoglobin molecule, is picked up by the blood cells in the lungs and unloaded in the tissues. Thus, the more hemoglobin molecules the RBCs contain, the more oxygen they will be able to transport. Normal blood contains 12 to 18 g of hemoglobin per 100 ml of blood. Hemoglobin content in men is slightly higher (13 to 18 g) than in women (12 to 16 g).

normally present in the plasma (clotting factors, or procoagulants) as well as some released by platelets and injured tissues. Basically hemostasis proceeds as follows **(Figure 22.8a)**: the injured tissues and platelets release **tissue factor (TF)** and **PF_3** respectively, which trigger the clotting mechanism, or cascade. Tissue factor and PF_3 interact with other blood protein clotting factors and calcium ions to form **prothrombin activator,** which in turn converts **prothrombin** (present in plasma) to **thrombin.** Thrombin then acts enzymatically to polymerize the soluble **fibrinogen** proteins (present in plasma) into insoluble **fibrin,** which forms a meshwork of strands that traps the RBCs and forms the basis of the clot (Figure 22.8b). Normally, blood removed from the body clots within 2 to 6 minutes.

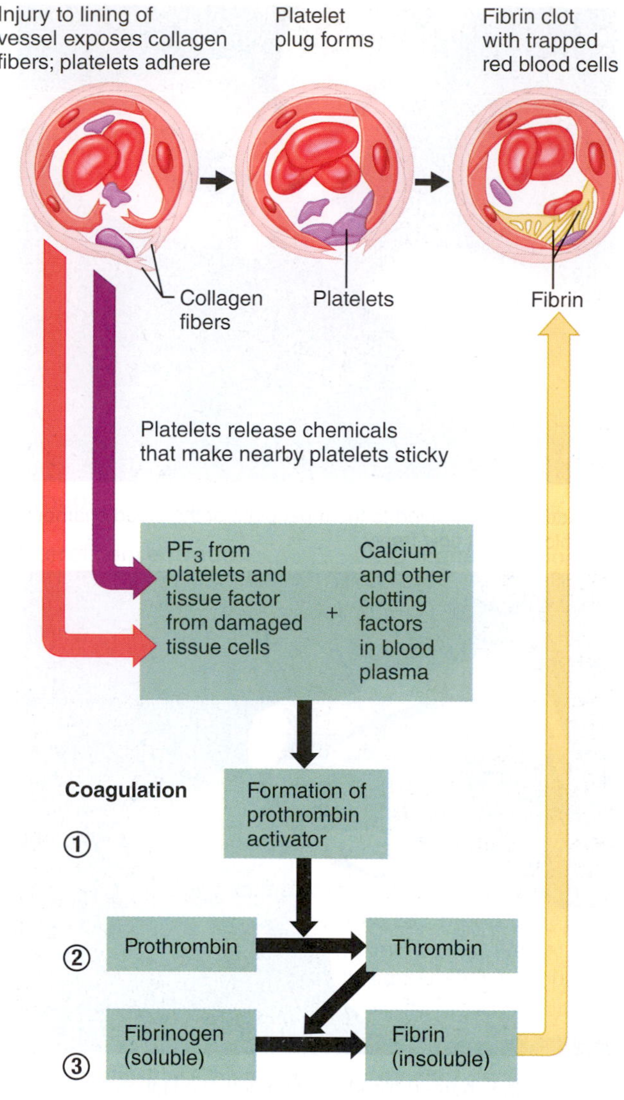

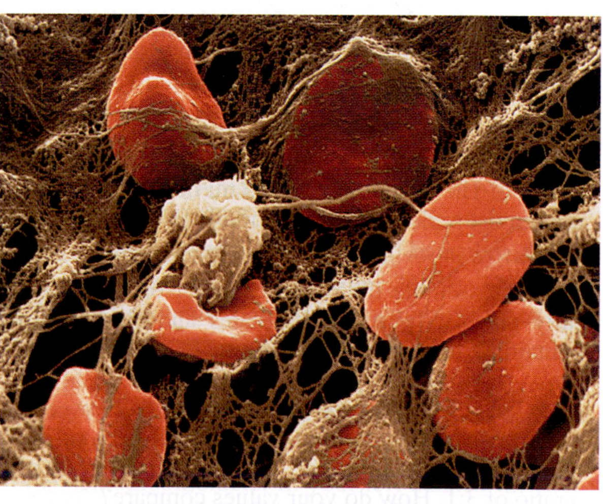

Figure 22.8 Events of hemostasis and blood clotting. (a) Simple schematic of events. Steps numbered 1–3 represent the major events of coagulation. **(b)** Photomicrograph of RBCs trapped in a fibrin mesh (2500×).

ACTIVITY 6

Determining Coagulation Time

1. Obtain a *nonheparinized* capillary tube, a timer (or watch), a lancet, cotton balls, a triangular file, and alcohol swabs.

2. Clean and prick the finger to produce a free flow of blood. Discard the lancet in the disposal container.

3. Place one end of the capillary tube in the blood drop, and hold the opposite end at a lower level to collect the sample.

4. Lay the capillary tube on a paper towel.

Record the time. ______________________

5. At 30-second intervals, make a small nick on the tube close to one end with the triangular file, and then carefully break the tube. Slowly separate the ends to see if a gel-like thread of fibrin spans the gap. When this occurs, record below and on the Test Data Sheet (page 374) the time for coagulation to occur. Are your results within the normal time range?

____

6. Put used supplies in the autoclave bag and broken capillary tubes into the sharps container. ■

Blood Typing

Blood typing is a system of blood classification based on the presence of specific glycoproteins on the outer surface of the RBC plasma membrane. Such proteins are called **antigens,** or **agglutinogens,** and are genetically determined. In many cases, these antigens are accompanied by plasma proteins, **antibodies,** or **agglutinins,** that react with RBCs bearing different antigens, causing them to be clumped, agglutinated, and eventually hemolyzed. It is because of this phenomenon that a person's blood must be carefully typed before a whole blood or packed cell transfusion.

Several blood typing systems exist, based on the various possible antigens, but the factors routinely typed for are antigens of the ABO and Rh blood groups, which are most commonly involved in transfusion reactions. Other blood factors, such as Kell, Lewis, M, and N, are not routinely typed for unless an individual will require multiple transfusions. The basis of the ABO typing is shown in the table **(Table 22.2)**.

Table 22.2 ABO Blood Typing

ABO blood type	Antigens present on RBC membranes	Antibodies present in plasma	% of U.S. population		
			White	Black	Asian
A	A	Anti-B	40	27	28
B	B	Anti-A	11	20	27
AB	A and B	None	4	4	5
O	Neither	Anti-A and anti-B	45	49	40

Individuals whose red blood cells carry the Rh antigen are Rh positive (approximately 85% of the U.S. population); those lacking the antigen are Rh negative. Unlike ABO blood groups, neither the blood of the Rh-positive (Rh^+) nor the blood of Rh-negative (Rh^-) individuals carries preformed anti-Rh antibodies. This is understandable in the case of the Rh-positive individual. However, Rh-negative persons who receive transfusions of Rh-positive blood become sensitized by the Rh antigens of the donor RBCs, and their systems begin to produce anti-Rh antibodies. On subsequent exposures to Rh-positive blood, typical transfusion reactions occur, resulting in the clumping and hemolysis of the donor blood cells.

Although the blood of dogs and other mammals does react with some of the human agglutinins (present in the antisera), the reaction is not as pronounced and varies with the animal blood used. Hence, the most accurate and predictable blood typing results are obtained with human blood. The artificial blood kit does not use any body fluids and produces results similar to but not identical to results for human blood.

ACTIVITY 7

Typing for ABO and Rh Blood Groups

Blood may be typed on glass slides or using blood test cards. Both methods are described next.

Typing Blood Using Glass Slides

1. Obtain two clean microscope slides, a wax marking pencil, anti-A, anti-B, and anti-Rh typing sera, toothpicks, lancets, alcohol swabs, a medicine dropper, and the Rh typing box.

2. Divide slide 1 into halves with the wax marking pencil. Label the lower left-hand corner "anti-A" and the lower right-hand corner "anti-B." Mark the bottom of slide 2 "anti-Rh."

3. Place one drop of anti-A serum on the *left* side of slide 1. Place one drop of anti-B serum on the *right* side of slide 1. Place one drop of anti-Rh serum in the center of slide 2.

4. If you are using your own blood, cleanse your finger with an alcohol swab, pierce the finger with a lancet, and wipe away the first drop of blood. Obtain 3 drops of freely flowing blood, placing one drop on each side of slide 1 and a drop on slide 2. Immediately dispose of the lancet in a designated disposal container.

If you are using instructor-provided animal blood or EDTA-treated red cells, use a medicine dropper to place one drop of blood on each side of slide 1 and a drop of blood on slide 2.

! 5. Quickly mix each blood-antiserum sample with a *fresh* toothpick. Then dispose of the toothpicks and used alcohol swab in the autoclave bag.

6. Place slide 2 on the Rh typing box, and rock it gently back and forth. (A slightly higher temperature is required for precise Rh typing than for ABO typing.)

7. After 2 minutes, observe all three blood samples for evidence of clumping. The agglutination that occurs in the positive test for the Rh factor is very fine and difficult to perceive; thus if there is any question, observe the slide under the microscope. Record your observations in the **Activity 7: Blood Typing** chart.

8. Interpret your ABO results (see the examples of each type in **Figure 22.9**). If clumping was observed on slide 2, you are Rh positive. If not, you are Rh negative.

9. Record your blood type in the Test Data Sheet (page 374).

10. Put the used slides in the bleach-containing bucket at the general supply area; put disposable supplies in the autoclave bag.

Using Blood Typing Cards

1. Obtain a blood typing card marked A, B, and Rh; dropper bottles of anti-A serum, anti-B serum, and anti-Rh serum; toothpicks; lancets; and alcohol swabs.

2. Place a drop of anti-A serum in the spot marked anti-A, place a drop of anti-B serum on the spot marked anti-B, and place a drop of anti-Rh serum on the spot marked anti-Rh (or anti-D).

Activity 7: Blood Typing

Result	Observed (+)	Not observed (–)
Presence of clumping with anti-A		
Presence of clumping with anti-B		
Presence of clumping with anti-Rh		

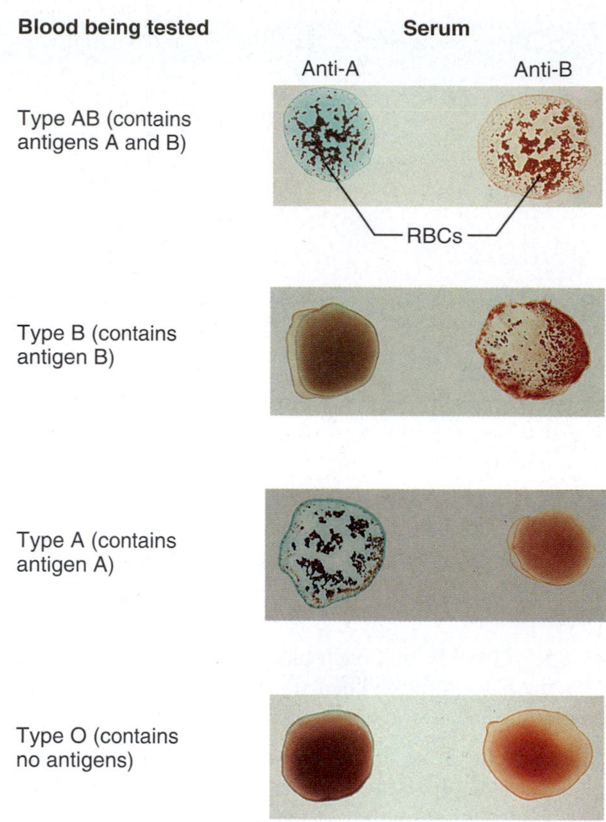

Figure 22.9 Blood typing of ABO blood types. When serum containing anti-A or anti-B antibodies (agglutinins) is added to a blood sample, agglutination will occur between the antibody and the corresponding antigen (agglutinogen A or B). As illustrated, agglutination occurs with both sera in blood group AB, with anti-B serum in blood group B, with anti-A serum in blood group A, and with neither serum in blood group O.

3. Carefully add a drop of blood to each of the spots marked "Blood" on the card. If you are using your own blood, refer to step 4 in the alternative instructions, "Typing Blood Using Glass Slides." Immediately discard the lancet in the designated disposal container.

4. Using a new toothpick for each test, mix the blood sample with the antibody. Dispose of the toothpicks appropriately.

5. Gently rock the card to allow the blood and antibodies to mix.

6. After 2 minutes, observe the card for evidence of clumping. The Rh clumping is very fine and may be difficult to observe. Record your observations in the **Activity 7: Blood Typing** chart. (Use Figure 22.9 to interpret your results).

7. Record your blood type in the Test Data Sheet (page 374), and discard the card in an autoclave bag. ■

ACTIVITY 8

Observing Demonstration Slides

Before continuing on to the cholesterol determination, take the time to look at the slides of *macrocytic hypochromic anemia, microcytic hypochromic anemia, sickle cell disease, lymphocytic leukemia* (chronic), and *eosinophilia* that have been put on demonstration by your instructor. Record your observations in the appropriate section of the Review Sheet for this exercise. You can refer to your notes, the text, and other references later to respond to questions about the blood pathologies represented on the slides. ■

Cholesterol Concentration in Plasma

Atherosclerosis is the disease process in which the body's blood vessels become increasingly blocked by plaques. By narrowing the arteries, the plaques can contribute to hypertensive heart disease. They also serve as starting points for the formation of blood clots (thrombi), which may break away and block smaller vessels farther downstream in the circulatory pathway causing heart attacks or strokes.

Ever since medical clinicians discovered that cholesterol is a major component of the smooth muscle plaques formed during atherosclerosis, it has had a bad press. Today, virtually no physical examination of an adult is considered complete until cholesterol levels are assessed along with other lifestyle risk factors. A normal value for plasma cholesterol in adults ranges from 130 to 200 mg per 100 ml of plasma; you will use blood to make such a determination.

Although the total plasma cholesterol concentration is valuable information, it may be misleading, particularly if a person's high-density lipoprotein (HDL) level is high and low-density lipoprotein (LDL) level is relatively low. Cholesterol, being water insoluble, is transported in the blood complexed to lipoproteins. In general, cholesterol bound into HDLs is destined to be degraded by the liver and then eliminated from the body, whereas that forming part of the LDLs is "traveling" to the body's tissue cells. When LDL levels are excessive, cholesterol is deposited in the blood vessel walls; hence, LDLs are considered to carry the "bad" cholesterol.

ACTIVITY 9

Measuring Plasma Cholesterol Concentration

1. Go to the appropriate supply area, and obtain a cholesterol test card and color scale, a lancet, and an alcohol swab.

2. Clean your fingertip with the alcohol swab, allow it to dry, then prick it with a lancet. Place a drop of blood on the test area of the card. Put the lancet in the designated disposal container.

3. After 3 minutes, remove the blood sample strip from the card and discard in the autoclave bag.

4. Analyze the underlying test spot, using the included color scale. Record the cholesterol level below and on the Test Data Sheet (page 374).

Cholesterol level ____________ mg/dl

5. Before leaving the laboratory, use the spray bottle of bleach solution to saturate a paper towel, and thoroughly wash down your laboratory bench. ■

Name ______________________________

Lab Time/Date ______________________

EXERCISE 22

REVIEW SHEET

Blood

Composition of Blood

1. What is the blood volume of an average-sized adult male? _____ liters

Of an average-sized adult female? ______ liters

2. What determines whether blood is bright red or a dull brick-red? ____

____

3. Use the key to identify the cell type(s) or blood elements that fit the following descriptive statements. (Some terms will be used more than once.)

Key:

a. red blood cell	d. basophil	g. lymphocyte
b. megakaryocyte	e. monocyte	h. formed elements
c. eosinophil	f. neutrophil	i. plasma

_______________ 1. most numerous leukocyte

_______________, _______________, and _______________ 2. granulocytes

_______________ 3. also called an erythrocyte; anucleate formed element

_______________, _______________ 4. actively phagocytic leukocytes

_______________, _______________ 5. agranulocytes

_______________ 6. precursor cell of platelets

_______________ 7. (a) through (g) are all examples of these

_______________ 8. number rises during parasite infections

_______________ 9. releases histamine; promotes inflammation

_______________ 10. many formed in lymphoid tissue

_______________ 11. transports oxygen

_______________ 12. primarily water, noncellular; the fluid matrix of blood

_______________ 13. increases in number during prolonged infections

_______________, _______________, _______________,

_______________, _______________ 14. also called white blood cells

4. List four classes of nutrients normally found in plasma. ______________________,

______________________, ______________________, and ______________________

Name two gases found in plasma. ______________________ and ______________________

Name three ions found in plasma. ______________________, ______________________, and ______________________

5. Describe the consistency and color of the plasma you observed in the laboratory. ______________________

____

6. What is the average life span of a red blood cell? How does its anucleate condition affect this life span?

____

____

7. From memory, describe the structural characteristics of each of the following blood cell types as accurately as possible, and note the percentage of each in the total white blood cell population.

eosinophils: ______________________

____

neutrophils: ______________________

____

lymphocytes: ______________________

____

basophils: ______________________

____

monocytes: ______________________

____

8. Correctly identify the blood disorders described in column A by matching them with selections from column B.

	Column A	Column B
______________	1. abnormal increase in the number of WBCs	a. anemia
______________	2. abnormal increase in the number of RBCs	b. leukocytosis
______________	3. condition of too few RBCs or of RBCs with hemoglobin deficiencies	c. leukopenia
		d. polycythemia
______________	4. abnormal decrease in the number of WBCs	

Hematologic Tests

9. Broadly speaking, why are hematologic studies of blood so important in diagnosing disease?

____

____

10. In the chart below, record information from the blood tests you read about or conducted. Complete the chart by recording values for healthy male adults and indicating the significance of high or low values for each test.

Test	Student test results	Normal values (healthy male adults)	Significance	
			High values	Low values
Total WBC count	No data			
Total RBC count	No data			
Hematocrit				
Hemoglobin determination				
Bleeding time	No data			
Coagulation time				

11. Why is a differential WBC count more valuable than a total WBC count when one is trying to pin down the specific source of pathology? ____

____

12. Define *hematocrit*. ____

13. If you had a high hematocrit, would you expect your hemoglobin determination to be high or low? ____________

Why? ____

____

14. What is an anticoagulant? ____

Name two anticoagulants used in conducting the hematologic tests. ____________________

and ____

What is the body's natural anticoagulant? ____

15. If your blood clumped with both anti-A and anti-B sera, your ABO blood type would be ____________

To which ABO blood groups could you give blood? ____

From which ABO donor types could you receive blood? ____

Which ABO blood type is most common? ____________________ Least common? ____________________

16. The blood of two patients has been typed for ABO blood type.

Typing results
Mr. Adams:

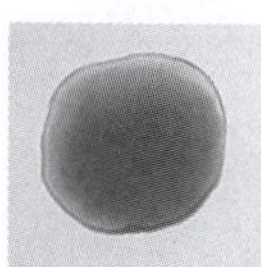

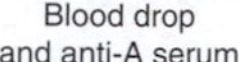

Blood drop
and anti-A serum

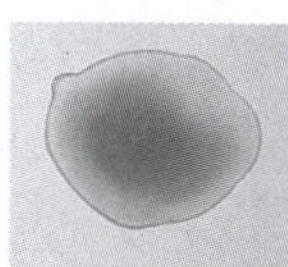

Blood drop
and anti-B serum

Typing results
Mr. Calhoon:

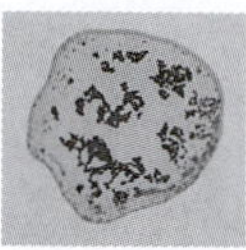

Blood drop
and anti-A serum

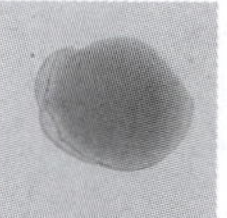

Blood drop
and anti-B serum

On the basis of these results, Mr. Adams has type ________ blood, and Mr. Calhoon has type ________ blood.

17. Record your observations of the five demonstration slides viewed.

a. Macrocytic hypochromic anemia: ____

b. Microcytic hypochromic anemia: ____

____

c. Sickle cell disease: ____

d. Lymphocytic leukemia (chronic): ____

____

e. Eosinophilia: ____

Which of the slides (a through e) above corresponds with the following conditions? (Consult your textbook or other reference source.)

______ 1. iron-deficient diet

______ 2. a type of bone marrow cancer

______ 3. genetic defect that causes hemoglobin to become sharp/spiky

______ 4. lack of vitamin B_{12}

______ 5. a tapeworm infestation in the body

______ 6. a bleeding ulcer

18. Provide the normal, or at least "desirable," range for plasma cholesterol concentration.

________ mg/100 ml

19. Describe the relationship between high blood cholesterol levels and cardiovascular diseases such as hypertension, heart attacks, and strokes.

____

____

____

____

23

EXERCISE

Anatomy of the Heart

MATERIALS

- ☐ X-ray image of the human thorax for observing the position of the heart in situ; X-ray viewing box
- ☐ Three-dimensional heart model and torso model or laboratory chart showing heart anatomy
- ☐ Red and blue pencils
- ☐ Highlighter
- ☐ Three-dimensional models of cardiac and skeletal muscle
- ☐ Compound microscope
- ☐ Prepared slides of cardiac muscle (l.s.)
- ☐ Preserved sheep heart, pericardial sacs intact or fresh hearts, pericardial sacs intact (if possible)
- ☐ Dissecting instruments and tray
- ☐ Pointed glass rods or blunt probes
- ☐ Disposable gloves
- ☐ Small plastic metric rulers
- ☐ Container for disposal of organic debris
- ☐ Laboratory detergent
- ☐ Spray bottle with 10% household bleach solution
- ☐ *Human Cardiovascular System: The Heart* video

OBJECTIVES

1. Describe the location of the heart.
2. Name and describe the covering and lining tissues of the heart.
3. Name and locate the major anatomical areas and structures of the heart when provided with an appropriate model, image, or dissected sheep heart, and describe the function of each.
4. Explain how the atrioventricular and semilunar valves operate.
5. Distinguish between blood vessels carrying oxygen-rich blood and those carrying carbon dioxide–rich blood, and describe the system used to color-code them in images.
6. Explain why the heart is called a double pump, and compare the pulmonary and systemic circuits.
7. Trace the pathway of blood through the heart.
8. Trace the functional blood supply of the heart, and name the associated blood vessels.
9. Describe the histology of cardiac muscle, and state the importance of its intercalated discs and the spiral arrangement of its cells.

PRE-LAB QUIZ

1. The heart is enclosed in a double-walled sac called the:
   a. apex b. mediastinum c. pericardium d. thorax
2. The heart is divided into __________ chambers.
   a. two b. three c. four d. five
3. What is the name of the two receiving chambers of the heart?
   ____
4. The left ventricle discharges blood into the __________, from which all systemic arteries of the body diverge to supply the body tissues.
   a. aorta c. pulmonary vein
   b. pulmonary artery d. vena cava
5. Circle True or False. Blood flows through the heart in one direction—from the atria to the ventricles.
6. Circle the correct underlined term. The right atrioventricular valve, or <u>tricuspid valve</u> / <u>mitral valve</u>, prevents backflow into the right atrium when the right ventricle is contracting.
7. Circle the correct underlined term. The heart serves as a double pump. The <u>right</u> / <u>left</u> side serves as the pulmonary circulation pump, shunting carbon dioxide–rich blood to the lungs.
8. The blood vessels that supply blood to the heart itself are the:
   a. aortas c. coronary arteries
   b. carotid arteries d. pulmonary trunks
9. Two microscopic features of cardiac cells that help distinguish them from other types of muscle cells are branching of the cells and:
   a. intercalated discs c. sarcolemma
   b. myosin fibers d. striations
10. Circle the correct underlined term. In the heart, the <u>left</u> / <u>right</u> ventricle has thicker walls and a basically circular cavity shape.

The major function of the **cardiovascular system** is transportation. Using blood as the transport vehicle, the system carries oxygen, digested foods, cell wastes, electrolytes, and many other substances vital to the body's homeostasis to and from the body cells. The system's propulsive force is the contracting heart, which can be compared to a muscular pump equipped with one-way valves. As the heart contracts, it forces blood into a closed system of large and small plumbing tubes (blood vessels) within which the blood is confined and circulated. This exercise focuses on the structure of the heart, or circulatory pump. (The anatomy of the blood vessels is considered separately in Exercise 24.)

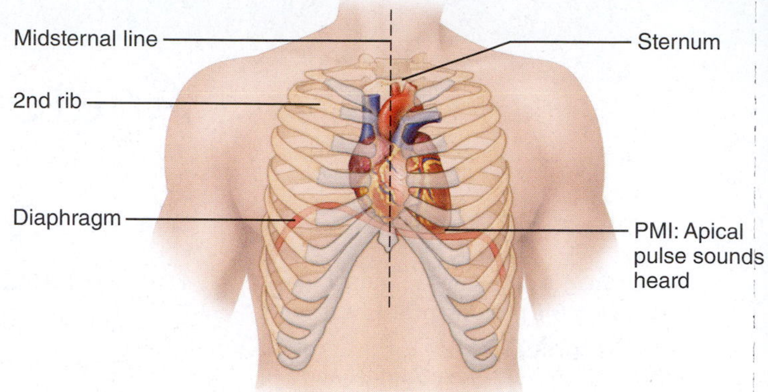

Figure 23.1 Location of the heart in the thorax. PMI is the point of the maximal intensity where the apical pulse is heard.

Gross Anatomy of the Human Heart

The **heart,** a cone-shaped organ approximately the size of a fist, is located within the mediastinum, or medial cavity, of the thorax. It is flanked laterally by the lungs, posteriorly by the vertebral column, and anteriorly by the sternum **(Figure 23.1)**. Its more pointed **apex** extends slightly to the left and rests on the diaphragm, approximately at the level of the fifth intercostal space. Its broader **base,** from which the great vessels emerge, lies beneath the second rib and points toward the right shoulder. In situ, the right ventricle of the heart forms most of its anterior surface.

The apical pulse may be heard in the fifth intercostal space at the point of maximal intensity (PMI).

☐ If an X-ray film of a human thorax is available, verify the relationships described above (otherwise, use Figure 23.1).

Check the box when you have completed this task.

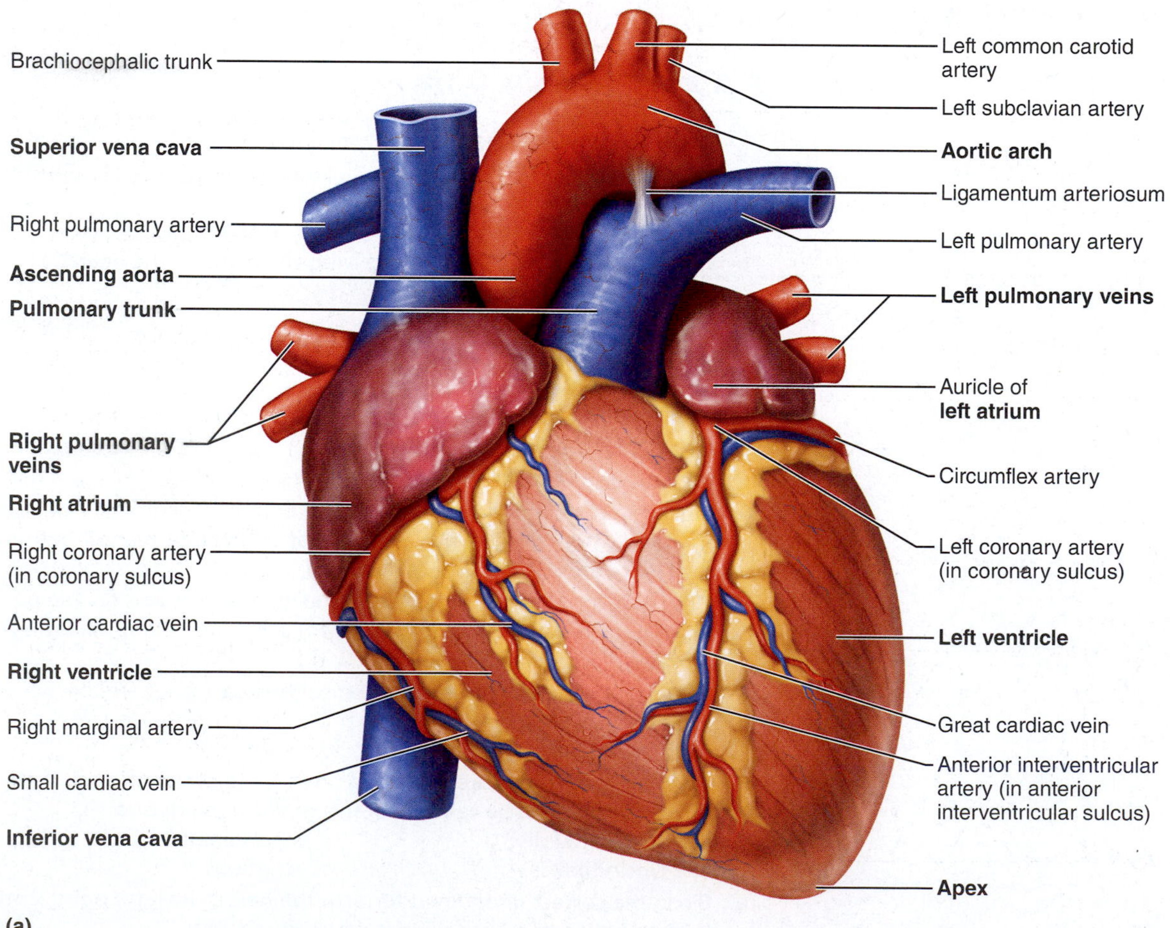

Figure 23.2 Gross anatomy of the human heart. (a) External anterior view.

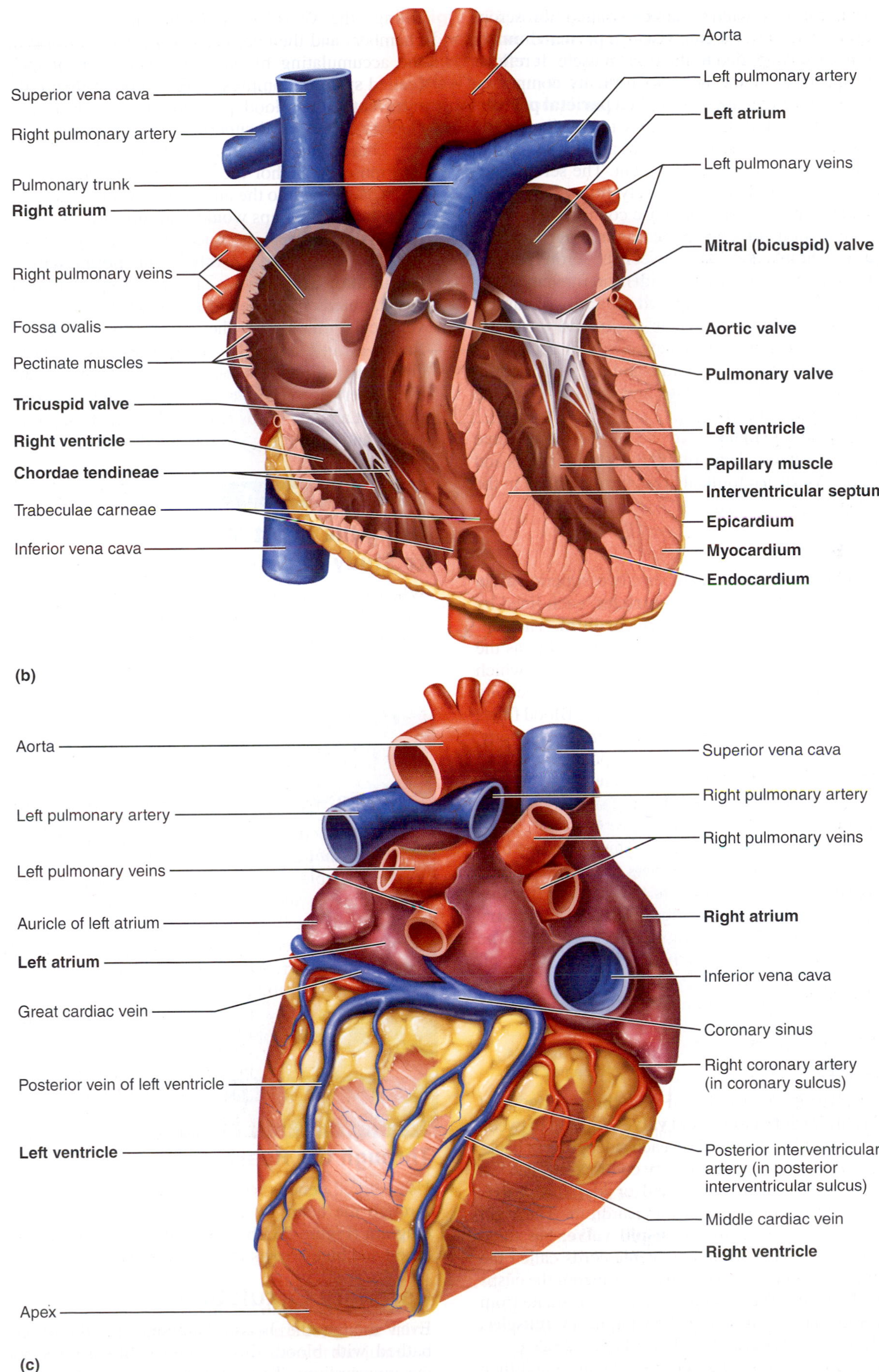
Aorta
Left pulmonary artery
Left atrium
Left pulmonary veins
Mitral (bicuspid) valve
Aortic valve
Pulmonary valve
Left ventricle
Papillary muscle
Interventricular septum
Epicardium
Myocardium
Endocardium
Superior vena cava
Right pulmonary artery
Pulmonary trunk
Right atrium
Right pulmonary veins
Fossa ovalis
Pectinate muscles
Tricuspid valve
Right ventricle
Chordae tendineae
Trabeculae carneae
Inferior vena cava
(b)
Aorta
Left pulmonary artery
Left pulmonary veins
Auricle of left atrium
Left atrium
Great cardiac vein
Posterior vein of left ventricle
Left ventricle
Apex
Superior vena cava
Right pulmonary artery
Right pulmonary veins
Right atrium
Inferior vena cava
Coronary sinus
Right coronary artery (in coronary sulcus)
Posterior interventricular artery (in posterior interventricular sulcus)
Middle cardiac vein
Right ventricle
(c)

Figure 23.2 *(continued)* **(b)** Frontal section. **(c)** External posterior view.

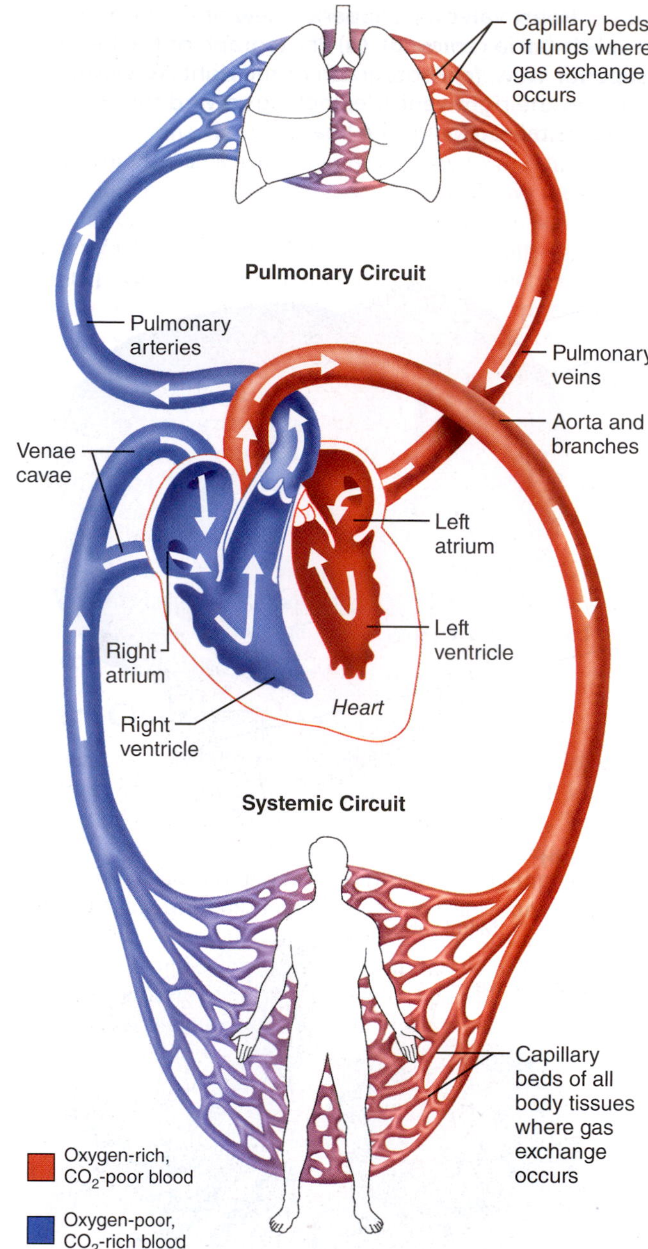

Figure 23.4 The systemic and pulmonary circuits. The heart is a double pump that serves two circulations. The right side of the heart pumps blood through the pulmonary circuit to the lungs and back to the left heart. (For simplicity, the actual number of two pulmonary arteries and four pulmonary veins has been reduced to one each.) The left side of the heart pumps blood via the systemic circuit to all body tissues and back to the right heart. Notice that blood flowing through the pulmonary circuit loses carbon dioxide (CO_2) and gains oxygen (O_2), as depicted by the color change from blue to red. Blood flowing through the systemic circuit loses oxygen and picks up carbon dioxide (red to blue color change).

the **anterior interventricular artery** (also called the *left anterior descending artery*) and the **circumflex artery.** The coronary arteries and their branches are compressed during systole and fill when the heart is relaxed.

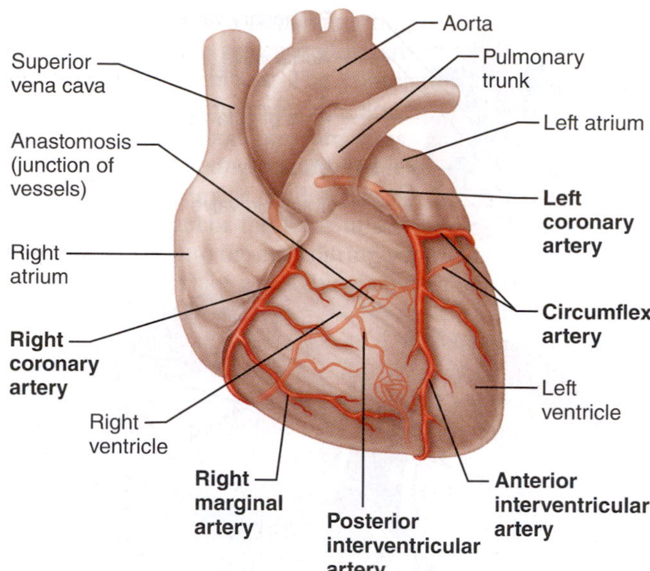

(a) The major coronary arteries

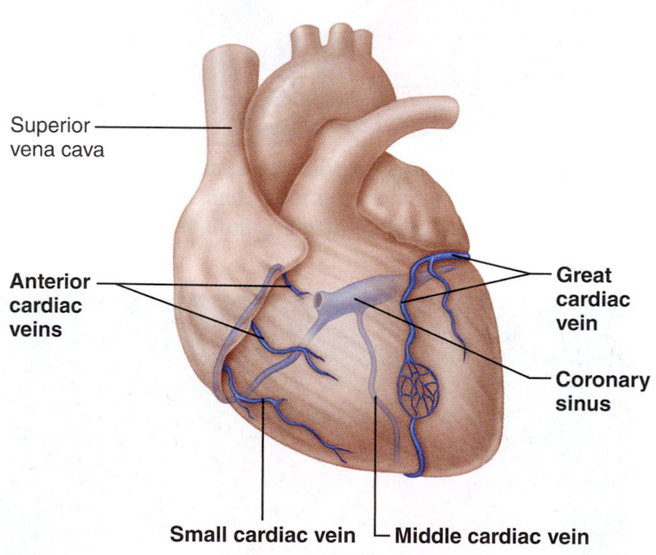

(b) The major cardiac veins

Figure 23.5 **Coronary circulation.**

The myocardium is largely drained by the **great, middle,** and **small cardiac veins,** which empty into the **coronary sinus.** The coronary sinus, in turn, empties into the right atrium. In addition, several **anterior cardiac veins** empty directly into the right atrium (Figure 23.5).

ACTIVITY 3

Using the Heart Model to Study Cardiac Circulation

1. Obtain a highlighter, and highlight all the names of the cardiac blood vessels in the external gross anatomy illustrations (Figure 23.2a and c). Note how arteries and veins travel together.

2. On a model of the heart, locate all the cardiac blood vessels shown in the coronary circulation illustration (Figure 23.5).

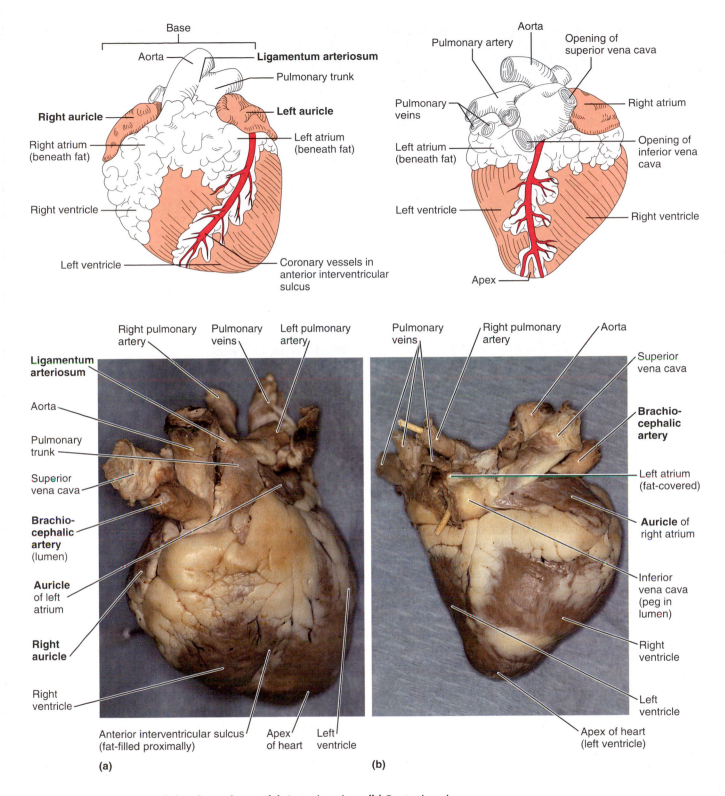

Figure 23.6 Anatomy of the sheep heart. (a) Anterior view. **(b)** Posterior view. Diagram views at top; photographs at bottom.

Use your finger to trace the pathway of blood from the right coronary artery to the lateral aspect of the right side of the heart and back to the right atrium. Name the arteries and veins along the pathway. Trace the pathway of blood from the left coronary artery to the anterior ventricular walls and back to the right atrium. Name the arteries and veins along the pathway. Note that there are multiple different pathways to distribute blood to these parts of the heart. ■

DISSECTION

The Sheep Heart

Dissecting a sheep heart is valuable because it is similar in size and structure to the human heart. Also, a dissection experience allows you to view structures in a way not possible with models and diagrams. (Refer to **Figure 23.6** as you proceed with the dissection.)

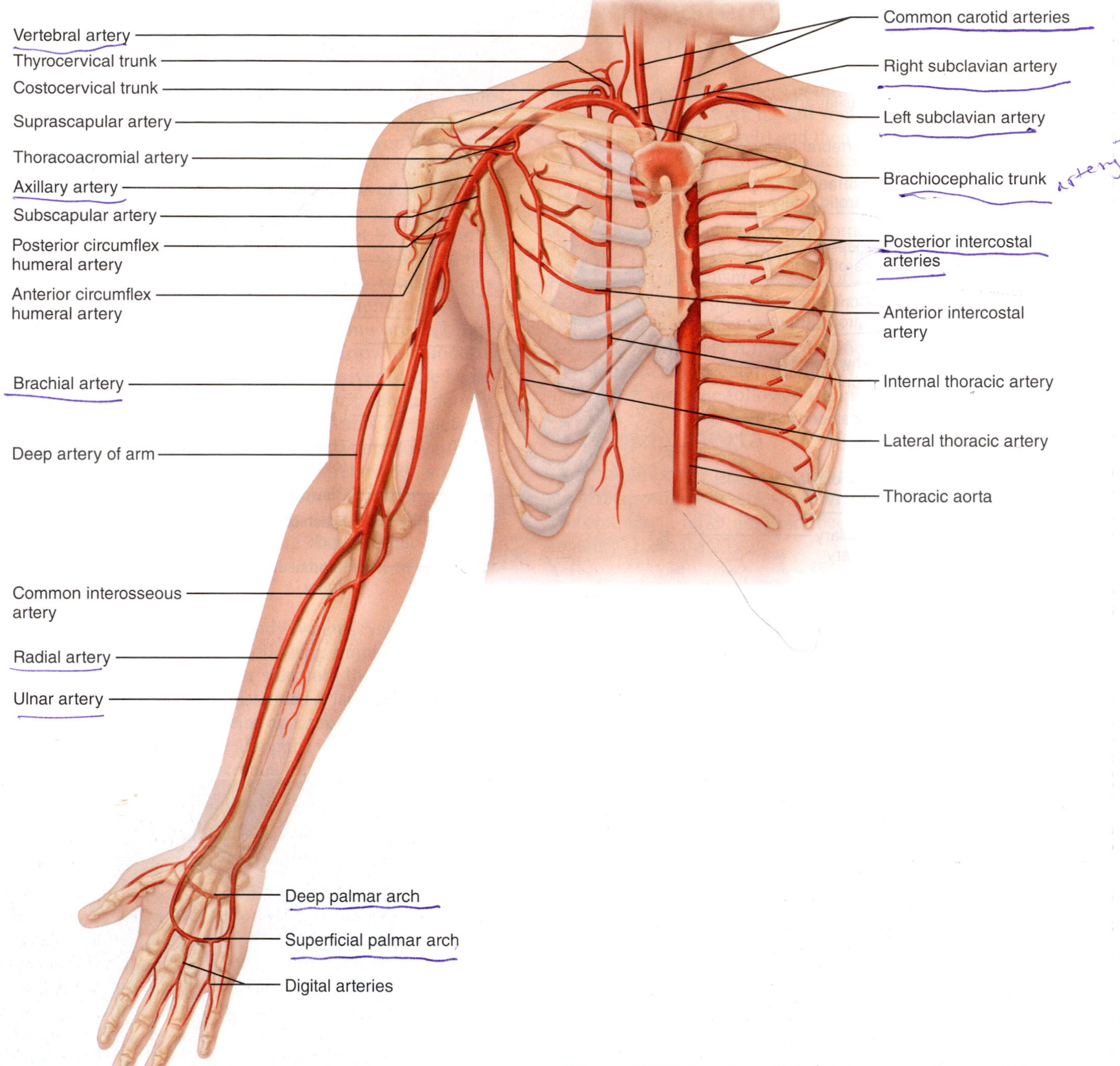

Figure 24.4 Arteries of the right upper limb and thorax.

Within the cranium, each internal carotid artery divides into **anterior** and **middle cerebral arteries,** which supply the bulk of the cerebrum. The right and left anterior cerebral arteries are connected by a short shunt called the **anterior communicating artery.** This shunt—along with shunts from each of the middle cerebral arteries, called the **posterior communicating arteries**—contributes to the formation of the **cerebral arterial circle** *(circle of Willis)*, an arterial anastomosis at the base of the brain surrounding the pituitary gland and the optic chiasma.

The paired **vertebral arteries** diverge from the subclavian arteries and pass superiorly through the foramina of the transverse process of the cervical vertebrae to enter the skull through the foramen magnum. Within the skull, the vertebral arteries unite to form a single **basilar artery,** which continues superiorly along the ventral aspect of the brain stem, giving off branches to the pons, cerebellum, and inner ear. At the base of the cerebrum, the basilar artery divides to form the **posterior cerebral arteries.** These supply portions of the temporal and occipital lobes of the cerebrum and complete the cerebral arterial circle posteriorly.

The uniting of the blood supply of the internal carotid arteries and the vertebral arteries via the cerebral arterial circle is a protective device that theoretically provides an

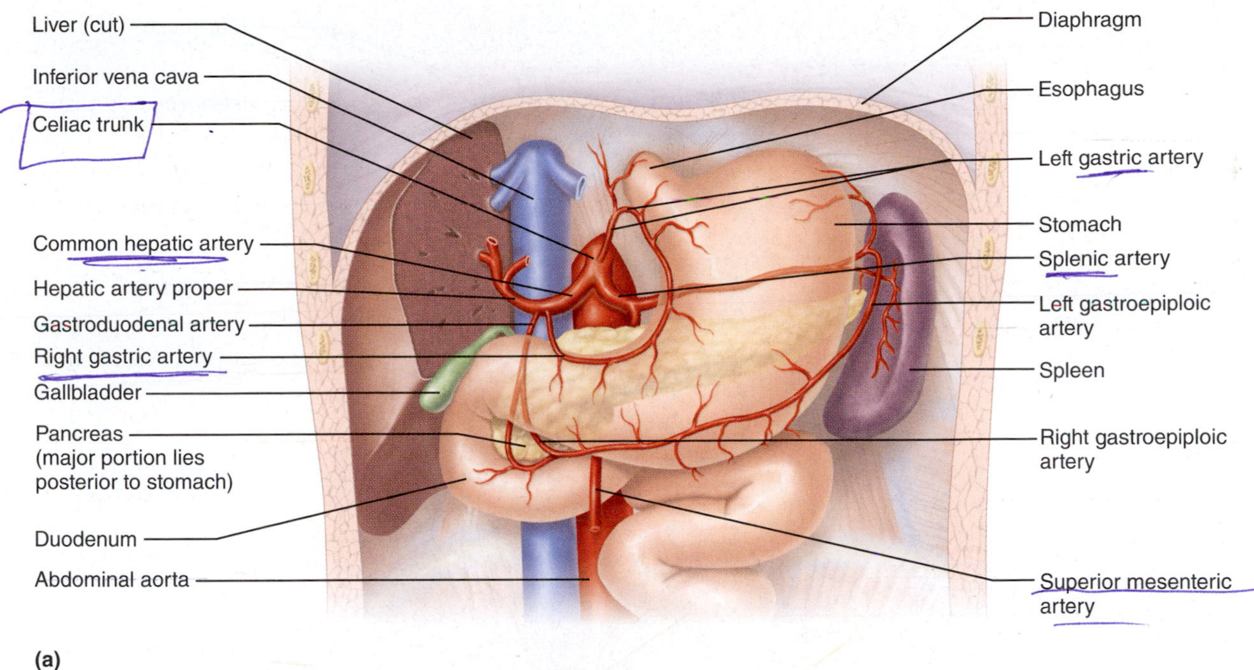

Figure 24.5 Arteries of the abdomen. (a) The celiac trunk and its major branches.

alternate set of pathways for blood to reach the brain tissue if an artery should become blocked or if blood flow is impaired anywhere in the system. In actuality, the communicating arteries are tiny, and in many cases the communicating system is not functional.

Arteries Serving the Thorax and Upper Limbs

As the **axillary artery** runs through the axilla, it gives off several branches to the chest wall and shoulder girdle **(Figure 24.4)**. These include the **thoracoacromial artery** (to shoulder and pectoral region), the **lateral thoracic artery** (lateral chest wall), the **subscapular artery** (to scapula and dorsal thorax), and the **anterior** and **posterior circumflex humeral arteries** (to the shoulder and the deltoid muscle). At the inferior edge of the teres major muscle, the axillary artery becomes the **brachial artery** as it enters the arm. The brachial artery gives off a deep branch, the **deep artery of the arm,** and as it nears the elbow it gives off several small branches. At the elbow, the brachial artery divides into the **radial** and **ulnar arteries,** which follow the same-named bones to supply the forearm and hand.

The **internal thoracic arteries** that arise from the subclavian arteries supply the mammary glands, most of the thorax wall, and anterior intercostal structures via their **anterior intercostal artery** branches. The first two pairs of **posterior intercostal arteries** arise from the costocervical trunk, noted above. The more inferior pairs arise from the thoracic aorta. Not shown in the figure (Figure 24.4) are the small arteries that serve the diaphragm *(phrenic arteries),* esophagus *(esophageal arteries),* bronchi *(bronchial arteries),* and other structures of the mediastinum (*mediastinal* and *pericardial arteries*).

Abdominal Aorta

Although several small branches of the descending aorta serve the thorax, the more major branches of the descending aorta are those serving the abdominal organs and ultimately the lower limbs **(Figure 24.5)**.

Arteries Serving Abdominal Organs

The **celiac trunk** (Figure 24.5a) is an unpaired artery that subdivides almost immediately into three branches: the **left gastric artery** supplying the stomach, the **splenic artery** supplying the spleen, and the **common hepatic artery,** which runs superiorly and gives off branches to the stomach **(right gastric artery)**, duodenum, and pancreas. Where the **gastroduodenal artery** branches off, the common hepatic artery becomes the **hepatic artery proper,** which serves the liver. The **right** and **left gastroepiploic arteries,** branches of the gastroduodenal and splenic arteries respectively, serve the greater curvature of the stomach.

The largest branch of the abdominal aorta, the **superior mesenteric artery** (Figure 24.5b and c), supplies most of the small intestine (via the intestinal arteries) and the first half of the large intestine (via the **ileocolic** and **colic arteries**). Flanking the superior mesenteric artery on the left and right are the **middle suprarenal arteries** that serve the adrenal glands, which sit atop the kidneys.

The paired **renal arteries** (Figure 24.5b) supply the kidneys, and the **gonadal arteries,** arising from the ventral aortic surface just below the renal arteries, run inferiorly to serve the gonads. They are called **ovarian arteries** in the female and **testicular arteries** in the male. Because these vessels must travel through the inguinal canal to supply the testes, they are considerably longer in the male than in the female.

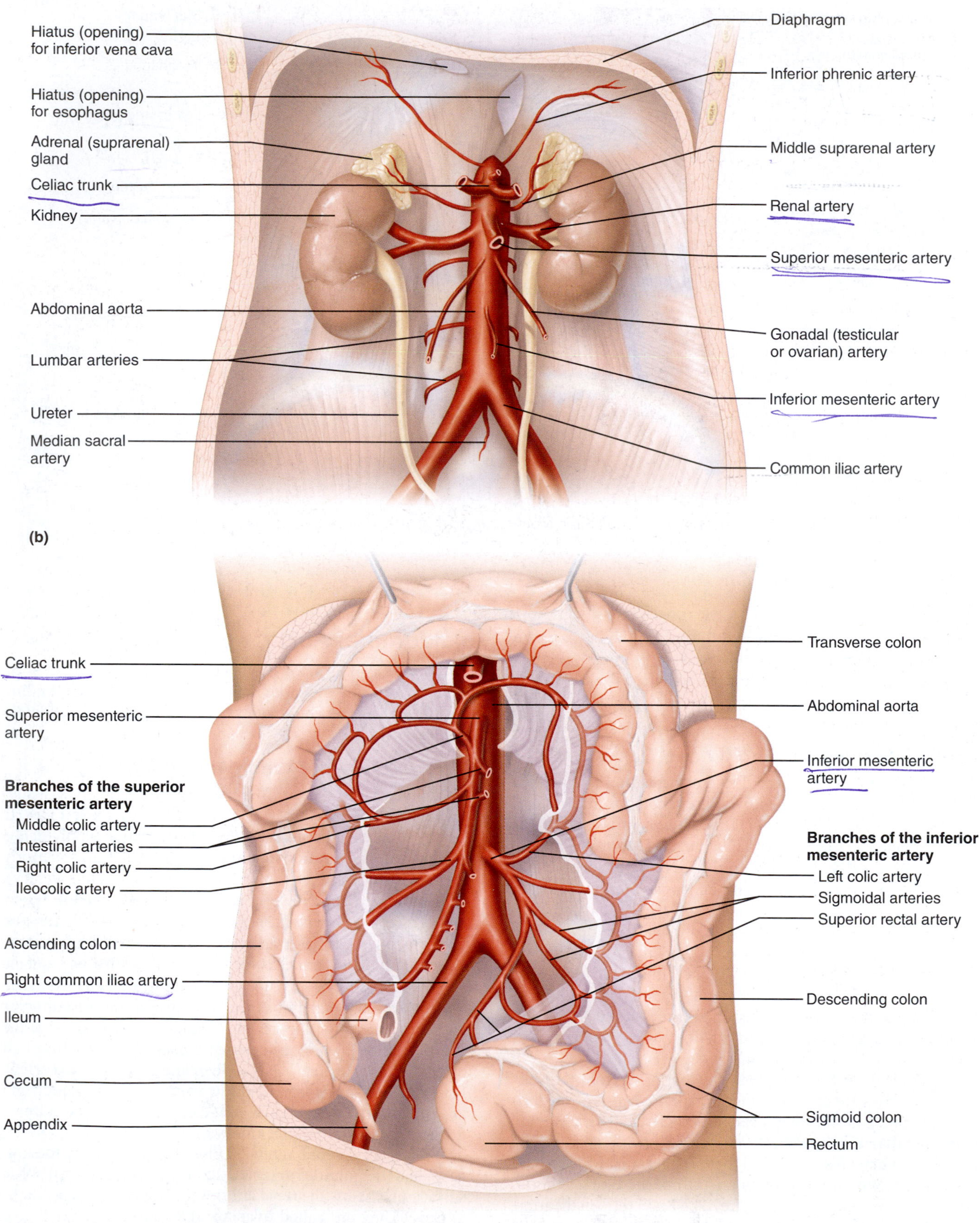

Figure 24.5 *(continued)* **Arteries of the abdomen. (b)** Major branches of the abdominal aorta. **(c)** Distribution of the superior and inferior mesenteric arteries, transverse colon pulled superiorly.

The final major branch of the abdominal aorta is the **inferior mesenteric artery** (Figure 24.5b and c), which supplies the distal half of the large intestine via several branches. Just below this, four pairs of **lumbar arteries** arise from the posterolateral surface of the aorta to supply the posterior abdominal wall (lumbar region).

In the pelvic region, the descending aorta divides into the two large **common iliac arteries,** which serve the pelvis, lower abdominal wall, and the lower limbs.

Arteries Serving the Lower Limbs

Each of the common iliac arteries extends for about 5 cm (2 inches) into the pelvis before it divides into the internal and external iliac arteries **(Figure 24.6)**. The **internal iliac artery** supplies the gluteal muscles via the **superior** and **inferior gluteal arteries** and the adductor muscles of the medial thigh via the **obturator artery,** as well as the external genitalia and perineum (via the *internal pudendal artery,* not illustrated).

The **external iliac artery** supplies the anterior abdominal wall and the lower limb. As it continues into the thigh, its name changes to **femoral artery.** Proximal branches of the femoral artery, the **circumflex femoral arteries,** supply the head and neck of the femur and the hamstring muscles. The femoral artery gives off a deep branch, the **deep artery of the thigh** (also called the *deep femoral artery*), which is the main supply to the thigh muscles (hamstrings, quadriceps, and adductors). In the knee region, the femoral artery briefly becomes the **popliteal artery;** its subdivisions—the **anterior** and **posterior tibial arteries**—supply the leg, ankle, and foot. The posterior tibial, which supplies flexor muscles, gives off one main branch, the **fibular artery,** that serves the lateral calf (fibular muscles), and then divides into the **lateral** and **medial plantar arteries** that supply blood to the sole of the foot. The anterior tibial artery supplies the extensor muscles and terminates with the **dorsalis pedis artery.** The dorsalis pedis supplies the dorsum of the foot and continues on as the **arcuate artery,** which issues the **dorsal metatarsal arteries** to the metatarsus of the foot. The dorsalis pedis is often palpated in patients with circulation problems of the leg to determine the circulatory efficiency to the limb as a whole.

☐ Palpate your own dorsalis pedis artery.

Check the box when you have completed this task.

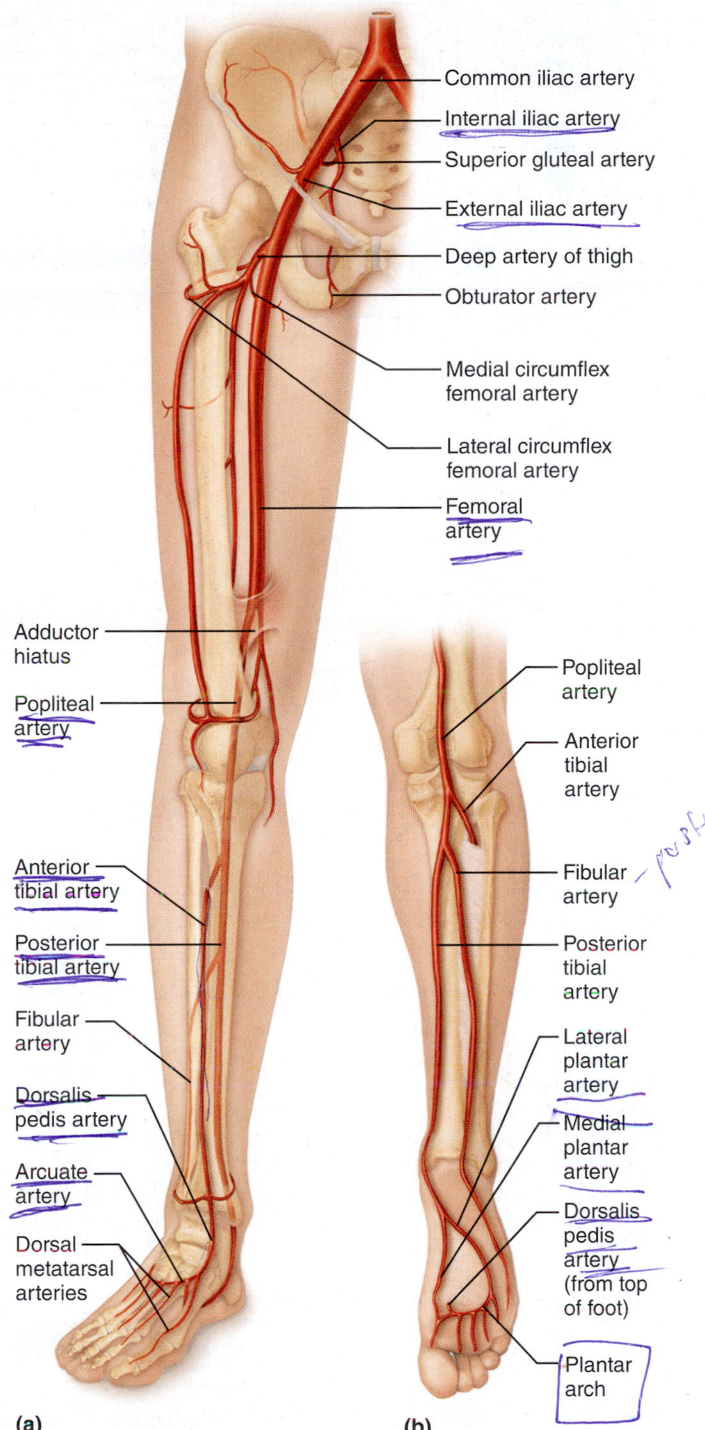

Figure 24.6 Arteries of the right pelvis and lower limb. **(a)** Anterior view. **(b)** Posterior view.

ACTIVITY 2

Locating Arteries on an Anatomical Chart or Model

Now that you have identified the arteries in the illustrations (Figures 24.2–24.6), attempt to locate and name them (without a reference) on a large anatomical chart or three-dimensional model of the vascular system. ■

Major Systemic Veins of the Body

Arteries are generally located in deep, well-protected body areas. However, many veins follow a more superficial course and are often easy to see and palpate on the body surface. Most deep veins parallel the course of the major arteries, and in many cases the naming of the veins and arteries is identical except for the designation of the vessels as veins. Whereas the major systemic arteries branch off the aorta, the veins tend to converge on the venae cavae, which enter the right atrium of the heart. Veins draining the head and upper extremities empty into the **superior vena cava,** and those draining the lower body empty into the **inferior vena cava.** The schematic **(Figure 24.7)** of the systemic veins and their relationship to the venae cave will get you started.

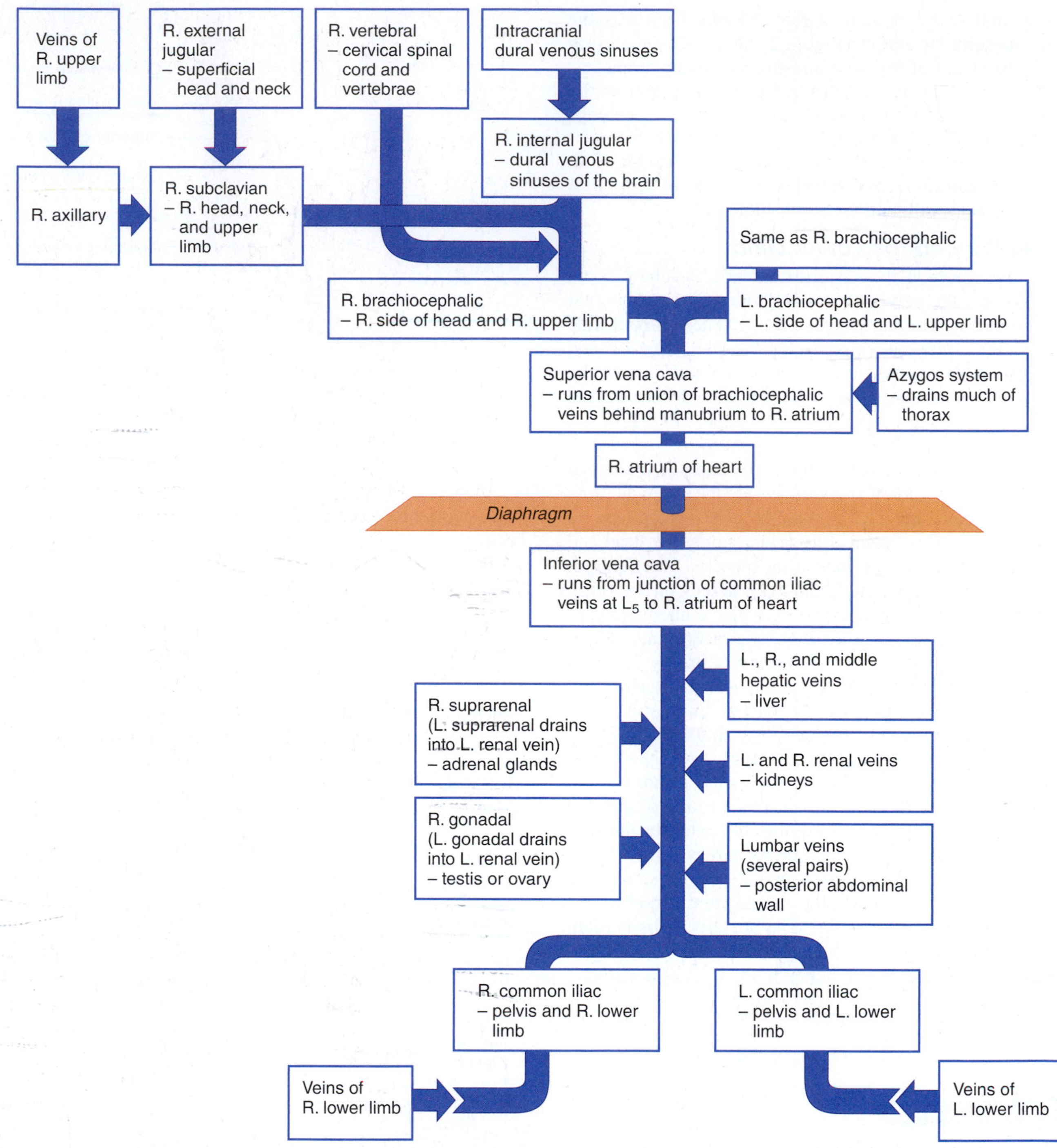

Figure 24.7 Schematic of systemic venous circulation. (R. = right, L. = left)

Veins Draining into the Inferior Vena Cava

The inferior vena cava, a much longer vessel than the superior vena cava, returns blood to the heart from all body regions below the diaphragm (see Figure 24.7). It begins in the lower abdominal region with the union of the paired **common iliac veins,** which drain venous blood from the legs and pelvis.

Veins of the Lower Limbs

Each common iliac vein is formed by the union of the **internal iliac vein,** draining the pelvis, and the **external iliac vein,** which receives venous blood from the lower limb **(Figure 24.8)**. Veins of the leg include the **anterior** and **posterior tibial veins,** which drain the calf and foot. The anterior tibial vein is a superior continuation of the **dorsalis pedis vein** of the foot. The posterior tibial vein is formed by the union of the

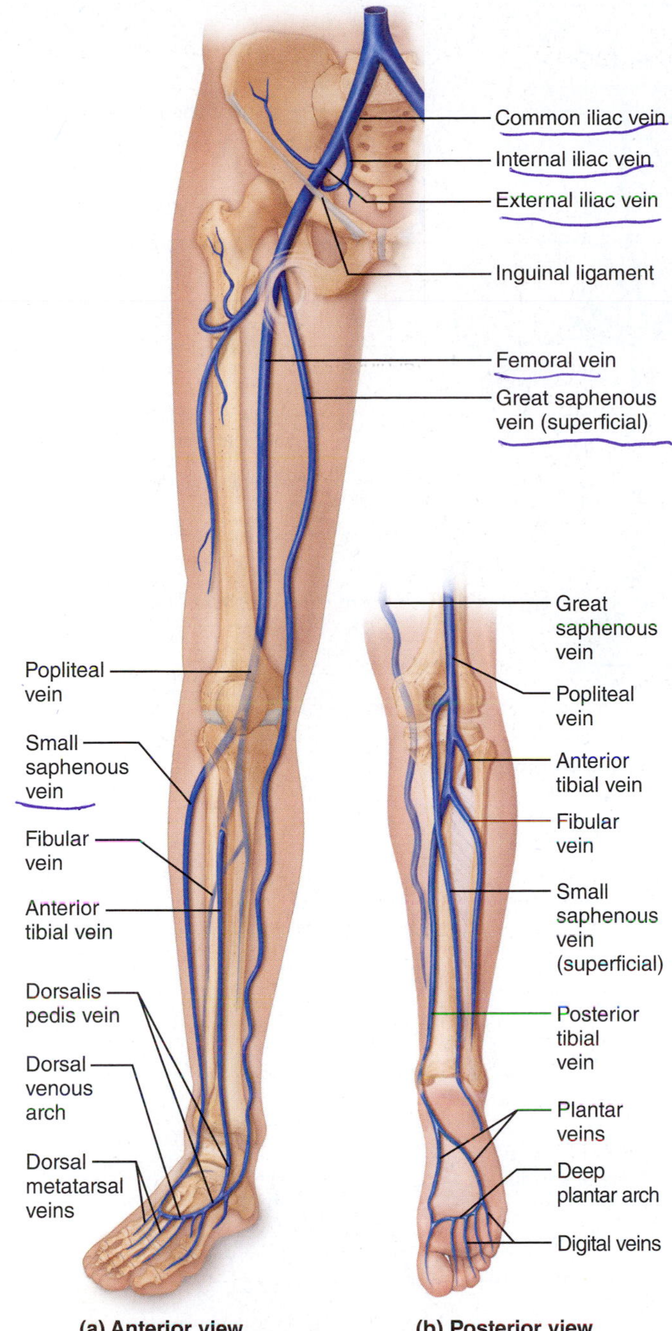

Figure 24.8 Veins of the right pelvis and lower limb. **(a)** Anterior view. **(b)** Posterior view.

medial and **lateral plantar veins** and ascends deep in the calf muscles. It receives the **fibular vein** in the calf and then joins with the anterior tibial vein at the knee to produce the **popliteal vein,** which crosses the back of the knee. The popliteal vein becomes the **femoral vein** in the thigh; the femoral vein in turn becomes the external iliac vein in the inguinal region.

The **great saphenous vein,** a superficial vein, is the longest vein in the body. Beginning in common with the **small saphenous vein** from the **dorsal venous arch,** it extends up the medial side of the leg, knee, and thigh to empty into the femoral vein. The small saphenous vein runs along the lateral aspect of the foot and through the calf muscle, which it drains, and then empties into the popliteal vein at the knee (Figure 24.8b).

Veins of the Abdomen

Moving superiorly in the abdominal cavity **(Figure 24.9)**, the inferior vena cava receives blood from the posterior abdominal wall via several pairs of **lumbar veins,** and from the right ovary or testis via the **right gonadal vein.** (The **left gonadal** [ovarian or testicular] **vein** drains into the left renal vein superiorly.) The paired **renal veins** drain the kidneys. Just above the right renal vein, the **right suprarenal vein** (receiving blood from the adrenal gland on the same side) drains into the inferior vena cava, but its partner, the **left suprarenal vein,** empties into the left renal vein inferiorly. The **hepatic veins** drain the liver. The unpaired veins draining the digestive tract organs empty into a special vessel, the **hepatic portal vein,** which carries blood to the liver to be processed before it enters the systemic venous system. (The hepatic portal system is discussed separately on pages 410–411.)

Veins Draining into the Superior Vena Cava

Veins draining into the superior vena cava are named from the superior vena cava distally, *but remember that the flow of blood is in the opposite direction.*

Veins of the Head and Neck

The **right** and **left brachiocephalic veins** drain the head, neck, and upper extremities and unite to form the superior vena cava **(Figure 24.10)**. Notice that although there is only one brachiocephalic artery, there are two brachiocephalic veins.

Branches of the brachiocephalic veins include the internal jugular, vertebral, and subclavian veins. The **internal jugular veins** are large veins that drain the superior sagittal sinus and other **dural venous sinuses.** As they move inferiorly, they receive blood from the head and neck via the **superficial temporal** and **facial veins.** The **vertebral veins** drain the posterior aspect of the head, including the cervical vertebrae and spinal cord. The **subclavian veins** receive venous blood from the upper extremity. The **external jugular vein** joins the subclavian vein near its origin to return the venous drainage of the extracranial (superficial) tissues of the head and neck.

Veins of the Upper Limb and Thorax

As the subclavian vein traverses the axilla, it becomes the **axillary vein** and then the **brachial vein** as it courses along the posterior aspect of the humerus **(Figure 24.11)**. The brachial vein is formed by the union of the deep **radial** and **ulnar veins** of the forearm. The superficial venous drainage of the arm includes the **cephalic vein,** which courses along the lateral aspect of the arm and empties into the axillary vein; the **basilic vein,** found on the medial aspect of the arm and entering the brachial vein; and the **median cubital vein,** which runs between the cephalic and basilic veins in the anterior aspect of the elbow (this vein is often the site of choice for removing blood for testing purposes). The **median antebrachial vein** lies between the radial and ulnar veins,

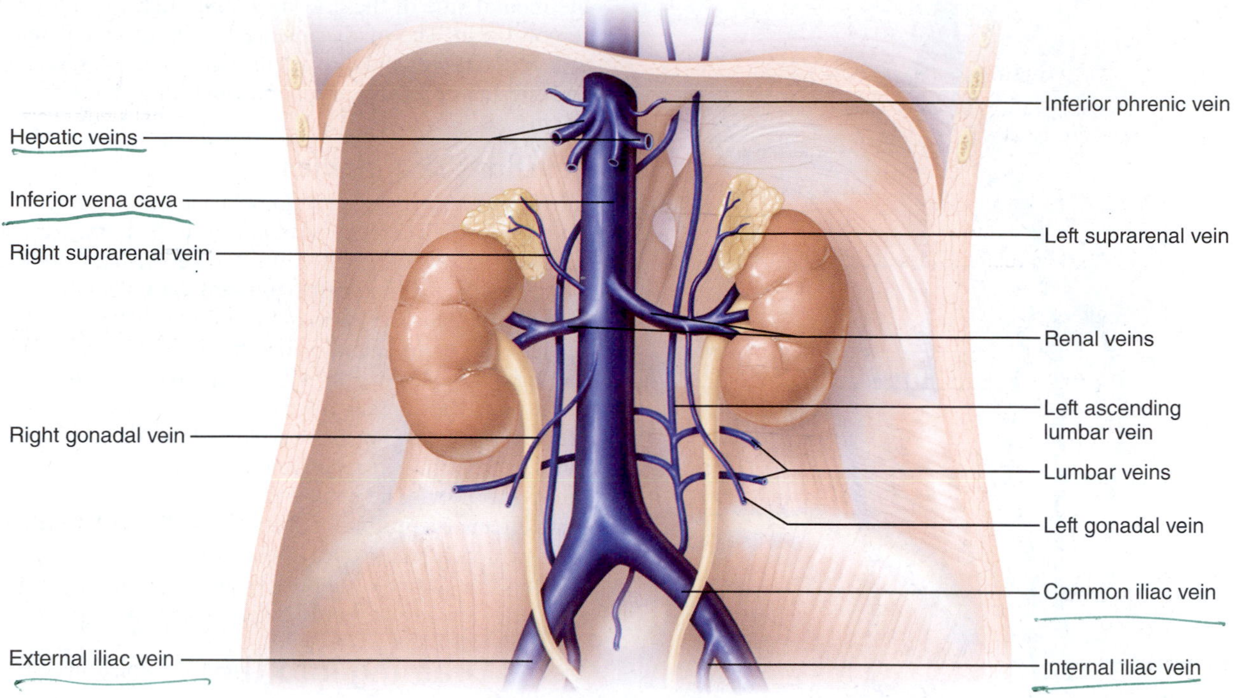

Figure 24.9 Venous drainage of abdominal organs not drained by the hepatic portal vein.

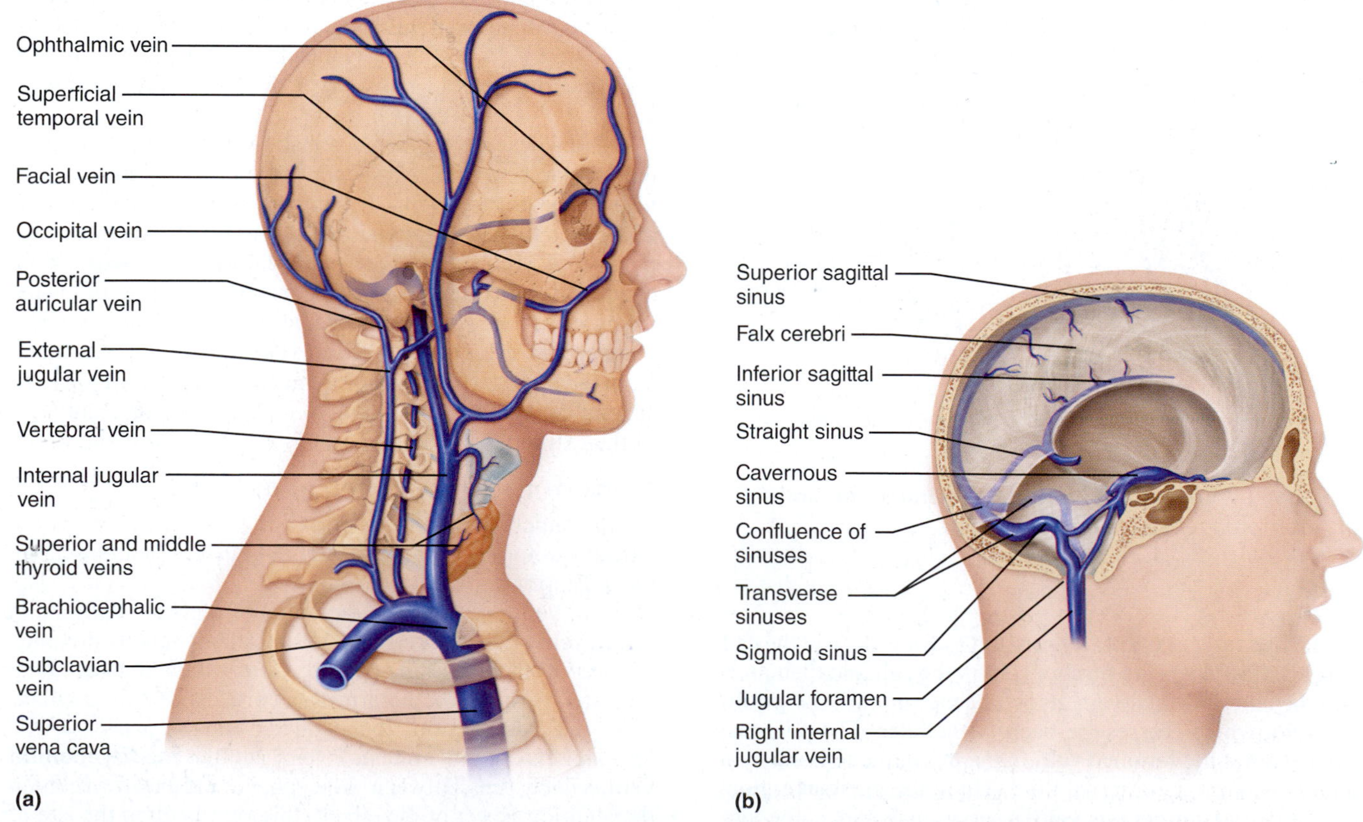

Figure 24.10 Venous drainage of the head, neck, and brain. (a) Veins of the head and neck, right superficial aspect. **(b)** Dural venous sinuses, right aspect.

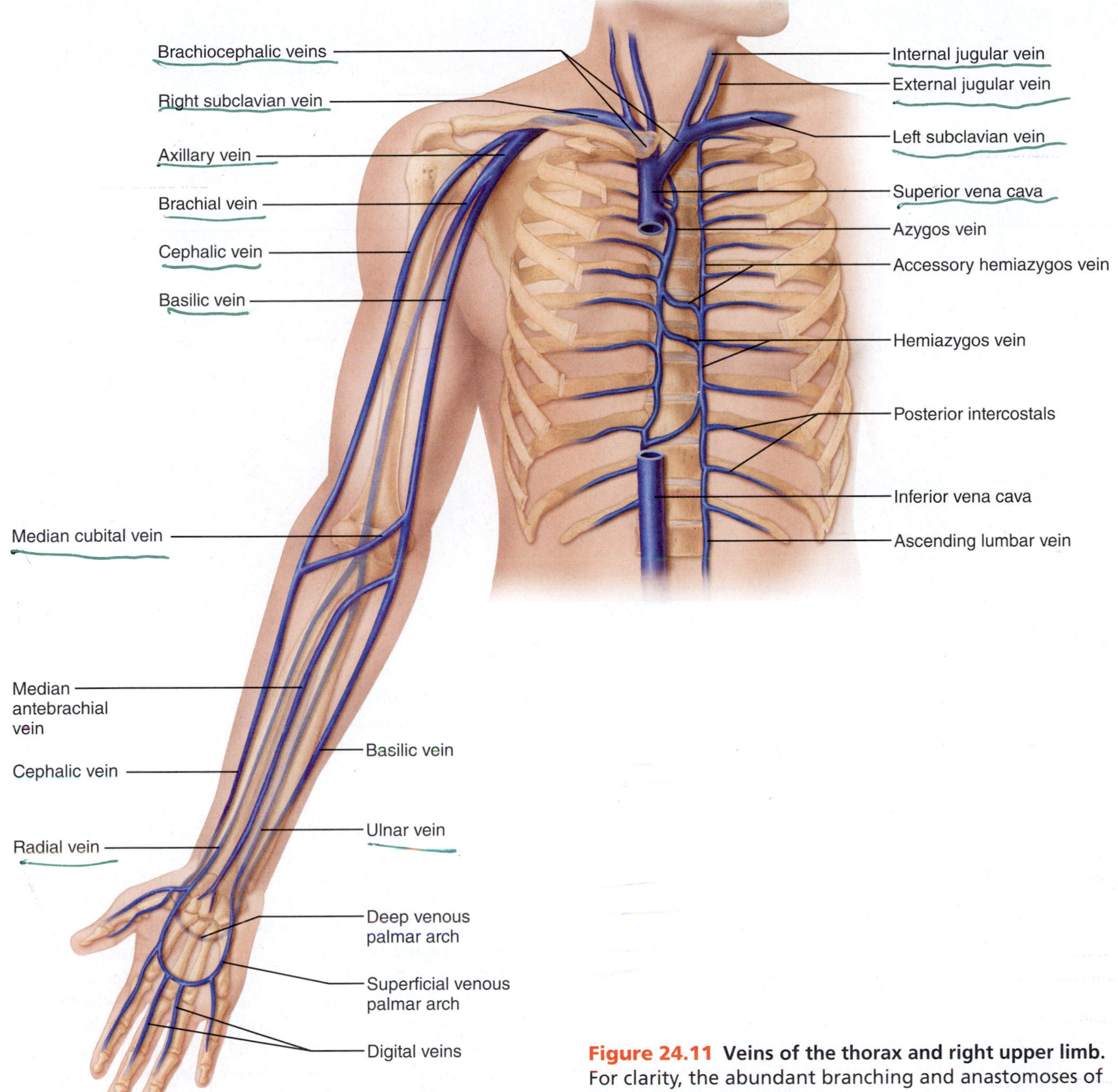

Figure 24.11 Veins of the thorax and right upper limb. For clarity, the abundant branching and anastomoses of these vessels are not shown.

and terminates variably by entering the cephalic or basilic vein at the elbow.

The **azygos system** (Figure 24.11) drains the intercostal muscles of the thorax and provides an accessory venous system to drain the abdominal wall. The **azygos vein,** which drains the right side of the thorax, enters the dorsal aspect of the superior vena cava immediately before that vessel enters the right atrium. Also part of the azygos system are the **hemiazygos** (a continuation of the **left ascending lumbar vein** of the abdomen) and the **accessory hemiazygos veins,** which together drain the left side of the thorax and empty into the azygos vein.

ACTIVITY 3

Identifying the Systemic Veins

Identify the important veins of the systemic circulation on the large anatomical chart or model without referring to the figures. ■

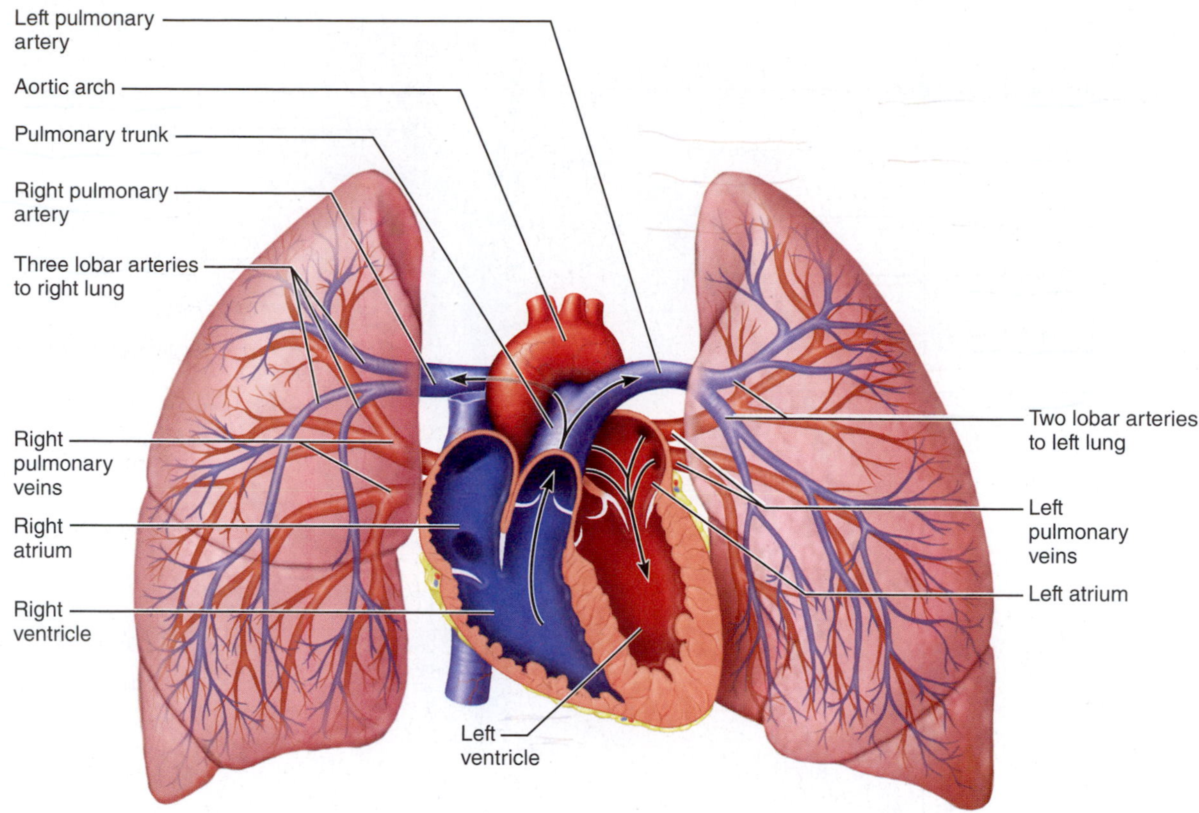

Figure 24.12 The pulmonary circulation. The pulmonary arterial system is shown in blue to indicate that the blood it carries is relatively oxygen-poor. The pulmonary venous drainage is shown in red to indicate that the blood it transports is oxygen-rich.

Special Circulations

Pulmonary Circulation

The pulmonary circulation (discussed previously in relation to heart anatomy on page 386) differs in many ways from systemic circulation because it does not serve the metabolic needs of the body tissues with which it is associated (in this case, lung tissue). It functions instead to bring the blood into close contact with the alveoli of the lungs to permit gas exchanges that rid the blood of excess carbon dioxide and replenish its supply of vital oxygen. The arteries of the pulmonary circulation are structurally much like veins, and they create a low-pressure bed in the lungs. (If the arterial pressure in the systemic circulation is 120/80, the pressure in the pulmonary artery is likely to be approximately 24/8.) The functional blood supply of the lungs is provided by the **bronchial arteries** (not shown), which diverge from the thoracic portion of the descending aorta.

ACTIVITY 4

Identifying Vessels of the Pulmonary Circulation

Find the vessels of the pulmonary circulation in the illustration **(Figure 24.12)** and on an anatomical chart (if one is available). ■

Pulmonary circulation begins with the large **pulmonary trunk,** which leaves the right ventricle and divides into the **right** and **left pulmonary arteries** about 5 cm (2 inches) above its origin. The right and left pulmonary arteries plunge into the lungs, where they subdivide into **lobar arteries** (three on the right and two on the left). The lobar arteries accompany the main bronchi into the lobes of the lungs and branch extensively within the lungs to form arterioles, which finally terminate in the capillary networks surrounding the alveolar sacs of the lungs. The respiratory gases diffuse across the walls of the alveoli and **pulmonary capillaries.** The pulmonary capillary beds are drained by venules, which converge to form sequentially larger veins and finally the four **pulmonary veins** (two leaving each lung), which return the blood to the left atrium of the heart.

Hepatic Portal Circulation

Blood vessels of the hepatic portal circulation drain the digestive viscera, spleen, and pancreas and deliver this blood to the liver for processing via the **hepatic portal vein (Figure 24.13)**. If a meal has recently been eaten, the hepatic portal blood will be nutrient-rich. The liver is the key body organ involved in maintaining proper sugar, fatty acid, and amino acid concentrations in the blood, and this system ensures that these substances pass through the liver before entering the systemic circulation. As blood percolates through the liver sinusoids, some of the nutrients are removed to be stored

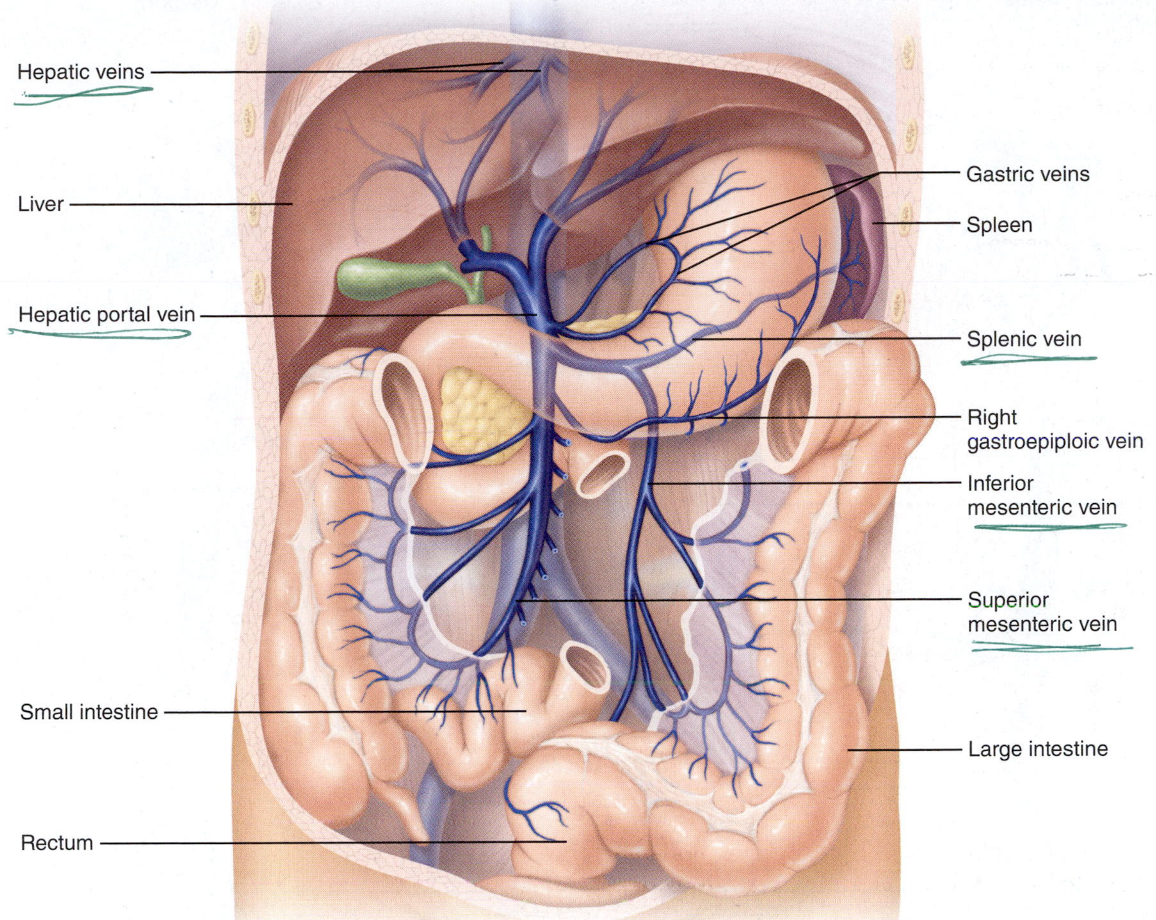

Figure 24.13 Hepatic portal circulation.

or processed in various ways for release to the general circulation. At the same time, the hepatocytes are detoxifying alcohol and other possibly harmful chemicals present in the blood, and the liver's macrophages are removing bacteria and other debris from the passing blood. The liver in turn is drained by the hepatic veins that enter the inferior vena cava.

ACTIVITY 5

Tracing the Hepatic Portal Circulation

Locate on the figure (Figure 24.13), and on an anatomical chart of the hepatic portal circulation (if available), the vessels named below in bold type. ■

The **inferior mesenteric vein,** draining the distal portions of the large intestine, joins the **splenic vein,** which drains the spleen and parts of the pancreas and stomach. The splenic vein and the **superior mesenteric vein,** which receives blood from the small intestine and the ascending and transverse colon, unite to form the hepatic portal vein. The **left gastric vein,** which drains the lesser curvature of the stomach, drains directly into the hepatic portal vein.

Fetal Circulation

In a developing fetus, the lungs and digestive system are not yet functional, and all nutrient, excretory, and gaseous exchanges occur through the placenta (**Figure 24.14a**). Nutrients and oxygen move across placental barriers from the mother's blood into fetal blood, and carbon dioxide and other metabolic wastes move from the fetal blood supply to the mother's blood.

ACTIVITY 6

Tracing the Pathway of Fetal Blood Flow

The pathway of fetal blood flow is indicated with arrows on the illustration (Figure 24.14a). Follow the flow of blood from the placenta to the fetal heart and back, noting the names of all specialized fetal circulatory structures. ■

Fetal blood travels through the umbilical cord, which contains three blood vessels: one large umbilical vein and two smaller umbilical arteries. The **umbilical vein** carries blood rich in nutrients and oxygen to the fetus; the **umbilical**

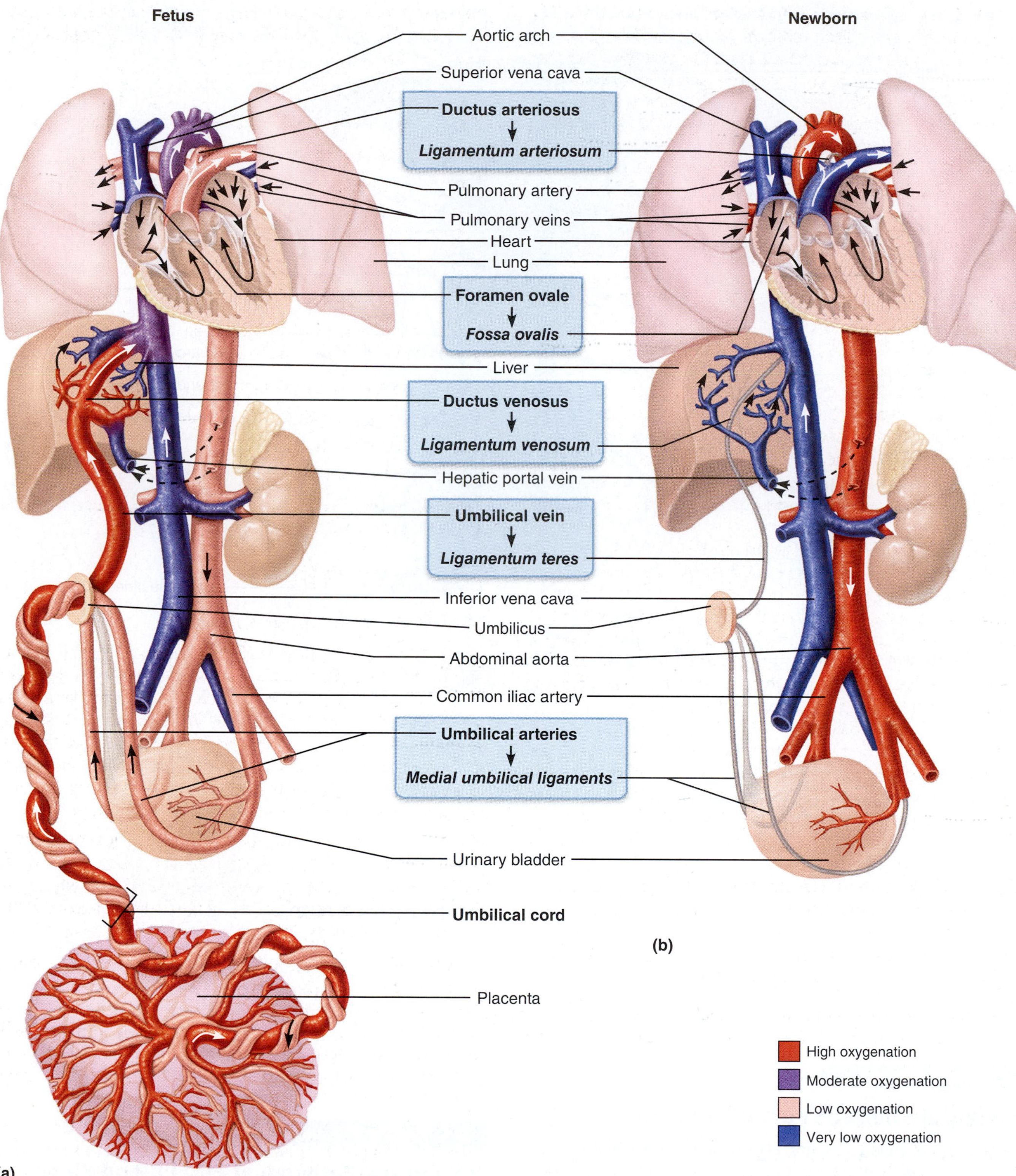

Figure 24.14 Circulation in fetus and newborn. Arrows indicate direction of blood flow. **(a)** Special adaptations for embryonic and fetal life. Arrows in the blue boxes go from the fetal structure to what it becomes after birth. The umbilical vein (red) carries oxygen- and nutrient-rich blood from the placenta to the fetus. The umbilical arteries (pink) carry waste-laden blood from the fetus to the placenta. **(b)** Changes in the cardiovascular system at birth. The umbilical vessels are occluded, as are the liver and lung bypasses (ductus venosus and arteriosus, and the foramen ovale).

arteries carry carbon dioxide and waste-laden blood from the fetus to the placenta. The umbilical arteries, which transport blood away from the fetal heart, meet the umbilical vein at the *umbilicus* (navel, or belly button) and wrap around the vein within the cord en route to their placental attachments. Newly oxygenated blood flows in the umbilical vein superiorly toward the fetal heart. Some of this blood perfuses the liver, but the larger proportion is ducted through the relatively nonfunctional liver to the inferior vena cava via a shunt vessel called the **ductus venosus,** which carries the blood to the right atrium of the heart.

Because fetal lungs are nonfunctional and collapsed, two shunting mechanisms ensure that blood almost entirely bypasses the lungs. Much of the blood entering the right atrium is shunted into the left atrium through the **foramen ovale,** a flaplike opening in the interatrial septum. The left ventricle then pumps the blood out the aorta to the systemic circulation. Blood that does enter the right ventricle and is pumped out of the pulmonary trunk encounters a second shunt, the **ductus arteriosus,** a short vessel connecting the pulmonary trunk and the aorta. Because the collapsed lungs present an extremely high-resistance pathway, blood more readily enters the systemic circulation through the ductus arteriosus.

The aorta carries blood to the tissues of the body; this blood ultimately finds its way back to the placenta via the umbilical arteries. The only fetal vessel that carries highly oxygenated blood is the umbilical vein. All other vessels contain varying degrees of oxygen-rich and oxygen-poor blood.

At birth, or shortly after, the foramen ovale closes and becomes the **fossa ovalis,** and the ductus arteriosus collapses and is converted to the fibrous **ligamentum arteriosum** (Figure 24.14b). Lack of blood flow through the umbilical vessels leads to their eventual obliteration, and the circulatory pattern becomes that of the adult. Remnants of the umbilical arteries persist as the **medial umbilical ligaments** on the inner surface of the anterior abdominal wall; the occluded umbilical vein becomes the **ligamentum teres** (or **round ligament** of the liver); and the ductus venosus becomes a fibrous band called the **ligamentum venosum** on the inferior surface of the liver.

DISSECTION AND IDENTIFICATION
The Blood Vessels of the Cat

If you have already opened your animal's ventral body cavity and identified many of its organs, begin this exercise with Activity 8. ■

ACTIVITY 7

Opening the Ventral Body Cavity and Preliminary Organ Identification

If the ventral body cavity has not yet been opened, do so now by following your instructor's instructions (see Exercise 21 on page 353).

Before you identify and trace the blood supply of the various organs of the cat, it can be helpful to do a preliminary identification of ventral body cavity organs shown in the photograph **(Figure 24.15)**. You will study the organ systems contained in the ventral cavity in later exercises, so the objective here is simply to identify the most important organs. Locate and identify the following body cavity organs (refer to Figure 24.15).

Thoracic Cavity Organs

Heart: In the mediastinum enclosed by the pericardium.

Lungs: Flanking the heart.

Thymus: Superior to and partially covering the heart (see Figure 21.5, page 355). The thymus is quite large in young cats but is largely replaced by fat as cats age.

Abdominal Cavity Organs

Liver: Posterior to the diaphragm.

- Lift the large, drapelike, fat-infiltrated greater omentum covering the abdominal organs to expose the following:

Stomach: Dorsally located and to the left side of the liver.

Spleen: A flattened, brown organ curving around the lateral aspect of the stomach.

Small intestine: Continuing posteriorly from the stomach.

Large intestine: Taking a U-shaped course around the small intestine and terminating in the rectum. ■

ACTIVITY 8

Identifying the Blood Vessels

1. Carefully clear away any thymus tissue or fat obscuring the heart and the large vessels associated with the heart. Before identifying the blood vessels, try to locate the *phrenic nerve* (from the cervical plexus), which innervates the diaphragm. The phrenic nerves lie ventral to the root of the lung on each side, as they pass to the diaphragm. Also attempt to locate the *vagus nerve* (cranial nerve X) passing laterally along the trachea and dorsal to the root of the lung.

2. Slit the parietal pericardium, and reflect it superiorly. Then, cut it away from its heart attachments. Review the structures of the heart. Notice its pointed inferior end (apex) and its broader superior portion. Identify the two *atria,* which appear darker than the inferior *ventricles.*

3. Identify the **aorta,** the largest artery in the body, issuing from the left ventricle. Also identify the *coronary arteries* in the sulcus on the ventral surface of the heart; these should be injected with red latex. To help you identify the blood vessels, the arteries of laboratory dissection specimens are injected with red latex; the veins are injected with blue latex. Exceptions to this will be noted as they are encountered.

4. Identify the two large venae cavae—the **superior** and **inferior venae cavae**—entering the right atrium. The superior vena cava is the largest dark-colored vessel entering the base of the heart. These vessels are called the *precava* and *postcava,* respectively, in the cat. The caval veins drain the same relative body areas as in humans. Also identify the **pulmonary trunk** (usually injected with blue latex) extending anteriorly from the right ventricle and the right and left pulmonary arteries. Trace the **pulmonary arteries** until they enter the lungs. Locate the **pulmonary veins** entering the left atrium and the ascending aorta arising from the left ventricle and running dorsal to the precava and to the left of the body midline.

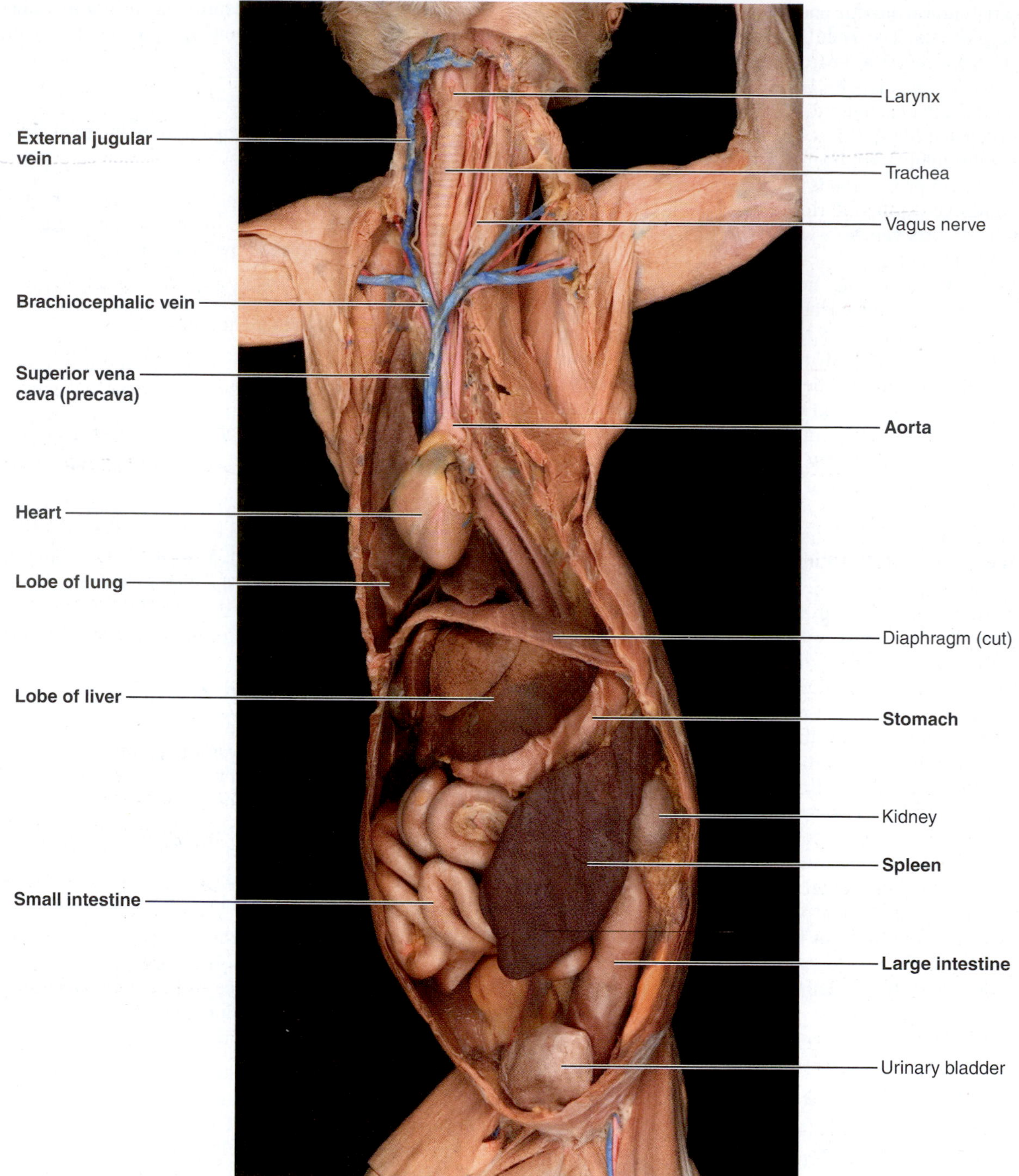

Figure 24.15 Ventral body cavity organs of the cat. (Greater omentum has been removed.)

Arteries of the Cat

Begin your dissection of the arterial system of the cat (refer to **Figure 24.16** and Figure 24.19).

1. Reidentify the aorta as it emerges from the left ventricle. As you observed in the dissection of the sheep heart, the first branches of the aorta are the **coronary arteries,** which supply the myocardium. The coronary arteries emerge from the base of the aorta and can be seen on the surface of the heart. Follow the aorta as it arches (aortic arch), and identify its major branches. In the cat, the aortic arch gives off two large vessels, the **brachiocephalic artery** and the **left subclavian artery.** The brachiocephalic artery has three major branches, the right subclavian artery and the right and left common carotid arteries. Note that in humans, the left common carotid artery and the left subclavian artery are direct branches off the aortic arch.

2. Follow the **right common carotid artery** along the right side of the trachea as it moves anteriorly, giving off branches

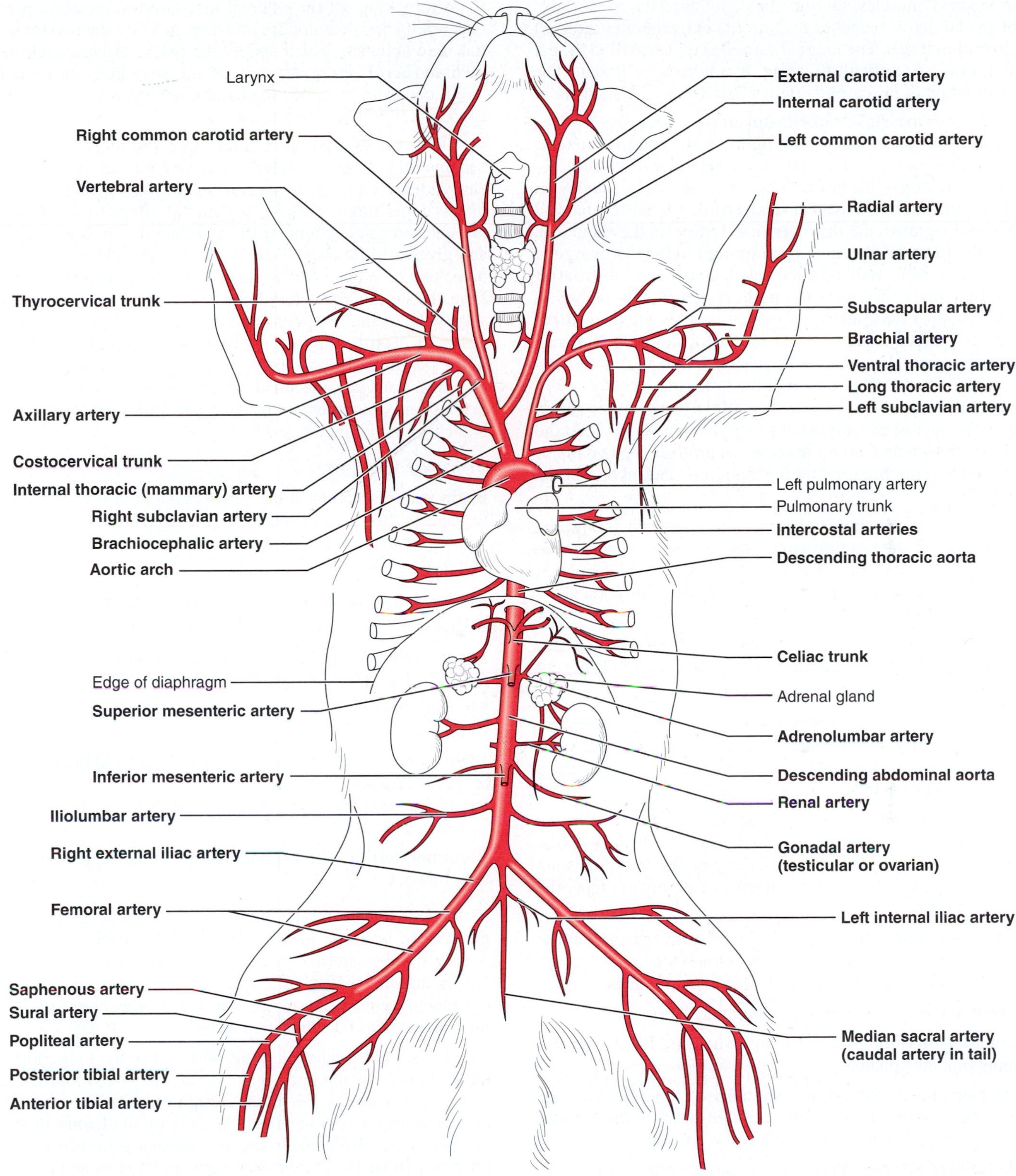
Larynx
Right common carotid artery
Vertebral artery
Thyrocervical trunk
Axillary artery
Costocervical trunk
Internal thoracic (mammary) artery
Right subclavian artery
Brachiocephalic artery
Aortic arch
External carotid artery
Internal carotid artery
Left common carotid artery
Radial artery
Ulnar artery
Subscapular artery
Brachial artery
Ventral thoracic artery
Long thoracic artery
Left subclavian artery
Left pulmonary artery
Pulmonary trunk
Intercostal arteries
Descending thoracic aorta
Celiac trunk
Edge of diaphragm
Superior mesenteric artery
Adrenal gland
Adrenolumbar artery
Descending abdominal aorta
Inferior mesenteric artery
Renal artery
Iliolumbar artery
Right external iliac artery
Gonadal artery
(testicular or ovarian)
Femoral artery
Left internal iliac artery
Saphenous artery
Sural artery
Popliteal artery
Posterior tibial artery
Anterior tibial artery
Median sacral artery
(caudal artery in tail)

Figure 24.16 Arterial system of the cat. (See also Figure 24.19 on page 420.)

to the neck muscles, thyroid gland, and trachea. At the level of the larynx, it branches to form the **external** and **internal carotid arteries.** The internal carotid is quite small in the cat and it may be difficult to locate. It may even be absent. The distribution of the carotid arteries parallels that in humans.

3. Follow the **right subclavian artery** laterally. It gives off four branches, the first being the tiny **vertebral artery,** which along with the internal carotid artery provides the arterial circulation of the brain. Other branches of the subclavian artery include the **costocervical trunk** (to the costal and cervical regions), the **thyrocervical trunk** (to the shoulder), and the **internal thoracic (mammary) artery** (serving the ventral thoracic wall). As the subclavian passes in front of the first rib, it becomes the **axillary artery.** Its branches, which supply the trunk and shoulder muscles, are the **ventral thoracic artery** (the pectoral muscles), the **long thoracic artery** (pectoral muscles and latissimus dorsi), and the subscapular artery (the trunk muscles). As the axillary artery enters the arm, it is called the **brachial artery,** and it travels with the median nerve down the length of the humerus. At the elbow, the brachial artery branches to produce the two major arteries serving the forearm and hand, the **radial** and **ulnar arteries.**

4. Return to the thorax, lift the left lung, and follow the course of the **descending aorta** through the thoracic cavity. The esophagus overlies it along its course. Notice the paired **intercostal arteries** that branch laterally from the aorta in the thoracic region.

5. Follow the aorta through the diaphragm into the abdominal cavity. Carefully pull the peritoneum away from its ventral surface, and identify the following vessels:

Celiac trunk: The first branch diverging from the aorta immediately as it enters the abdominal cavity; supplies the stomach, liver, gallbladder, pancreas, and spleen. (Trace as many of its branches to these organs as possible.)

Superior mesenteric artery: Immediately posterior to the celiac trunk; supplies the small intestine and most of the large intestine. (Spread the mesentery of the small intestine to observe the branches of this artery as they run to supply the small intestine.)

Adrenolumbar arteries: Paired arteries diverging from the aorta slightly posterior to the superior mesenteric artery; supply the muscles of the body wall and adrenal glands.

Renal arteries: Paired arteries supplying the kidneys.

Gonadal arteries (testicular or ovarian): Paired arteries supplying the gonads.

Inferior mesenteric artery: An unpaired thin vessel arising from the ventral surface of the aorta posterior to the gonadal arteries; supplies the second half of the large intestine.

Iliolumbar arteries: Paired, rather large arteries that supply the body musculature in the iliolumbar region.

External iliac arteries: Paired arteries that continue through the body wall and pass under the inguinal ligament to the hindlimb.

6. After giving off the external iliac arteries, the aorta persists briefly and then divides into three arteries: the two **internal iliac arteries,** which supply the pelvic viscera, and the **median sacral artery.** As the median sacral artery enters the tail, it comes to be called the **caudal artery.** Note that there is no common iliac artery in the cat.

7. Trace the external iliac artery into the thigh, where it becomes the **femoral artery.** The femoral artery is most easily identified in the *femoral triangle* at the medial surface of the upper thigh. Follow the femoral artery as it courses through the thigh (along with the femoral vein and nerve) and gives off branches to the thigh muscles. (These various branches are indicated on Figure 24.16.) As you approach the knee, the **saphenous artery** branches off the femoral artery to supply the medial portion of the leg. The femoral artery then descends deep to the knee to become the **popliteal artery** in the popliteal region. The popliteal artery in turn gives off two main branches, the **sural artery** and the **posterior tibial artery,** and continues as the **anterior tibial artery.** These branches supply the leg and foot.

Veins of the Cat

Begin your dissection of the venous system of the cat (refer to **Figure 24.17** and Figure 24.19). Keep in mind that the vessels are named for the region drained, not for the point of union with other veins.

1. Reidentify the **superior vena cava (precava)** as it enters the right atrium. Trace it anteriorly to identify veins that enter it:

Azygos vein: Passing directly into its dorsal surface; drains the thoracic intercostal muscles.

Internal thoracic (mammary) veins: Drain the chest and the abdominal walls.

Right vertebral vein: Drains the spinal cord and brain; usually enters right side of precava approximately at the level of the internal thoracic veins but may enter the brachiocephalic vein in your specimen.

Right and left brachiocephalic veins: Form the precava by their union.

2. Reflect the pectoral muscles, and trace the brachiocephalic vein laterally. Identify the two large veins that unite to form it—the external jugular vein and the subclavian vein. Notice that this differs from what is seen in humans, where the brachiocephalic veins are formed by the union of the internal jugular and subclavian veins.

3. Follow the **external jugular vein** as it courses anteriorly along the side of the neck to the point where it is joined on its medial surface by the **internal jugular vein.** The internal jugular veins are small and may be difficult to identify in the cat. Notice the difference in cat and human jugular veins. The internal jugular is considerably larger in humans and drains into the subclavian vein. In the cat, the external jugular is larger, and the interior jugular vein drains into it. Identify the *common carotid artery,* since it accompanies the internal jugular vein in this region. Also attempt to find the *sympathetic trunk,* which is located in the same area running lateral to the

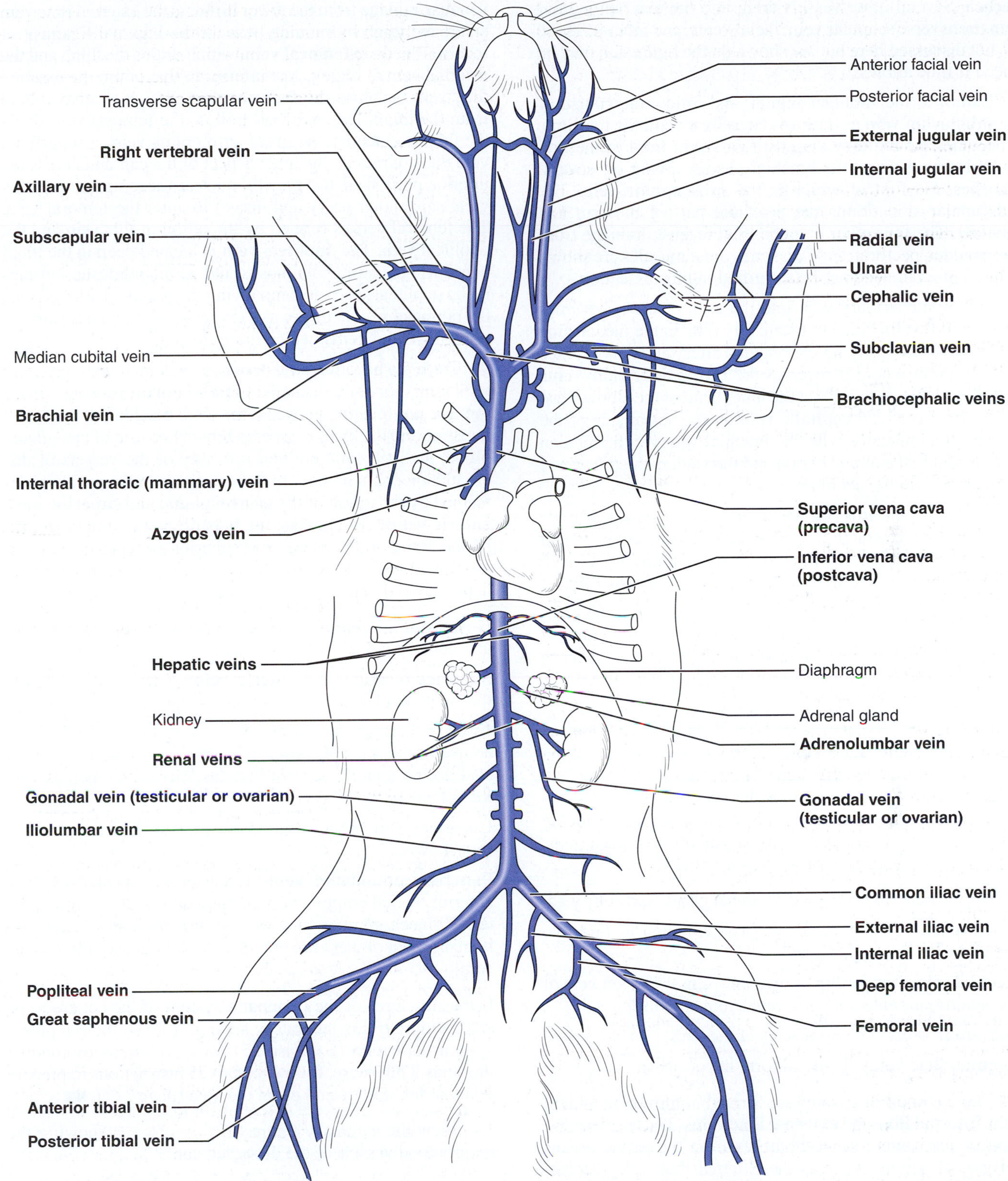
Anterior facial vein
Posterior facial vein
Transverse scapular vein
External jugular vein
Right vertebral vein
Internal jugular vein
Axillary vein
Subscapular vein
Radial vein
Ulnar vein
Cephalic vein
Subclavian vein
Median cubital vein
Brachiocephalic veins
Brachial vein
Internal thoracic (mammary) vein
Superior vena cava (precava)
Azygos vein
Inferior vena cava (postcava)
Hepatic veins
Diaphragm
Adrenal gland
Kidney
Adrenolumbar vein
Renal veins
Gonadal vein (testicular or ovarian)
Gonadal vein (testicular or ovarian)
Iliolumbar vein
Common iliac vein
External iliac vein
Internal iliac vein
Popliteal vein
Deep femoral vein
Great saphenous vein
Femoral vein
Anterior tibial vein
Posterior tibial vein

Figure 24.17 **Venous system of the cat.** (See also Figure 24.19 on page 420.)

trachea. Several other vessels drain into the external jugular vein (transverse scapular vein, facial veins, and others). These are not discussed here but are shown on the figure and may be traced if time allows.

4. Return to the shoulder region, and follow the course of the **subclavian vein** as it moves laterally toward the forelimb. It becomes the **axillary vein** as it passes in front of the first rib and runs through the brachial plexus, giving off several branches, the first of which is the **subscapular vein.** The subscapular vein drains the proximal part of the arm and shoulder. The four other branches that receive drainage from the shoulder, pectoral, and latissimus dorsi muscles are shown in the figure but need not be identified in this dissection.

5. Follow the axillary vein into the arm, where it becomes the **brachial vein.** You can locate this vein on the medial side of the arm accompanying the brachial artery and nerve. Trace it to the point where it receives the **radial** and **ulnar veins** (which drain the forelimb) at the inner bend of the elbow. Also locate the superficial **cephalic vein** on the dorsal side of the arm. It communicates with the brachial vein via the median cubital vein in the elbow region and then enters the transverse scapular vein in the shoulder.

6. Reidentify the **inferior vena cava (postcava),** and trace it to its passage through the diaphragm. Notice again as you follow its course that the intercostal veins drain into a much smaller vein lying dorsal to the postcava, the **azygos vein.**

7. Attempt to identify the **hepatic veins** entering the postcava from the liver. You may be able to see these if some of the anterior liver tissue is scraped away where the postcava enters the liver.

8. Displace the intestines to the left side of the body cavity, and proceed posteriorly to identify the following veins in order. All of these veins empty into the postcava and drain the organs served by the same-named arteries. In the cat, variations in the connections of the veins to be located are common, and in some cases the postcaval vein may be double below the level of the renal veins. If you observe deviations, call them to the attention of your instructor.

Adrenolumbar veins: From the adrenal glands and body wall.

Renal veins: From the kidneys (it is common to find two renal veins on the right side).

Gonadal veins (testicular or ovarian veins): The left vein of this venous pair enters the left renal vein anteriorly.

Iliolumbar veins: Drain muscles of the back.

Common iliac veins: Unite to form the postcava.

The common iliac veins are formed in turn by the union of the **internal iliac** and **external iliac veins.** The more medial internal iliac veins receive branches from the pelvic organs and gluteal region, whereas the external iliac vein receives venous drainage from the lower limb. As the external iliac vein enters the thigh by running beneath the inguinal ligament, it receives the **deep femoral vein,** which drains the thigh and the external genital region. Just inferior to that point, the external iliac vein becomes the **femoral vein,** which receives blood from the thigh, leg, and foot. Follow the femoral vein down the thigh to identify the **great saphenous vein,** a superficial vein that courses up the inner aspect of the calf and across the inferior portion of the gracilis muscle (accompanied by the great saphenous artery and nerve) to enter the femoral vein. The femoral vein is formed by the union of this vein and the popliteal vein. The **popliteal vein** is located deep in the thigh beneath the semimembranosus and semitendinosus muscles in the popliteal space accompanying the popliteal artery. Trace the popliteal vein to its point of division into the **posterior** and **anterior tibial veins,** which drain the leg.

9. Trace the hepatic portal drainage system in your cat (refer to **Figure 24.18**). Locate the **hepatic portal vein** by removing the peritoneum between the first portion of the small intestine and the liver. It appears brown because of coagulated blood, and it is unlikely that it or any of the vessels of this circulation contain latex. In the cat, the hepatic portal vein is formed by the union of the **gastrosplenic** and **superior mesenteric veins.** (In humans, the hepatic portal vein is formed by the union of the splenic and superior mesenteric veins.) If possible, locate the following vessels, which empty into the hepatic portal vein:

Gastrosplenic vein: Carries blood from the spleen and stomach; located dorsal to the stomach.

Superior (cranial) mesenteric vein: A large vein draining the small and large intestines and the pancreas.

Inferior (caudal) mesenteric vein (not shown): Parallels the course of the inferior mesenteric artery and empties into the superior mesenteric vein. In humans, this vessel merges with the splenic vein.

Coronary vein: Drains the lesser curvature of the stomach (not shown).

Pancreaticoduodenal veins (anterior and posterior): The anterior branch empties into the hepatic portal vein; the posterior branch empties into the superior mesenteric vein. (In humans, both of these are branches of the superior mesenteric vein.)

If the structures of the lymphatic system of the cat are to be studied during this laboratory session, turn to the lymphatic system dissection (Exercise 25) for instructions to conduct the study. Otherwise, follow the boxed instructions to prepare your cat for storage and clean the area (page 217). ■

View the summary figure **(Figure 24.19)** showing the major blood vessels of the dissected cat.

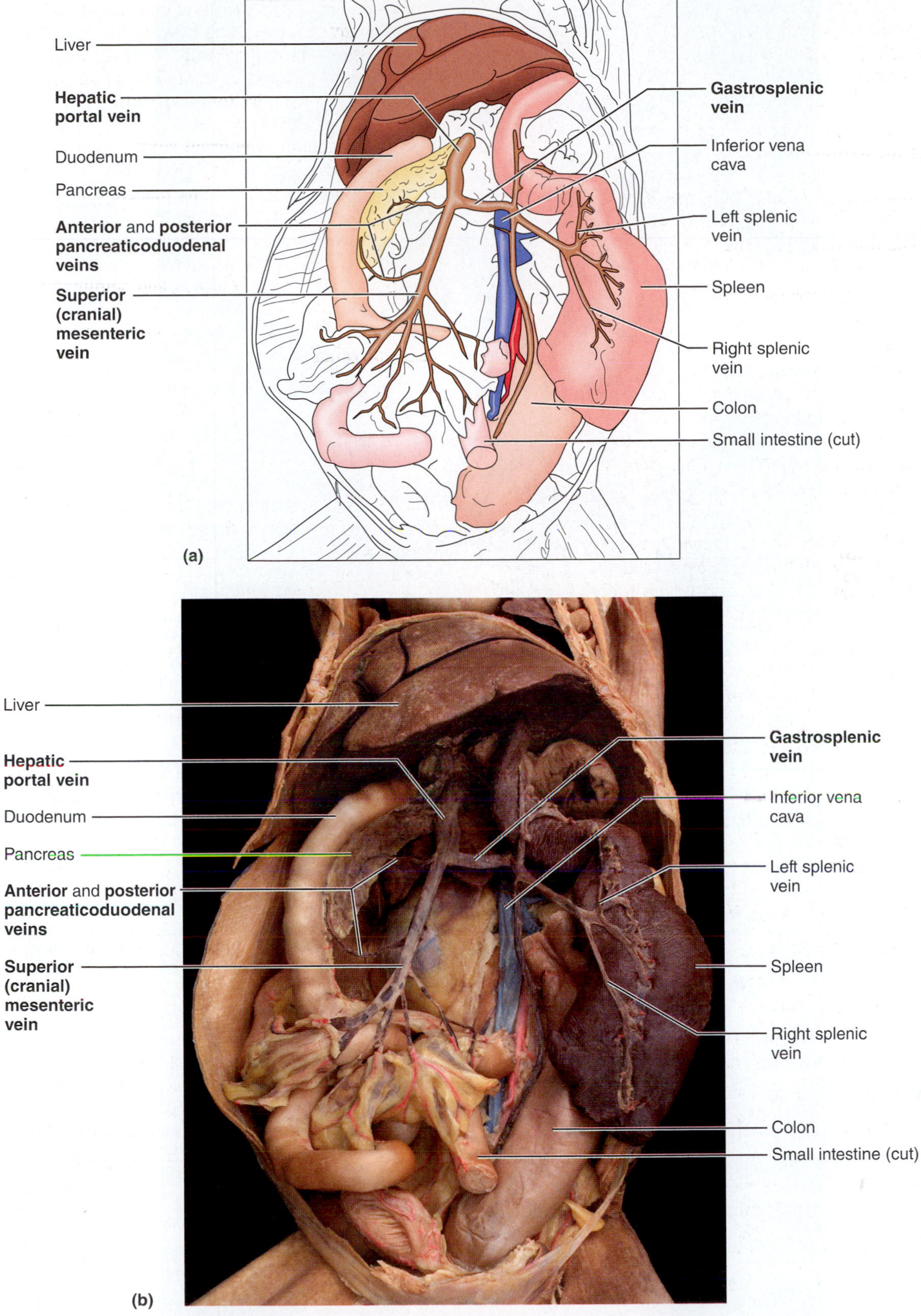

Figure 24.18 Hepatic portal circulation of the cat. (a) Diagram view. **(b)** Photograph of hepatic portal system of the cat. Intestines have been partially removed. The mesentery of the small intestine has been partially dissected to show the veins of the portal system.

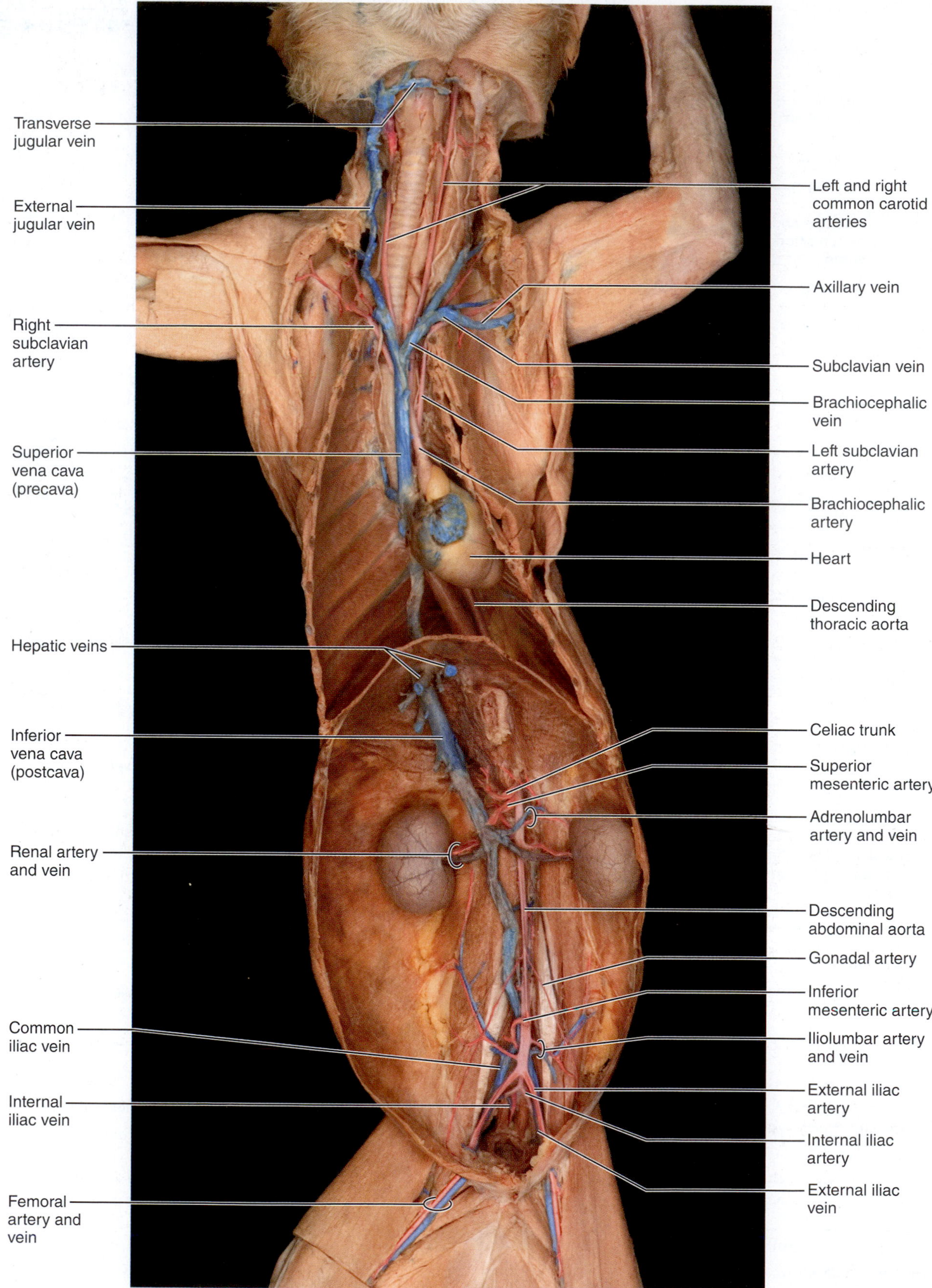

Figure 24.19 Cat dissected to reveal major blood vessels (summary figure).

Name ______________________________

Lab Time/Date ______________________

EXERCISE 24

REVIEW SHEET

Anatomy of Blood Vessels

Microscopic Structure of the Blood Vessels

1. Use the key choices to identify the blood vessel tunics described. (Responses may be used more than once.)

Key: a. tunica intima b. tunica media c. tunica externa

______________________ 1. innermost tunic

______________________ 2. bulky middle tunic; contains smooth muscle and elastin

______________________ 3. its smooth surface decreases resistance to blood flow

______________________ 4. tunic of capillaries

______________________, ______________________, ______________________ 5. tunic(s) of arteries and veins

______________________ 6. is especially thick in elastic arteries

______________________ 7. most superficial tunic

2. Servicing the capillaries is the essential function of the organs of the circulatory system. Explain this statement.

____

____

____

3. Cross-sectional views of an artery and of a vein are shown here. Identify each; and on the lines to the sides, note the structural details that enabled you to make these identifications:

______________________ (vessel type)

______________________ (a)

______________________ (b)

______________________ (vessel type)

______________________ (a)

______________________ (b)

4. Why are valves present in veins but *not* in arteries? ______________________________________

____

____

5. Name two events *occurring within the body* that aid in venous return.

______________________________________ and ______________________________________

6. Why are the walls of arteries proportionately thicker than those of the corresponding veins? ____________________

____

____

Major Systemic Arteries and Veins of the Body

7. Use the key on the right to identify the arteries or veins described on the left. Some terms are used more than once.

_________ 1. the arterial system has one of these; the venous system has two

_________ 2. these arteries supply the myocardium

_________, _________ 3. two paired arteries serving the brain

_________ 4. longest vein in the lower limb

_________ 5. artery on the dorsum of the foot checked after leg surgery

_________ 6. serves the posterior thigh

_________ 7. supplies the diaphragm

_________ 8. formed by the union of the radial and ulnar veins

_________, _________ 9. two superficial veins of the arm

_________ 10. artery serving the kidney

_________ 11. veins draining the liver

_________ 12. artery that supplies the distal half of the large intestine

_________ 13. drains the pelvic organs

_________ 14. what the external iliac artery becomes on entry into the thigh

_________ 15. artery that branches into radial and ulnar arteries

_________ 16. supplies most of the small intestine

_________ 17. join to form the inferior vena cava

_________ 18. an arterial trunk that has three major branches, which run to the liver, spleen, and stomach

_________ 19. major artery serving the tissues external to the skull

_________, _________, _________, _________ 20. four veins serving the leg

_________ 21. artery generally used to take the pulse at the wrist

Key:
a. anterior tibial
b. basilic
c. brachial
d. brachiocephalic
e. celiac trunk
f. cephalic
g. common carotid
h. common iliac
i. coronary
j. deep femoral
k. dorsalis pedis
l. external carotid
m. femoral
n. fibular
o. great saphenous
p. hepatic
q. inferior mesenteric
r. internal carotid
s. internal iliac
t. phrenic
u. posterior tibial
v. radial
w. renal
x. subclavian
y. superior mesenteric
z. vertebral

8. What two paired arteries enter the skull to supply the brain?

____________________ and ____________________

9. Branches of the paired arteries just named cooperate to form a ring of blood vessels encircling the pituitary gland, at the base of the brain. ____________________

What name is given to this communication network? ____________________

What is its function? ____________________

____

10. What portion of the brain is served by the anterior and middle cerebral arteries? ____________________

____

Both the anterior and middle cerebral arteries arise from the ____________________ arteries.

11. Trace the pathway of a drop of blood from the aorta to the left occipital lobe of the brain, noting all structures through which it flows. ____________________

____

____

____

12. The human arterial and venous systems are diagrammed on this page and the next. Identify all indicated blood vessels.

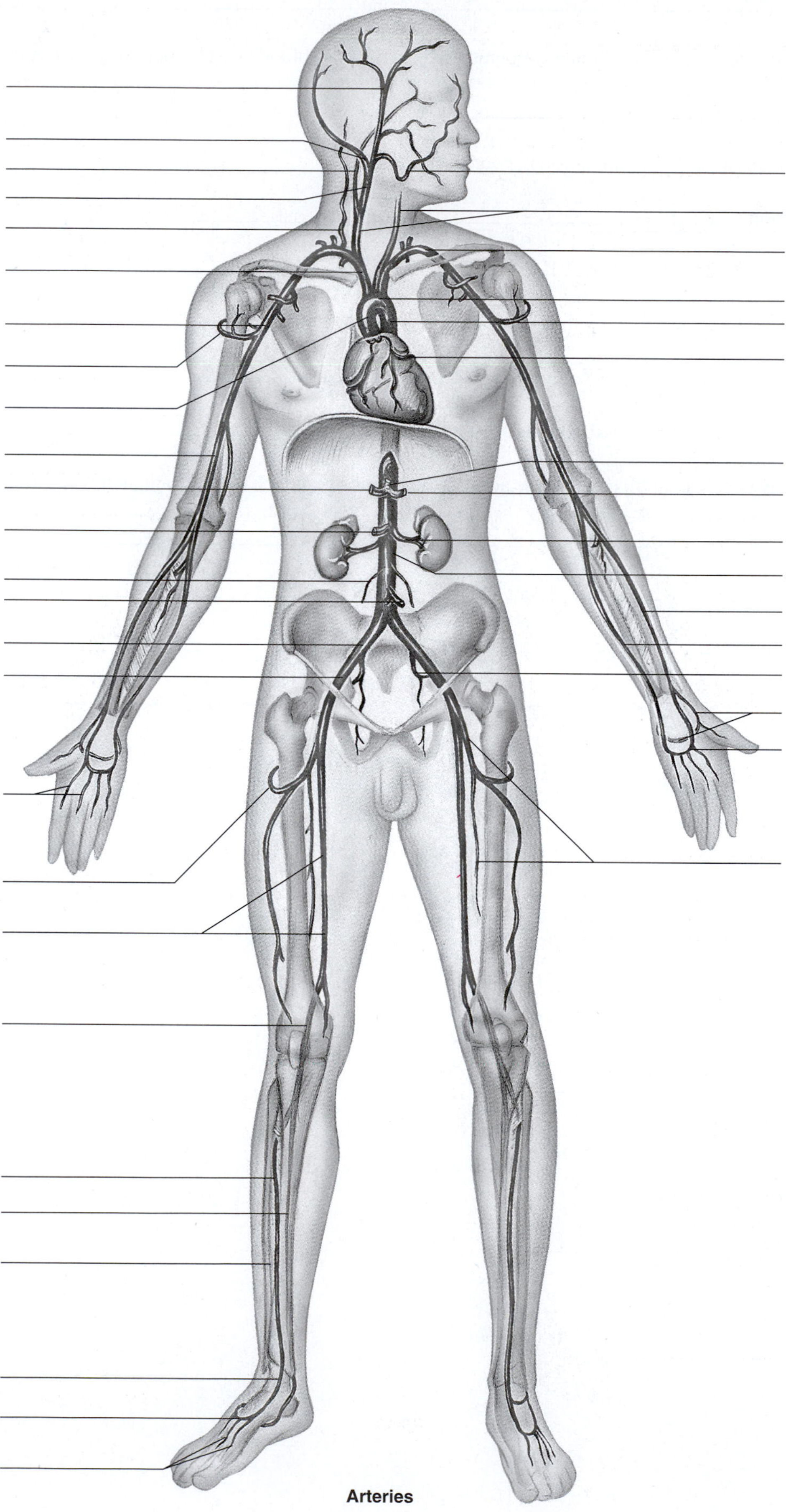

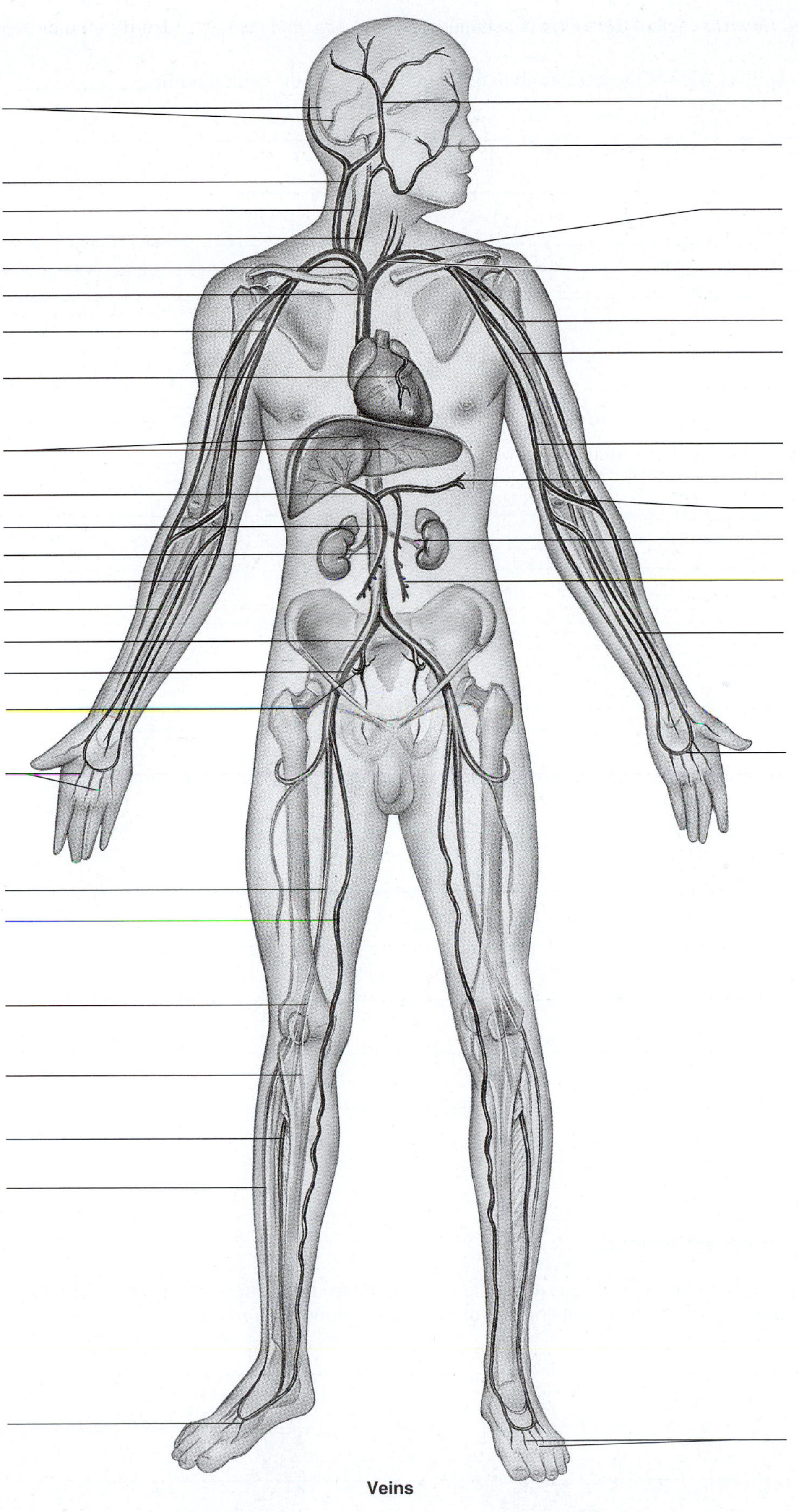

Veins

13. Trace the blood flow for each of the following situations.

a. From the capillary beds of the left thumb to the capillary beds of the right thumb: ______

b. From the mitral valve to the tricuspid valve by way of the great toe: ______

c. From the pulmonary vein to the pulmonary artery by way of the right side of the brain: ______

Pulmonary Circulation

14. Trace the pathway of a carbon dioxide gas molecule in the blood from the inferior vena cava until it leaves the bloodstream. Name all structures (vessels, heart chambers, and others) passed through en route.

______

15. Trace the pathway of an oxygen gas molecule from an alveolus of the lung to the right ventricle of the heart. Name all structures through which it passes. Circle the areas of gas exchange. ______________________________

16. Most arteries of the adult body carry oxygen-rich blood, and the veins carry oxygen-poor blood. How does this differ in the pulmonary arteries and veins? ______________________________

17. How do the arteries of the pulmonary circulation differ structurally from the systemic arteries? What condition is indicated by this anatomical difference? ______________________________

Hepatic Portal Circulation

18. What is the source of blood in the hepatic portal system? ______________________________

19. Why is this blood carried to the liver before it enters the systemic circulation? ______________________________

20. The hepatic portal vein is formed by the union of (a) __________, which drains the __________, __________, __________, __________; and (b) __________, which drains the __________ and __________. The __________ vein, which drains the lesser curvature of the stomach, empties directly into the hepatic portal vein.

21. Trace the flow of a drop of blood from the small intestine to the right atrium of the heart, noting all structures encountered or passed through on the way. ______________________________

Fetal Circulation

22. The failure of two of the fetal bypass structures to become obliterated after birth can cause congenital heart disease, in which the infant's blood is improperly oxygenated. Which two structures are these?

_______________ and _______________

23. For each of the following structures, first indicate its function in the fetus; and then note its fate (what happens to it or what it is converted to after birth). Circle the blood vessel that carries the most oxygen-rich blood.

Structure	Function in fetus	Fate
Umbilical artery		
Umbilical vein		
Ductus venosus		
Ductus arteriosus		
Foramen ovale		

24. What organ serves as a respiratory/digestive/excretory organ for the fetus? _______________

Dissection and Identification: The Blood Vessels of the Cat

25. What differences did you observe between the origin of the left common carotid arteries in the cat and in the human?

____

____

Between the origin of the internal and external iliac arteries? _______________

____

26. How do the relative sizes of the external and internal jugular veins differ in the human and the cat? _______________

____

____

27. In the cat, the inferior vena cava is called the _______________,

and the superior vena cava is referred to as the _______________.

28. Define the following terms.

ascending aorta: _______________

aortic arch: _______________

descending thoracic aorta: _______________

descending abdominal aorta: _______________

25

EXERCISE

The Lymphatic System and Immune Response

MATERIALS

- ☐ Large anatomical chart of the human lymphatic system
- ☐ Prepared slides of lymph node, spleen, and tonsil
- ☐ Compound microscope
- ☐ Dissection tray and instruments
- ☐ Animal specimen from previous dissections
- ☐ Embalming fluid
- ☐ Disposable gloves
- ☐ Organic debris container

OBJECTIVES

1. State the function of the lymphatic system, name its components, and compare its function to that of the blood vascular system.
2. Describe the formation and composition of lymph, and discuss how it is transported through the lymphatic vessels.
3. Relate immunological memory, specificity, and differentiation of self from nonself to immune function.
4. Differentiate between the roles of B cells and T cells in the immune response.
5. Describe the structure and function of lymph nodes, and indicate the location of T cells, B cells, and macrophages in a typical lymph node.
6. Describe the major micro-anatomical features of the spleen and tonsils.
7. Compare and contrast the lymphatic structures of the cat to those of the human.

PRE-LAB QUIZ

1. Circle True or False. The lymphatic system protects the body by removing foreign material such as bacteria from the lymphatic stream.
2. Lymph is excess:
   a. blood that has escaped from veins
   b. tissue fluid that has leaked out of capillaries
   c. tissue fluid that has escaped from arteries
3. Circle True or False. Collecting lymphatic vessels have three tunics and are equipped with valves like veins.
4. __________, which serve as filters for the lymphatic system, occur at various points along the lymphatic vessels.
   a. Glands
   b. Lymph nodes
   c. Valves
5. Circle True or False. The immune response is a systemic response that occurs when the body recognizes a substance as foreign and acts to destroy or neutralize it.
6. Three characteristics of the immune response are the ability to distinguish self from nonself, memory, and:
   a. autoimmunity b. specificity c. susceptibility
7. Circle the correct underlined term. B cells / T cells differentiate in the thymus.
8. Circle the correct underlined term. T cells mediate humoral / cellular immunity because they destroy cells infected with viruses and certain bacteria and parasites.
9. Circle True or False. Antibodies are produced by plasma cells in response to antigens and are found in all body secretions.
10. Circle True or False. The thymus contains both T and B cell–dependent regions.

The overall function of the lymphatic system is twofold: (1) It transports tissue fluid (lymph) to the blood vessels, and (2) it protects the body by removing foreign material such as bacteria from the lymphatic stream and by serving as a site for lymphocyte "policing" of body fluids and lymphocyte multiplication.

The Lymphatic System

The **lymphatic system** consists of a network of lymphatic vessels (lymphatics), lymphoid tissue, lymph nodes, and a number of other lymphoid organs, such as the tonsils, thymus, and spleen. We will focus on the lymphatic vessels and lymph nodes in this section. The white blood cells, which are the central actors in body immunity, are described later in this exercise.

Distribution and Function of Lymphatic Vessels and Lymph Nodes

As blood circulates through the body, the hydrostatic and osmotic pressures operating at the capillary beds result in fluid outflow at the arterial end of the bed and in fluid return at the venous end. However, not all of the lost fluid is returned to the bloodstream by this mechanism, and the fluid that lags behind in the tissue spaces must eventually return to the blood if the vascular system is to operate properly. (When fluid is not returned, it accumulates in the tissues, producing a condition called *edema.*) It is the microscopic, blind-ended **lymphatic capillaries** (**Figure 25.1a**), which branch through nearly all the tissues of the body, that pick up this leaked fluid (primarily water and a small amount of dissolved proteins) and carry it through successively larger vessels—**collecting lymphatic vessels** to **lymphatic trunks**—until the lymph finally returns to the blood vascular system through one of the two large ducts in the thoracic region (Figure 25.1b). The **right lymphatic duct,** present in some but not all individuals, drains lymph from the right upper extremity, head, and thorax delivered by the jugular, subclavian, and bronchomediastinal trunks. In individuals without a right lymphatic duct, those trunks open directly into veins of the neck. The large **thoracic duct** receives lymph from the rest of the body (see Figure 25.1c). In humans, both ducts empty the lymph into the venous circulation at the junction of the internal jugular vein and the subclavian vein, on their respective sides of the body. Notice that the lymphatic system, lacking both a contractile "heart" and arteries, is a one-way system; it carries lymph only toward the heart.

Like veins of the blood vascular system, the collecting lymphatic vessels have three tunics and are equipped with valves (**Figure 25.2**). However, lymphatics tend to be thinner walled, to have *more* valves, and to anastomose (form branching networks) more than veins. Because the lymphatic system is a pumpless system, lymph transport depends largely on the milking action of the skeletal muscles and on pressure changes within the thorax that occur during breathing.

As lymph is transported, it filters through bean-shaped **lymph nodes,** which cluster along the lymphatic vessels of the body. There are thousands of lymph nodes, but because they are usually embedded in connective tissue, they are not ordinarily seen. Within the lymph nodes are **macrophages,** phagocytes that destroy bacteria, cancer cells, and other foreign matter in the lymphatic stream, thus rendering many harmful substances or cells harmless before the lymph enters the bloodstream. Particularly large collections of lymph nodes are found in the inguinal, axillary, and cervical regions of the body. Although we are not usually aware of the filtering and protective nature of the lymph nodes, most of us have experienced "swollen glands" during an active infection. This swelling is a manifestation of the trapping function of the nodes.

Other lymphoid organs—the tonsils, thymus, and spleen (**Figure 25.3**)—resemble the lymph nodes histologically and house similar cell populations (lymphocytes and macrophages).

ACTIVITY 1

Identifying the Organs of the Lymphatic System

Study the large anatomical chart to observe the general plan of the lymphatic system. Notice the distribution of lymph nodes, various lymphatics, the lymphatic trunks, and the location of the right lymphatic duct and the thoracic duct. Also identify the **cisterna chyli,** the enlarged terminus of the thoracic duct that receives lymph from the digestive viscera. ■

The Immune Response

The **adaptive immune system** is a functional system that recognizes something as foreign and acts to destroy or neutralize it. This response is known as the **immune response.** It is a systemic response and is not restricted to the initial infection site. When operating effectively, the immune response protects us from bacterial and viral infections, bacterial toxins, and cancer. When it fails or malfunctions, the body is quickly devastated by pathogens or assaults by its own immune system.

Major Characteristics of the Immune Response

The most important characteristics of the immune response are its (1) **memory,** (2) **specificity,** and (3) **ability to differentiate self from nonself.** Not only does the immune system have a "memory" for previously encountered foreign antigens (the chicken pox virus, for example), but this memory is also remarkably accurate and highly specific.

An almost limitless variety of things are *antigens,* which are capable of provoking an immune response and reacting with the products of the response. Nearly all foreign proteins, many polysaccharides, bacteria and their toxins, viruses, mismatched RBCs, cancer cells, and many small molecules (haptens), when linked to our own body proteins, exhibit this capability. The cells that recognize antigens and initiate the immune response are lymphocytes, the second most numerous members of the leukocyte, or white blood cell (WBC), population. Each immunocompetent lymphocyte is virtually monospecific; that is, it has receptors on its surface allowing it to bind with only one or a few very similar antigens.

As a rule, our own proteins are tolerated, a fact that reflects the ability of the immune system to distinguish our own tissues (self) from foreign antigens (nonself). Nevertheless, an inability to recognize self can and does occasionally happen,

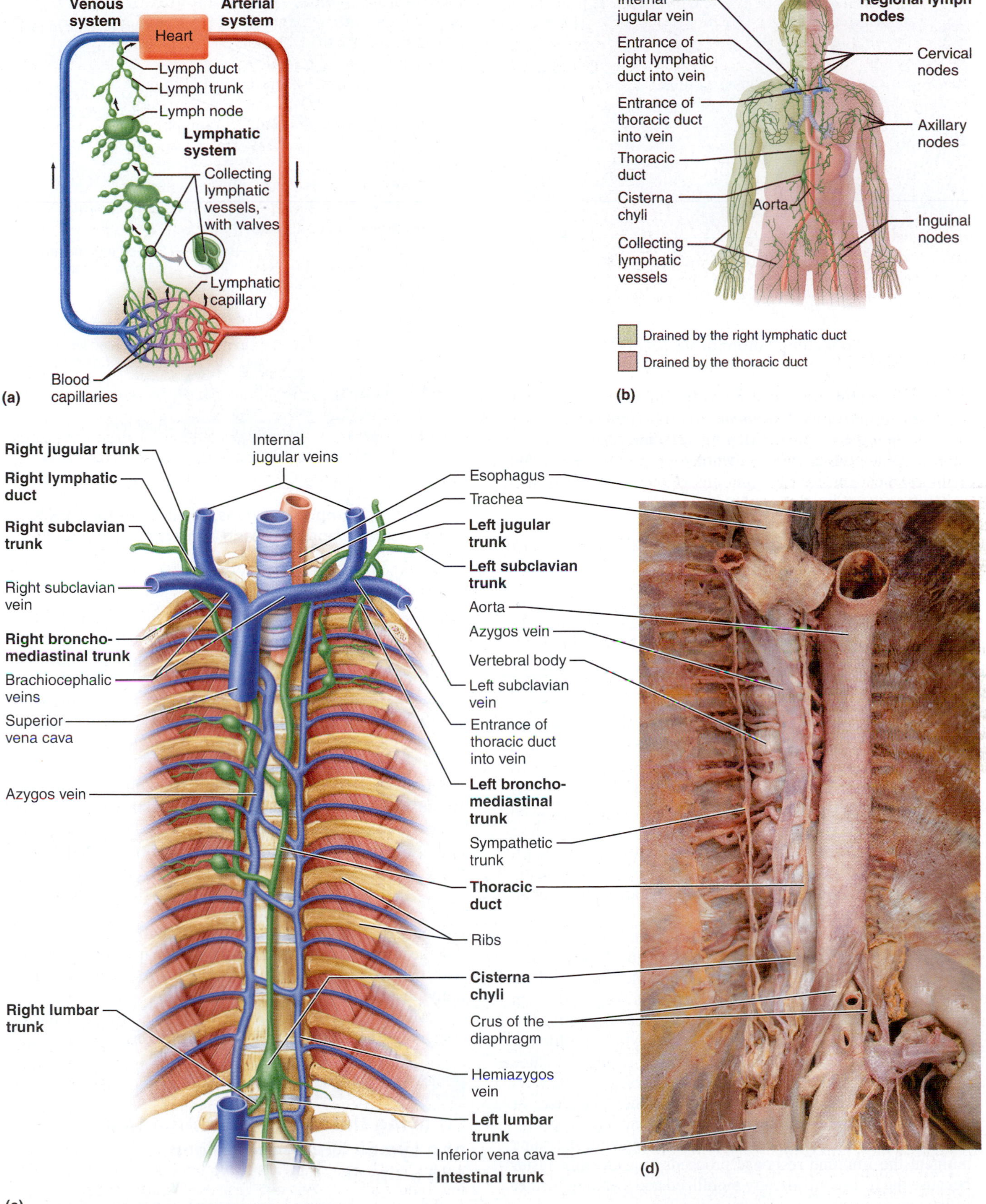

Figure 25.1 Lymphatic system. (a) Simplified scheme of the relationship of lymphatic vessels to blood vessels of the cardiovascular system. **(b)** Distribution of lymphatic vessels and lymph nodes. The green-shaded area represents body area drained by the right lymphatic duct. **(c)** Relationship of the major lymphatic trunks to the thoracic and right lymphatic ducts, and the entry points of the ducts into the subclavian veins. **(d)** Photograph showing the course of the thoracic duct.

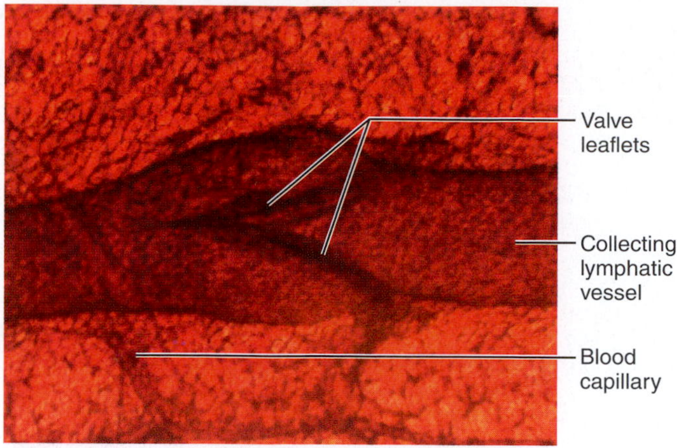

Figure 25.2 Collecting lymphatic vessel (120×).

and our own tissues are attacked by the immune system. This phenomenon is called *autoimmunity.* Autoimmune diseases include multiple sclerosis (MS), myasthenia gravis, Graves' disease, glomerulonephritis, rheumatoid arthritis (RA), and insulin-dependent diabetes mellitus (IDDM), which is also called type 1, or juvenile, diabetes.

Organs, Cells, and Cell Interactions of the Immune Response

The immune system utilizes as part of its arsenal the **lymphoid organs** and **lymphoid tissues,** including the thymus, lymph nodes, spleen, tonsils, appendix, and bone marrow. Of these, the thymus and bone marrow are considered to be the *primary lymphoid organs.* The others are *secondary lymphoid organs* and *tissues.*

The stem cells that give rise to the immune system arise in the bone marrow. Their subsequent differentiation into one of the two populations of immunocompetent lymphocytes occurs in the primary lymphoid organs. The **B cells** (B lymphocytes) differentiate in bone marrow, and the **T cells** (T lymphocytes) differentiate in the thymus. While in their "programming organs," the lymphocytes become *immunocompetent,* an event indicated by the appearance of specific cell-surface proteins that enable the lymphocytes to respond (by binding) to a particular antigen.

After differentiation, the B and T cells leave the bone marrow and thymus, respectively; enter the bloodstream; and travel to peripheral (secondary) lymphoid organs, where clonal selection occurs. **Clonal selection** is triggered when an antigen binds to the specific cell-surface receptors of a T or B cell. This event causes the lymphocyte to proliferate rapidly, forming a clone of like cells, all bearing the same antigen-specific receptors. Then, in the presence of certain regulatory signals, the progeny of the clone specialize, or differentiate—some forming memory cells and others becoming effector cells. Upon subsequent meetings with the same antigen, the immune response proceeds considerably faster because the troops are already mobilized and awaiting further orders, so to speak.

In the case of B cell clones, some become **memory B cells;** the others form antibody-producing **plasma cells.**

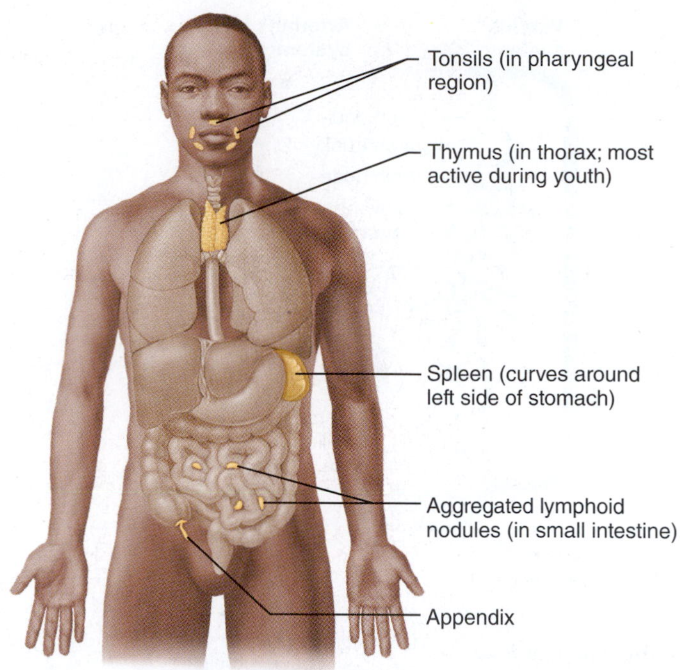

Figure 25.3 Lymphoid organs. Location of the tonsils, thymus, spleen, appendix, and aggregated lymphoid nodules.

Because the B cells act indirectly through the antibodies that their progeny release into the bloodstream (or other body fluids), they are said to provide **humoral immunity.** T cell clones are more diverse. Although all T cell clones also contain memory cells, some clones contain *cytotoxic T cells* (effector cells that directly attack virus-infected tissue cells). Others contain *helper T cells* (which help activate the B cells and cytotoxic T cells), and still others contain *regulatory T cells* (which can inhibit the immune response). Because certain T cells act directly to destroy cells infected with viruses, certain bacteria or parasites, and cancer cells, and to reject foreign grafts, T cells are said to mediate **cellular immunity.**

If T lymphocytes fail to differentiate in the thymus, both antibody and cell-mediated immune functions will be significantly depressed. Additionally, the observation that the thymus naturally shrinks with age has been correlated with the decline in immune function that occurs in elderly individuals. +

All lymphoid tissues except the thymus and bone marrow contain both T and B cell–dependent regions.

ACTIVITY 2

Studying the Microscopic Anatomy of a Lymph Node, the Spleen, and a Tonsil

1. Obtain a compound microscope and prepared slides of a lymph node, the spleen, and a tonsil. As you examine the lymph node slide, notice the following anatomical features,

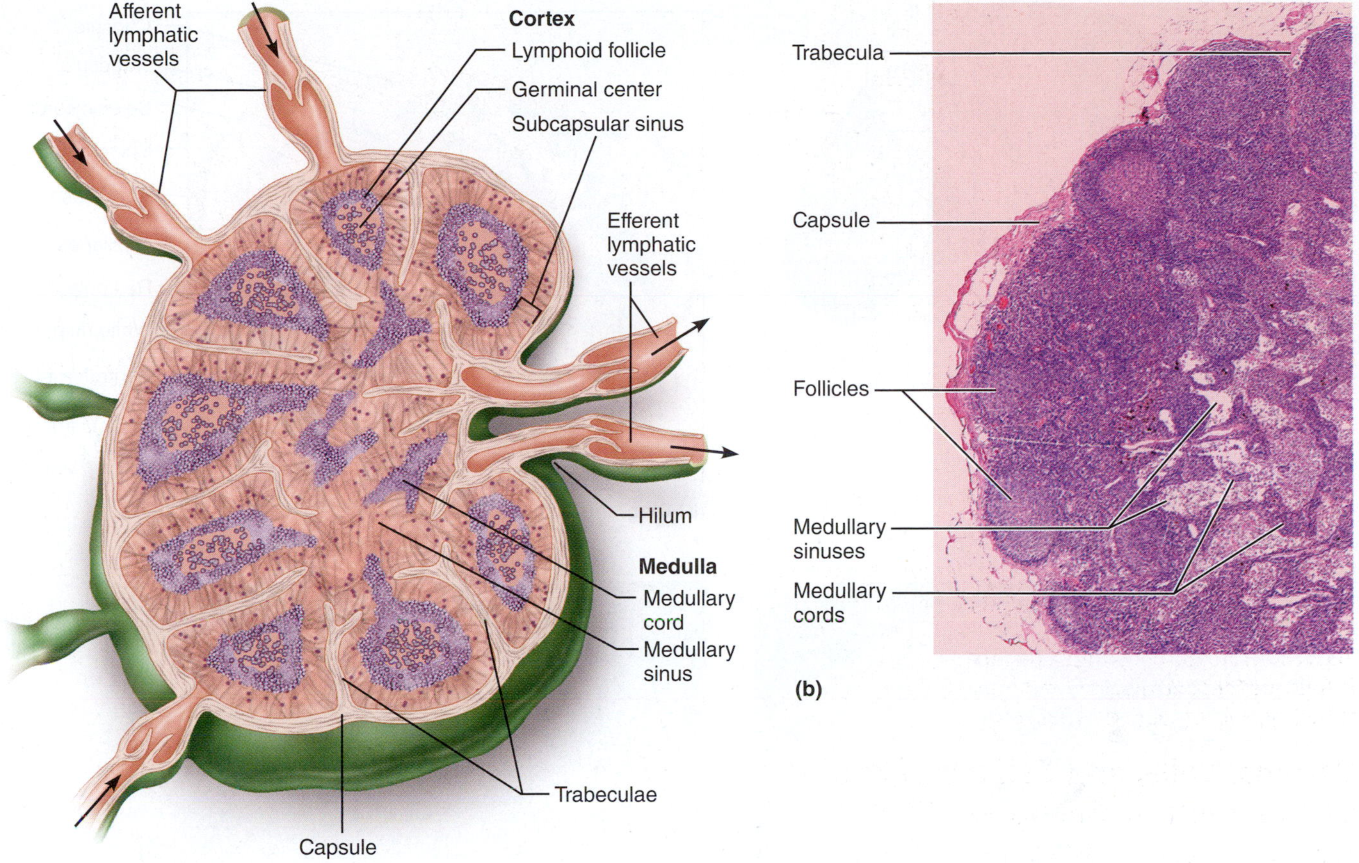

Figure 25.4 Structure of lymph node. (a) Longitudinal section of a lymph node, diagram view. Notice that the afferent vessels outnumber the efferent vessels, which slows the rate of lymph flow. The arrows indicate the direction of the lymph flow. **(b)** Photomicrograph of a part of a lymph node (70×).

which are depicted in the photomicrograph **(Figure 25.4b)**. The node is enclosed within a fibrous **capsule,** from which connective tissue septa (**trabeculae**) extend inward to divide the node into several compartments. Very fine strands of reticular connective tissue issue from the trabeculae, forming the stroma of the gland within which cells are found.

In the outer region of the node, the **cortex,** some of the cells are arranged in globular masses, referred to as germinal centers. The **germinal centers** contain rapidly dividing B cells. The rest of the cortical cells are primarily T cells that circulate continuously, moving from the blood into the node and then exiting from the node in the lymphatic stream.

In the internal portion of the node, the **medulla,** the cells are arranged in cordlike fashion. Most of the medullary cells are macrophages. Macrophages are important not only for their phagocytic function but also because they play an essential role in "presenting" the antigens to the T cells.

Lymph enters the node through a number of *afferent vessels,* circulates through *lymph sinuses* within the node, and leaves the node through *efferent vessels* at the **hilum.** Because each node has fewer efferent than afferent vessels, the lymph flow stagnates somewhat within the node. This allows time for generating an immune response and for the macrophages to remove debris from the lymph before it reenters the blood vascular system.

2. As you observe the slide of the spleen, look for the areas of lymphocytes suspended in reticular fibers, the **white pulp,** clustered around central arteries **(Figure 25.5)**. The remaining tissue in the spleen is the **red pulp,** which is composed of splenic sinusoids and areas of reticular tissue and macrophages called the **splenic cords.** The white pulp, composed primarily of lymphocytes, is responsible for the immune functions of the spleen. Macrophages remove worn-out red blood cells, debris, bacteria, viruses, and toxins from blood flowing through the sinuses of the red pulp.

3. As you examine the tonsil slide, notice the **follicles** containing **germinal centers** surrounded by scattered lymphocytes. The characteristic **tonsillar crypts** (invaginations of the mucosal epithelium) of the tonsils trap bacteria and other foreign material **(Figure 25.6)**. Eventually the bacteria work their way into the lymphoid tissue and are destroyed. ■

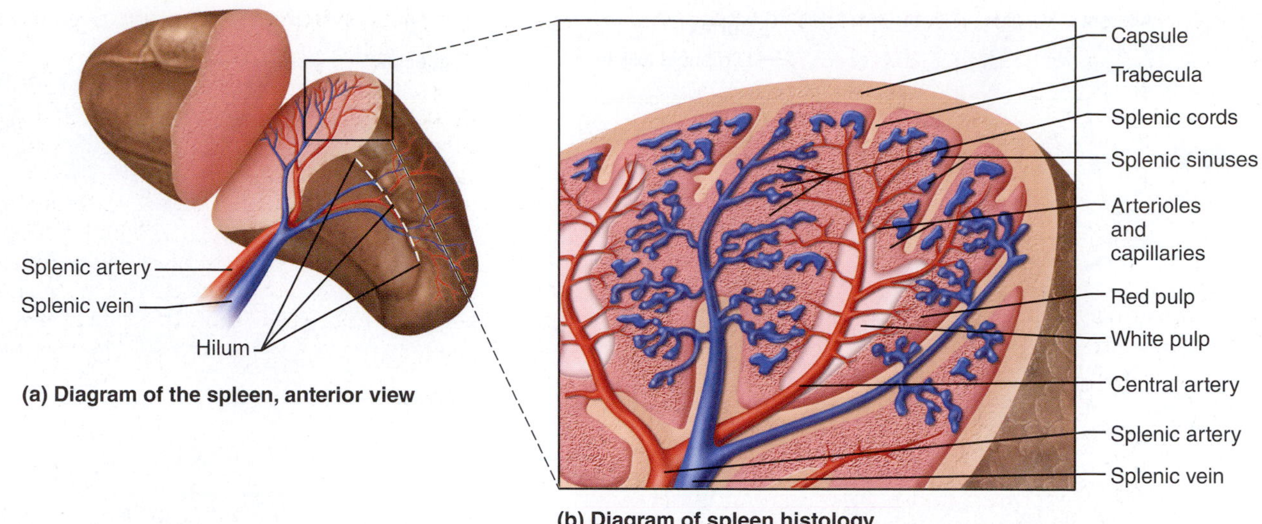

(a) Diagram of the spleen, anterior view

(b) Diagram of spleen histology

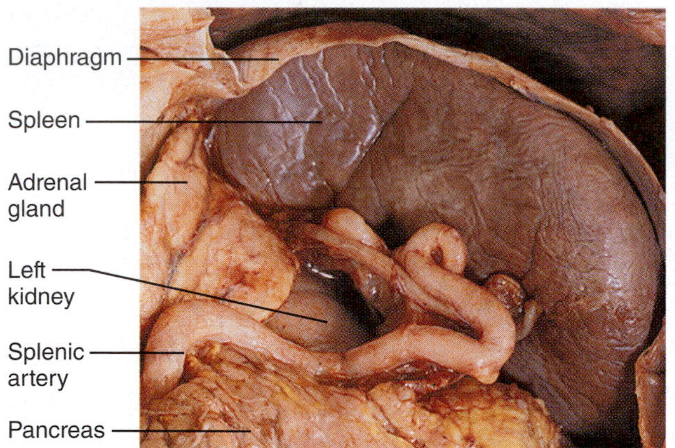

(c) Photograph of the spleen in its normal position in the abdominal cavity, anterior view

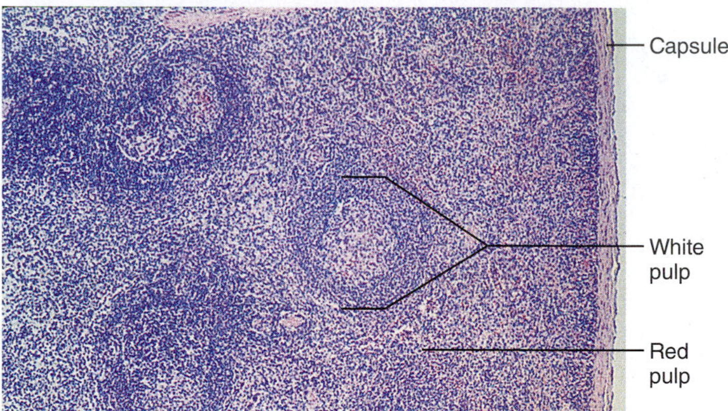

(d) Photomicrograph of spleen tissue (75×). The white pulp, a lymphoid tissue with many lymphocytes, is surrounded by red pulp containing abundant erythrocytes.

Figure 25.5 The spleen.

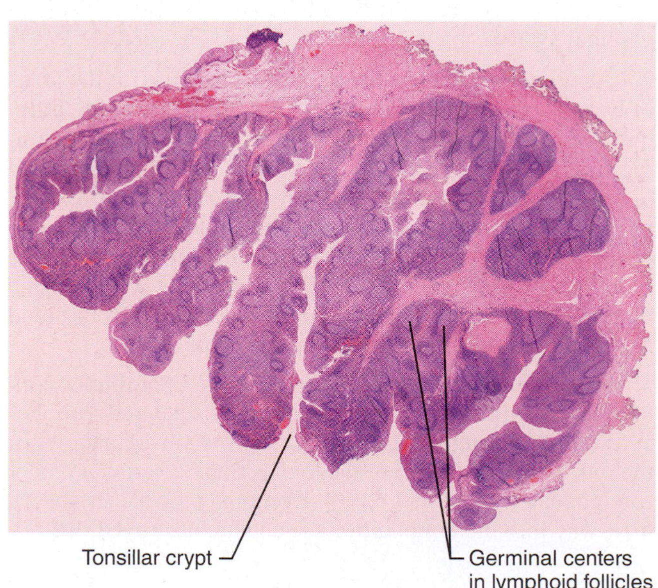

Figure 25.6 Histology of a palatine tonsil. The luminal surface is covered with epithelium that invaginates deeply to form crypts (10×).

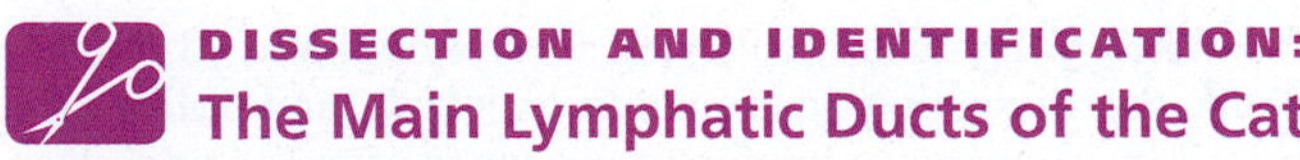

DISSECTION AND IDENTIFICATION: The Main Lymphatic Ducts of the Cat

ACTIVITY 3

Identifying the Main Lymphatic Ducts of the Cat

1. Don disposable gloves. Obtain your cat and a dissecting tray and instruments. Because lymphatic vessels are extremely thin-walled, it is difficult to locate them in a dissection unless the animal has been triply injected (with yellow or green latex for the lymphatic system). However, the large thoracic duct can be localized and identified.

2. Move the thoracic organs to the side to locate the **thoracic duct.** Typically it lies just to the left of the mid-dorsal line, abutting the dorsal aspect of the descending aorta. It is usually about the size of pencil lead and red-brown, with a segmented or beaded appearance caused by the valves within it. Trace it anteriorly to the site where it passes behind the left brachiocephalic vein and then bends and enters the venous system at the junction of the left subclavian and external jugular veins. If the veins are well injected, some of the blue latex may have slipped past the valves and entered the first portion of the thoracic duct.

GROUP CHALLENGE

Compare and Contrast Lymphoid Organs and Tissues

For each pair of lymphoid structures listed, describe ways in which they are similar and ways in which they differ. Remember to consider both structural and functional similarities and differences. Use your textbook or another appropriate reference as needed for comparing aggregated lymphoid nodules and the thymus to the other organs and tissues.

Lymphoid Pair	Similarities	Differences
Lymph node Spleen		
Lymph node Tonsil		
Aggregated lymphoid nodules Tonsils		
Tonsil Spleen		
Thymus Spleen		

3. While in this region, also attempt to identify the short **right lymphatic duct** draining into the right subclavian vein, and notice the collection of lymph nodes in the axillary region.

4. If the cat is triply injected, trace the thoracic duct posteriorly to identify the **cisterna chyli,** the saclike enlargement of its distal end. This structure, which receives fat-rich lymph from the intestine, begins at the level of the diaphragm and can be located posterior to the left kidney.

5. When you finish identifying these lymphatic structures, clean the dissecting instruments and tray, and follow the boxed instructions to prepare your cat for storage and clean the area (page 217). ■

Name ______________________________

Lab Time/Date ______________________

EXERCISE 25

REVIEW SHEET

The Lymphatic System and Immune Response

The Lymphatic System

1. Match the terms below with the correct letters on the diagram.

________ 1. axillary lymph nodes

________ 2. bone marrow

________ 3. cervical lymph nodes

________ 4. cisterna chyli

________ 5. inguinal lymph nodes

________ 6. lymphatic vessels

________ 7. aggregated lymphoid nodules (in small intestine)

________ 8. right lymphatic duct

________ 9. spleen

________ 10. thoracic duct

________ 11. thymus gland

________ 12. tonsils

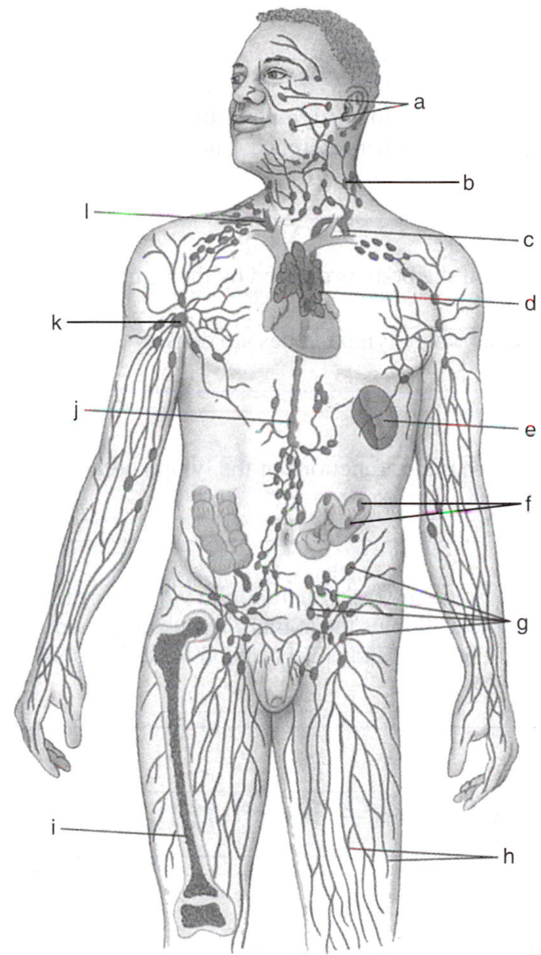

2. Explain why the lymphatic system is a one-way system, whereas the blood vascular system is a two-way system.

____

____

3. How do collecting lymphatic vessels resemble veins? ______________________________

How do lymphatic capillaries differ from blood capillaries? ______________________________

____

EXERCISE 26

Anatomy of the Respiratory System

MATERIALS

- ☐ Resin cast of the bronchial tree (if available)
- ☐ Human torso model
- ☐ Thoracic cavity structures model and/or chart of the respiratory system
- ☐ Larynx model (if available)
- ☐ Preserved inflatable lung preparation (obtained from a biological supply house) or fresh sheep pluck
- ☐ Source of compressed air
- ☐ 0.6 m (2-foot) length of laboratory rubber tubing
- ☐ Dissecting tray and instruments
- ☐ Disposable gloves
- ☐ Disposable autoclave bag
- ☐ Animal specimen from previous dissections
- ☐ Prepared slides of the following (if available): trachea (x.s.), lung tissue, both normal and pathological specimens (for example, sections taken from lung tissues exhibiting bronchitis, pneumonia, emphysema, or lung cancer)
- ☐ Compound and stereomicroscopes
- ☐ Organic debris container

OBJECTIVES

1. State the major functions of the respiratory system.
2. Define the following terms: *pulmonary ventilation, external respiration*, and *internal respiration*.
3. Identify the major respiratory system structures on models or appropriate images, and describe the function of each.
4. Describe the difference between the conducting and respiratory zones, and indicate which is referred to as *anatomical dead space.*
5. Name the serous membrane that encloses each lung, and describe its structure.
6. Demonstrate lung inflation in a fresh sheep pluck or preserved tissue specimen.
7. Recognize the histologic structure of the trachea and lung tissue microscopically or in an image, and describe the functions served by the observed structures.
8. Identify the major respiratory system organs in a dissected animal.

PRE-LAB QUIZ

1. The major role of the respiratory system is to:
   a. dispose of waste products in a solid form
   b. permit the flow of nutrients through the body
   c. supply the body with carbon dioxide and dispose of oxygen
   d. supply the body with oxygen and dispose of carbon dioxide
2. Circle True or False. Four processes—pulmonary ventilation, external respiration, transport of respiratory gases, and internal respiration—must all occur in order for the respiratory system to function fully.
3. The upper respiratory structures include the nose, the larynx, and the:
   a. epiglottis b. lungs c. pharynx d. trachea
4. Circle the correct underlined term. The thyroid cartilage / arytenoid cartilage is the largest and most prominent of the laryngeal cartilages.
5. Circle True or False. The epiglottis forms a lid over the larynx when we swallow food; it closes off the respiratory passageway to incoming food or drink.
6. Air flows from the larynx to the trachea, from which it enters the:
   a. left and right lungs c. pharynx
   b. left and right main bronchi d. segmental bronchi
7. Circle the correct underlined term. The lining of the trachea is pseudostratified ciliated columnar epithelium / transitional epithelium, which propels dust particles, bacteria, and other debris away from the lungs.
8. Circle True or False. All but the smallest branches of the bronchial tree have cartilaginous reinforcements in their walls.
9. ___________, tiny balloonlike expansions of the alveolar sacs, are composed of a single thin layer of squamous epithelium. They are the main structural and functional units of the lung and the actual sites of gas exchange.
10. Circle the correct underlined term. Fissures divide the lungs into lobes, three on the right and two / three on the left.

Body cells require an abundant and continuous supply of oxygen. As the cells use oxygen, they release carbon dioxide, a waste product that the body must get rid of. These oxygen-using cellular processes, collectively referred to as *cellular respiration,* are more appropriately described in conjunction with the topic of cellular metabolism. The major role of the **respiratory system,** our focus in this exercise, is to supply the body with oxygen and dispose of carbon dioxide. To fulfill this role, at least four distinct processes, collectively referred to as **respiration,** must occur:

Pulmonary ventilation: The tidelike movement of air into and out of the lungs so that the gases in the alveoli are continuously changed and refreshed. Also more simply called *ventilation,* or *breathing*.

External respiration: The gas exchange between the blood and the air-filled chambers of the lungs (oxygen loading and carbon dioxide unloading).

Transport of respiratory gases: The transport of respiratory gases between the lungs and tissue cells of the body accomplished by the cardiovascular system, using blood as the transport vehicle.

Internal respiration: Exchange of gases between systemic blood and tissue cells (oxygen unloading and carbon dioxide loading).

Only the first two processes are exclusive to the respiratory system, but all four must occur for the respiratory system to do its job. Hence, the respiratory and circulatory systems are irreversibly linked. If either system fails, cells begin to die from oxygen starvation and accumulation of carbon dioxide. Uncorrected, this soon causes death of the entire organism.

Upper Respiratory System Structures

The upper respiratory system structures—the nose, pharynx, and larynx—are described below and illustrated in the figure **(Figure 26.1)**. As you read through the descriptions, identify each structure in the figure.

Air generally passes into the respiratory tract through the **nostrils,** or **nares,** and enters the **nasal cavity,** which is divided by the **nasal septum.** It then flows posteriorly over three pairs of lobelike structures, the **inferior, superior,** and **middle nasal conchae,** which increase the air turbulence. As the air passes through the nasal cavity, it is also warmed, moistened, and filtered by the nasal mucosa. The air that flows directly beneath the superior part of the nasal cavity may chemically stimulate the olfactory receptors located in the mucosa of that region. The nasal cavity is surrounded by the **paranasal sinuses** in the frontal, sphenoid, ethmoid, and maxillary bones. These sinuses, named for the bones in which they are located, act as resonance chambers in speech and their mucosae, like that of the nasal cavity, warm and moisten the incoming air.

The nasal passages are separated from the oral cavity below by a partition composed anteriorly of the **hard palate** and posteriorly of the **soft palate.**

The genetic defect called **cleft palate** results from failure of the palatine bones and/or the palatine processes of the maxillary bones to fuse medially. It causes difficulty in breathing and oral cavity functions such as sucking, chewing, and speech. +

Of course, air may also enter the body via the mouth. From there it passes through the oral cavity to move into the pharynx posteriorly, where the oral and nasal cavities are joined temporarily.

Commonly called the *throat,* the funnel-shaped **pharynx** connects the nasal and oral cavities to the larynx (voice box) and esophagus inferiorly. It has three named parts (Figure 26.1):

1. The **nasopharynx** lies posterior to the nasal cavity and is continuous with it via the **posterior nasal aperture.** It lies above the soft palate; hence, it serves only as an air passage. High on its posterior wall is the *pharyngeal tonsil,* masses of lymphoid tissue that help to protect the respiratory passages from invading pathogens. The *pharyngotympanic (auditory) tubes,* which allow middle ear pressure to become equalized to atmospheric pressure, drain into the lateral aspects of the nasopharynx. The *tubal tonsils* surround the openings of these tubes into the nasopharynx (Figure 26.1b).

   Because the middle ear and nasopharyngeal mucosae are continuous, nasal infections may invade the middle ear cavity and cause **otitis media** (middle ear inflammation), which is difficult to treat. +

2. The **oropharynx** is continuous posteriorly with the oral cavity. Because it extends from the soft palate to the epiglottis of the larynx inferiorly, it serves as a common conduit for food and air. In its lateral walls are the *palatine tonsils.* The *lingual tonsil* covers the base of the tongue.

3. The **laryngopharynx,** like the oropharynx, accommodates both ingested food and air. It lies directly posterior to the upright epiglottis and extends to the larynx, where the common pathway divides into the respiratory and digestive channels. From the laryngopharynx, air enters the lower respiratory passageways by passing through the larynx and into the trachea below.

The **larynx (Figure 26.2)** consists of nine cartilages. The two most prominent are the large shield-shaped **thyroid cartilage,** whose anterior medial laryngeal prominence is commonly referred to as *Adam's apple,* and the inferiorly located, ring-shaped **cricoid cartilage,** whose widest dimension faces posteriorly. All the laryngeal cartilages are composed of hyaline cartilage except the flaplike **epiglottis,** a flexible elastic cartilage located superior to the opening of the larynx. The epiglottis, sometimes referred to as the "guardian of the airways," forms a lid over the larynx when we swallow. This closes off the respiratory passageways to incoming food or drink, which is routed into the posterior esophagus, or food chute.

☐ Palpate your larynx by placing your hand on the anterior neck surface approximately halfway down its length. Swallow. Can you feel the cartilaginous larynx rising?

Check the box when you have completed the task.

If anything other than air enters the larynx, a cough reflex attempts to expel the substance. Note that this reflex operates only when a person is conscious. Therefore, you should never try to feed or pour liquids down the throat of an unconscious person.

The mucous membrane of the larynx is thrown into two pairs of folds—the upper **vestibular folds,** also called the **false vocal cords,** and the lower **vocal folds,** or **true vocal cords,** which vibrate with expelled air for speech. The vocal cords are attached posterolaterally to the small triangular **arytenoid cartilages** by the *vocal ligaments.* The vocal folds and the slitlike passageway between them is called the **glottis.**

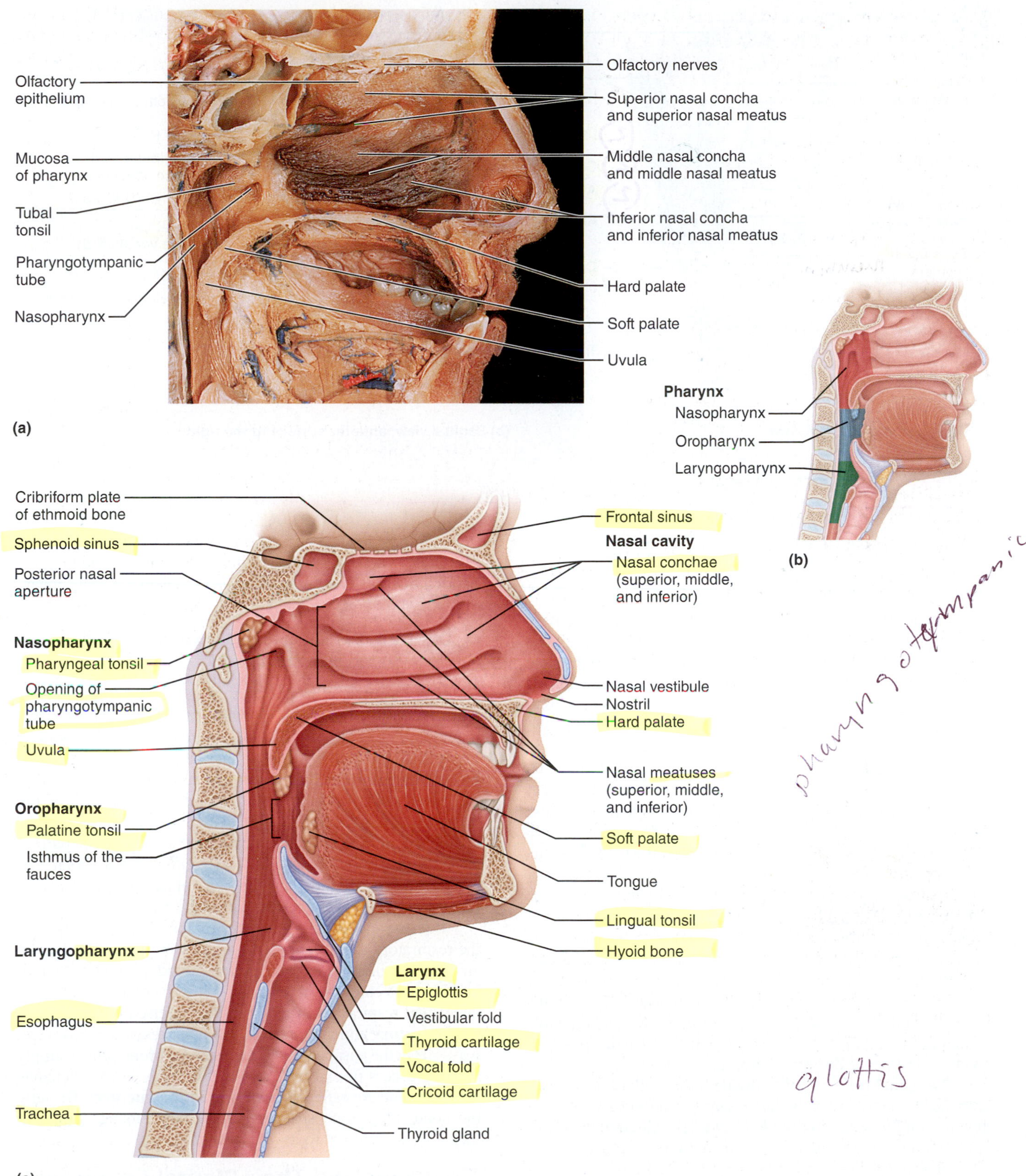

Figure 26.1 Structures of the upper respiratory tract (midsagittal section). **(a)** Photograph. **(b)** Regions of the pharynx. **(c)** Diagram view.

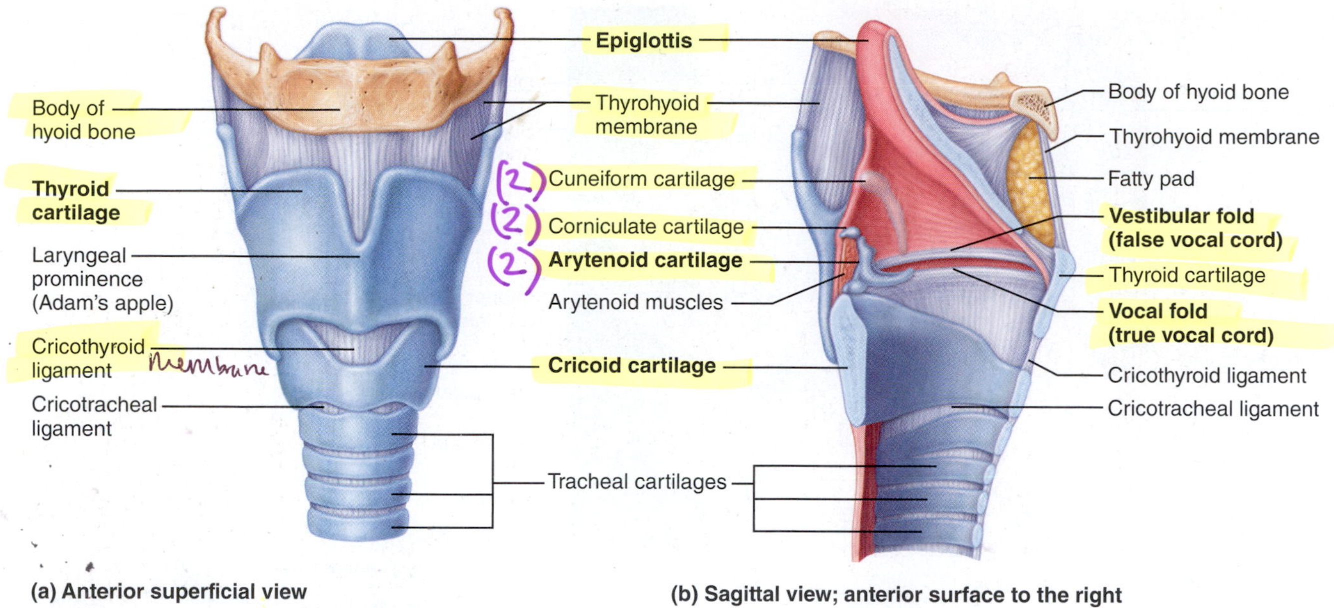

Figure 26.2 Structure of the larynx.

Lower Respiratory System Structures

Air entering the **trachea,** or windpipe, from the larynx travels down its length (about 11 cm or 4 inches) to the level of the *sternal angle* (or the disc between the fourth and fifth thoracic vertebrae). There the passageway divides into the right and left **main (primary) bronchi (Figure 26.3)**, which plunge into their respective lungs at an indented area called the **hilum** (see Figure 26.5c). The right main bronchus is wider, shorter, and more vertical than the left, and foreign objects that enter the respiratory passageways are more likely to become lodged in it.

The trachea is lined with a ciliated mucus-secreting, pseudostratified columnar epithelium, as are many of the other respiratory system passageways. The cilia propel mucus (produced by goblet cells) laden with dust particles, bacteria, and other debris away from the lungs and toward the throat, where it can be expectorated or swallowed. The walls of the trachea are reinforced with C-shaped cartilaginous rings, the incomplete portion located posteriorly. These C-shaped cartilages serve a double function. The incomplete parts allow the esophagus to expand anteriorly when a large food bolus is swallowed. The solid portions reinforce the trachea walls to maintain its open passageway regardless of the pressure changes that occur during breathing.

The main bronchi further divide into smaller and smaller branches—the lobar (secondary), segmental (tertiary), and on down—finally becoming the **bronchioles,** which have terminal branches called **respiratory bronchioles** (Figure 26.3b). All but the smallest branches have cartilaginous reinforcements in their walls, usually in the form of small plates of hyaline cartilage rather than cartilaginous rings. As the respiratory tubes get smaller and smaller, the relative amount of smooth muscle in their walls increases as the amount of cartilage declines and finally disappears. The complete layer of smooth muscle present in the bronchioles enables them to provide considerable resistance to airflow under certain conditions (asthma, hay fever, etc.). The continuous branching of the respiratory passageways in the lungs is often referred to as the **bronchial tree.** The comparison becomes much more meaningful if you observe a resin cast of the respiratory passages.

- Observe a resin cast of respiratory passages if one is available for observation in the laboratory.

The respiratory bronchioles in turn subdivide into several **alveolar ducts,** which terminate in alveolar sacs that rather resemble clusters of grapes. **Alveoli,** tiny balloonlike expansions along the alveolar sacs and occasionally found protruding from alveolar ducts and respiratory bronchioles, are composed of a single thin layer of squamous epithelium overlying a wispy basal lamina. The external surfaces of the alveoli are densely spiderwebbed with a network of pulmonary capillaries **(Figure 26.4)**. Together, the alveolar and capillary walls and their fused basement membranes form the **respiratory membrane,** also called the *blood air barrier*.

Because gas exchanges occur by simple diffusion across the respiratory membrane—oxygen passing from the alveolar air to the capillary blood and carbon dioxide leaving the capillary blood to enter the alveolar air—the alveolar sacs, alveolar ducts, and respiratory bronchioles are referred to collectively as **respiratory zone structures.** All other respiratory passageways (from the nasal cavity to the terminal bronchioles) simply serve as access or exit routes to and from these gas exchange chambers and are called **conducting zone structures.** Because the conducting zone structures have no exchange function, they are also referred to as *anatomical dead space.*

The Lungs and Their Pleural Coverings

The paired lungs are soft, spongy organs that occupy the entire thoracic cavity except for the *mediastinum,* which houses the heart, bronchi, esophagus, and other organs **(Figure 26.5)**. Each lung is connected to the mediastinum by a **root** containing its vascular and bronchial attachments. The structures of the root enter (or leave) the lung via a medial indentation

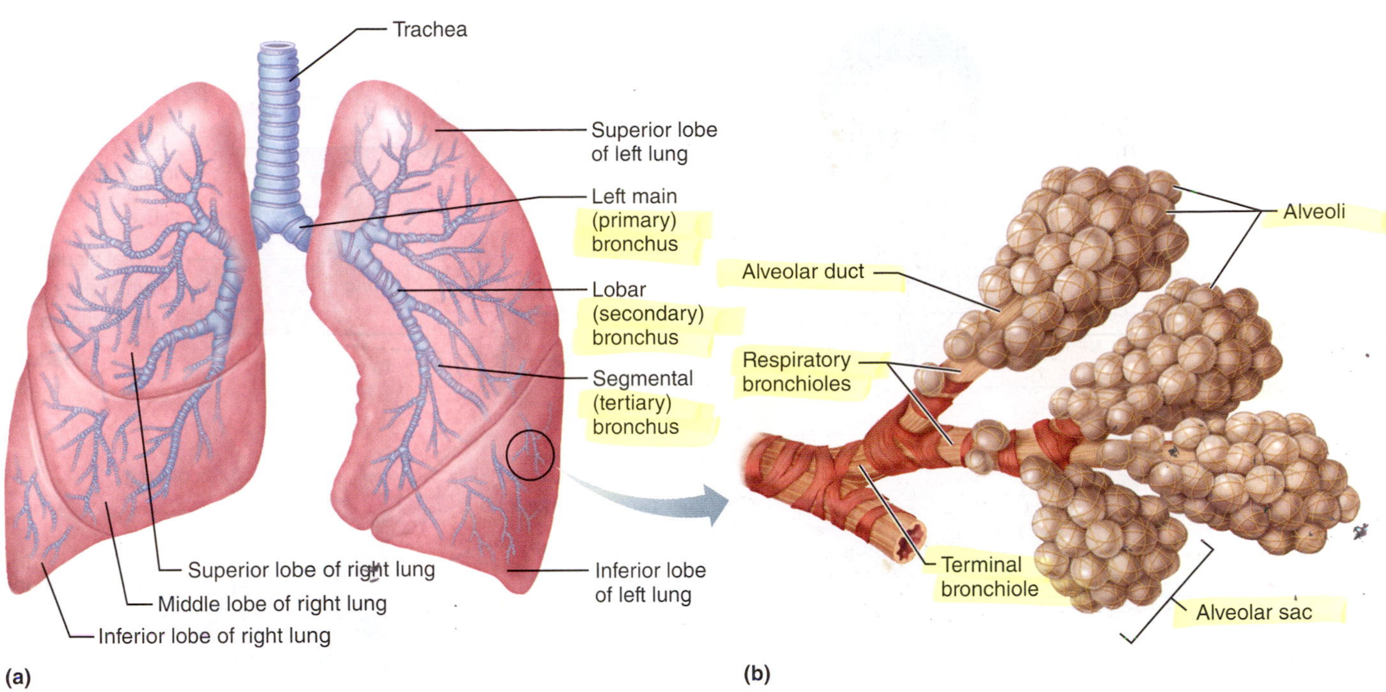

Figure 26.3 Structures of the lower respiratory tract. **(a)** Diagram view. **(b)** Enlarged view of alveoli.

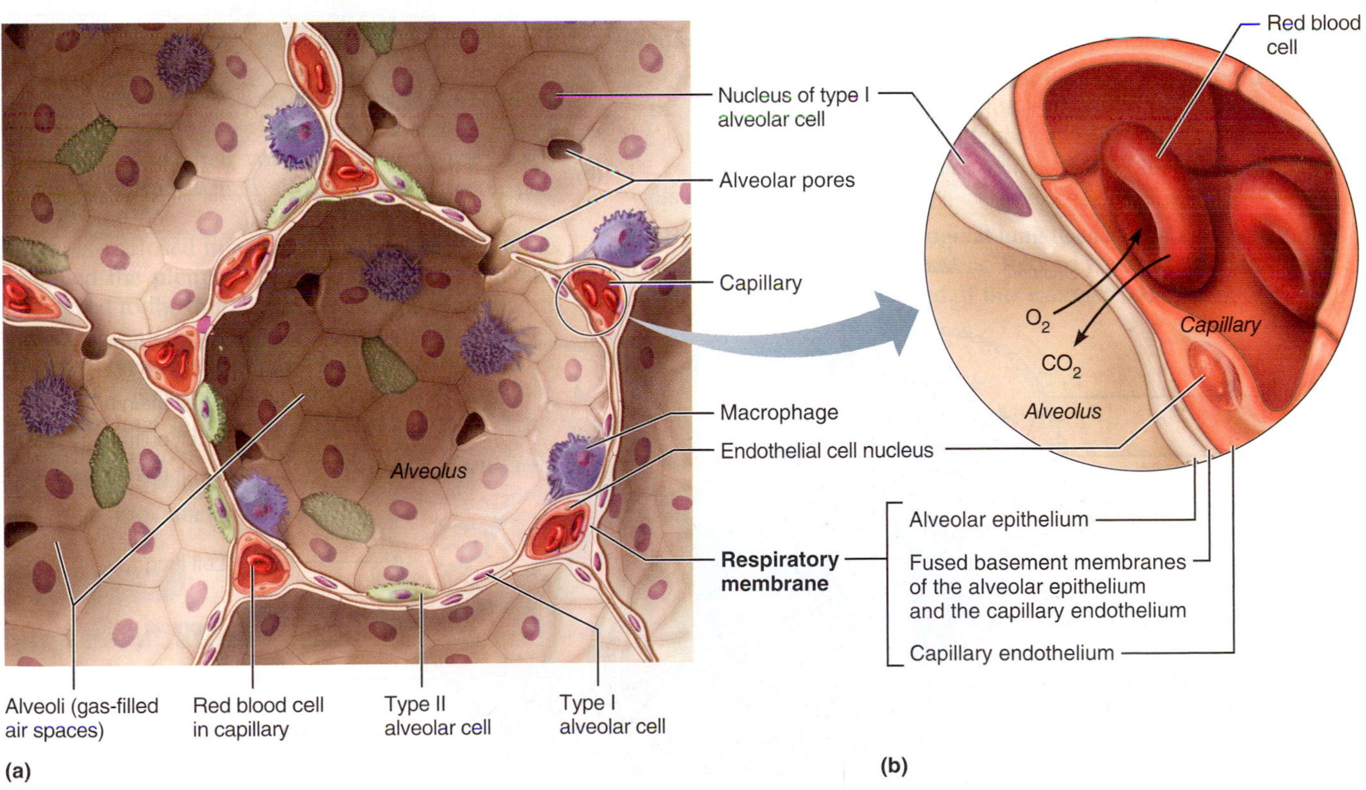

Figure 26.4 Relationship between the alveoli and pulmonary capillaries involved in gas exchange. Diagram views. **(a)** One alveolus surrounded by capillaries. **(b)** Enlargement of the respiratory membrane.

called the *hilum.* All structures distal to the main bronchi are found within the lung substance. A lung's **apex,** the narrower superior aspect, lies just deep to the clavicle, and its **base,** the inferior concave surface, rests on the diaphragm. Anterior, lateral, and posterior lung surfaces are in close contact with the ribs and, hence, are collectively called the **costal surface.**

4. Why is it important that the human trachea is reinforced with cartilaginous rings?

____

Why is it important that the rings are incomplete posteriorly? ____

____

5. What is the function of the pleural membranes? ____

____

____

6. Name two functions of the nasal cavity mucosa. ____

and ____

7. The following questions refer to the main bronchi.

Which is longer? ____________ Larger in diameter? Right More horizontal? Left

Which is more likely to trap a foreign object that has entered the respiratory passageways? Right

8. Appropriately label all structures provided with leader lines on the diagrams below.

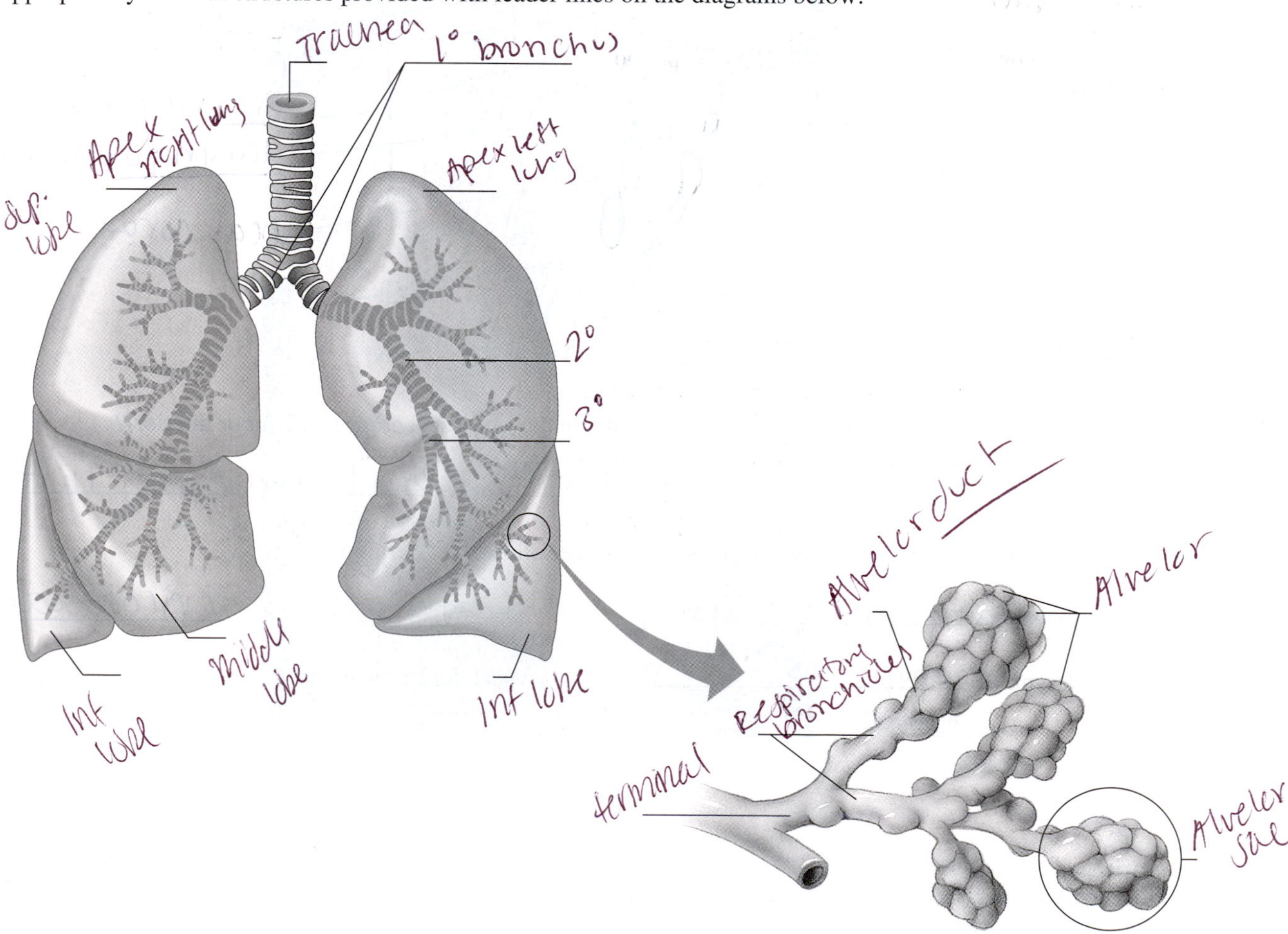

9. Trace a molecule of oxygen from the nostrils to the pulmonary capillaries of the lungs: Nostrils →

____

____

____

10. Match the terms in column B to the descriptions in column A. (Not all terms will be used.)

	Column A	Column B
n	1. connects the larynx to the main bronchi	a. alveolus
l	2. site of tonsils	b. bronchiole
E	3. food passageway posterior to the trachea	c. conchae
d	4. covers the glottis during swallowing of food	d. epiglottis
g	5. contains the vocal cords	e. esophagus
m	6. nerve that activates the diaphragm during inspiration	f. glottis
j	7. pleural layer lining the walls of the thorax	g. larynx
A	8. site from which oxygen enters the pulmonary blood	h. main bronchi
k	9. connects the middle ear to the nasopharynx	i. palate
f	10. contains opening between the vocal folds	j. parietal pleura
c	11. increase air turbulence in the nasal cavity	k. pharyngotympanic tube
i	12. separates the oral cavity from the nasal cavity	l. pharynx
		m. phrenic nerve
		n. trachea
		o. vagus nerve
		p. visceral pleura

11. Which portions of the respiratory system are referred to as *anatomical dead space?* ______________________

____

____

Why? ____

12. Define the following terms.

external respiration: ____

____

internal respiration: ____

____

Dissection and Identification: The Respiratory System of the Cat

21. Are the cartilaginous rings in the cat trachea complete or incomplete? ______________________

22. How does the number of lung lobes in the cat compare with the number in humans? ______________________

____

23. Describe the appearance of the bronchial tree in the cat lung. ______________________

____

____

24. Describe the appearance of lung tissue under the dissection microscope. ______________________

____

____

27

EXERCISE

Anatomy of the Digestive System

MATERIALS

- ☐ Dissectible torso model
- ☐ Anatomical chart of the human digestive system
- ☐ Prepared slides of the liver, pancreas, and mixed salivary glands; of longitudinal sections of the gastroesophageal junction and a tooth; and of cross sections of the stomach, duodenum, ileum, and large intestine
- ☐ Compound microscope
- ☐ Three-dimensional model of a villus (if available)
- ☐ Jaw model or human skull
- ☐ Three-dimensional model of liver lobules (if available)
- ☐ *Human Digestive System* video
- ☐ Dissection animal, tray, and instruments
- ☐ Bone cutters
- ☐ Disposable gloves
- ☐ Embalming fluid
- ☐ Hand lens
- ☐ Organic debris container

OBJECTIVES

1. State the overall function of the digestive system.
2. Describe the general histologic structure of the alimentary canal wall, and identify the following structures on an appropriate image of the wall: mucosa, submucosa, muscularis externa, and serosa or adventitia.
3. Identify on a model or image the organs of the alimentary canal, and name their subdivisions if any.
4. Describe the general function of each of the digestive system organs and structures.
5. List and explain the specializations of the structure of the stomach and small intestine that contribute to their functional roles.
6. Name and identify the accessory digestive organs, listing a function for each.
7. Describe the anatomy of the generalized tooth, and name the human deciduous and permanent teeth.
8. List the major enzymes or enzyme groups produced by the salivary glands, stomach, small intestine, and pancreas.
9. Recognize microscopically or in an image the histologic structure of the following organs: small intestine, tooth, liver, salivary glands, and stomach.
10. Identify on a dissected animal the organs that make up the alimentary canal and the accessory organs of the digestive system.

PRE-LAB QUIZ

1. The digestive system:
   a. eliminates undigested food
   b. provides the body with nutrients
   c. provides the body with water
   d. all of the above
2. Circle the correct underlined term. <u>Digestion</u> / <u>Absorption</u> occurs when small molecules pass through epithelial cells into the blood for distribution to the body cells.
3. The __________ abuts the lumen of the alimentary canal and consists of epithelium, lamina propria, and muscularis mucosae.
   a. mucosa b. serosa c. submucosa
4. Circle the correct underlined term. Approximately 25 cm long, the <u>esophagus</u> / <u>alimentary canal</u> conducts food from the pharynx to the stomach.
5. Wavelike contractions of the digestive tract that propel food along are called:
   a. digestion c. ingestion
   b. elimination d. peristalsis

(Text continues on next page.)

6. The __________ is located on the left side of the abdominal cavity and is hidden by the liver and diaphragm.
   a. gallbladder
   b. large intestine
   c. small intestine
   d. stomach
7. Circle True or False. Nearly all nutrient absorption occurs in the small intestine.
8. Circle the correct underlined term. The ascending colon / descending colon traverses down the left side of the abdominal cavity and becomes the sigmoid colon.
9. A tooth consists of two major regions, the crown and the:
   a. dentin
   b. enamel
   c. gingiva
   d. root
10. Located inferior to the diaphragm, the __________ is the largest gland in the body.
   a. gallbladder
   b. liver
   c. pancreas
   d. thymus

The **digestive system** provides the body with the nutrients, water, and electrolytes essential for health. The organs of this system ingest, digest, and absorb food and eliminate the undigested remains as feces.

The digestive system consists of a hollow tube extending from the mouth to the anus, into which various accessory organs or glands empty their secretions **(Figure 27.1)**. Food material within this tube, the **alimentary canal,** is technically outside the body because it has contact only with the cells lining the tract. For ingested food to become available to the body cells, it must first be broken down *physically* (by chewing or churning) and *chemically* (by enzymatic hydrolysis) into its smaller diffusible molecules—a process called **digestion.** The digested end products can then pass through the epithelial cells lining the tract into the blood for distribution to the body cells—a process called **absorption.** In one sense, the digestive tract can be viewed as a disassembly line, in which food is carried from one stage of its digestive processing to the next by muscular activity and its nutrients made available to the cells of the body en route.

The organs of the digestive system are traditionally separated into two major groups: the **alimentary canal,** or **gastrointestinal (GI) tract,** and the **accessory digestive organs.** The alimentary canal is approximately 9 meters long in a cadaver but is considerably shorter in a living person because of muscle tone. It consists of the mouth, pharynx, esophagus, stomach, and small and large intestines. The accessory structures include the teeth, which physically break down foods, and the salivary glands, gallbladder, liver, and pancreas, which secrete their products into the alimentary canal.

General Histologic Plan of the Alimentary Canal

From the esophagus to the anal canal, the basic structure of the alimentary canal is similar. So, it makes sense to begin our study by learning the features of this structure. As we study individual parts of the alimentary canal, we will note how this basic plan is modified to enable each subsequent organ to perform its unique digestive functions.

Essentially the alimentary canal wall has four basic **tunics** (layers). From the lumen outward, these are the *mucosa*, the *submucosa*, the *muscularis externa*, and either a *serosa* or *adventitia* **(Figure 27.2)**. Each of these tunics has a predominant tissue type and a specific function in the digestive process.

Mucosa (mucous membrane): The mucosa is the wet epithelial membrane abutting the alimentary canal lumen. It consists of a surface *epithelium* (in most cases, a simple columnar), a *lamina propria* (areolar connective tissue on which the epithelial layer rests), and a *muscularis mucosae* (a scant layer of smooth muscle fibers that enable local movements of the mucosa). The major functions of the mucosa are secretion (of enzymes, mucus, hormones, etc.), absorption of digested foodstuffs, and protection (against bacterial invasion). A particular mucosal region may be involved in one or all three functions.

Submucosa: Superficial to the mucosa, the submucosa is moderately dense connective tissue containing blood and lymphatic vessels, scattered lymph nodules, and nerve fibers. Its intrinsic nerve supply is called the *submucosal plexus.* Its vessels absorb and transport nutrients, and its elastic fibers help maintain the normal shape of each organ.

Muscularis externa: The muscularis externa, also simply called the *muscularis*, typically is a bilayer of smooth muscle, with the deeper layer running circularly and the superficial layer running longitudinally. This layer moves the contents of the canal along by segmentation and peristalsis. An important intrinsic nerve plexus, the *myenteric plexus*, is associated with this tunic; it is the major regulator of GI motility.

Serosa: The outermost covering of the intraperitoneal organs is the serosa, also called the *visceral peritoneum.* It consists of mesothelium associated with a thin layer of areolar connective tissue. The serosa reduces friction as the mobile digestive system organs work and slide across one another and the cavity walls. In the esophagus, which is *outside* the abdominopelvic cavity, the serosa is replaced by an **adventitia,** a layer of coarse fibrous connective tissue that binds the organ to surrounding tissues. The adventitia anchors and protects the surrounded organ.

Organs of the Alimentary Canal

ACTIVITY 1

Identifying Organs of the Alimentary Canal

The sequential pathway and fate of food as it passes through the alimentary canal organs are described in the next sections. Identify each structure in the digestive system illustration (Figure 27.1) and on the torso model or anatomical chart of the digestive system as you work. ■

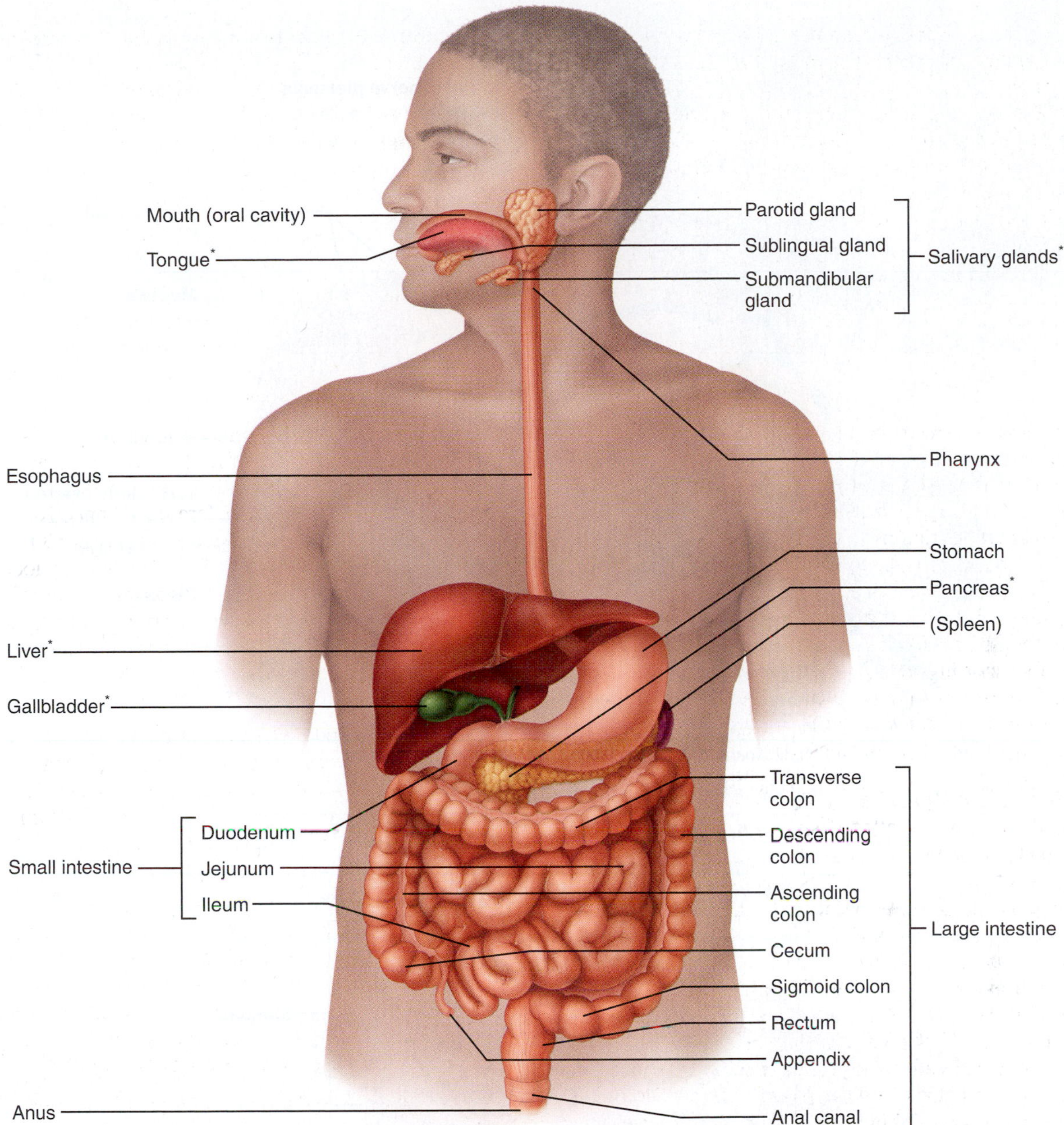

Figure 27.1 The human digestive system: alimentary tube and accessory organs. Organs with asterisks are accessory organs. Those without asterisks are alimentary canal organs (except the spleen, which is part of the lymphatic system).

Oral Cavity or Mouth

Food enters the digestive tract through the **oral cavity,** or **mouth (Figure 27.3)**. Within this mucous membrane–lined cavity are the gums, teeth, tongue, and openings of the ducts of the salivary glands. The **lips (labia)** protect the opening of the chamber anteriorly, the **cheeks** form its lateral walls, and the **palate,** its roof. The anterior portion of the palate is referred to as the **hard palate** because the palatine processes of the maxillae and the palatine bones underlie it. The posterior **soft palate** is a fibromuscular structure that is unsupported by bone. The **uvula,** a fingerlike projection of the soft palate, extends inferiorly from its posterior margin. The soft palate rises to close off the oral cavity from the nasal and pharyngeal passages during swallowing. The floor of the oral cavity is occupied by the muscular **tongue (Figure 27.4)**, which is largely supported by the *mylohyoid muscle* and attaches to the hyoid bone, mandible, styloid processes, and pharynx. A membrane called the **lingual frenulum** secures the inferior midline of the tongue to the floor of the mouth. The space between the lips and cheeks and the teeth and gums is the **oral vestibule;** the area that lies within the teeth and gums is the *oral cavity* proper. (The teeth and gums are discussed in more detail on pages 468–470.)

On each side of the mouth at its posterior end are masses of lymphoid tissue, the **palatine tonsils** (see Figure 27.3). Each lies in a concave area bounded anteriorly and posteriorly by membranes, the **palatoglossal arch** and the **palatopharyngeal**

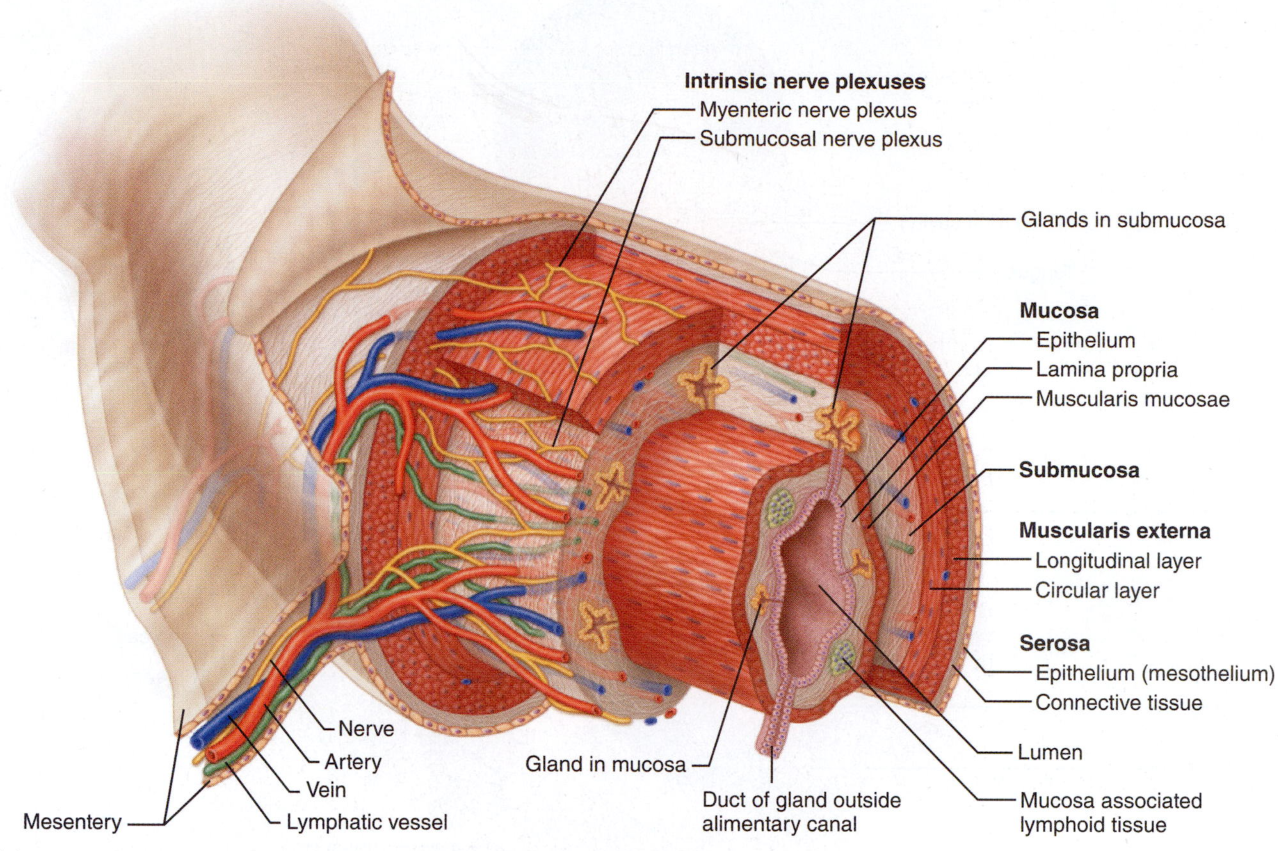

(a) Longitudinal and cross-sectional views through the small intestine

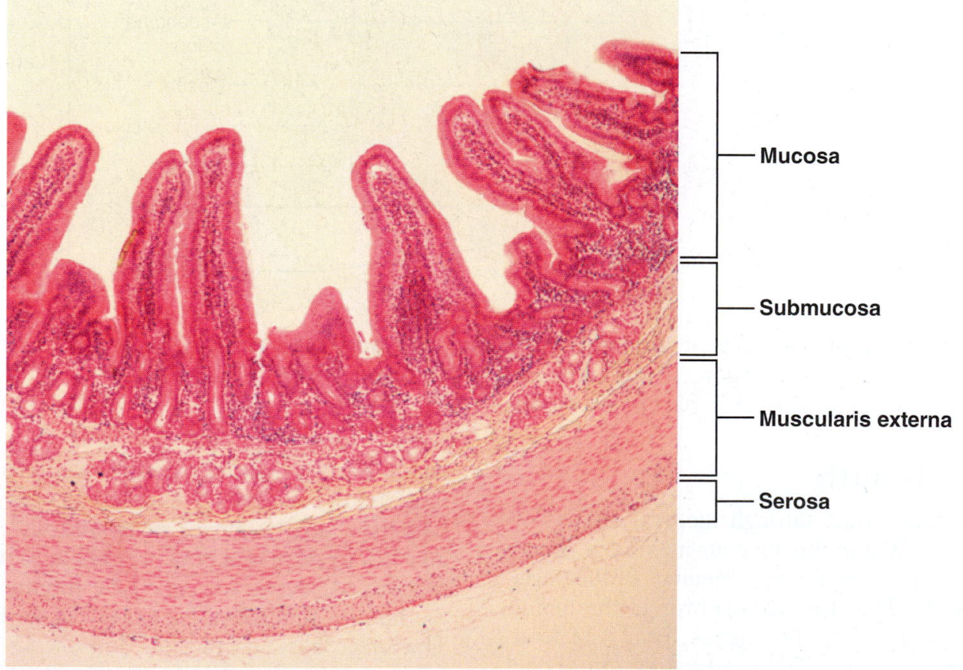

(b) Light micrograph cross section through the small intestine (90×)

Figure 27.2 **Basic structural pattern of the alimentary canal wall.**

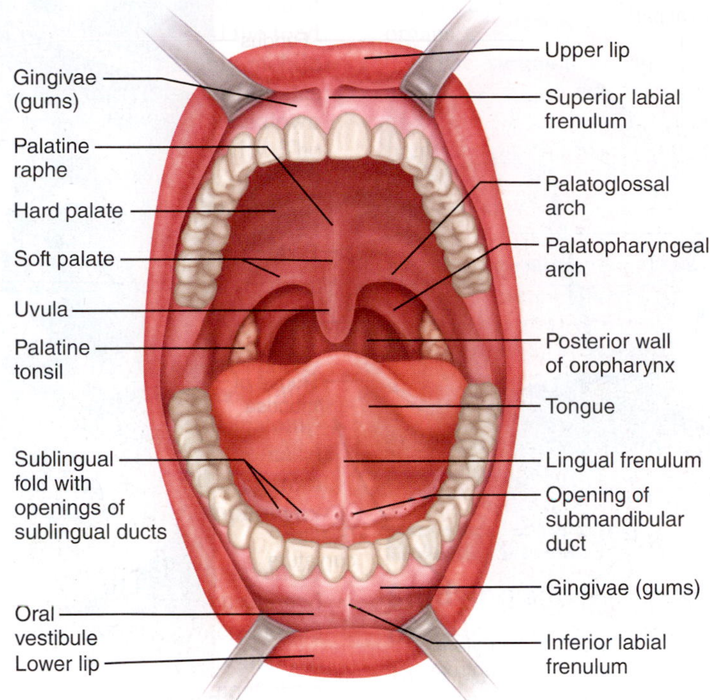

Figure 27.3 **Anterior view of the oral cavity.**

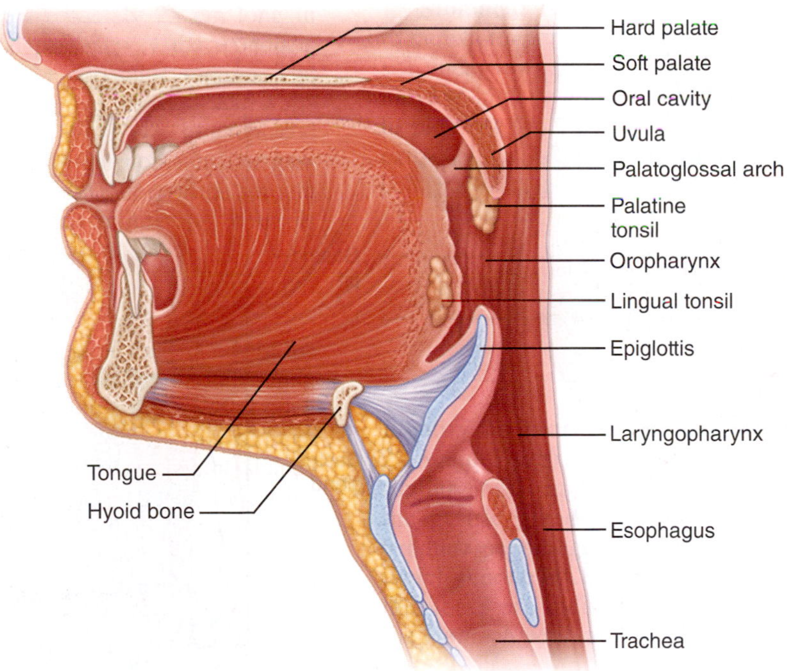

Figure 27.4 **Sagittal view of the head showing oral cavity and pharynx.**

arch, respectively. Another mass of lymphoid tissue, the **lingual tonsil** (see Figure 27.4), covers the base of the tongue, posterior to the oral cavity proper. The tonsils, in common with other lymphoid tissues, are part of the body's defense system. Very often in young children, the palatine tonsils become inflamed and enlarge, partially blocking the entrance to the pharynx posteriorly and making swallowing difficult and painful. This condition is called **tonsillitis.** +

Three pairs of salivary glands duct their secretion, saliva, into the oral cavity. One component of saliva, salivary amylase, begins the digestion of starchy foods within the oral cavity. (The salivary glands are discussed in more detail on page 470.)

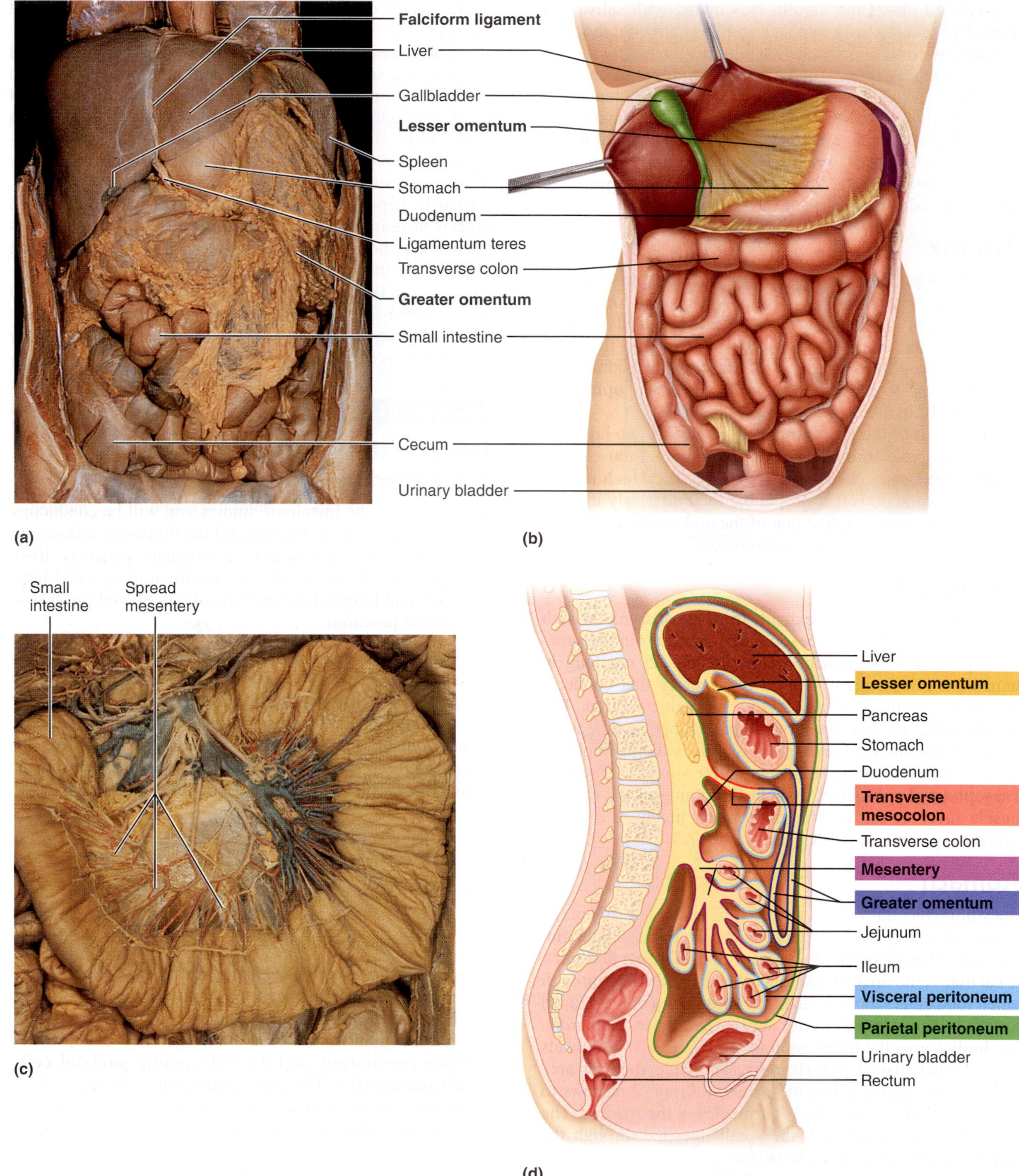

Figure 27.6 Peritoneal attachments of the abdominal organs. Superficial anterior views of the abdominal cavity. **(a)** Cadaver photograph with the greater omentum in place. **(b)** Diagram showing greater omentum removed and liver and gallbladder reflected superiorly. **(c)** Mesentery of the small intestine. **(d)** Sagittal view of a male torso. Mesentery labels appear in colored boxes.

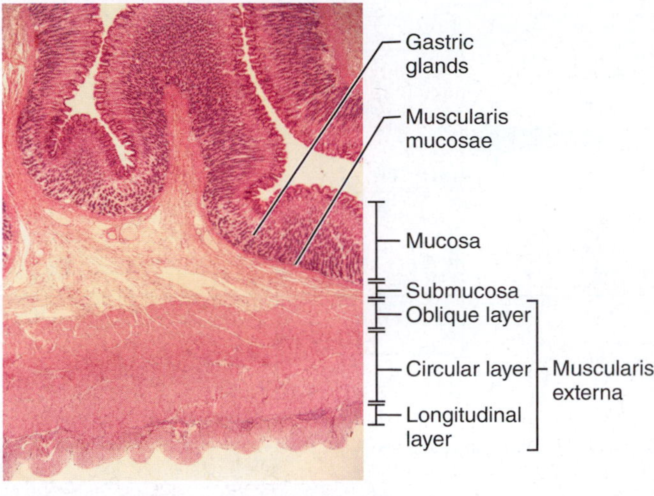

(a)

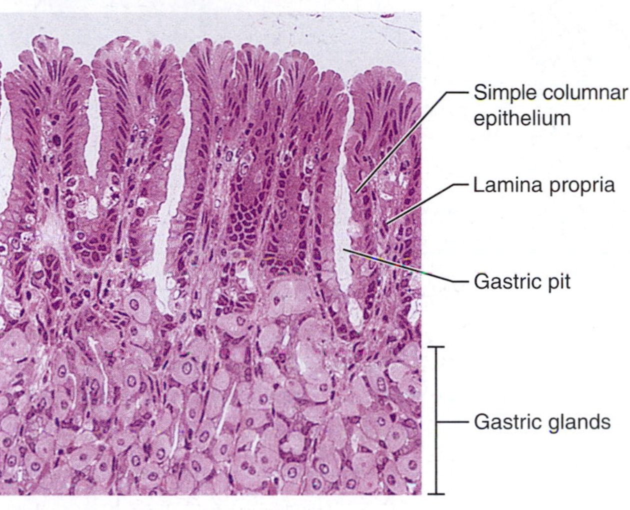

(b)

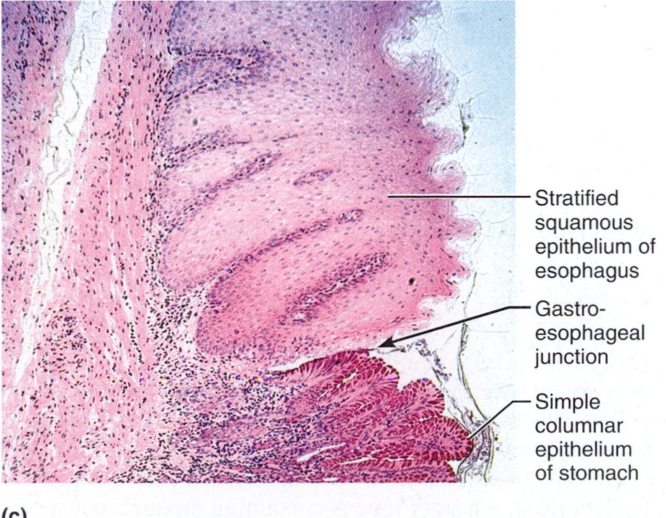

(c)

Figure 27.7 Histology of selected regions of the stomach and gastroesophageal junction. (a) Stomach wall (85×). **(b)** Gastric pits and glands (170×). **(c)** Gastroesophageal junction, longitudinal section (70×).

2. **Gastroesophageal junction:** Scan the slide under low power to locate the mucosal junction between the end of the esophagus and the beginning of the stomach, the gastroesophageal junction. Compare your observations to the photomicrograph (Figure 27.7c). What is the functional importance of the epithelial differences seen in the two organs?

____

____

____

Small Intestine

The **small intestine** is a convoluted tube, 6 to 7 m (about 20 feet) long in a cadaver but only about 2 m (6 feet) long during life because of its muscle tone. It extends from the pyloric sphincter to the ileocecal valve. The small intestine is suspended by a double layer of peritoneum, the fan-shaped **mesentery,** from the posterior abdominal wall (Figure 27.6), and it lies, framed laterally and superiorly by the large intestine, in the abdominal cavity. The small intestine has three subdivisions (Figure 27.1): (1) the **duodenum** extends from the pyloric sphincter for about 25 cm (10 inches) and curves around the head of the pancreas; most of the duodenum lies in a retroperitoneal position. (2) The **jejunum,** continuous with the duodenum, extends for 2.5 m (about 8 feet). Most of the jejunum occupies the umbilical region of the abdominal cavity. (3) The **ileum,** the terminal portion of the small intestine, is about 3.6 m (12 feet) long and joins the large intestine at the **ileocecal valve.** It is located inferiorly and somewhat to the right in the abdominal cavity, but its major portion lies in the hypogastric region.

Two types of enzymes complete digestion in the small intestine: **brush border enzymes,** which are hydrolytic enzymes bound to the microvilli of the columnar epithelial cells; and, more important, pancreatic enzymes, which are ducted into the duodenum largely via the **main pancreatic duct.** Bile (formed in the liver) also enters the duodenum via the **bile duct** in the same area. At the duodenum, the ducts join to form the bulblike **hepatopancreatic ampulla** and empty their products into the duodenal lumen through the **major duodenal papilla,** an orifice controlled by a muscular valve called the **hepatopancreatic sphincter.**

Nearly all nutrient absorption occurs in the small intestine, where three structural modifications increase the absorptive surface of the mucosa: the microvilli, villi, and circular folds (Figure 27.8). **Microvilli** are minute projections of the surface plasma membrane of the columnar epithelial lining cells of the mucosa. **Villi** are the finger-like projections of the mucosa tunic that give it a velvety appearance and texture. The **circular folds** are deep folds of the mucosa and submucosa layers that force chyme to spiral through the intestine, mixing it and slowing its progress. These structural modifications decrease in frequency and size toward the end of the small intestine. Any residue remaining undigested and unabsorbed at the terminus of the small intestine enters the large intestine through the ileocecal valve. In contrast, the amount of lymphoid tissue in the submucosa of the small intestine (especially the

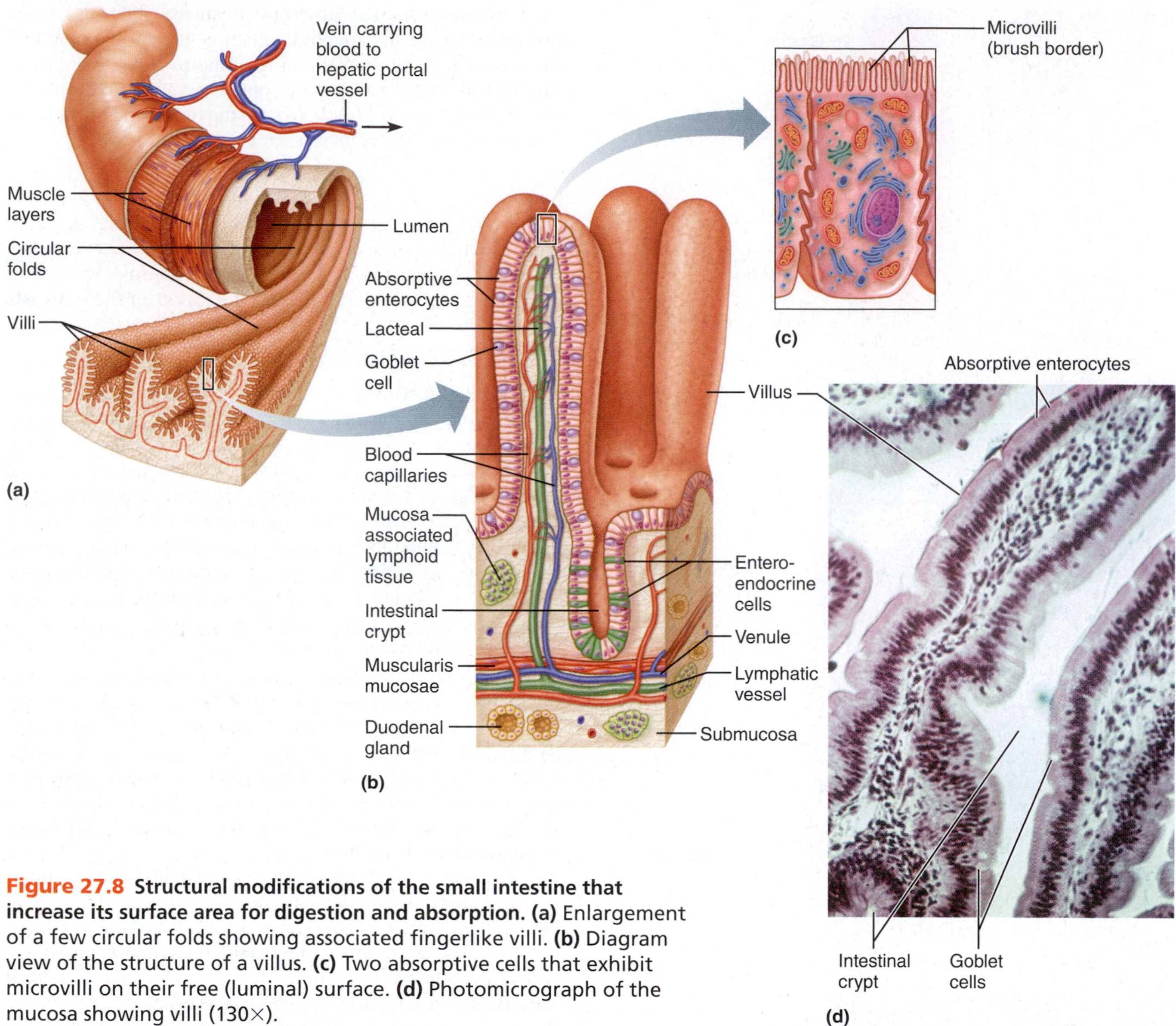

Figure 27.8 Structural modifications of the small intestine that increase its surface area for digestion and absorption. (a) Enlargement of a few circular folds showing associated fingerlike villi. **(b)** Diagram view of the structure of a villus. **(c)** Two absorptive cells that exhibit microvilli on their free (luminal) surface. **(d)** Photomicrograph of the mucosa showing villi (130×).

aggregated lymphoid nodules, also called *Peyer's patches,* Figure 27.9b) increases along the length of the small intestine and is very apparent in the ileum. This reflects the fact that the remaining undigested food residue contains large numbers of bacteria that must be prevented from entering the bloodstream.

ACTIVITY 3

Observing the Histologic Structure of the Small Intestine

1. **Duodenum:** Secure the slide of the duodenum to the microscope stage. Observe the tissue under low power to identify the four basic tunics of the intestinal wall—that is, the **mucosa** and its three sublayers, the **submucosa,** the **muscularis externa,** and the **serosa,** or *visceral peritoneum.* Consult the photomicrograph **(Figure 27.9a)** to help you identify the scattered mucus-producing **duodenal glands** in the submucosa.

What type of epithelium do you see here? ______________

____

Examine the large leaflike *villi,* which increase the surface area for absorption. Notice the scattered mucus-producing goblet cells in the epithelium of the villi. Note also the **intestinal crypts** (see also Figure 27.8), invaginated areas of the mucosa between the villi containing the cells that produce intestinal juice, a watery mucus-containing mixture that serves as a carrier fluid for absorption of nutrients from the chyme.

2. **Ileum:** The structure of the ileum resembles that of the duodenum, except that the villi are less elaborate because most of the absorption has occurred by the time that

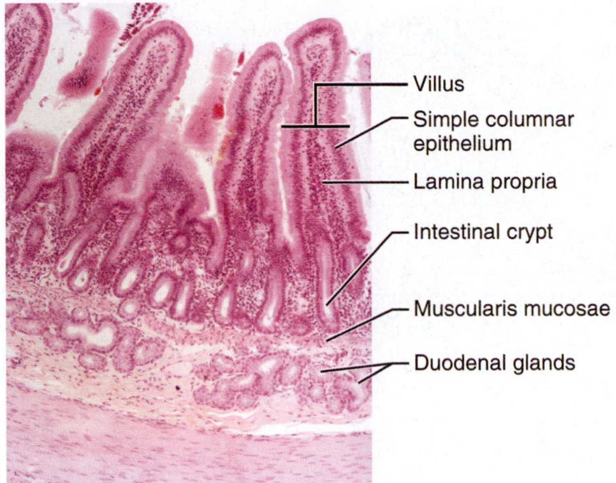

(a)

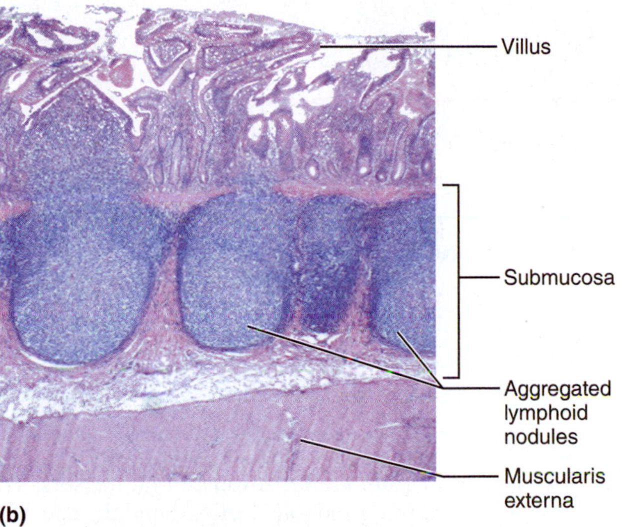

(b)

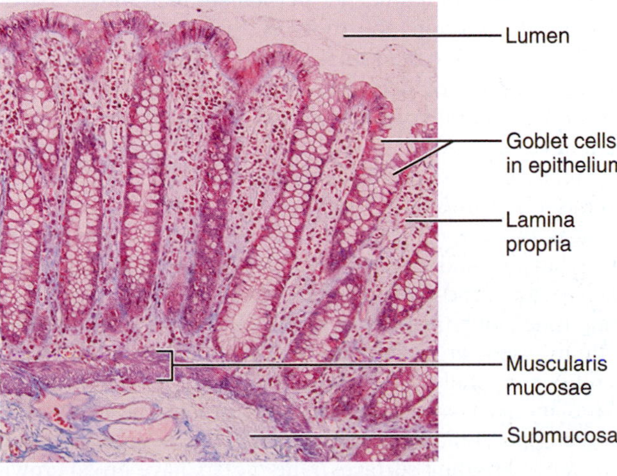

(c)

Figure 27.9 Histology of selected regions of the small and large intestines. Cross-sectional views. **(a)** Duodenum of the small intestine (180×). **(b)** Ileum of the small intestine (35×). **(c)** Large intestine (120×).

chyme reaches the ileum. Secure a slide of the ileum to the microscope stage for viewing. Observe the villi, and identify the four layers of the wall and the large, generally spherical aggregated lymphoid nodules (Figure 27.9b). What tissue type are aggregated lymphoid nodules?

---

3. If a villus model is available, identify the following cells or regions before continuing: absorptive epithelium, goblet cells, lamina propria, slips of the muscularis mucosae, capillary bed, and lacteal. If possible, also identify the intestinal crypts that lie between the villi. ■

Large Intestine

The **large intestine (Figure 27.10)** is about 1.5 m (5 feet) long and extends from the ileocecal valve to the anus. It encircles the small intestine on three sides and consists of the following subdivisions: **cecum, appendix, colon, rectum**, and **anal canal.**

The blind wormlike appendix, which hangs from the cecum, is a trouble spot in the large intestine. Because it is generally twisted, it provides an ideal location for bacteria to accumulate and multiply. Inflammation of the appendix, or appendicitis, is the result. +

The colon is divided into several distinct regions. The **ascending colon** travels up the right side of the abdominal cavity and makes a right-angle turn at the **right colic (hepatic) flexure** to cross the abdominal cavity as the **transverse colon.** It then turns at the **left colic (splenic) flexure** and continues down the left side of the abdominal cavity as the **descending colon,** where it takes an S-shaped course as the **sigmoid colon.** The sigmoid colon, rectum, and the anal canal lie in the pelvis anterior to the sacrum and thus are not considered abdominal cavity structures. Except for the transverse and sigmoid colons, which are secured to the dorsal body wall by mesocolons (see Figure 27.6), the colon is retroperitoneal.

The anal canal terminates in the **anus,** the opening to the exterior of the body. The anal canal has two sphincters, a voluntary *external anal sphincter* composed of skeletal muscle, and an involuntary *internal anal sphincter* composed of smooth muscle. The sphincters are normally closed except during defecation, when undigested food and bacteria are eliminated from the body as feces.

In the large intestine, the longitudinal layer of the muscularis externa is reduced to three longitudinal bands called the **teniae coli.** Because these bands are shorter than the rest of the wall of the large intestine, they cause the wall to pucker into small pocketlike sacs called **haustra.** Fat-filled pouches of visceral peritoneum, called *epiploic appendages,* hang from the colon's surface.

The major function of the large intestine is to consolidate and propel the unusable fecal matter toward the anus and eliminate it from the body. While it does that chore, it (1) provides a site where intestinal bacteria manufacture vitamins B and K; and (2) reclaims most of the remaining water from undigested food, thus conserving body water.

Watery stools, or **diarrhea,** result from any condition that rushes undigested food residue through the large intestine before it has had sufficient time to

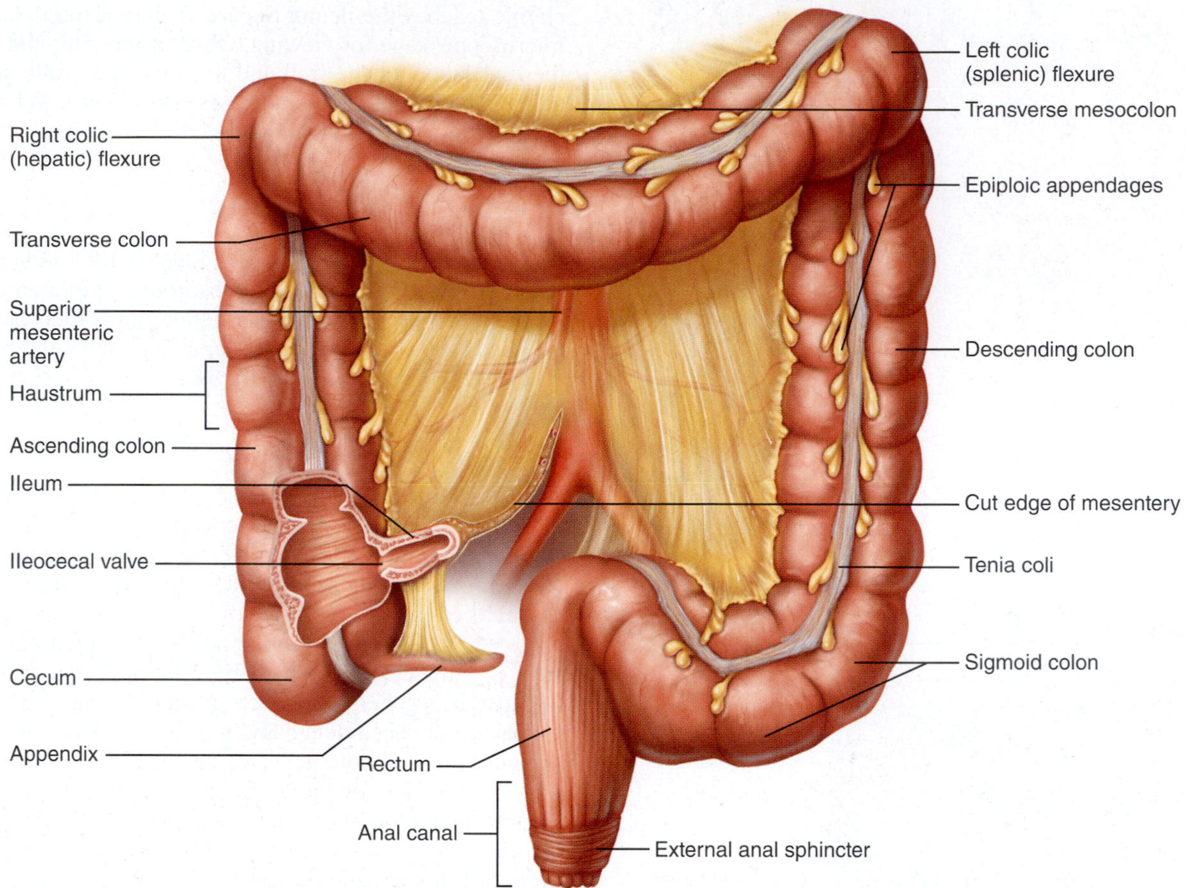

Figure 27.10 The large intestine. (Section of the cecum removed to show the ileocecal valve.)

absorb the water. Conversely, when food residue remains in the large intestine for extended periods, water is absorbed, and the stool becomes hard and difficult to pass, causing **constipation.** +

ACTIVITY 4

Examining the Histologic Structure of the Large Intestine

Secure a slide of the large intestine to the microscope stage for viewing. Observe the villi, and note the numerous goblet cells (Figure 27.9c). Why do you think the large intestine produces so much mucus?

____

____

Accessory Digestive Organs

Teeth

By the age of 21, two sets of teeth have developed **(Figure 27.11)**. The initial set, called the **deciduous** (or **milk**) **teeth,** normally appears between the ages of 6 months and 2½ years. The first of these to erupt are the lower central incisors. The child begins to shed the deciduous teeth around the age of 6, and a second set of teeth, the **permanent teeth,** gradually replaces them. As the deeper permanent teeth progressively enlarge and develop, the roots of the deciduous teeth are resorbed, leading to their final shedding. During years 6 to 12, the child has mixed dentition—both permanent and deciduous teeth. Generally, by the age of 12, all of the deciduous teeth have been shed.

Teeth are classified as **incisors, canines** *(eyeteeth),* **premolars** *(bicuspids),* and **molars.** Teeth names reflect differences in relative structure and function. The incisors are chisel-shaped and exert a shearing action used in biting. Canines are cone-shaped or fanglike, the latter description being much more applicable to the canines of animals whose teeth are used for the tearing of food. Incisors, canines, and premolars typically have single roots, though the first upper premolars may have two. The lower molars have two roots, but the upper molars usually have three. The premolars have two *cusps* (grinding surfaces); the molars have broad crowns with rounded cusps specialized for the fine grinding of food.

Dentition is described by means of a **dental formula,** which designates the numbers, types, and position of the teeth in one side of the jaw. (Because tooth arrangement is bilaterally symmetrical, it is necessary to designate only one side of the jaw.) The complete dental formula for the deciduous teeth

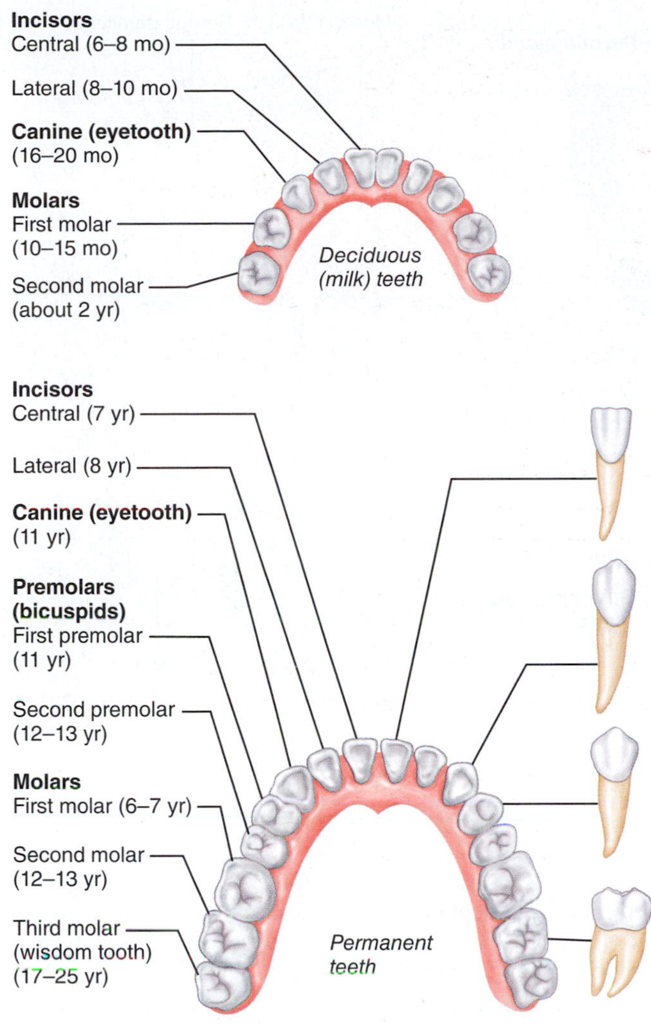

Figure 27.11 Human deciduous teeth and permanent teeth. (Approximate time of teeth eruption shown in parentheses.)

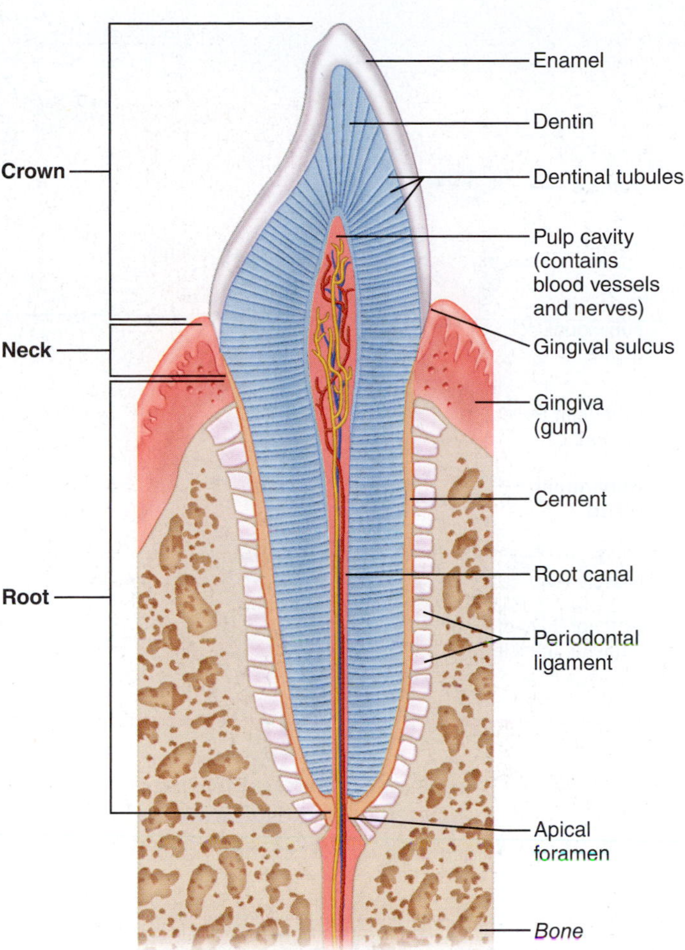

Figure 27.12 Longitudinal section of human canine tooth within its bony socket.

from the medial aspect of each jaw and proceeding posteriorly is as follows:

$$\frac{\text{Upper teeth: 2 incisors, 1 canine, 0 premolars, 2 molars}}{\text{Lower teeth: 2 incisors, 1 canine, 0 premolars, 2 molars}} \times 2$$

This formula is generally abbreviated to read as follows:

$$\frac{2,1,0,2}{2,1,0,2} \times 2 = 20 \text{ (number of deciduous teeth)}$$

The 32 permanent teeth are then described by the following dental formula:

$$\frac{2,1,2,3}{2,1,2,3} \times 2 = 32 \text{ (number of permanent teeth)}$$

Although 32 is designated as the normal number of permanent teeth, not everyone develops a full complement. In many people, the third molars, commonly called *wisdom teeth,* never erupt.

ACTIVITY 5

Identifying Types of Teeth

Identify the four types of teeth (incisors, canines, premolars, and molars) on the jaw model or human skull. ■

A tooth consists of two major regions, the *crown* and the *root.* A longitudinal section made through a tooth shows the following basic anatomical plan (**Figure 27.12**). The **crown** is the superior portion of the tooth. The portion of the crown visible above the **gingiva,** or **gum,** is referred to as the *clinical crown.* The entire area covered by **enamel** is called the *anatomical crown.* Enamel is the hardest substance in the body and is fairly brittle. It consists of 95% to 97% inorganic calcium salts (chiefly $CaPO_4$) and thus is heavily mineralized. The crevice between the end of the anatomical crown and the upper margin of the gingiva is referred to as the *gingival sulcus.*

That portion of the tooth embedded in the socket portion of the jaw is the **root,** and the root and crown are connected by a slight constriction, the **neck.** The outermost surface of the root is covered by **cement,** which is similar to bone in

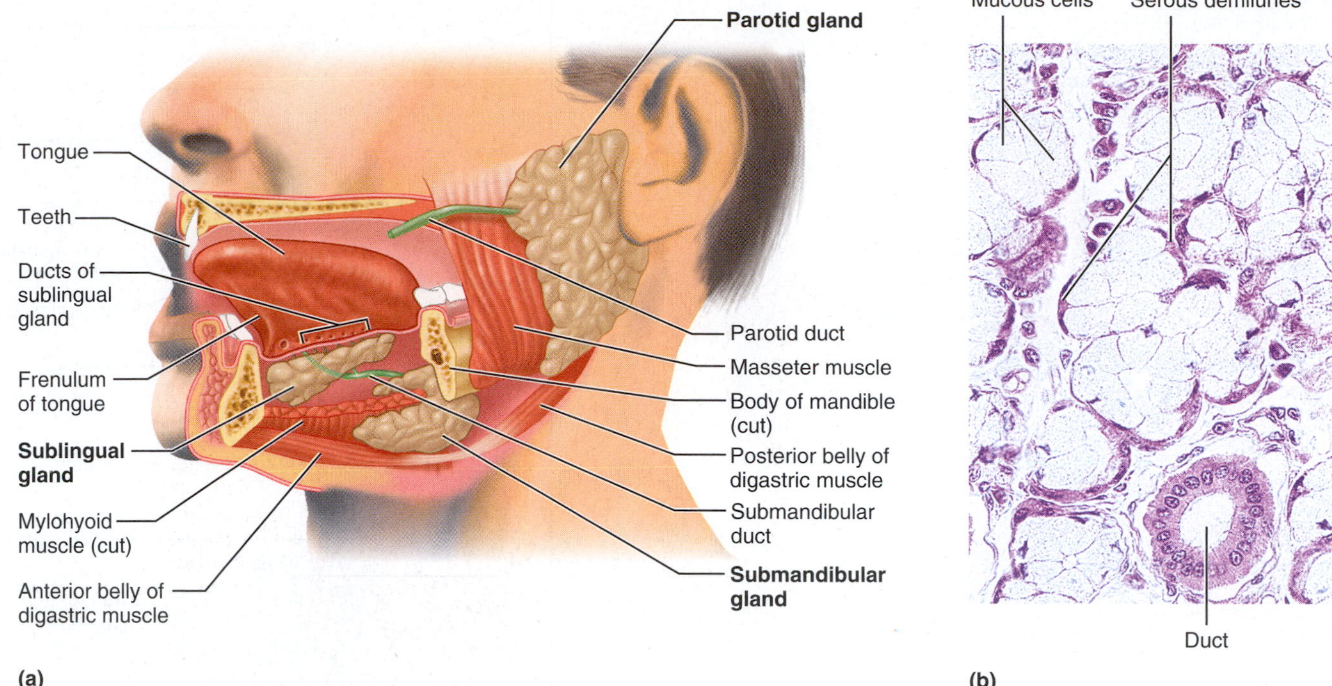

Figure 27.13 The salivary glands. (a) The parotid, submandibular, and sublingual salivary glands associated with the left aspect of the oral cavity. **(b)** Photomicrograph of the sublingual salivary gland (215×), which is a mixed salivary gland.

composition and less brittle than enamel. The cement attaches the tooth to the **periodontal ligament,** which holds the tooth in the tooth socket and exerts a cushioning effect. **Dentin,** which makes up the bulk of the tooth, is the bonelike material interior to the enamel and cement.

The **pulp cavity** occupies the central portion of the tooth. **Pulp,** connective tissue liberally supplied with blood vessels, nerves, and lymphatics, occupies this cavity and provides for tooth sensation and supplies nutrients to the tooth tissues. **Odontoblasts,** specialized cells that reside in the outer margins of the pulp cavity, produce the dentin. The pulp cavity extends into distal portions of the root and becomes the **root canal.** An opening at the root apex, the **apical foramen,** provides a route of entry into the tooth for blood vessels, nerves, and other structures from the tissues beneath.

ACTIVITY 6

Studying Microscopic Anatomy of the Tooth

Observe a slide of a longitudinal section of a tooth, and compare your observations with the structures detailed in the illustration (Figure 27.12). Identify as many of these structures as possible. ■

Salivary Glands

Three pairs of major **salivary glands** (**Figure 27.13**) empty their secretions into the oral cavity.

Parotid glands: Large glands located anterior to the ear and ducting into the mouth over the second upper molar through the parotid duct.

Submandibular glands: Located along the medial aspect of the mandibular body in the floor of the mouth, and ducting under the tongue to the base of the lingual frenulum.

Sublingual glands: Small glands located most anteriorly in the floor of the mouth and emptying under the tongue via several small ducts.

Food in the mouth and mechanical pressure (even chewing rubber bands or wax) stimulate the salivary glands to secrete saliva. Saliva consists primarily of a viscous glycoprotein called *mucin,* which moistens the food and helps to bind it together into a mass called a **bolus,** and a clear serous fluid containing the enzyme *salivary amylase.* Salivary amylase begins the digestion of starch, breaking it down into oligosaccharides and disaccharides. Parotid gland secretion is mainly serous, whereas the submandibular is a mixed gland that produces both mucin and serous components. The sublingual gland produces mostly mucin.

ACTIVITY 7

Examining Salivary Gland Tissue

Examine salivary gland tissue under low power and then high power to become familiar with the appearance of glandular tissue. Notice the clustered arrangement of the cells around their ducts. The cells are basically triangular, with their pointed ends facing the duct opening. If possible, differentiate mucus-producing cells, which look hollow or have a clear cytoplasm, from serous cells, which produce the clear, enzyme-containing fluid and have granules in their cytoplasm. The serous cells often form *demilunes* (caps) around the more central mucous cells. (Figure 27.13b may be helpful in this task.) ■

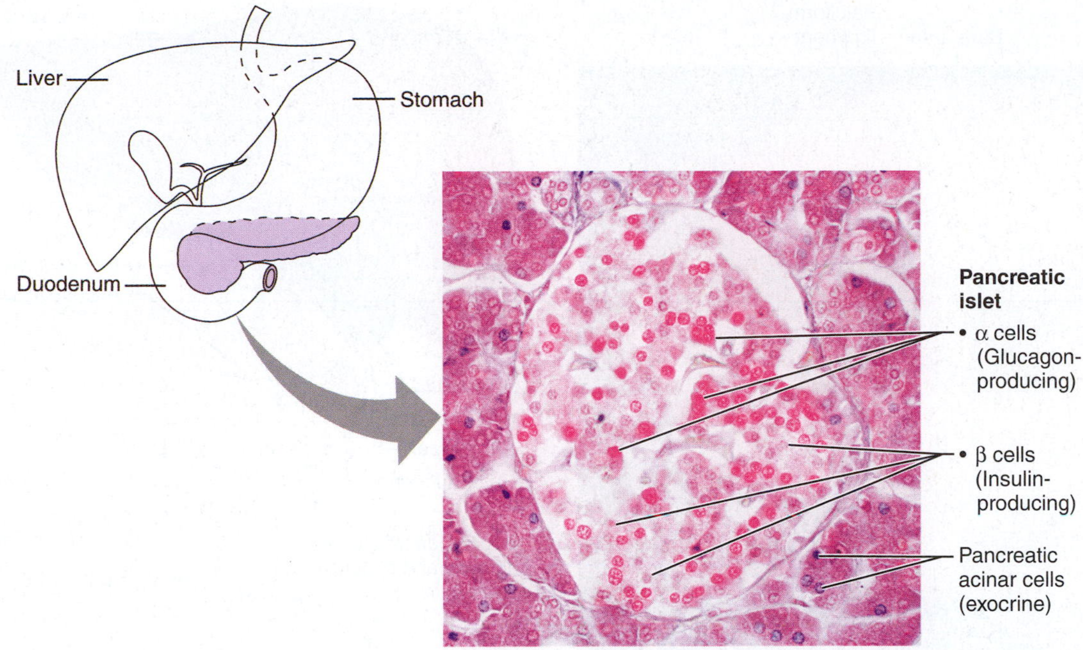

Figure 27.14 Histology of the pancreas. The pancreatic islet cells produce insulin and glucagon (hormones). The acinar cells synthesize digestive enzymes for "export" to the duodenum.

Pancreas

The **pancreas** is a soft, triangular gland that extends horizontally across the posterior abdominal wall from the spleen to the duodenum **(Figure 27.14)**. Like the duodenum, it is a retroperitoneal organ (see Figure 27.6). The pancreas has both an endocrine function, producing the hormones insulin and glucagon, and an exocrine function. Its exocrine secretion includes many hydrolytic enzymes, and is secreted into the duodenum through the pancreatic duct. Pancreatic juice is very alkaline. Its high concentration of bicarbonate ion (HCO_3^-) neutralizes the acidic chyme entering the duodenum from the stomach, enabling the pancreatic and intestinal enzymes to operate at their optimal pH, which is slightly alkaline.

ACTIVITY 8

Examining the Histology of the Pancreas

Observe pancreatic tissue under low power and then high power to distinguish between the lighter-staining, endocrine-producing clusters of cells called **pancreatic islets** and the deeper-staining **acinar cells,** which produce the hydrolytic enzymes and form the major portion of the pancreatic tissue (Figure 27.14). Notice the arrangement of the exocrine cells around their central ducts. If the tissue is differentially stained, you will also be able to identify specifically the lavender blue–stained insulin-secreting beta cells and the red-stained glucagon-secreting alpha cells of the pancreas. ■

Liver and Gallbladder

The **liver** (Figure 27.1 and **Figure 27.15**), the largest gland in the body, is located inferior to the diaphragm, more to the right than the left side of the body. As noted earlier, it hides the stomach from view in a superficial observation of abdominal contents. The human liver has four lobes and is suspended from the diaphragm and anterior abdominal wall by the **falciform ligament** (Figures 27.6a and 27.15).

The liver is one of the body's most important organs, and it performs many metabolic roles. However, its digestive function is to produce bile, which leaves the liver through the **common hepatic duct** and then enters the duodenum through the **bile duct (Figure 27.16)**. Bile has no enzymatic action but emulsifies fats by breaking up large fat particles into smaller ones, which creates a larger surface area to improve the efficiency of the enzyme lipase. Without bile, very little fat digestion or absorption occurs.

When digestive activity is not occurring in the digestive tract, bile backs up into the **cystic duct** and enters the **gallbladder,** a small green sac on the inferior surface of the liver. Bile is stored there until needed for the digestive process. While in the gallbladder, bile is concentrated by the removal of water and some ions. When fat-rich food enters the duodenum, a hormonal stimulus causes the gallbladder to contract, releasing the stored bile and making it available to the duodenum.

If the common hepatic or bile duct is blocked (for example, by wedged gallstones), bile is prevented from entering the small intestine, accumulates, and eventually backs up into the liver. This exerts pressure on the liver cells, and bile begins to enter the bloodstream. As the bile circulates through the body, the tissues become yellow, or jaundiced.

Blockage of the ducts is just one cause of jaundice. More often it results from actual liver problems such as **hepatitis,**

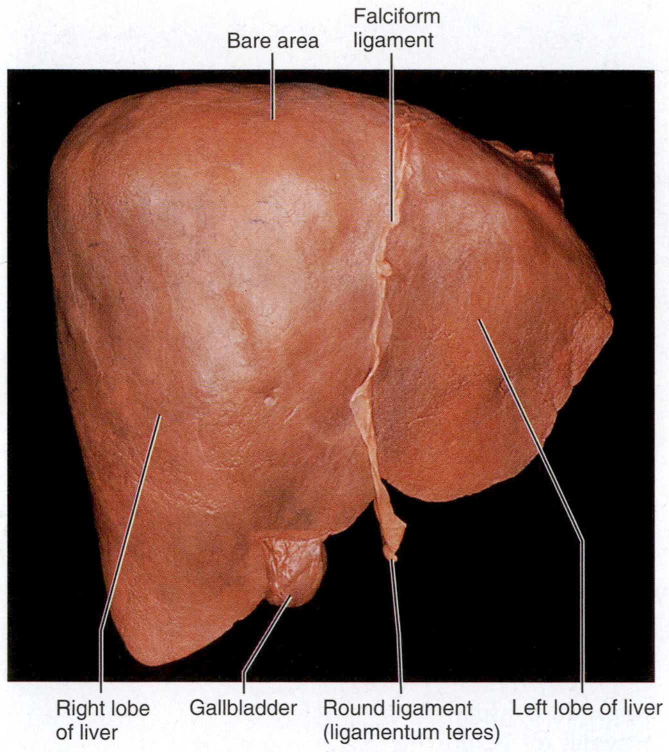

(a)

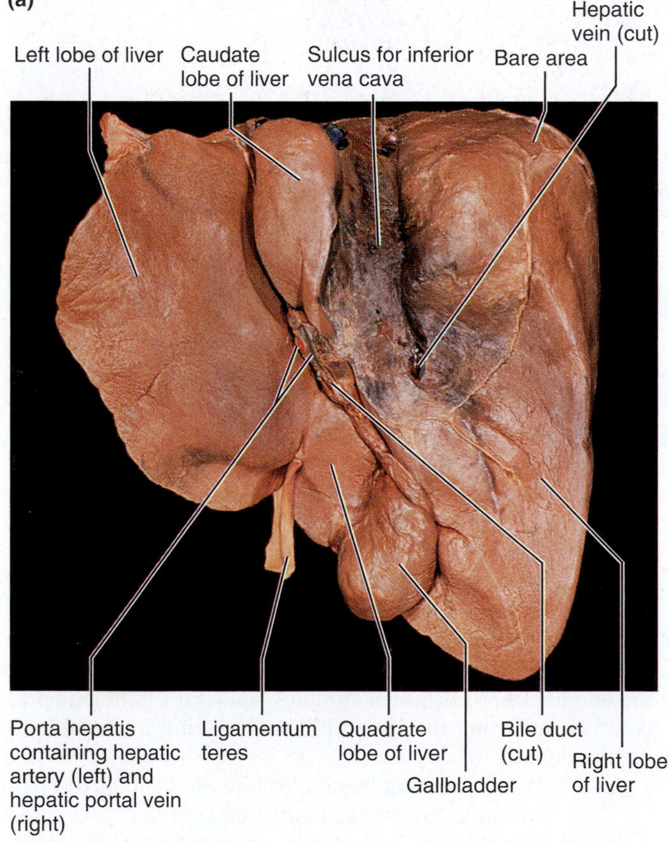

(b)

Figure 27.15 Gross anatomy of the human liver. **(a)** Anterior view. **(b)** Posteroinferior aspect. The four liver lobes are separated by a group of fissures in this view.

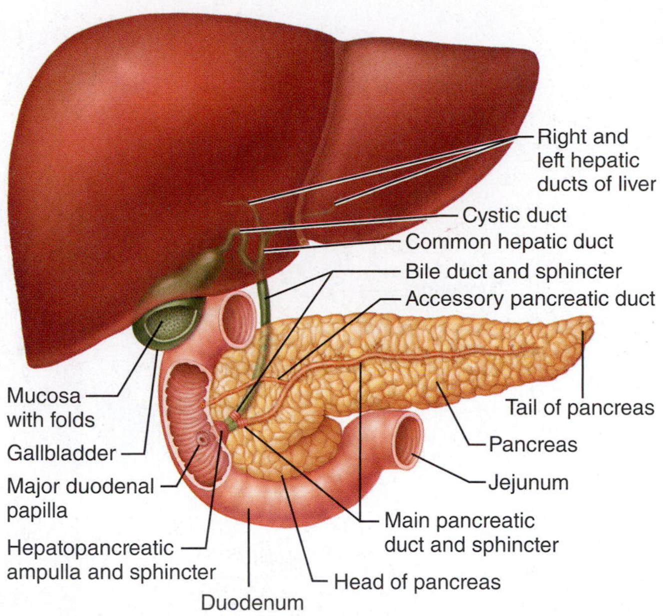

Figure 27.16 Ducts of accessory digestive organs.

which is any inflammation of the liver, or **cirrhosis,** a condition in which the liver is severely damaged and becomes hard and fibrous. Cirrhosis is prevalent in people who drink excessive alcohol for many years. ✚

As demonstrated by its highly organized anatomy, the liver **(Figure 27.17)** is very important in the initial processing of the nutrient-rich blood draining the digestive organs. Its structural and functional units are called **lobules.** Each lobule is a basically hexagonal structure consisting of cordlike arrays of **hepatocytes,** or *liver cells,* which radiate outward from a central vein running upward in the longitudinal axis of the lobule. At each of the six corners of the lobule is a **portal triad,** so named because three basic structures are always present there: a *portal arteriole* (a branch of the *hepatic artery,* the functional blood supply of the liver), a *portal venule* (a branch of the *hepatic portal vein* carrying nutrient-rich blood from the digestive viscera), and a *bile duct.* Between the hepatocytes are blood-filled spaces, or **sinusoids,** through which blood from the hepatic portal vein and hepatic artery percolates. **Stellate macrophages,** special phagocytic cells also called **hepatic macrophages,** line the sinusoids and remove debris such as bacteria from the blood as it flows past while the hepatocytes pick up oxygen and nutrients. Much of the glucose transported to the liver from the digestive system is stored as glycogen in the liver for later use, and amino acids are taken from the blood by the liver cells and utilized to make plasma proteins. The sinusoids empty into the central vein, and the blood ultimately drains from the liver via the *hepatic veins.*

Bile is continuously being made by the hepatocytes. It flows through tiny canals, the **bile canaliculi,** which run between adjacent hepatocytes toward the bile duct branches in the triad regions, where the bile eventually leaves the liver. Notice that the directions of blood and bile flow in the liver lobule are exactly opposite.

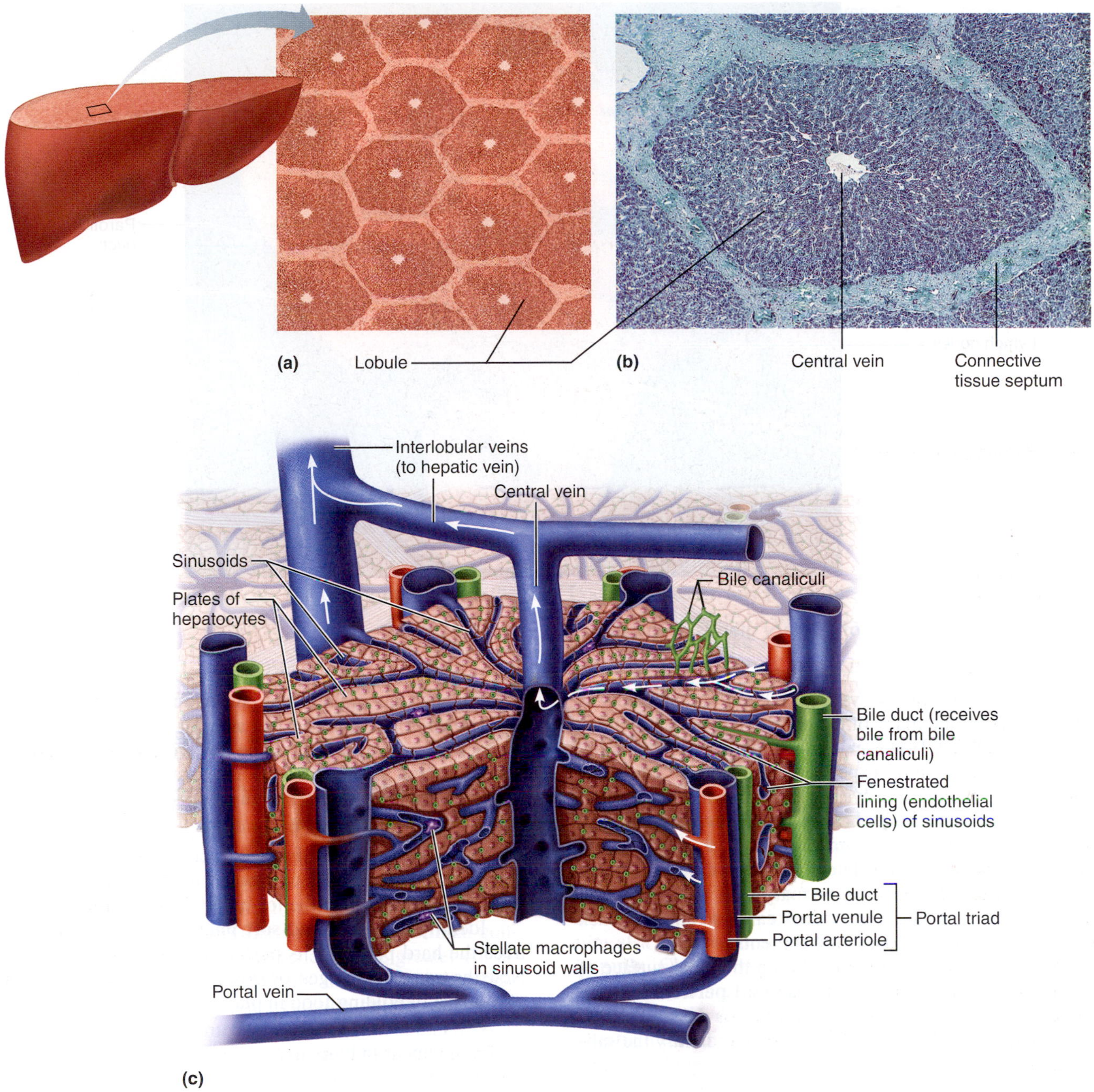

Figure 27.17 Microscopic anatomy of the liver. (a) Schematic view of the cut surface of the liver showing the hexagonal nature of its lobules. **(b)** Photomicrograph of one liver lobule (50×). **(c)** Enlarged three-dimensional diagram of one liver lobule. Arrows show direction of blood flow. Bile flows in the opposite direction toward the bile ducts.

ACTIVITY 9

Examining the Histology of the Liver

Examine a slide of liver tissue, and identify as many structural features as possible (see Figure 27.17). Also examine a three-dimensional model of liver lobules if this is available. ■

DISSECTION AND IDENTIFICATION

The Digestive System of the Cat

Don gloves, and obtain your cat and secure it to the dissecting tray, dorsal surface down. Obtain all necessary dissecting instruments. If you have completed the dissection of the

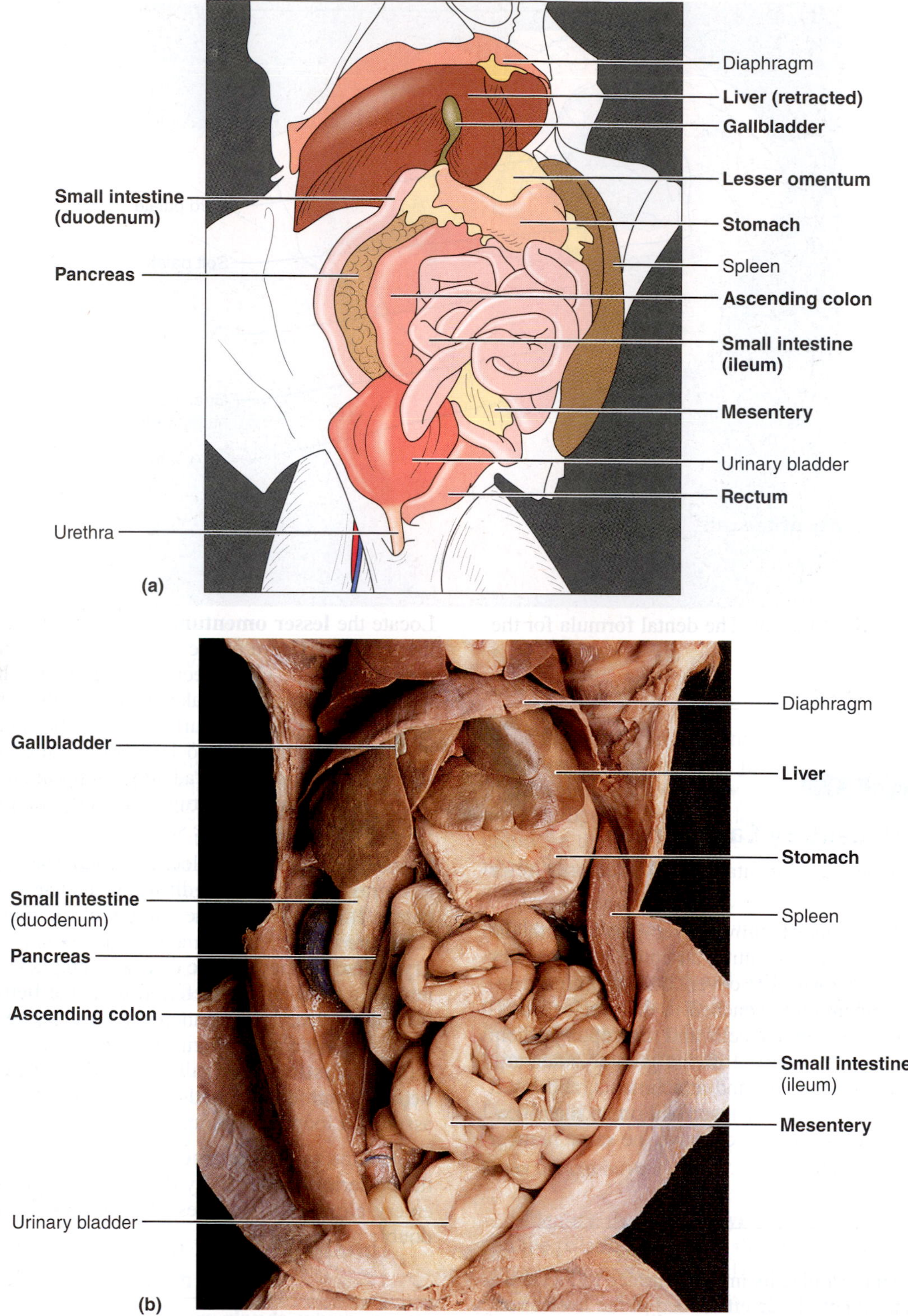

Figure 27.20 Digestive organs of the cat. (a) Diagram view. **(b)** Photograph. The greater omentum has been cut from its attachment to the stomach.

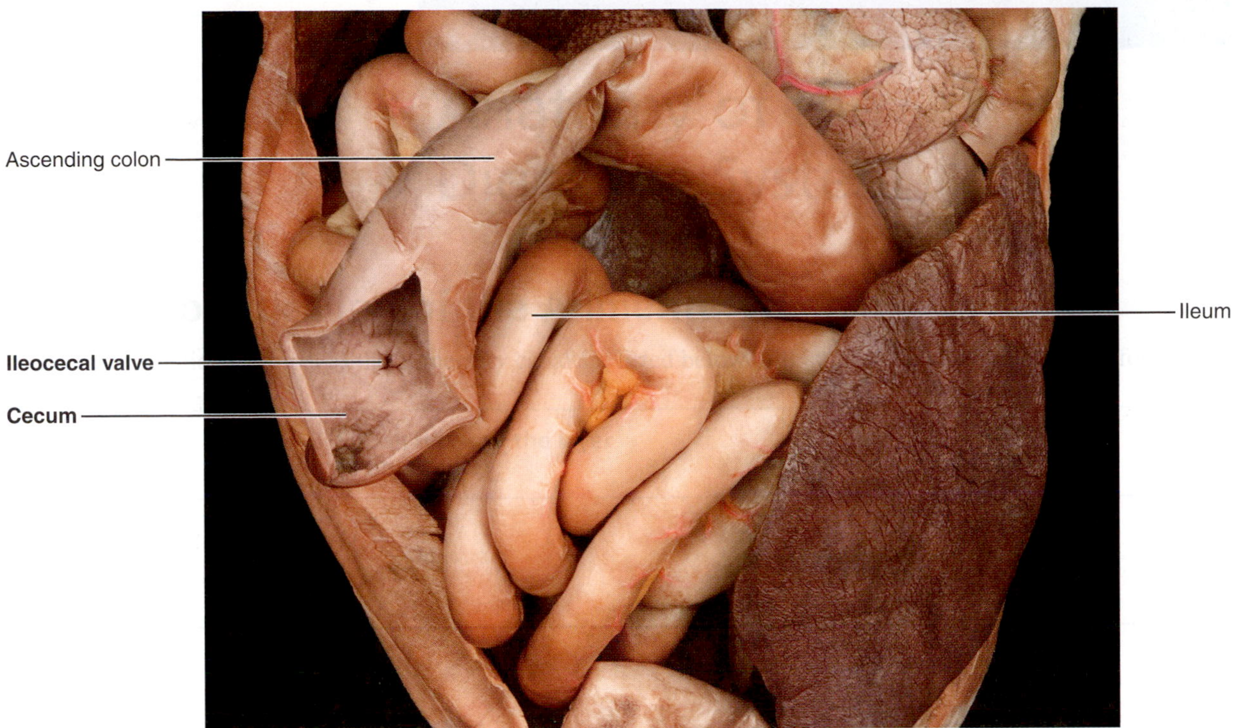

Figure 27.21 Ileocecal valve.

Trace the course of the small intestine from its proximal, (duodenal) end to its distal (ileal) end. Can you see any obvious differences in the external anatomy of the small intestine from one end to the other?

___

With a scalpel, slice open the distal portion of the ileum, and flush out the inner surface with water. Feel the inner surface with your fingertip. How does it feel?

___

Use a hand lens to see whether you can see any **villi** and to locate the areas of lymphoid tissue called **aggregated lymphoid nodules,** which appear as scattered white patches on the inner intestinal surface. (See also Figure 27.9b.)

Return to the duodenal end of the small intestine. Make an incision into the duodenum. As before, flush the surface with water, and feel the inner surface. Does it feel any different than the ileal mucosa?

___ If so, describe the difference. ___

___

Use the hand lens to observe the villi. What differences do you see in the villi in the two areas of the small intestine?

___

___

7. Make an incision into the junction between the ileum and cecum to locate the **ileocecal valve** (**Figure 27.21**). Observe the **cecum,** the initial expanded part of the large intestine. Lymph nodes may have to be removed from this area to observe it clearly. Does the cat have an appendix?

___

8. Identify the short ascending, transverse, and descending portions of the **colon** and the **mesocolon,** a membrane that attaches the colon to the posterior body wall. Trace the descending colon to the **rectum,** which penetrates the body wall, and identify the **anus.**

Identify the two portions of the peritoneum, the parietal peritoneum lining the abdominal wall (identified previously) and the visceral peritoneum, which is the outermost layer of the wall of the abdominal organs.

9. Before you leave the laboratory, follow the boxed instructions to prepare your cat for storage and clean the area (page 217).

6. Match the items in column B with the descriptive statements in column A. (The items in column B may be used more than once.)

Column A

_______ 1. structure that suspends the small intestine from the posterior body wall

_______ 2. fingerlike extensions of the intestinal mucosa that increase the surface area for absorption

_______ 3. large collections of lymphoid tissue found in the submucosa of the small intestine

_______ 4. deep folds of the mucosa and submucosa that extend completely or partially around the circumference of the small intestine

_______, _______ 5. regions that break down foodstuffs mechanically

_______ 6. mobile organ that manipulates food in the mouth and initiates swallowing

_______ 7. conduit for both air and food

_______, _______, _______ 8. three structures continuous with and representing modifications of the peritoneum

_______ 9. the gullet; no digestive or absorptive function

_______ 10. folds of the gastric mucosa

_______ 11. pocketlike sacs of the large intestine

_______ 12. projections of the plasma membrane of a mucosal epithelial cell

_______ 13. valve at the junction of the small and large intestines

_______ 14. primary region of food and water absorption

_______ 15. membrane securing the tongue to the floor of the mouth

_______ 16. absorbs water and forms feces

_______ 17. area between the teeth and lips/cheeks

_______ 18. wormlike sac that outpockets from the cecum

_______ 19. initiates protein digestion

_______ 20. structure attached to the lesser curvature of the stomach

_______ 21. organ immediately distal to the stomach

_______ 22. valve controlling food movement from the stomach into the duodenum

_______ 23. posterosuperior boundary of the oral cavity

_______ 24. location of the hepatopancreatic sphincter through which pancreatic secretions and bile pass

_______ 25. serous lining of the abdominal cavity wall

_______ 26. principal site for the synthesis of vitamin K by microorganisms

_______ 27. region containing two sphincters through which feces are expelled from the body

_______ 28. bone-supported anterosuperior boundary of the oral cavity

Column B

a. aggregated lymphoid nodules
b. anus
c. appendix
d. circular folds
e. esophagus
f. frenulum
g. greater omentum
h. hard palate
i. haustra
j. ileocecal valve
k. large intestine
l. lesser omentum
m. mesentery
n. microvilli
o. oral cavity
p. oral vestibule
q. parietal peritoneum
r. pharynx
s. pyloric valve
t. rugae
u. small intestine
v. soft palate
w. stomach
x. tongue
y. villi
z. visceral peritoneum

7. Correctly identify all organs depicted in the diagram below.

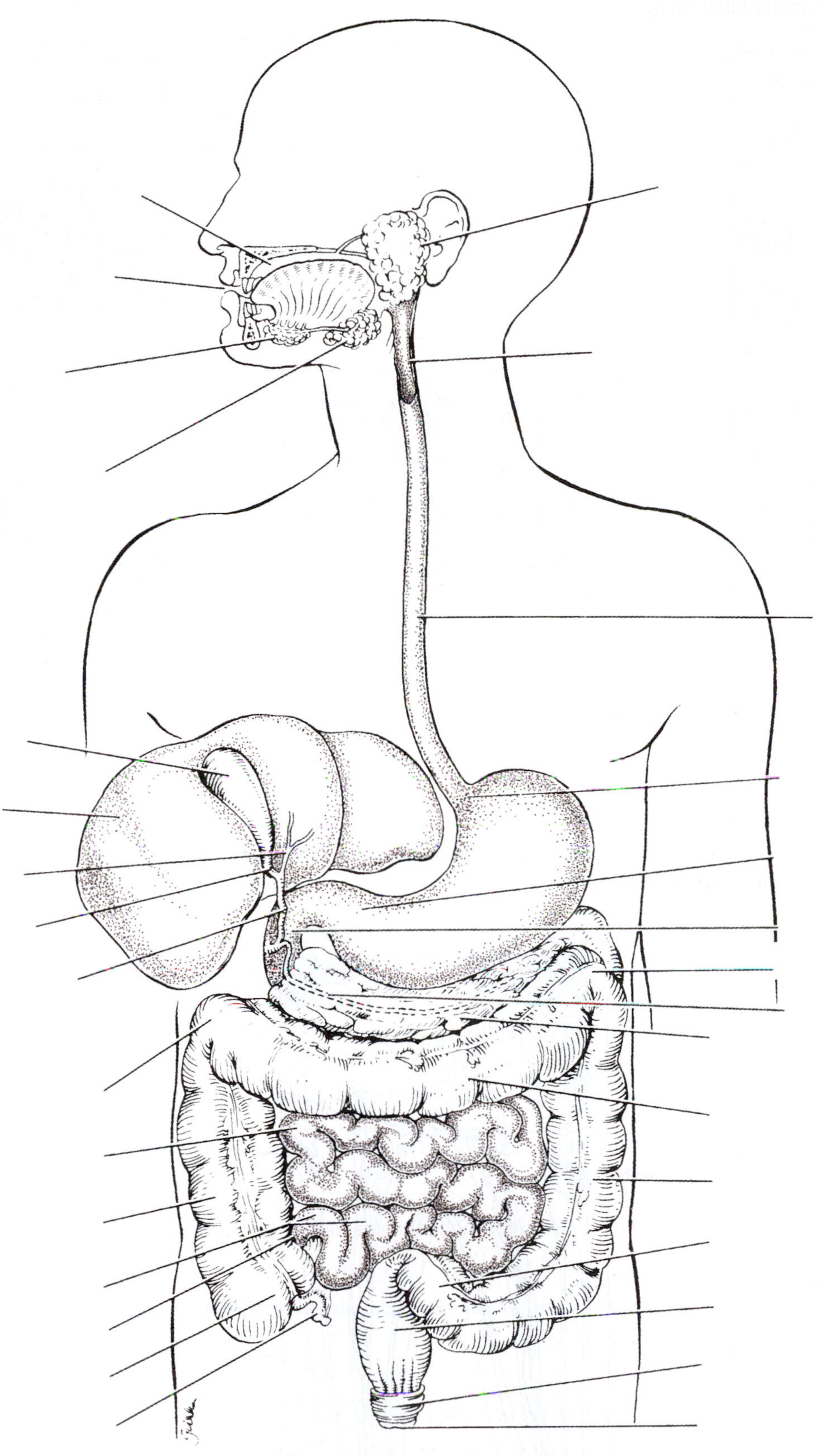

16. What is the role of the gallbladder? ______________________________

17. Name three structures always found in the portal triad regions of the liver. ______________________________,

______________________________, and ______________________________

18. Where would you expect to find the stellate macrophages of the liver? ______________________________

What is their function? ______________________________

19. Why is the liver so dark red in living animals? ______________________________

____

20. The pancreas has two major populations of secretory cells—those in the islets and the acinar cells. Which population serves the digestive process? ______________________________

Dissection and Identification: The Digestive System of the Cat

21. Several differences between cat and human digestive anatomy should have become apparent during the dissection. Note the pertinent differences between the human and the cat relative to the following structures:

Structure	Cat	Human
Tongue papillae		
Number of liver lobes		
Appendix		

28

EXERCISE

Anatomy of the Urinary System

MATERIALS

- ☐ Dissectible human torso model, three-dimensional model of the urinary system, and/or anatomical chart of the human urinary system
- ☐ Dissecting instruments and tray
- ☐ Pig or sheep kidney, doubly or triply injected
- ☐ Disposable gloves
- ☐ Animal specimen from previous dissections
- ☐ Embalming fluid
- ☐ Three-dimensional models of the cut kidney and of a nephron (if available)
- ☐ Compound microscope
- ☐ Prepared slides of section of kidney (l.s.) and of bladder (x.s.)
- ☐ Organic debris container

OBJECTIVES

1. List the functions of the urinary system.
2. Identify, on a model or image, the urinary system organs, and state the general function of each.
3. Compare the course and length of the urethra in males and females.
4. Identify these regions of the dissected kidney (longitudinal section): hilum, cortex, medulla, medullary pyramids, major and minor calyces, pelvis, renal columns, and fibrous and perirenal fat capsules.
5. Trace the blood supply of the kidney from the renal artery to the renal vein.
6. Define *nephron,* and describe its anatomy.
7. Define *glomerular filtration, tubular resorption,* and *tubular secretion,* and indicate the nephron areas involved in these processes.
8. Define *micturition,* and explain the differences in the control of the internal and external urethral sphincters.
9. Recognize the histologic structure of the kidney and ureter microscopically or in an image.
10. Identify on a dissection specimen the urinary system organs, and describe the general function of each.

PRE-LAB QUIZ

1. Circle the correct underlined term. In its excretory role, the urinary system is primarily concerned with the removal of <u>carbon-containing</u> / <u>nitrogenous</u> wastes from the body.
2. Which structure(s) perform(s) the excretory and homeostatic functions of the urinary system?
   a. kidneys
   b. ureters
   c. urinary bladder
   d. all of the above
3. Circle the correct underlined term. The <u>cortex</u> / <u>medulla</u> of the kidney is segregated into triangular regions with a striped appearance.
4. Circle the correct underlined terms. As the renal artery approaches a kidney, it is divided into branches known as the <u>segmental arteries</u> / <u>afferent arterioles</u>.
5. What do we call the anatomical units responsible for the formation of urine? ______________________
6. This knot of coiled capillaries, found in the kidneys, forms the filtrate. It is the:
   a. arteriole
   b. glomerulus
   c. podocyte
   d. tubule

(Text continues on next page.)

7. The section of the renal tubule closest to the glomerular capsule is the:
   a. collecting duct
   b. distal convoluted tubule
   c. nephron loop
   d. proximal convoluted tubule
8. Circle the correct underlined term. The afferent / efferent arteriole drains the glomerular capillary bed.
9. Circle True or False. During tubular resorption, components of the filtrate move from the bloodstream into the tubule.
10. Circle the correct underlined term. The internal / external urethral sphincter consists of skeletal muscle and is voluntarily controlled.

Metabolism of nutrients by the body produces wastes including carbon dioxide, nitrogenous wastes, and ammonia that must be eliminated from the body if normal function is to continue. Excretory processes involve multiple organ systems, with the **urinary system** primarily responsible for the removal of nitrogenous wastes from the body. In addition to this excretory function, the kidney maintains the electrolyte, acid-base, and fluid balances of the blood and is thus a major, if not *the* major, homeostatic organ of the body.

To perform its functions, the kidney acts first as a blood filter and then as a filtrate processor. It allows toxins, metabolic wastes, and excess ions to leave the body in the urine, while simultaneously retaining needed substances and returning them to the blood. Malfunction of the urinary system, particularly of the kidneys, leads to a failure in homeostasis which, unless corrected, is fatal.

Gross Anatomy of the Human Urinary System

The urinary system **(Figure 28.1)** consists of the paired kidneys and ureters and the single urinary bladder and urethra. The **kidneys** perform the functions described above and manufacture urine in the process. The remaining organs of the system provide temporary storage reservoirs or transportation channels for urine.

ACTIVITY 1

Identifying Urinary System Organs

Examine the human torso model, a large anatomical chart, or a three-dimensional model of the urinary system to locate and study the anatomy and relationships of the urinary organs.

1. Locate the paired kidneys on the dorsal body wall in the superior lumbar region. Notice that they are not positioned at exactly the same level. Because it is crowded by the liver, the right kidney is slightly lower than the left kidney. Three layers of support tissue surround each kidney. Beginning with the innermost layer they are (1) a transparent *fibrous capsule,* (2) a *perirenal fat capsule,* and (3) the fibrous *renal fascia* that holds the kidneys in place in a retroperitoneal position.

In cases of rapid weight loss or in very thin individuals, the fat capsule may be reduced or scarce in amount. In such a case, the kidneys are less securely anchored, so they may drop to a more inferior position in the abdominal cavity. This phenomenon is called **ptosis.** +

2. Observe the **renal arteries** as they diverge from the descending aorta and plunge into the indented medial region, called the **hilum,** of each kidney. Note also the **renal veins,** which drain the kidneys, and the two **ureters,** which carry urine from the kidneys, moving it by peristalsis to the bladder for temporary storage.

3. Locate the **urinary bladder,** and observe the point of entry of the two ureters into this organ. Also locate the single **urethra,** which drains the bladder. The triangular region of the bladder that is delineated by the openings of the ureters and the urethra is referred to as the **trigone (Figure 28.2)**.

4. Follow the course of the urethra to the body exterior. In the male, it is approximately 20 cm (8 inches) long, travels the length of the **penis,** and opens at its tip. The urethra is made up of three named regions—the *prostatic urethra,* the *intermediate part,* and *spongy urethra* (described in more detail in Exercise 29). The male urethra has a dual function. It carries urine to the body exterior, and it provides a passageway for semen ejaculation. Thus, in the male, the urethra is part of both the urinary and reproductive systems. In females, the urethra is short, approximately 4 cm (1½ inches) long (see Figure 28.2). There are no common urinary-reproductive pathways in the female, and the female's urethra serves only to transport urine to the body exterior. Its external opening, the **external urethral orifice,** lies anterior to the vaginal opening. ■

DISSECTION

Gross Internal Anatomy of the Pig or Sheep Kidney

1. In preparation for dissection, don gloves. Obtain a preserved sheep or pig kidney, dissecting tray, and instruments. Observe the kidney to identify the **fibrous capsule,** a smooth transparent membrane that adheres tightly to the external aspect of the kidney.

2. Find the ureter, renal vein, and renal artery at the hilum (indented) region. The renal vein has the thinnest wall and

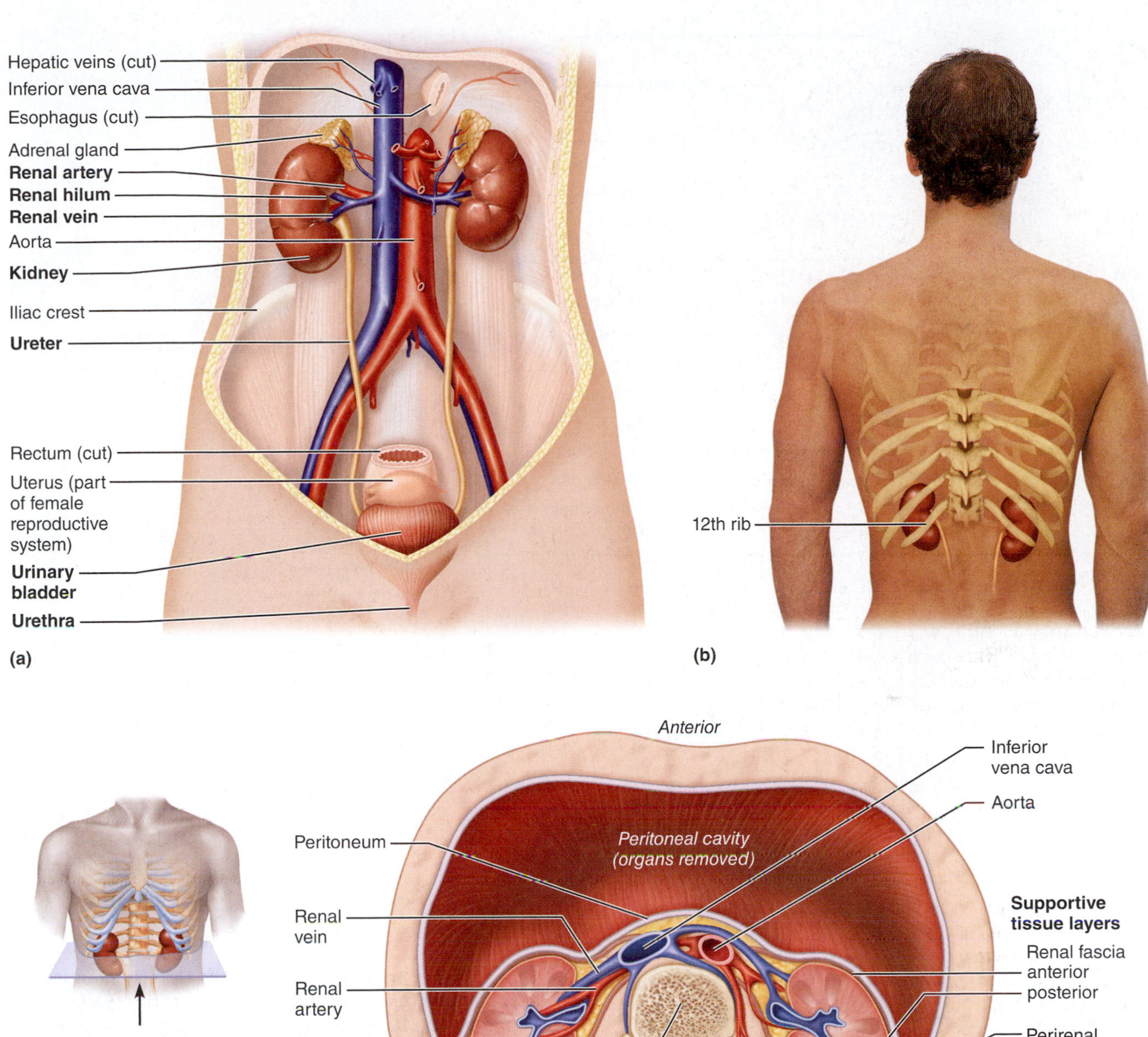

Figure 28.1 Organs of the urinary system. (a) Anterior view of the female urinary organs. Most unrelated abdominal organs have been removed. **(b)** Posterior in situ view showing the position of the kidneys relative to the twelfth ribs. **(c)** Cross section of the abdomen viewed from inferior direction. Note the retroperitoneal position and supportive tissue layers of the kidney.

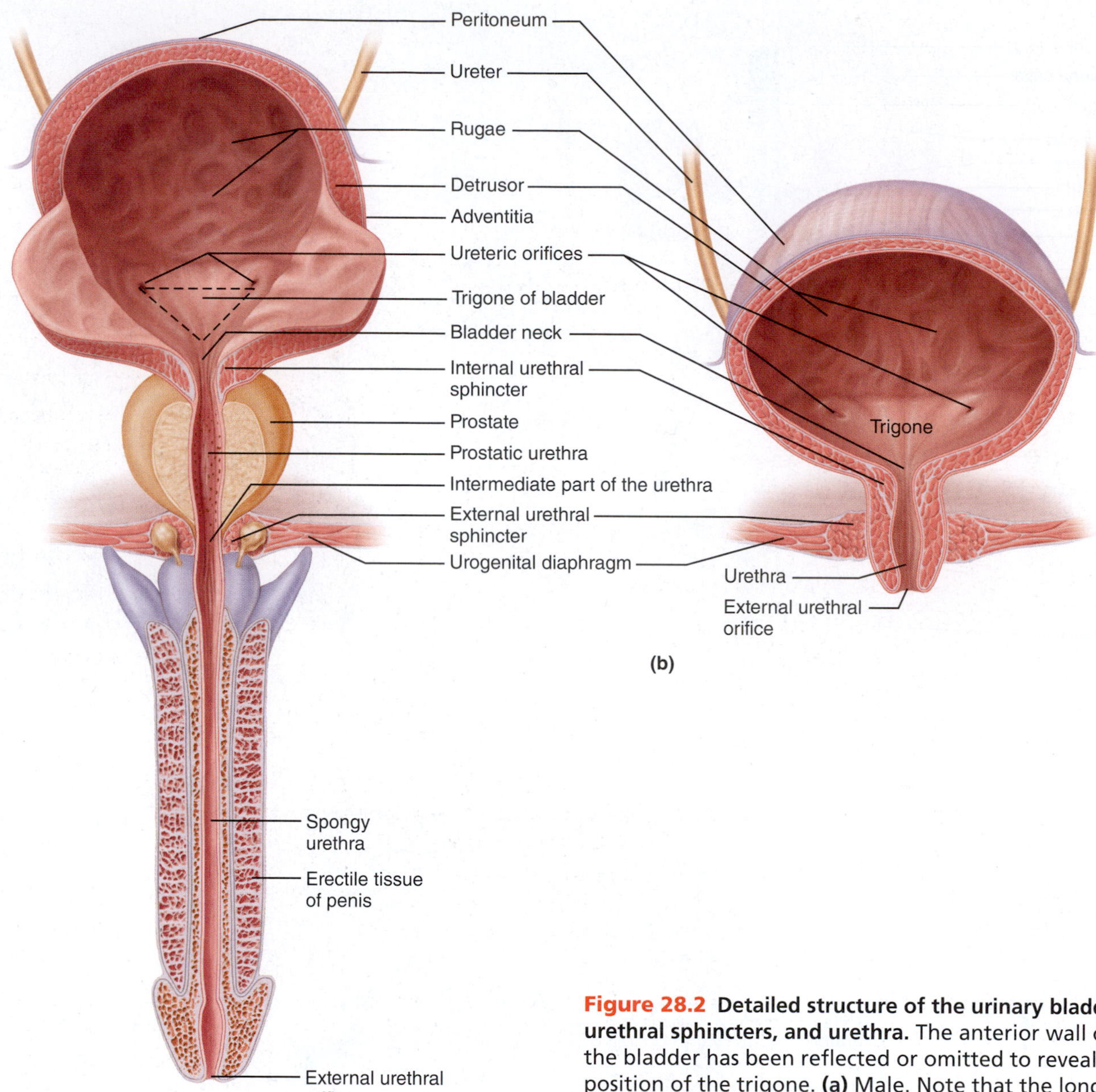

Figure 28.2 Detailed structure of the urinary bladder, urethral sphincters, and urethra. The anterior wall of the bladder has been reflected or omitted to reveal the position of the trigone. **(a)** Male. Note that the long male urethra has three regions: prostatic urethra, intermediate part, and spongy urethra. **(b)** Female.

will be collapsed. The ureter is the largest of these structures and has the thickest wall.

3. Section the kidney in the frontal plane, through the longitudinal axis and locate the anatomical areas described below and depicted in the illustration **(Figure 28.3)**.

Renal cortex: The superficial kidney region, which is lighter in color. If the kidney is doubly injected with latex, you will see a predominance of red and blue latex specks in this region indicating its rich vascular supply.

Medullary region: Deep to the cortex; a darker, reddish-brown color. The medulla is segregated into triangular regions that have a striped appearance—the **renal pyramids.** The base of each pyramid faces toward the cortex. Its more pointed *papilla,* or *apex,* points to the innermost kidney region.

Renal columns: Areas of tissue, more like the cortex in appearance, that dip inward between the pyramids separating them.

Renal pelvis: Extending inward from the hilum; a relatively flat, basinlike cavity that is continuous with the **ureter,** which exits from the hilum region. Fingerlike extensions of the pelvis should be visible. The larger, or primary, extensions are called the **major calyces** (singular: **calyx**); subdivisions of the major calyces are the **minor calyces.** Notice that the minor calyces terminate in cuplike areas that enclose the apexes of the medullary pyramids and collect urine draining from the pyramidal tips into the pelvis.

4. If the preserved kidney is doubly or triply injected, follow the renal blood supply from the renal artery to the *glomeruli.* The glomeruli appear as little red and blue specks in the cortex region.

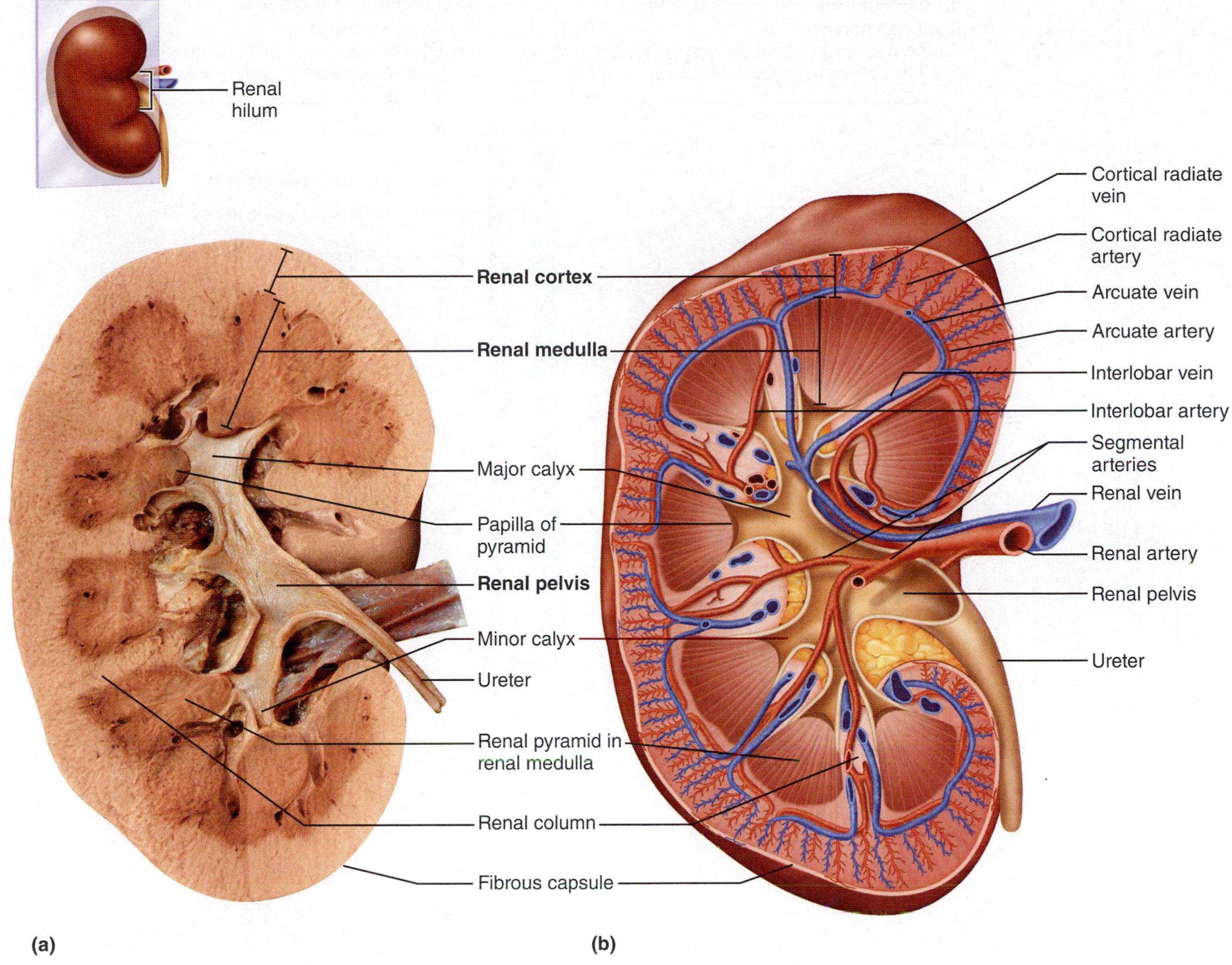

Figure 28.3 Internal anatomy of the kidney. Frontal section. **(a)** Photograph of a right kidney. **(b)** Diagram view showing the larger blood vessels supplying the kidney tissue.

Approximately a fourth of the total blood flow of the body is delivered to the kidneys each minute by the large **renal arteries.** As a renal artery approaches the kidney, it breaks up into branches called **segmental arteries,** which enter the hilum. Each segmental artery, in turn, divides into several **interlobar arteries,** which ascend toward the cortex in the renal columns. At the top of the medullary region, these arteries give off arching branches, the **arcuate arteries,** which curve over the bases of the medullary pyramids. Small **cortical radiate arteries** branch off the arcuate arteries and ascend into the cortex, giving off the individual **afferent arterioles (Figure 28.4)**, which lead to the capillary beds associated with the nephrons, the functional units of the kidney. Blood draining from the nephron capillary beds enters the **cortical radiate veins** and then drains through the **arcuate veins** and the **interlobar veins** to finally enter the **renal vein** in the pelvis region. There are no segmental veins.

Dispose of the kidney specimen as your instructor specifies. ■

Functional Microscopic Anatomy of the Kidney and Bladder

Kidney

Each kidney contains over a million **nephrons,** which are the structural and functional units responsible for filtering the blood and forming urine. The detailed structure and the relative positioning of the nephrons in the kidney is depicted in the illustration (Figure 28.4).

Each nephron consists of two major structures: a *renal corpuscle* and a *renal tubule*. All of the **renal corpuscles** are located in the cortex of the kidney. Each consists of a cuplike structure called the **glomerular** (or **Bowman's**) **capsule,** which contains a cluster of capillaries called the **glomerulus.** This special capillary bed resembles a ball of yarn. The capsule has a visceral layer of **podocytes** that have long branching *foot processes*. The foot processes interdigitate and cling

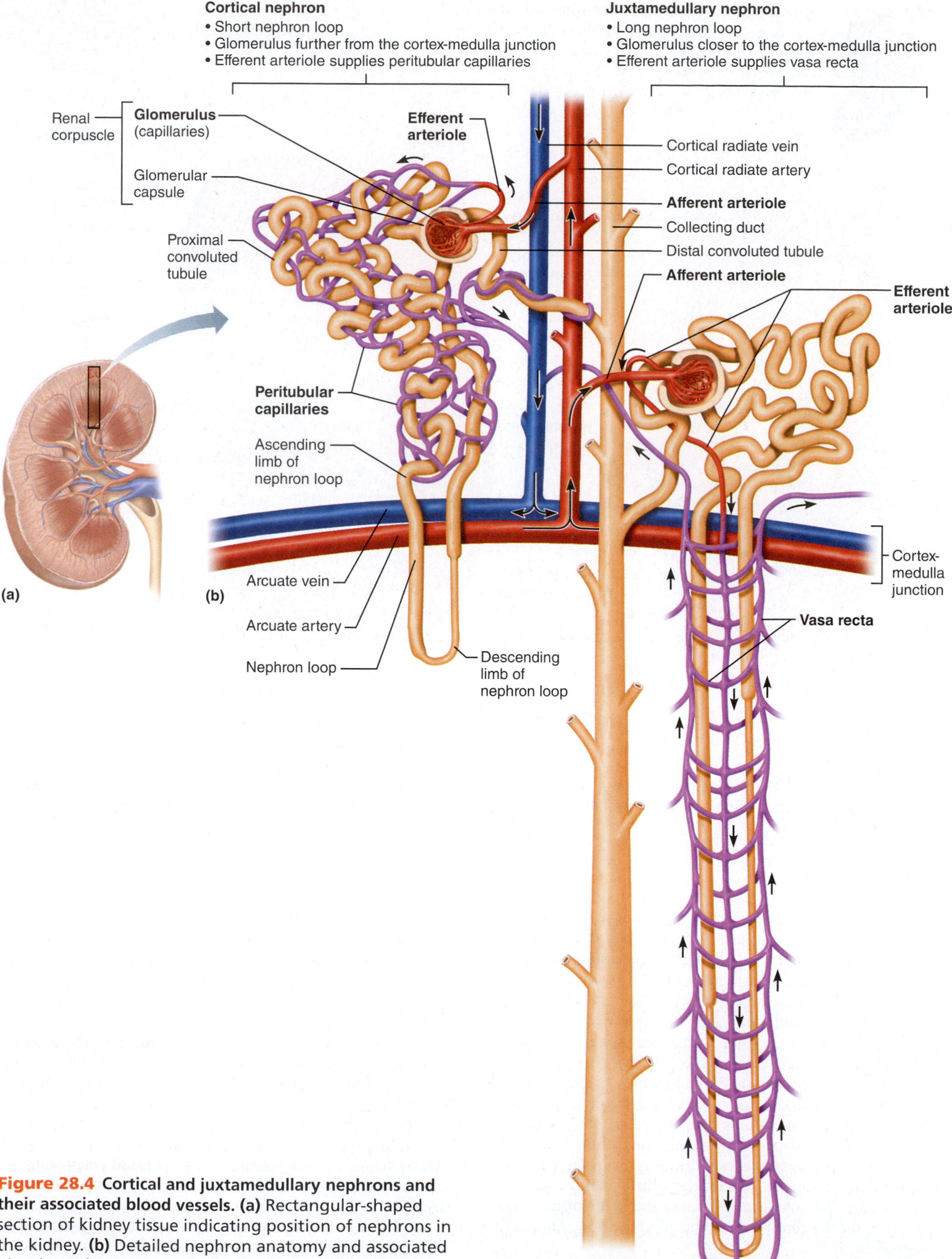

Figure 28.4 Cortical and juxtamedullary nephrons and their associated blood vessels. (a) Rectangular-shaped section of kidney tissue indicating position of nephrons in the kidney. **(b)** Detailed nephron anatomy and associated blood supply. Arrows indicate direction of blood flow.

to the basement membrane of the glomerulus, forming part of the filtration membrane. Blood in the glomerulus is filtered into the capsule to begin its journey along the renal tubule.

The **renal tubule** exits from the glomerular capsule to form the highly coiled **proximal convoluted tubule (PCT)** that then becomes the descending limb of the hairpin-like **nephron loop.** The ascending limb of the nephron loop leads to the coiled **distal convoluted tubule (DCT)** that empties into a collecting duct. A single layer of epithelial cells resting on a basement membrane forms the walls of both the renal tubule and the collecting duct. The cells of each part of the tubule and collecting duct are specialized to perform a particular function in processing filtrate. The cuboidal cells of the PCT have dense microvilli that greatly increase the surface area exposed to filtrate in the lumen. They are specialized for resorption. In contrast, the cuboidal cells of the DCT have far fewer microvilli, reflecting the decreased role of resorption in this part of the tubule.

There are two kinds of nephrons, cortical and juxtamedullary (see Figure 28.4). **Cortical nephrons** are most numerous, making up about 85% of nephrons. They are located almost entirely within the renal cortex, except for small parts of their nephron loops that dip into the renal medulla. The renal corpuscles of **juxtamedullary nephrons** are located deep in the cortex at the border with the medulla; their long nephron loops penetrate deeply into the medulla. Juxtamedullary nephrons play an important role in concentrating urine.

Filtrate leaving the renal tubule enters a **collecting duct,** where its volume and concentration can be modified. Each collecting duct receives filtrate from many nephrons. The ducts travel through the medullary pyramids, giving them a striped appearance, and then fuse near the renal pelvis. They deliver urine into the minor calyces via the papillae of the pyramids.

The function of the nephron depends on several unique features of the renal circulation **(Figure 28.5)**. The capillary vascular supply consists of two distinct capillary beds, the *glomerulus* and the *peritubular capillary bed.* Vessels leading to and from the glomerulus, the first capillary bed, are both glomerular arterioles: the **afferent arteriole** feeds the bed while the **efferent arteriole** drains it. The glomerular capillary is unique in the body. It is a high-pressure bed along its entire length. Its high pressure is a result of two major factors: (1) the bed is *fed and drained* by arterioles, and (2) the afferent feeder arteriole is larger in diameter than the efferent arteriole draining the bed. The high hydrostatic pressure created by these two anatomical features forces fluid and blood components smaller than proteins out of the glomerulus into the glomerular capsule. That is, it forms the filtrate, which is processed by the nephron tubule.

The **peritubular capillary bed** arises from the efferent arteriole draining the glomerulus. This set of capillaries clings intimately to the renal tubule and empties into the cortical radiate veins that leave the cortex. The peritubular capillaries are *low-pressure* porous capillaries adapted for absorption rather than filtration and readily take up the solutes and water resorbed from the filtrate by the tubule cells. Efferent arterioles that supply juxtaglomerular nephrons tend not to form peritubular capillaries. Instead, they form long, straight, highly interconnected vessels called **vasa recta** that run parallel and very close to the long nephron loops. This vasa recta is essential in concentrating urine.

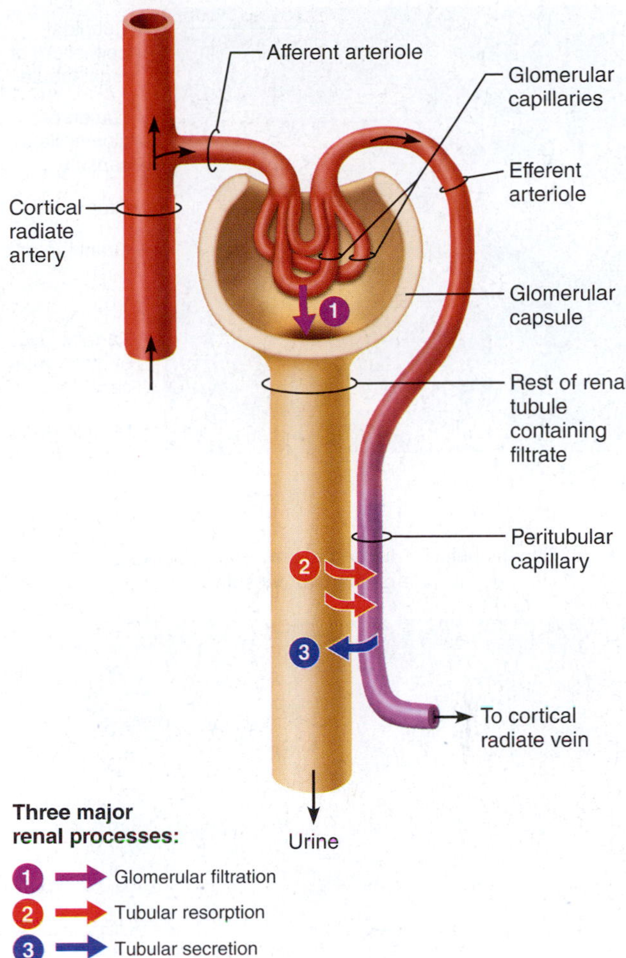

Figure 28.5 A schematic uncoiled nephron. A kidney actually has millions of nephrons acting in parallel. The three major mechanisms by which the kidneys adjust the composition of plasma are **(a)** glomerular filtration, **(b)** tubular resorption, and **(c)** tubular secretion. Black arrows show the path of blood flow through the renal microcirculation.

Each nephron also has a **juxtaglomerular complex (JGC)** **(Figure 28.6)** located where the most distal portion of the ascending limb of the nephron loop touches the afferent arteriole. Helping to form the JGC are (1) *granular cells,* also called *juxtaglomerular (JG) cells,* in the arteriole walls that sense blood pressure in the afferent arteriole; and (2) a group of columnar cells in the nephron loop called the *macula densa* that monitors sodium chloride (NaCl) concentration in the filtrate. The role of the JGC is to regulate the rate of filtration and systemic blood pressure.

Urine forms as a result of three processes: *filtration, resorption,* and *secretion* (see Figure 28.5). **Filtration,** the role of the glomerulus, is largely a passive process in which a portion of the blood passes from the glomerular capillary into the glomerular capsule. This filtrate then enters the proximal convoluted tubule, where tubular resorption and secretion begin. During **tubular resorption,** many of the filtrate components move through the tubule cells and return to the blood

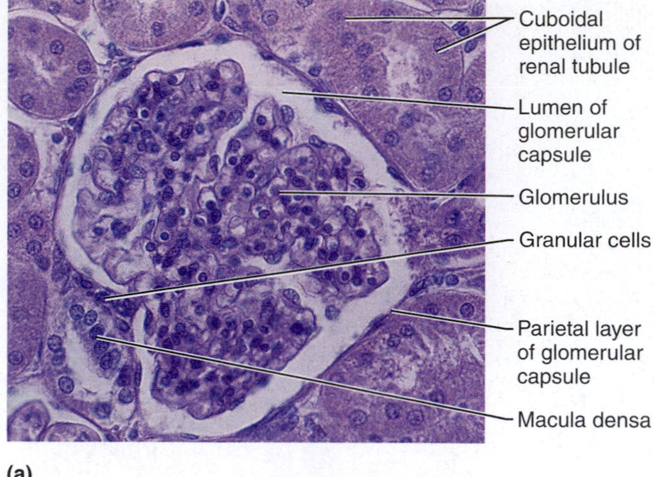

(a)

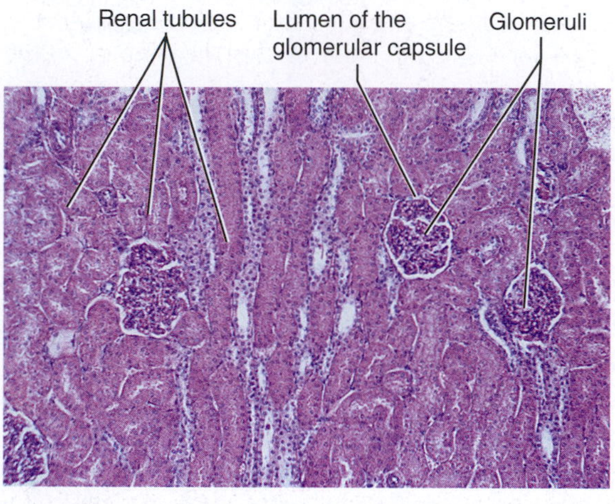

(b)

Figure 28.6 Microscopic structure of kidney tissue. **(a)** Detailed structure of the glomerulus (300×). **(b)** Low-power view of the renal cortex (75×).

in the peritubular capillaries. Some of this resorption is passive—such as that of water, which passes by osmosis—but the resorption of most substances depends on active transport processes and is highly selective. Which substances are resorbed at a particular time depends on the composition of the blood and the needs of the body at that time. Substances that are almost entirely resorbed from the filtrate include water, glucose, and amino acids. Various ions are selectively resorbed or allowed to go out in the urine according to what is required to maintain appropriate blood pH and electrolyte composition. Waste products, including urea, creatinine, uric acid, and drug metabolites, are resorbed to a much lesser degree or not at all. Most (75% to 80%) of tubular resorption occurs in the proximal convoluted tubule.

Tubular secretion is essentially the reverse process of tubular resorption. Substances such as hydrogen and potassium ions and creatinine move from the blood of the peritubular capillaries through the tubular cells into the filtrate to be disposed of in the urine. This process is particularly important for disposing of substances, such as drug metabolites, that are not already in the filtrate and as a device for controlling blood pH.

ACTIVITY 2

Studying Nephron Structure

1. Begin your study of nephron structure by identifying the glomerular capsule, proximal and distal convoluted tubule regions, and the nephron loop on a model of the nephron. Then, obtain a compound microscope and a prepared slide of kidney tissue to continue with the microscope study of the kidney.

2. Hold the longitudinal section of the kidney up to the light to identify the cortical and medullary areas. Then secure the slide on the microscope stage, and scan the slide under low power.

3. Move the slide so that you can see the cortical area. Identify a glomerulus, which appears as a ball of tightly packed material containing many small nuclei (Figure 28.6). It is usually surrounded by a vacant-appearing region (the *capsular space*) between the visceral and parietal layers of the glomerular capsule that surrounds it.

4. Notice that the renal tubules are cut at various angles. Try to differentiate the fuzzy cuboidal epithelium of the proximal convoluted tubule, which has dense microvilli, from the epithelium of the distal convoluted tubule with sparse microvilli. Also identify the thin-walled nephron loop. ■

Bladder

Although urine is produced by the kidney continuously, it is usually removed from the body only when voiding is convenient. In the meantime, the **urinary bladder,** which receives urine via the ureters and discharges it via the urethra, stores it temporarily.

Voiding, or **micturition,** is the act of emptying the bladder. Two sphincter muscles, or valves (see Figure 28.2), the **internal urethral sphincter** and the **external urethral sphincter,** control the outflow of urine from the bladder. Ordinarily, the bladder continues to collect urine until about 200 ml have accumulated, at which time the stretching of the bladder wall activates stretch receptors. Impulses transmitted to the central nervous system subsequently produce reflex contractions of the bladder wall through parasympathetic nervous system pathways via the pelvic splanchnic nerves. As contractions increase in force and frequency, stored urine is forced past the internal sphincter, which is a smooth muscle involuntary sphincter, into the superior part of the urethra. It is then that a person feels the urge to void. The inferior external sphincter consists of skeletal muscle and is voluntarily controlled. If it is not convenient to void, the opening of this sphincter can be inhibited. Conversely, if the time is convenient, the sphincter may be relaxed and the stored urine flushed from the body. If voiding is inhibited, the reflex contractions of the bladder cease temporarily, and urine continues to accumulate in the bladder. After another 200 to 300 ml of urine have been collected, the *micturition reflex* will again be initiated.

Lack of voluntary control over the external sphincter is referred to as **incontinence.** Incontinence is normal in children 2 years old or younger, because they have not yet gained control over the voluntary sphincter. In adults and older children, incontinence generally results from spinal cord injury, emotional problems, bladder irritability, or some other disorder of the urinary tract. ✚

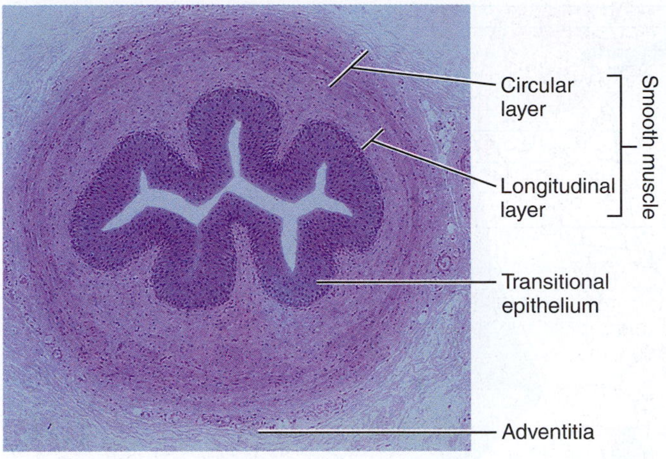

Figure 28.7 Structure of the ureter wall. Cross section of ureter (30×).

ACTIVITY 3

Studying Bladder Structure

1. Return the kidney slide to the supply area, and obtain a slide of bladder tissue. Scan the bladder tissue. Identify its three layers: mucosa, muscular layer, and fibrous adventitia.

2. Study the highly specialized transitional epithelium of the mucosa. The plump, transitional epithelial cells have the ability to slide over one another, thus decreasing the thickness of the mucosa layer as the bladder fills and stretches to accommodate the increased urine volume. Depending on the degree of stretching of the bladder, the mucosa may be three to eight cell layers thick. (Compare the transitional epithelium of the mucosa to that shown in Figure 5.3h, page 59.)

3. Examine the heavy muscular wall (detrusor), which consists of three irregularly arranged muscular layers. The innermost and outermost muscle layers are arranged longitudinally; the middle layer is arranged circularly. Attempt to differentiate the three muscle layers.

4. Compare the structure of the bladder wall you are observing to the structure of the ureter wall shown in the photomicrograph **(Figure 28.7)**. How are the two organs similar histologically?

____

____

What is/are the most obvious difference(s)?

____

______________________________ ■

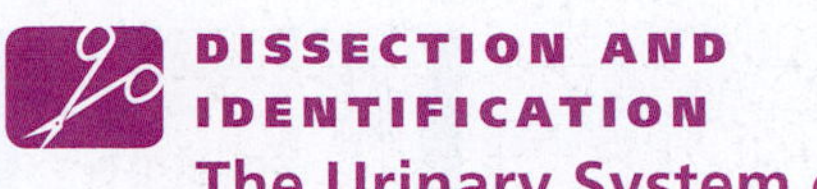

DISSECTION AND IDENTIFICATION

The Urinary System of the Cat

The structures of the reproductive and urinary systems are often considered together as the *urogenital system,* because they have common embryologic origins. However, the emphasis in this dissection is on identifying the structures of the urinary tract (**Figure 28.8** and **Figure 28.9**), with only a few references to contiguous reproductive structures. The anatomy of the reproductive system is studied in a separate exercise (the cat dissection section in Exercise 29). ■

ACTIVITY 4

Identifying Organs of the Urinary System

1. Don gloves. Obtain your dissection specimen, and place it ventral side up on the dissection tray. Reflect the abdominal viscera (in particular, the small intestine) to locate the kidneys high on the dorsal body wall. Note that the **kidneys** in the cat, as well as in the human, are retroperitoneal (behind the peritoneum).

2. Carefully remove the peritoneum, and clear away the bed of fat that invests the kidneys. Then locate the adrenal glands that lie superiorly and medial to the kidneys.

3. Identify the **renal artery** (red latex injected), the **renal vein** (blue latex injected), and the ureter at the hilum region of the kidney. You may find two renal veins leaving one kidney in the cat but not in humans.

4. To observe the gross internal anatomy of the kidney, slit the connective tissue *fibrous capsule* encasing a kidney, and peel it back. Make a midfrontal cut through the kidney, and examine one cut surface with a hand lens to identify the granular *cortex* and the central darker *medulla,* which will appear striated. Notice that the cat's renal medulla consists of just one pyramid, in contrast to the multipyramidal human kidney.

5. Trace the **ureters** to the **urinary bladder,** a smooth muscular sac located superiorly to the small intestine. If your cat is a female, be careful not to confuse the ureters with the uterine tubes, which lie superior to the bladder in the same general region (Figure 28.8). Observe the sites where the ureters enter the bladder. How would you describe the entrance point anatomically?

____

____

6. Cut through the bladder wall, and examine the region where the **urethra** exits to see whether you can discern any evidence of the *internal urethral sphincter.*

7. If your cat is a male, identify the prostate (part of the male reproductive system), which encircles the urethra distal to

(Text continues on page 496.)

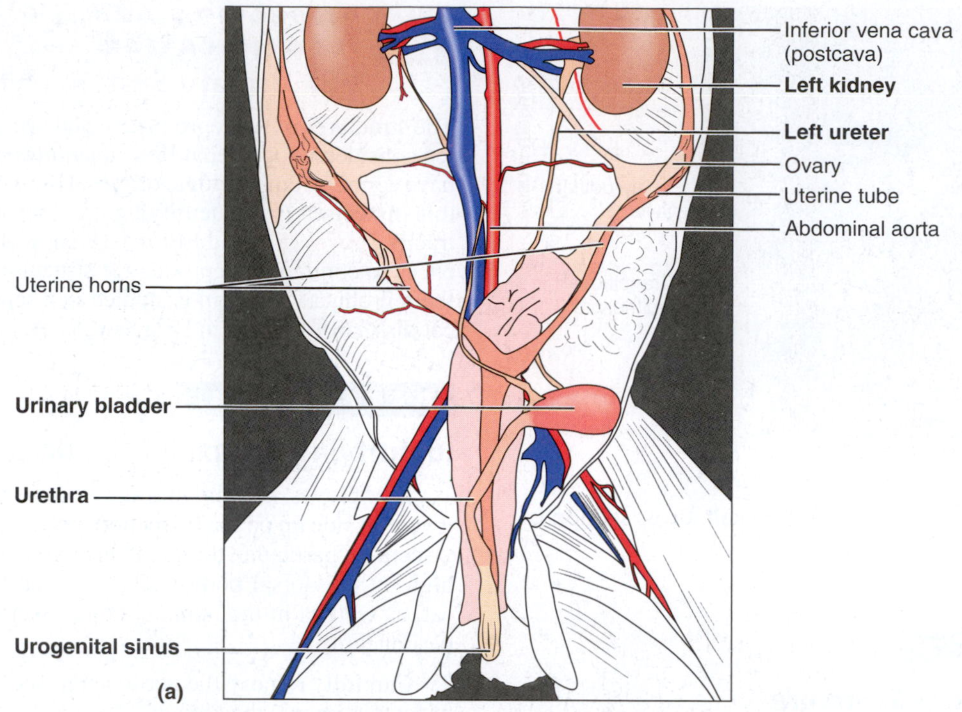

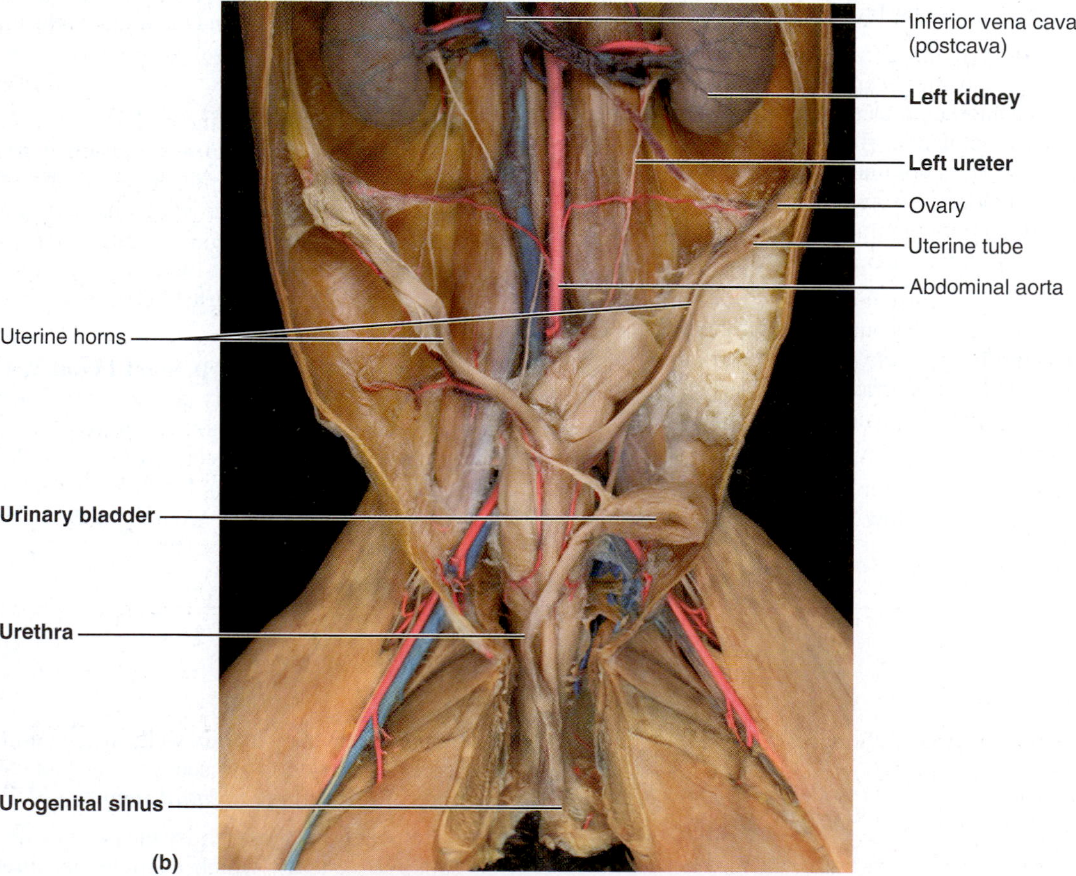

Figure 28.8 Urinary system of the female cat. (Reproductive structures are also indicated.) **(a)** Diagram view. **(b)** Photograph of female urogenital system.

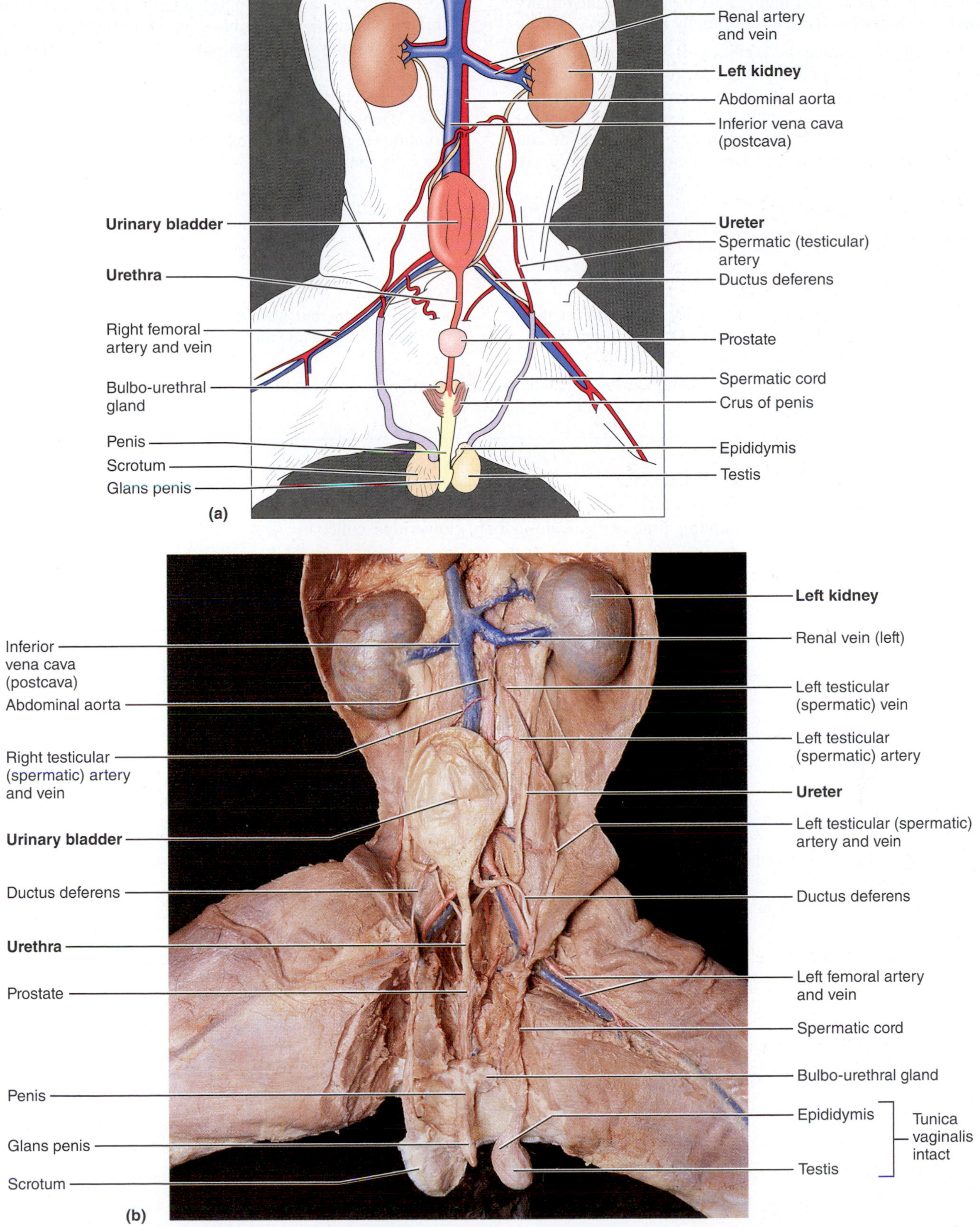

Figure 28.9 Urinary system of the male cat. (Reproductive structures are also indicated.) **(a)** Diagram view. **(b)** Photograph of male urogenital system.

the neck of the bladder (Figure 28.9). Notice that the urinary bladder is somewhat fixed in position by ligaments.

8. Using a probe, trace the urethra as it exits the bladder. In the male, the urethra goes through the prostate and enters the penis. In the female cat, the urethra terminates in the **urogenital sinus,** a common chamber into which both the vagina and the urethra empty. In the human female, the vagina and the urethra have separate external openings. Dissection to expose the urethra along its entire length should not be done at this time because of possible damage to the reproductive structures, which you may study in a separate exercise (Exercise 29).

9. To complete this exercise, observe a cat of the opposite sex. Before you leave the laboratory, follow the boxed instructions to prepare your cat for storage and clean the area (page 217). ■

GROUP CHALLENGE

Urinary System Sequencing

Arrange the following sets of urinary structures in the correct order for the flow of urine, filtrate, or blood. Work in small groups, but refrain from using a figure or other reference to determine the sequences.

1. renal pelvis, minor calyx, renal papilla, urinary bladder, ureter, major calyx, and urethra

____

____

2. distal convoluted tubule, ascending limb of the nephron loop, glomerulus, collecting duct, descending limb of the nephron loop, proximal convoluted tubule, and glomerular capsule

____

____

3. segmental artery, afferent arteriole, cortical radiate artery, glomerulus, renal artery, interlobar artery, and arcuate artery ____

____

4. arcuate vein, inferior vena cava, peritubular capillaries, renal vein, interlobar vein, cortical radiate vein, and efferent arteriole ____

____

■

Name ______________________

Lab Time/Date ______________________

EXERCISE 28

REVIEW SHEET

Anatomy of the Urinary System

Gross Anatomy of the Human Urinary System

1. Complete the following statements.

The kidney is referred to as an excretory organ because it excretes __1__ wastes. It is also a major homeostatic organ because it maintains the electrolyte, __2__, and __3__ balance of the blood.

Urine is continuously formed by the __4__ and is routed down the __5__ by the mechanism of __6__ to a storage organ called the __7__. Eventually, the urine is conducted to the body __8__ by the urethra. In the male, the urethra is __9__ centimeters long and transports both urine and __10__. The female urethra is __11__ centimeters long and transports only urine.

Voiding, or emptying the bladder, is called __12__. Voiding has both voluntary and involuntary components. The voluntary sphincter is the __13__ sphincter. An inability to control this sphincter is referred to as __14__.

1. ______________________
2. ______________________
3. ______________________
4. ______________________
5. ______________________
6. ______________________
7. ______________________
8. ______________________
9. ______________________
10. ______________________
11. ______________________
12. ______________________
13. ______________________
14. ______________________

2. What is the function of the fat cushion that surrounds the kidneys in life? ______________________

____

3. Define *ptosis*. ______________________

____

4. Why is incontinence a normal phenomenon in the child under 1½ to 2 years old? ______________________

____

In the adult what events may lead to incontinence? ______________________

____

____

9. Explain *why* the glomerulus is such a high-pressure capillary bed. ______

How does the high pressure help the glomerulus carry out its function of forming filtrate? ______

10. What structural modification of certain tubule cells enhances their ability to resorb substances from the filtrate?

______

11. Explain the mechanism of tubular secretion, and explain its importance in the urine formation process. ______

12. Compare and contrast the composition of blood plasma and glomerular filtrate. ______

13. Define *juxtaglomerular complex*. ______

14. Label the figure using the key letters of the correct terms.

Key: a. granular cells

b. cuboidal epithelium

c. macula densa

d. glomerular capsule (parietal layer)

e. ascending limb of nephron loop

15. What is important functionally about the specialized epithelium (transitional epithelium) in the bladder?

______

Dissection and Identification: The Urinary System of the Cat

16. How does the position of the kidneys in the cat differ from their position in humans? ______________________

____

____

____

____

17. How does the site of urethral emptying in the female cat differ from its termination point in the human female?

____

____

____

18. What gland encircles the neck of the bladder in the male? ______________________

Is this part of the urinary system? ______________________

What is its function? ______________________

____

____

5. The prostate, seminal glands, and bulbo-urethral glands produce __________, the liquid medium in which sperm leaves the body.
   a. seminal fluid c. urine
   b. testosterone d. water
6. Circle the correct underlined term. The interstitial endocrine cells / seminiferous tubules produce testosterone, the hormonal product of the testis.
7. The endocrine products of the ovaries are estrogen and:
   a. luteinizing hormone c. prolactin
   b. progesterone d. testosterone
8. The __________ is a pear-shaped organ that houses the embryo or fetus during its development.
   a. bladder c. uterus
   b. cervix d. vagina
9. Circle the correct underlined term. The endometrium / myometrium, or the thick mucosal lining of the uterus, has a superficial layer that sloughs off periodically.
10. Circle the correct underlined term. A developing egg is ejected from the ovary at the appropriate stage of maturity in an event known as menstruation / ovulation.

Most organ systems of the body function from the time they are formed to sustain the existing individual. However, the **reproductive system** begins its biological function, the perpetuation of the species, at puberty.

The essential organs of reproduction are the **gonads,** the testes and the ovaries, which produce the sex cells or **gametes** and the sex hormones. The reproductive role of the male is to manufacture sperm and to deliver them to the female reproductive tract. The female, in turn, produces eggs. If the time is suitable, the combination of sperm and egg produces a fertilized egg, which is the first cell of a new individual. Once fertilization has occurred, the female uterus provides a nurturing, protective environment in which the embryo, later called the *fetus*, develops until birth.

Gross Anatomy of the Human Male Reproductive System

The primary reproductive organs of the male are the **testes,** which produce sperm and the male sex hormones. All other reproductive structures are conduits or sources of secretions that help deliver the sperm safely to the body exterior or female reproductive tract.

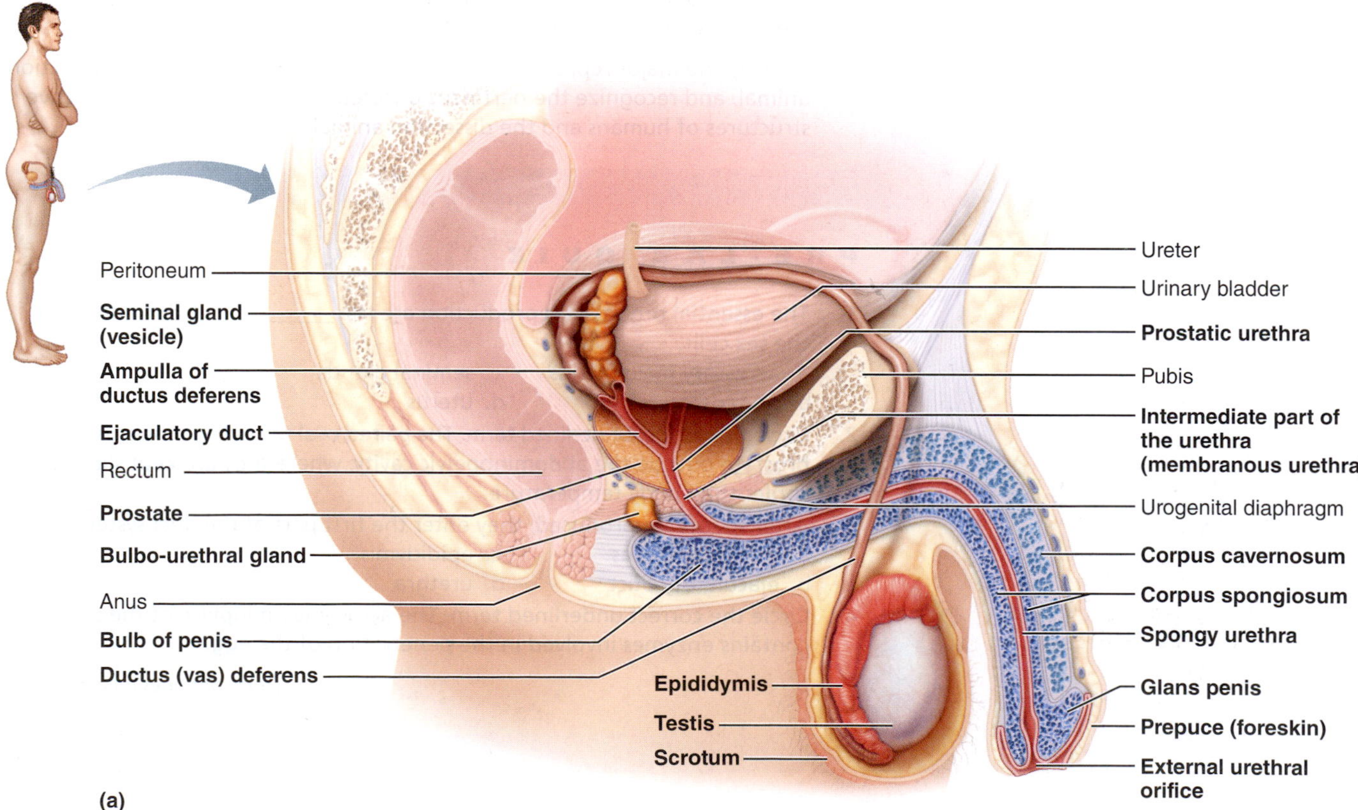

Figure 29.1 Reproductive organs of the human male. (a) Sagittal view.

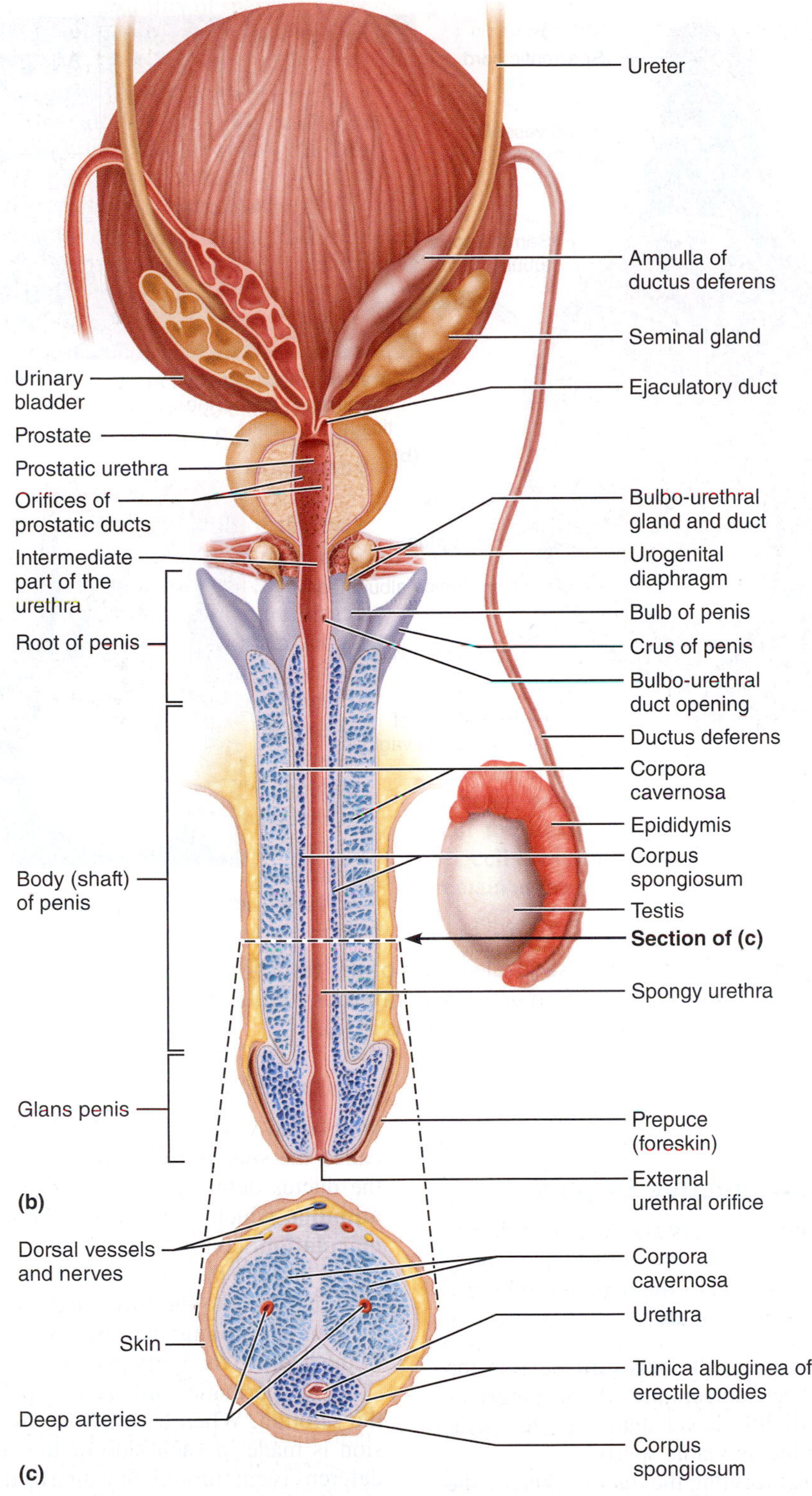

Figure 29.1 *(continued)* **Reproductive organs of the human male. (b)** Posterior view showing a longitudinal section of the penis. **(c)** Transverse section of the penis.

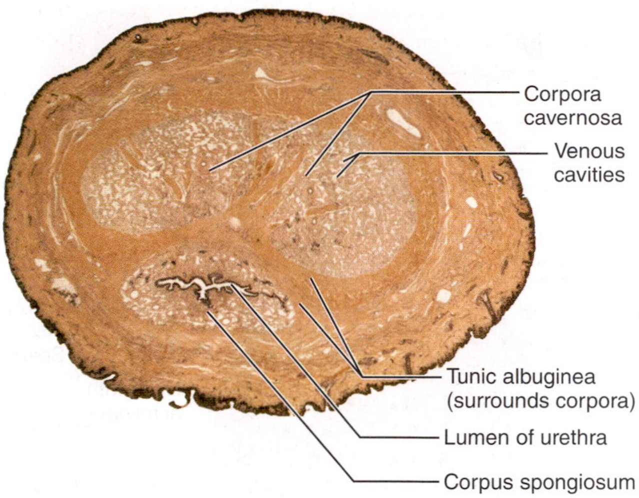

Figure 29.4 Transverse section of the penis (3×).

Penis

Obtain a cross section of the penis. Scan the tissue under low power to identify the urethra and the cavernous bodies. Compare your observations to the illustration (Figure 29.1c) and photomicrograph **(Figure 29.4)**. Observe the lumen of the urethra carefully. What type of epithelium do you see?

____

Explain the function of this type of epithelium.

____

____

____

Seminal Gland

Obtain a slide showing a cross-sectional view of the seminal gland. Examine the slide at low magnification to get an overall view of the highly folded mucosa of this gland. Switch to higher magnification, and notice that the folds of the gland protrude into the lumen where they divide further, giving the lumen a honeycomb look **(Figure 29.5)**. Notice that the loose connective tissue lamina propria is underlain by smooth muscle fibers—first a circular layer, and then a longitudinal layer. Identify the glandular secretion in the lumen, a viscous substance that is rich in fructose and prostaglandins.

Epididymis

Obtain a cross section of the epididymis. Notice the abundant tubule cross sections resulting from the fact that the coiling epididymis tubule has been cut through many times in the specimen. Using the photomicrographs (Figure 29.2b and **Figure 29.6**) as guides, look for sperm in the lumen of the tubule. Examine the composition of the tubule wall carefully. Identify the *stereocilia* of the pseudostratified columnar epithelial lining. These nonmotile microvilli absorb excess fluid and pass nutrients to the sperm in the lumen. Now identify the smooth muscle layer. What do you think the function of the smooth muscle is?

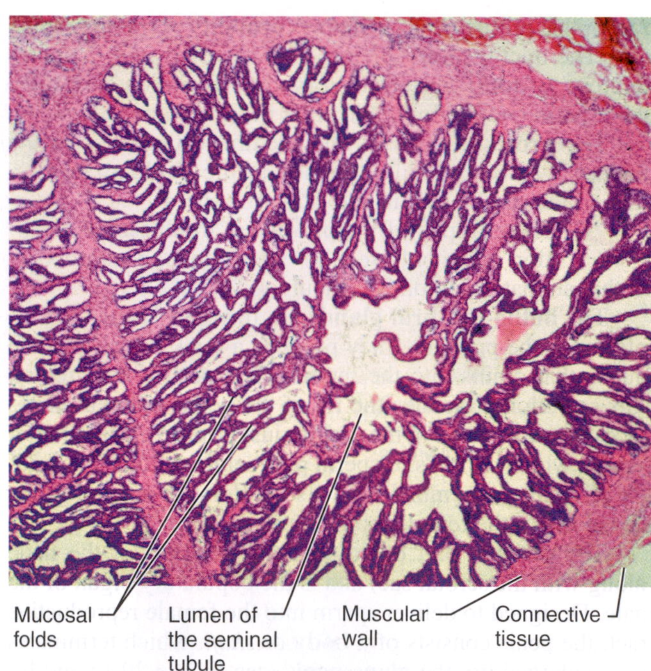

Figure 29.5 Histology of a seminal gland. Cross-sectional view showing the elaborate network of mucosal folds (12×). The glandular secretion is seen in the lumen.

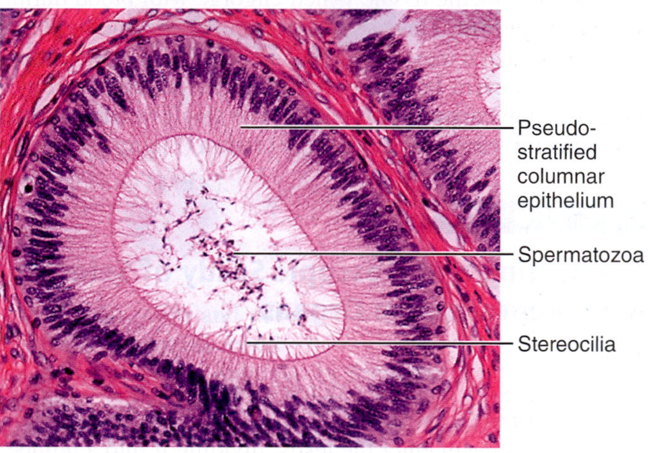

Figure 29.6 Cross section of epididymis (185×).

____

____

____

Gross Anatomy of the Human Female Reproductive System

The **ovaries** (female gonads) are the primary reproductive organs of the female. Like the testes of the male, the ovaries produce gametes (eggs, or ova) and also sex hormones (estrogens and progesterone). The other accessory structures

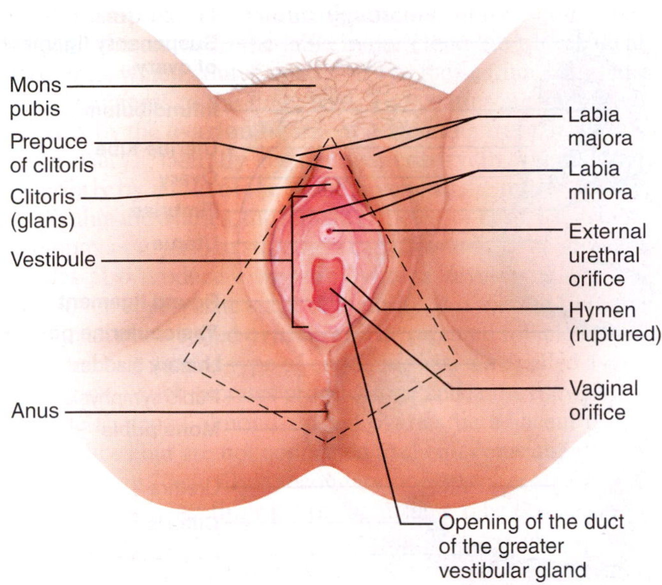

Figure 29.7 External genitalia of the human female. (The region enclosed by dashed lines is the perineum.)

of the female reproductive system transport, house, nurture, or otherwise serve the needs of the reproductive cells and/or the developing fetus.

The reproductive structures of the female are generally considered in terms of internal organs and external organs, or external genitalia.

ACTIVITY 3

Identifying Female Reproductive Organs

As you read the descriptions of these structures, locate them on the illustrations of external and internal female structures (**Figure 29.7** and **Figure 29.8**) and then on the female reproductive system model or large laboratory chart. ■

External Genitalia

The **external genitalia (vulva)** consist of the mons pubis, the labia majora and minora, the clitoris, the external urethral and vaginal orifices, the hymen, and the greater vestibular glands. The **mons pubis** is a rounded fatty eminence overlying the pubic symphysis. Running inferiorly and posteriorly from the mons pubis are two elongated, pigmented, hair-covered skin folds, the **labia majora,** which are homologous to the scrotum of the male. These enclose two smaller hair-free folds, the **labia minora.** (Terms indicating only one of the two folds in each case are *labium majus* and *minus,* respectively.) The labia minora, in turn, enclose a region called the **vestibule,** which contains many structures—the clitoris, most anteriorly, followed by the external urethral orifice and the vaginal orifice. The diamond-shaped region between the anterior end of the labial folds, the ischial tuberosities laterally, and the anus posteriorly is called the **perineum.**

The **clitoris** is a small protruding structure, homologous to the male penis. Like its counterpart, it is composed of highly sensitive, erectile tissue. It is hooded by skin folds of the anterior labia minora, referred to as the **prepuce of the clitoris.** The external urethral orifice, which lies posterior to the clitoris, is the outlet for the urinary system and has no reproductive function in the female. The vaginal opening is partially closed by a thin fold of mucous membrane called the **hymen** and is flanked by the pea-sized, mucus-secreting **greater vestibular glands.** These glands (see Figure 29.7 and Figure 29.8a) lubricate the distal end of the vagina during sexual intercourse and are homologous to the bulbo-urethral glands of males.

Internal Genitalia

The internal female organs include the vagina, uterus, uterine tubes, ovaries, and the ligaments and supporting structures that suspend these organs in the pelvic cavity (Figure 29.8). The **vagina** extends for approximately 10 cm (4 inches) from the vestibule to the uterus superiorly. It serves as a copulatory organ and birth canal and permits passage of the menstrual flow. The pear-shaped **uterus,** situated between the bladder and the rectum, is a muscular organ with its narrow end, the **cervix,** directed inferiorly. The major portion of the uterus is referred to as the **body;** its superior rounded region above the entrance of the uterine tubes is called the **fundus.** A fertilized egg is implanted in the uterus, which houses the embryo or fetus during its development.

In some cases, the fertilized egg may implant in a uterine tube or even on the abdominal viscera, creating an **ectopic pregnancy.** Such implantations are usually unsuccessful and may even endanger the mother's life because the uterine tubes cannot accommodate the increasing size of the fetus. +

The **endometrium,** the thick mucosal lining of the uterus, has a superficial **functional layer,** or **stratum functionalis,** that sloughs off periodically (about every 28 days) in response to cyclic changes in the levels of ovarian hormones in the woman's blood. This sloughing-off process, which is accompanied by bleeding, is referred to as **menstruation,** or **menses.** The deeper **basal layer,** or **stratum basalis,** forms a new functionalis after menstruation ends.

The **uterine,** or **fallopian, tubes** are about 10 cm (4 inches) long and extend from the ovaries in the peritoneal cavity to the superolateral region of the uterus. The distal ends of the tubes are funnel-shaped and have fingerlike projections called **fimbriae.** Unlike the male duct system, there is no actual contact between the female gonad and the initial part of the female duct system—the uterine tube.

Because of this open passageway between the female reproductive organs and the peritoneal cavity, reproductive system infections, such as gonorrhea and other **sexually transmitted infections (STIs),** also called **sexually transmitted diseases (STDs),** can cause widespread inflammations of the pelvic viscera, a condition called **pelvic inflammatory disease (PID).** +

The internal female organs are all retroperitoneal, except the ovaries. They are supported and suspended somewhat freely by ligamentous folds of peritoneum. The peritoneum takes an undulating course. From the pelvic cavity floor it moves superiorly over the top of the bladder and reflects over the anterior and posterior surfaces of the uterus, then over the rectum, and up the posterior body wall. The fold that encloses the uterine tubes and uterus and secures them to the lateral body walls is the **broad ligament** (Figure 29.8b). The part of the broad ligament specifically anchoring the uterus is called the **mesometrium** and that anchoring the uterine tubes,

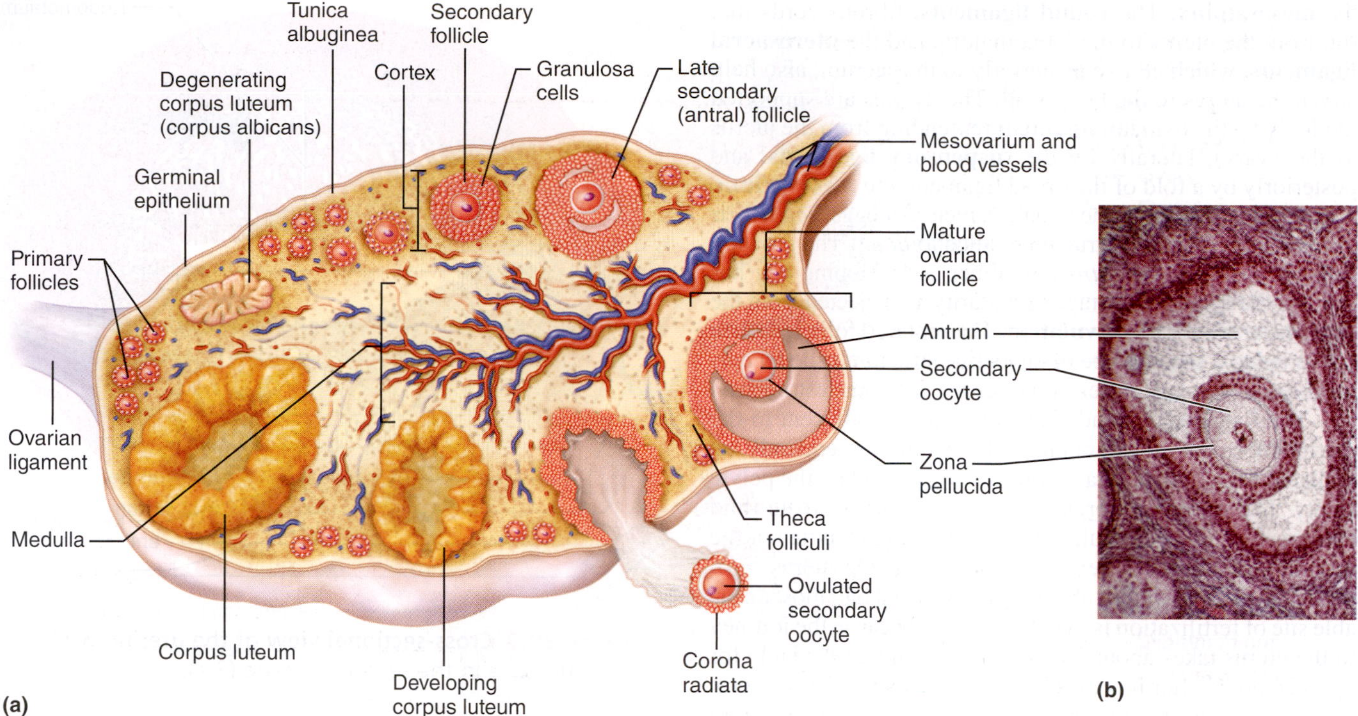

Figure 29.11 Anatomy of the human ovary. (a) The ovary has been sectioned to reveal the follicles in its interior. Note that not all the structures would appear in the ovary at the same time. **(b)** Photomicrograph of a mature ovarian follicle (115×).

Primary follicle: One or a few layers of cuboidal cells surrounding the large central developing ovum, or **oocyte,** the immature egg.

Secondary follicles: Follicles consisting of several layers of granulosa cells surrounding the central developing ovum and beginning to show evidence of fluid accumulation and **antrum** (central cavity) formation.

Mature ovarian follicle: At this stage of development, the follicle has a large antrum containing fluid produced by the granulosa cells. The developing ovum is pushed to one side of the follicle and is surrounded by a capsule of several layers of granulosa cells called the **corona radiata** (radiating crown). When the immature ovum (secondary oocyte) is released, it enters the uterine tubes with its corona radiata intact. The connective tissue adjacent to the mature follicle forms a capsule that encloses the follicle and is called the **theca folliculi.**

Corpus luteum: A solid glandular structure or a structure containing a scalloped lumen that develops from the ovulated follicle **(Figure 29.12)**. ■

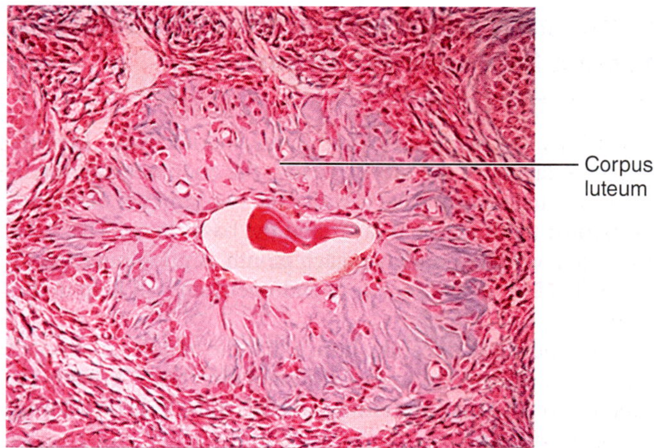

Figure 29.12 Glandular corpus luteum of an ovary. (170×).

The Mammary Glands

The **mammary glands** exist within the breasts in both sexes, but they normally have a reproduction-related function only in females. Because the function of the mammary glands is to produce milk to nourish the newborn infant, their importance is more closely associated with events that occur when reproduction has already been accomplished. Periodic stimulation by the female sex hormones, especially estrogens, increases the size of the female mammary glands at puberty. During this period, the duct system becomes more elaborate, and fat is deposited—fat deposition is the more important contributor to increased breast size.

The rounded, skin-covered mammary glands lie anterior to the pectoral muscles of the thorax, attached to them by connective tissue. Slightly below the center of each breast is a pigmented area, the **areola,** which surrounds a centrally protruding nipple **(Figure 29.13)**.

Internally, each mammary gland consists of 15 to 25 **lobes** that radiate around the nipple and are separated by fibrous connective tissue and adipose, or fatty, tissue. Within each lobe are smaller chambers called **lobules,** containing

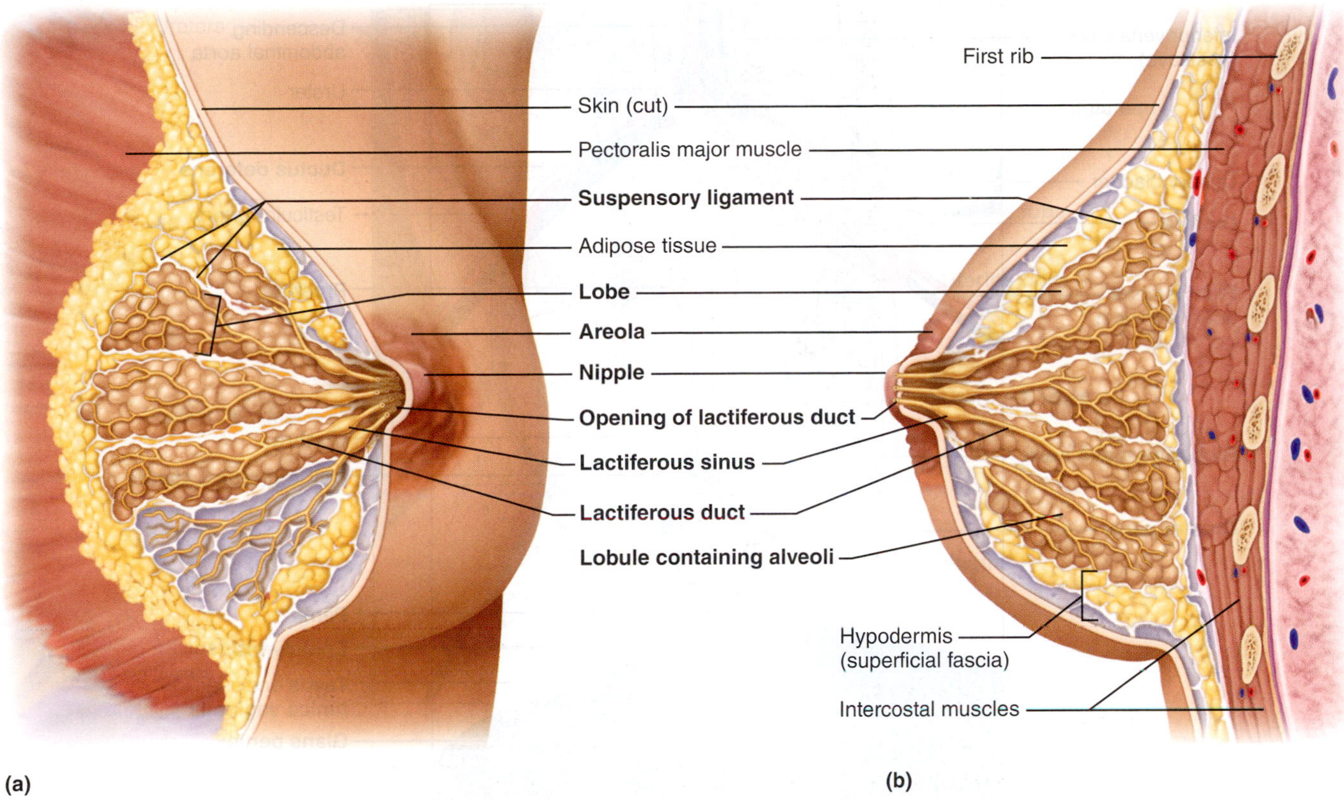

Figure 29.13 Anatomy of lactating mammary gland. (a) Anterior view of partially dissected breast. **(b)** Sagittal section of the breast.

the glandular **alveoli** that produce milk during lactation. The alveoli of each lobule pass the milk into a number of **lactiferous ducts,** which join to form an expanded storage chamber, the **lactiferous sinus,** as they approach the nipple. The sinuses open to the outside at the nipple.

DISSECTION AND IDENTIFICATION

The Reproductive System of the Cat

Don gloves, and obtain your cat, a dissection tray, and the necessary dissecting instruments. After you have completed the study of the reproductive structures of your specimen, observe a cat of the opposite sex. The following instructions assume that the abdominal cavity has been opened in previous dissection exercises. ■

ACTIVITY 5

Identifying Organs of the Male Reproductive System

Identify the male reproductive structures (refer to **Figure 29.14**).

1. Identify the **penis,** and notice the prepuce covering the glans. Carefully cut through the skin overlying the penis to expose the cavernous tissue beneath, then cross section the penis to see the relative positioning of the three cavernous bodies.

2. Identify the **scrotum,** and then carefully make a shallow incision through the scrotum to expose the **testes.** Notice that the scrotum is divided internally.

3. Lateral to the medial aspect of the scrotal sac, locate the **spermatic cord,** which contains the spermatic artery, vein, and nerve, as well as the ductus deferens, and follow it up through the inguinal canal into the abdominal cavity. It is not necessary to cut through the pubis; a slight tug on the spermatic cord in the scrotal sac region will reveal its position in the abdominal cavity. Carefully loosen the spermatic cord from the connective tissue investing it, and follow its course as it travels superiorly in the pelvic cavity. Then, follow the **ductus deferens** as it loops over the ureter and then courses posterior to the bladder and enters the prostate. Using bone cutters, carefully cut through the pubic symphysis to follow the urethra.

4. Notice that the **prostate,** an enlarged whitish mass abutting the urethra, is comparatively smaller in the cat than in the human, and it is more distal to the bladder. In the human, the prostate is immediately adjacent to the base of the bladder. Carefully slit open the prostate to follow the **ductus deferens** to the urethra, which exits from the bladder midline. The male cat urethra, like that of the human, serves as both a urinary and a sperm duct. In the human, the ductus deferens is joined by the duct of the seminal gland to form the ejaculatory duct, which enters the prostate. Seminal glands are not present in the cat.

5. Trace the **urethra** to the proximal ends of the cavernous tissues of the penis, each of which is anchored to the ischium by a band of connective tissue called the **crus** of the penis. The crus is covered ventrally by the ischiocavernosus muscle, and the **bulbo-urethral gland** lies beneath it (Figure 29.14a).

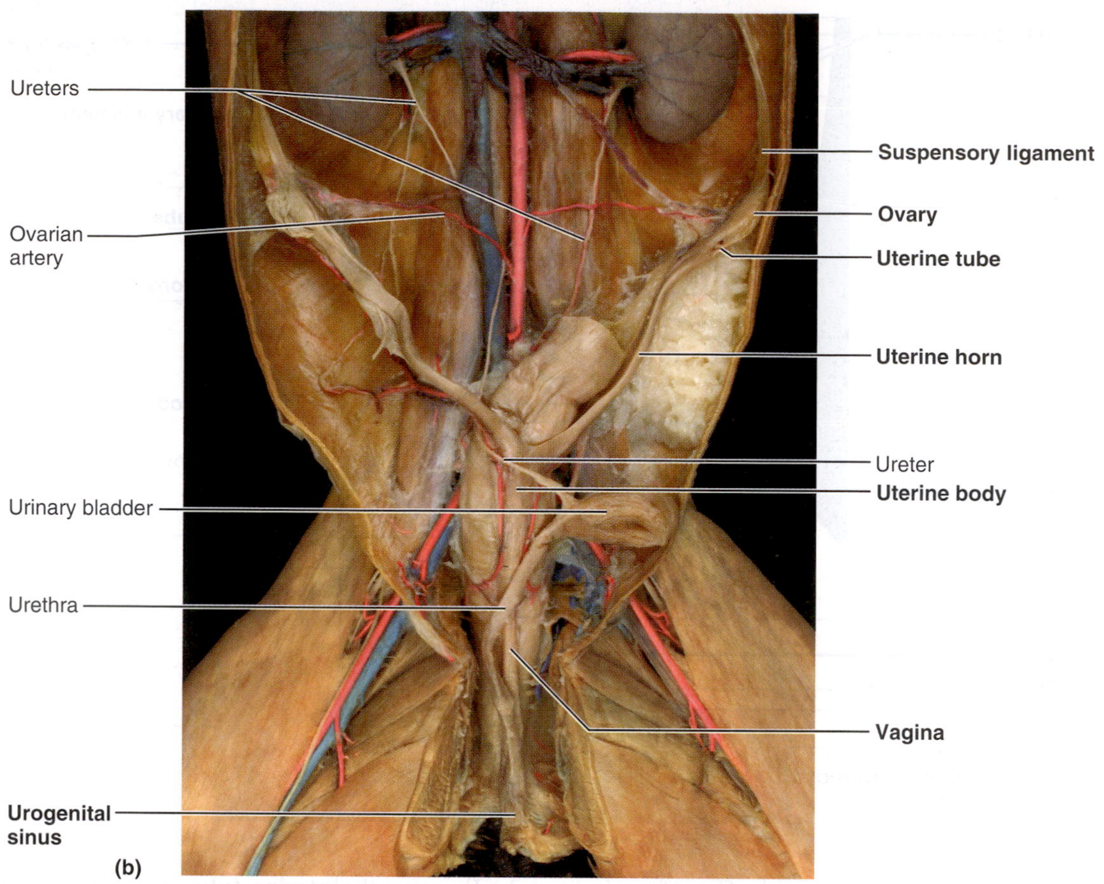

Figure 29.15 *(continued)* **Reproductive system of the female cat. (b)** Photograph.

muscular cervix of the uterus. Measure the distance between the urogenital sinus and the cervix. Approximately how long is the vagina of the cat?

____

7. To complete this exercise, observe a cat of the opposite sex.

8. Before you leave the laboratory, follow the boxed instructions to prepare your cat for storage and clean the area (page 217).

Name ______________________

Lab Time/Date ______________________

EXERCISE 29

REVIEW SHEET

Anatomy of the Reproductive System

Gross Anatomy of the Human Male Reproductive System

1. List the two principal functions of the testis: ______________________

   and ______________________

2. Identify all indicated structures or portions of structures on the male reproductive system diagram below.

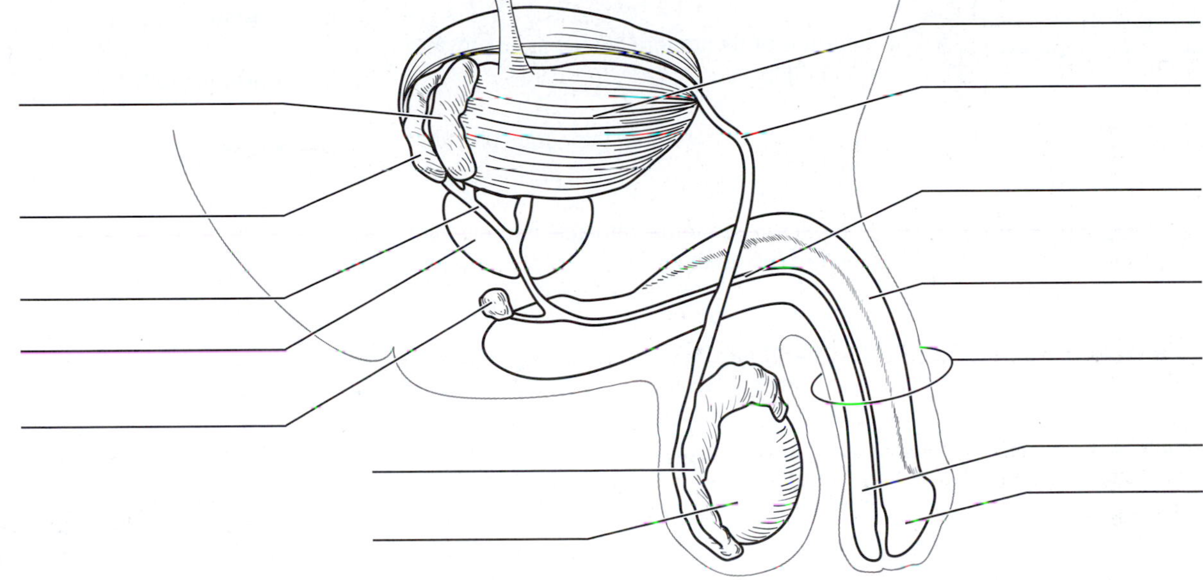

3. A common part of any physical examination of the male is palpation of the prostate. How is this accomplished? (Think!)

   ____

4. How might enlargement of the prostate interfere with urination or the reproductive ability of the male?

   ____

   ____

5. Why are the testes located in the scrotum rather than inside the ventral body cavity? ______________________

   ____

6. Match the terms in column B to the descriptive statements in column A. (Not all terms will be used.)

Column A

_______ 1. copulatory organ/penetrating device

_______ 2. site of sperm/androgen production

_______ 3. muscular passageway conveying sperm to the ejaculatory duct; in the spermatic cord

_______ 4. transports both sperm and urine

_______ 5. sperm maturation site

_______ 6. location of the testis in adult males

_______ 7. loose fold of skin encircling the glans penis

_______ 8. portion of the urethra between the prostate and the penis

_______ 9. empties a secretion into the prostatic urethra

_______ 10. empties a secretion into the intermediate part of the urethra

Column B

a. bulbo-urethral glands

b. ductus (vas) deferens

c. epididymis

d. glans penis

e. intermediate part of the urethra

f. penis

g. prepuce

h. prostate

i. prostatic urethra

j. scrotum

k. seminal gland

l. spongy urethra

m. testes

7. Describe the composition of semen, and name all structures contributing to its formation. _______

____

____

8. Of what importance is the fact that seminal fluid is alkaline? _______

____

9. What structures compose the spermatic cord? _______

____

Where is it located? _______

10. Using the following terms, trace the pathway of sperm from the testes to the urethra: rete testis, epididymis, seminiferous tubule, ductus deferens.

_______ → _______ → _______ → _______

Gross Anatomy of the Human Female Reproductive System

11. Name the structures composing the external genitalia, or vulva, of the female.

____

12. On the following diagram of a frontal section of a portion of the female reproductive system, identify all indicated structures.

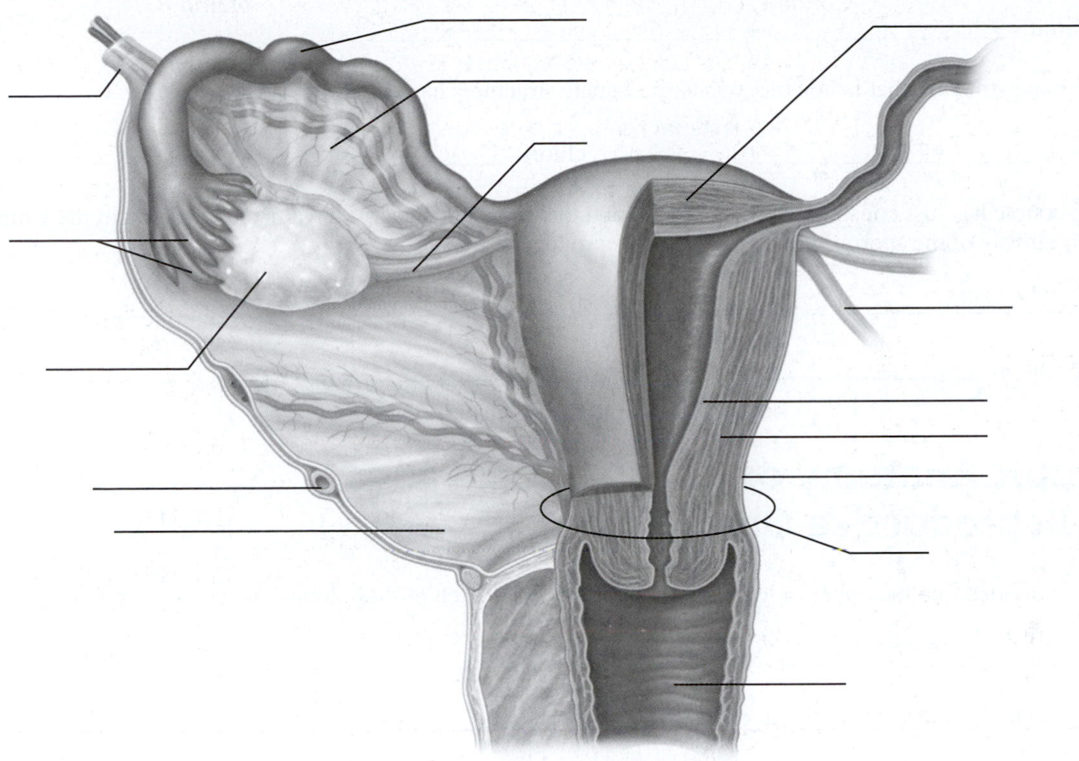

13. Identify the female reproductive system structures described below.

____________________ 1. site of fetal development

____________________ 2. copulatory canal

____________________ 3. egg typically fertilized here

____________________ 4. becomes erectile during sexual excitement

____________________ 5. duct extending from ovaries to the uterus

____________________ 6. partially closes the vaginal canal; a membrane

____________________ 7. produces oocytes, estrogens, and progesterone

____________________ 8. fingerlike ends of the uterine tube

14. Do any sperm enter the pelvic cavity of the female? Why or why not? ____________________

____

15. What is an ectopic pregnancy, and how can it happen? ____________________

____

____

____

16. Put the following vestibular-perineal structures in their proper order from the anterior to the posterior aspect: vaginal orifice, anus, external urethral opening, and clitoris.

Anterior limit: ____________________ → ____________________ → ____________________ → ____________________

17. Name the male structure that is homologous to the female structures named below.

labia majora ____________________ clitoris ____________________

18. Assume a couple has just consummated the sex act and that the male's sperm have been deposited in the woman's vagina. Trace the pathway of the sperm through the female reproductive tract.

____

19. Define *ovulation*. ____

Microscopic Anatomy of Selected Male and Female Reproductive Organs

20. The testis is divided into a number of lobes by connective tissue. Each of these lobes contains one to four ____________________ ____________________, which converge on a ____________________ which, in turn, empties into a tubular region at the testis mediastinum called the ____________________.

21. What is the function of the cavernous bodies seen in the male penis? ____

____

____

22. Describe the arrangement of the layers of smooth muscle in the seminal gland. ____

____

23. What is the function of the stereocilia exhibited by the epithelial cells of the mucosa of the epididymis? ____________________

____

24. Name the three layers of the uterine wall from the inside out.

____________________, ____________________, ____________________

Which of these is sloughed during menses? ____

Which contracts during childbirth? ____

25. Describe the epithelium found in the uterine tube. ____

____

26. On the diagram showing the sagittal section of the human testis, correctly identify all structures provided with leader lines.

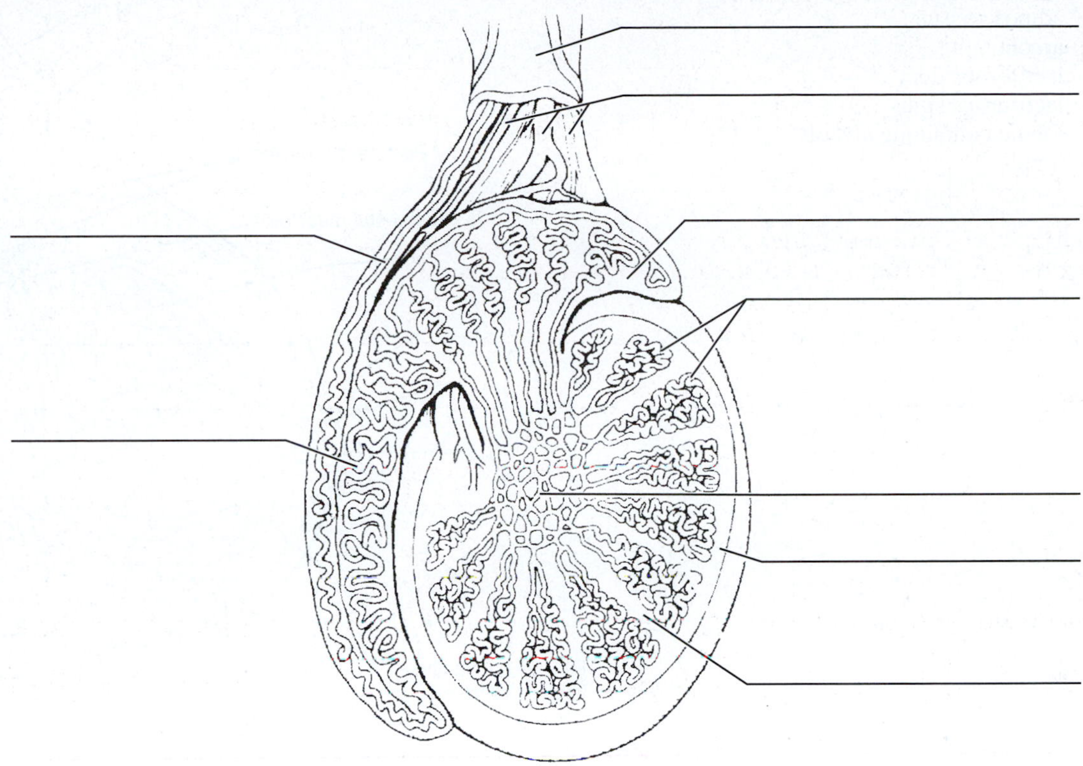

The Mammary Glands

27. Match the key term with the correct description.

		Key:
______________	1. gland that produces milk during lactation	a. alveolus
______________	2. subdivision of mammary lobe that contains alveoli	b. areola
______________	3. enlarged storage chamber for milk	c. lactiferous duct
______________	4. duct connecting alveoli to the storage chambers	d. lactiferous sinus
______________	5. pigmented area surrounding the nipple	e. lobule
______________	6. releases milk to the outside	f. nipple

28. Using the key terms, correctly identify the breast structures on the following diagram.

Key: a. adipose tissue
b. areola
c. lactiferous duct
d. lactiferous sinus
e. lobule containing alveoli
f. nipple

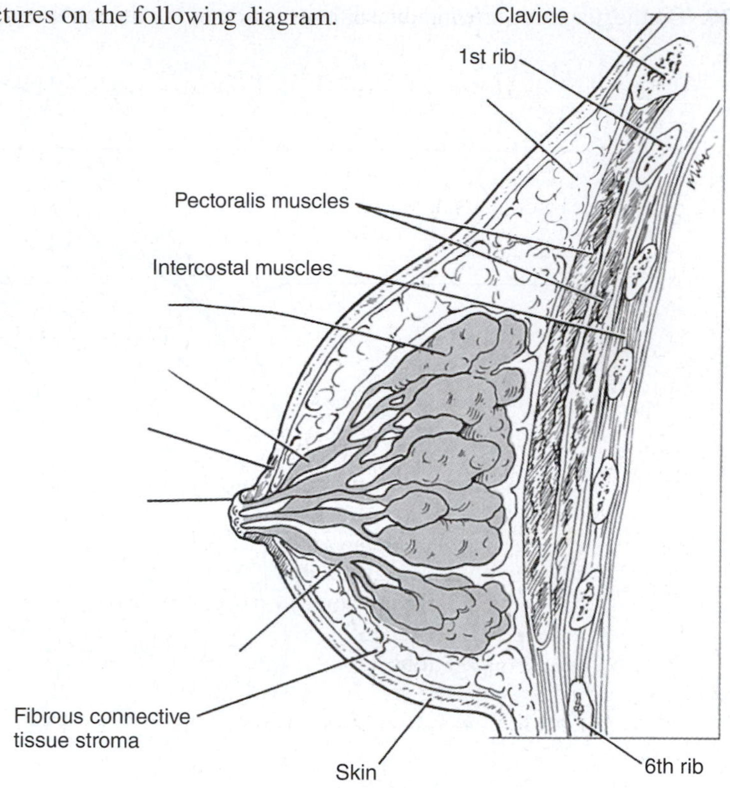

Clavicle
1st rib
Pectoralis muscles
Intercostal muscles
Fibrous connective tissue stroma
Skin
6th rib

29. Describe the procedure for self-examination of the breasts. (Men are not exempt from breast cancer, you know!) Use an appropriate reference as needed. ______________________________

____

____

____

Dissection and Identification: The Reproductive System of the Cat

30. The female cat has a ______________ uterus; that of the human female is ______________.

Explain the difference in structure of these two uterine types. ______________________________

____

____

31. What reproductive advantage is conferred by the feline uterine type?

____

____

32. Cite differences noted between the cat and the human relative to the following structures:

uterine tubes ______________________________

____

site of entry of ductus deferens into the urethra ______________________________

____

____

location of the prostate ______________________________

____

____

seminal glands ______________________________

____

____

urethral and vaginal openings in the female ______________________________

____

____

EXERCISE 30

Surface Anatomy Roundup

MATERIALS

- ☐ Articulated skeletons
- ☐ Three-dimensional models or charts of the skeletal muscles of the body
- ☐ Hand mirror
- ☐ Stethoscope
- ☐ Alcohol swabs
- ☐ Washable markers

OBJECTIVES

1. Define *surface anatomy*, and explain why it is an important field of study; define *palpation*.
2. Describe and palpate the major surface features of the cranium, face, and neck.
3. Describe the easily palpated bony and muscular landmarks of the back, and locate the vertebral spines on the living body.
4. List the bony surface landmarks of the thoracic cage, explain how they relate to the major soft organs of the thorax, and explain how to find the second to eleventh ribs.
5. Name and palpate the important surface features on the anterior abdominal wall, and explain how to palpate a full bladder.
6. Define and explain the following: *linea alba, umbilical hernia,* examination for an inguinal hernia, *linea semilunaris,* and *McBurney's point.*
7. Locate and palpate the main surface features of the upper limb.
8. Explain the significance of the cubital fossa, pulse points in the distal forearm, and the anatomical snuff box.
9. Describe and palpate the surface landmarks of the lower limb.
10. Explain exactly where to administer an injection in the gluteal region and in the other major sites of intramuscular injection.

PRE-LAB QUIZ

1. Why is it useful to study surface anatomy?
   a. You can easily locate deep muscle insertions.
   b. You can relate external surface landmarks to the location of internal organs.
   c. You can study cadavers more easily.
   d. You really can't learn that much by studying surface anatomy; it's a gimmick.
2. Circle the correct underlined term. Palpation / Dissection allows you to feel internal structures through the skin.
3. The epicranial aponeurosis binds to the subcutaneous tissue of the cranium to form the:
   a. mastoid process
   b. occipital protuberance
   c. true scalp
   d. xiphoid process
4. The ________ is the most prominent neck muscle and also the neck's most important landmark.
   a. buccinator
   b. epicranius
   c. masseter
   d. sternocleidomastoid
5. The three boundaries of the ________ are the trapezius medially, the latissimus dorsi inferiorly, and the scapula laterally.
   a. torso triangle
   b. triangle of ausculation
   c. triangle of back muscles
   d. McBurney's point
6. Circle True or False. The lungs do *not* fill the inferior region of the pleural cavity.

(Text continues on next page.)

inferiorly from the center of the axilla onto his lateral thoracic wall. This is the **midaxillary line** (Figure 30.6a). Now estimate the midpoint of his **clavicle,** and run a vertical line inferiorly from that point toward the groin. This is the **midclavicular line,** and it will pass about 1 cm medial to the nipple.

3. Next, feel along the V-shaped inferior edge of the rib cage, the **costal margin.** At the **infrasternal angle,** the superior angle of the costal margin, lies the **xiphisternal joint.** The heart lies on the diaphragm deep to the xiphisternal joint.

4. The thoracic cage provides many valuable landmarks for locating the vital organs of the thoracic and abdominal cavities. On the anterior thoracic wall, ribs 2–6 define the superior-to-inferior extent of the female breast, and the fourth intercostal space indicates the location of the **nipple** in men, children, and small-breasted women. The right costal margin runs across the anterior surface of the liver and gallbladder. Surgeons must be aware of the inferior margin of the *pleural cavities* because if they accidentally cut into one of these cavities, a lung collapses. The inferior pleural margin lies adjacent to vertebra T_{12} near the posterior midline (Figure 30.6b) and runs horizontally across the back to reach rib 10 at the midaxillary line. From there, the pleural margin ascends to rib 8 in the midclavicular line (Figure 30.6a) and to the level of the xiphisternal joint near the anterior midline. The *lungs* do not fill the inferior region of the pleural cavity. Instead, their inferior borders run at a level that is two ribs superior to the pleural margin, until they meet that margin near the xiphisternal joint.

5. Let's review the relationship of the *heart* of the thoracic cage. In essence, the superior right corner of the heart lies at the junction of the third rib and the sternum; the superior left corner lies at the second rib, near the sternum; the inferior left corner lies in the fifth intercostal space in the midclavicular line; and the inferior right corner lies at the sternal border of the sixth rib. You may wish to outline the heart on your chest or that of your lab partner by connecting the four corner points with a washable marker.

Muscles

The main superficial muscles of the anterior thoracic wall are the **pectoralis major** and the anterior slips of the **serratus anterior** (Figure 30.7).

- Palpate these two muscles on your chest. They both contract during push-ups, and you can confirm this by pushing yourself up from your desk with one arm while palpating the muscles with your opposite hand. ■

ACTIVITY 4

Palpating Landmarks of the Abdomen

Bony Landmarks

The anterior abdominal wall (Figure 30.7) extends inferiorly from the costal margin to an inferior boundary that is defined by several landmarks. Palpate these landmarks as they are described below.

1. **Iliac crest:** Locate the iliac crests by resting your hands on your hips.

2. **Anterior superior iliac spine:** Representing the most anterior point of the iliac crest, this spine is a prominent landmark. It can be palpated in everyone, even those who are overweight. Run your fingers anteriorly along the iliac crest to its end.

3. **Inguinal ligament:** The inguinal ligament, indicated by a groove on the skin of the groin, runs medially from the anterior superior iliac spine to the pubic tubercle of the pubis.

4. **Pubic crest:** You will have to press deeply to feel this crest on the pubis near the median **pubic symphysis.** The **pubic tubercle,** the most lateral point of the pubic crest, is easier to palpate, but you will still have to push deeply.

Inguinal hernias occur immediately superior to the inguinal ligament and may exit from a medial opening called the **superficial inguinal ring.** To locate this ring, one would palpate the pubic tubercle **(Figure 30.8)**. An inguinal hernia in a male can be detected by pushing into the superior inguinal ring. ✚

Muscles and Other Surface Features

The central landmark of the anterior abdominal wall is the *umbilicus* (navel). Running superiorly and inferiorly from the umbilicus is the **linea alba** (white line), represented in the skin of lean people by a vertical groove (see Figure 30.7). The linea alba is a tendinous seam that extends from the xiphoid process to the pubic symphysis, just medial to the rectus abdominis muscles (Table 12.3, page 192). The linea alba is a favored site for surgical entry into the abdominal cavity because the surgeon can make a long cut through this line with no muscle damage and minimal bleeding.

Several kinds of hernias involve the umbilicus and the linea alba. In an **acquired umbilical hernia,** the linea alba weakens until intestinal coils push through it just superior to the navel. The herniated coils form a bulge just deep to the skin.

Another type of umbilical hernia is a **congenital umbilical hernia,** present in some infants. This type of umbilical hernia is seen as a cherry-sized bulge deep to the skin of the navel that enlarges whenever the baby cries. Congenital umbilical hernias are usually harmless, and most correct themselves automatically before the child's second birthday. ✚

1. **McBurney's point** is the spot on the anterior abdominal skin that lies directly superficial to the base of the appendix (Figure 30.7). It is located one-third of the way along a line between the right anterior superior iliac spine and the umbilicus. Try to find it on your body.

McBurney's point is the most common site of incision in appendectomies, and it is often the place where the pain of appendicitis is experienced most acutely. Pain at McBurney's point after pressure is removed (rebound tenderness) can indicate appendicitis. This is not a *precise* method of diagnosis, however.

2. Flanking the linea alba are the vertical straplike **rectus abdominis** muscles (Figure 30.7). Feel these muscles contract just deep to your skin as you do a bent-knee sit-up (or as you bend forward after leaning back in your chair). In the skin of lean people, the lateral margin of each rectus muscle makes a groove known as the **linea semilunaris** (half-moon line). On your right side, estimate where your linea semilunaris crosses the costal margin of the rib cage. The *gallbladder* lies just deep to this spot, so this is the standard point of incision for gallbladder surgery. In muscular people, three horizontal grooves can be seen in the skin covering the rectus abdominis.

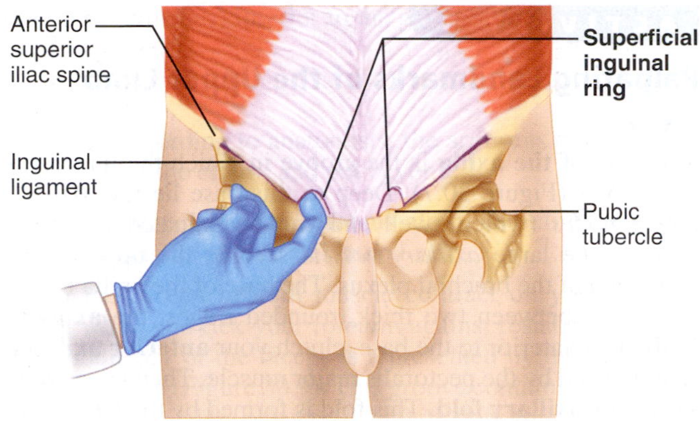

Figure 30.8 Clinical examination for an inguinal hernia in a male. The examiner palpates the patient's pubic tubercle, pushes superiorly to invaginate the scrotal skin into the superficial inguinal ring, and asks the patient to cough. If an inguinal hernia exists, it will push inferiorly and touch the examiner's fingertip.

These grooves represent the **tendinous intersections,** fibrous bands that subdivide the rectus muscle. Because of these subdivisions, each rectus abdominis muscle presents four distinct bulges. Try to identify these intersections on yourself or your partner.

3. The only other major muscles that can be seen or felt through the anterior abdominal wall are the lateral **external obliques.** Feel these muscles contract as you cough, strain, or raise your intra-abdominal pressure in some other way.

4. The anterior abdominal wall can be divided into four quadrants (Figure 1.8). A clinician listening to a patient's **bowel sounds** places the stethoscope over each of the four abdominal quadrants, one after another. Normal bowel sounds, which result as peristalsis moves air and fluid through the intestine, are high-pitched gurgles that occur every 5 to 15 seconds.

- Use the stethoscope to listen to your own or your partner's bowel sounds.

Abnormal bowel sounds can indicate intestinal disorders. Absence of bowel sounds indicates a halt in intestinal activity, which follows long-term obstruction of the intestine, surgical handling of the intestine, peritonitis, or other conditions. Loud tinkling or splashing sounds, by contrast, indicate an increase in intestinal activity. Such loud sounds may accompany gastroenteritis (inflammation and upset of the GI tract) or a partly obstructed intestine. +

The Pelvis and Perineum

The bony surface features of the *pelvis* are considered with the bony landmarks of the abdomen (page 532) and the gluteal region (page 536). Most *internal* pelvic organs are not palpable through the skin of the body surface. A full *bladder,* however, becomes firm and can be felt through the abdominal wall just superior to the pubic symphysis. A bladder that can be palpated more than a few centimeters above this symphysis is retaining urine and dangerously full, and it should be drained by catheterization.

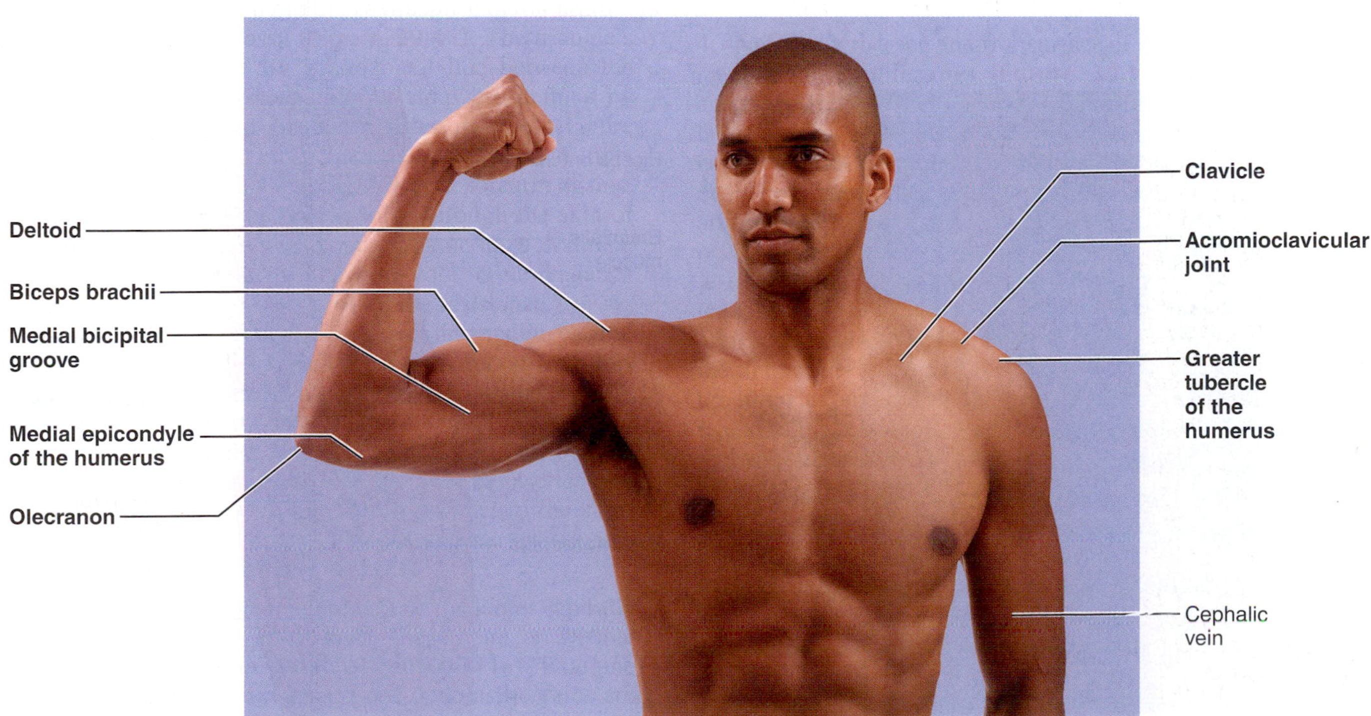

Figure 30.9 Shoulder and arm.

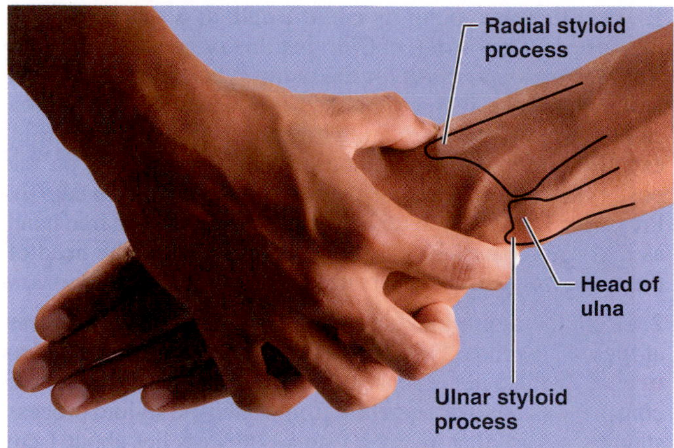

Figure 30.12 A way to locate the ulnar and radial styloid processes. The right hand is palpating the left hand in this picture. Note that the head of the ulna is not the same as the ulnar styloid process. The radial styloid process lies about 1 cm distal to the ulnar styloid process.

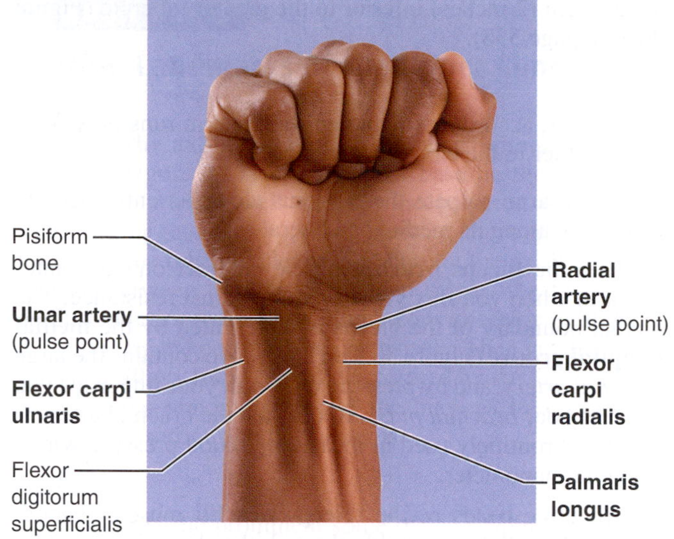

Figure 30.13 The anterior surface of the distal forearm and fist. The tendons of the flexor muscles guide the clinician to several sites for taking the pulse.

6. On the dorsum of your hand, observe the superficial veins just deep to the skin. This is the **dorsal venous network,** which drains superiorly into the cephalic vein. This venous network provides a site for drawing blood and inserting intravenous catheters and is preferred over the median cubital vein for these purposes. Next, extend your hand and fingers, and observe the tendons of the **extensor digitorum** muscle.

7. The anterior surface of the hand also contains some features of interest **(Figure 30.15)**. These features include the *epidermal ridges* (fingerprints) and many **flexion creases** in the skin. Grasp your **thenar eminence** (the bulge on the palm that contains the thumb muscles) and your **hypothenar eminence** (the bulge on the medial palm that contains muscles that move the little finger). ■

ACTIVITY 6

Palpating Landmarks of the Lower Limb

Gluteal Region

Dominating the gluteal region are the two *prominences* (cheeks) of the buttocks. These are formed by subcutaneous fat and by the thick **gluteus maximus** muscles **(Figure 30.16)**. The midline groove between the two prominences

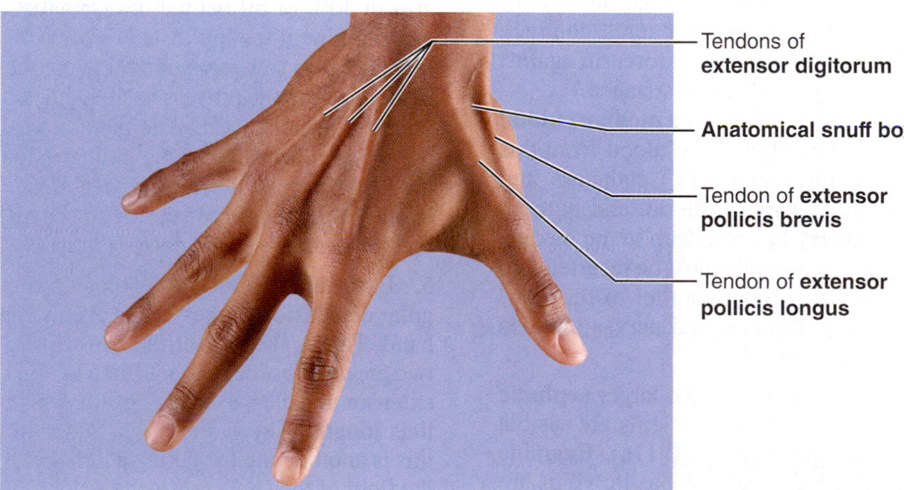

Figure 30.14 The dorsum of the hand. Note especially the anatomical snuff box and dorsal venous network.

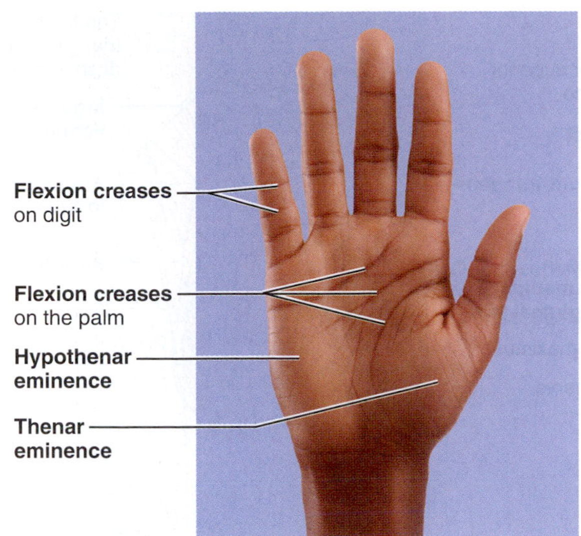

Figure 30.15 The palmar surface of the hand.

is called the **natal cleft** (*natal* = rump), or **gluteal cleft.** The inferior margin of each prominence is the horizontal **gluteal fold,** which roughly corresponds to the inferior margin of the gluteus maximus.

1. Try to palpate your **ischial tuberosity** just above the medial side of each gluteal fold (it will be easier to feel if you sit down or flex your thigh first). The ischial tuberosities are the robust inferior parts of the ischial bones, and they support the body's weight during sitting.

2. Next, palpate the **greater trochanter** of the femur on the lateral side of your hip (Figure 30.16). This trochanter lies just anterior to a hollow and about 10 cm (one hand's breadth, or 4 inches) inferior to the iliac crest. To confirm that you have found the greater trochanter, alternately flex and extend your thigh. Because this trochanter is the most superior point on the lateral femur, it moves with the femur as you perform this movement.

3. To palpate the sharp **posterior superior iliac spine** (Figure 30.16), locate your iliac crests again, and trace each to its most posterior point. You may have difficulty feeling this spine, but it is indicated by a distinct dimple in the skin that is easy to find. This dimple lies two to three fingerbreadths lateral to the midline of the back. The dimple also indicates the position of the *sacroiliac joint,* where the hip bone attaches to the sacrum of the spinal column. You can check *your* "dimples" out in the privacy of your home.

The gluteal region is a major site for administering intramuscular injections. When giving such injections, the health care provider must take extreme care to avoid piercing the major nerve that lies just deep to the gluteus maximus muscle. Can you guess what nerve this is?

This nerve is the thick *sciatic nerve,* which innervates much of the lower limb. Furthermore, the needle must avoid the gluteal nerves and gluteal blood vessels, which also lie deep to the gluteus maximus.

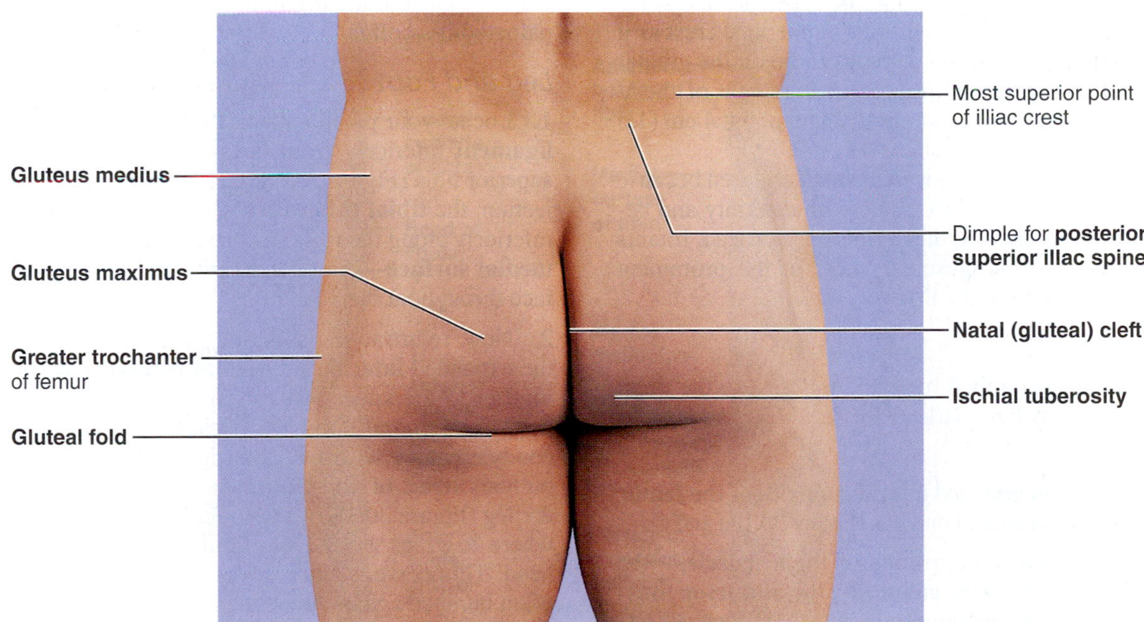

Figure 30.16 The gluteal region. The region extends from the iliac crests superiorly to the gluteal folds inferiorly. Therefore, it includes more than just the prominences of the buttock.

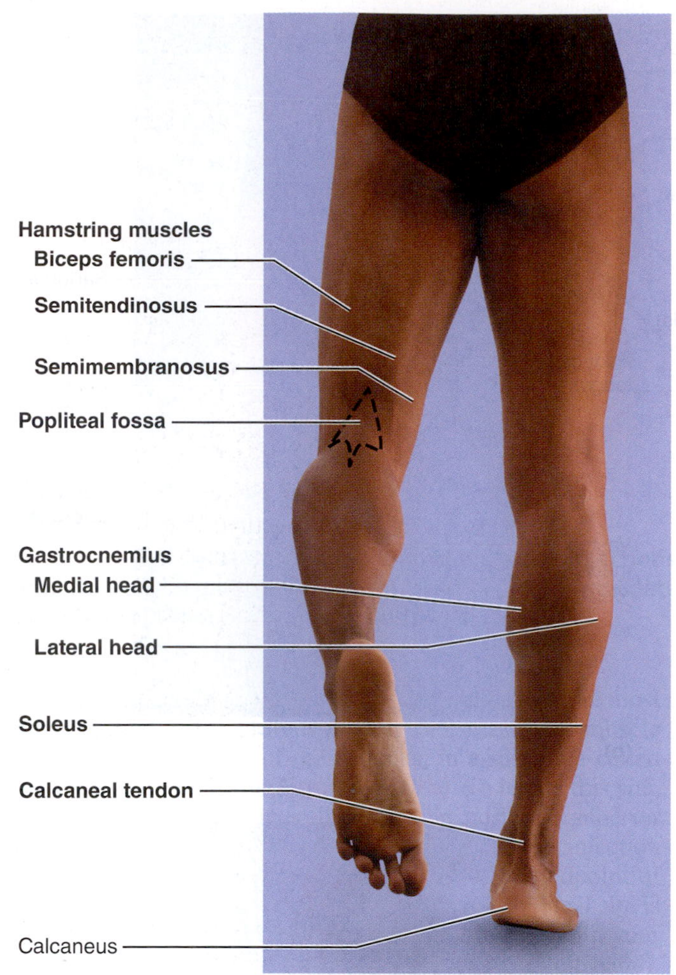

Figure 30.19 Posterior surface of the lower limb. Notice the diamond-shaped popliteal fossa posterior to the knee.

4. On the posterior aspect of the knee is a diamond-shaped hollow called the **popliteal fossa** (Figure 30.19). Palpate the large muscles that define the four borders of this fossa: the **biceps femoris** forming the superolateral border, the **semitendinosus** and **semimembranosus** defining the superomedial border, and the two heads of the **gastrocnemius** forming the inferior border. The main vessels to the leg, the *popliteal artery* and *vein,* lie deep within this fossa. To feel a popliteal pulse, flex your leg at the knee, and push your fingers firmly into the popliteal fossa. If a physician is unable to feel a patient's popliteal pulse, the femoral artery may be narrowed by atherosclerosis.

5. Observe the dorsum (superior surface) of your foot. You may see the superficial **dorsal venous arch** overlying the proximal part of the metatarsal bones (Figure 30.18d). This arch gives rise to both saphenous veins (the main superficial veins of the lower limb). Visible in lean people, the *great saphenous vein* ascends along the medial side of the entire limb (Figure 24.8, page 407). The *small saphenous vein* ascends through the center of the calf.

As you extend your toes, observe the tendons of the **extensor digitorum longus** and **extensor hallucis longus** muscles on the dorsum of the foot. Finally, place a finger on the extreme proximal part of the space between the first and second metatarsal bones. Here you should be able to feel the pulse of the **dorsalis pedis artery.** ■

Name ______________________________

Lab Time/Date ______________________

EXERCISE 30

Surface Anatomy Roundup

REVIEW SHEET

______________ 1. A blow to the cheek is most likely to break which superficial bone or bone part? (a) superciliary arches, (b) mastoid process, (c) zygomatic arch, (d) ramus of the mandible

______________ 2. Rebound tenderness (a) occurs in appendicitis, (b) is whiplash of the neck, (c) is a sore foot from playing basketball, (d) occurs when the larynx falls back into place after swallowing.

______________ 3. The anatomical snuff box (a) is in the nose, (b) contains the radial styloid process, (c) is defined by tendons of the flexor carpi radialis and palmaris longus, (d) cannot really hold snuff.

______________ 4. Some landmarks on the body surface can be seen or felt, but others are abstractions that you must construct by drawing imaginary lines. Which of the following pairs of structures is abstract and invisible? (a) umbilicus and costal margin, (b) anterior superior iliac spine and natal cleft, (c) linea alba and linea semilunaris, (d) McBurney's point and midaxillary line, (e) lacrimal fossa and sternocleidomastoid

______________ 5. Many pelvic organs can be palpated by placing a finger in the rectum or the vagina, but only one pelvic organ is readily palpated through the skin. This is the (a) nonpregnant uterus, (b) prostate, (c) full bladder, (d) ovaries, (e) rectum.

______________ 6. A muscle that contributes to the posterior axillary fold is the (a) pectoralis major, (b) latissimus dorsi, (c) trapezius, (d) infraspinatus, (e) pectoralis minor, (f) a and e.

______________ 7. Which of the following is *not* a pulse point? (a) anatomical snuff box, (b) inferior margin of mandible anterior to masseter muscle, (c) center of distal forearm at palmaris longus tendon, (d) medial bicipital groove on arm, (e) dorsum of foot between the first two metatarsals

______________ 8. Which pair of ribs inserts on the sternum at the sternal angle? (a) first, (b) second, (c) third, (d) fourth, (e) fifth

______________ 9. The inferior angle of the scapula is at the same level as the spinous process of which vertebra? (a) C_5, (b) C_7, (c) T_3, (d) T_7, (e) L_4

______________ 10. An important bony landmark that can be recognized by a distinct dimple in the skin is the (a) posterior superior iliac spine, (b) ulnar styloid process, (c) shaft of the radius, (d) acromion.

______________ 11. A nurse missed a patient's median cubital vein while trying to withdraw blood and then inserted the needle far too deeply into the cubital fossa. This error could cause any of the following problems, *except* this one: (a) paralysis of the ulnar nerve, (b) paralysis of the median nerve, (c) bruising the insertion tendon of the biceps brachii muscle, (d) spurting of blood from the brachial artery.

______________ 12. Which of these organs is almost impossible to study with surface anatomy techniques? (a) heart, (b) lungs, (c) brain, (d) nose

______________ 13. A preferred site for inserting an intravenous medication line into a blood vessel is the (a) medial bicipital groove on arm, (b) external carotid artery, (c) dorsal venous network of hand, (d) popliteal fossa.

______________ 14. One listens for bowel sounds with a stethoscope placed (a) on the four quadrants of the abdominal wall; (b) in the triangle of auscultation; (c) in the right and left midaxillary line, just superior to the iliac crests; (d) inside the patient's bowels (intestines), on the tip of an endoscope.

______________ 15. A stab wound in the posterior triangle of the neck could damage any of the following structures *except* the (a) accessory nerve, (b) phrenic nerve, (c) external jugular vein, (d) external carotid artery.

Credits

All illustrations are by Imagineering STA Media Services unless otherwise noted.

Illustrations

Exercise 1

1.5: Precision Graphics. 1.10: Adapted from Marieb and Mallatt, *Human Anatomy,* 3e, F1.10, © Benjamin Cummings, 2003.

Exercise 2

2.1: Precision Graphics.

Exercise 3

3.2–3.4: Precision Graphics.

Exercise 5

5.2: Precision Graphics.

Exercise 6

6.1: Tomo Narashima.

Exercise 8

8.1–8.3, 8.6, 8.8: Nadine Sokol.

Exercise 9

Table 9.1: Precision Graphics.

Exercise 11

11.4: Imagineering STA Media Services/Precision Graphics. 11.5: Electronic Publishing Services, Inc.

Exercise 12

12.18: Precision Graphics.

Exercise 13

13.1: Imagineering STA Media Services/Precision Graphics. 13.4: Precision Graphics. 13.8a: Charles Hoffman.

Exercise 14

14.4, 14.5, 14.7, 14.9: Electronic Publishing Services, Inc. 14.11, 14.13: Precision Graphics.

Exercise 15

15.10–15.12: Kristin Mount.

Exercise 17

17.1: Electronic Publishing Services, Inc.

Exercise 18

18.1: Shirley Bortoli. 18.4–18.6: Precision Graphics.

Exercise 19

19.1: Electronic Publishing Services, Inc./Precision Graphics. 19.2: Electronic Publishing Services, Inc. 19.6: Precision Graphics.

Exercise 20

20.1: Electronic Publishing Services, Inc.

Exercise 21

21.3: Precision Graphics. 21.4: Kristin Mount.

Exercise 22

22.2, 22.5, 22.6: Precision Graphics.

Exercise 23

23.1: Electronic Publishing Services, Inc./Precision Graphics. 23.2: Electronic Publishing Services, Inc. 23.6: Precision Graphics. 23.8: Barbara Cousins. 23.9: Electronic Publishing Services, Inc.

Exercise 24

24.2, 24.4–24.13: Electronic Publishing Services, Inc. 24.16, 24.17: Precision Graphics.

Exercise 26

26.2, 26.3, 26.5: Electronic Publishing Services, Inc. 26.9: Kristin Mount.

Exercise 27

27.1, 27.3, 27.4, 27.10, 27.16: Electronic Publishing Services, Inc. 27.17: Electronic Publishing Services, Inc./Precision Graphics. 27.20: Kristin Mount.

Exercise 28

28.1, 28.2: Electronic Publishing Services, Inc. 28.9a: Kristin Mount.

Exercise 29

29.2: Electronic Publishing Services, Inc. 29.14a: Kristin Mount.

Exercise 30

30.1–30.3, 30.5, 30.7, 30.9–30.19: Precision Graphics.

Photographs

Exercise 1

1.1: John Wilson White, Pearson Education. 1.3.1: Jenny Thomas, Pearson Education. 1.3a: Scott Camazine/Photo Researchers. 1.3b: James Cavallini/Photo Researchers. 1.3c: CNRI/Science Photo Library/Photo Researchers. 1.9a: John Wilson White, Pearson Education.

Exercise 2

2.1, 2.2, 2.3a, 2.4a, 2.5b, c: Elena Dorfman, Pearson Education. 2.3b, 2.4b, 2.5a, 2.6: From *A Stereoscopic Atlas of Human Anatomy* by David L. Bassett, M.D. 2.7: Arcady/Shutterstock.com.

Exercise 3

3.1: Leica Microsystems. 3.5: Victor Eroschenko, Pearson Education.

Exercise 4

4.1b: Don W. Fawcett/Science Source/Photo Researchers. 4.4, Review Sheet: William Karkow, Pearson Education.

Exercise 5

5.3a, 5.3d, 5.3g: William Karkow, Pearson Education. 5.3b, c, f, h: Allen Bell, Pearson Education. 5.3e: Nina Zanetti, Pearson Education. 5.5a, b, e, f, l: William Karkow, Pearson Education. 5.5c, g, i: Nina Zanetti, Pearson Education. 5.5d, j, k: Allen Bell, Pearson Education. 5.5h: Steve Downing, *PAL 2.0,* Pearson Education. 5.6: Biophoto Associates/Photo Researchers. 5.7a: Nina Zanetti, *PAL 1.0,* Pearson Education. 5.7b, c: William Karkow, Pearson Education.

Exercise 6

6.2a, 6.7b: William Karkow, Pearson Education. 6.3: Pearson Education. 6.4a, b: Lisa Lee, Pearson Education. 6.7a, Review Sheet: Marian Rice.

Exercise 7

7.3c: William Karkow, Pearson Education. 7.4: Deborah Vaughan. Review Sheet: Allen Bell, Pearson Education.

Exercise 8

8.5: Michael Wiley, Univ. of Toronto, Imagineering © Pearson Education. 8.6c: Elena Dorfman, Pearson Education. 8.7c: Ralph T. Hutchings, Pearson Education. 8.10a: Living Art Enterprises, LLC/Photo Researchers. 8.10b: BSIP/Photo Researchers. 8.17b: Karen Krabbenhoft, *PAL 3.0,* Pearson Education. 8.18c: Pearson Education. 8.19c, d: Larry Delay, *PAL 3.0,* Pearson Education.

Exercise 9

Table 9.1: From *A Stereoscopic Atlas of Human Anatomy* by David L. Bassett, M.D.

Exercise 10

10.2, 10.7c: From *A Stereoscopic Atlas of Human Anatomy* by David L. Bassett, M.D. 10.6: John Wilson White, Pearson Education. 10.8d, 10.10c: Karen Krabbenhoft, *PAL 3.0,* Pearson Education.

Exercise 11

11.1a: Marian Rice. 11.3: Victor Eroschenko, Pearson Education. 11.4b: William Karkow, Pearson Education. 11.6: Victor Eroschenko, Pearson Education.

Exercise 12

12.4b, 12.5a, 12.9a, 12.11f: Karen Krabbenhoft, *PAL 3.0,* Pearson Education. 12.8b: William Karkow, Pearson Education. 12.14b: From *A Stereoscopic Atlas of Human Anatomy* by David L. Bassett, M.D. 12.17c: CDC. 12.19–12.30: Shawn Miller (dissection) and Mark Nielsen (photography), Pearson Education.

Aorta *(continued)*
thoracic, 399*f*, 400–403, **400**, 402*f*
in cat, 415*f*, 420*f*
Aortic arch, 384*f*, 399*f*, 400–403, 410*f*
in cat, 414, 415*f*
Aortic impression, 446*f*
Aortic valve, 385*f*, **386**, 387*f*
in sheep, 390
Aperture selection dial (ophthalmoscope), 321*f*, **322**
Apex
of heart, **384**, 384*f*, 385*f*
in sheep, 389*f*, 391*f*
of lung, **445**, 446*f*
of nose, **527**, 527*f*
of renal pyramid (papilla), 488, 489*f*
Apical foramen, 469*f*, **470**
Apical pulse, 384, 384*f*
Apical surface, of epithelium, 54, 55*f*
Apocrine glands, **81**
Aponeuroses, **174**
epicranial, 187*f*, 527
palmar, 199*f*
Appendicitis/appendicitis pain, 467
at McBurney's point, 532
Appendicular (term), **2**, 3*f*
Appendicular skeleton, 92*f*, **93**, **133**–151. *See also specific region and specific bone*
pectoral girdle/upper limb, 92*f*, **134**–139, 134*f*, 135*f*, 136*f*, 137*f*, 138*f*
pelvic girdle/lower limb, 92*f*, **139**–144, 140*f*, 141*t*, 142*f*, 143*f*, 144*f*
Appendix, 432, 432*f*, 459*f*, **467**, 468*f*
Aqueous humor, 308*f*, **309**
Arachnoid granulations (villi), **261**, 262*f*, 263*f*
Arachnoid mater
brain, **261**, 262*f*, 263*f*
spinal cord, 280, 281*f*, 282*f*, 283
Arbor vitae, 258*f*, **259**, 261*f*
in sheep, 271*f*
Arches, in fingerprint, 85, 85*f*
Arches of foot, 144, 144*f*
Arch of vertebra, **117**, 117*f*
Arcuate arteries
of foot, **405**, 405*f*
renal, **489**, 489*f*, 490*f*
Arcuate line, **139**, 140*f*
Arcuate popliteal ligament, **159**, 162*f*
Arcuate veins, **489**, 489*f*, 490*f*
Areola (breast), **512**, 513*f*
Areolar connective tissue, **60**, 60*f*, 61, 62*f*
Arm. *See also* Forearm; Hand; Upper limb
bones of, 134–136, 136*f*
deep artery of, 402*f*, **403**
muscles of/muscles controlling, 184*f*, 185*f*, 192*f*, 198*f*
surface anatomy of, 533*f*, 534*f*, 535
Arm (microscope), 28*f*, **29**
Arrector pili muscle, 78*f*, 83*f*, **84**
Arteries, 398, 398*f*, 399–400, 399*f*, 400–405, 401*f*, 402*f*, 403–404*f*, 405*f*
in cat, 414–416, 415*f*, 420*f*
Arteriole(s), 398
nephron, 489, 490*f*, 491, 491*f*
portal, 472, 473*f*
Articular capsule, 156, 156*f*
Articular cartilage, **93**, 95*f*, **98**, 99*f*, 156, 156*f*
Articular discs, 156
Articular processes/facets, **117**, 117*f*, 118*f*, 119*f*, 119*t*, 120, 122*f*
Articular tubercle, **161**, 163*f*
Articulations, 93, 153–170, 154*f*, **155**. *See also* Joint(s)
Arytenoid cartilages, **442**, 444*f*
Ascending aorta, 384*f*, 399*f*, 400
in cat, 413
Ascending colon, 459*f*, **467**, 468*f*
in cat, 476*f*, 477, 477*f*
Ascending lumbar vein, 408*f*, **409**, 409*f*
Ascending (sensory) tracts, 283, 283*f*
Association areas of brain, 256*f*. *See also specific type*
Association neurons (interneurons), **244**
Association tracts, 257
Asters, 46*f*
Astigmatism/astigmatism testing, **318**, 320, 320*f*
Astrocytes, 240, 240*f*
Atherosclerosis, **378**
Atlantoaxial joint, 164*t*
Atlanto-occipital joint, 164*t*
Atlas (C_1), 117*f*, **118**
ATP
in active transport, 41
mitochondria producing, 44
Atria, cardiac, 384*f*, 385*f*, **386**, 388*f*
in cat, 413
in sheep, 389*f*, 390
Atrioventricular (AV) valves, **386**, 387*f*
Auditory association area, 256*f*
Auditory canal, external (external acoustic meatus), **106**, 107*f*, **328**, 328*f*, 526*f*, **527**
Auditory cortex, primary, 256*f*
Auditory ossicles, **328**, 328*f*
Auditory receptor (hair) cells, 329, 330–331, 330*f*, 331
Auditory (pharyngotympanic/eustachian) tube, **328**–329, 328*f*, 442, 443*f*
Auricle (pinna), 65*f*, **328**, 328*f*, **527**, 527*f*
Auricle(s), of heart, 384*f*, 385*f*
in sheep, 389*f*, **390**
Auricular nerve, greater, 285*f*, 286*t*
Auricular surface of ilium, **139**, 140*f*
Auricular vein, posterior, 408*f*
Auscultation, triangle of, 185*f*, **530**, 530*f*
Autoimmunity, 432
Autonomic nervous system (ANS), **254**, **299**–303, 300*f*, 301*f*
function of, 300
parasympathetic (craniosacral) division of, **300**, 300*f*
sympathetic (thoracolumbar) division of, **300**–**302**, 300*f*, 301*f*
Avascularity, epithelial, **54**
AV valves. *See* Atrioventricular (AV) valves
Axial (term), **2**, 3*f*
Axial skeleton, 92*f*, **93**, 105–131, **106**. *See also specific region and specific bone*
skull, 92*f*, **106**–115, 107*f*, 108*f*, 109*f*, 110–111*f*, 112*f*, 113*f*
fetal, 122, 123*f*
thoracic cage, 92*f*, **120**–122, 121*f*, 122*f*
vertebral column, 92*f*, **115**–120, 116*f*, 117*f*, 118*f*, 119*f*, 119*t*
Axilla, 531*f*, 534
Axillary (term), **2**, 3*f*
Axillary artery. *See* Axillary vessels
Axillary (lateral) border of scapula, 134, 135*f*
Axillary folds, 531*f*, **534**
Axillary lymph nodes, 431*f*, 534
Axillary nerve, 286*t*, 287*f*, **288**
in cat, 292*f*
Axillary vein. *See* Axillary vessels
Axillary vessels, 534
axillary artery, 399*f*, 401*f*, 402*f*, **403**
in cat, 292*f*, 415*f*, **416**
axillary vein, 406*f*, **407**, 409*f*
in cat, 417*f*, **418**, 420*f*
Axis (C_2), 117*f*, **118**
Axon(s), 68*f*, 241*f*, **242**, 243*f*, 245*f*, 246*f*
neuron classification and, 243, 244*f*
Axon collaterals, **242**
Axon hillock, 241*f*, **242**
Axon terminals (terminal boutons), **175**, 175*f*, 176*f*, 241*f*, **242**
Azygos system, 406*f*, **409**, 409*f*
Azygos vein, **409**, 409*f*, 431*f*
in cat, **416**, 417*f*, **418**

Back, surface anatomy (bones/muscles) of, 3*f*, 529–530, 530*f*
Balance tests, 334
Ball-and-socket joint, **157**, 157*f*
Bare area of liver, 472*f*
Basal epithelial cells (taste bud), **343**, 343*f*
Basal lamina, 54
Basal layer (stratum basalis) (endometrium), **509**
Basal nuclei (basal ganglia), 255*f*, **259**, 260*f*
Basal surface, of epithelium, 54, 55*f*
Base
of axilla, 531*f*, **534**
cranial, **106**
of heart, **384**
in sheep, 389*f*
of lung, **445**, 446*f*
of metacarpal, 138, 138*f*
Base (microscope), 28*f*, **29**
Basement membrane, **54**, 56*f*, 57*f*, 58*f*, 59*f*
Basilar artery, 401*f*, **402**
Basilar membrane, **330**, 330*f*, 331*f*
Basilar part of occipital bone, 107*f*
Basilic vein, **407**, 409*f*, 534*f*, **535**
Basophil cells, 356*f*, **357**
Basophils, 367*f*, 368*t*, **370**, 370*f*
B cells (B lymphocytes), 371, **432**
Beauchene skull, 106, 111*f*. *See also* Skull
Bedsores (decubitus ulcers), **81**, 81*f*
Belly button (umbilicus/navel), **411**, 412*f*, 532
Beta (β) cells, pancreatic, 356*f*, **357**, 471*f*
Biaxial joints/movement, 154*f*, 156, 157, 157*f*
Biceps (term), muscle name and, 182
Biceps brachii muscle, 183*f*, 184*f*, 192*f*, 197, 198*f*, 198*t*, 199*f*, 533*f*, 534*f*
in cat, 224*f*, **225**
tendon of, 199*f*, 534*f*
Biceps femoris muscle, 185*f*, 204, 207*f*, 208*t*, **540**, 540*f*
in cat, **226**–227, 226*f*
Bicipital groove (intertubercular sulcus), **136**, 136*f*
medial, 533*f*, **535**
Bicuspids (premolars), **468**, 469*f*
Bicuspid (mitral) valve, 385*f*, **386**, 387*f*
in sheep, 391
Bile, 465, 471, 472
Bile canaliculi, **472**, 473*f*
Bile duct, **465**, **471**, 472, 472*f*, 473*f*
in cat, **475**
Binocular vision, 320–321, 320*f*
Bipennate muscle structure, 183*f*
Bipolar cells, 309*f*, **310**
Bipolar neurons, **243**, 244*f*
Birth
changes in circulation at, 412*f*, 413
female pelvis and, 139, 141*t*
Bitter taste, 344
Blackheads, **81**
Bladder (urinary), **20**, 20*f*, 21*f*, 22*f*, 59*f*, 464*f*, **486**, 487*f*, 488*f*, **492**–493, 504*f*, 505*f*
in cat, 414*f*, 476*f*, **493**, 494*f*, 495*f*, 514*f*, 515*f*, 516*f*
functional microscopic anatomy of, **492**–493
palpable, 533
in rat, **20**, 20*f*, 21*f*
Blastula, 48
Bleeding time, **375**
Blind spot (optic disc), **307**, 308*f*, 317–318, 318*f*, 322*f*
in cow eye, 312*f*
Blood, **61**, 67*f*, 365–382
composition of, 366–371, 367*f*. *See also* Blood cells; Plasma
hematologic tests/analysis of, 371–378, 372*f*, 373*f*, 375*f*, 376*f*, 377*t*, 378*f*. *See also specific test*
oxygenation of, skin color reflecting, 81
precautions for handling, 366
Blood air barrier (respiratory membrane), **444**, 445*f*
Blood cells (formed elements), **366**, 367*f*, 368–371, 368*t*, 369*f*
tissues forming (hematopoietic tissue), **60**
Blood cholesterol, 378
Blood clotting, **375**–376, 376*f*
Blood smear, 368–369, 369*f*
Blood (hematologic) tests, 371–378, 372*f*, 373*f*, 375*f*, 376*f*, 377*t*, 378*f*. *See also specific test*
Blood type, 376–378, 377*t*, 378*f*
Blood vessel(s), 397–428. *See also* Circulation(s)
arteries, 398, 398*f*, 399–400, 399*f*, 400–405, 401*f*, 402*f*, 403–404*f*, 405*f*
in cat, 414–416, 415*f*, 420*f*
capillaries, 388*f*, 398, 398*f*, 400
in cat, 413–418, 415*f*, 417*f*, 419*f*, 420*f*
microscopic structure of, 398–400, 398*f*
veins, 398, 398*f*, 400, 405–409, 406*f*, 407*f*, 408*f*, 409*f*
in cat, 416–418, 417*f*, 419*f*, 420*f*
Body of bone
ilium, 140*f*
ischium, 140*f*
mandible, **111**, 112*f*
pubis, 140*f*
sphenoid, 109*f*, 113*f*
sternum, **120**, 121*f*, 530
vertebra, **117**, 117*f*, 118*f*, 119*f*, 119*t*, 120, 122*f*
Body cavities, 7–9, 7*f*, 8*f*, 9*f*, 10*f*
Body movements, 157–158, 158*f*, 159–160*f*
vertebrae and, 119*t*
Body orientation/direction, 6, 6*f*
Body of penis, 505*f*, **507**
Body planes/sections, 4*f*, **5**, 5*f*
Body of stomach, 462*f*, **463**
Body temperature. *See* Temperature
Body tube (microscope), **29**
Body of uterus, **509**, 510*f*
in cat, **515**, 515*f*, 516*f*
Bolus, **470**
Bone(s), **60**, **61**, 66*f*, 92–93, 92*f*, 93–98, 94*t*, 95*f*, 97*f*, 98*f*
chemical composition of, 95–96
classification of, 93
formation/growth/ossification of, 93, 95, **96**–98, 98*f*
gross anatomy of, 93–95, 95*f*
microscopic structure of, 96, 97*f*
Bone-conduction hearing, testing, 331–332, 332*f*
Bone markings, **93**, 94*t*
Bone marrow, 93–94, 95*f*, 432
Bone spurs, 165
Bone (osseous) tissue, **60**, **61**, 66*f*, 96, 97*f*

Bony labyrinth, **329**
Bony pelvis. *See* Pelvis
Bony thorax, 120, 530–532, 531f. *See also* Thoracic cage
Boutons, terminal (axon terminals), **175**, 175f, 176f, 241f, **242**
Bowel sounds, **533**
Bowman's (glomerular) capsule, **489**, 490f, 491f, 492f
Brachial (term), **2**, **3**, 3f
Brachial artery, 402f, **403**, 534f, **535**
  in cat, 292f, 415f, **416**
Brachialis muscle, 184f, 185f, 198f, 198t, 534f
  in cat, **223**, 224f
Brachial plexus, **285**–288, 285f, 286t, 287f, **529**, 529f
  in cat, 291, 292f
Brachial pulse, 535
Brachial vein, **407**, 409f
  in cat, 417f, **418**
Brachiocephalic trunk/artery, 384f, 399f, **400**, 401f, 402f
  in cat, **414**, 415f, 420f
  in sheep, 389f, **390**
Brachiocephalic veins, 406f, **407**, 408f, 409f, 431f
  in cat, 414f, **416**, 417f, 420f
Brachioradialis muscle, 184f, 185f, 198f, 198t, 199f, 201f, 534f, **535**
  in cat, **223**, 224f
  tendon of, 199f
Brachium, 134. *See also* Arm
Brain, 254–261
  arterial supply of, 400–403, 401f
  brain stem, 255f, **257**, 257f, 258f, 259, 267f
  cerebellum, 255f, **257**, 257f, 258f, 259, 262f, 267f
  cerebral hemispheres, **254**–256, 255–256f, 257–259, 258f, 267f
  cerebrospinal fluid and, **261**, 263f
  diencephalon, 255f, **256**–257, 257f, 258f, **259**
  embryonic development of, 254, 255f
  meninges of, **261**, 262f
  sheep, dissection of, 267–272, 268–269f, 270f, 271f, 272f
  veins draining, 407, 408f
Brain stem, 255f, **257**, 257f, 258f, 259, 267f
Brain vesicles, 255f
Breast, 512–513, 513f
Breathing, **442**. *See also under Respiratory*
  muscles of, 191t, 193f
Brevis (term), muscle name and, 182
Bridge of nose, **527**, 527f
Broad ligament, **509**, 510f
  in cat, 515
Broca's area, **256**, 256f
Bronchi, **18**, 444, 445f, 446f
  in cat, **449**
  in rat, **18**
Bronchial arteries, 399f, 403, **410**
Bronchial cartilages, **98**, 99f
Bronchial tree, **444**, 445f
Bronchioles, **444**, 445f, 447, 448f
Bronchomediastinal trunks, 431f
Brow ridges (superciliary arches), 526f, **527**
Brush border, 57f, 465, 466f
Brush border enzymes, **465**
Buccal (term), 2
Buccinator muscle, 186t, 187f, 188f
Bulbar conjunctiva, 306, 306f
Bulbo-urethral glands, 21f, 505f, **507**
  in cat, 495f, **513**, 514f
  in rat, 21f
Bulbous corpuscles, **246**, 247f
Bursa(e), 156
Buttocks, 226, 536, 537f
Calcaneal (term), **3**, 3f
Calcaneal tendon, 185f, 211f, 213f, 540f
  in cat, 230f
Calcaneus (heel bone), 143, 144, 144f, 540f
Calcification zone, bone growth/ossification and, 98f
Calcitonin, **351**
Calcitriol, 352
Calcium
  in bone, 95
  calcitonin in regulation of, 351
  parathyroid hormone in regulation of, 352
Calvaria, **106**
Canaliculi
  bile, **472**, 473f
  in bone, **96**, 97f
  lacrimal, **305**, 306f
Cancellous (spongy) bone, **93**, 95f, 96, 97f
Canines (eye teeth), **468**, 469f
Capillaries, 388f, 398, 398f, 400
  glomerular, 489–491, 490f, 491, 491f, 492
  lymphatic, **430**, 431f
  peritubular, 490f
  pulmonary, 388f, **410**, 445f
Capitate, 137, 138f
Capitulum, **136**, 136f
Capsular space, renal, 492
Capsule
  kidney, 486, 487f, 489f
    in cat, 493
  lymph node, **433**, 433f
  spleen, 434f
Carbohydrate digestion, by salivary amylase, 470
Cardia (cardial part of stomach), 462f, **463**
Cardiac circulation, 386–389, 388f
Cardiac impression, 446f
Cardiac muscle, **68**, 69f, 392, 392f. *See also* Heart
Cardiac notch, 446f, **447**
Cardiac skeleton, 386, 387f
Cardiac valves, 385f, 386, 387f
Cardiac veins, 384f, 385f, 388, 388f
Cardiac ventricles, 384f, 385f, **386**, 388f
  in cat, 413
  right versus left, 391, 392f
  in sheep, 389f, 390
Cardial part of stomach (cardia), 462f, **463**
Cardinal (lateral cervical) ligaments, 510f
Cardiovascular system, 16t, **384**. *See also* Blood vessel(s); Heart
Carotene, 81
Carotid artery, **529**, 529f
  common, 384f, 399f, **400**, 401f, 402f, 528
    in cat, **414**–416, 415f, 416, 420f, 448
  external, 399f, **400**, 401f, **528**
    in cat, 415f, **416**
  internal, 399f, **400**, 401f
    in cat, 415f, **416**
Carotid canal, **106**, 107f
Carotid pulse, 528, 529
Carpal (term), **2**, 3f
Carpals, 92f, **137**–138, 138f
Carpometacarpal joints, 164t
Carpus, **137**. *See also* Wrist
Cartilage(s)/cartilage tissue, **61**, 65f, 92, 98, 99f. *See also specific type*
Cartilaginous joints, 154f, **155**, 155f, 164–165t
Caruncle, lacrimal, **306**, 306f
Cataracts, **309**
Cat dissection
  blood vessels, 413–418, 415f, 417f, 419f, 420f
  digestive system, 354, 354f, 473–477, 474f, 475f, 476f, 477f
  endocrine system, 353–354, 353f, 354f, 355f
  lymphatic ducts, 434–435
  muscles, 215–230
    forelimb, 222–225, 224f
    hindlimb, 225–230, 226f, 227f, 229f, 230f
    neck and trunk, 217–222
      abdominal wall, 218f, 219–220, 219f
      anterior neck, 217–218, 217f
      deep laterodorsal trunk and neck, 221–222, 222f, 223f
      superficial chest, 218–219, 218f
      superficial shoulder/dorsal trunk and neck, 220–221, 220f
  preparation/incisions for, 215–216, 216f
  reproductive systems, 513–516, 514f, 515–516f
  respiratory system, 448–450, 449f, 450f
  spinal nerves, 291–292, 292f, 293f, 294f
  urinary system, 493–496, 494f, 495f
  ventral body cavity, 353–354, 353f, 354f, 413, 414f
Cauda equina, **280**, 281f, 285f
  in cat, 294f
Caudal (term), **6**, 6f
Caudal ("saddle block") anesthesia, 120, 280
Caudal artery, in cat, 415f, **416**
Caudal mesenteric vein, in cat, **418**
Caudate nucleus, **259**, 260f
  in sheep, 272f
Caudofemoralis muscle, in cat, **226**, 226f
Cavernous sinus, 408f
C (parafollicular) cells, **357**
Cecum, **20**, 20f, 459f, 464f, **467**, 468f
  in cat, **477**, 477f
  in rat, **20**
Celiac ganglia, 302
Celiac trunk, 399f, **403**, 403f, 404f
  in cat, 415f, **416**, 420f
Cell(s), **15**, **40**
  anatomy of, 40–44, 40f, 41f, 42f, 43t
  cytoplasm/organelles of, **40**, 40f, **41**–44, 42f, 43t
  differences/similarities in, 44–45
  division/life cycle of, **45**–48, 46–47f
  membrane of. *See* Plasma membrane
  microscopic examination of, 33–34, 33f, 34f, 40, 40f
  nucleus of, **40**, 40f, 42f
  permeability of, 40–41
Cell body, neuron, 67, 68f, **240**, 244f, 245f, 246f
Cell division, **45**–48, 46–47f
Cell junctions, epithelia, 54
Cell processes, neuron, 67, 68f, 241f, 242, 243, 246f
Cellular immunity, **432**
Cellular respiration, 44, 442
Cement (of tooth), **469**–470, 469f
Centimeter, 32t
Central artery
  retinal, 308f, 322f
  of spleen, 434f
Central (Haversian) canal, 66f, **96**, 97f
Central canal of spinal cord, 255f, **280**, 282f
Central nervous system (CNS), **254**. *See also* Brain; Spinal cord
  supporting cells in, 240, 240f
Central processes, neuron, 243, 244f, 246f
Central sulcus, **254**, 255f, 256f
Central vein
  of liver, 472, 473f
  retinal, 308f, 322f
Centrioles, 42f, 43t, **44**, 46f
Centromere, 46f
Centrosome/centrosome matrix, 42f, 44, 46f, 47f
Cephalad/cranial (term), **6**, 6f
Cephalic (term), **2**, **3**, 3f
Cephalic vein, **407**, 409f, 533f, 534f, **535**
  in cat, 417f, **418**
Cerebellar peduncles, in sheep, 270
Cerebellum, 255f, **257**, 257f, 258f, 259, 261f, 262f, 267f
  development of, 254, 255f
  in sheep, 269f, 270, 270f, 271f, 272
Cerebral aqueduct, 255f, 258f, **259**, 263f
  in sheep, 271, 271f
Cerebral arterial circle (circle of Willis), 401f, **402**
Cerebral arteries, 401f, **402**
Cerebral cortex/cerebral gray matter, 255f, **256**, 257, 260f
Cerebral fissures, **254**, 255–256f
  in sheep, 267, 267f, 270
Cerebral hemispheres (cerebrum), **254**–256, 255–256f, 257–259, 258f, 267f
  embryonic development of, 254, 255f
  in sheep, 269f, 270, 270f, 271f
Cerebral peduncles, **257**, 257f, 259
  in sheep, 268, 268f, 269f, 271, 271f
Cerebral white matter, 255f, **256**, 257–259, 260f
Cerebrospinal fluid, **261**, 263f
  analysis of, 280
Cerebrum. *See* Cerebral hemispheres
Ceruminous glands, **328**
Cervical (term), **2**, 3f
Cervical curvature, 116, 116f
Cervical enlargement, 281f, 285f
Cervical (cardinal) ligaments, lateral, 510f
Cervical lymph nodes, 431f, 528
Cervical nerve, transverse, 285f, 286t
Cervical plexus/cervical spinal nerves, 281f, 284, **285**, 285f, 286t, **529**, 529f
Cervical vertebrae, 115, 116f, 117f, 118–119, 118f, 119t
Cervix, **509**, 510f
Cheekbone. *See* Zygomatic bone
Cheeks, **459**
Cheeks (gluteal prominences), 536, 537f
Chemoreceptors, **342**. *See also* Olfaction/olfactory receptors; Taste
"Chenille stick" mitosis, 48
Chest muscles, in cat, 218–219, 218f
Chewing (mastication), 463
  muscles of, 188f, 188t
Chief cells, gastric, 462f, **463**
Childbirth, female pelvis and, 139, 141t
Cholesterol
  blood/plasma concentration of, 378
  in plasma membrane, 40, 41f
Cholinergic fibers, **300**
Chondroblasts, 98
Chondrocytes, 65f, 66f
Chordae tendinae, 385f, **386**, 387f
  in sheep, 391, 391f
Choroid, **307**, 308f, 309f
  pigmented coat of, in cow/sheep eye, **311**, 312f
Choroid plexus, 258f, **259**, **261**, 263f
Chromatids, 46f
Chromatin, **40**, 40f, 42f, 46f
Chromatophilic substance (Nissl bodies), 241f, **242**
Chromophobes, 356f, **357**
Chromosome(s), **40**, 46f, 47f
Chyme, **463**
Cilia
  olfactory, **342**, 342f
  tracheal, 57f, 444
Ciliary body, **307**, 308f
  in cow/sheep eye, **310**, 312f

Ciliary glands, **306**
Ciliary muscles, 307, 308*f*, 309
Ciliary processes, **307**, 308*f*, **309**
Ciliary zonule, 308*f*, **309**
Circle of Willis (cerebral arterial circle), 401*f*, **402**
Circular folds, **465**, 466*f*
Circular muscle structure, 183*f*
Circulation(s)
coronary/cardiac, 386–389, 388*f*
fetal, 411–413, 412*f*
hepatic portal, 410–411, 411*f*
in cat, 418, 419*f*
pulmonary, **386**, 388*f*, 410, 410*f*
systemic, **386**, 388*f*
arterial, 398, 398*f*, 399–400, 399*f*, 400–405, 401*f*, 402*f*, 403–404*f*, 405*f*
in cat, 414–416, 415*f*, 420*f*
venous, 398, 398*f*, 400, 405–409, 406*f*, 407*f*, 408*f*, 409*f*
in cat, 416–418, 417*f*, 419*f*, 420*f*
Circumcision, 507
Circumduction (body movement), **158**, 160*f*
Circumferential lamellae, 96, 97*f*
Circumflex artery, 384*f*, **388**, 388*f*
Circumflex femoral arteries, **405**, 405*f*
Circumflex humeral arteries, 402*f*, **403**
Cirrhosis, **472**
Cisterna chyli, **430**, 431*f*
in cat, **435**
Clavicle (collarbone), 92*f*, **134**, 134*f*, 135*f*, 138, 528*f*, 529*f*, 531*f*, **532**, 533*f*, **534**
Clavicular head, of sternocleidomastoid muscle, **528**, 528*f*
Clavicular notch, 121*f*
Clavodeltoid (clavobrachialis) muscle, in cat, 220*f*, **221**, 224*f*
Clavotrapezius muscle, in cat, 217*f*, 220*f*, **221**
Cleavage furrow, 47*f*
Cleft palate, **442**
Cleidomastoid muscle, in cat, **217**, 218
Clinical crown of tooth, 469
Clitoris, **509**, 509*f*, 510*f*
Clonal selection, **432**
Clotting (blood), **375**–376, 376*f*
cm. *See* Centimeter
CNS. *See* Central nervous system
Coagulation/coagulation time, **375**–376, 376*f*
Coarse adjustment knob (microscope), 28*f*, **29**
Coccygeal spinal nerve, 281*f*, 285*f*
Coccyx, 115, 116*f*, 119*f*, **120**, 140*f*, 529
in male versus female, 141*t*
Cochlea, **329**, 329*f*, 330–331, 330*f*
Cochlear duct, **329**, 329*f*, 330*f*, 331*f*
Cochlear nerve, **329**, 329*f*, 330*f*, 331*f*
Cock's comb (crista galli), **108**, 108*f*, 109*f*
Colic artery, **403**, 404*f*
Colic flexures, 467, 468*f*
Collagen fibers, 60–61, **60**, 60*f*, 62*f*, 63*f*, 64*f*, 66*f*
in bone, 95
Collarbone. *See* Clavicle
Collateral (prevertebral) ganglion, 301*f*, **302**
Collateral ligaments, of knee (fibular and tibial), **159**, 162*f*
Collaterals, axon, **242**
Collecting ducts, 490*f*, 491, **491**
Collecting lymphatic vessels, **430**, 431*f*, 432*f*
Colles' fracture, **535**
Colliculi, inferior and superior, **257**, 258*f*, 311*f*
in sheep, 270*f*
Colloid, in thyroid gland follicles, 355, 356*f*
Colon, 459*f*, 464*f*, **467**, 468*f*. *See also* Large intestine
in cat, 476*f*, **477**, 477*f*
Color blindness, 320
Color vision, cones in, 310, 320
Columnar epithelium, **54**, 55*f*
pseudostratified, **54**, 57*f*
simple, 57*f*
stratified, 59*f*
Commissures, 258*f*, 259
anterior and posterior, 258*f*
gray, **280**, 282*f*
lateral and medial, **306**
Common carotid artery, 384*f*, 399*f*, **400**, 401*f*, 402*f*, 528
in cat, **414**–416, 415*f*, 416, 420*f*, 448, 450*f*
Common fibular nerve, **288**, 290*f*, 291*t*, 538
in cat, **292**, 294*f*
Common hepatic artery, **403**, 403*f*
Common hepatic duct, **471**, 472*f*
in cat, **475**
Common iliac arteries, 399*f*, 404*f*, **405**, 405*f*
Common iliac veins, **406**, 406*f*, 407*f*, 408*f*
in cat, 417*f*, **418**, 420*f*
Common interosseous artery, 402*f*
Common tendinous ring, extrinsic eye muscles, 307*f*
Communicating arteries, 401*f*, **402**
Compact bone, **93**, 95*f*, 96, 97*f*
Compound microscope, **27**, 28*f*. *See also* Microscope(s)
Concentric lamellae, **96**
Conchae, nasal (nasal turbinates)
inferior, 110*f*, **113**, **442**, 443*f*
middle, 109*f*, 110*f*, **111**, **442**, 443*f*
superior, **111**, **442**, 443*f*
Condenser (microscope), 28*f*, **29**
Conducting zone structures, **444**
Conduction/conductive deafness, testing for, 331–332, 332*f*
Conductivity, neuron, 67
Condylar joint, **156**, 157*f*
Condylar process, 107*f*, **111**, 112*f*, 527
Condyle (bone marking), 94*t*. *See also* Lateral condyle; Medial condyle
Cones, 307, 309*f*, **310**
color blindness and, 320
Congenital umbilical hernia, **532**
Conjunctiva, **306**, 306*f*
Conjunctival sac, 306*f*
Conjunctivitis, **306**
Connective tissue, **60**–67, 60*f*, 61–67*f*. *See also specific type*
embryonic (mesenchyme), 61, 61*f*
epithelia supported by, **54**, 56*f*, 58*f*, 59*f*
as nerve covering, 245, 246*f*
of skeletal muscle, 174, 175*f*
Connective tissue fibers, 60, 60*f*, 61*f*, 62*f*, 63*f*, 64*f*, 66*f*
Connective tissue proper, **61**, 62–64*f*
Conoid tubercle, **134**, 135*f*
Constipation, **468**
Constrictor muscles, of pharynx, 463
Conus medullaris, **280**, 281*f*
Convergent muscle structure, 183*f*
Coracoacromial ligament, 166*f*
Coracobrachialis muscle, 192*f*, 198*f*
in cat, **225**
Coracohumeral ligament, **161**, 166*f*
Coracoid process, **134**, 135*f*
Cords, brachial plexus, 287*f*, 288
Cornea, **306**, 306*f*, 308*f*
in cow/sheep eye, 310, **311**, 312*f*
Corniculate cartilage, 444*f*
Cornua (horns)
hyoid bone, **113**, 113*f*, 115
uterine
in cat, 494*f*, **515**, 515*f*, 516*f*
in rat, 21*f*, 22
Coronal (frontal) plane/section, 4*f*, **5**, 5*f*
Coronal suture, **106**, 107*f*
Corona radiata (cerebral projection fibers), **259**
Corona radiata (oocyte), **512**, 512*f*
Coronary arteries, 384*f*, **387**, 388*f*, 399*f*, **400**
in cat, 413, **414**
Coronary circulation, 386–389, 388*f*
Coronary sinus, 385*f*, 386, **388**, 388*f*
in sheep, **391**, 391*f*
Coronary sulcus, **387**
Coronary vein, in cat, **418**
Coronoid fossa, **136**, 136*f*
Coronoid process
of mandible, 107*f*, **111**, 112*f*
of ulna, 136*f*, **137**, 137*f*
Corpora cavernosa, 504*f*, 505*f*, **507**, 508*f*
Corpora quadrigemina, **257**, 258*f*, 259
in sheep, 270, 271, 271*f*
Corpus albicans, 512*f*
Corpus callosum, 258*f*, **259**, 260*f*, 311*f*
in sheep, 271*f*
Corpus luteum, 511, **512**, 512*f*
Corpus spongiosum, 505*f*, **507**, 508*f*
Corrugator supercilii muscle, 186*t*, 187*f*
Cortex
adrenal, **352**, 356*f*, 357
cerebellar, 259, 261*f*
cerebral, 255*f*, **256**, 257, 260*f*
of hair, 83, 83*f*
kidney, **488**, 489*f*, 492, 492*f*
in cat, 493
in pig/sheep, **488**
lymph node, **433**, 433*f*
ovarian, 512*f*
Cortical nephrons, 490*f*, **491**
Cortical radiate arteries, **489**, 489*f*, 490*f*, 491*f*
Cortical radiate veins, **489**, 489*f*, 490*f*, 491*f*
Corticosteroids, **352**
Corticosterone, 352
Cortisol (hydrocortisone), 352
Cortisone, 352
Costal cartilages, 65*f*, **98**, 99*f*, 121*f*, 122*f*
Costal facets/demifacets, 118*f*, 120, 122*f*
Costal groove, 122*f*
Costal margin, 121*f*, 531*f*, **532**
Costal surface, **445**
Costocervical trunk, **400**, 401*f*, 402*f*
in cat, 415*f*, **416**
Costovertebral joint, 164*t*
Cough reflex, 442
Cow eye, dissection of, 310–311, 312*f*
Coxal (term), **2**, 3*f*
Coxal bones (ossa coxae/hip bones), **139**, 140*f*
Coxal (hip) joint, **158**, 161*f*, 165*t*
Cranial/cephalad (term), **6**, 6*f*
Cranial base, **106**
Cranial cavity, **7**, 7*f*
Cranial fossae, **106**, 108*f*
Cranial mesenteric vein, in cat, **418**, 419*f*
Cranial nerves, **262**–**267**, 264–265*t*, 266*f*
Cranial sutures, 106, 155
Craniosacral (parasympathetic) division of autonomic nervous system, **300**, 300*f*
function of, 300
Cranium, 92*f*, 106–111, **106**, 107*f*, 108*f*, 109*f*, 110–111*f*, 113*f*
surface anatomy of, 526*f*, 527, 527*f*
Crest (bone marking), 94*t*
Cribriform foramina, 108*f*, 110
Cribriform plates, 108*f*, **110**, 443*f*
Cricoid cartilage, 99*f*, **442**, 443*f*, 444*f*, **528**, 528*f*
in cat, **448**, 450*f*
Cricothyroid ligament, 444*f*, **528**, 528*f*
Cricotracheal ligament, 444*f*
Crista(e), mitochondrial, 43*t*, 44
Crista ampullaris, 329*f*, 333–334, **333**, 333*f*, 334*f*
Crista galli, **108**, 108*f*, 109*f*
Cross section, 4*f*, **5**, 5*f*
Crown of tooth, **469**, 469*f*
Cruciate ligaments, **159**, 162*f*
Crural (term), **2**, 3*f*
Crus of penis, 505*f*
in cat, 495*f*, **513**, 514*f*
Crypts
intestinal, **466**, 466*f*, 467*f*
tonsillar, **433**, 435*f*
CSF. *See* Cerebrospinal fluid
Cubital fossa, 534*f*, **535**
Cubital vein, median, **407**, 409*f*, 534*f*, **535**
in cat, 417*f*
Cuboid, 144*f*
Cuboidal epithelium, **54**, 55*f*
simple, 56*f*
stratified, 58*f*
Cuneiform cartilage, 444*f*
Cuneiforms, 144*f*
Curvatures of spine, 116, 116*f*
Curvatures of stomach, 462*f*, 463
Cusp(s), of tooth, 468
Cutaneous femoral nerve
anterior, 288, 289*f*
lateral, 288*t*, 289*f*
posterior, 290*f*, 291*t*
Cutaneous glands, 81–82, 82*f*
Cutaneous maximus muscle, in cat, **216**
Cutaneous membrane. *See* Skin
Cutaneous nerves, 285*f*
brachial plexus, 285*f*, 286*t*, 287*f*
lateral, in cat, 293*f*
Cutaneous receptors, 78*f*, 79, 81, 246–247, 247*f*, 248*f*
Cuticle
hair, 83, 83*f*
nail (eponychium), **82**, 83*f*
Cyanosis, 81
Cystic duct, **471**, 472*f*
in cat, **475**
Cytokinesis, **45**, 47*f*
Cytoplasm, **40**, 40*f*, 41–44, 42*f*
division of (cytokinesis), **45**, 47*f*
Cytoplasmic inclusions, **44**
Cytoplasmic organelles, 40, 40*f*, **41**–44, 42*f*, 43*t*
Cytoskeletal elements/cytoskeleton, 41*f*, 42*f*, 43*t*, **44**
Cytosol, **41**, 42*f*
Cytotoxic T cells, 432

Daughter cells, 45
Daughter chromosomes, 47*f*
Daughter nuclei, 45
DCT. *See* Distal convoluted tubule
Dead space, anatomical, 444
Deafness, 331. *See also* Hearing
testing for, 331–332, 332*f*
Deciduous (milk) teeth, **468**, 469*f*
Decubitus ulcers/bedsores, **81**, 81*f*
Decussation of pyramids, **257**, 257*f*
Deep/internal (term), **6**
Deep artery of arm, 402*f*, **403**
Deep artery of thigh (deep femoral artery), **405**, 405*f*
Deep fascia, **174**
Deep femoral vein, in cat, 417*f*, **418**
Deep fibular nerve, 290*f*
Deep palmar arch, 402*f*
Deep plantar arch, 407*f*
Deep venous palmar arch, 409*f*
Deltoid muscles, 183*f*, 184*f*, 185*f*, 191, 191*t*, 192*f*, 195*f*, 198*f*, 530*f*, 533*f*, **534**–535, 538*f*

in cat, 220*f*, **221**, 224*f*
for intramuscular injection, 534–535, 538*f*
Deltoid tuberosity, **136**, 136*f*
Demilunes, in salivary glands, 470, 470*f*
Dendrite(s), 68*f*, 241*f*, **242**, 245*f*
neuron classification and, 243, 244*f*
Dendritic cells, epidermal (Langerhans' cells), **79**, 80*f*
Dens, 117*f*, **118**
Dense connective tissue, **60**, 64*f*
Dental formula, **468**–469
for cat, 475
Denticulate ligaments, 281*f*
Dentin, 469*f*, **470**
Dentinal tubules, 469*f*
Dentition (teeth), 468–470, 469*f*
in cat, **475**
Depressor anguli oris muscle, 186*t*, 187*f*
Depressor labii inferioris muscle, 186*t*, 187*f*
Depth of field (microscope), **33**
Depth perception, **320**–321
using microscope, 33
Dermal papillae, 78*f*, **79**, 83*f*, **84**
Dermal plexus, 78*f*, 79
Dermis, 63*f*, 78, 78*f*, 79–81, 80*f*, 84*f*
Dermography, 85–86, 85*f*. *See also* Fingerprints
Descending aorta, **20**, 21*f*
in cat, 415*f*, **416**, 420*f*
in rat, **20**
Descending colon, 459*f*, **467**, 468*f*
in cat, 477
Descending (motor) tracts, 283, 283*f*
Desmosomes, in skin, 80*f*
Detrusor, 488*f*, 493
Diabetes insipidus, **351**
Diabetes mellitus, **352**
Diapedesis, **370**
Diaphragm, 7*f*, **18**, 19*f*, 191*t*, 193*f*, 446*f*, **447**
in cat, 354*f*, 414*f*, **449**, 449*f*, 450*f*
in rat, **18**, 19*f*
Diaphysis, **93**, 95*f*
Diarrhea, **467**–468
Diarthroses, **155**. *See also* Synovial joints
Diastole, **386**
Diencephalon, 255*f*, **256**–257, 257*f*, 258*f*, **259**
Differential white blood cell count, **371**–372, 372*f*
Digastric muscle, 190*f*, 190*t*
in cat, **217**, 217*f*
Digestion, **458**
chemical (enzymatic action), 458, 463
in large intestine, 467–468
liver in, 471
in mouth, 461–463
pancreatic enzymes in, 465, 471
physical, 458
salivary gland secretions in, 461, 463, 470
in small intestine, 465, 466*f*
in stomach, 463
Digestive system, 16*t*, 457–484, **458**, 459*f*. *See also specific organ*
accessory organs of, **458**, 459*f*
alimentary canal (GI tract), **458**–468, 459*f*. *See also specific organ*
in cat, 354, 354*f*, 475–477, 476*f*, 477*f*
histological plan of, 458, 460*f*
in cat, 354, 354*f*, 473–477, 474*f*, 475*f*, 476*f*, 477*f*
Digit(s). *See* Finger(s); Toe(s)
Digital (term), **2**, 3*f*
Digital arteries, 402*f*
Digital veins
in fingers, 409*f*
in toes, 407*f*
Dilator pupillae, **307**
Diopter window (ophthalmoscope), **321**
Dislocations, **165**
Distal (term), **6**, 6*f*
Distal convoluted tubule, 490*f*, **491**, 492
Distal radioulnar joint, 137*f*, 164*t*
Divisions, brachial plexus, 287*f*, 288
DNA, 40
replication of, 45, 46*f*
Dorsal (term), **3**, 3*f*, **6**, 6*f*
Dorsal body cavity, 7, 7*f*
Dorsal funiculus, 282*f*, **283**, 284*f*
Dorsal (posterior) horns, 263*f*, **280**, 282*f*, 284, 284*f*
Dorsal interossei muscles
of foot, 214*f*, 214*t*
of hand, 202*t*, 203*f*
Dorsalis pedis artery, **405**, 405*f*, 539*f*, **540**
pulse at, 539*f*, 540, **540**
Dorsalis pedis vein, **406**, 407*f*
Dorsal median sulcus, **280**, 281*f*, 282*f*, 284*f*
Dorsal metatarsal arteries, **405**, 405*f*
Dorsal metatarsal veins, 407*f*
Dorsal rami, **284**, 285*f*, 301*f*
Dorsal root, **280**, 281*f*, 282*f*, 284, 301*f*
Dorsal root ganglion, 244, 244*f*, 245*f*, **280**, 282*f*, 284, 301*f*
Dorsal scapular nerve, 286*t*, 287*f*
Dorsal venous arch, **407**, 407*f*, 539*f*, **540**
Dorsal venous network, **536**
Dorsiflexion (body movement), **158**, 160*f*
Dorsum (term), 6
Dorsum of foot, muscles on, 209*f*, 212*t*
Dorsum nasi, **527**, 527*f*
Duct system
female reproductive, 509
male reproductive, 504–505*f*, 506, 506*f*
mammary gland, 513, 513*f*
Ductus arteriosus, 412*f*, **413**
in sheep, **390**
Ductus (vas) deferens, **21**, 21*f*, **22**, 22*f*, 504*f*, 505*f*, **506**, 506*f*
in cat, 495*f*, **513**, 514*f*
in rat, **21**, 21*f*
Ductus venosus, 412*f*, **413**
Duke method, 375
Duodenal glands, **466**, 466*f*, 467*f*
Duodenal papilla, major, **465**, 472*f*
Duodenum, 459*f*, 462*f*, 464*f*, **465**, **466**, 467*f*, 472*f*
in cat, 476*f*, 477
Dural venous sinuses, 406*f*, **407**, 408*f*
Dura mater
brain, **261**, 262*f*, 263*f*
spinal cord, 280, 281*f*, 282*f*, 283
Dwarfism, 350
Dynamic equilibrium, **333**

Ear, 327–340
cartilage in, 65*f*, 98, 99*f*, 328
equilibrium and, 333–335, 333*f*, 334*f*
gross anatomy of, 328–330, 328*f*, 329*f*, 330*f*
hearing and, 330–333, 331*f*, 332*f*
microscopic anatomy of
of equilibrium apparatus, 333–334, 334*f*
of spiral organ, 330–331, 330*f*, 331*f*
otoscopic examination of, 329–330
surface anatomy of, 527, 527*f*
Eardrum (tympanic membrane), **328**, 328*f*
Earlobe (lobule of ear), **328**, 328*f*, 526*f*, **527**
Eccrine glands (merocrine sweat glands), 78*f*, **81**, 82*f*
Ectopic pregnancy, **509**
Edema, 430
Efferent arteriole, 490*f*, **491**, 491*f*
Efferent (motor) nerves, **245**, **254**
Efferent (motor) neurons, 241*f*, 242, **244**, 246*f*
Efferent vessels, lymphatic, 433, 433*f*
Eggs (ova), 352, 504, 511
fertilized, 504
implantation of, 509
Ejaculation, **506**
Ejaculatory duct, 504*f*, 505*f*, **506**
Elastic arteries, 399
Elastic cartilage, 65*f*, **98**, 99*f*
Elastic connective tissue, 64*f*
Elastic fibers, **60**, 60*f*, 62*f*, 64*f*
in skin, 79
Elastic (titin) filaments, 172*f*
Elbow/elbow joint, 136*f*, 154*f*, 164*t*
surface anatomy of, 138–139, 534*f*, 535
Embryonic connective tissue (mesenchyme), 61, 61*f*
Emmetropic eye, **318**, 319*f*
Emulsification, by bile, 471
Enamel (tooth), **469**, 469*f*
Encapsulated nerve endings, 246, 247*f*
Encephalitis, **261**
Endocardium, 385*f*, **386**
Endochondral ossification, **96**–98, 98*f*
Endocrine system/glands, 16*t*, **54**, 55*f*, 349–363, **350**. *See also specific gland and hormone*
in cat, 353–354, 353*f*, 354*f*, 355*f*
epithelial cells forming, **54**, 55*f*
gross anatomy/basic function of, 350–353, 351*f*, 353*f*
microscopic anatomy of, 355–357, 356*f*
Endolymph, **329**, 330*f*, 333, 333*f*
Endometrium, **509**, 510*f*, **511**, 511*f*
Endomysium, **174**, 175*f*
Endoneurium, **245**, 246*f*
Endoplasmic reticulum (ER), 40*f*, **42**, 43*t*
rough, 41, **42**, 42*f*, 43*t*
smooth, **42**, 42*f*, 43*t*
muscle cell (sarcoplasmic reticulum), **173**, 173*f*
Endosteum, **94**–95, 95*f*, 97*f*
Endothelium, blood vessel, 398*f*, 399
Energy
for active transport, 41
mitochondria producing, 44
Enterocytes, 466*f*
Enteroendocrine cells, 462*f*
Envelope, nuclear, **40**, 40*f*, 42*f*, 46*f*
Enzyme(s). *See also specific type*
in digestion, 463
Eosinophilia, 378
Eosinophils, 367*f*, 368*t*, 369*f*, **370**, 370*f*
Ependymal cells, 240, 240*f*
Epicardium (visceral pericardium), 8*f*, 385*f*, **386**
in sheep, 390
Epicondyle (bone marking), 94*t*. *See also* Lateral epicondyle; Medial epicondyle
Epicranial aponeurosis, 187*f*, 527
Epicranius muscle, 183, 184*f*, 185*f*, 186*t*, 187*f*, 527, 527*f*
Epidermal dendritic cells (Langerhans' cells), **79**, 80*f*
Epidermal/friction ridges (fingerprints), 79, 85–86, 85*f*, 536
Epidermis, 78, 78*f*, 79, 80*f*, 84*f*
Epididymis, 504*f*, **506**, 506*f*, 508, 508*f*
in cat, 495*f*, 514*f*, **515**
Epidural space, 282*f*
Epigastric region, **9**, 9*f*
Epiglottis, 99*f*, **442**, 443*f*, 444*f*, 461*f*
in cat, **448**, 449*f*, 450*f*, **474**
Epimysium, **174**, 175*f*
Epinephrine, **352**
Epineurium, **245**, 246*f*
Epiphyseal lines, **93**, 95*f*
Epiphyseal (growth) plate, **93**, 95, 155, 155*f*
Epiphysis, **93**, 95*f*
Epiploic appendages, 467, 468*f*
Epithalamus, 255*f*, 258*f*, **259**
Epithelial cells, cheek, 33, 34*f*
Epithelial tactile (Merkel) cells, **79**, 80*f*, 246
Epithelial tactile complexes (Merkel discs), **246**
Epithelial tissues/epithelium, **54**–55, 55*f*, 56–59*f*. *See also specific type*
Epitrochlearis muscle, in cat, 224*f*, **225**
Eponychium, **82**, 83*f*
Equator (spindle), 47*f*
Equilibrium, 333–335, 333*f*, 334*f*. *See also* Ear
dynamic, **333**
static, **333**
ER. *See* Endoplasmic reticulum
Erection, **507**
Erector spinae muscles, 196*f*, 197*t*, **529**, 530*f*
Erythrocyte(s) (red blood cells/RBCs), 67*f*, **366**, 367*f*, 368*t*, 369–370, 369*f*
Erythrocyte (red blood cell) antigens, blood typing and, **376**, 377*t*, 378*f*
Erythrocyte (red blood cell) count, total, **371**
Esophageal arteries, 399*f*, 403
Esophageal opening, in cat, **474**
Esophagus, **18**, 58*f*, 459*f*, 461*f*, 462*f*, **463**, 465*f*
in rat, **18**
Estrogen(s), **352**, 511
adrenal, 352
Ethmoidal air cells (sinuses), 109*f*, 113, 114*f*, 442
Ethmoidal labyrinths, 109*f*, **110**
Ethmoid bone, 107*f*, **108**–111, 108*f*, 109*f*, 110*f*, 111*f*, 113*f*
Eustachian tube. *See* Pharyngotympanic (auditory/eustachian) tube
Eversion (body movement), **158**, 160*f*
Excitability, neuron, 67
Exocrine glands, **54**, 55*f*
epithelial cells forming, **54**, 55*f*
Exocytosis, 42*f*
Extension (body movement), **158**, 159*f*
Extensor carpi radialis brevis muscle, 200*t*, 201*f*
in cat, 224*f*, **225**
tendon of, 201*f*
Extensor carpi radialis longus muscle, 185*f*, 199*f*, 200*t*, 201*f*
in cat, 224*f*, **225**
tendon of, 201*f*
Extensor carpi ulnaris muscle, 185*f*, 200*t*, 201*f*
in cat, 224*f*, **225**
Extensor digiti minimi muscle, 201*f*
Extensor digitorum brevis muscle, 209*f*, 212*t*
Extensor digitorum communis muscle, in cat, 224*f*, **225**
Extensor digitorum lateralis muscle, in cat, 224*f*, **225**
Extensor digitorum longus muscle, 183*f*, 184*f*, 209*f*, 210*t*, 539*f*, **540**
in cat, 227*f*, **228**
tendons of, 539*f*, **540**
Extensor digitorum muscle, 185*f*, 197, 200*t*, 201*f*, **536**
tendons of, 201*f*, 536*f*
Extensor expansion, 201*f*
Extensor hallucis brevis muscle, 209*f*
Extensor hallucis longus muscle, 209*f*, 210*t*, 539*f*, **540**
tendon of, 539*f*, **540**
Extensor indicis muscle, 201*f*

Extensor muscles of upper limb, 535
Extensor pollicis longus and brevis muscles, 200*t*, 201*f*, **535**
tendons of, 535, 536*f*
Extensor retinacula, superior and inferior, 209*f*
External/superficial (term), **6**
External acoustic (auditory) meatus/canal, **106**, 107*f*, **328**, 328*f*, 526*f*, **527**
External anal sphincter, 467
External carotid artery, 399*f*, **400**, 401*f*, **528**
in cat, 415*f*, **416**
External (outer) ear, **328**, 328*f*
cartilage in, 65*f*, 98, 99*f*, 328
External genitalia
of female (vulva), 21*f*, 22, 22*f*, **509**, 509*f*
in cat, **515**, 515–516*f*
in rat, 21*f*, 22
of male, 21, 21*f*, 22, 22*f*, 507
in rat, 21, 21*f*
External iliac artery, **405**, 405*f*
in cat, 415*f*, **416**, 420*f*
External iliac vein, **406**, 407*f*, 408*f*
in cat, 417*f*, **418**, 420*f*
External intercostal muscles, 191*t*, 193*f*
External jugular vein, 406*f*, **407**, 408*f*, 409*f*, 528, **529**, 529*f*
in cat, 217, 217*f*, 292*f*, 414*f*, **416**, 417*f*, 420*f*, 450*f*, 474*f*
External nares, in cat, 449*f*, 475*f*
External oblique muscle, 184*f*, 192*t*, 193*f*, 194*f*, 531*f*, **533**
in cat, 218*f*, 219*f*, **220**, 222*f*
External occipital crest, 107*f*, **108**, 110*f*
External occipital protuberance, 107*f*, **108**, 110*f*, 115, 526*f*, **527**
External os, 510*f*
External respiration, **442**
External urethral orifice, 21*f*, 486, 488*f*, **492**, 504*f*, 505*f*, 509, 509*f*, 510*f*
in rat, 21*f*
External urethral sphincter, 488*f*
Exteroceptors, **245**
Extracellular matrix, **60**, 60*f*, 65*f*
Extrafusal muscle fibers, 248*f*
Extrinsic eye muscles, **306**, 307*f*
in cow, 312*f*
Eye, 305–310, 306*f*, 307*f*, 308*f*, 309*f*. *See also under Visual and* Vision
cow/sheep, dissection of, 310–311, 312*f*
emmetropic, **318**, 319*f*
external anatomy/accessory structures of, 305–306, 306*f*, 307*f*
extrinsic muscles of, **306**, 307*f*
in cow, 312*f*
hyperopic, **318**, 319*f*
internal anatomy of, 306–310, 308*f*, 309*f*
myopic, **318**, 319*f*
ophthalmoscopic examination of, 321–322, 321*f*, 322*f*
retinal anatomy and, 309*f*, 310
Eyebrow, 306*f*
Eyelashes, **306**, 306*f*
Eyelids, **306**, 306*f*
Eye muscles, extrinsic, **306**, 307*f*
in cow, 312*f*
Eyepiece/ocular (microscope), **29**
Eye teeth (canines), **468**, 469*f*

Face. *See also under Facial*
bones of, 92*f*, **106**, 111–113, 112*f*
muscles of, 184*f*, 186*t*, 187*f*, **527**
surface anatomy of, 526*f*, 527, 527*f*
Facet (bone marking), 94*t*
Facial artery, **400**, 401*f*, 526*f*, **527**
pulse at, 526*f*, 527
Facial bones, 92*f*, **106**, 111–113, 112*f*
Facial expression, muscles controlling, 184*f*, 186*t*, 187*f*, **527**
Facial nerve (cranial nerve VII), 265*t*, 266*f*, 329*f*, 343
in sheep, 268, 269*f*
Facial vein, **407**, 408*f*
in cat, 417*f*, 450*f*, 474*f*
Falciform ligament, 464*f*, **471**, 472*f*
in cat, **475**
Fallen arches (flat feet), 144
Fallopian (uterine) tubes, **22**, 22*f*, **509**, 510*f*, 511, 511*f*
in cat, 494*f*, **515**, 515*f*, 516*f*
False pelvis, **139**
False ribs, 121, 121*f*
False vocal cords (vestibular folds), **442**, 443*f*, 444*f*
in cat, 448
Falx cerebelli, **261**, 262*f*
Falx cerebri, **261**, 262*f*
Farsightedness (hyperopia), **318**, 319*f*
Fascia
deep, **174**
superficial (hypodermis), **78**, 78*f*
Fascia lata (iliotibial band), in cat, **225**
Fascicle(s)/fasciculus
muscle, **174**, 175*f*
arrangements of/muscle structure and, 182, 183*f*
muscle name and, 182
nerve, **245**, 246*f*
Fat capsule, perirenal, 486, 487*f*
Fat cell(s), 60*f*, 62*f*, 67
Fat (adipose tissue), **60**, 62*f*, 67, 78*f*
Fauces, isthmus of, 443*f*
Feet. *See* Foot
Female
gonads in, 352, 353*f*. *See also* Ovary/ovaries
pelvis in, male pelvis compared with, 139, 141*t*
reproductive system in, 16*t*, 21*f*, 22, 22*f*, 504, 508–513, 509*f*, 510*f*, 511*f*, 512*f*, 513*f*
in cat, 515–516, 515–516*f*
in human, 22, 22*f*, 504, 508–513, 509*f*, 510*f*, 511*f*, 512*f*, 513*f*
in rat, 21*f*, 22
urethra in, 486, 488*f*, 510*f*
in cat, 494*f*, 496, 515, 515*f*, 516*f*
urinary system in, 486, 487*f*, 488*f*
in cat, 494*f*
Femoral (term), **2**, **3**, 3*f*
Femoral arteries, **405**, 405*f*, 538
in cat, 293*f*, 415*f*, **416**, 420*f*
circumflex, **405**, 405*f*
Femoral cutaneous nerve
anterior, 288, 289*f*
lateral, 288*t*, 289*f*
posterior, 290*f*, 291*t*
Femoral nerve, **288**, 288*t*, 289*f*
in cat, **291**–292, 293*f*, 294*f*
Femoral pulse, 538
Femoral triangle, **538**, 538*f*, 539*f*
in cat, **228**, 291–292, 416
Femoral vein, **407**, 407*f*, 538
in cat, 293*f*, 417*f*, **418**, 420*f*
Femoropatellar joint, **159**, 162*f*, 165*t*
Femur, 92*f*, **142**–143, 142*f*, 144, 539*f*
ligament of head of, **158**, 161*f*
Fertilization/fertilized egg, 504, 511
implantation and, 509
Fetal circulation, 411–413, 412*f*
Fetal skull, 122, 123*f*
Fetus, 504
circulation in, 411–413, 412*f*
Fibers, muscle. *See* Muscle fibers
Fibrin/fibrin mesh (clot), **376**, 376*f*
Fibrinogen, **376**, 376*f*
Fibroblast(s), 60*f*, 61, 62*f*, 63*f*, 64*f*
Fibrocartilage, 66*f*, **98**, 99*f*
Fibrous capsule of kidney, 486, 487*f*, 489*f*
in cat, 493
in pig/sheep, **486**
Fibrous joints, 154*f*, **155**, 164–165*t*
Fibrous layer of eye, **306**
Fibrous layer of synovial joint, 156, 156*f*
Fibrous pericardium, **386**
in sheep, 390
Fibrous sheath, of hair, 83*f*, 84
Fibula, 92*f*, **143**, 143*f*, 538, 539*f*
Fibular (term), **2**, 3*f*
Fibular artery, **405**, 405*f*
Fibular collateral ligament, **159**, 162*f*
Fibularis (peroneus) brevis muscle, 208*t*, 209*f*, 211*f*
Fibularis (peroneus) longus muscle, 184*f*, 185*f*, 208*t*, 209*f*, 211*f*, 213*f*
tendon of, 214*f*
Fibularis (peroneus) muscles, in cat, 227*f*, **228**
Fibularis (peroneus) tertius muscle, 209*f*, 210*t*
Fibular nerve
common, **288**, 290*f*, 291*t*, 538
in cat, **292**, 294*f*
deep and superficial, 290*f*
Fibular notch, 143*f*
Fibular retinaculum, 209*f*
Fibular vein, **407**, 407*f*
Field (microscope), **30**, 32–33, 32*t*
"Fight-or-flight" response, 300. *See also* Sympathetic (thoracolumbar) division of autonomic nervous system
Filiform papillae, in cat, 474
Filter switch (ophthalmoscope), 321*f*, **322**
Filtrate, 491
Filtration, **491**
glomerular, **491**
Filum terminale, **280**, 281*f*
Fimbriae, **509**, 510*f*
Fine adjustment knob (microscope), 28*f*, **29**
Finger(s)
bones of (phalanges), 92*f*, **138**, 138*f*
forearm muscles acting on, 199–201*f*, 199–200*t*
intrinsic hand muscles acting on, 202*t*, 203*f*
joints of, 164*t*
Fingerprints, 79, 85–86, 85*f*, 536
Fissure(s) (bone marking), 94*t*
Fissure(s) (cerebral), **254**, 255–256*f*
in sheep, 267, 267*f*, 270
Fissure(s) (lung), 446*f*, 447
Fixators (fixation muscles), **182**
Flat bones, **93**
Flat feet (fallen arches), 144
Flexion (body movement), **158**, 159*f*
plantar, **158**, 160*f*
Flexion creases, **536**, 537*f*
Flexor accessorius (quadratus plantae) muscle, 212*t*, 213*f*
Flexor carpi radialis muscle, 184*f*, 199*f*, 199*t*, 534*f*, **535**, 536*f*
in cat, 224*f*, **225**
tendon of, 199*f*, 203*f*, 535, 536*f*
Flexor carpi ulnaris muscle, 185*f*, 199*f*, 200*t*, 201*f*, **535**, 536*f*
in cat, 224*f*, **225**
tendon of, 203*f*, 536*f*
Flexor digiti minimi brevis muscle
of foot, 213*f*, 214*f*, 214*t*
of hand, 202*t*, 203*f*
Flexor digitorum brevis muscle, 212*t*, 213*f*
Flexor digitorum longus muscle, 210*t*, 211*f*
in cat, **230**, 230*f*
tendon of, 213*f*
Flexor digitorum profundus muscle, 199*f*, 200*t*
tendon of, 199*f*, 203*f*
Flexor digitorum superficialis muscle, 199*f*, 200*t*, 536*f*
tendon of, 199*f*, 203*f*, 536*f*
Flexor hallucis brevis muscle, 212*t*, 213*f*, 214*f*
Flexor hallucis longus muscle, 209*f*, 210*t*, 211*f*
in cat, **230**
tendon of, 213*f*, 214*f*
Flexor muscles, of upper limb, 197, 535
Flexor pollicis brevis muscle, 202*t*, 203*f*
Flexor pollicis longus muscle, 199*f*, 200*t*
tendon of, 199*f*, 203*f*
Flexor retinaculum, 199*f*, 203*f*, 213*f*
Floating (vertebral) ribs, 121, 121*f*
Flocculonodular lobe, 259
Floor of orbit, 113*f*
Flower spray endings, 248*f*
Fluid mosaic model of plasma membrane, 40, 41*f*
Foliate papillae, **343**, 343*f*, 344*f*
Follicle(s)
hair, 78*f*, 83*f*, **84**, 84*f*
lymphoid, 433*f*
ovarian, 511, 512, 512*f*
thyroid gland, **255**
of tonsils, **433**, 435*f*
Follicle-stimulating hormone (FSH), **350**, 351*f*
Fontanelles, **122**, 123*f*
Food, digestion of, **458**. *See also* Digestion
Foot. *See also* Lower limb
arches of, 144, 144*f*
bones of, 143–144, 144*f*
movements of, 158, 160*f*
muscles acting on/intrinsic muscles of, 204, 208–210*t*, 209*f*, 211*f*, 212–214*t*, 213–214*f*
surface anatomy of, 3*f*, 539*f*, 540, 540*f*
Footdrop, **288**
Foot processes, 489–491
Foramen (bone marking), 94*t*
Foramen lacerum, **106**, 107*f*, 108*f*
Foramen magnum, 107*f*, **108**, 108*f*
Foramen ovale (fetal heart), 412*f*, **413**
in sheep, **391**
Foramen ovale (skull), 107*f*, **108**, 108*f*, 109*f*
Foramen rotundum, **108**, 108*f*, 109*f*
Foramen spinosum, 107*f*, **108**, 108*f*, 109*f*
Forearm. *See also* Upper limb
bones of, 136–137, 137*f*
muscles of/muscles controlling, 184*f*, 185*f*, 198*f*, 198*t*, 199–201*f*, 199–200*t*
surface anatomy of, 535, 536*f*
Forebrain, **254**, 255*f*
Forelimb blood vessels, in cat, 292*f*
Forelimb muscles, in cat, 222–225, 224*f*
Forelimb nerves, in cat, 291, 292*f*
Foreskin (prepuce), 504*f*, 505*f*, **507**
in cat, **513**
Formed elements of blood (blood cells), **366**, 367*f*, 368–371, 368*t*, 369*f*
tissues forming (hematopoietic tissue), **60**
Fornix, 258*f*, **259**
in sheep, 270, 271*f*, 272*f*
Fornix (vaginal), 510*f*
Fossa (bone marking), 94*t*
Fossa ovalis, 385*f*, 412*f*, **413**
in sheep, **391**, 391*f*
Fovea capitis, **142**, 142*f*, **158**
Fovea centralis, 308*f*, **309**
Freckle(s), 79
Free edge of nail, **82**, 83*f*
Free (nonencapsulated) nerve endings, in skin, 78*f*, 79, 81, **246**, 247*f*
Frequency range of hearing, testing, 332
Friction/epidermal ridges (fingerprints), 79, 85–86, 85*f*, 536

Frontal (term), **2**, 3*f*
Frontal belly of epicranius muscle, 183, 184*f*, 186*t*, 187*f*, **527**, 527*f*
Frontal bone, **106**, 107*f*, 108*f*, 110*f*, 111*f*, 113*f*
in fetal skull, 123*f*
Frontal eye field, 256*f*
Frontal (anterior) fontanelle, 122, 123*f*
Frontal lobe, **254**, 255*f*, 257*f*
in sheep, 269, 271*f*
Frontal (coronal) plane/section, 4*f*, **5**, 5*f*
Frontal process, 112*f*, 113*f*
Frontal sinuses, 113, 114*f*, 442, 443*f*
Frontal suture, 123*f*
Frontonasal suture, 110*f*
FSH. *See* Follicle-stimulating hormone
Functional layer (stratum functionalis) (endometrium), **509**
Fundus
of eye, **307**
ophthalmoscopic examination of, 321–322, 321*f*, 322*f*
of stomach, 462*f*, **463**
of uterus, **509**, 510*f*
Fungiform papillae, **343**, 343*f*
Funiculi (dorsal/lateral/ventral), 282*f*, **283**, 284*f*
"Funny bone." *See* Medial epicondyle, of humerus
Fusiform muscle structure, 183*f*

Gallbladder, 459*f*, 464*f*, **471**, 472*f*, 532
in cat, **475**, 476*f*
Gamete(s), **504**
Ganglia (ganglion), **240**, 246*f*
dorsal root, 244, 244*f*, 245*f*
Ganglion cells of retina, 309*f*, **310**
Gas exchange, 444, 445*f*
external respiration, **442**
internal respiration, **442**
Gas (respiratory) transport, **442**
Gastric arteries, **403**, 403*f*
Gastric glands, 462*f*, **463**, 465*f*
microscopic structure of, 462*f*, 463
Gastric pits, 462*f*, 463, 465*f*
Gastric veins, **411**, 411*f*
Gastrocnemius muscle, 184*f*, 185*f*, 204, 207*f*, 209*f*, 210*t*, 211*f*, **540**, 540*f*
in cat, 226*f*, **227**–228, 227*f*, 230*f*
tendon of, 211*f*
Gastroduodenal artery, **403**, 403*f*
Gastroepiploic arteries, **403**, 403*f*
Gastroepiploic vein, 411*f*
Gastroesophageal junction, **465**, 465*f*
Gastroesophageal sphincter, **463**
Gastrointestinal (GI) tract (alimentary canal), **458**–468, 459*f*. *See also specific organ*
in cat, 354, 354*f*, 475–477, 476*f*
histology of, 458, 460*f*
Gastrosplenic vein, in cat, **418**, 419*f*
Gene(s), 40
General sensory receptors, 245–247, **246**, 247*f*, 248*f*. *See also under Sensory*
Geniculate nucleus, lateral, **310**, 311*f*
Genital (pubic) (term), **3**, 3*f*
Genitalia
external
of female (vulva), 21*f*, 22, 22*f*, **509**, 509*f*
in cat, **515**, 515–516*f*
in rat, 21*f*, 22
of male, 21, 21*f*, 22, 22*f*, 507
in rat, 21, 21*f*
internal, of female, 509–511, 510*f*
Genitofemoral nerve, 288*t*, 289*f*
Germinal centers
lymph node, **433**, 433*f*
of tonsils, **433**, 435*f*
Germinal epithelium (ovary), **511**, 512*f*
GH. *See* Growth hormone
Gigantism, 350
Gingiva (gums), 461*f*, **469**
Gingival sulcus, 469, 469*f*
Glabella, **106**, 110*f*
Gland(s). *See also specific type*
epithelial cells forming, 54, 55*f*
Glans penis, 504*f*, 505*f*, **507**
in cat, 495*f*, 514*f*
Glassy membrane, 83*f*, 84
Glaucoma, **309**
Glenohumeral (shoulder) joint, 154*f*, **161**, 164*t*, 166*f*
Glenohumeral ligaments, **161**, 166*f*
Glenoid cavity, **134**, 135*f*, 166*f*
Glenoid labrum, **161**, 166*f*
Glial cells (neuroglia), **67**, 68*f*, **240**, 240*f*
Globus pallidus, **259**, 260*f*
Glomerular (Bowman's) capsule, **489**, 490*f*, 491*f*, 492*f*
Glomerular filtration, **491**
Glomerulus (glomeruli), 488, **489**–491, 490*f*, 491, 491*f*, 492, 492*f*
Glossopharyngeal nerve (cranial nerve IX), 265*t*, 266*f*, 343
in sheep, 269*f*, 270
Glottis, **442**
Glucagon, **352**
Glucocorticoids, **352**
Glucose, insulin affecting, 352
Gluteal (term), **3**, 3*f*
Gluteal arteries, **405**, 405*f*
Gluteal (natal) cleft, **537**, 537*f*
Gluteal fold, **537**, 537*f*
Gluteal lines, 140*f*
Gluteal nerves, inferior and superior, 290*f*, 291*t*
Gluteal prominences, 536, 537*f*
Gluteal region, 536–538, 537*f*, 538*f*
for intramuscular injection, 537–538, 538*f*
Gluteal tuberosity, **142**, 142*f*
Gluteus maximus muscle, 185*f*, 204, 207*f*, 208*t*, **536**–537, 537*f*, 538*f*
in cat, **226**, 226*f*
Gluteus medius muscle, 185*f*, 207*f*, 208*t*, 537*f*, **538**, 538*f*
in cat, **226**, 226*f*
for intramuscular injection, **538**, 538*f*
Gluteus minimus muscle, 208*t*
Glycocalyx (carbohydrate side chains), 40, 41*f*
Glycolipids, in plasma membrane, 41*f*
Glycoproteins, in plasma membrane, 41*f*
Goblet cells
large intestine, 467*f*
small intestine, 57*f*, 466, 466*f*
tracheal, 57*f*, 447
Golgi apparatus, **42**–43, 42*f*, 43*t*
Gomphosis, **155**
Gonad(s), **352**, **504**. *See also* Ovary; Testis
in cat, **354**, 355*f*
Gonadal arteries, 399*f*, **403**, 404*f*
in cat, 415*f*, **416**, 420*f*
Gonadal veins, 406*f*, **407**, 408*f*
in cat, 417*f*, **418**
Gonadocorticoids (sex hormones), **352**
Gonadotropins, **350**, 351*f*
Gracilis muscle, 184*f*, 205*f*, 206*t*, 207*f*
in cat, **228**, 229*f*
Granular (juxtaglomerular) cells, 491, 492*f*
Granulocytes, 368*t*, **370**, 370*f*
Granulosa cells, 512*f*
Gravitational pull, in equilibrium, 333, 334*f*
Gray commissure, **280**, 282*f*
Gray matter
cerebellar, 259
cerebral, 255*f*, **256**, 257, 260*f*
of spinal cord, **280**, 282*f*
Gray ramus communicans, 301*f*, **302**
Great cardiac vein, 384*f*, 385*f*, **388**, 388*f*
Greater auricular nerve, 285*f*, 286*t*
Greater curvature of stomach, 462*f*, **463**
Greater omentum, **20**, 20*f*, **463**, 464*f*
in cat, 354*f*, 475
in rat, **20**
Greater sciatic notch, **139**, 140*f*
Greater trochanter, **142**, 142*f*, 144, **537**, 537*f*
Greater tubercle, **134**–136, 136*f*, 533*f*, **534**
Greater vestibular glands, **509**, 509*f*, 510*f*
Greater wings of sphenoid, 107*f*, **108**, 108*f*, 109*f*, 110*f*, 113*f*, 115
Great saphenous vein, **407**, 407*f*, 540
in cat, 417*f*, **418**
Groove (bone marking), 94*t*
Gross anatomy, **1**
Ground substance, 60–61, **60**, 60*f*, 61*f*
Growth (ossification) centers
in fetal skull, **122**, 123*f*
primary, 98
Growth hormone (GH), **350**, 351*f*
Growth (epiphyseal) plate, **93**, 95, 155, 155*f*
Gums (gingiva), 461*f*, **469**
Gustatory cortex, 256*f*
Gustatory epithelial cells, **343**, 343*f*
Gustatory hairs, **343**, 343*f*
Gyri (gyrus), **254**, 255*f*

Hair(s), **83**–84, 83*f*
Hair bulb, 83*f*, **84**
Hair cells
in equilibrium, 333, **333**, 333*f*
in hearing, 329, 330–331, 330*f*, 331
Hair follicle, 78*f*, 83*f*, **84**, 84*f*
Hair follicle receptor (root hair plexus), 78*f*, 81, **246**, 247*f*
Hair matrix, 83*f*
Hair root, 78*f*, **83**, 83*f*
Hair shaft, 78*f*, **83**, 83*f*, 84*f*
Hallux (term), **2**, 3*f*
Hamate, 138, 138*f*
Hammer (malleus), **328**, 328*f*
Hamstring muscles, 185*f*, 207*f*, 208*t*, **538**, 540*f*
in cat, **226**–227, 226*f*
Hand (manus), **2**, **3**, 3*f*. *See also* Upper limb
bones of, 137–138, 138*f*
forearm muscles acting on, 199–201*f*, 199–200*t*
intrinsic muscles of, 202*t*, 203*f*
surface anatomy of, 3*f*, 535–536, 536*f*, 537*f*
Haptens, 430
Hard palate, 17, 107*f*, **442**, 443*f*, **459**, 461*f*
in cat, 449*f*, **474**, 475*f*
in rat, 17
Haustra, **467**, 468*f*
Haversian (central) canal, 66*f*, **96**, 97*f*
Haversian system (osteon), **96**, 97*f*
Hb. *See* Hemoglobin
HDL. *See* High-density lipoprotein
Head
arteries of, 399*f*, 400, 401*f*
muscles of, 182–183, 184*f*, 185*f*, 186–188*t*, 187–188*f*
surface anatomy of, 3*f*, 526*f*, 527, 527*f*
veins of, 406*f*, 407, 408*f*
Head of bone, 94*t*
of femur, 139, 142, 142*f*
ligament of, **158**, 161*f*
of fibula, 143*f*, **538**, 539*f*
of humerus, 134, 136*f*
of metacarpals, 138, 138*f*
of radius, 136–137, 136*f*, 137*f*, 534*f*, **535**
of ulna, **137**, 137*f*, 534*f*, **535**, 536*f*
Head (microscope), 28*f*, **29**
Head of sperm, 507, 507*f*
Hearing, 330–333, 331*f*, 332*f*. *See also* Ear
cortical areas in, 256, 256*f*
Hearing loss, 331
testing for, 331–332, 332*f*
Heart, **18**, 19*f*, 383–396, **384**, 532. *See also under Cardiac*
anatomy of, 383–396
gross (human), 384–386, 384*f*, 384–385*f*, 387*f*
microscopic (cardiac muscle), **68**, 69*f*, 392, 392*f*
in sheep, 389–392, 389*f*, 391*f*, 392*f*
blood supply of, 386–389, 388*f*
in cat, **354**, 354*f*, **413**, 414*f*, 450*f*
chambers of, 384*f*, 385*f*, 386
fibrous skeleton of, 386, 387*f*
location of, 384, 384*f*, 446*f*, 532
in rat, **18**, 19*f*
in sheep, 389–392, 389*f*, 391*f*, 392*f*
Heart valves, 385*f*, 386, 387*f*
Heat, bone affected by, 96
Heel bone (calcaneus), 143, 144, 144*f*, 540*f*
Helix of ear, 328*f*, 526*f*, **527**
Helper T cells, 432
Hematocrit, **372**–373, 373*f*
Hematologic tests, 371–378, 372*f*, 373*f*, 375*f*, 376*f*, 377*t*, 378*f*. *See also specific test*
Hematopoietic stem cells, 370
Hematopoietic tissue, **60**
Heme, 373
Hemiazygos vein, **409**, 409*f*, 431*f*
Hemoglobin, 373
Hemoglobin concentration, 373–375, 375*f*
Hemoglobinometer, 374–375, 375*f*
Hemostasis, 375–376, 376*f*
Hepatic artery, 472, 472*f*
common, **403**, 403*f*
Hepatic artery proper, **403**, 403*f*
Hepatic duct, 472*f*
common, **471**, 472*f*
in cat, **475**
Hepatic (right colic) flexure, **467**, 468*f*
Hepatic (stellate) macrophages, **472**, 473*f*
Hepatic portal circulation, 410–411, 411*f*
in cat, 418, 419*f*
Hepatic portal vein, **407**, **410**, 411*f*, 472, 472*f*, 473*f*
in cat, **418**, 419*f*
Hepatic veins, 406*f*, **407**, 408*f*, 411*f*, 472*f*
in cat, 417*f*, **418**, 420*f*
Hepatitis, **471**–472
Hepatocytes (liver cells), **472**, 473*f*
Hepatopancreatic ampulla, **465**, 472*f*
in cat, **475**
Hepatopancreatic sphincter, **465**
Hernia
inguinal, **532**, 533*f*
umbilical, 532
Herniated disc, **116**
High-density lipoprotein (HDL), 378
High-power lens (microscope), **29**
Hilum
lung, **444**, 445, 446*f*
lymph node, **433**, 433*f*
renal, **486**, 487*f*
spleen, 434*f*
Hindbrain, **254**, 255*f*, 259
in sheep, 268

Hindlimb muscles, in cat, 225–230, 226*f*, 227*f*, 229*f*, 230*f*
Hindlimb nerves, in cat, 291–292, 293*f*, 294*f*
Hinge joint, **156**, 157*f*
Hip bones (coxal bones/ossa coxae), **139**, 140*f*
Hip fracture, 142
Hip (pelvic) girdle, **139**. *See also* Hip joint; Pelvis
bones of, 92*f*, **139**–142, 140*f*, 141*t*
surface anatomy of, 144
Hip joint, **158**, 161*f*, 165*t*
Hip muscles, 185*f*, 207*f*
in cat, 225–228, 226*f*
Hirsutism, **352**
Histamine, 370
Histology, **54**
Horizontal cells, retinal, 309*f*, 310
Horizontal fissure (lung), 446*f*
Horizontal plate of ethmoid bone, 110
Horizontal plate of palatine bone, 107*f*, 110
Hormone(s), **350**. *See also specific hormone and* Endocrine system/glands
Horns (cornua)
hyoid bone, **113**, 113*f*, 115
uterine
in cat, 494*f*, **515**, 515*f*, 516*f*
in rat, 21*f*, 22
Human(s)
reproductive systems in. *See also* Reproductive system
female, 22, 22*f*, 504, 508–513, 509*f*, 510*f*, 511*f*, 512*f*, 513*f*
male, 22, 22*f*, 504, 504–508, 504–505*f*, 506*f*, 507*f*, 508*f*
torso/organ system identification and, 23–24, 23*f*
Humeral arteries, circumflex, 402*f*, **403**
Humeral ligament, transverse, 166*f*
Humerus, 92*f*, 95*f*, **134**–136, 136*f*, 138
muscles of acting on forearm, 198*f*, 198*t*
Humoral immunity, **432**. *See also* Antibodies
Hyaline cartilage, 65*f*, **98**, 99*f*, 156, 156*f*
tracheal. *See* Tracheal cartilages
"Hyaline cartilage" bones, ossification of, 96–98, 98*f*
Hydrocephalus, **261**
Hydrochloric acid
bone affected by, 96
in digestion, 463
Hydrocortisone (cortisol), 352
Hydrolases, acid, 43
Hymen, **509**, 509*f*, 510*f*
Hyoid bone, **113**, 113*f*, 115, 443*f*, 444*f*, 461*f*, **528**, 528*f*
in cat, 448, 449*f*
Hyperextension (body movement), 158, 159*f*
Hyperopia (farsightedness), **318**, 319*f*
Hypertrophic zone, bone growth/ossification and, 98*f*
Hypochondriac regions, **9**, 9*f*
Hypochromic anemia, macrocytic and microcytic, 378
Hypodermis (superficial fascia), **78**, 78*f*
Hypogastric ganglia, inferior, 302
Hypogastric (pubic) region, **9**, 9*f*
Hypoglossal canal, **108**, 108*f*
Hypoglossal nerve (cranial nerve XII), 265*t*, 266*f*, 285*f*
in sheep, 269*f*, 270
Hypoglycemia, **352**
Hyponychium, **82**
Hypophyseal arteries, inferior and superior, 351*f*
Hypophyseal fossa, 108*f*, 109*f*
Hypophyseal portal system, **350**, 351*f*
Hypophysis (pituitary gland), **257**, 257*f*, 258*f*, 259, **350**–351, 351*f*, 353*f*
anatomy/basic function of, 350–351, 351*f*
microscopic anatomy of, 356*f*, 357
in sheep, 271*f*
Hypotension, orthostatic, **300**
Hypothalamo-hypophyseal tract, 351, 351*f*
Hypothalamus, 255*f*, 258*f*, **259**, 353*f*
pituitary gland relationships and, 350, 351*f*
in sheep, 271, 272*f*
Hypothenar eminence, **536**, 537*f*
Hypothenar muscles, 202*t*
H zone, 172*f*, 173*f*

I bands, 172*f*, 173, 173*f*
Ileocecal valve, **465**, 468*f*
in cat, **477**, 477*f*
Ileocolic artery, **403**, 404*f*
Ileum, 459*f*, 464*f*, **465**, **466**–467, 467*f*, 468*f*
in cat, 476*f*, 477, 477*f*
Iliac arteries
common, 399*f*, 404*f*, **405**, 405*f*
internal and external, **405**, 405*f*
in cat, 415*f*, **416**, 420*f*
Iliac crest, **139**, 140*f*, 144, **529**, 530*f*, 531*f*, **532**
Iliac fossa, **139**, 140*f*
Iliac (inguinal) regions, **9**, 9*f*
Iliac spines
anterior inferior, 140*f*
anterior superior, **139**, 140*f*, 144, 531*f*, **532**, 533*f*
posterior inferior, 140*f*
posterior superior, **139**, 140*f*, 144, **537**, 537*f*
Iliacus muscle, 205*f*, 206*t*
Iliac veins
common, **406**, 406*f*, 407*f*, 408*f*
in cat, 417*f*, **418**, 420*f*
external and internal, **406**, 407*f*, 408*f*
in cat, 417*f*, **418**, 420*f*
Iliocostalis muscles, 196*f*, 197*t*
Iliofemoral ligament, **158**, 161*f*
Iliohypogastric nerve, 288*t*, 289*f*
Ilioinguinal nerve, 288*t*, 289*f*
Iliolumbar arteries, in cat, 415*f*, **416**, 420*f*
Iliolumbar veins, in cat, 417*f*, **418**, 420*f*
Iliopsoas muscle, 184*f*, 205*f*, 206*t*
in cat, **228**, 229*f*
Iliotibial band (fascia lata), in cat, **225**
Iliotibial tract, 185*f*, 207*f*, 208*t*
Ilium, **139**, 140*f*
Image (microscope), real and virtual, **29**, 29*f*
Immune response/immunity, **430**–432
deficient, 432
Immune system, 15, 16*t*. *See also* Immune response; Lymphatic system
adaptive, **430**
Immunocompetence, 432
Immunodeficiency, 432
Implantation, 509
Incisive fossa, 107*f*, **112**
Incisors, **468**, 469*f*
Inclusions, cytoplasmic, **44**
Incontinence, **492**
Incus (anvil), **328**, 328*f*
Indirect pathway, 259
Inferior (term), **6**, 6*f*
Inferior angle of scapula, 134, 135*f*, **529**, 530*f*
Inferior articular process/facet, **117**, 117*f*, 118*f*, 119*t*, 120
Inferior colliculi, **257**, 258*f*
in sheep, 270*f*
Inferior extensor retinacula, 209*f*
Inferior gluteal artery, **405**
Inferior gluteal line, 140*f*
Inferior gluteal nerve, 290*f*, 291*t*
Inferior horns, 260*f*, 263*f*
Inferior hypogastric ganglia, 302
Inferior hypophyseal artery, 351*f*
Inferior mesenteric artery, 399*f*, 404*f*, **405**
in cat, 415*f*, **416**, 420*f*
Inferior mesenteric ganglia, 302
Inferior mesenteric vein, **411**, 411*f*
in cat, **418**
Inferior nasal conchae/turbinates, 110*f*, **113**, **442**, 443*f*
Inferior nasal meatus, 443*f*
Inferior nuchal line, 107*f*, 110*f*
Inferior oblique muscle of eye, 307*f*
Inferior orbital fissure, 110*f*, 113*f*
Inferior phrenic artery, 399*f*, 404*f*
Inferior phrenic vein, 408*f*
Inferior pubic ramus, 140*f*
Inferior rectus muscle of eye, 307*f*
Inferior sagittal sinus, 408*f*
Inferior tibiofibular joint, 143*f*, 165*t*
Inferior vena cava, **21**, 21*f*, 384*f*, 385*f*, **386**, 403*f*, **405**, 406*f*, 408*f*, 409*f*, 431*f*
in cat (postcava), 293*f*, **413**, 417*f*, **418**, 419*f*, 420*f*
in rat, **21**, 21*f*
in sheep, 389*f*, 390, 391*f*
veins draining into, 406–407, 406*f*, 407*f*, 408*f*
Inferior vertebral notch, 118*f*
Infraglenoid tubercle, 135*f*
Infrahyoid muscles, **529**, 529*f*
Infraorbital foramen, 107*f*, 110*f*, **112**, 112*f*, 113*f*, 115
Infraorbital groove, 113*f*
Infrapatellar bursa, 162*f*
Infrapatellar fat pad, 162*f*
Infraspinatus muscle, 161, 185*f*, 194*t*, 195*f*, 198*f*
in cat, **222**
Infraspinous fossa, 134, 135*f*
Infrasternal angle, 531*f*, **532**
Infundibulum (pituitary), **259**, **350**, 351*f*
in sheep, 268, 268*f*, 269*f*, 271, 271*f*
Infundibulum (uterine tube), 510*f*
Inguinal (term), **2**, 3*f*
Inguinal hernia, **532**, 533*f*
Inguinal ligament, 139, 193*f*, 194*f*, 531*f*, **532**, 533*f*, **538**, 538*f*
Inguinal lymph nodes, 431*f*, 538
Inguinal (iliac) regions, **9**, 9*f*
Inguinal ring, superficial, **532**, 533*f*
Initial segment, axon, 241*f*, 242
Inner (internal) ear, 328*f*, **329**
equilibrium and, 333–335, 333*f*, 334*f*
hearing and, 330–333, 331*f*, 332*f*
Insertion (muscle), **157**, 158*f*, 174
muscle name and, 182
Insula, **254**
Insulin, **352**
Integral proteins, in plasma membrane, 41*f*
Integument/integumentary system, 16*t*, 77–90, **78**, 78*f*. *See also* Skin
Interatrial septum, **386**
Intercalated discs, **68**, 69*f*, **392**, 392*f*
Intercarpal joints, 164*t*
Intercondylar eminence, **143**, 143*f*
Intercondylar fossa, **142**, 142*f*
Intercostal arteries
anterior, 402*f*, **403**
in cat, 415*f*, **416**
posterior, 399*f*, 402*f*, **403**
Intercostal muscles, 184*f*, 191*t*, 193*f*, 446*f*
Intercostal nerves, **284**, 285*f*
Intercostal spaces, 121*f*
Intercostal veins, 409*f*
in cat, 418
Interlobar arteries, **489**, 489*f*
Interlobar veins, **489**, 489*f*
Interlobular veins, in liver, 473*f*
Intermaxillary suture, 107*f*
Intermediate filaments, 42*f*, 43*t*, **44**
Intermediate mass (interthalamic adhesion), 258*f*, **259**
in sheep, 271, 271*f*, 272*f*
Intermediate part of urethra, 486, 488*f*, 504*f*, 505*f*, **506**
Internal/deep (term), **6**
Internal acoustic meatus, **106**, 108*f*
Internal anal sphincter, 467
Internal capsule, **259**
Internal carotid artery, 399*f*, **400**, 401*f*
in cat, 415*f*, **416**
Internal (inner) ear, 328*f*, **329**
equilibrium and, 333–335, 333*f*, 334*f*
hearing and, 330–333, 331*f*, 332*f*
Internal genitalia, in female, 509–511, 510*f*
Internal iliac artery, **405**, 405*f*
in cat, 415*f*, **416**, 420*f*
Internal iliac vein, **406**, 407*f*, 408*f*
in cat, 417*f*, **418**, 420*f*
Internal intercostal muscles, 191*t*, 193*f*
Internal jugular vein, 406*f*, **407**, 408*f*, 409*f*, 431*f*, 528, 529*f*
in cat, **416**, 417*f*, 448
Internal oblique muscle, 184*f*, 192*t*, 193*f*, 194*f*
in cat, 219*f*, **220**
Internal os, 510*f*
Internal pudendal artery, 405
Internal respiration, **442**
Internal thoracic artery, 401*f*, 402*f*, **403**
in cat, 415*f*, **416**
Internal thoracic (mammary) vein, in cat, **416**, 417*f*
Internal urethral sphincter, 488*f*, **492**
in cat, 493
Interneurons (association neurons), **244**
Interoceptors, **245**
Interossei muscles
of foot, 214*f*, 214*t*
dorsal, 214*f*, 214*t*
plantar, 214*f*, 214*t*
of hand, 201*f*, 202*t*, 203*f*
dorsal, 202*t*, 203*f*
palmar, 202*t*, 203*f*
Interosseous artery, common, 402*f*
Interosseous membrane, 137*f*, 143*f*
Interphalangeal joints
of fingers, 164*t*
of toes, 165*t*
Interphase, **45**, 46*f*
Interstitial endocrine cells (testis), 506*f*, **507**
Interstitial lamellae, 96, 97*f*
Intertarsal joint, 165*t*
Interthalamic adhesion (intermediate mass), 258*f*, **259**
in sheep, 271, 271*f*, 272*f*
Intertrochanteric crest, **142**, 142*f*
Intertrochanteric line, **142**, 142*f*
Intertubercular sulcus (bicipital groove), **136**, 136*f*
medial, 533*f*, **535**
Interventricular arteries
anterior, 384*f*, **388**, 388*f*
posterior, 385*f*, **387**, 388*f*
Interventricular foramen, 258*f*, **259**, **261**, 263*f*
in sheep, 271
Interventricular septum, 385*f*, **386**, 387*f*
Intervertebral discs, **98**, 99*f*, **115**–116, 116*f*, 154*f*, 164*t*
herniated, **116**
Intervertebral foramina, 116*f*, **117**, 280
Intervertebral joints, 155, 164*t*
Intestinal arteries, 404*f*
Intestinal crypts, **466**, 466*f*, 467*f*
Intestinal trunks, 431*f*

Intrafusal muscle fibers, **247**, 248*f*
Intramuscular injections, 537–538, 538*f*
Intrinsic muscles of foot, 209*f*, 212–214*t*, 213–214*f*
Intrinsic muscles of hand, 202*t*, 203*f*
Intrinsic nerve plexuses, in alimentary canal, 458, 460*f*
Inversion (body movement), **158**, 160*f*
Involuntary nervous system, 254, 299. *See also* Autonomic nervous system
Ion channels, in neuromuscular junction, 175, 175*f*
Iris, **307**, 308*f*
  in cow/sheep eye, **311**
Iris diaphragm lever (microscope), 28*f*, **29**
Irregular bones, **93**
Irregular connective tissue, 61
Ischial ramus, 140*f*
Ischial spine, **139**, 140*f*
  in male versus female, 141*t*, 142
Ischial tuberosity, **139**, 140*f*, **537**, 537*f*
Ischiofemoral ligament, **158**
Ischium, **139**, 140*f*
Ishihara's color plates, 320
Islets of Langerhans (pancreatic islets), 356*f*, **357**, **471**, 471*f*
Isthmus
  of fauces, 443*f*
  of thyroid gland, 528
  of uterine tube, 510*f*
Ivy method, 375

Jaundice, **81**, 471
Jejunum, 459*f*, 464*f*, **465**, 472*f*
JGC. *See* Juxtaglomerular complex
JG cells. *See* Juxtaglomerular (granular) cells
Joint(s), 93, 153–170, 154*f*, **155**, 164–165*t*. *See also specific joint*
  cartilaginous, 154*f*, **155**, 155*f*, 164–165*t*
  disorders of, 165
  fibrous, 154*f*, **155**, 164–165*t*
  synovial, 154*f*, **155**–161, 156*f*, 157*f*, 158*f*, 159–160*f*, 164–165*t*
Joint (synovial) cavities, 156, 156*f*
Jugular foramen, **106**, 107*f*, 108*f*
Jugular notch, **120**, 121*f*, 134, **528**, 528*f*, 531*f*
Jugular trunks, 431*f*
Jugular vein, **529**
  external, 406*f*, **407**, 408*f*, 409*f*, 528, **529**, 529*f*
    in cat, 217, 217*f*, 292*f*, 414*f*, **416**, 417*f*, 420*f*, 450*f*, 474*f*
  internal, 406*f*, **407**, 408*f*, 409*f*, 431*f*, 528, 529*f*
    in cat, **416**, 417*f*, 448
  transverse, in cat, 420*f*
Juxtaglomerular (granular) cells, 491, 492*f*
Juxtaglomerular complex, **491**
Juxtamedullary nephrons, 490*f*, **491**

Keratin, **79**
Keratinocytes, **79**, 80*f*
Keratohyalin granules, 79
Kidney(s), **20**, 21*f*, **486**, 487*f*. *See also under Renal*
  in cat, **354**, 355*f*, 414*f*, **493**, 494*f*, 495*f*
  functional microscopic anatomy of, 489–492, 490*f*, 491*f*, 492*f*
  gross anatomy of, 486–489, 487*f*, 489*f*, 490*f*
  in pig/sheep, 486–489
  in rat, **20**, 21*f*
Kinetochore(s), 46*f*
Kinetochore microtubules, 46*f*
Kinocilia, in equilibrium, 333*f*, 334*f*
Knee/knee joint, **159**–160, 162*f*, 165*t*, 538, 539*f*, 540
  bones of, 142*f*
Kneecap. *See* Patella
Knuckles (metacarpophalangeal joints), 139, 164*t*
Kyphosis, 116, 116*f*

Labia (lips), **459**, 461*f*
Labial frenulum, 461*f*
Labia majora, **509**, 509*f*, 510*f*
  in cat, **515**
Labia minora, **509**, 509*f*, 510*f*
Labrum
  acetabular, **158**, 161*f*
  glenoid, **161**, 166*f*
Labyrinth, 328*f*. *See also* Internal (inner) ear
  bony, **329**
  membranous, **329**, 333*f*
Lacrimal apparatus, **305**, 306*f*
Lacrimal bone, 107*f*, 110*f*, **113**, 113*f*
Lacrimal canaliculi, **305**, 306*f*
Lacrimal caruncle, **306**, 306*f*
Lacrimal fossa, 107*f*, **113**, **527**, 527*f*
Lacrimal glands, **305**, 306*f*
Lacrimal puncta, **305**, 306*f*
Lacrimal sac, **305**
Lacteal, 466*f*
Lactiferous ducts, **513**, 513*f*
Lactiferous sinus, **513**, 513*f*
Lacunae
  in bone, 60, 66*f*, **96**, 97*f*
  connective tissue matrix, 60, 65*f*, 66*f*
Lambdoid suture, 107*f*, **108**, 110*f*
Lamella(e), in bone, 66*f*, **96**, 97*f*
  circumferential, 96, 97*f*
  concentric, **96**
  interstitial, 96, 97*f*
Lamellar corpuscles, 78*f*, **79**, 81, **246**, 247, 247*f*
Lamellar granules, 79
Lamina(e), vertebral, 117*f*
Lamina propria
  alimentary canal, 458, 460*f*
    gastric, 465*f*
    large intestine, 467*f*
    small intestine, 467*f*
  bronchiole, 448*f*
  tracheal, 447*f*
Landmarks, anatomical, 2–5, 3*f*. *See also* Surface anatomy
Langerhans, islets of (pancreatic islets), 356*f*, **357**, **471**, 471*f*
Langerhans' cells (dendritic cells), **79**, 80*f*
Language, cortical areas in, 256, 256*f*
Large intestine, **20**, 20*f*, 459*f*, **467**–468, 467*f*, 468*f*
  in cat, **354**, 354*f*, **413**, 414*f*
  histology of, 467, 467*f*, 468
  in rat, **20**, 20*f*
Laryngeal cartilages, **98**, 99*f*, 442
Laryngeal prominence (Adam's apple), 442, 444*f*, **528**, 528*f*
Laryngopharynx, **442**, 443*f*, **463**
  in cat, 449*f*, 475*f*
Larynx, 99*f*, **442**, 443*f*, 444*f*
  in cat, 414*f*, **448**, 449*f*, 450*f*
Lateral (term), **6**
Lateral angle of scapula, 134, 135*f*
Lateral apertures, 259, 263*f*
Lateral (axillary) border of scapula, 134, 135*f*
Lateral circumflex femoral artery, 405*f*
Lateral commissure, **306**, 306*f*
Lateral condyle
  of femur, **142**, 142*f*, 144, **538**, 539*f*
  of tibia, **143**, 143*f*, 144, **538**, 539*f*
Lateral cutaneous nerve, 285*f*
  in cat, 293*f*
Lateral (acromial) end of clavicle, 134, 135*f*
Lateral epicondyle
  of femur, **142**, 142*f*
  of humerus, **136**, 136*f*, 138, 534*f*, **535**
Lateral femoral cutaneous nerve, 288*t*, 289*f*
Lateral fornix, 510*f*
Lateral funiculus, 282*f*, **283**, 284*f*
Lateral geniculate nucleus, **310**, 311*f*
Lateral horn, **280**, 282*f*, 301*f*
Lateral ligament (temporomandibular joint), **161**, 163*f*
Lateral longitudinal arch of foot, 144, 144*f*
Lateral malleolus, **143**, 143*f*, 144, **538**, 539*f*
Lateral masses, of vertebra, 117*f*
Lateral meniscus of knee, 162*f*
Lateral muscle compartment, lower limb muscles and, 208*t*, 209*f*, 211*f*
Lateral patellar retinaculum, **159**, 162*f*
Lateral pectoral nerve, 286*t*, 287*f*
Lateral plantar artery, **405**, 405*f*
Lateral plantar vein, **407**
Lateral pterygoid muscle, 188*f*, 188*t*
Lateral rectus muscle of eye, 307*f*
Lateral sacral crest, 119*f*
Lateral sulcus, **254**, 255*f*
Lateral supracondylar line, 142*f*
Lateral supracondylar ridge, 136*f*
Lateral thoracic artery, 402*f*, **403**
Lateral wall of orbit, 113*f*
Latissimus dorsi muscle, 185*f*, 191, 194*t*, 195*f*, 198*f*, **529**, 530*f*
  in cat, 218*f*, 220*f*, **221**
LDL. *See* Low-density lipoprotein
Left, anatomical, **8**
Left anterior descending (anterior interventricular) artery, 384*f*, **388**, 388*f*
Left ascending lumbar vein, 408*f*, **409**, 409*f*
Left atrium, 384*f*, 385*f*, 388*f*
  in sheep, 389*f*
Left brachiocephalic vein, 406*f*, **407**
  in cat, **416**
Left colic (splenic) flexure, **467**, 468*f*
Left common carotid artery, 384*f*, 399*f*, **400**
  in cat, 415*f*, 448, 450*f*
Left coronary artery, 384*f*, 387–388, 388*f*, 399*f*, **400**
Left gastric artery, **403**, 403*f*
Left gastric vein, **411**
Left gastroepiploic artery, **403**, 403*f*
Left gonadal vein, 406*f*, **407**, 408*f*
Left pulmonary artery, 384*f*, 385*f*, **410**, 410*f*, 446*f*
  in cat, 415*f*
  in sheep, 389*f*
Left pulmonary vein, 384*f*, 385*f*, 410*f*, 446*f*
Left subclavian artery, 384*f*, 399*f*, **400**, 402*f*
  in cat, **414**, 415*f*, 420*f*
Left subclavian vein, 409*f*, 431*f*
  in cat, 450*f*
Left suprarenal vein, 406*f*, **407**, 408*f*
Left upper and lower abdominopelvic quadrants, 8, 8*f*
Left ventricle, 384*f*, 385*f*, 388*f*
  right ventricle compared with, 391, 392*f*
  in sheep, 389*f*, 390, 391, 392*f*
Leg. *See also* Lower limb
  bones of, 143, 143*f*
  muscles of/muscles acting on, 184*f*, 185*f*, 204, 205*f*, 206*t*, 208*t*, 209*f*, 211*f*
    in cat, 227–228, 227*f*, 230*f*
  surface anatomy of, 538, 539*f*
  veins of, 406–407
Lens of eye, 308*f*, **309**
  in cow/sheep eye, **311**, 312*f*
  opacification of (cataracts), **309**
Lens selection disc (ophthalmoscope), **321**, 321*f*
Lesser curvature of stomach, 462*f*, **463**
Lesser occipital nerve, 285*f*, 286*t*
Lesser omentum, **463**, 464*f*
  in cat, **475**, 476*f*
Lesser sciatic notch, **139**, 140*f*
Lesser trochanter, **142**, 142*f*
Lesser tubercle, **134**–136, 136*f*
Lesser wings of sphenoid, **108**, 108*f*, 109*f*, 113*f*
Leukemia, **371**
  lymphocytic, 378
Leukocyte(s) (white blood cells/WBCs), 67*f*, **366**, 367*f*, 368*t*, 369*f*, 370–371, 370*f*
Leukocyte count
  differential, **371**–372, 372*f*
  total, **371**
Leukocytosis, **371**
Leukopenia, **371**
Levator labii superioris muscle, 186*t*, 187*f*
Levator palpebrae superioris muscle, 306*f*
Levator scapulae muscle, 195*f*, 195*t*
Levator scapulae ventralis muscle, in cat, 220*f*, **221**
LH. *See* Luteinizing hormone
Life cycle, cell, 45–48, 46–47*f*
Ligament(s), 156, 156*f*. *See also specific type*
Ligament of head of femur (ligamentum teres), **158**, 161*f*
Ligamentum arteriosum, 384*f*, 412*f*, **413**
  in sheep, 389*f*, **390**
Ligamentum nuchae, 196*f*
Ligamentum teres (ligament of head of femur), **158**, 161*f*
Ligamentum teres (round ligament), 412*f*, **413**, 464*f*, 472*f*
  in cat, 475
Ligamentum venosum, 412*f*, **413**
Light pathway, in vision, 309*f*
Light refraction in eye, 318–320, 318*f*, 319*f*
Line (bone marking), 94*t*
Linea alba, 192*t*, 193*f*, 531*f*, **532**
  in cat, **220**
Linea aspera, **142**, 142*f*
Linea semilunaris, 531*f*, **532**
Lingual artery, 401*f*
Lingual frenulum, **459**, 461*f*, 470*f*
  in cat, **474**
Lingual tonsil, 442, 443*f*, **461**, 461*f*
Lipids/lipid bilayer, in plasma membrane, 40, 41*f*
Lipoproteins, 378
Lips (labia), **459**, 461*f*
Liver, 19*f*, **20**, 20*f*, 459*f*, 464*f*, **471**–473, 472*f*, 473*f*. *See also under Hepatic*
  in cat, **354**, 354*f*, **413**, 414*f*, **475**, 476*f*
  in rat, 19*f*, **20**, 20*f*
  round ligament of (ligamentum teres), 412*f*, **413**, 464*f*, 472*f*
    in cat, 475
Liver cells (hepatocytes), **472**, 473*f*
Lobar arteries, of lung, **410**, 410*f*
Lobar (secondary) bronchi, 444, 445*f*
Lobe(s)
  of cerebral hemispheres, 254, 255–256*f*, 257*f*
  of lung, 445*f*, 446*f*, **447**
    in cat, 414*f*, 449, 449*f*
  of mammary gland, **512**, 513*f*
  of pituitary gland. *See* Adenohypophysis; Neurohypophysis
  of thyroid gland, 528

SR. *See* Sarcoplasmic reticulum
Stage (microscope), 28f, **29**
Stapes (stirrup), **328**, 328f, 329f, 331f
Starch digestion, by salivary amylase, 470
Static equilibrium, **333**
Stellate (hepatic) macrophages, **472**, 473f
Stem cells, 432
hematopoietic, 370
Stereocilia
epididymal, 508, 508f
in equilibrium, 333f, 334f
in spiral organ, 330, 330f
Sternal (term), **3**, 3f
Sternal angle, **120**, 121f, 444, **530**, 531f
Sternal (medial) end of clavicle, 134, 135f
Sternal head, of sternocleidomastoid muscle, **528**, 528f
Sternal puncture, 120
Sternoclavicular joint, 134, 164t
Sternocleidomastoid muscle, 184f, 185f, 187f, 189f, 189t, 190f, 192f, **528**, 528f, 529f
Sternocostal joints, 164t
Sternohyoid muscle, 184f, 189f, 190f, 190t
in cat, 217f, **218**
Sternomastoid muscle, in cat, **217**, 217f
Sternothyroid muscle, 189f, 190t
in cat, **218**
Sternum, 92f, **120**, 121f, 530
Stirrup (stapes), **328**, 328f, 329f
STI/STDs. *See* Sexually transmitted infections/diseases
Stomach, **20**, 20f, 459f, 462f, **463**, 464f, 465f. *See also under Gastric*
in cat, **354**, 354f, **413**, 414f, **475**, 476f
histology of, 462f, 463, 465f
in rat, **20**, 20f
Straight sinus, 262f, 408f
Stratified epithelium, **54**, 55f
columnar, 59f
cuboidal, 58f
squamous, 58f
Stratum basale (stratum germinativum) (skin), **79**, 80f, 84f
Stratum basalis (basal layer) (endometrium), **509**
Stratum corneum (skin), **79**, 80f, 84f
Stratum functionalis (functional layer) (endometrium), **509**
Stratum granulosum (skin), **79**, 80f, 84f
Stratum lucidum (skin), **79**, 84f
Stratum spinosum (skin), **79**, 80f, 84f
Stretch receptors, in bladder wall, 492
Striated muscle, 173. *See also* Skeletal muscle(s)
Striations, muscle tissue, 68, 69f, 173. *See also* Skeletal muscle(s)
Striatum, **259**, 260f
Stria vascularis, 330f
Sty, **306**
Stylohyoid muscle, 190f, 190t
Styloid process
of radius, 137f, 139, **535**, 536f
of temporal bone, **106**, 107f
of ulna, **137**, 137f, 139, **535**, 536f
Stylomastoid foramen, **106**, 107f
Subarachnoid space
brain, 261, 262f, 263f
spinal cord, 282f
Subcapsular sinus, lymph node, 433f
Subclavian artery, 384f, 399f, **400**, 401f, 402f, **528**, 528f, **529**, 529f
in cat, **414**, 415f, **416**, 420f
Subclavian trunks, 431f
Subclavian veins, 406f, **407**, 408f, 409f, 431f
in cat, 417f, **418**, 420f, 450f
Subclavius muscle, 192f
nerve to, 287f
Subdural space
brain, **261**, 262f
spinal cord, 282f
Sublingual glands, 459f, **470**, 470f
in cat, **474**
Submandibular duct, 461f, 470, 470f
Submandibular glands, 459f, **470**, 470f, **529**, 529f
in cat, **474**, 474f
Submucosa
alimentary canal, **458**, 460f
duodenal, **466**
gastric, 465f
ileum, 467f
large intestine, 467f
tracheal, 447f
Submucosal plexus, 458, 460f
Subpapillary plexus, 78f, 79
Subscapular artery, 402f, **403**
in cat, 292f, 415f, 416
Subscapular fossa, 134, 135f
Subscapularis muscle, 161, 192f
in cat, **221**, 222f
Subscapular nerve, 286t, 287f
in cat, 292f
Subscapular vein, in cat, 417f, **418**
Substage light (microscope), 28f, **29**
Sudoriferous (sweat) glands, 81, 82, 82f
apocrine, **81**
merocrine (eccrine glands), 78f, **81**, 82f
Sulci (sulcus), **254**, 255f
Superciliary arches, 526f, **527**
Superficial/external (term), **6**
Superficial fascia of skin (hypodermis), **78**, 78f
Superficial fibular nerve, 290f
Superficial inguinal ring, **532**, 533f
Superficial palmar arch, 402f
Superficial temporal artery, **400**, 401f, 526f, **527**
Superficial temporal vein, **407**, 408f
Superficial transverse ligament of palm, 199f
Superficial venous palmar arch, 409f
Superior (term), **6**, 6f
Superior angle of scapula, 134, 135f
Superior articular process/facet, **117**, 117f, 118f, 119f, 119t, 120
Superior border of scapula, 134, 135f
Superior colliculi, **257**, 258f, 311f
in sheep, 270f
Superior costal facet, 118f, 122f
Superior extensor retinacula, 209f
Superior gluteal artery, **405**, 405f
Superior gluteal nerve, 290f, 291t
Superior hypophyseal artery, 351f
Superior mesenteric artery, 399f, **403**, 403f, 404f
in cat, 415f, **416**, 420f
Superior mesenteric ganglia, 302
Superior mesenteric vein, **411**, 411f
in cat, **418**, 419f
Superior nasal conchae/turbinates, **111**, **442**, 443f
Superior nasal meatus, 443f
Superior nuchal line, 107f, 110f, 527
Superior oblique muscle of eye, 307f
Superior orbital fissure, **108**, 109f, 110f, 113f
Superior pubic ramus, 140f
Superior rectal artery, 404f
Superior rectus muscle of eye, 307f
Superior sagittal sinus, **261**, 262f, 263f, 408f
Superior thyroid artery, 401f
Superior tibiofibular joint, 143f, 165t
Superior vena cava, 19f, 384f, 385f, **386**, 388f, **405**, 406f, 408f, 409f, 431f
in cat (precava), **413**, 414f, **416**, 417f, 420f, 450f
in sheep, 389f, 390
veins draining into, 406f, 407–409, 408f, 409f
Supination (body movement), **158**, 160f
Supinator muscle, 199f, 200t, 201f
Supporting cells
in cochlea, 330f
in crista ampullaris, 333f
nervous tissue (neuroglia/glial cells), **67**, 68f, **240**, 240f
olfactory epithelium, 342, 342f
Suprachiasmatic nucleus, 311f
Supraclavicular nerves, 285f, 286t
Supracondylar lines, 142f
Supracondylar ridges, 136f
Supracristal line, **529**, 530f
Supraglenoid tubercle, 135f
Suprahyoid muscles, **529**, 529f
Supraorbital foramen (notch), **106**, 110f, 113f, 115
Supraorbital margin, 110f
Suprapatellar bursa, 162f
Suprarenal arteries, 399f, **403**, 404f
Suprarenal glands. *See* Adrenal (suprarenal) glands
Suprarenal veins, 406f, **407**, 408f
Suprascapular artery, 402f
Suprascapular nerve, 286t, 287f
Suprascapular notch, **134**, 135f
Supraspinatus muscle, 161, 195f, 195t, 198f
in cat, **222**, 223f
Supraspinous fossa, 134, 135f
Sural (term), 3f, **5**
Sural artery, in cat, 415f, **416**
Sural nerve, 290f, 291t
Surface anatomy, 2–5, 2f, 3f, 525–541, **526**
of abdomen, 3f, 532–533, 533f
of head, 3f, 526f, 527, 527f
of lower limb, 3f, 144, 536–540, 537f, 538f, 539f, 540f
of neck, 3f, 528–529, 528f, 529f
terminology and, 2–5, 3f
of trunk, 529–532, 530f, 531f
of upper limb, 3f, 138–139, 533f, 534–536, 534f, 536f, 537f
Surgical neck of humerus, **134**, 136f
Suspensory ligament
of breast, 513f
of eye (ciliary zonule), 308f, **309**
of ovary, 510f, **511**
in cat, 515f, 516f
Sustentocytes (Sertoli cells), 506f, **507**
Sutural bones, **93**, 110f
Sutures, 154f, **155**
cranial, 106, 155
Sweat (sudoriferous) glands, 81, 82, 82f
apocrine, **81**
merocrine (eccrine glands), 78f, **81**, 82f
Sweat pores, 78f, 81
Sweet taste, 344
Sympathetic (thoracolumbar) division of autonomic nervous system, **300**–302, 300f, 301f
function of, 300
Sympathetic trunk ganglion/chain/paravertebral ganglion, 285f, **300**, 301f, 302
in cat, 416–418
Symphyses, 154f, **155**
Synapses, neural, 241f, **242**
Synaptic cleft, **175**, 175f, 241f, **242**
Synaptic potential, 242
Synaptic vesicles, 175f, 241f, 242
Synarthroses, **155**
Synchondroses, 154f, **155**, 155f
Syndesmoses, 154f, **155**
Synergists (muscle), **182**
Synovial (joint) cavities, **9**, 10f, 156, 156f
Synovial fluid, 156, 156f
Synovial joints, 154f, **155**–161, 156f, 157f, 164–165t. *See also specific joint*
movements at, 157, 158f, 159–160f
types of, 156–157, 157f
Synovial membranes, 156, 156f
Systemic circulation, **386**, 388f
arterial, 398, 398f, 399–400, 399f, 400–405, 401f, 402f, 403–404f, 405f
venous, 398, 398f, 400, 405–409, 406f, 407f, 408f, 409f
Systole, **386**

T_3 (triiodothyronine), **351**
T_4 (thyroxine), **351**
Tactile corpuscles, **79**, **246**, 247, 247f
Tactile epithelial (Merkel) cells, **79**, 80f, 246
Tactile epithelial complexes (Merkel discs), **246**
Tail of sperm, 507, 507f
Tallquist method, 374
Talus, 143, 144f
Tapetum lucidum, **311**, 312f
Target organs, **350**
Tarsal (term), **3**, 3f
Tarsal bones, 92f, **143**, 144f
Tarsal glands, **306**
Tarsal plate, 306f
Tarsometatarsal joint, 165t
Taste, 343–344, 343f, 344f
cortical areas in, 256f
smell/texture/temperature and, 344, 345
Taste buds, **343**–344, 343f, 344f
Taste pore, **343**, 343f
T cells (T lymphocytes), 371, **432**
thymus in development of, 352, 432
Tears (lacrimal secretion), 305–306
Tectorial membrane, **330**, 330f
Teeth, 468–470, 469f, 470f
in cat, **475**
Telencephalon, 255f
Telophase, **45**, 47f
Temperature
skin in regulation of, 79
sweat glands in regulation of, 81
taste and, 345
Temperature receptors, 246, 247f
Temporal artery, superficial, **400**, 401f, 526f, **527**
Temporal bone, **106**, 107f, 108f, 110f, 111f
in fetal skull, 123f
Temporalis muscle, 183, 184f, 187f, 188f, 188t
Temporal lobe, **254**, 255f, 257f
Temporal vein, superficial, **407**, 408f
Temporomandibular joint (TMJ), 112f, 115, **161**, 163f, 164t, 526f, **527**
Tendinous intersections, 193f, 220, 531f, **533**
Tendon(s), 64f, **174**
Tendon organs, **247**, 248f
Teniae coli, **467**, 468f
Tensor fasciae latae muscle, 184f, 205f, 206t
in cat, **226**, 226f, 229f
Tentorium cerebelli, **261**, 262f, 263f
Teres major muscle, 185f, 195f, 195t, 198f, 530f
Teres minor muscle, 161, 194t, 195f, 198f
Terminal boutons (axon terminals), **175**, 175f, 176f, 241f, **242**
Terminal bronchioles, 445f
Terminal cisterns, **173**, 173f
Terminal web, 44
Terminology (anatomical), 1–14
Tertiary (segmental) bronchi, 444, 445f
Testicular artery, 399f, **403**, 404f
in cat (spermatic artery), 415f, **416**, 495f, 513, 514f
Testicular (gonadal) vein, **407**
in cat, 417f, **418**
Testis/testes, **21**, 21f, **22**, 22f, 352, 353f, **504**, 504f, 505f, 506, 506f, 507
in cat, **354**, 355f, 495f, **513**, 514f, 515
in rat, **21**, 21f

Testosterone, **352**
Tetany, **352**
Texture, taste affected by, 345
TF. *See* Tissue factor
TH. *See* Thyroid hormone
Thalamic nuclei, in sheep, 272*f*
Thalamus, 255*f*, 258*f*, **259**, 260*f*
  in sheep, 271
Theca folliculi, **512**, 512*f*
Thenar eminence, **536**, 537*f*
Thenar muscles of thumb, 199*f*, 202*t*
Thick (myosin) filaments, 172*f*, **173**
Thick skin, 79, 84, 84*f*
Thigh. *See also* Lower limb
  bones of, 142–143, 142*f*
  deep artery of, **405**, 405*f*
  muscles of/muscles acting on, 184*f*, 185*f*, 204, 205*f*, 206*t*, 207*f*, 208*t*, 538, 539*f*
    in cat, 225–228, 226*f*, 229*f*
  surface anatomy of, 538, 539*f*
Thin (actin) filaments, 44, 172*f*, **173**
Thin skin, 79, 80*f*, 82, 84, 84*f*
Thoracic (term), **3**, 3*f*
Thoracic aorta, 399*f*, 400–403, **400**, 402*f*
  in cat, 415*f*, 420*f*
Thoracic artery
  internal, 401*f*, 402*f*, **403**
    in cat, 415*f*, **416**
  lateral, 402*f*, **403**
  long, in cat, 415*f*, **416**
  ventral, in cat, 415*f*, **416**
Thoracic cage, 92*f*, **120**–122, 121*f*, 122*f*, 530–532, 531*f*
Thoracic cavity, **7**, 7*f*, 18–20, 19*f*
  in cat, 354, 354*f*, 355*f*, 413, 414*f*, 449
  in rat, 18–20, 19*f*
  relationships of organs in, 446*f*
Thoracic curvature, 116, 116*f*
Thoracic duct, **430**, 431*f*
  in cat, **434**
Thoracic nerve, long, 286*t*
Thoracic spinal nerves, 281*f*, 284, 285*f*
Thoracic splanchnic nerves, 301*f*
Thoracic vein, internal (mammary), in cat, **416**, 417*f*
Thoracic vertebrae, 115, 116*f*, 118*f*, 119*t*, 120
Thoracoacromial artery, 402*f*, **403**
Thoracodorsal nerve, 287*f*
Thoracolumbar (sympathetic) division of autonomic nervous system, **300**–302, 300*f*, 301*f*
  function of, 300
Thorax
  arteries of, 402*f*, 403
  bony, 120, 530–532, 531*f*. *See also* Thoracic cage
  heart in, 384, 384*f*
  muscles of, 184*f*, 191*t*, 192*f*, 193*f*, 194–195*t*, 195*f*, 531*f*, 532
    in cat, 218–219, 218*f*, 222*f*, 223*f*
  surface anatomy of, 530–532, 531*f*
  veins of, 407–409, 409*f*
Three-dimensional vision (depth perception), **320**–321
Throat (pharynx), **442**, 443*f*, 459*f*
  in cat, 449*f*
Thrombin, **376**, 376*f*
Thumb (pollex), **3**, 3*f*
  bones of, 138
  thenar muscles of, 199*f*, 202*t*
Thumb joint (carpometacarpal joint), 164*t*
Thymopoietins, **352**
Thymosins, **352**
Thymulin, **352**
Thymus, **18**, 19*f*, 352, 353*f*, 432, 432*f*, 446*f*
  in cat, **354**, 354*f*, 355*f*, **413**, 449*f*, 450*f*
  in rat, **18**, 19*f*
Thyrocervical trunk, **400**, 401*f*, 402*f*
  in cat, 415*f*, **416**
Thyroglobulin, **355**
Thyrohyoid membrane, 444*f*
Thyrohyoid muscle, 190*f*, 190*t*
Thyroid artery, superior, 401*f*
Thyroid cartilage, 99*f*, **442**, 443*f*, 444*f*, 528, 528*f*
  in cat, **448**, 449*f*, 450*f*
Thyroid gland, 351–352, 353*f*, 355–357, 356*f*
  in cat, **354**, 355*f*, **448**, 449*f*, 450*f*
  isthmus of, 528
  lobes of, 528
Thyroid hormone (TH), **351**
Thyroid-stimulating hormone (TSH/thyrotropin), **350**, 351*f*
Thyroid veins, 408*f*
Thyrotropin. *See* Thyroid-stimulating hormone
Thyroxine (T_4), **351**
Tibia (shinbone), **143**, 143*f*, 538, 539*f*
Tibial arteries
  anterior, **405**, 405*f*
    in cat, 415*f*, **416**
  posterior, **405**, 405*f*, 538
    in cat, 415*f*, **416**
    pulse at, 538
Tibial collateral ligament, **159**, 162*f*
Tibialis anterior muscle, 184*f*, 204, 209*f*, 210*t*, 539*f*
  in cat, 227*f*, **228**, 230*f*
  tendon of, 213*f*
Tibialis posterior muscle, 210*t*, 211*f*
  in cat, **230**
  tendon of, 211*f*, 213*f*
Tibial nerve, **288**, 290*f*, 291*t*
  in cat, **292**, 294*f*
Tibial tuberosity, **143**, 143*f*, 144, **538**, 539*f*
Tibial veins, anterior and posterior, **406**–407, 407*f*
  in cat, 417*f*, **418**
Tibiofemoral joint, **159**, 165*t*
Tibiofibular joints, 143*f*, 165*t*
Tissue(s), **15**, **54**. *See also specific type*
  classification of, 53–76
  connective, **60**–67, 60*f*, 61–67*f*
  epithelial (epithelium), **54**–55, 55*f*, 56–59*f*
  muscle, **67**–68, 69–70*f*
  nervous, **67**, 68*f*
Tissue factor (TF), **376**, 376*f*
Titin (elastic) filaments, 172*f*
TM. *See* Total magnification
TMJ. *See* Temporomandibular joint
Toe(s)
  bones of (phalanges), 92*f*
  foot muscles controlling movement of, 209*f*, 212–214*t*, 213–214*f*
  joints of, 165*t*
Tongue, 443*f*, **459**, 459*f*, 461*f*, 470*f*
  in cat, **474**
  taste buds on, 343, 343*f*, 344*f*
Tonsil(s), 432, 432*f*, 433, 435*f*, 442, 461
  lingual, 442, 443*f*, **461**, 461*f*
  palatine, 435*f*, 442, 443*f*, **459**, 461, 461*f*
    in cat, 474
  pharyngeal, 432*f*, 442, 443*f*
  tubal, 442, 443*f*
Tonsillar crypts, **433**, 435*f*
Tonsillitis, **461**
Tooth. *See* Teeth
Total blood counts, **371**
Total magnification (TM), microscope, **29**–30
Total red blood cell count, **371**
Total white blood cell count, **371**
Touch receptors, in skin, 78*f*, 79, 246
Trabeculae
  bone, 93, **96**
  lymph node, **433**, 433*f*
  spleen, 434*f*
Trabeculae carneae, 385*f*
  in sheep, **391**
Trachea, **18**, 19*f*, 57*f*, 99*f*, 443*f*, **444**, 445*f*, 446*f*, 447, 447*f*, 461*f*
  in cat, **354**, 354*f*, 355*f*, 414*f*, **448**, 449*f*, 450*f*
  in rat, **18**, 19*f*
Tracheal cartilages, **98**, 99*f*, 444, 444*f*, 447*f*
  in cat, 448, 450*f*
Trachealis muscle, 447*f*
Tracheal wall, 447, 447*f*
Tracts, **242**
  spinal cord, **283**, 283*f*
Transitional epithelium, **54**, 59*f*
  of bladder, 493
Transport
  cell, active and passive processes in, 41
  of respiratory gases, **442**
Transverse (term), muscle name and, 182
Transverse arch of foot, 144, 144*f*
Transverse cerebral fissure, 255*f*, 267*f*
Transverse cervical nerve, 285*f*, 286*t*
Transverse colon, 459*f*, 464*f*, **467**, 468*f*
  in cat, 477
Transverse costal facet, 118*f*, 122*f*
Transverse foramen, in vertebra, 117*f*, 118*f*, 119
Transverse humeral ligament, 166*f*
Transverse jugular vein, in cat, 420*f*
Transverse ligament of palm, superficial, 199*f*
Transverse mesocolon, 464*f*, 468*f*
Transverse plane, 4*f*, **5**
Transverse process, 116*f*, **117**, 117*f*, 118*f*, 119, 119*t*, 120
Transverse scapular vein, in cat, 417*f*
Transverse sinus, 262*f*, 408*f*
Transverse (T) tubules, **173**, 173*f*
Transversus abdominis muscle, 184*f*, 192*t*, 193*f*, 194*f*
  in cat, 219*f*, **220**
Trapezium, 137, 138*f*
Trapezius muscles, 184*f*, 185*f*, 187*f*, 191, 194*t*, 195*f*, **529**, 529*f*, 530*f*
  in cat, 220*f*, **221**, 223*f*
Trapezoid, 137, 138*f*
Trapezoid line, 135*f*
Traveling (sound) waves, 330–332, 332*f*
Triads, **173**, 173*f*
Triangle of auscultation, 185*f*, **530**, 530*f*
Triceps (term), muscle name and, 182
Triceps brachii muscle, 184*f*, 185*f*, 197, 198*f*, 198*t*, 534*f*, **535**
  in cat, **222**–223, 224*f*
Triceps surae muscle, 210*t*
  in cat, **227**
Tricuspid valve, 385*f*, **386**, 387*f*
  in sheep, 390, 391*f*
Trigeminal nerve (cranial nerve V), 264*t*, 266*f*
  in sheep, 268, 268*f*, 269*f*
Trigone, **486**, 488*f*
Triiodothyronine (T_3), **351**
Triquetrum, 137, 138*f*
Trochanter (bone marking), 94*t*
  greater and lesser, **142**, 144
Trochlea
  of humerus, **136**, 136*f*
  of talus, 144*f*
Trochlear nerve (cranial nerve IV), 264*t*, 266*f*
  in sheep, 268, 269*f*
Trochlear notch, **137**, 137*f*
Tropic hormones, pituitary, **350**, 351*f*
True pelvis, **139**
True (vertebrosternal) ribs, 121, 121*f*, 122*f*
True vocal cords (vocal folds), **442**, 443*f*, 444*f*
  in cat, 448, 450*f*
Trunk. *See also specific region*
  muscles of, 183–191, 191–192*t*, 192*f*, 193–194*f*, 194–197*t*, 195–196*f*
    in cat, 217–222
      deep, 221–222, 222*f*, 223*f*
      superficial, 220–221, 220*f*
  surface anatomy of, 529–532, 530*f*, 531*f*
Trunks
  brachial plexus, 287*f*, 288
  lymphatic, **430**, 431*f*
TSH. *See* Thyroid-stimulating hormone
T (transverse) tubules, **173**, 173*f*
Tubal tonsils, 442, 443*f*
Tubercle (bone marking), 94*t*
Tuberosity (bone marking), 94*t*
Tubular resorption, **491**–492
Tubular secretion, **492**
Tubulins, 44
Tunic(s)
  alimentary canal, **458**, 460*f*
  blood vessel, 398–400, 398*f*
Tunica albuginea
  of ovary, 512*f*
  of testis, 505*f*, 506*f*, **507**, 508*f*
Tunica externa, 398*f*, **399**
Tunica intima, 398*f*, **399**
Tunica media, 398*f*, **399**
Tunica vaginalis, 506*f*
  in cat, 495*f*, 514*f*, **515**
Tuning fork tests, of hearing, 331–332, 332*f*
Turbinates, nasal (nasal conchae)
  inferior, 110*f*, **113**, **442**, 443*f*
  middle, 109*f*, 110*f*, **111**, **442**, 443*f*
  superior, **111**, **442**, 443*f*
Turk's saddle (sella turcica), **108**, 108*f*, 109*f*
Tympanic cavity, **328**
Tympanic membrane (eardrum), **328**, 328*f*
Tympanic part of temporal bone, **106**
Type I alveolar cell, 445*f*
Type II alveolar cell, 445*f*
Type A blood, 377*t*, 378*f*
Type AB blood, 377*t*, 378*f*
Type B blood, 377*t*, 378*f*
Type lines, in fingerprint, 85
Type O blood, 377*t*, 378*f*

Ulcers, decubitus (bedsores), **81**, 81*f*
Ulna, 92*f*, 136, 136*f*, **137**, 137*f*, 138–139, 138*f*, 535, 536*f*
Ulnar artery, 402*f*, **403**, **535**, 536*f*
  in cat, 415*f*, **416**
Ulnar head, **137**, 137*f*, 534*f*, **535**, 536*f*
Ulnar nerve, 136, 286*t*, 287*f*, **288**
  in cat, **291**, 292*f*
Ulnar notch, **137**, 137*f*
Ulnar pulse, 535, 536*f*
Ulnar styloid process, **137**, 137*f*, 139, **535**, 536*f*
Ulnar vein, **407**, 409*f*
  in cat, 417*f*, **418**
Umami (taste), 344
Umbilical (term), **3**, 3*f*
Umbilical arteries, **411**–413
Umbilical cord, 411–413, 412*f*
Umbilical hernia, acquired, **532**
Umbilical ligaments, medial, 412*f*, **413**
Umbilical region, **9**, 9*f*
Umbilical vein, **411**, 412*f*
Umbilicus (navel), **411**, 412*f*, 532
Uncus, **256**, 258*f*
Uniaxial joints/movement, 154*f*, 156, 157*f*
Unipennate muscle structure, 183*f*
Unipolar neurons, **243**, 244*f*
Upper limb. *See also* Arm; Forearm; Hand
  arteries of, 399*f*, 402*f*, 403
  blood vessels of, in cat, 292*f*

Upper limb *(continued)*
bones of, 92*f*, 134–139, 134*f*, 135*f*, 136*f*, 137*f*, 138*f*
muscles of, 197, 198*f*, 198*t*, 199–201*f*, 199–200*t*, 202*t*, 203*f*
in cat, 222–225, 224*f*
nerves of/brachial plexus, **285**–288, 285*f*, 286*t*, 287*f*, **529**, 529*f*
in cat, 291, 292*f*
surface anatomy of, 3*f*, 138–139, 533*f*, 534–536, 534*f*, 536*f*, 537*f*
veins of, 406*f*, 407–409, 409*f*
Upper respiratory system, 442, 443*f*, 444*f*
Ureter(s), **20**, 21*f*, 22*f*, **486**, 487*f*, **488**, 488*f*, 489*f*, 493*f*, 504*f*, 505*f*, 510*f*
in cat, **493**, 494*f*, 495*f*, 514*f*, 515*f*, 516*f*
in pig/sheep, **488**
in rat, **20**, 21*f*
Urethra, 22, 59*f*, **486**, 487*f*, 488*f*, 504*f*, 505*f*, 506, 510*f*
in cat, 476*f*, **493**, 494*f*, 495*f*, 496, **513**, 514*f*, 515*f*, 516*f*
in rat, 21
Urethral orifice, external, 21*f*, 486, 488*f*, **492**, 504*f*, 505*f*, 509, 509*f*, 510*f*
in rat, 21*f*
Urethral sphincters, 488*f*, 492
in cat, 493
Urinary bladder, **20**, 20*f*, 21*f*, 22*f*, 464*f*, **486**, 487*f*, 488*f*, **492**–493, 505*f*
in cat, 414*f*, 476*f*, **493**, 494*f*, 495*f*, 514*f*, 515*f*, 516*f*
functional microscopic anatomy of, **492**–493
palpable, 533
in rat, **20**, 20*f*, 21*f*
Urinary system, 16*t*, 485–501, **486**, 487*f*, 488*f*. *See also specific organ*
in cat, 493–496, 494*f*, 495*f*
functional microscopic anatomy of kidney/bladder and, 489–493, 490*f*, 491*f*, 492*f*
gross anatomy of, 486–489, 487*f*, 488*f*, 489*f*, 490*f*
in pig/sheep, 486–489
Urination (voiding/micturition), **492**
Urine, formation of, 491–492, 491*f*
Urogenital diaphragm, 488*f*, 504*f*, 505*f*, 510*f*
Urogenital sinus, in cat, 494*f*, **496**, **515**, 515*f*, 516*f*
Urogenital system, in cat, 493
Uterine body, **509**, 510*f*
in cat, **515**, 515*f*, 516*f*
Uterine horns
in cat, 494*f*, **515**, 515*f*, 516*f*
in rat, 21*f*, 22
Uterine (Fallopian) tube(s), **22**, 22*f*, **509**, 510*f*, 511, 511*f*
in cat, 494*f*, **515**, 515*f*, 516*f*
Uterine wall, 511, 511*f*
Uterosacral ligaments, 510*f*, **511**
Uterus, 21*f*, **22**, 22*f*, **509**, 510*f*, 511, 511*f*
in cat, 515, 515*f*
in rat, 21*f*, **22**
wall of, 511, 511*f*
Utricle, 329*f*, **333**, 334*f*
Uvea, 307
Uvula, 443*f*, **459**, 461*f*

Vagina, 21*f*, **22**, 22*f*, **509**, 510*f*
in cat, 515–516, **515**, 515*f*, 516*f*
in rat, 21*f*, **22**
Vaginal orifice, 21*f*, **22**, 22*f*, 509*f*
in rat, 21*f*, **22**
Vagus nerve (cranial nerve X), 265*t*, 266*f*, 343
in cat, 413, 414*f*, 448
in sheep, 269*f*, 270
Vallate papillae, **343**, 343*f*
Valves
of collecting lymphatic vessels, 430, 431*f*, 432*f*
of heart, 385*f*, 386, 387*f*
venous, 398*f*, 400
Vasa recta, 490*f*, **491**
Vasa vasorum, 398*f*, **399**
Vascular layer of eye, **307**
Vas (ductus) deferens, **21**, 21*f*, **22**, 22*f*, 504*f*, 505*f*, **506**, 506*f*
in cat, 495*f*, **513**, 514*f*
in rat, **21**, 21*f*
Vasectomy, 506
Vastus intermedius muscle, 205*f*, 206*t*
in cat, **228**
Vastus lateralis muscle, 184*f*, 205*f*, 206*t*, **538**, 538*f*, 539*f*
in cat, **228**, 229*f*
for intramuscular injections, **538**, 538*f*
Vastus medialis muscle, 184*f*, 205*f*, 206*t*, 539*f*
in cat, **228**, 229*f*
Veins, 398, 398*f*, 400, 405–409, 406*f*, 407*f*, 408*f*, 409*f*
in cat, 416–418, 417*f*, 419*f*, 420*f*
Venae cavae, **386**, 388*f*
inferior, **21**, 21*f*, 384*f*, 385*f*, **386**, 403*f*, **405**, 406*f*, 408*f*, 409*f*, 431*f*
in cat (postcava), 293*f*, **413**, 417*f*, **418**, 419*f*, 420*f*
in rat, **21**, 21*f*
in sheep, 389*f*, 390
veins draining into, 406–407, 406*f*, 407*f*, 408*f*
superior, 19*f*, 384*f*, 385*f*, **386**, 388*f*, **405**, 406*f*, 409*f*, 431*f*
in cat (precava), **413**, 414*f*, **416**, 417*f*, 420*f*, 450*f*
in sheep, 389*f*
veins draining into, 406*f*, 407–409, 408*f*, 409*f*
Venous valves, 398*f*, 400
Venter (term), 6
Ventilation (breathing), **442**. *See also under Respiratory*
muscles of, 191*t*, 193*f*
Ventral (term), 3*f*, **6**, 6*f*
Ventral body cavity, 7–9, 7*f*, 8*f*, 9*f*, 17–24
in cat, 353–354, 353*f*, 354*f*, 413, 414*f*
on human torso model, 23–24, 23*f*
in rat
examining, 18–23, 19*f*, 20*f*, 21*f*, 22*f*
opening, 17–18, 17–18*f*, 18*f*
serous membranes of, **7**–8, 8*f*
Ventral funiculus, 282*f*, **283**, 284*f*
Ventral gluteal site, for intramuscular injections, **538**, 538*f*
Ventral (anterior) horns, 260*f*, 263*f*, **280**, 282*f*, 284, 284*f*
Ventral median fissure, **280**, 282*f*, 284*f*
Ventral rami, **284**, 285*f*, 287*f*, 289*f*, 290*f*, 301*f*
Ventral root, **280**, 282*f*, 284, 301*f*
Ventral thoracic artery, in cat, 415*f*, **416**
Ventricles
brain, **254**, 255*f*, 258*f*, 259
cerebrospinal fluid circulation and, 261, 263*f*
in sheep, 270, 271, 271*f*, 272*f*
cardiac, 384*f*, 385*f*, **386**, 388*f*
in cat, 413
right versus left, 391, 392*f*
in sheep, 389*f*, 390
Venule(s), 398
portal, 472, 473*f*
Vermis, **259**, 261*f*
Vertebra(e), 92*f*, **115**–120, 116*f*, 117*f*, 118*f*, 119*t*
cervical, 115, 116*f*, 117*f*, 118–119, 118*f*, 119*t*
in coccyx, 115, 116*f*, 119*f*, 120
lumbar, 115, 116*f*, 118*f*, 119*t*, 120
relationship of to ribs, 120, 121
in sacrum, 115, 116*f*, 119*f*, 120
structure of, 117, 117*f*, 118*f*, 119*f*, 119*t*, 120
thoracic, 115, 116*f*, 118*f*, 119*t*, 120
Vertebral (term), 3*f*, **5**
Vertebral arch, **117**, 117*f*
Vertebral artery, 399*f*, **400**, 401*f*, **402**, 402*f*
in cat, 415*f*, **416**
Vertebral (medial) border of scapula, 134, 135*f*, **529**, 530*f*
Vertebral (spinal) cavity, **7**, 7*f*
Vertebral column, 92*f*, **115**–120, 116*f*, 117*f*, 118*f*, 119*f*, 119*t*
muscles associated with, 195–197*t*, 196*f*
Vertebral foramen, **117**, 117*f*, 118*f*, 119, 119*t*, 120
Vertebral (floating) ribs, 121, 121*f*
Vertebral veins, 406*f*, **407**, 408*f*
in cat, **416**, 417*f*
Vertebra prominens, 116*f*, 119, 528
Vertebrochondral ribs, 121, 121*f*
Vertebrosternal (true) ribs, 121, 121*f*, 122*f*
Vesicouterine pouch, 510*f*
Vestibular apparatus, **333**
Vestibular folds (false vocal cords), **442**, 443*f*, 444*f*
in cat, 448
Vestibular ganglion, 329*f*
Vestibular glands, greater, **509**, 509*f*, 510*f*
Vestibular membrane, **330**, 330*f*
Vestibular nerve, 329*f*, 333*f*
Vestibule
of ear, **329**, 329*f*, 333
nasal, 443*f*
oral, **459**, 461*f*
of vagina, **509**, 509*f*
Vestibulocochlear nerve (cranial nerve VIII), 265*t*, 266*f*
in sheep, 269*f*, 270
Viewing window (ophthalmoscope), **321**, 321*f*
Villi
arachnoid (arachnoid granulations), **261**, 262*f*, 263*f*
intestinal, **465**, 466, 466*f*, 467*f*
in cat, **477**
Vinegar, bone affected by, 96
Virtual image, **29**, 29*f*
Visceral layer of serous membrane (visceral serosa), **7**
Visceral (smooth) muscle, **68**, 70*f*
Visceral pericardium (epicardium), 8*f*, 385*f*, **386**
in sheep, 390
Visceral peritoneum, 8*f*, 458, 464*f*
in cat, 477
in duodenum, **466**
Visceral pleura, 8*f*, 446*f*, **447**
Visceral serosa (visceral layer of serous membrane), **7**
Vision. *See also under Visual and* Eye
binocular, 320–321, 320*f*
color, 320
cortical areas in, 256, 256*f*, 310, 311*f*
equilibrium and, 335
tests/experiments and, 317–325. *See also specific test*
visual system anatomy and, 305–316. *See also* Visual system anatomy
Visual acuity, **319**–320
Visual association area, 256*f*
Visual cortex, primary, 256*f*, **310**, 311*f*
Visual fields, 311*f*, 320, 320*f*
Visual pathways, 310, 311*f*
Visual system anatomy, 305–316
in cow/sheep, 310–311, 312*f*
external eye/accessory structures, 305–306, 306*f*, 307*f*
internal eye, 306–310, 308*f*, 309*f*
pathways to brain, 310, 311*f*
retinal, 309*f*, 310
Visual tests, 317–325. *See also specific test*
Vitreous humor/vitreous body, 308*f*, **309**
in cow/sheep, 310, 312*f*
Vocal cords
false (vestibular folds), **442**, 443*f*, 444*f*
in cat, 448
true (vocal folds), **442**, 443*f*, 444*f*
in cat, 448, 450*f*
Vocal folds (true vocal cords), **442**, 443*f*, 444*f*
in cat, 448, 450*f*
Vocal ligaments, 442
Voiding (micturition), **492**
Volkmann's (perforating) canals, **96**, 97*f*
Voluntary muscle, 173. *See also* Skeletal muscle(s)
Voluntary (somatic) nervous system, **254**, **299**
Vomer, 107*f*, 110*f*, **113**
Vulva (external genitalia), 21*f*, 22, 22*f*, **509**, 509*f*
in cat, **515**, 515–516*f*
in rat, 21*f*, 22

WBCs. *See* White blood cell(s)
Weber test, 331, 332*f*
Wernicke's area, **256**, 256*f*
Wet mount, 33–34, 33*f*, 34*f*
White blood cell(s) (leukocytes/WBCs), 67*f*, **366**, 367*f*, 368*t*, 369*f*, 370–371, 370*f*
White blood cell count
differential, **371**–372, 372*f*
total, **371**
White columns, 282*f*, **283**
White matter
cerebellar, 259, 261*f*
cerebral, 255*f*, **256**, 257–259, 260*f*
of spinal cord, **280**–283, 282*f*
White pulp, **433**, 434*f*
White ramus communicans, **300**, 301*f*
Whorls, in fingerprint, 85, 85*f*
Willis, circle of (cerebral arterial circle), 401*f*, **402**
Wisdom teeth, 469, 469*f*
Working distance (microscope), **30**, 31*f*
Wrist/wrist joint, 154*f*, 164*t*, 535, 536*f*
bones of, 137–138, 137*f*, 138*f*
fracture of (Colles' fracture), **535**
muscles acting on, 199–201*f*, 199–200*t*

Xiphihumeralis muscle, in cat, 218*f*, **219**
Xiphisternal joint, **120**, 121*f*, 531*f*, **532**
Xiphoid process, **120**, 121*f*, **530**, 531*f*

Yellow marrow, **93**, 95*f*

Z discs, 172*f*, 173, 173*f*
Zona fasciculata, 356*f*, **357**
Zona glomerulosa, 356*f*, **357**
Zona pellucida, 512*f*
Zona reticularis, 356*f*, **357**
Zygomatic arch, 106, 115, 526*f*, **527**
Zygomatic bone, 106, 107*f*, 110*f*, 111*f*, **112**–113, 113*f*, 115
Zygomatic process, **106**, 107*f*, 112*f*, 113*f*
Zygomaticus muscles, 183, 184*f*, 186*t*, 187*f*

DISCARD